Statistical Multisource-Multitarget Information Fusion

For a listing of recent titles in the *Artech House Information Warfare Library*,
turn to the back of this book.

Statistical Multisource-Multitarget Information Fusion

Ronald P. S. Mahler

ARTECH HOUSE
BOSTON | LONDON
artechhouse.com

Library of Congress Cataloging-in-Publication Data
A catalog record for this book is available from the U.S. Library of Congress.

British Library Cataloguing in Publication Data
A catalogue record for this book is available from the British Library.

ISBN 13: 978-1-59693-092-6

Cover design by Yekaterina Ratner

685 Canton Street
Norwood, MA 02062

10 9 8 7 6 5 4 3 2 1

To I. R. Goodman

Contents

Preface **xxiii**

Acknowledgments **xxv**

Chapter 1 Introduction to the Book 1
- 1.1 What Is the Purpose of This Book? 1
- 1.2 Major Challenges in Information Fusion 7
- 1.3 Why Random Sets—or FISST? 8
  - 1.3.1 Why Isn't Multitarget Filtering Straightforward? 9
  - 1.3.2 Beyond Heuristics 10
  - 1.3.3 How Do Single-Target and Multitarget Statistics Differ? 11
  - 1.3.4 How Do Conventional and Ambiguous Data Differ? 11
  - 1.3.5 What Is Formal Bayes Modeling? 13
  - 1.3.6 How Is Ambiguous Information Modeled? 13
  - 1.3.7 What Is Multisource-Multitarget Formal Modeling? 14
- 1.4 Random Sets in Information Fusion 15
  - 1.4.1 Statistics of Multiobject Systems 15
  - 1.4.2 Statistics of Expert Systems 16
  - 1.4.3 Finite Set Statistics 17
- 1.5 Organization of the Book 17
  - 1.5.1 Part I: Unified Single-Target Multisource Integration 17
  - 1.5.2 Part II: Unified Multitarget-Multisource Integration 20
  - 1.5.3 Part III: Approximate Multitarget Filtering 21
  - 1.5.4 Appendixes 22

I Unified Single-Target Multisource Integration 23

Chapter 2 Single-Target Filtering 25
2.1 Introduction to the Chapter 25
2.1.1 Summary of Major Lessons Learned 26
2.1.2 Organization of the Chapter 27
2.2 The Kalman Filter 27
2.2.1 Kalman Filter Initialization 28
2.2.2 Kalman Filter Predictor 28
2.2.3 Kalman Filter Corrector 29
2.2.4 Derivation of the Kalman Filter 30
2.2.5 Measurement Fusion Using the Kalman Filter 32
2.2.6 Constant-Gain Kalman Filters 32
2.3 Bayes Formulation of the Kalman Filter 33
2.3.1 Some Mathematical Preliminaries 34
2.3.2 Bayes Formulation of the KF: Predictor 35
2.3.3 Bayes Formulation of the KF: Corrector 37
2.3.4 Bayes Formulation of the KF: Estimation 40
2.4 The Single-Target Bayes Filter 42
2.4.1 Single-Target Bayes Filter: An Illustration 43
2.4.2 Relationship Between the Bayes and Kalman Filters 45
2.4.3 Single-Target Bayes Filter: Modeling 51
2.4.4 Single-Target Bayes Filter: Formal Bayes Modeling 56
2.4.5 Single-Target Bayes Filter: Initialization 61
2.4.6 Single-Target Bayes Filter: Predictor 61
2.4.7 Single-Target Bayes Filter: Corrector 62
2.4.8 Single-Target Bayes Filter: State Estimation 63
2.4.9 Single-Target Bayes Filter: Error Estimation 64
2.4.10 Single-Target Bayes Filter: Data Fusion 67
2.4.11 Single-Target Bayes Filter: Computation 68
2.5 Single-Target Bayes Filter: Implementation 70
2.5.1 Taylor Series Approximation: The EKF 71
2.5.2 Gaussian-Mixture Approximation 72
2.5.3 Sequential Monte Carlo Approximation 79
2.6 Chapter Exercises 87

Chapter 3 General Data Modeling 89
3.1 Introduction to the Chapter 89
3.1.1 Summary of Major Lessons Learned 91

3.1.2 Organization of the Chapter 91
3.2 Issues in Modeling Uncertainty 92
3.3 Issues in Modeling Uncertainty in Data 94
3.4 Examples 97
3.4.1 Random, Slightly Imprecise Measurements 97
3.4.2 Imprecise, Slightly Random Measurements 101
3.4.3 Nonrandom Vague Measurements 102
3.4.4 Nonrandom Uncertain Measurements 103
3.4.5 Ambiguity Versus Randomness 106
3.5 The Core Bayesian Approach 109
3.5.1 Formal Bayes Modeling in General 109
3.5.2 The Bayes Filter in General 110
3.5.3 Bayes Combination Operators 111
3.5.4 Bayes-Invariant Measurement Conversion 113
3.6 Formal Modeling of Generalized Data 114
3.7 Chapter Exercise 117

Chapter 4 Random Set Uncertainty Representations 119
4.1 Introduction to the Chapter 119
4.1.1 Summary of Major Lessons Learned 119
4.1.2 Organization of the Chapter 120
4.2 Universes, Events, and the Logic of Events 120
4.3 Fuzzy Set Theory 121
4.3.1 Fuzzy Logics 122
4.3.2 Random Set Representation of Fuzzy Events 123
4.3.3 Finite-Level Fuzzy Sets 126
4.3.4 Copula Fuzzy Logics 129
4.3.5 General Random Set Representations of Fuzzy Sets 131
4.4 Generalized Fuzzy Set Theory 133
4.4.1 Random Set Representation of Generalized Fuzzy Events 134
4.5 Dempster-Shafer Theory 134
4.5.1 Dempster's Combination 136
4.5.2 "Zadeh's Paradox" and Its Misinterpretation 138
4.5.3 Converting b.m.a.s to Probability Distributions 141
4.5.4 Random Set Representation of Uncertain Events 143
4.6 Fuzzy Dempster-Shafer Theory 144
4.6.1 Random Set Representation of Fuzzy DS Evidence 145

4.7	Inference Rules	147
4.7.1	What Are Rules?	147
4.7.2	Combining Rules Using Conditional Event Algebra	148
4.7.3	Random Set Representation of First-Order Rules	150
4.7.4	Random Set Representation of Composite Rules	151
4.7.5	Random Set Representation of Second-Order Rules	152
4.8	Is Bayes Subsumed by Other Theories?	152
4.9	Chapter Exercises	154
Chapter 5	UGA Measurements	157
5.1	Introduction to the Chapter	157
5.1.1	Notation	158
5.1.2	Summary of Major Lessons Learned	159
5.1.3	Organization of the Chapter	161
5.2	What Is a UGA Measurement?	162
5.2.1	Modeling UGA Measurements	162
5.2.2	Modeling the Generation of UGA Measurements	164
5.3	Likelihoods for UGA Measurements	164
5.3.1	Special Case: Θ Is Statistical	165
5.3.2	Special Case: Θ Is Fuzzy	166
5.3.3	Special Case: Θ Is Generalized Fuzzy	169
5.3.4	Special Case: Θ Is Discrete/Dempster-Shafer	171
5.3.5	Special Case: Θ Is Fuzzy Dempster-Shafer	173
5.3.6	Special Case: Θ Is a First-Order Fuzzy Rule	174
5.3.7	Special Case: Θ Is a Composite Fuzzy Rule	179
5.3.8	Special Case: Θ Is a Second-Order Fuzzy Rule	180
5.4	Bayes Unification of UGA Fusion	181
5.4.1	Bayes Unification of UGA Fusion Using Normalized and Unnormalized Dempster's Combinations	185
5.4.2	Bayes Unification of UGA Fusion Using Normalized and Unnormalized Fuzzy Dempster's Combinations	186
5.4.3	Bayes Unification of UGA Fusion Using Copula Fuzzy Conjunctions	186
5.4.4	Bayes Unification of UGA Rule-Firing	187
5.4.5	If $\mathfrak{Z}_0$ Is Finite, Then Generalized Likelihoods Are Strict Likelihoods	188

5.4.6 Bayes-Invariant Conversions Between UGA Measurements 189
5.5 Modeling Other Kinds of Uncertainty 194
5.5.1 Modeling Unknown Statistical Dependencies 195
5.5.2 Modeling Unknown Target Types 196
5.6 The Kalman Evidential Filter (KEF) 199
5.6.1 Definitions 204
5.6.2 KEF Predictor 205
5.6.3 KEF Corrector (Fuzzy DS Measurements) 205
5.6.4 KEF Corrector (Conventional Measurements) 207
5.6.5 KEF State Estimation 208
5.6.6 KEF Compared to Gaussian-Mixture and Kalman Filters 208
5.7 Chapter Exercises 209

Chapter 6 AGA Measurements 211
6.1 Introduction to the Chapter 211
6.1.1 Summary of Major Lessons Learned 212
6.1.2 Organization of the Chapter 213
6.2 AGA Measurements Defined 213
6.3 Likelihoods for AGA Measurements 214
6.3.1 Special Case: Θ and $\Sigma_{\mathbf{x}}$ Are Fuzzy 215
6.3.2 Special Case: Θ and $\Sigma_{\mathbf{x}}$ Are Generalized Fuzzy 219
6.3.3 Special Case: Θ and $\Sigma_{\mathbf{x}}$ Are Dempster-Shafer 219
6.3.4 Special Case: Θ and $\Sigma_{\mathbf{x}}$ Are Fuzzy DS 220
6.4 Filtering with Fuzzy AGA Measurements 221
6.5 Example: Filtering with Poor Data 222
6.5.1 A Robust-Bayes Classifier 223
6.5.2 Simulation 1: More Imprecise, More Random 225
6.5.3 Simulation 2: Less Imprecise, Less Random 225
6.5.4 Interpretation of the Results 232
6.6 Unmodeled Target Types 232
6.7 Example: Target ID Using Link INT Data 238
6.7.1 Robust-Bayes Classifier 240
6.7.2 "Pseudodata" Simulation Results 243
6.7.3 "LONEWOLF-98" Simulation Results 243
6.8 Example: Unmodeled Target Types 244
6.9 Chapter Exercises 245

Chapter 7 AGU Measurements 249
7.1 Introduction to the Chapter 249
7.1.1 Summary of Major Lessons Learned 250
7.1.2 Why Not Robust Statistics? 250
7.1.3 Organization of the Chapter 251
7.2 Random Set Models of UGA Measurements 252
7.2.1 Random Error Bars 252
7.2.2 Random Error Bars: Joint Likelihoods 252
7.3 Likelihoods for AGU Measurements 254
7.4 Fuzzy Models of AGU Measurements 255
7.5 Robust ATR Using SAR Data 260
7.5.1 Summary of Methodology 264
7.5.2 Experimental Ground Rules 266
7.5.3 Summary of Experimental Results 268

Chapter 8 Generalized State-Estimates 271
8.1 Introduction to the Chapter 271
8.1.1 Summary of Major Lessons Learned 273
8.1.2 Organization of the Chapter 274
8.2 What Is a Generalized State-Estimate? 274
8.3 What Is a UGA DS State-Estimate? 275
8.4 Posterior Distributions and State-Estimates 277
8.4.1 The Likelihood of a DS State-Estimate 278
8.4.2 Posterior Distribution Conditioned on a DS State-Estimate 278
8.4.3 Posterior Distributions and Pignistic Probability 279
8.5 Unification of State-Estimate Fusion Using Modified Dempster's Combination 280
8.6 Bayes-Invariant Transformation 280
8.7 Extension to Fuzzy DS State-Estimates 281
8.8 Chapter Exercises 285

Chapter 9 Finite-Set Measurements 287
9.1 Introduction to the Chapter 287
9.1.1 Summary of Major Lessons Learned 287
9.1.2 Organization of the Chapter 288
9.2 Examples of Finite-Set Measurements 288
9.2.1 Ground-to-Air Radar Detection Measurements 288
9.2.2 Air-to-Ground Doppler Detection Measurements 291

9.2.3 Extended-Target Detection Measurements 292
9.2.4 Features Extracted from Images 292
9.2.5 Human-Mediated Features 292
9.2.6 General Finite-Set Measurements 293
9.3 Modeling Finite-Set Measurements? 293
9.3.1 Formal Modeling of Finite-Set Measurements 293
9.3.2 Multiobject Integrals 297
9.3.3 Finite-Set Measurement Models 299
9.3.4 True Likelihoods for Finite-Set Measurements 302
9.3.5 Constructive Likelihood Functions 302
9.4 Chapter Exercises 303

II Unified Multitarget-Multisource Integration 305

Chapter 10 Conventional Multitarget Filtering 307
10.1 Introduction to the Chapter 307
10.1.1 Summary of Major Lessons Learned 308
10.1.2 Organization of the Chapter 311
10.2 Standard Multitarget Models 311
10.2.1 Standard Multitarget Measurement Model 311
10.2.2 Standard Multitarget Motion Model 313
10.3 Measurement-to-Track Association 315
10.3.1 Distance Between Measurements and Tracks 315
10.4 Single-Hypothesis Correlation (SHC) 319
10.4.1 SHC: No Missed Detections, No False Alarms 319
10.4.2 SHC: Missed Detections and False Alarms 320
10.5 Multihypothesis Correlation (MHC) 321
10.5.1 Elements of MHC 323
10.5.2 MHC: No Missed Detections or False Alarms 326
10.5.3 MHC: False Alarms, No Missed Detections 329
10.5.4 MHC: Missed Detections and False Alarms 332
10.6 Composite-Hypothesis Correlation (CHC) 335
10.6.1 Elements of CHC 335
10.6.2 CHC: No Missed Detections or False Alarms 337
10.6.3 CHC: Probabilistic Data Association (PDA) 337
10.6.4 CHC: Missed Detections, False Alarms 338
10.7 Conventional Filtering: Limitations 338
10.7.1 Real-Time Performance 338

10.7.2 Is a Hypothesis Actually a State Variable? 340
10.8 MHC with Fuzzy DS Measurements 341

Chapter 11 Multitarget Calculus 343
11.1 Introduction to the Chapter 343
11.1.1 Transform Methods in Conventional Statistics 344
11.1.2 Transform Methods in Multitarget Statistics 345
11.1.3 Summary of Major Lessons Learned 346
11.1.4 Organization of the Chapter 348
11.2 Random Finite Sets 348
11.3 Fundamental Statistical Descriptors 356
11.3.1 Multitarget Calculus—Why? 357
11.3.2 Belief-Mass Functions 359
11.3.3 Multiobject Density Functions and Set Integrals 360
11.3.4 Important Multiobject Probability Distributions 364
11.3.5 Probability-Generating Functionals (p.g.fl.s) 370
11.4 Functional Derivatives and Set Derivatives 375
11.4.1 Functional Derivatives 375
11.4.2 Set Derivatives 380
11.5 Key Multiobject-Calculus Formulas 383
11.5.1 Fundamental Theorem of Multiobject Calculus 384
11.5.2 Radon-Nikodým Theorems 385
11.5.3 Fundamental Convolution Formula 385
11.6 Basic Differentiation Rules 386
11.7 Chapter Exercises 394

Chapter 12 Multitarget Likelihood Functions 399
12.1 Introduction to the Chapter 399
12.1.1 Summary of Major Lessons Learned 401
12.1.2 Organization of the Chapter 402
12.2 Multitarget State and Measurement Spaces 403
12.2.1 Multitarget State Spaces 403
12.2.2 Multisensor State Spaces 406
12.2.3 Single-Sensor, Multitarget Measurement Spaces 407
12.2.4 Multisensor-Multitarget Measurement Spaces 408
12.3 The Standard Measurement Model 408
12.3.1 Measurement Equation for the Standard Model 411
12.3.2 Case I: No Target Is Present 412
12.3.3 Case II: One Target Is Present 414

12.3.4 Case III: No Missed Detections or False Alarms 416
12.3.5 Case IV: Missed Detections, No False Alarms 418
12.3.6 Case V: Missed Detections and False Alarms 420
12.3.7 p.g.fl.s for the Standard Measurement Model 421
12.4 Relationship with MHC 422
12.5 State-Dependent False Alarms 424
12.5.1 p.g.fl. for State-Dependent False Alarms 426
12.6 Transmission Drop-Outs 426
12.6.1 p.g.fl. for Transmission Drop-Outs 427
12.7 Extended Targets 427
12.7.1 Single Extended Target 428
12.7.2 Multiple Extended Targets 430
12.7.3 Poisson Approximation 431
12.8 Unresolved Targets 432
12.8.1 Point Target Clusters 434
12.8.2 Single-Cluster Likelihoods 435
12.8.3 Multicluster Likelihoods 442
12.8.4 Continuity of Multicluster Likelihoods 444
12.9 Multisource Measurement Models 445
12.9.1 Conventional Measurements 445
12.9.2 Generalized Measurements 447
12.10 A Model for Bearing-Only Measurements 448
12.10.1 Multitarget Measurement Model 450
12.10.2 Belief-Mass Function 451
12.10.3 Multitarget Likelihood Function 452
12.11 A Model for Data-Cluster Extraction 452
12.11.1 Finite-Mixture Models 453
12.11.2 A Likelihood for Finite-Mixture Modeling 456
12.11.3 Extraction of Soft Data Classes 457
12.12 Chapter Exercises 458

Chapter 13 Multitarget Markov Densities 461
13.1 Introduction to the Chapter 461
13.1.1 Summary of Major Lessons Learned 465
13.1.2 Organization of the Chapter 466
13.2 "Standard" Multitarget Motion Model 466
13.2.1 Case I: At Most One Target Is Present 469
13.2.2 Case II: No Target Death or Birth 470

13.2.3 Case III: Target Death, No Birth 471
13.2.4 Case IV: Target Death and Birth 471
13.2.5 Case V: Target Death and Birth with Spawning 472
13.2.6 p.g.fl.s for the Standard Motion Model 473
13.3 Extended Targets 474
13.4 Unresolved Targets 475
13.4.1 Intuitive Dynamic Behavior of Point Clusters 475
13.4.2 Markov Densities for Single Point Clusters 476
13.4.3 Markov Densities for Multiple Point Clusters 477
13.5 Coordinated Multitarget Motion 478
13.5.1 Simple Virtual Leader-Follower 478
13.5.2 General Virtual Leader-Follower 481
13.6 Chapter Exercises 482

Chapter 14 The Multitarget Bayes Filter 483
14.1 Introduction to the Chapter 483
14.1.1 Summary of Major Lessons Learned 484
14.1.2 Organization of the Chapter 486
14.2 Multitarget Bayes Filter: Initialization 486
14.2.1 Initialization: Multitarget Poisson Process 486
14.2.2 Initialization: Target Number Known 487
14.3 Multitarget Bayes Filter: Predictor 487
14.3.1 Predictor: No Target Birth or Death 489
14.4 Multitarget Bayes Filter: Corrector 490
14.4.1 Conventional Measurements 490
14.4.2 Generalized Measurements 493
14.4.3 Unified Multitarget-Multisource Integration 493
14.5 Multitarget Bayes Filter: State Estimation 494
14.5.1 The Failure of the Classical State Estimators 494
14.5.2 Marginal Multitarget (MaM) Estimator 497
14.5.3 Joint Multitarget (JoM) Estimator 498
14.5.4 JoM and MaM Estimators Compared 501
14.5.5 Computational Issues 504
14.5.6 State Estimation and Track Labeling 505
14.6 Multitarget Bayes Filter: Error Estimation 509
14.6.1 Target Number RMS Deviation 509
14.6.2 Track Covariances 509
14.6.3 Global Mean Deviation 510

14.6.4 Information Measures of Multitarget Dispersion 512
14.7 The JoTT Filter 514
14.7.1 JoTT Filter: Models 516
14.7.2 JoTT Filter: Initialization 518
14.7.3 JoTT Filter: Predictor 519
14.7.4 JoTT Filter: Corrector 520
14.7.5 JoTT Filter: Estimation 520
14.7.6 JoTT Filter: Error Estimation 523
14.7.7 SMC Implementation of JoTT Filter 523
14.8 The p.g.fl. Multitarget Bayes Filter 528
14.8.1 The p.g.fl. Multitarget Predictor 528
14.8.2 The p.g.fl. Multitarget Corrector 530
14.9 Target Prioritization 531
14.9.1 Tactical Importance Functions (TIFs) 533
14.9.2 The p.g.fl. for a TIF 533
14.9.3 The Multitarget Posterior for a TIF 535
14.10 Chapter Exercises 537

III Approximate Multitarget Filtering 539

Chapter 15 Multitarget Particle Approximation 541
15.1 Introduction to the Chapter 541
15.1.1 Summary of Major Lessons Learned 542
15.1.2 Organization of the Chapter 543
15.2 The Multitarget Filter: Computation 543
15.2.1 Fixed-Grid Approximation 544
15.2.2 SMC Approximation 545
15.2.3 When Is the Multitarget Filter Appropriate? 546
15.2.4 Implementations of the Multitarget Filter 547
15.3 Multitarget Particle Systems 551
15.4 M-SMC Filter Initialization 554
15.4.1 Target Number is Known 554
15.4.2 Null Multitarget Prior 555
15.4.3 Poisson Multitarget Prior 555
15.5 M-SMC Filter Predictor 556
15.5.1 Persisting and Disappearing Targets 557
15.5.2 Appearing Targets 558
15.6 M-SMC Filter Corrector 560

15.7 M-SMC Filter State and Error Estimation 561
15.7.1 PHD-Based State and Error Estimation 561
15.7.2 Global Mean Deviation 562
15.7.3 Track Labeling for the Multitarget SMC Filter 563

Chapter 16 Multitarget-Moment Approximation 565
16.1 Introduction to the Chapter 565
16.1.1 Single-Target Moment-Statistic Filters 566
16.1.2 First-Order Multitarget-Moment Filtering 568
16.1.3 Second-Order Multitarget-Moment Filtering 572
16.1.4 Summary of Major Lessons Learned 574
16.1.5 Organization of the Chapter 575
16.2 The Probability Hypothesis Density (PHD) 576
16.2.1 First-Order Multitarget Moments 576
16.2.2 PHD as a Continuous Fuzzy Membership Function 579
16.2.3 PHDs and Multitarget Calculus 580
16.2.4 Examples of PHDs 583
16.2.5 Higher-Order Multitarget Moments 586
16.3 The PHD Filter 587
16.3.1 PHD Filter Initialization 587
16.3.2 PHD Filter Predictor 587
16.3.3 PHD Filter Corrector 590
16.3.4 PHD Filter State and Error Estimation 595
16.3.5 Target ID and the PHD Filter 599
16.4 Physical Interpretation of PHD Filter 599
16.4.1 Physical Interpretation of PHD Predictor 600
16.4.2 Physical Interpretation of PHD Corrector 603
16.5 Implementing the PHD Filter 609
16.5.1 Survey of PHD Filter Implementations 610
16.5.2 SMC-PHD Approximation 615
16.5.3 GM-PHD Approximation 623
16.6 Limitations of the PHD Filter 631
16.7 The Cardinalized PHD (CPHD) Filter 632
16.7.1 CPHD Filter Initialization 633
16.7.2 CPHD Filter Predictor 634
16.7.3 CPHD Filter Single-Sensor Corrector 636
16.7.4 CPHD Filter State and Error Estimation 639
16.7.5 Computational Complexity of the CPHD Filter 640

16.7.6 CPHD and JoTT Filters Compared 641
16.8 Physical Interpretation of CPHD Filter 642
16.9 Implementing the CPHD Filter 642
16.9.1 Survey of CPHD Filter Implementations 643
16.9.2 Particle Approximation (SMC-CPHD) 644
16.9.3 Gaussian-Mixture Approximation (GM-CPHD) 646
16.10 Deriving the PHD and CPHD Filters 649
16.10.1 Derivation of PHD and CPHD Predictors 650
16.10.2 Derivation of PHD and CPHD Correctors 651
16.11 Partial Second-Order Filters? 652
16.12 Chapter Exercise 653

Chapter 17 Multi-Bernoulli Approximation 655
17.1 Introduction to the Chapter 655
17.1.1 p.g.fl.-Based Multitarget Approximation 655
17.1.2 Why Multitarget Multi-Bernoulli Processes? 657
17.1.3 The Multitarget Multi-Bernoulli Filter 657
17.1.4 The Para-Gaussian Filter 658
17.1.5 Summary of Major Lessons Learned 659
17.1.6 Organization of the Chapter 660
17.2 Multitarget Multi-Bernoulli Filter 660
17.2.1 MeMBer Filter Initialization 661
17.2.2 MeMBer Filter Predictor 661
17.2.3 MeMBer Filter Corrector 662
17.2.4 MeMBer Filter Pruning and Merging 665
17.2.5 MeMBer Filter State and Error Estimation 666
17.2.6 Relationship with the Moreland-Challa Filter 667
17.3 Para-Gaussian Filter 668
17.3.1 Para-Gaussian Filter Initialization 669
17.3.2 Para-Gaussian Filter Predictor 669
17.3.3 Para-Gaussian Filter Corrector 671
17.3.4 Para-Gaussian Filter Pruning and Merging 673
17.3.5 Para-Gaussian Filter State and Error Estimation 675
17.4 MeMBer Filter Derivation 675
17.4.1 Derivation of the MeMBer Filter Predictor 675
17.4.2 Derivation of the MeMBer Filter Corrector 677
17.5 Chapter Exercise 682

Appendix A Glossary of Notation 683

A.1 Transparent Notational System 683
A.2 General Mathematics 684
A.3 Set Theory 685
A.4 Fuzzy Logic and Dempster-Shafer Theory 686
A.5 Probability and Statistics 687
A.6 Random Sets 689
A.7 Multitarget Calculus 690
A.8 Finite-Set Statistics 691
A.9 Generalized Measurements 692

Appendix B Dirac Delta Functions 693

Appendix C Gradient Derivatives 695
C.1 Relationship with Partial Derivatives 696
C.2 Multidimensional Taylor Series 696
C.3 Multidimensional Extrema 696

Appendix D Fundamental Gaussian Identity 699

Appendix E Finite Point Processes 705
E.1 Mathematical Representations of Multiplicity 705
E.2 Random Point Processes 707
E.3 Point Processes Versus Random Finite Sets 708

Appendix F FISST and Probability Theory 711
F.1 Multiobject Probability Theory 711
F.2 Belief-Mass Functions Versus Probability Measures 713
F.3 Set Integrals Versus Measure Theoretic Integrals 714
F.4 Set Derivatives Versus Radon-Nikodým Derivatives 715

Appendix G Mathematical Proofs 717
G.1 Likelihoods for First-Order Fuzzy Rules 717
G.2 Likelihoods for Composite Rules 718
G.3 Likelihoods for Second-Order Fuzzy Rules 720
G.4 Unification of DS Combinations 721
G.5 Unification of Rule-Firing 722
G.6 Generalized Likelihoods: $\mathfrak{Z}_0$ Is Finite 723
G.7 NOTA for Fuzzy DS Measurements 724
G.8 KEF Predictor 726

G.9 KEF Corrector (Fuzzy DS Measurements) 729
G.10 Likelihoods for AGA Fuzzy Measurements 732
G.11 Likelihoods for AGA Generalized Fuzzy Measurements 733
G.12 Likelihoods for AGA Fuzzy DS Measurements 734
G.13 Interval Argsup Formula 735
G.14 Consonance of the Random State Set $\hat{\Gamma}_{\mathbf{z}}$ 736
G.15 Sufficient Statistics and Modified Combination 737
G.16 Transformation Invariance 738
G.17 MHT Hypothesis Probabilities 739
G.18 Likelihood for Standard Measurement Model 742
G.19 p.g.fl. for Standard Measurement Model 745
G.20 Multisensor Multitarget Likelihoods 747
G.21 Continuity of Likelihoods for Unresolved Targets 749
G.22 Association for Fuzzy Dempster-Shafer 751
G.23 JoTT Filter Predictor 753
G.24 JoTT Filter Corrector 755
G.25 p.g.fl. Form of the Multitarget Corrector 757
G.26 Induced Particle Approximation of PHD 758
G.27 PHD Counting Property 760
G.28 GM-PHD Filter Predictor 761
G.29 GM-PHD Filter Corrector 763
G.30 Exact PHD Corrector 765
G.31 GM-CPHD Filter Predictor 767
G.32 GM-CPHD Filter Corrector 768
G.33 MeMBer Filter Target Number 771
G.34 Para-Gaussian Filter Predictor 773
G.35 Para-Gaussian Filter Corrector 774

Appendix H Solutions to Exercises 777

References **821**

About the Author **837**

Index **839**

Preface

A number of excellent books on standard information fusion techniques are already in existence, the most well known of which are the following:

- R. Antony, *Principles of Data Fusion Automation* [7];
- Y. Bar-Shalom and X.-R. Li, *Estimation and Tracking: Principles, Techniques, and Software*, [11];
- S. Blackman and S. Popoli, *Design and Analysis of Modern Tracking Systems* [18];
- D. Hall, *Mathematical Techniques in Multisensor Data Fusion* [82];
- D. Hall and J. Llinas (eds.), *Handbook of Multisensor Data Fusion* [83];
- E. Waltz and J. Llinas, *Multisensor Data Fusion* [245].

The purpose of this book is not to reprise what has already been adequately addressed. Rather, it is a textbook style introduction to a fundamentally new, seamlessly unified, and fully statistical approach to information fusion.

The emergence of unconventional defense and security challenges has greatly increased the need for fusing and exploiting unconventional and highly disparate forms of information. Conventional data is supplied by sensors and can often be modeled and characterized statistically. Nontraditional information tends to involve target identity and often requires human mediation. Typical examples include attributes extracted from images by human operators; features extracted from signatures by digital signal processors or by human analysts; textual or verbal natural language statements; and inference rules drawn from knowledge bases.

Numerous expert systems approaches have been proposed to address such problems. The great majority of these approaches bear no obvious relationship with the most mature subdiscipline of information fusion: single-target and multitarget detection and tracking theory. As a consequence it has often been unclear how to develop systematic and integrated solutions to many real-world challenges.

This book is the result of a decade long effort on the author's part to address such challenges. The fundamental methodology I propose is conceptually parsimonious. It consists of a systematic and novel utilization of *formal Bayes modeling* and the *recursive Bayes filter*. It provides techniques for modeling uncertainties due to randomness or ignorance, propagating these uncertainties through time, and extracting estimates of desired quantities (as well as measures of reliability of those estimates) that optimally reflect the influence of inherent system uncertainties.

The process just described is well known. What makes my version of it unique is my systematic application of it to multitarget information, and to unconventional information in both single-target and multitarget problems. One of its consequences is a seamlessly unified statistical approach to multitarget-multisource integration. This seamless unification includes the following:

- A unified theory of measurements, both single-target and multitarget;
- Unified mathematical representation of uncertainty, including randomness, imprecision, vagueness, ambiguity, and contingency;
- A unified single-target and multitarget modeling methodology based on generalized likelihood functions;
- A unification of much of expert systems theory, including fuzzy, Bayes, Dempster-Shafer, and rule-based techniques;
- Unified and optimal single-target and multitarget detection and estimation;
- Unified and optimal fusion of disparate information;
- A systematic multitarget calculus for devising new approximations.

It is my hope that readers will find the book informative, useful, thought provoking, occasionally provocative, and—perhaps—even a bit exciting.

Acknowledgments

The basic research reported in this book was supported by Lockheed Martin internal research and development funding in 1993 and 1994. Since then it has been supported by the U.S. Army Research Office (ARO) under contracts DAAH04-94-C-0011 and DAAG55-98-C-0039 and by the U.S. Air Force Office of Scientific Research (AFOSR) under contract F49620-01-C-0031. The content does not necessarily reflect the position or policy of either the U.S. government or Lockheed Martin. No official endorsement should be inferred in either case.

I thank the following individuals for their help, encouragement, and support during the past decade: Dr. William Sander of the Army Research Office (retired); Stanton Musick of the Air Force Research Laboratory/SNAT (retired); Dr. Mark Alford of the Air Force Research Laboratory/IFEA; Craig Poling and Charles Mills of Lockheed Martin; Drs. Adel El-Fallah, Alex Zatezalo, and Raman Mehra of Scientific Systems Co., Inc.; Dr. Ravi Ravichandran, now of BAE Systems; Dr. Ivan Kadar of Interlink Systems Sciences; Professor David Hall of Pennsylvania State University; and Professor James Llinas of the State University of New York at Buffalo.

I gratefully acknowledge the ongoing original research and thought-provoking correspondence of Professor Ba-Ngu Vo of the University of Melbourne (Australia). The research conducted by Professor Vo and his students has resulted in basic new computational and theoretical techniques. His communications have helped clarify many issues.

I thank Dr. Adel El-Fallah of Scientific Systems Co. for his assistance in preparing the camera ready copy of the manuscript.

Finally, I acknowledge my profound debt to the groundbreaking research of Dr. I.R. Goodman of the U.S. Navy SPAWAR Systems Center. It was Dr.

Goodman who, beginning in late 1993, exposed me to the potentially revolutionary implications of random set techniques in information fusion.

The manuscript for this book was produced using Version 4.0 of MacKichan Software's *Scientific WorkPlace*. The task of writing the book would have been vastly more difficult without it. Figures were prepared using *Scientific WorkPlace* and *Microsoft PowerPoint*. Figures 1.1, 3.3, and 12.2 include clip art downloaded from the *Microsoft PowerPoint* online library of clip art.

Chapter 1

Introduction to the Book

1.1 WHAT IS THE PURPOSE OF THIS BOOK?

The subject of this book is *finite set statistics* (FISST) [70, 134, 144, 145, 148], a recently developed theory that *unifies much of information fusion under a single probabilistic—in fact, Bayesian—paradigm*. It does so by directly generalizing the "statistics 101" formalism that most signal processing practitioners learn as undergraduates. Since its introduction in 1994, FISST has addressed an increasingly comprehensive expanse of information fusion, including multitarget-multisource integration (MSI), also known as level 1 fusion; expert systems theory; sensor management for level 1 fusion, including management of dispersed mobile sensors; group target detection, tracking, and classification; robust automatic target recognition; and scientific performance evaluation. This research program is portrayed schematically in Figure 1.1.

FISST has attracted much international interest in a relatively short time. FISST-based research efforts are in progress in at least a dozen nations. FISST-based approaches and algorithms are being or have been investigated under more than a dozen basic and applied R&D contracts from U.S. Department of Defense agencies such as the Army Research Office (ARO), the Air Force Office of Scientific Research (AFOSR), the Navy SPAWAR Systems Center, the Missile Defense Agency (MDA), the Army Missile Research and Defense Command (AMRDEC), the Defense Advanced Research Projects Administration (DARPA), and four different sites of the Air Force Research Laboratory (AFRL).

The following constitute the most important topics of this book.

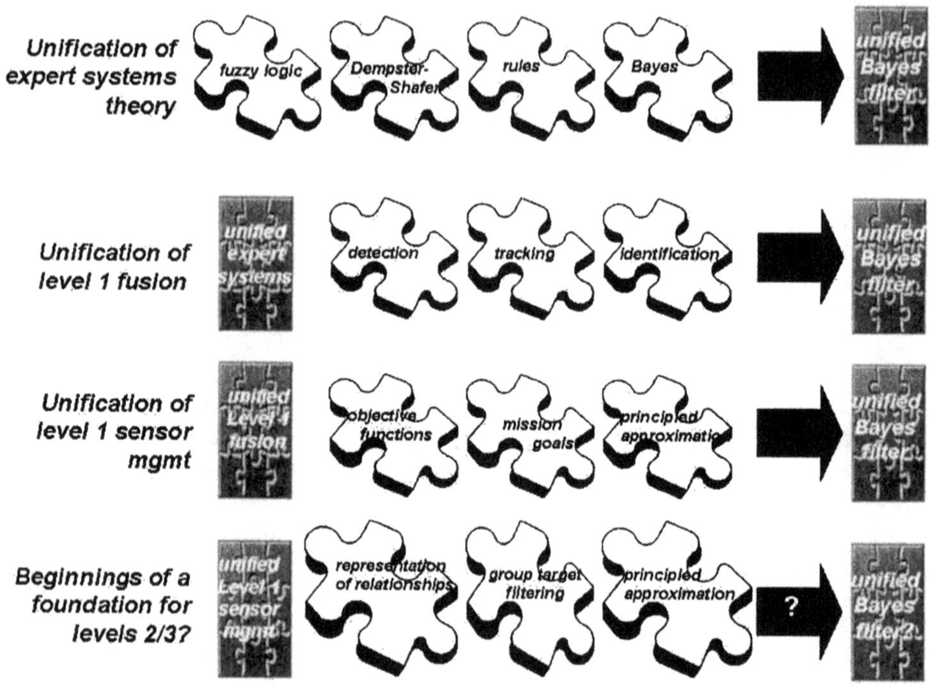

Figure 1.1 The FISST research program. Unification of expert systems, and of generalized data, sets the stage for a unification of multisource integration. This sets the stage for a unification of sensor and platform management. This may, ultimately, set the stage for a unification of all of the data fusion levels.

I. Single-Target Multisource Integration (Single-Target MSI):

- Systematic modeling and processing of unconventional evidence—rules, features, attributes, natural language statements—essentially in the same way as radar returns;
- Systematic modeling of evidential conflict—including a Bayesian resolution of Zadeh's paradox;
- Systematic employment of expert systems techniques—Bayes, fuzzy logic, Dempster-Shafer, rule-based—in a probabilistic and optimal manner;
- Systematic common ground for these techniques based on a common mathematical framework;
- A systematic methodology for converting data from one uncertainty representation to another without loss of relevant information;
- A systematic methodology for implementing these models and techniques despite their inherent nonlinearity;
- New single-target filter implementation approaches:
    - Kalman evidential (KEF), Section 5.6;
    - Joint target detection & tracking (JoTT), Section 14.7.

II. Multitarget-Multisource Integration (Multitarget MSI):

- Systematic incorporation of nontraditional data into traditional multitarget techniques such as multihypothesis correlation (MHC);
- Explicit incorporation of target prioritization and tactical significance into models;
- Comprehensive modeling of multisource-multitarget information sources, including sensor fields of view, missed detections, false alarms, and clutter;
- Comprehensive modeling of multitarget dynamicism, including target appearance, disappearance, spawning, and coordinated motion;
- New multitarget filter implementation approaches:
    - SMC probability hypothesis density (SMC-PHD), Section 16.5.2;
    - Gaussian-mixture PHD (GM-PHD), Section 16.5.3;
    - SMC cardinalized PHD (SMC-CPHD), Section 16.9.2;

- Gaussian-mixture CPHD (GM-CPHD), Section 16.9.3;
- Para-Gaussian, Section 17.3.

III. Multitarget Statistics:

- Identification of the three fundamental statistical descriptors of a multitarget system, and what they are good for:
  - Multitarget probability densities, Section 11.3.3;
  - Belief-mass functions, Section 11.3.2;
  - Probability-generating functionals (p.g.fl.s), Section 11.3.5.
- A multitarget integral and differential calculus, and its systematic employment to develop fundamental new techniques:
  - Multitarget set integrals, Section 11.3.3;
  - Multitarget set derivatives, Section 11.4.2;
  - Multitarget functional derivatives, Section 11.4.1.

The purpose of this book is to make the basic concepts, techniques, and algorithms of FISST accessible to a broader segment of the signal processing community than has heretofore been the case. Until now, the primary entry points into FISST have been the book *Mathematics of Data Fusion*, coauthored with I.R. Goodman and H.T. Nguyen [70]; Chapter 14 of the *Handbook of Multisensor Data Fusion* [145]; the Lockheed Martin Technical Monograph *An Introduction to Multisource-Multitarget Statistics and Its Applications* [134]; and the invited *IEEE Aerospace and Electronics Magazine* paper, "'Statistics 101' for Multisensor, Multitarget Data Fusion" [148]. The first presumes working familiarity with measure theoretic probability theory and is not readily penetrable by most engineers or scientists. The second is a condensed version of the third, which is a 114-page broad brush overview, which, as such, was never intended to serve as an introduction to the subject. The fourth is a short, elementary, and even more broad brush tutorial.

This book will, it is hoped, fill the gap insofar as level 1 information fusion (also known as multisource integration) is concerned. Its intended audiences are the following: (1) bachelor's degree signal processing engineers and graduate students; (2) researchers, graduate students, advanced undergraduates, and practitioners in information fusion, expert system theory, and single-target and multitarget tracking; and (3) computer scientists, physicists, and mathematicians interested in these subjects.

No less than with many other more familiar subjects—Fourier theory, wavelet theory, the Kalman filter—FISST arises out of applied mathematics rather than computer science. Indeed, a central aspect of FISST is its emphasis on one of the most critical and difficult—as well as one of the more neglected—steps in information fusion algorithm development:

- *Careful specification of concrete real-world statistical models, on the basis of which practical implementation can proceed.*

Central to this emphasis is my employment of what, to most readers, will be an unfamiliar new statistical concept:

- The *random set*.

Nevertheless, I have made every effort to make the material accessible to its intended audiences. The emphasis of the book is on the derivation of concrete techniques and formulas that can be applied to concrete problems. Consequently, the introduction to every chapter includes a section detailing the most practically significant concepts, formulas, and results of that chapter.

Much material is devoted to illustrative problems of practical interest: single-target tracking using both conventional and Dempster-Shafer evidence (Section 5.6); robust target identification using human mediated intelligence (INT) datalink attributes (Section 6.7); target identification in the presence of unmodeled target types (Section 6.6); automatic target recognition (ATR) of motionless ground targets using synthetic aperture radar (SAR) (Section 7.5); and joint detection and tracking of single targets in arbitrary false alarm backgrounds (Section 14.7).

I make extensive use of graphical illustrations, illustrative simple examples, and exercises of varying degrees of difficulty (with solutions included in Appendix H). I have avoided unnecessarily cluttering the exposition with mathematical proofs or other extended mathematical derivations. These have been relegated to appendixes when it has not been possible or desirable to direct the more theoretically engaged reader to other publications.

Because I use a "transparent" system of notation (see Appendix A.1), the reader will usually be able to infer the meaning of mathematical symbols at a glance. I also employ a *crawl-walk-run* style of exposition. I begin with more familiar concepts and techniques and then build upon them to introduce more complex ones. Thus, the book is divided into three parts that address increasingly more complex aspects of information fusion.

Part I is devoted to *unified single-target, multisource integration*. It begins with the Kalman filter and progresses to the Bayesian formulation of the Kalman filter. From there, it takes up the single-target Bayes recursive filter and its most familiar practical implementations. This lays the groundwork for the remaining chapters of Part I.

FISST is, in part, a *comprehensive theory of measurements*. Thus Part I is primarily devoted to deriving measurement models and generalized likelihood functions for "nontraditional" measurements, such as attributes, features, natural language statements, and rules. Once such likelihoods have been specified, very disparate forms of information ("from rules to radar," so to speak) can be processed using the Bayes recursive filter.

Part I concludes the treatment of single-target problems with an introduction to detection type measurements and the concept of a *random finite measurement set*. This sets the stage for Part II.

Part II is devoted to *unified multitarget-multisource integration*. Once again, I begin with more familiar concepts and techniques and build upon them. I begin with a summary exposition of the basic concepts of conventional multitarget processing: "hard" and "soft" measurement to track association, single-target and multihypothesis tracking, and joint probabilistic data association (JPDA).

As already noted, FISST is in part a systematic theory of measurements. The first chapters of Part II extend this theory of measurements to multitarget situations. I show how to construct multisource-multitarget measurement models (and multitarget motion models). From these, I show how to construct the corresponding "true" multisource-multitarget likelihood functions and multitarget Markov transition densities. This provides the foundation for the primary subject of Part II, the multisource-multitarget recursive Bayes filter.

Part III is devoted to approaches for real-time implementation of the multitarget Bayes filter. Three approximation techniques are considered: multitarget sequential Monte Carlo, multitarget first moment, and multitarget multi-Bernoulli. Recent developments due to researchers throughout the world will be summarized.[1]

I assume that the reader is familiar with undergraduate level probability theory and the Kalman filter, though I review both. It would be helpful, but is not essential, for the reader to have some familiarity with multitarget tracking techniques such as data association and multihypothesis correlation. It would be helpful, though once again not essential, for the reader to have some familiarity with standard expert systems approaches such as Bayesian probability theory, fuzzy logic, and Dempster-Shafer theory.

1 The author has endeavored to describe the work of other researchers as accurately as possible. Any errors in these descriptions should be ascribed to me and not to those whose work is being summarized.

In the remainder of this chapter I ask, and answer, the following questions:

- What are the major challenges that have impeded a unified approach to information fusion?
- Why isn't multitarget filtering straightforward?
- How do multitarget and single-target statistics differ?
- How do conventional and "ambiguous" data differ?
- What is formal Bayes modeling?
- How is ambiguous data modeled?
- What is multisource-multitarget formal modeling?
- Why random sets, and why FISST in particular?
- What are the antecedents of random set theory in information fusion?
- What detailed topics will the three parts of this book cover?

1.2 MAJOR CHALLENGES IN INFORMATION FUSION

Three primary challenges have inhibited the maturation of general information fusion as a unified, systematic, scientifically founded discipline.

- The first is the *highly disparate and ambiguous forms that information can have*. Many kinds of data, such as that supplied by tracking radars, can be adequately described in statistical form. On the other hand, statistically uncharacterizable real-world variations make other kinds of data, such as synthetic aperture radar (SAR) images, difficult to model. It has been even more unclear how still other forms of data—natural-language statements, features extracted from signatures, rules drawn from knowledge bases—might be mathematically modeled, let alone statistically processed. Numerous expert systems approaches—fuzzy set theory, the Dempster-Shafer theory, rule-based inference, and so on—have been proposed to address such problems. A major goal of this book is to unify and integrate the most familiar expert systems fusion techniques (Dempster's combination, fuzzy conjunction, rule firing) within a Bayesian paradigm.

- The second obstacle is the fact that multisource-multitarget systems introduce a major complication. Such systems are comprised of *randomly varying numbers of randomly varying objects of various kinds*: randomly varying collections of targets, randomly varying collections of sensors and sensor carrying platforms, and randomly varying observation sets collected by those sensors. A rigorous mathematical foundation for stochastic multiobject problems—*point process theory* [36, 214, 220]—has been in existence for decades. However, this theory has traditionally been formulated with the requirements of pure mathematicians rather than engineers and scientists in mind. A second major goal of the book is to elaborate a more "practitioner friendly" version of point process theory that can be readily applied to multisource-multitarget problems.

- The third obstacle is the most crucial of all. The daunting combinatorial complexity of multisensor-multitarget systems (see Section 15.2) guarantees that no systematic unification of information fusion will be of practical interest unless it incorporates an equally *systematic calculus for devising principled algorithmic approximation strategies*. The third major goal of this book is to describe principled new approximation techniques potentially applicable to real-time problems.

1.3 WHY RANDOM SETS—OR FISST?

The point of FISST is not that multitarget problems can be formulated in terms of random sets or point processes. It is, rather, that random set techniques provide a systematic toolbox of explicit, rigorous, and general procedures that address many difficulties—those involving ambiguous evidence, unification, and computation, especially. A major purpose of this book is to describe this toolbox in enough detail to make it available to practitioners for real-world, real-time application.

The following points will be central to the discussion which follows, in this section and in the rest of the book:

- The fact that multisensor-multitarget problems can be formulated using random sets—or any other particular mathematical formalism—is, in and of itself, of limited practical interest;

- If a formulation requires increased mathematical complexity, this should result in increased practical advantage, especially computationally;

- Conversely, if a formulation is so heedlessly or vaguely formulated that it repeatedly leads to blunders in practice, it is simplistic—not simple;
- The point of approximation is to tractably preserve as much application realism as possible—not to achieve tractability under simplistic assumptions.

1.3.1 Why Isn't Multitarget Filtering Straightforward?

The statisticians J.C. Naylor and A.F.M. Smith have remarked, "The implementation of Bayesian inference procedures can be made to appear deceptively simple" [176, p. 214]. In this section I argue that, as a consequence, "computational logjams" await the unwary.

1.3.1.1 Computability, 1: The Single-Sensor, Single-Target Bayes Filter

The single-sensor, single-target recursive Bayes filter—see (2.81) and (2.82)—is far more computationally demanding than conventional techniques such as the extended Kalman filter (EKF). Much recent research has been devoted to its approximate real-time implementation. Caution is in order, however. Fixed-grid implementation, for example, is easily understood and straightforward to implement. It is based on discretization of the measurement and state spaces into finite numbers of "cells." However, in actuality, computational tractability using such techniques can be achieved usually only for "toy" problems—see Section 2.4.11.

Matters can be improved if we use a principled approximation technique such as sequential Monte Carlo (SMC) approximation (also known as particle systems; see Section 2.5.3). However, even SMC based algorithms will be computationally demanding in those applications for which SMC methods are appropriate—which is to say, those where traditional methods fail.

A more subtle issue is the fact that, in such approaches, modeling and implementation have been irrevocably intertwined. The models are *heuristic contrivances that are specific to a particular implementation technique*. Such approaches also neglect the problem of modeling and processing nontraditional information such as natural language statements and rules.

1.3.1.2 Computability, 2: The Single-Sensor, Multitarget Bayes Filter

The multitarget Bayes filter—see (14.14), (14.50), and (14.51)—is far more computationally challenging than its single-sensor, single-target special case, and so even more powerful approximation strategies are required. Much research is being devoted to devising such strategies, but once again caution is in order. As just one example, multitarget fixed-grid implementation is certainly "straightforward." However, even with drastic and highly unrealistic simplifications, it is also *inherently intractable even when using the simplest possible model of targets moving between discrete locations in one dimension*—see Section 15.2.1.

Once again, matters can be improved if we employ multitarget SMC techniques (see Section 15.2.2)—but only up to a point. Such methods will be computationally demanding under realistic conditions, such as lower SNR, in which conventional methods such as multihypothesis tracking fail and where, therefore, SMC methods are appropriate. So one must ask: In which applications will SMC approximation be *both tractable and appropriate*? What does one do when it is *appropriate but intractable*?

Once again, a more subtle issue is the fact that the multitarget models are heuristic contrivances specific to some particular implementation technique.

1.3.2 Beyond Heuristics

Deeper insight is required if such difficulties are to be surmounted. This is precisely what FISST strives to achieve. We need to presume as much realism as possible and then try to devise approximations that are potentially tractable despite this fact.

Furthermore, we need a formal modeling methodology for ambiguous evidence and for multitarget problems that, like the already accepted one for single-target problems, is *nonheuristic* and *implementation-independent*. That is, we need a modeling approach that results in general mathematical formulas for Markov densities and likelihood functions, constructed from general statistical models. Once one has chosen a specific implementation technique, these formulas can be carefully reformulated in terms of this technique.

This will ensure that the statistical assumptions underlying the original models are being implemented as faithfully as possible. This in turn helps ensure that the conditions under which the implemented filter will be approximately optimal are explicitly specified. Algorithm behavior should also tend to be more explicable, because assumptions and important design decisions in both the models and the approximations have been carefully parsed into a systematic, disciplined chain of reasoning.

1.3.3 How Do Single-Target and Multitarget Statistics Differ?

FISST is "practitioner friendly" in that it is *geometric* (i.e., treats multitarget systems as visualizable images; see Section 11.2); and in that it *directly generalizes the Bayes "Statistics 101" formalism that most signal processing practitioners already understand*—including formal Bayes-statistical modeling methods.

However, these methods do not necessarily generalize in a straightforward manner [126, 132, 134, 145, 148]. The following are examples of how multisensor-multitarget statistics differs from single-sensor, single-target statistics (differences that FISST specifically addresses):

- The standard Bayes optimal state estimators are not defined in general.
- Neither are such familiar concepts as expected value, least squares optimization, and Shannon entropy.
- Other concepts, such as miss distance, require major reworking (see [87] or Section 14.6.3).
- In the absence of FISST no explicit, general, and systematic techniques exist for modeling multisensor-multitarget problems and then transforming these models into Bayesian form.

1.3.4 How Do Conventional and Ambiguous Data Differ?

The modeling of observations as vectors is ubiquitous to the point that it is commonplace to think of a vector as itself "data." However, this is not the case. A vector is actually a *mathematical abstraction* that serves as a representation of some real-world entity called a "datum." Voltages are modeled as real or complex numbers, signatures as functions, features as integers, and so on. The only uncertainty in such data is that associated with randomness, and likelihood functions are interpreted as the uncertainty models for such data.

Unfortunately, conventional reasoning does not suffice for data types such as rules, natural language statements, human-extracted attributes or, more generally, any information involving some kind of human mediation. Information of this kind introduces two additional forms of uncertainty:

- Ignorance of the actual underlying statistical phenomenology;
- The ambiguities inherent in the process of constructing actionable mathematical models.

Comprehensive data modeling requires four steps. In conventional practice the first, second, and fourth steps are usually taken for granted, so that only the third step remains and ends up viewed as the complete data model. The four steps are:

- Creating the mathematized abstractions that represent observations;
- Modeling any ambiguities inherent to this act of abstraction;
- Modeling the generation of observations;
- Modeling any ambiguities caused by our ignorance of the actual data generation phenomenologies.

This realization leads, in turn, to fundamental distinctions between different kinds of data:

- *Measurements* versus *state-estimates* (i.e., opinions about what is being observed versus deductions about target state based on measurements that are not passed on to us);
- *Unambiguously generated unambiguous (UGU) measurements* (i.e., conventional measurements);
- *Unambiguously generated ambiguous (UGA) measurements* (i.e., the relationship between states and measurements is precisely understood, but there is considerable uncertainty about what is being observed and how to model it);
- *Ambiguously generated ambiguous (AGA) measurements* (i.e., there is considerable uncertainty about the relationship between states and measurements, as well as what is being observed);
- *Ambiguously generated unambiguous (AGU) measurements* (i.e., conventional measurements in which the process by which states generate measurements is imperfectly understood).

To deal with information of the last three kinds, we must expand our palette of mathematical data-modeling abstractions beyond numbers, vectors, functions, and the like. We must be willing to embrace a much more general data-modeling abstraction called a *random (closed) set*.

1.3.5 What Is Formal Bayes Modeling?

Formal Bayes modeling has become the accepted standard in single-sensor, single-target research and development. In it, we begin with an implementation-independent *formal measurement model* that describes sensor behavior; and an implementation-independent *formal motion model* that models interim target behavior. Using ordinary undergraduate calculus, one can derive concrete formulas for the *true likelihood density* and *true Markov density* (i.e., the probability densities that faithfully reflect the original models).[2]

Given this, we can address single-sensor, single-target tracking, detection, and identification problems using the Bayes recursive filter.

1.3.6 How Is Ambiguous Information Modeled?

One of the basic goals of FISST is to extend formal Bayes modeling in a rigorous fashion to nontraditional information such as features and attributes, natural language statements, and rules. A second basic goal is to accomplish this extension in such a fashion that it seamlessly incorporates many familiar data fusion techniques, including Dempster's combination, fuzzy conjunction, and rule-based inference. We would like to have general and systematic procedures for the following:

- Constructing measurement models for generalized measurements (i.e., measurements that are ambiguous either in their definition, their generation, or both);
- Transforming a generalized measurement model into a corresponding generalized likelihood function (i.e., one that faithfully reflects the generalized measurement model while, at the same time, hedging against uncertainties in the modeling process);
- Constructing generalized measurement models for generalized measurements (i.e., measurements that are ambiguous either in their definition, their generation, or both);
- Transforming a generalized measurement model into a corresponding generalized likelihood function (i.e., one that faithfully reflects the generalized

2 In practice, these formulas can be looked up in a textbook and so we need never actually bother with their formal derivation. However, in the multitarget or generalized-data cases, until now no textbook has existed that allows such an easy exit.

measurement model while, at the same time, hedging against uncertainties in the modeling process);

- Demonstrating that, under certain conditions, many familiar expert system data fusion techniques are rigorously defensible within a Bayesian paradigm.

If we shirk such issues we unnecessarily relinquish the data processing of nontraditional information to heuristics. We also abandon any attempt to reach a deeper understanding of the nature of nontraditional data and of the methods necessary to fuse such data.

1.3.7 What Is Multisource-Multitarget Formal Modeling?

A third basic goal of FISST is to rigorously extend formal Bayes modeling to multisensor-multitarget problems. We would like to have general and systematic procedures for the following.

- Constructing a *multitarget measurement model* for any given sensor(s);
- Constructing a *multitarget motion model;*
- Transforming the multitarget measurement model into the corresponding *true multitarget likelihood function* (i.e., the one that faithfully reflects the multitarget sensor model);
- Transforming the multitarget motion model into the corresponding *true multitarget Markov density* (i.e., the one that faithfully reflects the multitarget motion model).

Such procedures should be inherently *implementation-independent*, in that they produce *general mathematical formulas* for these probability densities. Implementations that preserve the statistical assumptions underlying the original models can then be derived from these formulas.

If we shirk such issues we fail to grasp that there is an inherent problem. How do we know that the multitarget likelihood density is not a heuristic contrivance or not erroneously constructed? Any boast of "optimality" is hollow if the likelihood function models the wrong sensors. Similar comments apply to the Markov density function.

Once in possession of these basics we can address multisensor-multitarget tracking, detection, and identification problems using the multitarget Bayes filter. However, even here we encounter complications because, in general, the integrals in multitarget generalizations of those equations must sum over not only all possible target states but also over all possible numbers of targets. This leads to the concept of a *set integral*.

Last, but not least, we would like:

- *Multitarget Bayes optimal state estimators* that allow us to determine the number and states of the targets.

Here, however, we encounter an unexpected difficulty: the naïve generalizations of the usual single-target Bayes optimal estimators do not exist in general. We must devise new estimators and show that they are well behaved.

1.4 RANDOM SETS IN INFORMATION FUSION

This section sketches the antecedents of FISST. At a purely mathematical level, FISST is a synthesis of three separate though intertwined strands of research:

- Random measure theory, employed as a foundation for the study of randomly varying populations;
- Random set theory, employed as a foundation for the study of randomly varying geometrical shapes;
- Random set theory, employed as a potential foundation for expert systems theory.

I summarize these strands and describe how they have been integrated to produce FISST.

1.4.1 Statistics of Multiobject Systems

Point process theory arose historically as the mathematical theory of *randomly varying time-arrival phenomena* such as queues. For example, the number of customers waiting in line for a bank teller varies randomly with time. The Poisson distribution $\pi(n) = e^{-\lambda}\lambda^n/n!$ is the simplest model of random processes of this type. It states that the probability that there will be n customers in a queue at any given time is $\pi(n)$ where λ is the average length of the queue. However, what if the customers could be anywhere in the building? Their randomness would include not only number but position.

Such a physically dispersed queue is an example of a *population process* or *multidimensional point process* [36, 214, 220]. Moyal [169] introduced the first theory of population processes over 40 years ago. Noting that population processes could be regarded as randomly varying sets, he instead chose a different mathematical formulation: random counting measures.[3] By the time of Fisher's survey in 1972 [59], random measure theory had become the usual language for point processes among mathematicians.[4]

Other mathematicians, meanwhile, were concerned about a completely different problem. In ordinary signal processing one models physical phenomena as "signals" obscured by "noise," the former to be extracted from the latter by optimal filters. However, how can we construct such filters if the signal is some geometric shape and the noise is a geometrically structured clutter process? Mathéron devised the first systematic theory of random (closed) subsets in 1975 to address such "stochastic geometry" problems [12, 160, 165, 220, 166].

It was subsequently shown that random locally finite subsets provide an essentially equivalent mathematical foundation for point processes—see, for example, Ripley in 1976 [195] and Baudin in 1986 [13]. This point is now widely understood [220, pp. 100-102].

1.4.2 Statistics of Expert Systems

Beginning in the 1970s, a number of researchers began uncovering close connections between random set theory and many aspects of expert system theory. Orlov [180, 181] and Höhle [90] demonstrated relationships between random sets and fuzzy set theory, though the most systematic work in this direction is due to Goodman [68]. In 1978, Nguyen connected random set theory with the Dempster-Shafer theory of evidence [179]. Dubois and Prade have published a number of papers relating random sets with expert system theory [46, 47]. In the late 1990s, the author proposed random sets as a means of probabilistically modeling rule-based evidence [146, 147].[5]

By the early 1990s, some researchers were proposing random set theory as a unifying foundation for expert systems theory [113, 84, 112, 191]. Such publications included books by Goodman and Nguyen in 1985 [71] and by Kruse, Schwecke and Heinsohn in 1991 [114].

3 See Appendix E for a summary of the relationship between FISST and conventional point process theory.

4 For a more comprehensive history, see [37, pp. 1-18].

5 The author was also principal organizer of a 1996 international workshop, the purpose of which was to bring together the different communities of random set researchers in statistics, image processing, expert systems, and information fusion [75].

1.4.3 Finite Set Statistics

Inspired by the pioneering work of I.R. Goodman and H.T. Nguyen [71], in 1994 the author proposed random set theory as a theoretically rigorous way of integrating point process theory with expert systems theory—thereby producing a unified and purely probabilistic foundation for much of multisensor-multitarget information fusion [154]. Most importantly, we noted that this synthesis could be recast as a direct generalization of the "Statistics 101" formalism that most signal processing practitioners learn as undergraduates.

In this book we report on those aspects of FISST that are pertinent to multitarget-multisource integration (level 1 fusion). Results of our research in level 2 fusion (situation assessment), level 3 fusion (threat assessment), and level 4 fusion (sensor and platform management) are topics for another day and, perhaps, another book.

1.5 ORGANIZATION OF THE BOOK

I describe the contents of the book in more detail. Figure 1.2 is a "road map" that depicts the relationships between the chapters.

1.5.1 Part I: Unified Single-Target Multisource Integration

The purpose of Part I is to build up, piece by piece, a comprehensive and unified Bayesian approach to multisource, single-target information fusion. Chapter 2 sets the stage for the rest of the book by introducing, within the context of single-target tracking, two concepts that are fundamental to everything that follows: *formal statistical modeling* and the *recursive Bayes nonlinear filter*.

I employ a crawl-walk-run style of exposition, beginning with the familiar Kalman filter (crawl); progressing to the Bayesian formulation of the Kalman filter (walk); and from there to the general Bayes filter (run). Chapter 3 is devoted to a relatively informal, intuition-oriented discussion of comprehensive data modeling. There we discuss the problem of modeling uncertainty and, in particular, modeling uncertainty in data. I present five simple examples that illustrate our data-modeling approach. In this chapter I also introduce the major concepts of the book:

- General formal Bayes modeling;
- The general Bayes filter;

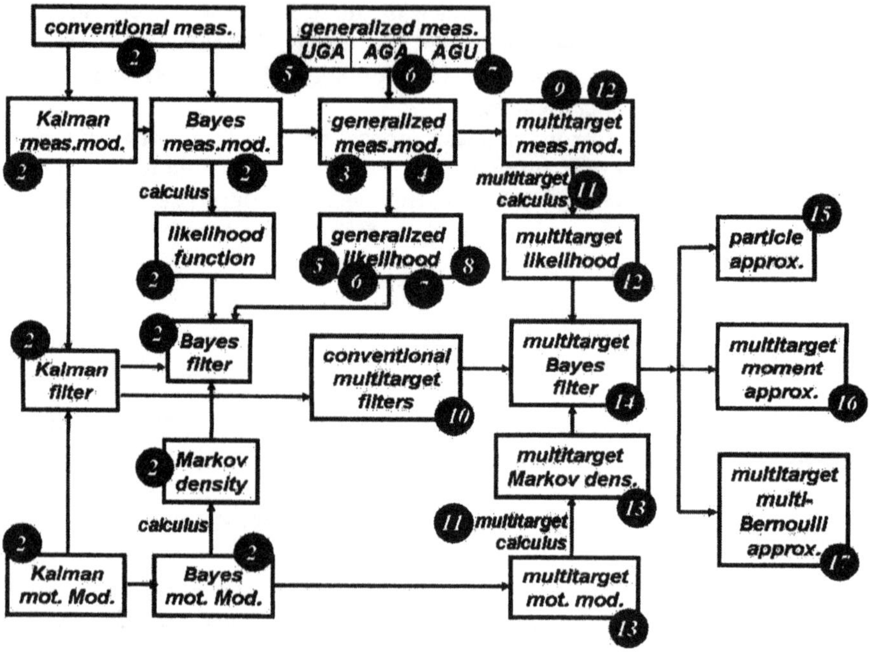

Figure 1.2 The structure of the book. Boxes indicate the major subjects to be covered. The numbered black disks indicate which chapters cover which subjects. Chapter 2 sets the stage by introducing the book's important concepts in the context of single-target filtering. Chapters 3-8 describe how single-target filtering can be extended to include very general forms of information. Chapter 10 sets the stage for the rest of the book by summarizing the primary conventional multitarget filtering methods (and how they can be extended to nontraditional information). Chapters 11-14 develop the foundations of multitarget Bayes filtering. Chapters 15, 16, and 17 describe approximate implementation approaches.

- The distinction between *state-estimate data* and *measurement data*;
- The four types of measurements that will concern us throughout Part I: *unambiguously generated unambiguous* (UGU); *unambiguously generated ambiguous* (UGA); *ambiguously generated ambiguous* (AGA); and *ambiguously generated unambiguous* (AGU).

The mathematical core of my modeling approach is covered in Chapter 4: the representation of data uncertainty using *random closed subsets of measurement or state space*. There I introduce random set representations of several uncertainty modeling formalisms: (1) *fuzzy logic*; (2) *generalized fuzzy logic*; (3) *Dempster-Shafer theory*; (4) *fuzzy Dempster-Shafer theory*; and (5) *first- and second-order fuzzy rules*.

I then turn to the detailed data-modeling problem. The purpose of Chapter 5 is to (1) define and model UGA measurements; (2) define the concept of a generalized likelihood function for a UGA measurement; and (3) construct concrete formulas for these likelihoods in specific cases: fuzzy, generalized fuzzy, Dempster-Shafer, and so on. I also present my seamless *Bayesian unification of UGA measurement fusion*. I show that the following measurement fusion methods are equivalent to measurement fusion using Bayes' rule: copula type fuzzy conjunctions; normalized and unnormalized Dempster's combination; and firing and partial firing of rules. I also show how to convert uncertainty representations from one to the other (e.g., fuzzy to probabilistic and vice versa) without losing estimation-relevant information.

The chapter concludes with a concrete application: target tracking with conventional and fuzzy Dempster-Shafer measurements using the *Kalman evidential filter* (KEF). In the process, it is shown that data update using Dempster's combination is equivalent to Bayes' rule, in the same sense that data update using the Kalman corrector equations is equivalent to Bayes' rule.

AGA measurements are defined and modeled in Chapter 6, along with their generalized likelihoods and derivation of concrete formulas for these likelihoods in various special cases. The chapter concludes with two applications: robust target identification using human mediated INT attributes transmitted over datalink, and robust target identification in the presence of unmodeled target types.

AGU measurements, the subject of Chapter 7, require a random set representation formalism that differs from that employed in Chapters 5 and 6: the *random error bar* or its special case, the *fuzzy error bar*. This chapter includes the following application: robust automatic target recognition (ATR) of motionless ground targets using synthetic aperture radar (SAR) intensity images.

In Chapter 8, I take initial steps toward addressing the fusion of information consisting not of generalized measurements but, rather, of *generalized estimates of target state*. The chapter includes a unification result analogous to those presented in Chapter 5—namely, that fusing independent Dempster-Shafer state-estimates using "modified Dempster's combination" is equivalent to fusing them using Bayes' rule alone.

Part I concludes, in Chapter 9, with an introduction to measurements that take the form of *finite sets of conventional measurements*. This chapter serves as an entry point into the theory and statistics of multitarget systems covered in Part II.

1.5.2 Part II: Unified Multitarget-Multisource Integration

The purpose of Part II is to do for multisource-multitarget information fusion what Part I did for multisource, single-target information fusion: build up, piece by piece, a comprehensive and unified Bayesian approach.

In Chapter 10, I initiate my discussion with an overview of the most familiar concepts in multitarget detection and tracking: "hard" measurement to track association and single- and multihypothesis tracking and "soft" measurement to track association and joint probabilistic data association (JPDA).

The mathematical core of my approach is described in Chapter 11. This includes fundamental statistical descriptors—multiobject density functions, belief-mass functions, and probability-generating functionals (p.g.fl.s)—and a multiobject differential and integral calculus necessary to apply them in practical problems. I then turn to the problem of formal Bayes modeling in multisource-multitarget problems.

In Chapter 12, I describe the process of constructing formal multisource-multitarget statistical measurement models, and of the "true" multisource-multitarget likelihood functions that faithfully encapsulate them. My discussion begins, in Section 12.2, with a detailed discussion of multisensor-multitarget measurement spaces and of multitarget state spaces. The chapter concentrates on the "standard" multitarget measurement model: detection type measurements extracted from signatures.

I consider a number of variants and generalizations of the standard model: transmission dropouts (Section 12.6), state-dependent false alarms (Section 12.5), extended targets (Section 12.7), and unresolved targets (Section 12.8). I derive concrete formulas for the "true" multitarget likelihood functions corresponding to all of these cases. I also consider the multisensor case and the case of nontraditional information sources (Section 12.9).

Chapter 12 concludes with two "nonstandard" models. First, a measurement model for asynchronously reporting bearings-only sensors, due to Vihola (Section 12.10). Second, a measurement model that addresses the problem of Bayes optimal extraction of the number and shapes of "soft" data clusters from a time series of dynamically evolving measurement sets (Section 12.11).

Chapter 13 is devoted to constructing formal multitarget motion models and, from them, "true" multitarget Markov transition density functions. In Section 13.2 I propose a "standard" multitarget measurement model that is mathematically analogous to the standard multitarget measurement model of Chapter 12. The standard model encompasses statistically independent target motion, target appearance, target spawning, and target disappearance. I then extend the standard model to address extended targets (Section 13.3) and unresolved targets (Section 13.4). Chapter 13 concludes in Section 13.5 with two simple multitarget *coordinated-motion* models: virtual leader-follower with and without target appearance and disappearance.

The material in Chapters 12 and 13 provide the theoretical foundation for all multisource-multitarget detection, tracking, and classification problems: the multisource-multitarget recursive Bayes filter (Chapter 14).

The bulk of Chapter 14 addresses the individual steps of the filtering process: initialization (Section 14.2), prediction (Section 14.3), correction or data-update (Section 14.4), state estimation (Section 14.5), and error estimation (Section 14.6). Because multitarget state estimation turns out to be unexpectedly tricky (Section 14.5.1), I introduce two multitarget state estimators: the marginal multitarget (MaM) estimator (Section 14.5.2) and the joint multitarget (JoM) estimator (Section 14.5.3).

I also describe a special case of the recursive Bayes filter, the joint target-detection and tracking (JoTT) filter, in which it is assumed that target number is at most one.[6]

1.5.3 Part III: Approximate Multitarget Filtering

The question of potentially practical implementation of the multitarget Bayes filter occupies the remainder of the book. Chapter 15 begins with a discussion of computational issues associated with the multitarget Bayes filter. In Section 15.2 I point out that this filter will often be computationally intractable in those applications for which it is *appropriate* (i.e., those applications in which more conventional multitarget detection and tracking approaches fail).

6 This generalizes the integrated probabilistic data association (IPDA) filter of [172].

With this as a backdrop, Chapter 15 goes on to summarize multitarget sequential Monte Carlo (also known as particle) approximation.

Chapter 16 addresses the concept of *multitarget first-moment* approximation and two of its consequences: the *probability hypothesis density* (PHD) *filter* (Section 16.2) and the *cardinalized probability hypothesis density* (CPHD) *filter* (Section 16.7). Thus far, the PHD and CPHD filters have been implemented with reasonable success by many researchers throughout the world. This work is summarized.

These implementations typically employ sequential Monte Carlo approximations (Section 16.5.2). Vo and his associates, however, have developed a Gaussian-mixture approximation which, under suitable assumptions, provides fast closed-form solutions of the PHD and CPHD filter equations (Section 16.5.3).

The final chapter, Chapter 17, is devoted to a more speculative implementation technique: *multitarget multi-Bernoulli approximation.* It is based on the observation that the two set integrals in the multitarget Bayes filter can be solved in closed form when this approximation is used. When combined with Gaussian-mixture techniques, multi-Bernoulli approximation leads to a potentially tractable multitarget filter, the *para-Gaussian filter.*

1.5.4 Appendixes

Various mathematical details and tangential matters have been relegated to the appendixes, as follows:

- Appendix A: Glossary of notation;
- Appendix B: Identities involving the Dirac delta function;
- Appendix C: The gradient derivative of a vector transformation;
- Appendix D: The fundamental identity for Gaussian distributions;
- Appendix E: Point processes versus random sets;
- Appendix F: FISST versus measure theoretic probability;
- Appendix G: Mathematical proofs;
- Appendix H: Solutions to the chapter exercises.

Part I

Unified Single-Target Multisource Integration

Chapter 2

Single-Target Filtering

2.1 INTRODUCTION TO THE CHAPTER

The purpose of this chapter is to set the stage for the rest of the book by introducing, within the familiar context of single-target tracking, two concepts that are fundamental to everything that follows:

- *Formal statistical modeling*;
- The *recursive Bayes nonlinear filter*.

To make these concepts as transparent as possible, I follow a crawl-walk-run style of exposition. Many readers will have already had some degree of exposure to the Kalman filter, and in particular to the sensor and motion models upon which it is based. Since the Kalman filter is a special case of the Bayes filter, and since its models are special cases of the models used in the Bayes filter, it provides a convenient starting point.

I then build upon the Kalman filter through a reformulation of it that leads directly to the general Bayes filter, and to general single-sensor, single-target measurement models and general single-target motion models. Once we have a good understanding of target tracking based on formal Bayes modeling and the Bayes filter, in later chapters I will introduce the more general measurement models that are the primary subject of Part I.

This chapter also sets the stage for general formal Bayes modeling and general Bayes filtering, as introduced in Section 3.5.

2.1.1 Summary of Major Lessons Learned

These are the major points that the reader will have absorbed by chapter's end:

- The Kalman filter can be reformulated within a fully Bayesian viewpoint (Section 2.3).
- The Kalman filter corrector step is actually a special case of Bayes' rule (Section 2.3).
- A Bayesian viewpoint provides an intuitive understanding of what the Kalman filter is and the conditions under which it will succeed or fail (Section 2.4.2).
- The single-target recursive Bayes filter, (2.81) and (2.82), is potentially very powerful, capable of addressing applications that would otherwise appear to defy solution (Section 2.4.1).
- It also can be made to appear *deceptively simple*, especially where implementation is concerned (Section 2.4.11).
- Its practical real-time implementation requires principled approximation approaches (Sections 2.5.2 and 2.5.3).
- It provides a rigorous basis for multisensor, single-target data fusion, provided that intersensor correlations can be effectively modeled (Section 2.4.10).
- Nevertheless, the following question should be central: When is the Bayes filter *inappropriate*—that is, when can the application at hand be adequately addressed using conventional techniques (Section 2.4.11)?
- In heuristic modeling, the models tend to be ad hoc contrivances, dictated as much by constraints imposed by the implementation technique as by real-world phenomenology (Section 2.4.4).
- Formal modeling begins with *implementation-free statistical models*, to which principled approximation techniques can then be applied (Section 2.4.4).
- One of the basic purposes of formal modeling is to derive likelihood functions and Markov densities that are *true*—that is, that faithfully reflect the measurement and motion models, respectively (Section 2.4.4).

- One of the consequences of formal modeling is that algorithmic behavior should be more explicable, because important assumptions in both modeling and design have been parsed into a systematic, disciplined chain of reasoning (Section 2.4.4).

2.1.2 Organization of the Chapter

I begin by reviewing the Kalman filter in Section 2.2. This includes a summary of the Kalman filter predictor and corrector steps, their derivation using least-squares techniques, and fusion of data using the corrector equations. The Bayesian formulation of the Kalman filter is described in Section 2.3 in terms of predictor and corrector densities and state estimation.

The general single-target Bayes filter is introduced in Section 2.4, beginning with a simple illustration of the performance of this filter. This section includes the following topics: modeling issues for the Bayes filter; formal Bayes modeling; the Bayes filter initialization, prediction, correction, and estimation steps; fusion of measurements using the Bayes filter corrector; computational issues; and three commonly employed real-time implementation approaches: the extended Kalman filter (EKF), the Gaussian-mixture (GM) filter, and sequential Monte Carlo (SMC) or particle system techniques. Exercises for the chapter can be found in Section 2.6.

2.2 THE KALMAN FILTER

The Kalman filter is by far the most familiar and widely used algorithm employed in information fusion applications. It is within this familiar context that I introduce some of the basic concepts upon which this book depends. Assume that through time step k the sensor has collected a time sequence $Z^k : \mathbf{z}_1, ..., \mathbf{z}_k$ of observation-vectors. At each time step k and on the basis of the currently available measurements Z^k, we want to determine two things:

- An estimate $\mathbf{x}_{k|k}$ of the state vector of the target;
- An estimate of the likely error associated with $\mathbf{x}_{k|k}$.

The state vector $\mathbf{x}_{k|k}$ contains the information about the target (e.g., position and velocity) that we wish to know. The measure of error in $\mathbf{x}_{k|k}$ is the *error covariance matrix* $P_{k|k}$, and it measures error in the following sense. For each integer $k > 0$, the inequality

$$\left(\mathbf{x} - \mathbf{x}_{k|k}\right)^T P_{k|k}^{-1} \left(\mathbf{x} - \mathbf{x}_{k|k}\right) \leq 1 \tag{2.1}$$

defines the hyperellipsoid of all states $\mathbf{x}$ that are located within one sigma of $\mathbf{x}_{k|k}$, and is therefore a multidimensional generalization of the concept of an error bar.

The Kalman filter shows us how to recursively propagate $\mathbf{x}_{k|k}$ and $P_{k|k}$ through time, thus resulting in a time sequence

$$(\mathbf{x}_{0|0}, P_{0|0}) \to (\mathbf{x}_{1|0}, P_{1|0}) \to (\mathbf{x}_{1|1}, P_{1|1}) \to ... \tag{2.2}$$

$$\to (\mathbf{x}_{k|k}, P_{k|k}) \to (\mathbf{x}_{k+1|k}, P_{k+1|k}) \to (\mathbf{x}_{k+1|k+1}, P_{k+1|k+1}) \tag{2.3}$$

$$\to ... \tag{2.4}$$

In what follows I describe this time evolution in more detail. The basic Kalman filter steps—initialization, prediction, and correction—are summarized in Sections 2.2.1, 2.2.2, and 2.2.3. The derivation of the Kalman filter predictor and corrector using least-squares is summarized in Section 2.2.4. Multisensor, single-target data fusion using the Kalman filter is described in Section 2.2.5. A simplified form of the Kalman filter, the constant-gain Kalman filter, is discussed in Section 2.2.6.

2.2.1 Kalman Filter Initialization

One must first *initialize* the filter by choosing an initial guess $\mathbf{x}_{0|0}$ about the target state and a guess $P_{0|0}$ about the degree of certainty in this state. If $\mathbf{x}_{0|0}$ and $P_{0|0}$ are poorly chosen, the Kalman filter will not be able to acquire the target and will diverge.

2.2.2 Kalman Filter Predictor

At recursive time step $k+1$ we collect a new observation $\mathbf{z}_{k+1}$. Since the target has moved during the interval between time steps k and $k+1$, we must account for the uncertainty created by this motion. This is accomplished using a *formal statistical motion model* of the form

$$\mathbf{X}_{k+1} = F_k \mathbf{x} + \mathbf{V}_k. \tag{2.5}$$

This formal motion model consists of two parts. The first is a *deterministic* motion model, $\mathbf{x}_{k+1} = F_k \mathbf{x}$, , which is essentially a guess about how the target will move during the interval between time steps k and $k+1$. For example, we may guess that the target will move in a straight line with constant-velocity, or that the target will move as a ballistic object. The deterministic model states that the target will have state $\mathbf{x}_{k+1}$ at time step $k+1$ if it had state $\mathbf{x}$ at time step k. The matrix F_k is called the *state transition matrix.*

Since actual target motion is unknown and $\mathbf{x}_{k+1} = F_k\mathbf{x}$ is only a guess, the actual $\mathbf{x}_{k+1}$ will usually be not $F_k\mathbf{x}$ but, rather, some perturbation $F_k\mathbf{x} + \Delta\mathbf{X}$ of it. We assume that this perturbation, the *plant noise*, is a zero-mean Gaussian random vector $\Delta\mathbf{X} = \mathbf{V}_k$ with covariance Q_k. If actual target motion deviates significantly from the deterministic model, we can often hedge against this mismatch by making $\mathbf{V}_k$ noisier. The fact that F_k, $\mathbf{V}_k$ can be chosen differently at each time step permits additional flexibility. We can, for example, attempt to detect target maneuvers and account for them by choosing F_k suitably.

Given this, the Kalman filter *predictor equations* tell us that at time step $k+1$, the most likely target state will be

$$\mathbf{x}_{k+1|k} = F_k\mathbf{x}_{k|k} \tag{2.6}$$

and that the possible error in this prediction is measured by the predicted error covariance matrix

$$P_{k+1|k} = F_k P_{k|k} F_k^T + Q_k. \tag{2.7}$$

2.2.3 Kalman Filter Corrector

Having extrapolated the target state to the time of the next measurement at time step $k+1$, we need to correct this prediction using the actual observation $\mathbf{z}_{k+1}$. To do this we must have a model of how the sensor generates observations. This is accomplished using a *measurement model* of the form

$$\mathbf{Z}_{k+1} = H_{k+1}\mathbf{x} + \mathbf{W}_{k+1}. \tag{2.8}$$

Like the motion model, it consists of two parts. The first is a deterministic sensor *state-to-measurement transform model*, $\mathbf{z} = H_{k+1}\mathbf{x}$, that states that if a target has state $\mathbf{x}$ at time step $k+1$, then it will generate observation $\mathbf{z}$. The $M \times N$ matrix H_{k+1} captures the fact that the sensor cannot observe the entire target state $\mathbf{x}$ but, rather, only some incomplete and transformed view of it. In addition, because of internal noise the sensor will actually collect not $H_{k+1}\mathbf{x}$ but instead some random perturbation $H_{k+1}\mathbf{x}+\Delta\mathbf{Z}$. We assume that $\Delta\mathbf{Z} = \mathbf{W}_{k+1}$ is a zero-mean Gaussian random vector with covariance matrix R_{k+1}.

Given this, the Kalman filter *corrector equations* tell us that the best estimate of the target state is

$$\mathbf{x}_{k+1|k+1} = \mathbf{x}_{k+1|k} + K_{k+1}\left(\mathbf{z}_{k+1} - H_{k+1}\mathbf{x}_{k+1|k}\right) \tag{2.9}$$

and that the possible error in this estimate is measured by the corrected error covariance matrix

$$P_{k+1|k+1} = \left(I - K_{k+1}H_{k+1}\right)P_{k+1|k}. \tag{2.10}$$

Here, the $N \times M$ Kalman gain matrix K_{k+1} is

$$K_{k+1} = P_{k+1|k}H_{k+1}^T\left(H_{k+1}P_{k+1|k}H_{k+1}^T + R_{k+1}\right)^{-1}. \tag{2.11}$$

The *innovation* $\mathbf{s}_{k+1} = \mathbf{z}_{k+1} - H_{k+1}\mathbf{x}_{k+1|k}$ indicates the degree to which the actual measurement $\mathbf{z}_{k+1}$ differs from the predicted measurement $H_{k+1}\mathbf{x}_{k+1|k}$, and K_{k+1} determines the degree to which the predicted state $\mathbf{x}_{k+1|k}$ should be corrected to reflect this deviation.

2.2.4 Derivation of the Kalman Filter

The Kalman filter equations are most commonly derived in a bottom-up fashion as the solution to some linear least-squares estimation problem. In this section I briefly sketch the basic ideas behind this approach.

2.2.4.1 Derivation of Kalman Filter Corrector

For the corrector step,

$$\begin{aligned} \varepsilon_{k+1|k+1}^2(\mathbf{x}) &= (\mathbf{x} - \mathbf{x}_{k+1|k})^T P_{k+1|k}^{-1}(\mathbf{x} - \mathbf{x}_{k+1|k}) & (2.12) \\ &\quad +(\mathbf{z}_{k+1} - H_{k+1}\mathbf{x})^T R_{k+1}^{-1}(\mathbf{z}_{k+1} - H_{k+1}\mathbf{x}) & (2.13) \end{aligned}$$

is the weighted average of two errors: the square error for the predicted estimate $\mathbf{x}_{k+1k}$; and the square error associated with the new observation $\mathbf{z}_{k+1}$. We are to find the value of $\mathbf{x}$ that minimizes the joint error. Using matrix completion of the square, (2.12) can be rewritten as

$$\begin{aligned} &\varepsilon_{k+1|k+1}^2(\mathbf{x}) = (\mathbf{x} - \mathbf{x}_{k+1|k+1})^T P_{k+1|k+1}^{-1}(\mathbf{x} - \mathbf{x}_{k+1|k+1}) & (2.14) \\ &+(\mathbf{z}_{k+1} - H_{k+1}\mathbf{x}_{k+1|k})^T C_{k+1}^{-1}(\mathbf{z}_{k+1} - H_{k+1}\mathbf{x}_{k+1|k}) & (2.15) \end{aligned}$$

where

$$P^{-1}_{k+1|k+1} \triangleq P^{-1}_{k+1|k} + H^T_{k+1} R^{-1}_{k+1} H_{k+1} \quad (2.16)$$

$$P^{-1}_{k+1|k+1} \mathbf{x}_{k+1|k+1} \triangleq P^{-1}_{k+1|k} \mathbf{x}_{k+1|k} + H^T_{k+1} R^{-1}_{k+1} \mathbf{z}_{k+1} \quad (2.17)$$

$$C^{-1}_{k+1} \triangleq R^{-1}_{k+1} + H_{k+1} P^{-1}_{k+1|k} H^T_{k+1}. \quad (2.18)$$

Since the second term of (2.14) is constant it follows that $\mathbf{x}_{k+1|k+1}$ produces the smallest square error and $P_{k+1|k+1}$ is the error covariance of the estimate. Note that $P^{-1}_{k+1|k+1}$ is symmetric and positive-definite, since

$$\mathbf{e}^T P^{-1}_{k+1|k+1} \mathbf{e} = \mathbf{e}^T P^{-1}_{k+1|k} \mathbf{e} + \mathbf{e}^T H^T_{k+1} R^{-1}_{k+1} H_{k+1} \mathbf{e} \geq \mathbf{e}^T P^{-1}_{k+1|k} \mathbf{e} > 0 \quad (2.19)$$

for all $\mathbf{e}$. Application of standard matrix identities [61] then leads to the usual corrector equations (2.9) and (2.11) [64, p. 111].

2.2.4.2 Derivation of Kalman Filter Predictor

As for the predictor step, suppose that the target had state $\mathbf{x}'$ at time step k. Then

$$\varepsilon^2_{k+1|k}(\mathbf{x}|\mathbf{x}') = (\mathbf{x} - F_k \mathbf{x}')^T Q_k^{-1} (\mathbf{x} - F_k \mathbf{x}') + (\mathbf{x}' - \mathbf{x}_{k|k})^T P^{-1}_{k|k} (\mathbf{x}' - \mathbf{x}_{k|k}) \quad (2.20)$$

is the weighted average of two errors: the square error for the estimate $\mathbf{x}_{k|k}$ from the previous time step; and the square error associated with the predicted target state $F_k \mathbf{x}'$. We want to find that value of $\mathbf{x}$ that minimizes $\varepsilon^2_{k+1|k}(\mathbf{x}|\mathbf{x}')$. Matrix completion of the square leads to

$$(\mathbf{x} - F_k \mathbf{x}')^T Q_k^{-1} (\mathbf{x} - F_k \mathbf{x}') + (\mathbf{x}' - \mathbf{x}_{k|k})^T P^{-1}_{k|k} (\mathbf{x}' - \mathbf{x}_{k|k}) \quad (2.21)$$

$$= (\mathbf{x} - F_k \mathbf{x}_{k|k})^T P^{-1}_{k+1|k} (\mathbf{x} - F_k \mathbf{x}_{k|k}) + (\mathbf{x}' - \mathbf{e})^T E^{-1} (\mathbf{x}' - \mathbf{e}) \quad (2.22)$$

where

$$E^{-1} \triangleq P^{-1}_{k|k} + F_k^T Q_k^{-1} F_k \quad (2.23)$$

$$E^{-1}\mathbf{e} \triangleq P^{-1}_{k|k} \mathbf{x}_{k|k} + F_k^T Q_k^{-1} \mathbf{x} \quad (2.24)$$

$$P_{k+1|k} \triangleq Q_k + F_k^T P_{k|k} F_k. \quad (2.25)$$

Thus $\mathbf{x} = \mathbf{x}_{k+1|k} \triangleq F_k \mathbf{x}_{k|k}$ minimizes $\varepsilon^2_{k+1|k}(\mathbf{x}|\mathbf{x}')$ with corresponding error covariance $P_{k+1|k}$.

2.2.5 Measurement Fusion Using the Kalman Filter

Since a different model $H_{k+1}, \mathbf{W}_{k+1}$ can be chosen at each time step, the Kalman filter provides a way of fusing data provided by different sensors. For example, if *statistically independent* observations $\overset{1}{\mathbf{z}}_{k+1}$ and $\overset{2}{\mathbf{z}}_{k+1}$ of the target are collected from two sensors at the same time, one can first process $\overset{1}{\mathbf{z}}_{k+1}$ using the corrector equations for the model $\overset{1}{\mathbf{Z}} = \overset{1}{H}_{k+1}\mathbf{x} + \overset{1}{\mathbf{W}}_{k+1}$; and then process $\overset{2}{\mathbf{z}}_{k+1}$ using the corrector equations for the model $\overset{2}{\mathbf{Z}} = \overset{2}{H}_{k+1}\mathbf{x} + \overset{2}{\mathbf{W}}_{k+1}$.

If the sensors are not statistically independent, then fusion of their data requires construction of a joint measurement model

$$\begin{pmatrix} \overset{1}{\mathbf{Z}} \\ \overset{2}{\mathbf{Z}} \end{pmatrix} = \begin{pmatrix} \overset{1}{H}_{k+1} \\ \overset{2}{H}_{k+1} \end{pmatrix} \mathbf{x} + \begin{pmatrix} \overset{1}{\mathbf{W}}_{k+1} \\ \overset{2}{\mathbf{W}}_{k+1} \end{pmatrix} \tag{2.26}$$

for the joint observation $(\overset{1}{\mathbf{z}}^T, \overset{2}{\mathbf{z}}^T)^T$, where the joint correlation matrix of $\overset{1}{\mathbf{W}}_{k+1}$ and $\overset{2}{\mathbf{W}}_{k+1}$ is chosen to model the statistical dependence of the two sensors. This increases the size of the measurement space, with the consequence that computational burden for the matrix inversion for the Kalman gain, (2.11), can increase substantially. Another version of the Kalman filter, called the *information Kalman filter* [203, p. 28], is often used in such situations since in this formulation the gain is independent of the dimensionality of the joint measurement space.

2.2.6 Constant-Gain Kalman Filters

The Kalman gain matrix K_{k+1} can be computationally expensive because of the matrix inversion in (2.11). One remedy is to assume that the gain matrix is constant, $K_k = K$ for all $k = 0, 1, ...$ In this case (2.9)-(2.10) reduce to

$$\mathbf{x}_{k+1|k+1} = \mathbf{x}_{k+1|k} + K\left(\mathbf{z}_{k+1} - H_{k+1}\mathbf{x}_{k+1|k}\right) \tag{2.27}$$

and the filter propagates only the state-estimate,

$$\mathbf{x}_{0|0} \to \mathbf{x}_{1|0} \to \mathbf{x}_{1|1} \to \ldots \to \mathbf{x}_{k|k} \to \mathbf{x}_{k+1|k} \to \mathbf{x}_{k+1|k+1} \to \ldots \tag{2.28}$$

Constant-gain Kalman filters will obviously perform less effectively than the Kalman filter. They work best under conditions in which targets do not undergo rapid maneuvers and in which sensor noise is small (i.e., the matrix norms $\|Q_k\|$ and $\|R_{k+1}\|$ are small).[1]

Such conditions are usually valid in air traffic control applications. Commercial aircraft follow straight-line paths occasionally punctuated by shallow maneuvers. They also carry transponders, which greatly increase their observability. The most well known constant-gain Kalman filter, the *alpha-beta filter* [82, pp. 123-125], is widely used in air traffic control applications.

Let $\mathbf{x} = (\mathbf{p}, \mathbf{v})^T$ where $\mathbf{p}$ is vector position and $\mathbf{v}$ is vector velocity. Assume that the sensor measures positions; that target motion is constant-velocity; and that position and velocity are uncorrelated. In this case

$$F_k = \begin{pmatrix} I_p & \Delta t \cdot I_v \\ 0 & I_v \end{pmatrix}, \qquad H_{k+1} = \begin{pmatrix} I_x & 0 \\ 0 & 0 \end{pmatrix} \tag{2.29}$$

where Δt is the time interval between measurements and where I_p and I_v are identity matrices. The alpha-beta filter has the form

$$\mathbf{p}_{k+1|k} = \mathbf{p}_{k|k} + \Delta t \cdot \mathbf{v}_{k|k} \tag{2.30}$$

$$\mathbf{v}_{k+1|k} = \mathbf{v}_{k|k} \tag{2.31}$$

$$\mathbf{p}_{k+1|k+1} = \mathbf{p}_{k+1|k} + \alpha \cdot \left(\mathbf{z}_{k+1} - \mathbf{p}_{k+1|k}\right) \tag{2.32}$$

$$\mathbf{v}_{k+1|k+1} = \mathbf{v}_{k+1|k} + \frac{\beta}{\Delta t} \cdot \left(\mathbf{z}_{k+1} - \mathbf{v}_{k|k}\right) \tag{2.33}$$

where the constants α and β are suitably chosen gain parameters.

2.3 BAYES FORMULATION OF THE KALMAN FILTER

Linear least-squares estimation formulations of the Kalman filter are too restrictive for the primary purpose of this book, which is to develop a rigorous and useful filtering theory for much more general problems such as nontraditional information problems and multitarget problems.

Equally importantly, such formulations are not very informative from the point of view of practical intuition. Least-squares approaches are popular not

1 The norm of a square matrix Q is defined as

$$\|Q\|^2 \triangleq \sum_{i,j}^{n} Q_{i,j}^2.$$

because there is something especially insightful or theoretically privileged about them, but because they are *mathematically tractable*. Other formulations, such as minimizing the total absolute error rather than the sum of squares error, are no less plausible but are difficult to work with mathematically.

What real basis is there, then, for the often-claimed *optimality* of the Kalman filter, given that the theoretical basis for this optimality appears to be predicated on expediency? More generally, what can bottom-up least-squares approaches tell us about when Kalman filters work, and why? I return to these questions in Section 2.4.2.

We must reexamine the Kalman filter from a different, *top-down Bayesian perspective*, first introduced into the engineering literature by Ho and Lee in 1964 [86]. The basic approach is to reformulate the Kalman filtering problem, and the Kalman predictor and corrector equations in particular, entirely in terms of probability density functions. This reformulation leads to filtering equations—the recursive Bayes filter equations—that greatly generalize the Kalman filter. This new filter will provide us with a solid foundation for single-target and multitarget information fusion problems much more complicated than those that can be addressed by the Kalman filter and its variants.

The Bayesian formulation of the Kalman filter will also provide insight into the questions just posed. It will lead us to conclude that least-squares is not actually an artifact of expediency. A Bayesian formulation also provides us with a framework within which we can understand why the Kalman filter often fails to track maneuvering targets, and why such misperformance can often be corrected by increasing plant noise.

After introducing some mathematical prerequisites in Section 2.3.1, I derive Bayesian versions of the Kalman predictor, corrector, and estimator equations in Sections 2.3.2, 2.3.3, and 2.3.4.

2.3.1 Some Mathematical Preliminaries

First recall that a general Gaussian distribution has the form

$$N_C(\mathbf{x}-\mathbf{c}) \triangleq \frac{1}{\sqrt{\det 2\pi C}} \cdot \exp\left(-\frac{1}{2}(\mathbf{x}-\mathbf{c})^T C^{-1}(\mathbf{x}-\mathbf{c})\right) \tag{2.34}$$

where $\mathbf{c}$ is the mean value and where C is a positive-definite (symmetric) matrix. It is a probability density function, since $\int N_C(\mathbf{c})d\mathbf{c} = 1$. This density function faithfully encapsulates all information contained in the pair $C, \mathbf{c}$ without introducing extraneous information. This is because

$$\int \mathbf{x} \cdot N_C(\mathbf{x} - \mathbf{c})d\mathbf{x} = \mathbf{c}, \qquad \int \mathbf{x}\mathbf{x}^T \cdot N_C(\mathbf{x} - \mathbf{c})d\mathbf{x} = C. \tag{2.35}$$

where in general 'C^T' denotes the transpose of the matrix C.

In what follows and throughout the book we will require the following algebraic identity, which is proved in Appendix D. Let $N \geq M$ and let P be an $N \times N$ positive-definite matrix; R an $M \times M$ positive-definite matrix; and H an $M \times N$ matrix. Then

$$N_R(\mathbf{r} - H\mathbf{x}) \cdot N_P(\mathbf{x} - \mathbf{p}) = N_{R+HPH^T}(\mathbf{r} - H\mathbf{p}) \cdot N_E(\mathbf{x} - \mathbf{e}) \tag{2.36}$$

where the N-vector $\mathbf{e}$ and the $N \times N$ positive-definite matrix E are defined by

$$E^{-1} \triangleq P^{-1} + H^T R^{-1} H, \qquad E^{-1}\mathbf{e} \triangleq P^{-1}\mathbf{p} + H^T R^{-1}\mathbf{r}. \tag{2.37}$$

Note that everything is well defined, since E is symmetric and positive-definite:

$$(E^{-1})^T = (P^{-1})^T + (H^T R^{-1} H)^T \tag{2.38}$$

$$= (P^T)^{-1} + H^T R^{-1} (H^T)^T \tag{2.39}$$

$$= P^{-1} + H^T R^{-1} H = E^{-1} \tag{2.40}$$

$$\mathbf{x}^T E^{-1} \mathbf{x} = \mathbf{x}^T P^{-1} \mathbf{x} + \mathbf{x}^T H^T R^{-1} H \mathbf{x} \geq \mathbf{x}^T P^{-1} \mathbf{x} > 0. \tag{2.41}$$

2.3.2 Bayes Formulation of the KF: Predictor

From the motion model (2.5), define a probability density function—the *Markov transition density*—that encapsulates the information contained in this model:

$$f_{k+1|k}(\mathbf{x}|\mathbf{x}') \triangleq N_{Q_k}(\mathbf{x} - F_k\mathbf{x}'). \tag{2.42}$$

Note that $f_{k+1|k}(\mathbf{x}|\mathbf{x}')$ is large when $\mathbf{x}$ is near the predicted state $F_k\mathbf{x}'$ and small when it is far from $F_k\mathbf{x}'$.

Likewise, define a probability density function that encapsulates the information contained in the Kalman state and covariance estimates $\mathbf{x}_{k|k}, P_{k|k}$ at time step k:

$$f_{k|k}(\mathbf{x}|Z^k) \triangleq N_{P_{k|k}}(\mathbf{x} - \mathbf{x}_{k|k}). \tag{2.43}$$

Again, $f_{k|k}(\mathbf{x}|Z^k)$ is large when the current target state $\mathbf{x}$ is near $\mathbf{x}_{k|k}$ and small when it is far from $\mathbf{x}_{k|k}$.

Since $f_{k|k}(\mathbf{x}'|Z^k)$ is the prior probability (density) that the target was in state $\mathbf{x}'$, the product $f_{k+1|k}(\mathbf{x}|\mathbf{x}') \cdot f_{k|k}(\mathbf{x}'|Z^k)$ is the probability (density) that the target will have state $\mathbf{x}$ at time step $k+1$ given that it had state $\mathbf{x}'$ at time step k. Integrating over all prior states $\mathbf{x}'$, we get the total probability (density) $f_{k+1|k}(\mathbf{x})$ that the target will have state $\mathbf{x}$ at time step $k+1$:

$$f_{k+1|k}(\mathbf{x}) \triangleq \int f_{k+1|k}(\mathbf{x}|\mathbf{x}') \cdot f_{k|k}(\mathbf{x}'|Z^k) d\mathbf{x}' \tag{2.44}$$

$$= \int N_{Q_k}(\mathbf{x} - F_k\mathbf{x}') \cdot N_{P_{k|k}}(\mathbf{x}' - \mathbf{x}_{k|k}) d\mathbf{x}'. \tag{2.45}$$

This is a probability density function, since

$$\int f_{k+1|k}(\mathbf{x}) d\mathbf{x} = \int \left(\int f_{k+1|k}(\mathbf{x}|\mathbf{x}') d\mathbf{x} \right) \cdot f_{k|k}(\mathbf{x}'|Z^k) d\mathbf{x}' \tag{2.46}$$

$$= \int f_{k|k}(\mathbf{x}'|Z^k) d\mathbf{x}' = 1. \tag{2.47}$$

Note that

$$f_{k+1|k}(\mathbf{x}) = \int N_{Q_k + F_k P_{k|k} F_k^T}(\mathbf{x} - F_k \mathbf{x}_{k|k}) \cdot N_E(\mathbf{x}' - \mathbf{e}) d\mathbf{x}' \tag{2.48}$$

$$= N_{Q_k + F_k P_{k|k} F_k^T}(\mathbf{x} - F_k \mathbf{x}_{k|k}) \cdot \int N_E(\mathbf{x}' - \mathbf{e}) d\mathbf{x}' \tag{2.49}$$

$$= N_{Q_k + F_k P_{k|k} F_k^T}(\mathbf{x} - F_k \mathbf{x}_{k|k}) \tag{2.50}$$

$$= N_{P_{k+1|k}}(\mathbf{x} - \mathbf{x}_{k+1|k}) \tag{2.51}$$

$$= f_{k+1|k}(\mathbf{x}|Z^k). \tag{2.52}$$

Equation (2.49) results from an application of (2.36) with $P = P_{k|k}$, $R = Q_k$, and $H = F_k$. Equation (2.50) results from the fact that indefinite integrals are translation invariant and that $N_E(\mathbf{x})$ is a probability density function.

In other words, the integral produces the same result that would be expected from application of the Kalman predictor equations, (2.6 and 2.7). Equation (2.44) is therefore

- *A mathematically equivalent formulation of the Kalman predictor step, expressed entirely in terms of probability density functions.*

Example 1 *I use a simple one-dimensional problem to illustrate the Bayes-Kalman filter predictor. Assume that* $x_{k|k} = -2$ *and* $P_{k|k} = 16$*, so that*

$$f_{k|k}(x) = N_{16}(x+2) \tag{2.53}$$

$$= \frac{1}{\sqrt{2\pi}\cdot 4} \cdot \exp\left(-\frac{(x+2)^2}{2\cdot 16}\right). \tag{2.54}$$

Further suppose that $F_k = -2$ *and* $Q_k = 4$*, so that the Markov transition density is*

$$f_{k+1|k}(x|x') = N_4(x+2x') \tag{2.55}$$

$$= \frac{1}{\sqrt{2\pi}\cdot 2} \cdot \exp\left(-\frac{(x+2x')^2}{2\cdot 4}\right). \tag{2.56}$$

It follows that $x_{k+1|k} = F_k x_{k|k} = (-2)\cdot(-2) = 4$ *and* $P_{k+1|k} = Q_k + F_k P_{k|k} F_k^T = 4 + 2\cdot 16\cdot 2 = 68$*. Thus the predicted posterior density is*

$$f_{k+1|k}(x) = N_{68}(x-4) \tag{2.57}$$

$$= \frac{1}{\sqrt{2\pi}\cdot 68} \cdot \exp\left(-\frac{(x-4)^2}{2\cdot 68}\right). \tag{2.58}$$

The transition from $f_{k|k}(x)$ *to* $f_{k+1|k}(x)$ *is shown in Figure 2.1.*

2.3.3 Bayes Formulation of the KF: Corrector

From the measurement model (2.8) define a probability density function—a *likelihood function*—that encapsulates the information contained in the model:

$$f_{k+1}(\mathbf{z}|\mathbf{x}) \triangleq N_{R_{k+1}}(\mathbf{z} - H_{k+1}\mathbf{x}). \tag{2.59}$$

Likewise, define a probability density function that encapsulates the information contained in the Kalman state and covariance estimates $\mathbf{x}_{k+1|k}$, $P_{k+1|k}$ at time step $k+1$:

$$f_{k+1|k}(\mathbf{x}|Z^k) \triangleq N_{P_{k+1|k}}(\mathbf{x} - \mathbf{x}_{k+1|k}). \tag{2.60}$$

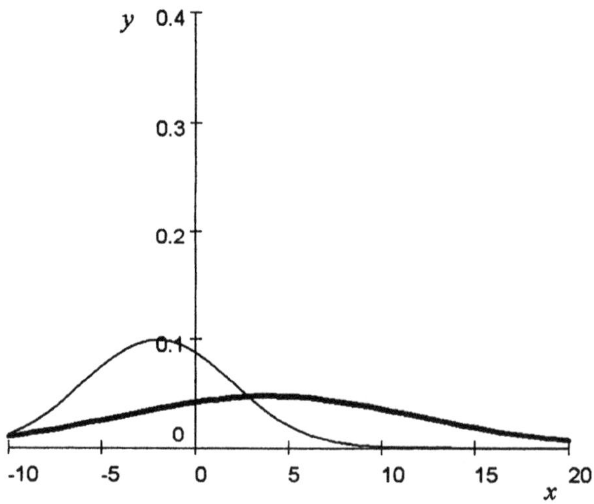

Figure 2.1 The motion-transition from the posterior at time step k (the curve on the left) to the predicted posterior at time step $k+1$ (the curve on the right). At time step k, the target is poorly localized. Because of the uncertainty introduced during the time between data collections, the uncertainty in target location has only increased.

Next, consider the posterior density function constructed using Bayes' rule:

$$f_{k+1|k+1}(\mathbf{x}) \triangleq \frac{f_{k+1}(\mathbf{z}_{k+1}|\mathbf{x}) \cdot f_{k+1|k}(\mathbf{x}|Z^k)}{\int f_{k+1}(\mathbf{z}_{k+1}|\mathbf{y}) \cdot f_{k+1|k}(\mathbf{y}|Z^k)d\mathbf{y}} \tag{2.61}$$

$$= \frac{N_{R_{k+1}}(\mathbf{z}_{k+1} - H_{k+1}\mathbf{x}) \cdot N_{P_{k+1|k}}(\mathbf{x} - \mathbf{x}_{k+1|k})}{\int N_{R_{k+1}}(\mathbf{z}_{k+1} - H_{k+1}\mathbf{y}) \cdot N_{P_{k+1|k}}(\mathbf{y} - \mathbf{x}_{k+1|k})d\mathbf{y}}. \tag{2.62}$$

The density $f_{k+1|k}(\mathbf{x}|Z^k)$ encapsulates all *a priori* information up until collection of $\mathbf{z}_{k+1}$. The density $f_{k+1}(\mathbf{z}_{k+1}|\mathbf{x})$ is largest for those $\mathbf{x}$ such that $H_{k+1}\mathbf{x}$ is closest to $\mathbf{z}_{k+1}$, and smallest for those $\mathbf{x}$ such that $H_{k+1}\mathbf{x}$ is distant from $\mathbf{z}_{k+1}$. Consequently, the product $f_{k+1}(\mathbf{z}_{k+1}|\mathbf{x}) \cdot f_{k+1|k}(\mathbf{x}|Z^k)$ adjusts the prior $f_{k+1|k}(\mathbf{x}|Z^k)$ using the new data, increasing its value at $\mathbf{x}$ if $\mathbf{z}_{k+1}$ favors $\mathbf{x}$ and decreasing it otherwise.

Now note that

$$\begin{aligned} f_{k+1|k+1}(\mathbf{x}) &= \frac{N_{D_{k+1}}(\mathbf{z}_{k+1} - H_{k+1}\mathbf{x}_{k+1|k})N_E(\mathbf{x} - \mathbf{e})}{N_{D_{k+1}}(\mathbf{z}_{k+1} - H_{k+1}\mathbf{x}_{k+1|k}) \int N_E(\mathbf{y} - \mathbf{e})d\mathbf{y}} && (2.63) \\ &= N_E(\mathbf{x} - \mathbf{e}) && (2.64) \\ &= N_{P_{k+1|k+1}}(\mathbf{x} - \mathbf{x}_{k+1|k+1}) && (2.65) \\ &= f_{k+1|k+1}(\mathbf{x}|Z^{k+1}). && (2.66) \end{aligned}$$

Equation (2.63) results from an application of (2.36) with $P = P_{k+1|k}$, $R = R_{k+1}$, and $H = H_{k+1}$; and in which

$$\begin{aligned} P^{-1}_{k+1|k+1} &= E^{-1} \triangleq P^{-1}_{k+1|k} + H^T_{k+1}R^{-1}_{k+1}H_{k+1} && (2.67) \\ P^{-1}_{k+1|k+1}\mathbf{x}_{k+1|k+1} &= E^{-1}\mathbf{e} \triangleq P^{-1}_{k+1|k}\mathbf{x}_{k+1|k} + H^T_{k+1}R^{-1}_{k+1}\mathbf{z}_{k+1} && (2.68) \\ D_{k+1} &\triangleq R_{k+1} + H_{k+1}P_{k+1|k}H^T_{k+1}. && (2.69) \end{aligned}$$

Equation (2.65) results from (2.16).

Equation (2.61) is therefore

- *A mathematically equivalent formulation of the Kalman corrector step, expressed entirely in terms of probability density functions.*

Example 2 *From Example 1 we know that the predicted posterior is*

$$f_{k+1|k}(x) = N_{68}(x-4) = \frac{1}{\sqrt{2\pi \cdot 66}} \cdot \exp\left(-\frac{(x-4)^2}{2 \cdot 68}\right). \tag{2.70}$$

Suppose that $R_{k+1} = 1$ *and* $H_{k+1} = 1$ *and* $z_{k+1} = 6$. *Then the likelihood function is*

$$f_{k+1}(z_{k+1}|x) = N_1(6 - x) \tag{2.71}$$

$$= \frac{1}{\sqrt{2\pi \cdot 1}} \cdot \exp\left(-\frac{(6-x)^2}{2 \cdot 1}\right) \tag{2.72}$$

Then

$$P^{-1}_{k+1|k+1} = P^{-1}_{k+1|k} + H^T_{k+1} R^{-1}_{k+1} H_{k+1} = 68^{-1} + 1 \cdot (1)^{-1} \cdot 1 = 1.0147 \tag{2.73}$$

and

$$P^{-1}_{k+1|k+1} x_{k+1|k+1} = P^{-1}_{k+1|k} x_{k+1|k} + H^T_{k+1} R^{-1}_{k+1} H_{k+1} \tag{2.74}$$

$$= 68^{-1} \cdot 4 + 1 \cdot (1)^{-1} \cdot 6 = 6.0588. \tag{2.75}$$

Thus $P_{k+1|k+1} = 0.98551$ *and* $x_{k+1|k+1} = 0.98551 \cdot 6.0588 = 5.971$. *The data-updated posterior distribution is*

$$f_{k+1|k+1}(x) = N_{0.98551}(x - 5.971) \tag{2.76}$$

$$= \frac{1}{\sqrt{2\pi \cdot 0.98551}} \cdot \exp\left(-\frac{(x-5.971)^2}{2 \cdot 0.98551}\right). \tag{2.77}$$

The transition from $f_{k+1|k}(x)$ *to* $f_{k+1|k+1}(x)$ *is pictured in Figure 2.2.*

2.3.4 Bayes Formulation of the KF: Estimation

In the Bayesian formulation of the corrector step, the end result is a probability distribution

$$f_{k+1|k+1}(\mathbf{x}|Z^{k+1}) = N_{P_{k+1|k+1}}(\mathbf{x} - \mathbf{x}_{k+1|k+1}). \tag{2.78}$$

However, what we actually require are estimates of the state and error covariance. These quantities can be extracted in closed form as

$$\int \mathbf{x} \cdot N_{P_{k+1|k+1}}(\mathbf{x} - \mathbf{x}_{k+1|k+1}) d\mathbf{x} = \mathbf{x}_{k+1|k+1} \tag{2.79}$$

$$\int \mathbf{x}\mathbf{x}^T \cdot N_{P_{k+1|k+1}}(\mathbf{x} - \mathbf{x}_{k+1|k+1}) d\mathbf{x} = P_{k+1|k+1}. \tag{2.80}$$

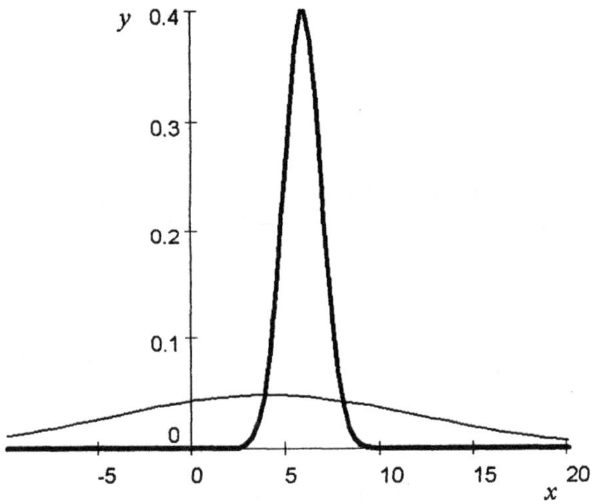

Figure 2.2 The data-updated transition from the predicted posterior at time step $k+1$ (bottom curve) to the data-upated posterior (top curve). The predicted target is localized at $x = 4$, but poorly so. Because sensor accuracy is good, the measurement-collection $z_{k+1} = 6$ results in a more accurate and localized position estimate of $x = 5.971$.

2.4 THE SINGLE-TARGET BAYES FILTER

In summary, thus far I have shown that the Kalman predictor and corrector steps are special cases of the formulas

$$f_{k+1|k}(\mathbf{x}|Z^k) = \int f_{k+1|k}(\mathbf{x}|\mathbf{x}') \cdot f_{k|k}(\mathbf{x}'|Z^k)d\mathbf{x}' \tag{2.81}$$

$$f_{k+1|k+1}(\mathbf{x}|Z^{k+1}) = \frac{f_{k+1}(\mathbf{z}_{k+1}|\mathbf{x}) \cdot f_{k+1|k}(\mathbf{x}|Z^k)}{f_{k+1}(\mathbf{z}_{k+1}|Z^k)} \tag{2.82}$$

where $Z^k : \mathbf{z}_1, ..., \mathbf{z}_k$ is the time sequence of collected observations at time step k; and where

$$f_{k+1}(\mathbf{z}_{k+1}|Z^k) \triangleq \int f_{k+1}(\mathbf{z}_{k+1}|\mathbf{x}) \cdot f_{k+1|k}(\mathbf{x}|Z^k)d\mathbf{x} \tag{2.83}$$

is the Bayes normalization factor. Equations (2.81) and (2.82) define the *single-sensor, single-target Bayes recursive filter* (see [11, 38, 99, 216]). It propagates through time a sequence

$$f_{0|0}(\mathbf{x}|Z^0) \rightarrow f_{1|0}(\mathbf{x}|Z^0) \rightarrow f_{1|1}(\mathbf{x}|Z^1) \rightarrow ... \tag{2.84}$$

$$\rightarrow f_{k|k}(\mathbf{x}|Z^k) \rightarrow f_{k+1|k}(\mathbf{x}|Z^k) \rightarrow f_{k+1|k+1}(\mathbf{x}|Z^{k+1}) \tag{2.85}$$

$$\rightarrow \ldots \tag{2.86}$$

of *Bayes posterior probability distributions* (also known as "filtering densities"). It is the theoretical foundation for single-target fusion, detection, tracking, and identification. Its generalizations, to be introduced in later chapters, will form the foundation for single-target information fusion using generalized information, and for multisensor-multitarget fusion, detection, tracking, and identification.

Though not considered in this volume, its further generalizations are the basis for unified approaches to sensor management [139, 149, 150], group target detection and tracking [130, 131], and so on.

The following sections describe the single-target Bayes filter in more detail. I begin in Section 2.4.1 with a simple one-dimensional target detection and tracking problem that illustrates the potential power of the Bayes filter. This example will also lead us to a better understanding of the behavior of the Kalman and general Bayes filters in Section 2.4.2. Modeling issues, and the accepted approach for addressing them, *formal Bayes modeling*, are described in Sections 2.4.3 and 2.4.4, respectively.

The Bayes filter initialization, predictor, and corrector steps are addressed in Sections 2.4.5, 2.4.6, and 2.4.7, respectively. Determination of state estimates, and of the uncertainty in those estimates, are described in Sections 2.4.8 and 2.4.9, respectively. Fusion of data from different sensors is addressed in Section 2.4.10. Section 2.4.11 describes computational issues surrounding the Bayes filter.

In Sections 2.5.1-2.5.3, I sketch some commonly used real-time implementation approaches: Taylor series, Gaussian-mixture, and sequential Monte Carlo (also known as particle systems).

2.4.1 Single-Target Bayes Filter: An Illustration

In this section I request the forbearance of the reader. Many concepts employed here—false detections (false alarms), missed detections, and so on—will not be formally introduced until Chapter 9.

The problem to be addressed is depicted in Figure 2.3.[2] A particle moves around a circular "racetrack," the perimeter of which has unit length. Its motion is uniform but dithered by Brownian-motion perturbations. Its state is $(x, v)^T$, where x is position and v is velocity. At midscenario (the time of the 250th observation-collection), the target abruptly reverses direction.

The target is interrogated by a position-observing sensor, the likelihood function of which is a non-Gaussian "witch's hat" function (see Figure 2.3). During each scan the sensor collects an observation from the target 93% of the time. During each scan, the sensor also collects an average of 40 *false detections* (see Section 9.2.1). The false detections are uniformly distributed spatially over the racetrack and are Poisson-distributed in time. (That is, the probability that there will be n false detections at any time is given by the Poisson distribution $p(n) = e^{-40}40^n/n!$) Using techniques to be introduced in Part II, a measurement model is constructed from this information and the true likelihood function $f_{k+1}(Z|\mathbf{x})$ constructed from that. The assumed motion model for the target is constant-velocity: $\varphi(x, v)^T = (x + \Delta t \cdot v, v)^T$.

Figure 2.4 shows a time sequence of observations. The horizontal axis is observation number and the vertical axis is position on the racetrack. The target-generated observation track is essentially undetectable by the naked eye, despite the fact we have an advantage that the filter does not: being able to examine the entire scenario all at once. Nevertheless, the Bayes filter must determine if a target is present, find it, and then track it. It is not provided with an estimate of initial target position, or any indication that the target actually exists. (That is, the

2 This material first appeared in [156, p. 227].

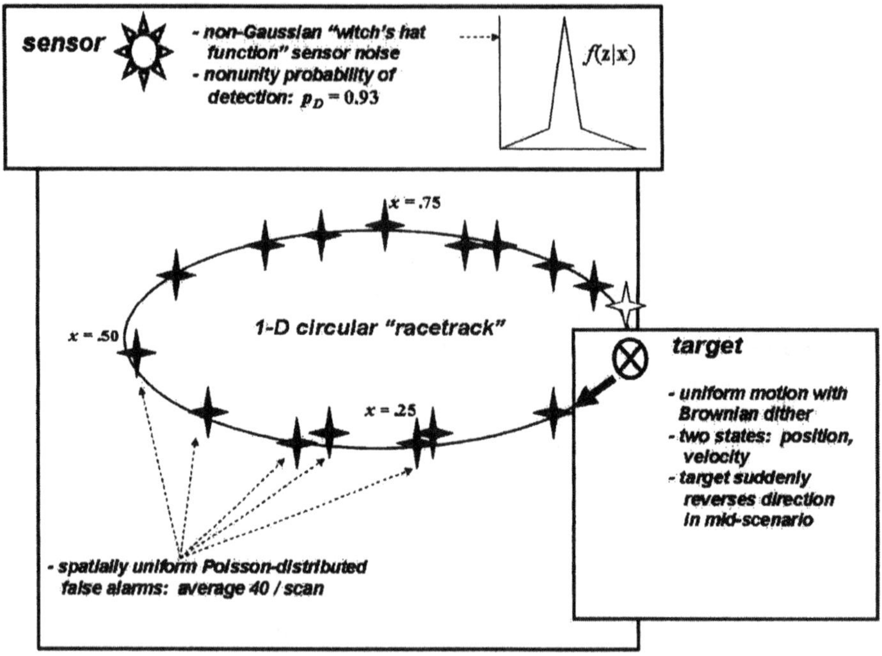

Figure 2.3 A single target moves around a circular "racetrack" with uniform motion perturbed by a small Brownian dither. The target is observed by a sensor, the noise statistics of which are governed by a non-Gaussian "witch's hat" distribution. The target is not detected 7% of the time. An average of 40 false measurements are collected per scan. (© 2004 IEEE. Reprinted from [148] with permission.)

initial distribution $f_{0|0}(\mathbf{x})$ is chosen to be uniform.) After the abrupt maneuver at midscenario, the filter must reacquire and relock on the target.

Figure 2.5 shows the target trajectory (solid line) and position estimates produced by the Bayes filter (dots). The filter finds and locks onto the target after about 20 scans, though becoming momentarily confused afterwards. After midscenario, about 20 observations are required before the filter detects the abrupt maneuver and another 20 before it reacquires and relocks. The delays arise from the bias created by the constant-velocity motion model, which must be overcome by accumulation of countervailing data.

Though the scenario is an artificially simple one it clearly demonstrates that the general Bayes filter is, at least in principle, capable of addressing practical problems that would initially appear to defy solution.[3]

2.4.2 Relationship Between the Bayes and Kalman Filters

At the beginning of Section 2.3, I questioned the pertinence of least-squares theoretical formulations of the Kalman filter, given that they appeared to be based on expediency. I questioned their value for practical intuition for the same reason. We are in a better position to address such questions because of the Bayesian formulation of the Kalman filter explicated in Section 2.3 and because of the example just presented in Section 2.4.1.

First, it is now clear that the least-squares objective function is not a mere mathematical convenience. Rather, it is central to the theory of the Kalman filter. It arises inevitably when one does the following:

- Stipulates that the sensor has linear-Gaussian statistics;
- Asks the question: What special form must the Markov density $f_{k+1|k}(\mathbf{x}|\mathbf{x}')$ and the posterior distributions $f_{k|k}(\mathbf{x}|Z^k)$ and $f_{k+1|k}(\mathbf{x}|Z^k)$ have so that the Bayes filter equations (2.81) and (2.82) can be solved in closed form?

The answer—that these distributions must also be linear-Gaussian and therefore also definable in terms of quadratics—is what makes least-squares optimization intrinsic to the Kalman filter.

Second, the Bayesian formulation of the Kalman filter provides an intuitive basis for understanding why the Kalman filter works when it does, and why it does not work when it does not.

3 I revisit this problem in Section 14.7.

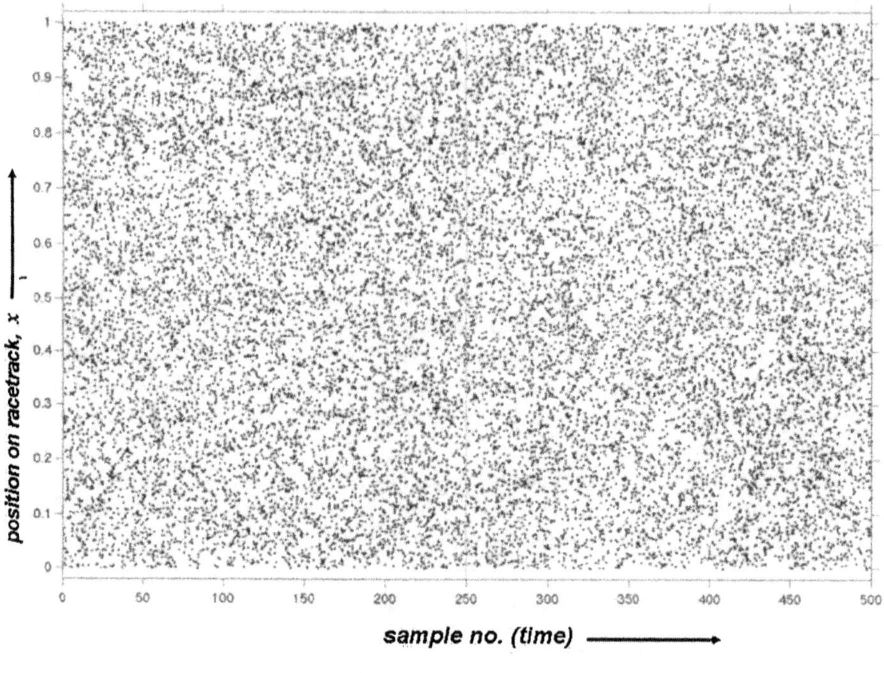

Figure 2.4 The measurements—both target detections and false detections—collected by the sensor during the entire scenario. The target path is essentially invisible to the human eye, even though we have the advantage of pattern recognition throughout the entire scenario. (© 2004 IEEE. Reprinted from [148] with permission.)

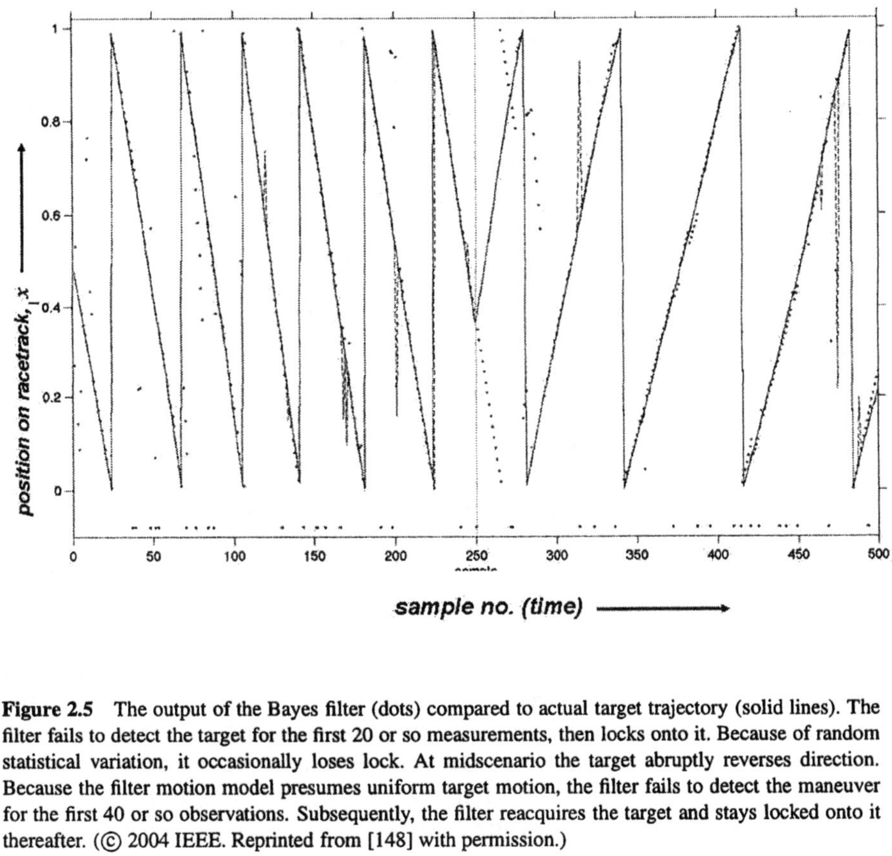

Figure 2.5 The output of the Bayes filter (dots) compared to actual target trajectory (solid lines). The filter fails to detect the target for the first 20 or so measurements, then locks onto it. Because of random statistical variation, it occasionally loses lock. At midscenario the target abruptly reverses direction. Because the filter motion model presumes uniform target motion, the filter fails to detect the maneuver for the first 40 or so observations. Subsequently, the filter reacquires the target and stays locked onto it thereafter. (© 2004 IEEE. Reprinted from [148] with permission.)

The Kalman filter is based on the assumption that sensor and target behaviors are benign enough that the single-target system can be adequately explained by posterior distributions $f_{k|k}(\mathbf{x}|Z^k)$ that are essentially unimodal and not too skewed. When this assumption is valid, the Kalman filter performs adequately. When this assumption is violated—as occurs with rapid target maneuvers or low signal-to-noise ratio—it will fail.

Consider rapid target maneuvers first. Assume a one-dimensional problem like that of Section 2.4.1, with target state $(x, v_x)^T$ and constant-velocity motion model $(x, v_x)^T \mapsto (x + v_x \Delta t, v_x)^T$ where Δt is the time interval between measurements. Suppose that, much as in Section 2.4.1, the target moves with constant speed and reverses direction but, now, not instantaneously. Rather, it slows down until it comes to a stop and then, after the reversal, regains its former speed. I assume that Δt is small enough that several measurements can be collected during the slowdown, direction reversal, and speedup.

Figure 2.6 illustrates typical Kalman filter performance under such conditions and compares it to the corresponding behavior of the Bayes filter. As the maneuver commences, the Bayes posterior develops a minor second mode, built up from newly collected measurements during the maneuver. The major mode continues in the positive direction because of the constant-velocity motion model. The Kalman filter, which attempts to approximate the actual posterior with a Gaussian distribution, overshoots the actual trajectory; and its error ellipses increase in size. This reflects increasing uncertainty due to the increasing discrepancy between measurements and state-estimates. Because the data rate is high enough relative to the rapidity of the maneuver, the two Bayes filter modes remain cojoined and can thus still be roughly approximated by a Gaussian even as the formerly minor mode becomes dominant. As measurements accumulate, the minor mode disappears and the new dominant mode—as well as the Kalman filter—locks back onto the actual target position.[4]

Figure 2.7 illustrates the result of repeating this experiment, now assuming that significantly fewer measurements are collected during the maneuver. In this case the two modes of the posterior are sufficiently separated that the posterior can no longer be well approximated by a Gaussian distribution. In this case the Kalman filter will diverge but the full Bayes filter will not. This explains the behavior of the Bayes filter in the example of Section 2.4.1.

4 Many approaches have been devised for making the Kalman filter respond more robustly to maneuvers. While many are ad hoc, in recent years more principled approaches—most notably, interacting multiple model (IMM) techniques—have achieved popularity.

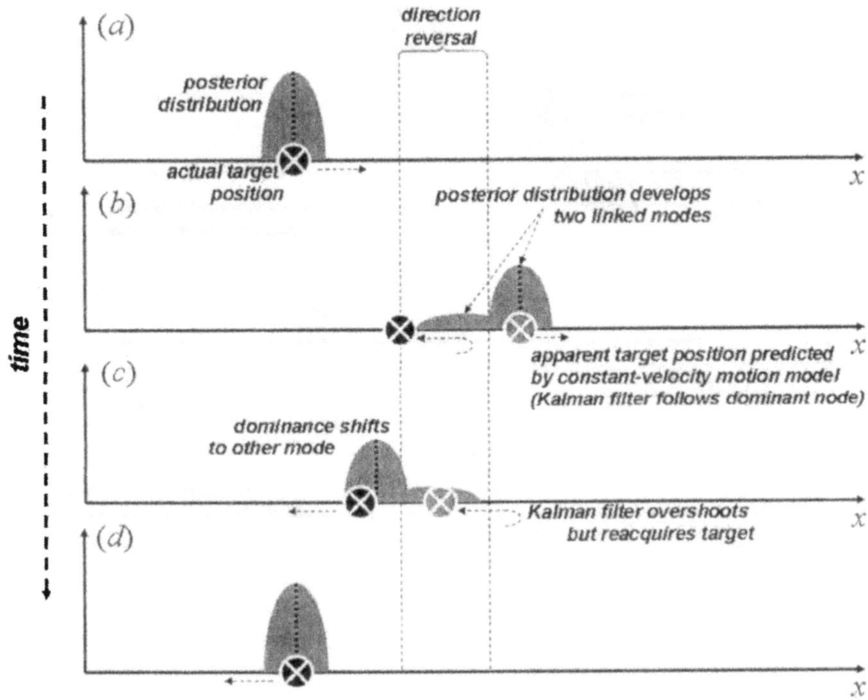

Figure 2.6 A schematic illustration of the time evolution of the Bayes posterior distribution for a one-dimensional target that gradually reverses direction. (a) The mode of the posterior closely tracks actual target position. (b) Following the maneuver, the posterior develops two modes. The dominant mode is determined by the constant-velocity motion model and the minor mode by the measurements, but jointly the two modes can be roughly approximated as a single-mode Gaussian distribution. (c) As measurements accumulate, the previously minor mode becomes dominant and more closely tracks the actual target position. (d) The formerly dominant mode vanishes, and the new dominant mode closely tracks the target.

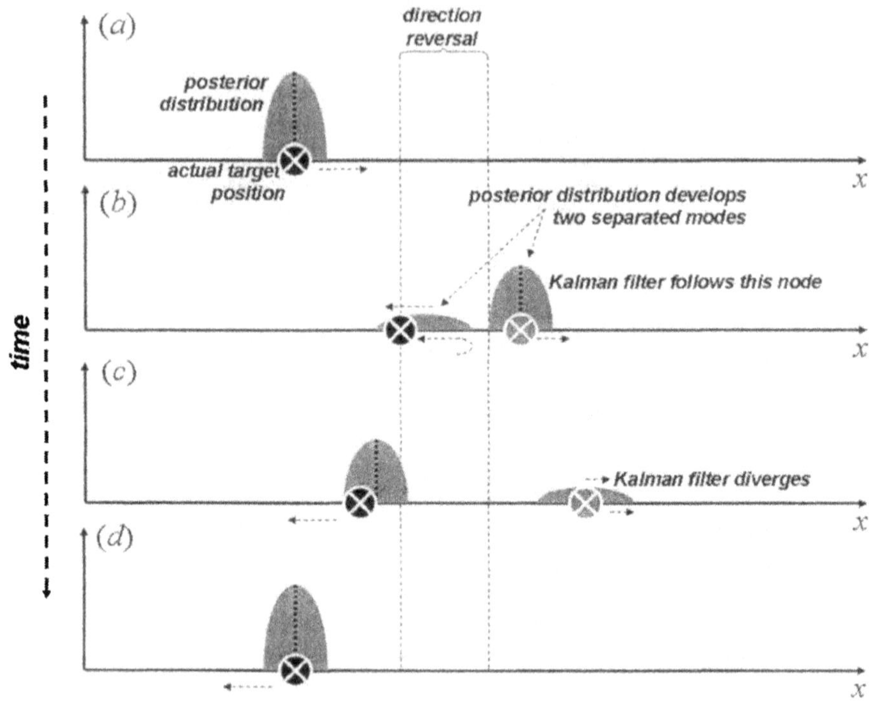

Figure 2.7 The figure schematically illustrates the time evolution of the Bayes posterior distribution for a one-dimensional target that rapidly reverses direction. (a) The mode of the posterior closely tracks actual target position. (b) Following the maneuver, the posterior develops two modes. The two modes are significantly separated and can no longer be readily approximated as a Gaussian. (c) The previously minor mode becomes dominant, but the Kalman filter has been "fooled" into following the previously dominant—but now minor—mode. (d) As the formerly dominant mode vanishes, the Kalman filter diverges.

The same reasoning explains the Kalman filter's behavior in low-SNR scenarios. Once again assume a one-dimensional problem with state $(x, v_x)^T$. Let measurements be pixelized image intensity vectors and assume a measurement model of the form $z_x = x + W_x$ where W_x is zero-mean noise. If noise in nontarget pixels is small, then $f_{k|k}(x|Z^k)$ will have a single dominant mode—corresponding to the target state—and a large number of very small, randomly shifting minor modes induced by noise in nontarget pixels. This is pictured in Figure 2.8(a). As noise power increases these minor modes increase in size until they begin to account for an increasingly greater percentage of the probability mass of $f_{k|k}(x|Z^k)$. At a certain point this percentage is so high that $f_{k|k}(x|Z^k)$ is highly multimodal, as pictured in Figure 2.8(b). The Bayes filter can still determine an accurate estimate of the target state, but only after recursively processing a sufficiently large number of observations. The smaller the SNR, the more measurements that must be collected to achieve this.

2.4.3 Single-Target Bayes Filter: Modeling

The Bayes filtering equations (2.81) and (2.82) provide a potentially powerful basis for single-sensor, single-target problems. However, we should be concerned at this point. As will be shown in Sections 2.4.6 and 2.4.7, these equations are straightforward consequences of two elementary principles of probability theory: the total probability theorem and Bayes' rule. What this means is that I have developed a foundation for single-target information fusion consisting of simple mathematical abstractions. In particular, I have solved the problem by using two apparently magical quantities:

- The density function $f_{k+1|k}(\mathbf{x}|\mathbf{x}')$ (the *Markov transition density* of the target);
- $L_{k+1,\mathbf{z}}(\mathbf{x}) \triangleq f_{k+1}(\mathbf{z}|\mathbf{x})$ (the *sensor likelihood function*).

Theoretically speaking, both are conditional probability density functions. However, this bare fact merely transforms magic into equally unhelpful "mathemagic." The real questions that must be answered are these:

- What explicit, general procedures allow us to derive concrete formulas for $f_{k+1|k}(\mathbf{x}|\mathbf{x}')$ and $L_{k+1,\mathbf{z}}(\mathbf{x})$ in any specific practical situation?

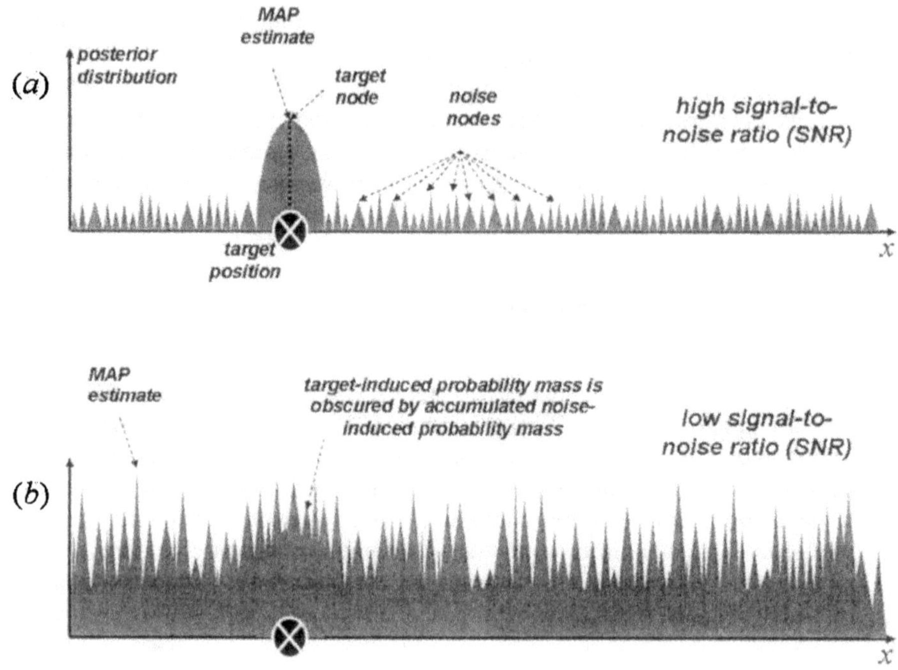

Figure 2.8 The posterior distribution of a motionless target obscured in a noisy background. (a) If noise power is small, the target state is easily extracted from the posterior using a MAP estimator. (b) If noise power is large, the target-generated mode is buried in a large number of noise-generated modes.

- How do we know that these formulas are "true," in the sense that $L_{k+1,\mathbf{z}}(\mathbf{x})$ faithfully reflects the actual behavior of the sensor and that $f_{k+1|k}(\mathbf{x}|\mathbf{x}')$ faithfully reflects our underlying assumptions about interim target motion?

As a specific example:

- How do we know that the density functions of (2.42) and (2.59) faithfully encapsulate all of the information contained in the Kalman measurement and motion models of (2.5) and (2.8), respectively—but without inadvertently incorporating information extraneous to the problem?
- What does "faithfully encapsulate" even mean in precise terms?
- Stated more informally: How do we know that $f_{k+1|k}(\mathbf{x}|\mathbf{x}')$ and $L_{k+1,\mathbf{z}}(\mathbf{x})$ are not ad hoc contrivances or have not been erroneously constructed?

Answering these questions incorrectly can have significant implications in practice. For example, if the likelihood $L_{k+1,\mathbf{z}}(\mathbf{x})$ is not true, then any claim of "optimality," Bayes or otherwise, is hollow, because it applies only to whatever sensor is actually modeled by the incorrect likelihood. In particular, if $L_{k+1,\mathbf{z}}(\mathbf{x})$ is too inaccurately modeled then an algorithm will "waste" data trying (and perhaps failing) to overcome the mismatch with reality.

More generally, the real algorithm-design issue is sidestepped:

- What does one do when $L_{k+1,\mathbf{z}}(\mathbf{x})$ cannot be well-characterized—as, for example, may be the case with synthetic aperture radar (SAR)and high range-resolution radar (HRRR) signatures?

Analogous comments apply to the Markov density $f_{k+1|k}(\mathbf{x}|\mathbf{x}')$. The more accurately that it models interim target motion, the more effectively the Bayes filter will do its job. Otherwise, a certain amount of data must be expended in overcoming poor motion-model selection. This leads to the following question:

- How do we ensure that $f_{k+1|k}(\mathbf{x}|\mathbf{x}')$ faithfully reflects target motion if we have modeled it correctly?

2.4.3.1 Heuristic Modeling

The answer to these questions, *formal Bayes modeling*, is to be contrasted with *heuristic modeling*. Heuristic modeling takes a stance exactly opposite to the one advocated in this book. It can be summarized in the following statement:

- *It is not very important how one models an application—all one needs are some internal algorithm parameters to adjust.*

The very existence of this standpoint *has everything to do with computers.* Computers have been in many respects a boon to developers of information fusion algorithms. Many techniques that once would have been impossible to implement in real time are now potentially viable in at least some applications. (The Bayes filter is but one instance.)

However, in other respects computers have had a detrimental effect on the course of algorithm research and development. The easy availability of inexpensive computing power and the existence of efficient high-level computer languages has led to a greater emphasis on computer science and to a lessened emphasis on classical physics. This has resulted in what might be called an "implement before think" attitude toward algorithm development. This includes a tendency toward ignoring the single most critical and difficult step in algorithm development:

- *Careful specification of concrete real-world statistical models, on the basis of which practical implementation can proceed.*

That is, instead of beginning with models and moving toward an algorithmic approach and from there to a computer implementation technique, one does the opposite. One begins with some specific computer implementation technique and an algorithmic approach based on it. As a result, both the models and the approach typically end up as ad hoc contrivances, the form of which is dictated as much by constraints imposed by the implementation technique as by real-world phenomenology.

Algorithm behavior may be difficult to diagnose because of hidden assumptions and ad hoc design choices—the "lumpy pillow syndrome." That is, let an algorithm be analogized as a pillow and various kinds of algorithmic misbehavior as its "lumps." Trial-and-error adjustment of one parameter may smooth out one lump, but cause misbehavior elsewhere. So one varies another parameter by trial-and-error to remove the new lump, resulting in still another lump. Additional trial-and-error leads to even more lumps and even more parameter adjusting.

Any experienced algorithm implementer will be all too familiar with the result: confusion and make-do "fixes" engendered by the fact that too many parameters have been varied for too many poorly understood reasons.

Challa, Evans, and Musicki have succinctly summarized one of the primary viewpoints of this book [22, p. 437]:

> In practice, an intelligent combination of probability theory using Bayes' theorem and ad hoc logic is used to solve tracking problems. The ad hoc aspects of practical tracking algorithms are there to limit the complexity of the probability theory-based solutions. One of the key things that is worth noting is that the success of Bayesian solution depends on the models used. Hence, before applying Bayes' theorem to practical problems, a significant effort must be spent on modeling.

Indeed, most of the chapters of this book are devoted to the careful development of a general, unified theory of data modeling for the following:

- Unconventional measurements (Chapters 3-7);
- Unconventional state-estimates (Chapter 8);
- Multitarget measurements (Chapters 9, 12);
- Multitarget motion (Chapter 13).

2.4.3.2 Formal Modeling

Formal Bayes modeling has become the accepted "gold standard" for modeling in single-target tracking applications, especially in research and development. Formal modeling is what takes the filter equations (2.81) and (2.82) out of the realm of mathematical abstraction and connects them with the real world. The end products of formal Bayes modeling are general, *implementation-free* mathematical formulas for Markov densities and likelihood functions, constructed from general statistical models. Once one has chosen a specific implementation technique these formulas can then be carefully reformulated in terms of this technique.

This helps ensure that the statistical assumptions underlying the original models are being implemented as faithfully as possible. This in turn helps ensure that the conditions under which the implemented filter will be approximately optimal are explicitly specified. Algorithm behavior should also tend to be more explicable, because assumptions and important design decisions in both the model and the approximations have been carefully parsed into a systematic, disciplined chain of reasoning.

Formal modeling helps avoid the "lumpy pillow syndrome," because it clarifies which parameters are important, to what degree, and how they are interrelated. Stated differently:

- It reduces to a more manageable size the algorithmic solution space that must be searched.

2.4.4 Single-Target Bayes Filter: Formal Bayes Modeling

Bayes filtering is based on the so-called *dynamic state space model* [108, p. 2592]. This model consists of the following six steps (to be described in more detail in the sections that follow):

1. Mathematically modeling the states of the single-target system;
2. Mathematically modeling the measurements collected by the sensor;
3. Devising a statistical motion model for the target;
4. Devising a statistical measurement model for the sensor;
5. Constructing a true Markov transition density from the motion model;
6. Constructing a true likelihood function from the measurement model.

2.4.4.1 The States of a Single-Target System

Suppose that a single sensor collects data from a single, moving target. Thinking like a physicist, we first precisely model the physical "states" that our "single-particle system"—the target—could be in. The state model is typically (but not always) a vector, such as

$$\mathbf{x} = (x, y, z, v_x, v_y, v_z, a_x, a_y, a_z, c)^T, \tag{2.87}$$

that contains pertinent target descriptors such as position x, y, z, velocity v_x, v_y, v_z, acceleration a_x, a_y, a_z, target identity, type or label c, and so on.

The *state space* is the set $\mathfrak{X}_0$ of all target states. In most information fusion applications it is one of the following.

- A Euclidean vector space, $\mathfrak{X}_0 = \mathbb{R}^N$.
- A finite set, $\mathfrak{X}_0 = C$.
- A hybrid space, $\mathfrak{X}_0 = \mathbb{R}^N \times C$, where '$\times$' denotes Cartesian product.

2.4.4.2 Conventional Single-Sensor Measurements of a Single-Target System

We must precisely model the observation that the sensor collects. Measurements are frequently (but not always) modeled as vectors. For example, the sensor may observe only target position in spherical coordinates, in which case observations will be vectors, such as $\mathbf{z} = (\rho, \theta, \phi)^T$. The *measurement space* is the set $\mathfrak{Z}_0$ of all measurements, and is usually Euclidean, $\mathfrak{X}_0 = \mathbb{R}^M$, finite, $\mathfrak{X}_0 = D$, or a hybrid, $\mathfrak{X}_0 = \mathbb{R}^M \times D$.

Vector representations are rich enough to model a wide variety of sensor types. I discuss some of the most important: signatures, images and frames, scans, and detections/reports.

Generally speaking, a *signature* is a scalar- or vector-valued function $s(t)$ of time over some time interval $t_0 \leq t \leq t_1$. The most familiar example is the radio frequency (RF) intensity output of a rotating-antenna radar as it sweeps out a single 2π radian "scan." If the scan is divided into a number of measurement "bins," it can be modeled as a scan-vector $\mathbf{z}$.

High range-resolution radar (HRRR) [194] provides another example. An HRRR transmits a very narrow range pulse, which is back-propagated as it is reflected from a target. Time-of-arrival information at the radar receiver is converted to ranges, resulting in an intensity-versus-range signature, which is conventionally discretized into range-bins and represented as a vector.

The intensity-pixels in a synthetic aperture radar (SAR) intensity image can similarly be represented as the elements of an image vector.

More generally, any of these RF outputs could be in in-phase and quadrature (I&Q) form. In this case the signature is complex-valued, and the modulus and phase can be discretized into bins and represented as a vector.

A *detection* results from applying some threshold to an intensity signature. In the case of a phased-array radar, for example, a detection could be any $\mathbf{z} = (\rho, \theta, \phi)$ such that $s(\rho, \theta, \phi) > \tau$ for a suitable threshold τ.[5]

2.4.4.3 Conventional Single-Target Motion Models

I introduced the concept of formal statistical motion models $\mathbf{X}_{k+1} = F_k \mathbf{x} + \mathbf{V}_k$ for the Kalman filter in Section 2.2.2. Such models are *linear* and *additive*. Formal modeling can encompass far more general models. The most obvious

5 In general, the collection $\mathbf{z}_1, ..., \mathbf{z}_m$ of detections extracted from an intensity signature will vary in number m. As we shall see in Chapter 9, such data is most conveniently modeled as a finite set $Z = \{\mathbf{z}_1, ..., \mathbf{z}_m\}$.

generalizations are nonlinear-additive models of the form

$$\mathbf{X}_{k+1} = \varphi_k(\mathbf{x}) + \mathbf{V}_k \tag{2.88}$$

where $\varphi_k(\mathbf{x})$ is a (in general) nonlinear vector-valued function of $\mathbf{x}$. Most generally, we can assume fully nonlinear models of the form

$$\mathbf{X}_{k+1} = \varphi_k(\mathbf{x}, \mathbf{V}_k) \tag{2.89}$$

where $\varphi_k(\mathbf{x}, \mathbf{x}')$ is nonlinear in both $\mathbf{x}$ and $\mathbf{x}'$. For the purposes of deriving Markov densities, we must also assume that $\varphi_{k,\mathbf{x}}(\mathbf{x}') \triangleq \varphi_k(\mathbf{x}, \mathbf{x}')$ is a family of nonsingular vector transformations indexed by $\mathbf{x}$. In this case we can also write $\varphi_{k,\mathbf{x}}^{-1}(\mathbf{X}_{k+1}) = \mathbf{V}_k$.

2.4.4.4 Conventional Single-Sensor, Single-Target Measurement Models

Essentially identical comments apply to the linear-additive statistical measurement models $\mathbf{Z}_{k+1} = H_{k+1}\mathbf{x} + \mathbf{W}_{k+1}$ that we introduced for the Kalman filter in Section 2.2.3. These can be generalized to nonlinear-additive models

$$\mathbf{Z}_{k+1} = \eta_{k+1}(\mathbf{x}) + \mathbf{W}_{k+1} \tag{2.90}$$

and from there to fully nonlinear models

$$\mathbf{Z}_{k+1} = \eta_{k+1}(\mathbf{x}, \mathbf{W}_{k+1}). \tag{2.91}$$

Once again, it must be assumed that $\eta_{k+1,\mathbf{x}}(\mathbf{z}) \triangleq \eta_{k+1}(\mathbf{x}, \mathbf{z})$ is a family of nonsingular transformations indexed by $\mathbf{x}$, so that we can also write $\eta_{k+1,\mathbf{x}}^{-1}(\mathbf{Z}_{k+1}) = \mathbf{W}_{k+1}$.

2.4.4.5 True Markov Transition Densities

In practice, the true Markov density $f_{k+1|k}(\mathbf{x}|\mathbf{x}')$ corresponding to the additive motion model $\mathbf{X}_{k+1} = \varphi_k(\mathbf{x}) + \mathbf{V}_k$ can be simply obtained from standard reference texts such as [11, 27]. It is essentially identical in form to (2.42):

$$f_{k+1|k}(\mathbf{x}|\mathbf{x}') = N_{Q_k}(\mathbf{x} - \varphi_k(\mathbf{x})). \tag{2.92}$$

However,

- *How do we actually know that this is the Markov density that faithfully encapsulates the information in the model* $\mathbf{X}_{k+1} = \varphi_k(\mathbf{x}) + \mathbf{V}_k$?

This is not a purely theoretical question. Indeed, answering it will be central to my approach to formally modeling more nontraditional sources of information such as human operators, human observers, feature extraction algorithms, and rule-bases. Consequently, it is necessary to understand how it is addressed in the more familiar context of the single-target problem.

I begin with the following well known fact from probability theory. For any region S of state space $\mathfrak{X}_0$,

$$\Pr(\mathbf{X}_{k+1} \in S|\mathbf{x}') = \int_S f_{\mathbf{X}_{k+1}}(\mathbf{x}|\mathbf{x}')d\mathbf{x} \tag{2.93}$$

where the Markov transition density $f_{k+1|k}(\mathbf{x}|\mathbf{x}') \triangleq f_{\mathbf{X}_{k+1}}(\mathbf{x})$ is the probability density function of $\mathbf{X}_{k+1}$. Then

$$\begin{aligned}
\int_S f_{k+1|k}(\mathbf{x}|\mathbf{x}')d\mathbf{x} &= \Pr(\mathbf{X}_{k+1} \in S|\mathbf{x}') && (2.94)\\
&= \Pr(F_k\mathbf{x}' + \mathbf{V}_k \in S|\mathbf{x}') && (2.95)\\
&= \Pr(\mathbf{V}_k \in S - F_k\mathbf{x}') && (2.96)\\
&= \int_{S-F_k\mathbf{x}'} f_{\mathbf{V}_k}(\mathbf{x})d\mathbf{x} && (2.97)\\
&= \int \mathbf{1}_{S-F_k\mathbf{x}'}(\mathbf{x}) \cdot f_{\mathbf{V}_k}(\mathbf{x})d\mathbf{x} && (2.98)\\
&= \int \mathbf{1}_S(\mathbf{x} + F_k\mathbf{x}') \cdot f_{\mathbf{V}_k}(\mathbf{x})d\mathbf{x} && (2.99)\\
&= \int \mathbf{1}_S(\mathbf{x}) \cdot f_{\mathbf{V}_k}(\mathbf{x} - F_k\mathbf{x}')d\mathbf{x} && (2.100)\\
&= \int_S f_{\mathbf{V}_k}(\mathbf{x} - F_k\mathbf{x}')d\mathbf{x} && (2.101)
\end{aligned}$$

where

$$S - F_k\mathbf{x}' \triangleq \{\mathbf{x} - F_k\mathbf{x}' |\ \mathbf{x} \in S\} \tag{2.102}$$

and where (2.100) follows from the translation invariance of the indefinite integral. Since (2.94)-(2.101) are true for all S, it follows that

$$f_{k+1|k}(\mathbf{x}|\mathbf{x}') = f_{\mathbf{V}_k}(\mathbf{x} - F_k\mathbf{x}') \tag{2.103}$$

almost everywhere.

The same analysis can be applied to general nonlinear models $\mathbf{X}_{k+1} = \varphi_k(\mathbf{x}, \mathbf{V}_k)$, though with greater complication. The true Markov transition density for the general nonlinear motion model is

$$\int_S f_{k+1|k}(\mathbf{x}|\mathbf{x}')d\mathbf{x} = \Pr(\mathbf{X}_{k+1} \in S|\mathbf{x}') \tag{2.104}$$

$$= \Pr(\varphi_{k,\mathbf{x}'}(\mathbf{V}_k) \in S|\mathbf{x}') \tag{2.105}$$

$$= \Pr(\mathbf{V}_k \in \varphi_{k,\mathbf{x}'}^{-1}(S)) \tag{2.106}$$

$$= \int \mathbf{1}_{\varphi_{k,\mathbf{x}'}^{-1}(S)}(\mathbf{x}) \cdot f_{\mathbf{V}_k}(\mathbf{x})d\mathbf{x} \tag{2.107}$$

$$= \int \mathbf{1}_S(\varphi_{k,\mathbf{x}}(\mathbf{x}')) \cdot f_{\mathbf{V}_k}(\mathbf{x})d\mathbf{x} \tag{2.108}$$

$$= \int \mathbf{1}_S(\mathbf{y}) \cdot f_{\mathbf{V}_k}(\varphi_{k,\mathbf{x}'}^{-1}(\mathbf{y})) \cdot J_{\varphi_{k,\mathbf{x}'}^{-1}}(\mathbf{y})d\mathbf{y} \tag{2.109}$$

$$= \int_S f_{\mathbf{V}_k}(\varphi_{k,\mathbf{x}'}^{-1}(\mathbf{y})) \cdot J_{\varphi_{k,\mathbf{x}'}^{-1}}(\mathbf{y})d\mathbf{y} \tag{2.110}$$

where $J_{\varphi_{k,\mathbf{x}'}^{-1}}(\mathbf{y})$ is the Jacobian determinant of the transformation $\varphi_{k,\mathbf{x}'}^{-1}$. Equation (2.109) is a consequence of the integral change of variables formula [67, p. 104]

$$\int f(\mathbf{x})d\mathbf{x} = \int f(T(\mathbf{y})) \cdot J_T(\mathbf{y})d\mathbf{y}. \tag{2.111}$$

Thus

$$f_{k+1|k}(\mathbf{x}|\mathbf{x}') = f_{\mathbf{V}_k}(\varphi_{k,\mathbf{x}'}^{-1}(\mathbf{x})) \cdot J_{\varphi_{k,\mathbf{x}'}^{-1}}(\mathbf{x}). \tag{2.112}$$

2.4.4.6 True Likelihood Functions

The discussion of the previous section is applicable essentially without alteration to measurement models. Using the same derivation as in the previous section, the true likelihood function $f_{k+1}(\mathbf{z}|\mathbf{x})$ corresponding to the additive measurement model $\mathbf{Z}_{k+1} = \eta_{k+1}(\mathbf{x}) + \mathbf{W}_{k+1}$ is easily shown to be

$$f_{k+1}(\mathbf{z}|\mathbf{x}) = N_{R_{k+1}}(\mathbf{z} - \eta_{k+1}(\mathbf{x})). \tag{2.113}$$

The same is true for the likelihood function of the general nonlinear measurement model $\mathbf{Z}_{k+1} = \eta_{k+1}(\mathbf{x}, \mathbf{W}_{k+1})$:

$$f_{k+1}(\mathbf{z}|\mathbf{x}) = f_{\mathbf{W}_{k+1}}(\eta_{k+1,\mathbf{x}}^{-1}(\mathbf{z})) \cdot J_{\eta_{k+1,\mathbf{x}}^{-1}}(\mathbf{z}). \tag{2.114}$$

The details are left to the reader as Exercise 1.

2.4.5 Single-Target Bayes Filter: Initialization

The Bayes filter recursion must begin with time step $k = 0$. This means that the filter must be initialized with an *initial distribution*

$$f_{0|0}(\mathbf{x}|Z^0) = f_0(\mathbf{x}). \tag{2.115}$$

This specifies, on the basis of whatever a priori information is available, that $f_0(\mathbf{x})$ is the probability (density) that $\mathbf{x}$ is the initial state of the target.

If we have relatively good a priori information, $f_0(\mathbf{x})$ might have a sharp, well defined peak at some initial state $\mathbf{x}_0$, or it could be multimodal, indicating several possible initial target states.

If we have very poor a priori information, $f_0(\mathbf{x})$ can be chosen to be a uniform distribution in some bounded domain S_0. In this case it is the task of the Bayes filter to both acquire (detect) the target and track it thereafter.

2.4.6 Single-Target Bayes Filter: Predictor

The Bayes filter predictor equation, (2.81), is a special case of the *total probability theorem.* Assume that the state-transition density has the form $f_{k+1|k}(\mathbf{x}|\mathbf{x}', Z^k)$. That is, target state at time step $k+1$ can depend on the previous data-collection history Z^k as well as on the previous target state $\mathbf{x}$. Given this,

$$\int f_{k+1|k}(\mathbf{x}|\mathbf{x}', Z^k) \cdot f_{k|k}(\mathbf{x}'|Z^k)d\mathbf{x}' \tag{2.116}$$

$$= \int \frac{f_{k+1|k}(\mathbf{x}, \mathbf{x}', Z^k)}{f_{k+1|k}(\mathbf{x}', Z^k)} \cdot \frac{f_{k|k}(\mathbf{x}', Z^k)}{f_{k|k}(Z^k)} d\mathbf{x}' \tag{2.117}$$

$$= \frac{\int f_{k+1|k}(\mathbf{x}, \mathbf{x}', Z^k)d\mathbf{x}'}{f_{k|k}(Z^k)} \tag{2.118}$$

$$= \frac{f_{k+1|k}(\mathbf{x}, Z^k)}{f_{k|k}(Z^k)} \tag{2.119}$$

$$= f_{k+1|k}(\mathbf{x}|Z^k) \tag{2.120}$$

where (2.119) follows from the total probability theorem. In summary,

$$f_{k+1|k}(\mathbf{x}|Z^k) = \int f_{k+1|k}(\mathbf{x}|\mathbf{x}', Z^k) \cdot f_{k|k}(\mathbf{x}'|Z^k) d\mathbf{x}'. \tag{2.121}$$

The more familiar Bayes filter predictor equation results from assuming that current state is independent of observation history:

$$f_{k+1|k}(\mathbf{x}|\mathbf{x}', Z^k) = f_{k+1|k}(\mathbf{x}|\mathbf{x}'). \tag{2.122}$$

2.4.7 Single-Target Bayes Filter: Corrector

The Bayes filter corrector equation, (2.81), is a special case of *Bayes' rule*. To see this, assume that the likelihood has the form $f_{k+1}(\mathbf{z}|\mathbf{x}, Z^k)$. That is, the generation of current observations depends on the previous data-collection history Z^k as well as current target state $\mathbf{x}$. Given this,

$$\begin{aligned}
& \frac{f_{k+1}(\mathbf{z}_{k+1}|\mathbf{x}, Z^k) \cdot f_{k+1|k}(\mathbf{x}|Z^k)}{f_{k+1}(\mathbf{z}_{k+1}|Z^k)} && (2.123) \\
= & \frac{\frac{f_{k+1}(\mathbf{z}_{k+1}, \mathbf{x}, Z^k)}{f_{k+1}(\mathbf{x}, Z^k)} \cdot \frac{f_{k+1|k}(\mathbf{x}, Z^k)}{f_{k+1|k}(Z^k)}}{\frac{f_{k+1}(\mathbf{z}_{k+1}, Z^k)}{f_{k+1}(Z^k)}} && (2.124) \\
= & \frac{f_{k+1}(\mathbf{z}_{k+1}, \mathbf{x}, Z^k)}{f_{k+1}(\mathbf{z}_{k+1}, Z^k)} && (2.125) \\
= & f_{k+1}(\mathbf{x}|\mathbf{z}_{k+1}, Z^k) && (2.126) \\
= & f_{k+1|k+1}(\mathbf{x}|Z^{k+1}) && (2.127)
\end{aligned}$$

or in summary,

$$f_{k+1|k+1}(\mathbf{x}|Z^{k+1}) = \frac{f_{k+1}(\mathbf{z}_{k+1}|\mathbf{x}, Z^k) \cdot f_{k+1|k}(\mathbf{x}|Z^k)}{f_{k+1}(\mathbf{z}_{k+1}|Z^k)}. \tag{2.128}$$

Taking integrals of both sides, we find that

$$f_{k+1}(\mathbf{z}_{k+1}|Z^k) = \int f_{k+1}(\mathbf{z}_{k+1}|\mathbf{x}, Z^k) \cdot f_{k+1|k}(\mathbf{x}|Z^k) d\mathbf{x}. \tag{2.129}$$

The usual Bayes filter corrector equation results from assuming that current observations are independent of all previous measurements, for example, that

$$f_{k+1}(\mathbf{z}|\mathbf{x}, Z^k) = f_{k+1}(\mathbf{z}|\mathbf{x}). \tag{2.130}$$

2.4.8 Single-Target Bayes Filter: State Estimation

The Bayes posterior distribution $f_{k+1|k+1}(\mathbf{x}|Z^{k+1})$ encapsulates everything we know about the target state, based on current observations and a priori information. It is not useful unless we can extract the information about the target that we really want: position, velocity, identity, and so on. This is the purpose of a *Bayes-optimal state estimator* [231, pp. 54-63]. In this section I briefly review this concept.

2.4.8.1 Bayes-Optimal State Estimators

A *state estimator* is a family of functions $\hat{\mathbf{x}}_m(Z_m)$ indexed by $m \geq 1$, the values of which are state vectors and the arguments of which are sets of measurement vectors $Z_m = \{\mathbf{z}_1, ..., \mathbf{z}_m\}$. Let $C(\mathbf{x}, \mathbf{y}) \geq 0$ be a *cost function* defined on states $\mathbf{x}$, $\mathbf{y}$, meaning that $C(\mathbf{x}, \mathbf{y}) = C(\mathbf{y}, \mathbf{x})$ and $C(\mathbf{x}, \mathbf{y}) = 0$ implies $\mathbf{x} = \mathbf{y}$. The *posterior cost* of $\hat{\mathbf{x}}$, given the measurement set Z_m, is the cost averaged with respect to the posterior distribution:

$$\bar{C}_m(\hat{\mathbf{x}}|Z_m) = \int C(\mathbf{x}, \hat{\mathbf{x}}_m(Z_m)) \cdot f_{k|k}(\mathbf{x}|Z^k, Z_m)d\mathbf{x}. \tag{2.131}$$

The *Bayes risk* is the expected value of the posterior cost with respect to all possible observation-sequences:

$$R_m(\hat{\mathbf{x}}) = \mathrm{E}[\bar{C}_m(\hat{\mathbf{x}}|Z_m)] = \int \bar{C}_m(\hat{\mathbf{x}}|Z_m) \cdot f_k(Z_m)dZ_m. \tag{2.132}$$

The state estimator $\hat{\mathbf{x}}$ is *Bayes-optimal* if it minimizes the Bayes risk for all m.

2.4.8.2 Familiar Bayes-Optimal State Estimators: EAP and MAP

The most well known examples are the expected a posteriori (EAP) estimator (also known as the *posterior expectation*)

$$\hat{\mathbf{x}}^{\mathrm{EAP}}_{k|k} \triangleq \int \mathbf{x} \cdot f_{k|k}(\mathbf{x}|Z^k)d\mathbf{x} \tag{2.133}$$

and the maximum a posteriori (MAP) estimator

$$\hat{\mathbf{x}}^{\mathrm{MAP}}_{k|k} \triangleq \arg\sup_{\mathbf{x}} \ f_{k|k}(\mathbf{x}|Z^k). \tag{2.134}$$

The EAP estimator results from choosing $C(\mathbf{x}, \mathbf{y}) = \|\mathbf{x} - \mathbf{y}\|^2$. The MAP estimator results from choosing the "infinitesimal notch" cost function defined by $C(\mathbf{x}, \mathbf{y}) = 1$ except when $\|\mathbf{x} - \mathbf{y}\| \leq \varepsilon$ for some arbitrarily small $\varepsilon > 0$, in which case $C(\mathbf{x}, \mathbf{y}) = 0$. See [231, pp. 54-63] for details.

2.4.8.3 Maximum Likelihood Estimator (MLE)

From (2.134) and Bayes' rule it follows that

$$\hat{\mathbf{x}}^{\text{MAP}}_{k+1|k+1} = \arg\sup_{\mathbf{x}} \; f_{k+1}(\mathbf{z}_{k+1}|\mathbf{x}) \cdot f_{k+1|k}(\mathbf{x}|Z^k). \tag{2.135}$$

Assume that the prior distribution $f_{k+1|k}(\mathbf{x}|Z^k)$ is uniform and therefore constant. Then this last equation reduces to the familiar maximum likelihood estimator (MLE):

$$\hat{\mathbf{x}}^{\text{MLE}}_{k+1|k+1} \triangleq \arg\sup_{\mathbf{x}} \; f_{k+1}(\mathbf{z}_{k+1}|\mathbf{x}). \tag{2.136}$$

The MLE is Bayes-optimal only when the prior is uniform. Since this is a typically unlikely eventuality, the MLE is in general *not Bayes-optimal.*

Remark 1 (Practical State Estimators) *State estimators must be selected with care. If they have unrecognized inefficiencies, then data will be unnecessarily "wasted" in trying to overcome them—though not necessarily with success. For example, the EAP estimator plays an important part in optimal filtering theory, but will behave poorly in applications in which the posteriors* $f_{k|k}(\mathbf{x}|Z^k)$ *are significantly multimodal. The MAP estimator is generally useful for practical application.*

2.4.9 Single-Target Bayes Filter: Error Estimation

A state-estimate $\hat{\mathbf{x}}_{k|k}$ provides us the most likely state of the target, given all current information. We also need to determine the possible error associated with this estimate. For example, $\hat{\mathbf{x}}_{k|k}$ is precisely known if $f_{k|k}(\mathbf{x}|Z^k)$ is unimodal and very "peaky" at $\hat{\mathbf{x}}_{k|k}$. It is inconclusively known if $f_{k|k}(\mathbf{x}|Z^k)$ has other significant modes. It is imprecisely known if $f_{k|k}(\mathbf{x}|Z^k)$ is very flat near $\hat{\mathbf{x}}_{k|k}$. Thus we require measures of the degree of error associated with $\hat{\mathbf{x}}_{k|k}$.

2.4.9.1 Covariance Matrices

The most familiar measure of the possible error in $\hat{\mathbf{x}}_{k|k}$ is the *covariance matrix*

$$C_{k|k} \triangleq \int (\mathbf{x}-\hat{\mathbf{x}})(\mathbf{x}-\hat{\mathbf{x}})^T \cdot f_{k|k}(\mathbf{x}|Z^k)d\mathbf{x} \tag{2.137}$$

$$= \int \mathbf{x}\mathbf{x}^T \cdot f_{k|k}(\mathbf{x}|Z^k)d\mathbf{x} - \hat{\mathbf{x}}\hat{\mathbf{x}}^T \tag{2.138}$$

where I have abbreviated $\hat{\mathbf{x}} \overset{\text{abbr}}{=} \hat{\mathbf{x}}_{k|k}$. The *one sigma containment region* is the hyperellipsoid defined by all $\mathbf{x}$, such that

$$(\mathbf{x}-\hat{\mathbf{x}})^T C_{k|k}^{-1}(\mathbf{x}-\hat{\mathbf{x}}) \leq 1. \tag{2.139}$$

Example 3 *The covariance matrix of a two-dimensional state estimate* $\hat{\mathbf{x}} = (\hat{x},\hat{y})^T$ *is*

$$\begin{pmatrix} C_{11} & C_{12} \\ C_{21} & C_{22} \end{pmatrix} \tag{2.140}$$

$$= \int \begin{pmatrix} x-\hat{x} \\ y-\hat{y} \end{pmatrix} \begin{pmatrix} x-\hat{x} & y-\hat{y} \end{pmatrix} f_{k|k}(x,y|Z^k)dxdy \tag{2.141}$$

$$= \int \begin{pmatrix} (x-\hat{x})^2 & (x-\hat{x})(y-\hat{y}) \\ (x-\hat{x})(y-\hat{y}) & (y-\hat{y})^2 \end{pmatrix} \tag{2.142}$$

$$\cdot f_{k|k}(x,y|Z^k)dxdy. \tag{2.143}$$

From this it follows that

$$C_{11} = \int (x-\hat{x})^2 f_{k|k}(x,y|Z^k)dxdy \tag{2.144}$$

$$C_{21} = C_{12} = \int (x-\hat{x})(y-\hat{y}) f_{k|k}(x,y|Z^k)dxdy \tag{2.145}$$

$$C_{22} = \int (y-\hat{y})^2 f_{k|k}(x,y|Z^k)dxdy. \tag{2.146}$$

2.4.9.2 Information Measures of Uncertainty

The covariance matrix is defined only when the state space is Euclidean, $\mathfrak{X}_0 = \mathbb{R}^N$. If $\mathfrak{X}_0$ is finite or hybrid, other measures of statistical dispersion must be used.

Entropy: The most common is the *entropy*, defined by

$$\varepsilon_{k|k} \triangleq -\int f_{k|k}(\mathbf{x}|Z^k) \cdot \log f_{k|k}(\mathbf{x}|Z^k) d\mathbf{x} \quad (2.147)$$

Strictly speaking, the entropy is not well defined if $\mathfrak{X}_0$ has continuous state variables with units of measurement. This is because $\log x$ must be a function of a unitless variable x, but $f_{k|k}(\mathbf{x}|Z^k)$ will not be unitless in general. The conundrum is easily resolved by noting that the entropy is actually a special case of another, and well defined, information-theoretic concept.

Kullback-Leibler discrimination: The Kullback-Leibler discrimination or cross-entropy (of a probability density $f(\mathbf{x})$ as compared to a reference probability density $f_0(\mathbf{x})$) is

$$K(f; f_0) \triangleq \int f(\mathbf{x}) \cdot \log \left(\frac{f(\mathbf{x})}{f_0(\mathbf{x})} \right) d\mathbf{x}. \quad (2.148)$$

It has the property that $K(f; f_0) \geq 0$ and that $K(f; f_0) = 0$, if and only if $f = f_0$ (almost everywhere).

Assume that $f(\mathbf{x})$ and $f_0(\mathbf{x})$ are defined on some bounded domain D of (hyper)volume $V = |D|$ and that $f_0(\mathbf{x}) = V^{-1}$ is the corresponding uniform distribution on D. Then it is left to the reader as Exercise 2 to show that (2.148) reduces to

$$K(f_{k|k}; f_0) = -\varepsilon_{k|k} + \log V. \quad (2.149)$$

The more $f_{k|k}$ resembles the uniform distribution f_0, the smaller the value of $K(f_{k|k}; f_0)$ and thus the larger the value of the entropy $\varepsilon_{k|k}$. Consequently:

- Entropy is a special case of Kullback-Leibler discrimination, combined with an abuse of notation—$\log f(\mathbf{x})$—in the event that $\mathbf{x}$ has a continuous state variable with units of measurement.

Central entropy: Equation (2.147) is a measure of the *total* or *overall* dispersion of the posterior distribution $f_{k|k}(\mathbf{x}|Z^k)$. That is, it is a measure of the degree to which it differs from a uniform distribution. The information-theoretic analog of a central moment is the *central entropy*.

Suppose that $\hat{\mathbf{x}}_{k|k}$ is a state-estimate. Let E be a small neighborhood of (hyper)volume $\varepsilon = |E|$ containing $\hat{\mathbf{x}}_{k|k}$. The central entropy is defined as

$$\kappa_{k|k} \triangleq -\log\left(\varepsilon \cdot f_{k|k}(\hat{\mathbf{x}}_{k|k}|Z^k)\right). \quad (2.150)$$

The "peakier" (less dispersed) that $f_{k|k}(\mathbf{x}|Z^k)$ is at $\mathbf{x} = \hat{\mathbf{x}}_{k|k}$, the smaller the value of $\kappa_{k|k}$. Note that the size ε of E must be small enough that $\varepsilon \cdot f_{k|k}(\hat{\mathbf{x}}|Z^k) \leq 1$, and thus that $\kappa_{k|k} \geq 0$.

The central entropy is a consequence of the following approximation. Let $u_{k|k}(\mathbf{x})$ be the uniform distribution defined by $u_{k|k}(\mathbf{x}) \triangleq \varepsilon^{-1}$ if $\mathbf{x} \in E$, and $u_{k|k}(\mathbf{x}) \triangleq 0$ otherwise.[6] Then

$$K(u_{k|k}; f_{k|k}) = \int u_{k|k}(\mathbf{x}) \cdot \log\left(\frac{u_{k|k}(\mathbf{x})}{f_{k|k}(\mathbf{x}|Z^k)}\right) d\mathbf{x} \tag{2.151}$$

$$= \int_E \varepsilon^{-1} \cdot \log\left(\frac{\varepsilon^{-1}}{f_{k|k}(\mathbf{x}|Z^k)}\right) d\mathbf{x} \tag{2.152}$$

$$\cong \varepsilon^{-1} \cdot \log\left(\frac{\varepsilon^{-1}}{f_{k|k}(\hat{\mathbf{x}}_{k|k}|Z^k)}\right) \cdot \varepsilon \tag{2.153}$$

$$= -\log\left(\varepsilon \cdot f_{k|k}(\hat{\mathbf{x}}_{k|k}|Z^k)\right). \tag{2.154}$$

2.4.10 Single-Target Bayes Filter: Data Fusion

Suppose that two statistically independent sensors with respective likelihood functions $\overset{1}{f}_{k+1}(\overset{1}{\mathbf{z}}|\mathbf{x})$ and $\overset{2}{f}_{k+1}(\overset{2}{\mathbf{z}}|\mathbf{x})$ collect observations $\overset{1}{\mathbf{z}}_{k+1}$ and $\overset{2}{\mathbf{z}}_{k+1}$, respectively. These measurements can be fused by applying the Bayes corrector equation sequentially, first for $\overset{1}{\mathbf{z}}_{k+1}$ and then for $\overset{2}{\mathbf{z}}_{k+1}$:

$$f_{k+1|k+1}(\mathbf{x}|Z^k, \overset{1}{\mathbf{z}}_{k+1}) \propto \overset{1}{f}_{k+1}(\overset{1}{\mathbf{z}}_{k+1}|\mathbf{x}) \cdot f_{k+1|k}(\mathbf{x}|Z^k) \tag{2.155}$$

$$f_{k+1|k+1}(\mathbf{x}|Z^k, \overset{1}{\mathbf{z}}_{k+1}, \overset{2}{\mathbf{z}}_{k+1}) \propto \overset{2}{f}_{k+1}(\overset{2}{\mathbf{z}}_{k+1}|\mathbf{x}) \tag{2.156}$$

$$\cdot f_{k+1|k+1}(\mathbf{x}|Z^k, \overset{1}{\mathbf{z}}_{k+1}) \tag{2.157}$$

$$\propto \overset{1}{f}_{k+1}(\overset{1}{\mathbf{z}}_{k+1}|\mathbf{x}) \cdot \overset{2}{f}_{k+1}(\overset{2}{\mathbf{z}}_{k+1}|\mathbf{x}) \tag{2.158}$$

$$\cdot f_{k+1|k}(\mathbf{x}|Z^k). \tag{2.159}$$

This is equivalent to applying the corrector equation a single time, using the joint likelihood function

$$\overset{12}{f}_{k+1}(\overset{1}{\mathbf{z}}_{k+1}, \overset{2}{\mathbf{z}}_{k+1}|\mathbf{x}) \triangleq \overset{1}{f}_{k+1}(\overset{1}{\mathbf{z}}_{k+1}|\mathbf{x}) \cdot \overset{2}{f}_{k+1}(\overset{2}{\mathbf{z}}_{k+1}|\mathbf{x}). \tag{2.160}$$

If the sensors are not independent then measurements can be fused, provided that the joint likelihood function for the joint sensor can be constructed.

6 This is an approximation of the Dirac delta function $\delta_{\hat{\mathbf{x}}_{k|k}}(\mathbf{x})$.

2.4.11 Single-Target Bayes Filter: Computation

As I remarked in the introduction to this chapter, the single most important challenge in Bayes filtering is *real-time computability*. To better understand why this is so, let us examine two approximation techniques: one that is usually intractable, *fixed-grid discretization*, and one that is potentially tractable, *particle system discretization*. For a more systematic discussion of computational issues in nonlinear filtering, see the excellent tutorial article by Daum [38].

2.4.11.1 Fixed-Grid Discretization

In fixed-grid approximation, one chooses fixed bounded regions of the measurement space $\mathfrak{Z}_0 = \mathbb{R}^M$ and the state space $\mathfrak{X}_0 = \mathbb{R}^N$. One then discretizes these regions into collections of $\mu = \mu_0^M$ measurement cells and $\nu = \nu_0^N$ state cells, where μ_0 and ν_0 are the respective numbers of single-dimensional cells. Target and sensor constraints are modeled by heuristically specifying Markov transitions $f_{k+1|k}(x|x')$ from each state cell x' to all others x; and sensor noise by specifying the likelihood $f_{k+1}(z|x)$ that the sensor will collect an observation-cell z if the target is present at a given state cell x. The Bayes filter equations, (2.81) and (2.82), are then implemented on a cell-by-cell basis. The target state-estimate at time step k is chosen to be that cell that maximizes the discretized posterior $f_{k|k}(x|Z^k)$.

However, as the statisticians J.C. Naylor and A.F.M. Smith have remarked [176, p. 214]:

- "The implementation of Bayesian inference procedures can be made to appear deceptively simple."

This is indeed the case here.

It is easy to derive an approximate estimate of the computational load of the fixed-grid filter. Let a be the number of operations required to compute $f_{k+1|k}(x|x')$ for fixed x, x'. Then from (2.81), computation of

$$f_{k+1k}(x|Z^k) = \sum_{x'} f_{k+1|k}(x|x') \cdot f_{k+1k}(x'|Z^k) \tag{2.161}$$

requires at least νa operations for each x. Thus computation of $f_{k+1k}(\cdot|Z^k)$ requires at least $\nu^2 a$ operations.

Next let b be the number of operations required to compute $f_{k+1}(z|x)$ for fixed z, x. Then from (2.82), computation of

$$f_{k+1|k+1}(x|Z^{k+1}) \propto f_{k+1}(z_{k+1}|x) \cdot f_{k+1|k}(x'|Z^k) \tag{2.162}$$

for each x requires at least b operations. So, computation of $f_{k+1|k+1}(\cdot|Z^{k+1})$ requires at least νb operations.

Each recursive cycle of (2.81) and (2.82) consequently requires at least $\nu^2 a + \nu b$ operations. Since this is quadratic in the number ν of unknowns, the fixed-grid filter is computationally problematic for real-time application. As Arulampalam et al have written, "As the dimensionality of the state space increases, the computational cost of the approach...increases dramatically" [8, p. 177].

Computational tractability, in the sense of linearity in the number of unknowns, can be achieved essentially only for "toy" problems. For example, $\nu^2 a + \nu b$ reduces to $(1 + b)\nu$ if we assume that states transition only to immediately neighboring cells and to 2ν if we further assume binary observations (i.e., $z = 0$ or $z = 1$).

2.4.11.2 Particle System Discretization

Matters can be improved if we use a principled approximation technique, such as particle systems (see Section 2.5.3). The number ν of state cells is replaced by the generally smaller number π of particles and $\nu^2 a + \nu b$ becomes $(a + b)\pi$—or $(a + 1)\pi$, if binary observations are assumed. The particle system approximation is thus linear in the number π of unknowns. However, π will be large—and often prohibitively large—in those applications for which particle methods are appropriate to begin with (e.g., in applications where traditional methods fail).

For example, in low-SNR scenarios such as the one described in Section 2.4.1, posterior distributions will be highly multimodal long before they become unimodal. During the early phases of the filtering process a large—and perhaps prohibitively large—number of particles will be required if the posteriors are to be sufficiently well approximated. Large numbers of particles will also be required if rapid maneuvers are to be detected, if target motion is sharply constrained by terrain constraints, and so on.

2.4.11.3 When Is the Bayes Filter Appropriate?

For decades the extended Kalman filter (EKF), see Section 2.5.1, has been the "workhorse" technique for real-time implementation of the single-target recursive

Bayes filter, (2.81) and (2.82). It is versatile and robust and often surprisingly so. When used with techniques such as boot-strapping and covariance gating, it is capable of detecting targets and tracking them in moderately small SNRs [11]. It can be made more robust to strong target maneuvers by increasing plant noise.

Nevertheless, the EKF still has inherent limitations. When SNR decreases to the point that false alarms or clutter returns cannot be effectively rejected via gating, it will prove incapable of tracking targets, let alone detecting them. When target motions are limited by extreme nonlinearities such as terrain constraints, it will behave poorly—for example, by allowing tanks to vault cliffs and submarines to tunnel inland. Similar problems occur if sensor nonlinearities are severe (e.g., those inherent to change of coordinate frames).

It is in applications such as these—where even the best EKF-based algorithms may perform poorly—that the Bayes filter becomes potentially important.

In recent years some tracking practitioners have, because of inappropriate application of nonlinear filtering techniques to benign sensing environments, reached untenable assessments of their real-world practicality. Some, for example, have concluded that single-target particle filters are generically impractical because the EKF works perfectly well with much smaller computational load. However, if SNR is so large and nonlinearities are so moderate that an EKF suffices for this application, there is no need for any nonlinear filtering approach to begin with.

The following questions should therefore be fundamental to any application of the general Bayes filter to a real-world problem:

- *Is the Bayes filter appropriate or inappropriate for the application at hand?*
- *That is, can the application be adequately addressed using conventional techniques because, for example, SNR is small enough or target and sensor models are linear enough?*

2.5 SINGLE-TARGET BAYES FILTER: IMPLEMENTATION

The sections that follow summarize a few of the most common real-time approximations of the Bayes filter. I begin with a short explication of the EKF in Section 2.5.1. Another long-standing and often effective technique, the Gaussian-mixture filter, is summarized in Section 2.5.2. Sequential Monte Carlo (SMC) filters, also known as particle filters or particle system filters, have attracted exponentially expanding interest in the research and development community since 2000. These are described in Section 2.5.3.

In each case I emphasize the role of formal Bayes modeling (Section 2.4.4). I begin with specific models, use them to define the specific form of the Bayes filter equations (2.81) and (2.82), and then specialize these equations using the specific implementation technique to derive filtering equations for that technique.

2.5.1 Taylor Series Approximation: The EKF

I cannot attempt more than a brief summary of the EKF here. For more details see [11, pp. 28-30] or [27, pp. 108, 111]. My discussion is also mathematically more general than is usually the case. It presumes familiarity with the concept of the *gradient derivative of a vector transformation.* Readers unfamiliar with general gradient derivatives may wish to consult Appendix C.

Let us be given the motion model $\mathbf{X}_{k+1} = \varphi_k(\mathbf{x}) + \mathbf{V}_k$ and measurement model $\mathbf{Z}_{k+1} = \eta_{k+1}(\mathbf{x}) + \mathbf{W}_{k+1}$ introduced in Section 2.4.4. The EKF is based on first-order Taylor series approximations of the nonlinear functions φ_k and η_{k+1}.

2.5.1.1 EKF Predictor

Assume that, from time step k, we have a state-estimate $\mathbf{x}_{k|k}$ and error covariance matrix $P_{k|k}$. Expand φ_k to first order around $\mathbf{x}_{k|k}$,

$$\varphi_k(\mathbf{x}) - \varphi_k(\mathbf{x}_{k|k}) \cong F_k(\mathbf{x} - \mathbf{x}_{k|k}) \tag{2.163}$$

where the state-transition matrix F_k is defined by

$$F_k\mathbf{x} \triangleq \frac{\partial \varphi_k}{\partial \mathbf{x}}(\mathbf{x}_{k|k}); \tag{2.164}$$

where the indicated derivative is a gradient derivative, and where the approximation is valid for $\mathbf{x}$ near $\mathbf{x}_{k|k}$. The EKF predictor equations are

$$\mathbf{x}_{k+1|k} = \varphi_k(\mathbf{x}_{k|k}) \tag{2.165}$$

$$P_{k+1|k} = F_k P_{k|k} F_k^T + Q_k. \tag{2.166}$$

2.5.1.2 EKF Corrector

Assume that we have a predicted state-estimate $\mathbf{x}_{k+1|k}$ and error covariance matrix $P_{k+1|k}$. Expand η_{k+1} to first order around $\mathbf{x}_{k+1|k}$,

$$\eta_{k+1}(\mathbf{x}) - \eta_{k+1}(\mathbf{x}_{k+1|k}) \cong H_{k+1}(\mathbf{x} - \mathbf{x}_{k+1|k}) \tag{2.167}$$

where the observation matrix H_{k+1} is defined by

$$H_{k+1}\mathbf{x} \triangleq \frac{\partial \eta_{k+1}}{\partial \mathbf{x}}(\mathbf{x}_{k+1|k}) \tag{2.168}$$

and where the approximation is valid for $\mathbf{x}$ near $\mathbf{x}_{k+1|k}$. Given this, the EKF corrector equations are, expressed for convenience in information form,

$$P_{k+1|k+1}^{-1}\mathbf{x}_{k+1|k+1} = P_{k+1|k}^{-1}\mathbf{x}_{k+1|k} + H_{k+1}^T R_{k+1}^{-1} \tilde{\mathbf{z}}_{k+1} \tag{2.169}$$

$$P_{k+1|k+1}^{-1} = P_{k+1|k}^{-1} + H_{k+1}^T R_{k+1}^{-1} H_{k+1} \tag{2.170}$$

where the "adjusted observation" is $\tilde{\mathbf{z}}_{k+1} \triangleq \mathbf{z}_{k+1} + H\mathbf{x}_{k+1|k} - \eta_{k+1}(\mathbf{x}_{k+1|k})$. Alternatively, expressed in standard Kalman filter form,

$$\mathbf{x}_{k+1|k+1} = \mathbf{x}_{k+1|k} + K_{k+1}(\tilde{\mathbf{z}}_{k+1} - H_{k+1}\mathbf{x}_{k+1|k}) \tag{2.171}$$

$$= \mathbf{x}_{k+1|k} + K_{k+1}(\mathbf{z}_{k+1} - \eta_{k+1}(\mathbf{x}_{k+1|k})) \tag{2.172}$$

$$P_{k+1|k+1}^{-1} = (I - K_{k+1}H_{k+1})\, P_{k+1|k} \tag{2.173}$$

$$K_{k+1} = P_{k+1|k}H_{k+1}^T \left(H_{k+1}P_{k+1|k}H_{k+1}^T + R_{k+1}\right)^{-1}. \tag{2.174}$$

2.5.2 Gaussian-Mixture Approximation

The Gaussian-mixture filter (GMF) is a generalization of the Kalman filter in which Gaussian distributions are replaced by Gaussian-mixture distributions. I cannot attempt more than a brief summary of SMC approximation here. For more details, see [217]. (See also [98] for a related technique.)

In this approach it is assumed that the Markov transition density and likelihood function are nonlinear, but not so nonlinear that they cannot be well approximated as weighted sums of Gaussian densities (also known as Gaussian mixtures):

$$f_{k+1|k}(\mathbf{x}|\mathbf{x}') = \sum_{j=1}^{T_k} \tau_k^j \cdot N_{Q_k^j}(\mathbf{x} - F_k^j\mathbf{x}') \tag{2.175}$$

$$f_{k+1}(\mathbf{z}|\mathbf{x}) = \sum_{j=1}^{L_{k+1}} \lambda_k^j \cdot N_{R_k^j}(\mathbf{z} - H_k^j\mathbf{x}) \tag{2.176}$$

with $\tau_k^j \geq 0$, $\lambda_k^j \geq 0$, $\sum_{j=1}^T \tau_k^j = 1$, and $\sum_{j=1}^L \lambda_k^j = 1$. The Gaussian-mixture filter results from assuming that the posterior distributions $f_{k|k}(\mathbf{x}|Z^k)$ and

$f_{k+1|k}(\mathbf{x}|Z^k)$ are also Gaussian mixtures:

$$f_{k|k}(\mathbf{x}|Z^k) = \sum_{i=1}^{n_{k|k}} w^i_{k|k} \cdot N_{P^i_{k|k}}(\mathbf{x} - \mathbf{x}^i_{k|k}) \quad (2.177)$$

$$f_{k+1|k}(\mathbf{x}|Z^k) = \sum_{i=1}^{n_{k+1|k}} w^i_{k+1|k} \cdot N_{P^i_{k+1|k}}(\mathbf{x} - \mathbf{x}^i_{k+1|k}). \quad (2.178)$$

In the Gaussian-mixture filter, the posterior components $(w^i_{k|k}, \mathbf{x}^i_{k|k}, P^i_{k|k})$ are propagated through time using the Bayes recursive filter equations:

$$(w^i_{0|0}, \mathbf{x}^i_{0|0}, P^i_{0|0})_{i=1,..,n_{0|0}} \quad (2.179)$$

$$\rightarrow (w^i_{1|0}, \mathbf{x}^i_{1|0}, P^i_{1|0})_{i=1,..,n_{1|0}} \quad (2.180)$$

$$\rightarrow (w^i_{1|1}, \mathbf{x}^i_{1|1}, P^i_{1|1})_{i=1,...,n_{1|1}} \rightarrow \cdots \quad (2.181)$$

$$\rightarrow (w^i_{k|k}, \mathbf{x}^i_{k|k}, P^i_{k|k})_{i=1,...,n_{k|k}} \quad (2.182)$$

$$\rightarrow (w^i_{k+1|k}, \mathbf{x}^i_{k+1|k}, P^i_{k+1|k})_{i=1,...,n_{k+1|k}} \quad (2.183)$$

$$\rightarrow (w^i_{k+1|k+1}, \mathbf{x}^i_{k+1|k+1}, P^i_{k+1|k+1})_{i=1,...,n_{k+1|k+1}} \quad (2.184)$$

$$\rightarrow \cdots \quad (2.185)$$

Example 4 *Figure 2.9 depicts a one-dimensional posterior distribution $f_{k|k}(x)$ that is a mixture of four equally weighted Gaussian components:*

$$f_{k|k}(x) = \frac{N_{16}(x+2) + N_1(x-2) + N_4(x-6) + N_9(x-13)}{4}. \quad (2.186)$$

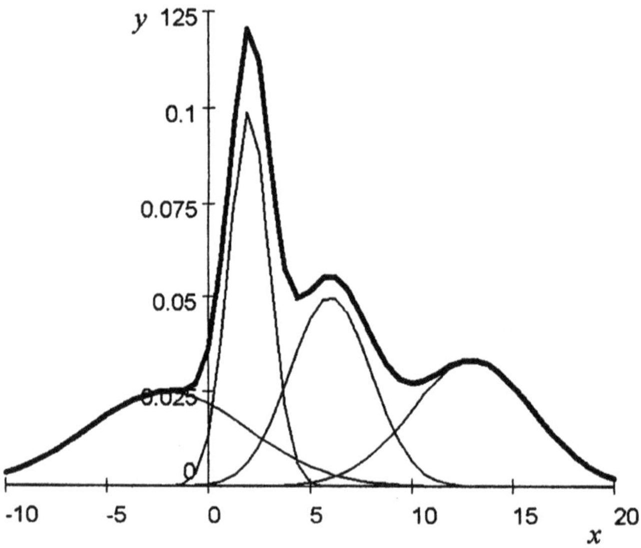

Figure 2.9 A one-dimensional posterior distribution $f_{k|k}(x)$ (thick curve), which is a mixture of four equally weighted Gaussian components (thin curves).

2.5.2.1 GMF Predictor

From (2.81) the predictor equation becomes

$$f_{k+1|k}(\mathbf{x}|Z^k) = \int f_{k+1|k}(\mathbf{x}|\mathbf{x}') \cdot f_{k|k}(\mathbf{x}'|Z^k)d\mathbf{x}' \quad (2.187)$$

$$= \sum_{j=1}^{T_k}\sum_{i=1}^{n_{k|k}} \tau_k^j \cdot w_{k|k}^i \cdot \int N_{Q_k^j}(\mathbf{x} - F_k^j\mathbf{x}') \quad (2.188)$$

$$\cdot N_{P_{k|k}^i}(\mathbf{x}' - \mathbf{x}_{k|k}^i)d\mathbf{x} \quad (2.189)$$

$$= \sum_{j=1}^{T_k}\sum_{i=1}^{n_{k|k}} \tau_k^j \cdot w_{k|k}^i \quad (2.190)$$

$$\cdot \int N_{Q_k^j + F_k^i P_{k|k}^i (F_k^i)^T}(\mathbf{x} - F_k^j \mathbf{x}_{k|k}^i) \quad (2.191)$$

$$\cdot N_{E_{k|k}^{i,j}}(\mathbf{x} - \mathbf{e}_{k|k}^{i,j})d\mathbf{x} \quad (2.192)$$

$$= \sum_{j=1}^{T_k}\sum_{i=1}^{n_{k|k}} \tau_k^j \cdot w_{k|k}^i \quad (2.193)$$

$$\cdot N_{Q_k^j + F_k^i P_{k|k}^i (F_k^i)^T}(\mathbf{x} - F_k^j \mathbf{x}_{k|k}^i) \quad (2.194)$$

where (2.190) follows from the fundamental Gaussian identity, (2.36) or (D.1).

Consequently, $f_{k+1|k}(\mathbf{x}|Z^k)$ is a Gaussian mixture with $n_{k+1|k} = T_k \cdot n_{k|k}$ components $N_{Q_k^j + F_k^i P_{k|k}^i (F_k^i)^T}(\mathbf{x} - F_k^j \mathbf{x}_{k|k}^i)$ and weights $w_{k+1|k}^{i,j} = w_{k|k}^i \cdot \tau_k^j$. Note that if $T_k = 1$ and $\tau_k^1 = 1$ then $n_{k+1|k} = n_{k|k}$ and $w_{k+1|k}^{i,1} = w_{k|k}^i$.

Example 5 *Let* $f_{k+1|k}(x|x') = N_4(x - 2x')$ *be the Markov transition density. We apply it to the prior posterior* $f_{k|k}(x)$ *of (2.186). Using the Kalman predictor equations we predict each of the components of* $f_{k|k}(x)$ *in turn. It follows that the predicted posterior has the form*

$$f_{k+1|k}(x) = \frac{N_{68}(x+4) + N_{20}(x-4) + N_{20}(x-12) + N_{40}(x-26)}{4}. \quad (2.195)$$

The graph of this function is shown in Figure 2.10.

2.5.2.2 GMF Corrector

The Bayes filter corrector step, (2.82), likewise results in a Gaussian mixture with $n_{k+1|k+1} = L_{k+1} \cdot n_{k+1|k}$ components:

$$f_{k+1|k+1}(\mathbf{x}|Z^{k+1}) = \sum_{j=1}^{L_{k+1}} \sum_{i=1}^{n_{k+1|k}} w^{i,j}_{k+1|k+1} \cdot N_{E^{i,j}_{k+1|k}}(\mathbf{x} - \mathbf{e}^{i,j}_{k+1|k}) \tag{2.196}$$

where

$$(E^{i,j}_{k+1|k})^{-1}\mathbf{e}^{i,j}_{k+1|k+1} = (P^i_{k+1|k})^{-1}\mathbf{x}^i_{k+1|k} \tag{2.197}$$

$$+(H^j_{k+1})^T(R^j_{k+1})^{-1}\mathbf{z}_{k+1} \tag{2.198}$$

$$(E^{i,j}_{k+1|k})^{-1} = (P^i_{k+1|k})^{-1} + (H^j_{k+1})^T(R^j_{k+1})^{-1}H^j_{k+1} \tag{2.199}$$

and

$$w^{i,j}_{k+1|k+1} \tag{2.200}$$

$$= \frac{\lambda^j_k \cdot w^i_{k+1|k} \cdot N_{C^{j,i}_k}(\mathbf{z}_{k+1} - H^j_k\mathbf{x}^i_{k+1|k})}{\sum_{e=1}^{L_{k+1}} \sum_{l=1}^{n_{k+1|k}} \lambda^e_k \cdot w^l_{k+1|k} \cdot N_{C^{j,i}_k}(\mathbf{z}_{k+1} - H^e_k\mathbf{x}^l_{k+1|k})} \tag{2.201}$$

where

$$C^{j,i}_k \triangleq R^j_k + H^j_k P^i_{k+1|k}(H^j_k)^T. \tag{2.202}$$

The derivation of these formulas is left to the reader as Exercise 3. Note that if $L_{k+1} = 1$ and $\lambda^1_k = 1$ then $n_{k+1|k+1} = n_{k+1|k}$ and $w^{i,1}_{k+1|k+1} = w^i_{k+1|k+1}$.

The number of Gaussian components in $f_{k|k}(\mathbf{x}|Z^k)$ and $f_{k+1|k}(\mathbf{x}|Z^k)$ increases combinatorially as time progresses. If computational tractability is to be achieved, various techniques must be used to merge similar components and prune ones with small weights [217].

Example 6 *Let $f_{k+1}(z_{k+1}|x') = N_1(6-x)$ be the likelihood function. We apply it to the predicted density $f_{k=1|k}(x)$ of (2.195). Using the Kalman corrector equations we predict each of the components of $f_{k+1|k}(x)$ in turn. It follows that the data-updated posterior has the form*

$$f_{k+1|k}(x) = \frac{\left(\begin{array}{c} N_{0.98}(x-5.85) + N_{0\ 95}(x-5.7) \\ +N_{0.95}(x-5.88) + N_{0.97}(x-6.22) \end{array}\right)}{4}. \tag{2.203}$$

The graph of this function is shown in Figure 2.11.

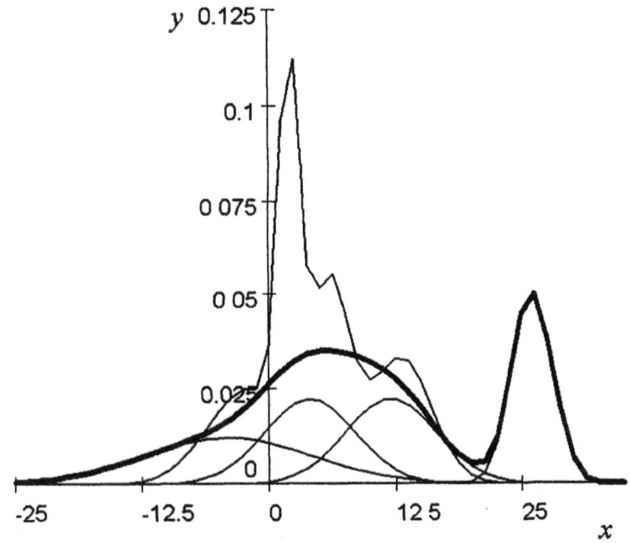

Figure 2.10 The Gaussian mixture distribution $f_{k|k}(x)$ (light curve), the predicted density $f_{k+1|k}(x)$ (dark curve), and the components of $f_{k+1|k}(x)$.

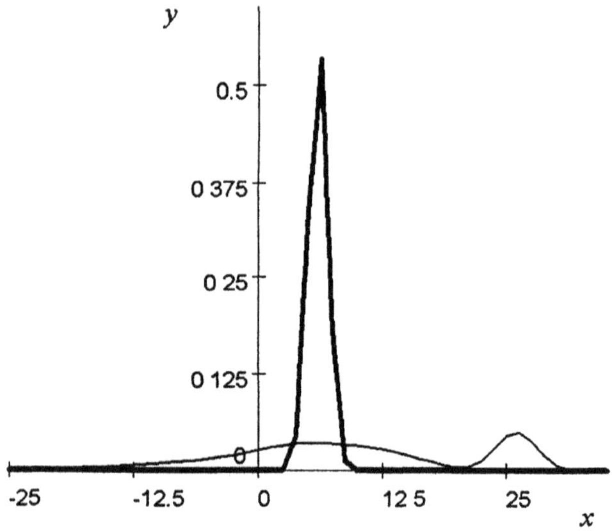

Figure 2.11 The predicted Gaussian mixture distribution $f_{k+1|k}(x)$ (light curve) and the data-updated density $f_{k+1|k+1}(x)$ (dark curve)

2.5.2.3 GMF State Estimation

It is left to the reader as Exercise 4 to show that the EAP estimator of (2.133) reduces to:

$$\hat{\mathbf{x}}^{\text{EAP}}_{k+1|k+1} = \int \mathbf{x} \cdot f_{k+1|k+1}(\mathbf{x}|Z^{k+1}) d\mathbf{x} \quad (2.204)$$

$$= \sum_{j=1}^{L_{k+1}} \sum_{i=1}^{n_{k+1|k}} w^{i,j}_{k+1|k+1} \cdot \mathbf{e}^{i,j}_{k+1|k}. \quad (2.205)$$

where the notation is as in (2.196).

When $f_{k+1|k+1}(\mathbf{x}|Z^{k+1})$ is significantly multimodal, the EAP estimator will not behave well and the MAP estimator of (2.134) should be used instead. This requires determining the absolute maximum of a weighted sum of Gaussian distributions. As a general rule this requires numerical techniques. Approximately, however, after pruning and merging of the components of $f_{k+1|k+1}(\mathbf{x}|Z^{k+1})$ one can choose

$$\hat{\mathbf{x}}^{\text{MAP}}_{k+1|k+1} \cong \mathbf{e}^{i_0,j_0}_{k+1|k} \quad (2.206)$$

where $\mathbf{e}^{i_0,j_0}_{k+1|k}$ corresponds to the largest mixture coefficient $w^{i_0,j_0}_{k+1|k+1}$. This approximation will be especially accurate if the $\mathbf{e}^{i,j}_{k+1|k}$ are sufficiently well separated from each other.

2.5.3 Sequential Monte Carlo Approximation

A seminal paper by Gordon [74] is generally credited with popularizing particle system methods in single-target tracking. I cannot attempt more than a brief and simplified overview of sequential Monte Carlo (SMC) approximation here.[7] For more details, see [6, 8, 43, 115, 196].[8] My discussion draws from Kadirkamanthan et al. [102, p. 2393] and Arulampalam et al. [8].

A *particle system approximation* of the posterior distribution $f_{k|k}(\mathbf{x}|Z^k)$ is a collection $\mathbf{x}^1_{k|k}, ...\mathbf{x}^{\nu}_{k|k}$ of state vectors (the "particles") and positive "importance

7 Target filtering using SMC approximation is also known as particle, particle systems, or condensation filtering.

8 For a tutorial on particle filtering with an emphasis on application to wireless communications, see Djurić et al. [41]. See also the "Special Issue on Monte Carlo Methods for Statistical Signal Processing,"*IEEE Trans. Sign. Proc* , Vol. 50, No. 2, 2002. See also Kotecha and Djurić [108, 109] for a related technique called "Gaussian-sum particle filtering," that integrates the Gaussian-mixture and SMC approximations.

weights" $w_{k|k}^1, ... w_{k|k}^\nu$ with $\sum_{i=1}^\nu w_{k|k}^i = 1$, such that

$$\int \theta(\mathbf{x}) \cdot f_{k|k}(\mathbf{x}|Z^k) d\mathbf{x} \cong \sum_{i=1}^{\nu} w_{k|k}^i \cdot \theta(\mathbf{x}_{k|k}^i) \tag{2.207}$$

for any unitless function $\theta(\mathbf{x})$ of $\mathbf{x}$. The following convergence property must be satisfied:[9]

$$\lim_{\nu \to \infty} \sum_{i=1}^{\nu} w_{k|k}^i \cdot \theta(\mathbf{x}_{k|k}^i) = \int \theta(\mathbf{x}) \cdot f_{k|k}(\mathbf{x}|Z^k) d\mathbf{x}. \tag{2.208}$$

The particles are interpreted as random samples drawn from the posterior distribution $f_{k|k}(\mathbf{x}|Z^k)$:

$$\mathbf{x}_{k|k}^1, ... \mathbf{x}_{k|k}^\nu \sim f_{k|k}(\cdot|Z^k). \tag{2.209}$$

A common abuse of notation is to abbreviate (2.207) as

$$f_{k|k}(\mathbf{x}|Z^k) \cong \sum_{i=1}^{\nu} w_{k|k}^i \cdot \delta_{\mathbf{x}_{k|k}^i}(\mathbf{x}) \tag{2.210}$$

where $\delta_{\mathbf{x}_i}(\mathbf{x})$ denotes the Dirac delta density concentrated at $\mathbf{x}_i$. Caution is in order, however, since (2.210) has no rigorous mathematical basis and can lead to erroneous results if understood otherwise. Equation (2.207) rather than (2.210) should always be used in derivations.

The importance weights are often assumed to be equal: $w_{k|k}^i = 1/\nu$ for all i, and this is what we presume hereafter. Equal weighting is consistent with the interpretation of $\mathbf{x}_{k|k}^1, ... \mathbf{x}_{k|k}^\nu$ as random samples drawn from $f_{k|k}(\mathbf{x}|Z^k)$. That is, far more of the $\mathbf{x}_{k|k}^i$ will be drawn from regions of state space where $f_{k|k}(\mathbf{x}|Z^k)$ is large than from regions where it is small. Thus the sampling process induces an implicit weighting.

The SMC filter propagates the time sequence of particles:

$$\{\mathbf{x}_{0|0}^i\}_{0 \leq i \leq \nu} \rightarrow \{\mathbf{x}_{1|0}^i\}_{0 \leq i \leq \nu} \rightarrow \{\mathbf{x}_{1|1}^i\}_{0 \leq i \leq \nu} \rightarrow \cdots \tag{2.211}$$

$$\rightarrow \{\mathbf{x}_{k|k}^i\}_{0 \leq i \leq \nu} \rightarrow \{\mathbf{x}_{k+1|k}^i\}_{0 \leq i \leq \nu} \tag{2.212}$$

$$\rightarrow \{\mathbf{x}_{k+1|k+1}^i\}_{0 \leq i \leq \nu} \rightarrow \cdots \tag{2.213}$$

where $\{\mathbf{x}_{k|k}^i\}_{0 \leq i \leq \nu} \overset{\text{abbr.}}{=} \{\mathbf{x}_{k|k}^1, ..., \mathbf{x}_{k|k}^\nu\}$. The concept is illustrated in Figure 2.12.

9 An SMC filter that does not satisfy this property is, strictly speaking, not actually an SMC filter.

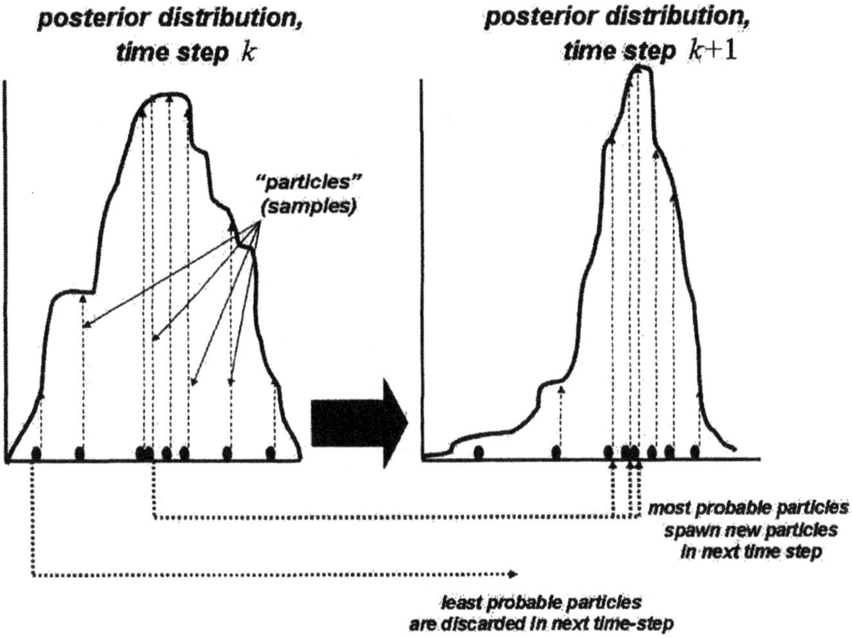

Figure 2.12 The basic concept of sequential Monte Carlo (particle systems) approximation of the recursive Bayes filter. Particles represent random samples drawn from posterior distributions. Particles are supposed to be more densely located where posterior probability is greatest. The idea is to propagate particles from time step to time step, so that this assumption remains valid.

2.5.3.1 SMC Filter Initialization

If an initial distribution $f_{0|0}(\mathbf{x})$ is available, then the filter is initialized by drawing ν samples from it:

$$\mathbf{x}_{0|0}^{1}, ..., \mathbf{x}_{0|0}^{\nu} \sim f_{0|0}(\cdot). \tag{2.214}$$

If nothing is known about the target state, then $f_{0|0}(\mathbf{x})$ could be chosen to be uniform in the region of interest in state space.

Alternatively, an "implicit initial distribution" is assumed, meaning that ν particles are selected on the basis of intuition and convenience. This can be constructed, for example, from the first observation.

2.5.3.2 SMC Filter Predictor

Assume that the Bayes posterior at the previous time step has been approximated by a particle system:

$$f_{k|k}(\mathbf{x}|Z^k) \cong \frac{1}{\nu}\sum_{i=1}^{\nu} \delta_{\mathbf{x}_{k|k}^{i}}(\mathbf{x}). \tag{2.215}$$

Assume further that for each $\mathbf{x}_{k|k}^{i}$, τ random samples have been drawn from the Markov transition density $f_{k+1|k}(\mathbf{x}|\mathbf{x}_{k|k}^{i}, Z^k)$, that is,[10]

$$\mathbf{x}_{k+1|k}^{1,i}, ..., \mathbf{x}_{k+1|k}^{\tau,i} \sim f_{k+1|k}(\cdot|\mathbf{x}_{k|k}^{i}, Z^k). \tag{2.216}$$

Thus we can approximate the Markov densities as

$$f_{k+1|k}(\mathbf{x}|\mathbf{x}_{k|k}^{i}) \cong \frac{1}{\tau}\sum_{j=1}^{\tau} \delta_{\mathbf{x}_{k+1|k}^{j,i}}(\mathbf{x}). \tag{2.217}$$

10 Here we have assumed the somewhat more general form $f_{k+1|k}(\mathbf{x}|\mathbf{x}', Z^k)$ rather than the usual form $f_{k+1|k}(\mathbf{x}|\mathbf{x}')$. See (3.54) and (3.55).

Substituting (2.215) and (2.217) into the Bayes filter predictor (2.81), we get

$$\int \theta(\mathbf{x}) \cdot f_{k+1|k}(\mathbf{x}|Z^k)dx \cong \int\int \theta(\mathbf{x}) \cdot f_{k+1|k}(\mathbf{x}|\mathbf{x}') \tag{2.218}$$

$$\cdot f_{k|k}(\mathbf{x}'|Z^k)d\mathbf{x}'dx \tag{2.219}$$

$$= \frac{1}{\nu}\sum_{i=1}^{\nu} \int \theta(\mathbf{x}) \cdot f_{k+1|k}(\mathbf{x}|\mathbf{x}^i_{k|k})dx \tag{2.220}$$

$$= \frac{1}{\tau\nu}\sum_{j=1}^{\tau}\sum_{i=1}^{\nu} \theta(\mathbf{x}^{j,i}_{k+1|k}). \tag{2.221}$$

If $\tau > 1$ then the number of particles would increase as $\tau^k \nu$ for $k = 1, 2, ...$ The usual practice is to set $\tau = 1$—that is, limit the draw to a single sample:

$$\mathbf{x}^i_{k+1|k} \sim f_{k+1|k}(\cdot|\mathbf{x}^i_{k|k}, Z^k). \tag{2.222}$$

In this case

$$\int \theta(\mathbf{x}) \cdot f_{k+1|k}(\mathbf{x}|Z^k)d\mathbf{x} \cong \frac{1}{\nu}\sum_{i=1}^{\nu} \theta(\mathbf{x}^i_{k+1|k}). \tag{2.223}$$

The predicted particle system is, therefore, $\mathbf{x}^1_{k+1|k}, ..., \mathbf{x}^\nu_{k+1|k}$.

This is the simplest approach to *importance sampling*. That is, the *importance sampling density* (*proposal density*) has been chosen to be the so-called "dynamic prior" $f_{k+1|k}(\mathbf{x}|\mathbf{x}')$ [74]. Implementations using the dynamic prior are simple and intuitively appealing, and therefore also popular. However, they also tend to require a large number ν of samples.

One can use more general importance sampling densities that result in improved performance [8, p. 178]. Arulampalam et al. describe the choice of a good importance sampling density as a "crucial design step" [8, p. 179]. Even more forthrightly, Daum notes that "The key ingredient that makes particle filters work for high-dimensional problems is a good proposal density..." [38, p. 62].

2.5.3.3 SMC Filter Corrector

Assume that

$$f_{k+1|k}(\mathbf{x}|Z^k) \cong \frac{1}{\nu}\sum_{i=1}^{\nu} \delta_{\mathbf{x}^i_{k+1|k}}(\mathbf{x}). \tag{2.224}$$

Substituting (2.224) into the Bayes normalization factor, (2.83), yields:

$$f_{k+1}(\mathbf{z}_{k+1}|Z^k) = \int f_{k+1}(\mathbf{z}_{k+1}|\mathbf{x}) \cdot f_{k+1|k}(\mathbf{x}|Z^k)d\mathbf{x} \tag{2.225}$$

$$\cong \frac{1}{\nu}\sum_{i=1}^{\nu} f_{k+1}(\mathbf{z}_{k+1}|\mathbf{x}^i_{k+1|k}). \tag{2.226}$$

The Bayes corrector equation, (2.82), thereby reduces to

$$\int \theta(\mathbf{x}) \cdot f_{k+1|k+1}(\mathbf{x}|Z^{k+1})d\mathbf{x} \tag{2.227}$$

$$= \frac{\int \theta(\mathbf{x}) \cdot f_{k+1}(\mathbf{z}_{k+1}|\mathbf{x}) \cdot f_{k+1|k}(\mathbf{x}|Z^k)d\mathbf{x}}{f_{k+1}(\mathbf{z}_{k+1}|Z^k)} \tag{2.228}$$

$$\cong \frac{\frac{1}{\nu}\sum_{i=1}^{\nu} \theta(\mathbf{x}^i_{k+1|k}) \cdot f_{k+1}(\mathbf{z}_{k+1}|\mathbf{x}^i_{k+1|k})}{\frac{1}{\nu}\sum_{e=1}^{\nu} f_{k+1}(\mathbf{z}_{k+1}|\mathbf{x}^e_{k+1|k})} \tag{2.229}$$

$$= \sum_{i=1}^{\nu} w^i_{k+1|k+1} \cdot \theta(\mathbf{x}^i_{k+1|k+1}) \tag{2.230}$$

where $\mathbf{x}^i_{k+1|k+1} \triangleq \mathbf{x}^i_{k+1|k}$ and where

$$w^i_{k+1|k+1} \triangleq \frac{f_{k+1}(\mathbf{z}_{k+1}|\mathbf{x}^i_{k+1|k})}{\sum_{e=1}^{\nu} f_{k+1}(\mathbf{z}_{k+1}|\mathbf{x}^e_{k+1|k})}. \tag{2.231}$$

Since the particles now have unequal weights, we must replace them with new particles that have equal weights but that reflect the influence of the $w^i_{k+1|k+1}$. One wishes to eliminate particles with small weights and duplicate particles with large weights. Intuitively speaking, the goal is to replace each $\mathbf{x}^i_{k+1|k+1}$ with $w^i_{k+1|k+1} \cdot \nu$ copies of itself, resulting in a new posterior particle system $\tilde{\mathbf{x}}^1_{k+1|k+1}, ..., \tilde{\mathbf{x}}^\nu_{k+1|k+1}$ and a new approximation

$$\int \theta(\mathbf{x}) \cdot f_{k+1|k+1}(\mathbf{x}|Z^{k+1})d\mathbf{x} \cong \frac{1}{\nu}\sum_{i=1}^{\nu} \theta(\tilde{\mathbf{x}}^i_{k+1|k+1}). \tag{2.232}$$

This is accomplished using various *resampling* techniques: see [8, p. 179] or [20, 42, 197]. Resampling introduces random variation into the copying process by sampling from the discrete distribution

$$f_{k+1|k+1}(\mathbf{x}) = \sum_{i=1}^{\nu} w^i_{k+1|k+1} \cdot \delta_{\mathbf{x}^i_{k+1|k+1}}(\mathbf{x}). \tag{2.233}$$

Methods include order statistics resampling, stratified resampling, residual sampling, and systematic resampling.

Here I illustrate the concept via the most easily understood approach: multinomial resampling [35, 42]. Abbreviate $\tilde{w}_i \overset{\text{abbr}}{=} w^i_{k+1|k+1}$ for all $i = 1, ..., \nu$. The multinomial distribution

$$\mu(i_1, ..., i_\nu) = \frac{\nu!}{i_1! \cdots i_\nu!} \cdot \tilde{w}_1^{i_1} \cdots \tilde{w}_\nu^{i_\nu} \tag{2.234}$$

is a probability distribution on all ν-tuples $(i_1, ..., i_\nu)$ of nonnegative integers for which $i_1 + ... + i_\nu = \nu$. Draw a random sample $(e_1, ..., e_\nu) \sim \mu(\cdot)$ from this distribution. If $e_1 = 0$ then eliminate the particle $\mathbf{x}^1_{k+1|k+1}$. Otherwise, make e_1 identical copies of $\mathbf{x}^1_{k+1|k+1}$. Repeat this process for $i = 2, ..., \nu$. Since $e_1 + ... + e_\nu = \nu$ we end up with ν equally weighted new particles $\tilde{\mathbf{x}}^1_{k+1|k+1}, ..., \tilde{\mathbf{x}}^\nu_{k+1|k+1}$, the distribution of which corresponds to the original unequal weighting $w^1_{k+1|k+1}, ..., w^\nu_{k+1|k+1}$.

Because resampling creates identical copies of particles, it contributes to "particle impoverishment." After a few recursive steps all but a relatively small number of particles (or at the extreme, a single particle) will have negligible weights. This problem is especially severe when $f_{k+1|k}(\mathbf{x}|\mathbf{x}^i_{k|k})$ can be approximated by a Dirac delta density (i.e., when process noise $\mathbf{V}_k$ is very small or zero), since then $\mathbf{x}^i_{k+1|k} \cong \mathbf{x}^i_{k|k}$ for all i.

Various techniques have been proposed to ameliorate particle impoverishment by increasing the statistical diversity of the resampled particles. Many of these have the effect of increasing statistical diversity excessively. In *roughening*, copies of particles are randomly jittered so that they are no longer identical. *Resample-move* or *rejuvenation* is a theoretically rigorous but computationally expensive form of roughening, in which Markov Chain Monte Carlo (MCMC) techniques are used to create the jitter. In *regularization* or *smoothing* [8, 174], particles are diversified by drawing equally weighted, interpolated random samples

$$\tilde{\mathbf{x}}^1_{k+1|k+1}, ..., \tilde{\mathbf{x}}^\nu_{k+1|k+1} \sim \bar{f}_{k+1|k+1}(\cdot) \tag{2.235}$$

from the smoothed distribution

$$\bar{f}_{k+1|k+1}(\mathbf{x}) = \sum_{i=1}^{\nu} w^i_{k+1|k+1} \cdot K(\mathbf{x} - \mathbf{x}^i_{k+1|k+1}) \tag{2.236}$$

where $K(\mathbf{x})$ is an Epanechnikov smoothing kernel.

2.5.3.4 SMC Filter State Estimation

Equation (2.133) reduces to the following formula for the EAP estimate:

$$\hat{\mathbf{x}}^{\text{EAP}}_{k+1|k+1} = \int \mathbf{x} \cdot f_{k+1|k+1}(\mathbf{x}|Z^{k+1}) d\mathbf{x} \tag{2.237}$$

$$\cong \frac{1}{\nu} \sum_{i=1}^{\nu} \mathbf{x}^i_{k+1|k+1}. \tag{2.238}$$

For SMC filters MAP estimation, as defined in (2.134), is conceptually more difficult. One frequently used technique is to apply Dempster's *expectation-maximization* (EM) *algorithm* [15, 162]. First, the number N of significant modes in $f_{k+1|k+1}(\mathbf{x}|Z^{k+1})$ is estimated. Then one assumes that $f_{k+1|k+1}(\mathbf{x}|Z^{k+1})$ can be approximated as a weighted sum of Gaussians,

$$f_{k+1|k+1}(\mathbf{x}|Z^{k+1}) \cong \sum_{i=1}^{N} w_i \cdot N_{C_i}(\mathbf{x} - \mathbf{x}_i). \tag{2.239}$$

The EM algorithm is used to estimate the unknown parameters w_i, C_i, and $\mathbf{x}_i$. The component of the sum with the largest w_i corresponds to the dominant mode. Its mean, $\mathbf{x}_i$, is chosen as the MAP state-estimate. The corresponding covariance C_i is chosen as the uncertainty in that estimate.

2.5.3.5 SMC Filter Error Estimation

Given a state-estimate $\hat{\mathbf{x}} \overset{\text{abbr}}{=} \hat{\mathbf{x}}_{k+1|k+1}$, (2.137) reduces to the following formula for the covariance matrix:

$$C_{k|k} = \int (\mathbf{x} - \hat{\mathbf{x}})(\mathbf{x} - \hat{\mathbf{x}})^T \cdot f_{k|k}(\mathbf{x}|Z^k) d\mathbf{x} \tag{2.240}$$

$$\cong \frac{1}{\nu} \sum_{i=1}^{\nu} (\mathbf{x}^i_{k+1|k+1} - \hat{\mathbf{x}})(\mathbf{x}^i_{k+1|k+1} - \hat{\mathbf{x}})^T. \tag{2.241}$$

2.6 CHAPTER EXERCISES

Exercise 1 *Prove (2.114), the formula for the likelihood corresponding to a general nonlinear measurement model.*

Exercise 2 *Show that (2.149) is true:* $K(f_{k|k}; f_0) = \log V - \varepsilon_{k|k}$. *(That is, entropy is a special case of Kullback-Leibler discrimination.)*

Exercise 3 *Prove the corrector equations for the Gaussian sum filter, (2.196)-(2.200).*

Exercise 4 *Prove the formula for the expected a posteriori (EAP) estimator for the Gaussian sum filter, (2.204).*

Chapter 3

General Data Modeling

3.1 INTRODUCTION TO THE CHAPTER

One of the most challenging aspects of information fusion has been the highly disparate and ambiguous forms that information can have. Many kinds of data, such as that supplied by tracking radars, can be described in statistical form. However, statistically uncharacterizable real-world variations make the modeling of other kinds of data, such as synthetic aperture radar (SAR) images, highly problematical.

It has been even more unclear how still other forms of data—natural-language statements, features extracted from signatures, rules drawn from knowledge bases—might be mathematically modeled and processed. Numerous expert systems approaches have been proposed to address such problems. However, their burgeoning number and variety have generated much confusion and controversy.

Nowhere has this been more evident than in the Dempster-Shafer theory of evidence. In recent years, Dempster's familiar rule for combining evidence (see Section 4.5) has been increasingly criticized for its supposed deficiencies in properly resolving conflicts between evidence. Much of this criticism has arisen from "Zadeh's paradox," which I address in Section 4.5.2. These controversies have accelerated the proliferation of alternative or generalized Dempster's rules—of which there are now a literally infinite number [40, 48, 60, 97, 100, 4, 94, 117, 128, 129, 171, 206, 212, 221, 223, 251, 256, 257, 258].[1]

1 Full disclosure: The author has contributed to the confusion by introducing the infinite family of "conditioned" and "modified" Dempster's combinations [128, 129, 60].

Dempster-Shafer theory is not alone in this respect, since there is also a literal infinitude of fuzzy logics (see [70, p. 458] and [80, 164]) and of rule-based inference logics.[2]

My stance toward this state of affairs is similar to that recently expressed by R. Haenni in a paper addressing the proliferation of alternative Dempster's rules [81]:

- *Confusion is inevitable if one fails to securely attach evidential representation and fusion formalisms to explicit, concrete, and carefully designed models.*

What underlies much of the confusion and controversy, I believe, has been a failure to apply and adhere to concrete formal modeling of the kind taken for granted in single-target tracking (see Chapter 2). This chapter extends formal Bayes modeling for conventional measurements, as described in Section 2.4.4, to unconventional measurements and other kinds of data. It is here that I introduce the two types of data and the four types of measurements that will concern us throughout Part I:

- Two types of data: state-estimates versus measurements;
- Four types of measurements: unambiguously generated unambiguous (UGU) measurements; unambiguously generated ambiguous (UGA) measurements; ambiguously generated ambiguous (AGA) measurements; and ambiguously generated unambiguous (AGU) measurements.

UGU measurements are simply conventional measurements of the kind discussed in Chapter 2.

AGU measurements include conventional measurements, such as synthetic aperture radar (SAR) or high range-resolution radar (HRRR), whose likelihood functions are ambiguously defined because of statistically uncharacterizable real-world variations.

UGA measurements resemble conventional measurements in that their relationship to target state is precisely known, but differ in that there is ambiguity regarding what is actually being observed. Examples include attributes or features extracted by humans or digital signal processors from signatures; natural-language statements; and rules.

AGA measurements are the same as UGA measurements except that not only the measurements themselves but also their relationship to target state is ambiguous.

2 For example, the FGM fuzzy logics are infinite in number—see (4.57) and (4.58). The FGM family is only one of many other infinite families of fuzzy logics.

Examples include attributes, features, natural-language statements, and rules, for which sensor models must be constructed from human-mediated knowledge bases.

3.1.1 Summary of Major Lessons Learned

The following are the major points that the reader will encounter while reading this chapter:

- Confusion is guaranteed if uncertainty representations are not based on concrete, systematic, real-world models (Section 3.2);
- Conventional data modeling neglects, or takes for granted, three of the four steps of systematic data modeling (Section 3.3);
- The distinctions between randomness, imprecision, and uncertainty, and the basic methodology for modeling them (Sections 3.4-3.4.5);
- Formal Bayes modeling in general (Section 3.5.1);
- The recursive Bayes filter in full generality (Section 3.5.2);
- The concept of a Bayes combination operator—that is, a measurement fusion method that is equivalent to Bayes' rule (Section 3.5.3);
- The concept of Bayes-invariant conversion of uncertainty representations (Section 3.5.4);
- The concept of a generalized measurement;
- The concepts of UGA, AGA, and AGU measurements (Section 3.6);
- The concept of a generalized likelihood function for a UGA measurement; see (3.14).

3.1.2 Organization of the Chapter

The chapter is organized as follows. Sections 3.2 and 3.3 summarize the crucial issues for the modeling of uncertainty, and for the modeling of uncertainty in data in particular. Section 3.4 illustrates my data-modeling approach with simple examples, as follows:

- Extending the additive measurement model $\mathbf{Z} = \eta_{k+1}(\mathbf{x}) + \mathbf{W}_{k+1}$ to include a small degree of imprecision (Section 3.4.1);

- Imprecise measurements with a small degree of randomness (Section 3.4.2);
- Deterministic "vague" (also known as "fuzzy") measurements (Section 3.4.3);
- Deterministic "uncertain" (also known as Dempster-Shafer) measurements (Section 3.4.4);
- Comparison of, and contrast between, randomness and uncertainty (Section 3.4.5).

In Section 3.5, I describe formal Bayes modeling and the recursive Bayes filter in general. This section also introduces the concepts of Bayes combination operator and Bayes-invariant data conversion. I sketch my approach for extending formal modeling of UGU measurements to UGA and AGA measurements in Section 3.6. Exercises for the chapter are in Section 3.7.

3.2 ISSUES IN MODELING UNCERTAINTY

Evidential conflict, and the uncertainties that it engenders, can have many causes and each cause will typically have its own phenomenology. The following questions are thus fundamental:

- What real-world phenomena are producing evidential conflict?
- How do we model these phenomena even if we do not understand them in detail?
- What techniques can be applied based on these models?

Consider the following instances:

- *Uncertainty due to multiple targets.* Suppose that two targets are present but we treat all evidence as though it were generated by a single target. Measurements originating with the two sources will exhibit considerable conflict, thereby generating greater confusion and uncertainty than usual. The proper resolution of the problem is not to invent a data fusion rule that will "properly resolve" the conflict. Rather, it is to create a *two target measurement model* that properly reflects the underlying phenomenology, and to devise methods (such as measurement-to-track association) based on this model;
- *Uncertainty due to false measurements.* Evidence that does not originate with any target, such as false alarms or clutter observations, will likewise produce

great uncertainty. Once again, the proper resolution of the problem is not to invent some new fusion rule but to devise a more general measurement model that accounts for the underlying false alarm or clutter phenomenology, coupled with a suitable clutter-rejection approach based on the model;

- *Uncertainty due to unreliability.* Data sources can be unreliable, delivering biased evidence. Such sources are essentially no different than temporally or spatially misregistered conventional sensors. The proper resolution of the problem is to model the bias, and develop bias estimation and removal approaches based on the model;
- *Uncertainty due to deception.* Data sources that deliberately deceive are much like conventional sensors confronted by spoofing techniques. The phenomenology of deception should be modeled, and countermeasure techniques constructed from these models;
- *Uncertainty due to evasive action.* Rapidly maneuvering targets generate conflicting evidence that can result in great uncertainty—unless one models variability of maneuver and uses these models to develop adaptive motion modeling techniques such as IMM trackers;
- *Uncertainty due to unknown correlations.* Spatially and temporally correlated sources can generate uncertainty and confusion (specifically, unwarranted certainty in state-estimates) if they are processed as though statistically independent. The proper resolution is to model the correlation in some way and incorporate these models into the reasoning process.

We will address some of these types of uncertainty later in the book. The primary goal of this chapter, however, is to address those forms of uncertainty that affect the modeling of data and especially the modeling of measurements. This includes uncertainties attributable to the combined effects of randomness and ignorance, whether in the modeling of data itself or in the modeling of its generation. Uncertainties due to ignorance include the following.

- *Imprecision*;
- *Vagueness* (imprecisely characterized imprecision);
- *Uncertainty* (imprecisely characterized vagueness);
- *Contingency* (as with rules).

3.3 ISSUES IN MODELING UNCERTAINTY IN DATA

The modeling of observations as vectors $\mathbf{z}$ in some Euclidean space $\mathbb{R}^M$ is ubiquitous to the point that it is commonplace to think of $\mathbf{z}$ as itself the "data." However, this is not actually the case:

- $\mathbf{z}$ is actually a *mathematical abstraction, which serves as a representation of some real-world entity called a "datum."*

The following are examples of actual data that occur in the real world: a voltage; a radio frequency (RF) intensity-signature; an RF in-phase and quadrature (I&Q) signature; a feature extracted from a signature by a digital signal processor; an attribute extracted from an image by a human operator; a natural-language statement supplied by a human observer; a rule drawn from a knowledge base; and so on.

All of these measurement types are mathematically meaningless—which is to say, we cannot do anything algorithmic with them—unless we first construct mathematical abstractions that model them.

Thus voltages are commonly modeled as real numbers. Intensity signatures are modeled as real-valued functions or, when discretized into bins, as vectors. I&Q signatures are modeled as complex-valued functions or as complex vectors. Features are commonly modeled using integers, real numbers, feature vectors, and so on.

For these kinds of data, relatively little ambiguity adheres to the representation of a given datum by its associated model. The only uncertainty in such data is that associated with the randomness of the generation of measurements by targets. This uncertainty is modeled by the likelihood function $L_{\mathbf{z}}(\mathbf{x}) \triangleq f_{k+1}(\mathbf{z}|\mathbf{x})$. Thus it is conventional to think of the $\mathbf{z}$ in $f_{k+1}(\mathbf{z}|\mathbf{x})$ as a "datum" and of $f_{k+1}(\mathbf{z}|\mathbf{x})$ as the full encapsulation of its uncertainty model.

Conventional reasoning of this sort will not, unfortunately, suffice for data types such as rules, natural-language statements, human-extracted attributes, or more generally, any information involving some kind of human mediation.

In reality, $\mathbf{z}$ is a model $\mathbf{z}_D$ of some real-world datum D, and the likelihood actually has the form $f_{k+1}(D|\mathbf{x}) = f_{k+1}(\mathbf{z}_D|\mathbf{x})$.

Consider, for example, a natural-language report supplied by a human observer, such as

$$D = \text{'The target is near sector five.'} \tag{3.1}$$

Two additional kinds of ambiguity other than randomness impede the modeling of this datum.

The first kind is due to ignorance. The observer will make random errors because of factors such as fatigue, excessively high data rates, deficiencies in training, and deficiencies in ability. In principle, one could conduct a statistical analysis of the observer to determine a "likelihood function" that models his/her data generation process. In practice, such an analysis is rarely feasible. This fact introduces a nonstatistical component to the uncertainty associated with D, and we must find some way to model that uncertainty to mitigate its contaminating effects.

The second kind of uncertainty is caused by the ambiguities associated with constructing an actionable mathematical model of D. How do we model "fuzzy" and context-dependent concepts such as "near," for example?

A complete data model for D must have the form $f_{k+1}(\Theta_D|\mathbf{x})$ where the following are true:

- Θ_D is a mathematical model of both D and the uncertainties associated with constructing Θ_D;
- $f_{k+1}(\cdot|\mathbf{x})$ is a mathematical model of both the process by which Θ_D is generated given $\mathbf{x}$ and the uncertainties associated with constructing $f_{k+1}(\cdot|\mathbf{x})$.

In summary, *comprehensive data-modeling* requires a *unified, systematic, and theoretically defensible procedure* for accomplishing the following four steps (see Figure 3.1):

1. Creating the mathematized abstractions that represent individual physical-world observations;
2. Some approach for modeling any ambiguities that may be inherent in this act of abstraction;
3. Creating the random variable that, by selecting among the possible mathematized abstractions, models data generation;
4. Some approach for modeling any ambiguities caused by gaps in our understanding of how data generation occurs.

In conventional applications, steps 1, 2, and 4 are usually taken for granted, so that only the third step remains and ends up being described as the complete data model. If we are to process general and not just conventional information sources, however, we must address the other three steps.

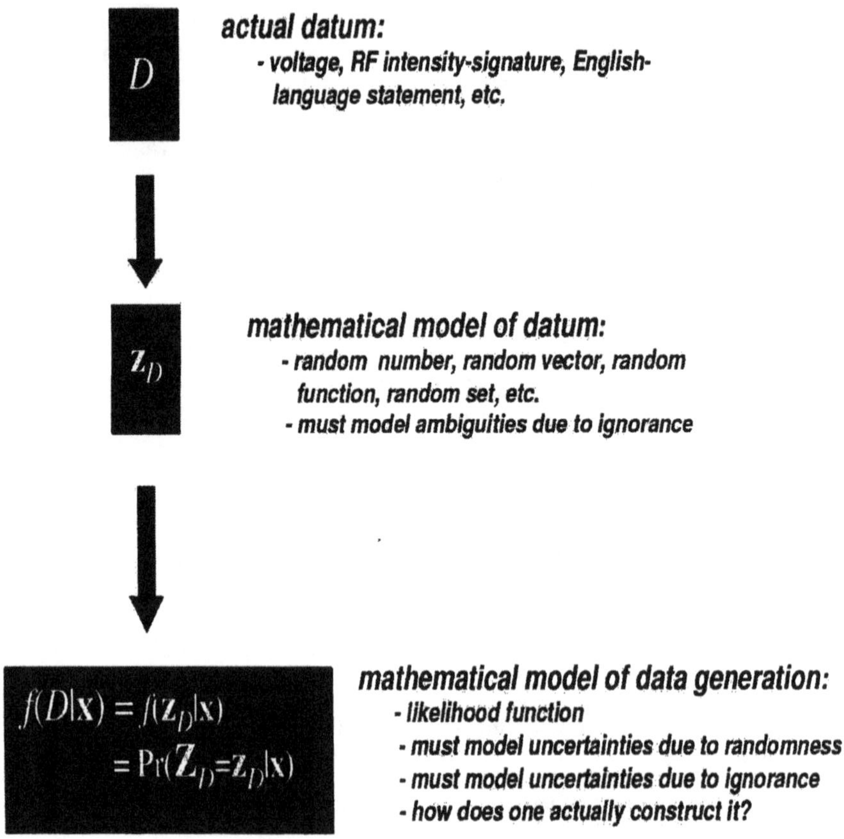

Figure 3.1 A schematic illustration of systematic data modeling. Actual data (i.e., observed real-world phenomena) must first be converted to mathematical models. Additional models must be constructed to describe not only how the mathematical data is generated, but also any ambiguities inherent in the two modeling steps themselves.

Our analysis thus far still does not completely address the data-modeling problem. We have implicitly assumed that information always takes the form of a *measurement.* Broadly speaking, a measurement is:

- An opinion ventured by an information source (a sensor, a human expert) regarding what has or has not been observed.

There are forms of information that are not measurements. Many radars, for example, do not supply measurements but, rather, feed these measurements into a Kalman filter and supply the filter's state-estimates instead. Unlike measurement information, which can often be assumed to be statistically independent from time step to time step, this kind of information is inherently time correlated. Another example is an a posteriori opinion supplied by a human observer about the state of a target, based on information that the witness has observed but does not choose to share with us.

Our data-modeling methodology must be general enough to encompass these types of data as well. To accomplish our goals, we must expand our palette of mathematical data-modeling abstractions beyond familiar ones such as real and complex numbers, vectors, real- or complex-valued functions, and so on.

- *We must be willing to embrace a much more general modeling abstraction called a random set.*

A menagerie of random set representations are portrayed in Figure 3.2. The meaning of the figure will be explained in detail in the examples that follow.

3.4 EXAMPLES

3.4.1 Random, Slightly Imprecise Measurements

I begin with a simple example that illustrates my basic data-modeling methodology. Even conventional sensors can exhibit nonrandom forms of uncertainty. Consider, for example, readings taken from a digital voltmeter. An analog voltmeter returns a voltage-value that is an instantiation of a randomly varying voltage. A digital voltmeter, on the other hand, discretizes voltage into a series of intervals and returns one of these intervals as a reading. Because of noise, the returned interval may exhibit a certain amount of random variation. Thus the uncertainty associated with a digital voltmeter includes *imprecision* as well as randomness.

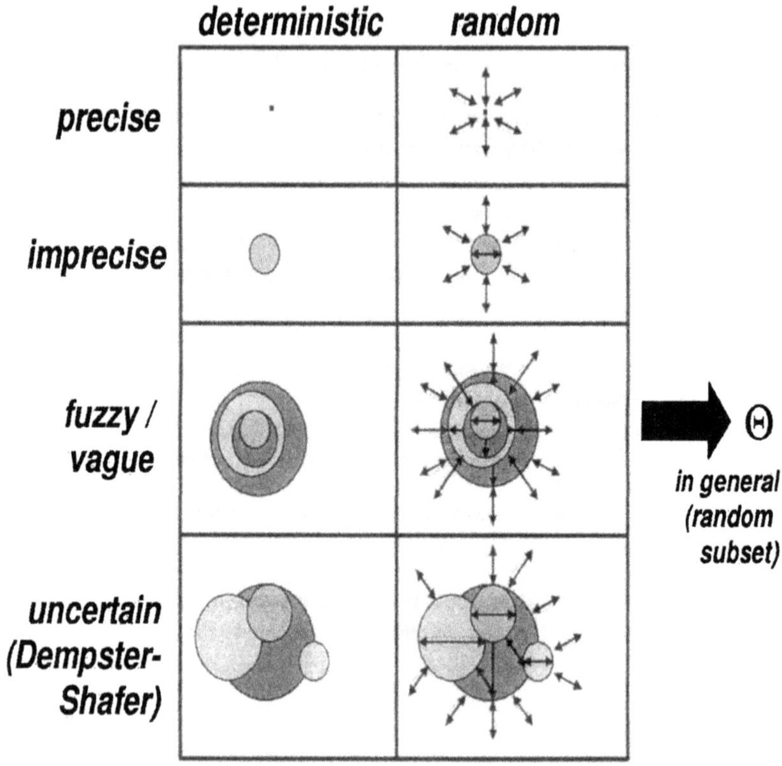

Figure 3.2 A menagerie of measurement types that model different kinds of ambiguity. Deterministic but increasingly more ambiguous types are portrayed from top to bottom in the left column: precise, imprecise, vague, and uncertain. The right column portrays the respective types when one or more parameters (diameter, centroid, number of components) have been randomized. All eight pictured types are special cases of a general concept: a *random subset of measurement space.* Stated differently: random set models encompass complex forms of uncertainty, whether due to randomness, ignorance, or both simultaneously.

With this in mind we now turn to our example. Recall the conventional nonlinear-additive measurement model $\mathbf{Z} = \eta_{k+1}(\mathbf{x}) + \mathbf{W}_{k+1}$ introduced in Section 2.4.7, where $\mathbf{W}_{k+1}$ is a zero-mean random noise vector with probability density $f_{\mathbf{W}_{k+1}}(\mathbf{z})$. Recall that the corresponding likelihood function is $f_{k+1}(\mathbf{z}|\mathbf{x}) = f_{\mathbf{W}_{k+1}}(\mathbf{z} - \eta_{k+1}(\mathbf{x}))$. Let an observation $\mathbf{z}$ be collected. We begin by segregating states from measurements:

$$\mathbf{z} - \mathbf{W}_{k+1} = \eta_{k+1}(\mathbf{x}). \tag{3.2}$$

Intuitively speaking, $\mathbf{z} - \mathbf{W}_{k+1}$ encapsulates all of the uncertainty associated with collection of $\mathbf{z}$ (in this case, just randomness), and $f_{k+1}(\mathbf{z}|\mathbf{x})$ is the probability (density) that $\mathbf{z} - \mathbf{W}_{k+1}$ agrees with $\eta_{k+1}(\mathbf{x})$.

It is possible that the value $\mathbf{z}$ may not be known with complete precision. To model this additional uncertainty, let $E_{\mathbf{z}} \subseteq \mathfrak{Z}_0$ be a closed hypersphere of very small fixed radius centered at $\mathbf{z}$ and that therefore has constant (hyper)volume $|E_{\mathbf{z}}| = \varepsilon$. Replace the precise random observation $\eta_{k+1}(\mathbf{x}) + \mathbf{W}_{k+1}$ with the imprecise random observation $E_{\mathbf{Z}} = E_{\eta_{k+1}(\mathbf{x})+\mathbf{W}_{k+1}}$. Thus any collected $\mathbf{z}$ is known only to within some degree of precision: $\mathbf{z} \in E_{\mathbf{Z}}$ rather than $\mathbf{Z} = \mathbf{z}$. The random closed subset

$$\Theta_{\mathbf{z}} \triangleq E_{\mathbf{z}-\mathbf{W}_{k+1}} \tag{3.3}$$

models the uncertainty associated with collection of $\mathbf{z}$ (in this case, randomness *and* imprecision). The probability that it agrees with $\eta_{k+1}(\mathbf{x})$ is

$$\begin{aligned}
f_{k+1}(\Theta_{\mathbf{z}}|\mathbf{x}) &\triangleq \Pr(\mathbf{z} \in E_{\mathbf{Z}}|\mathbf{x}) = \Pr(\eta_{k+1}(\mathbf{x}) \in \Theta_{\mathbf{z}}|\mathbf{x}) && (3.4)\\
&= \Pr(\mathbf{W}_{k+1} \in E_{\mathbf{z}-\eta_{k+1}(\mathbf{x})}) && (3.5)\\
&= \int_{E_{\mathbf{z}-\eta_{k+1}(\mathbf{x})}} f_{\mathbf{W}_{k+1}}(\mathbf{w})d\mathbf{w} && (3.6)\\
&\cong f_{\mathbf{W}_{k+1}}(\mathbf{z} - \eta_{k+1}(\mathbf{x})) \cdot |E_{\mathbf{z}-\eta_{k+1}(\mathbf{x})}| && (3.7)\\
&= f_{\mathbf{W}_{k+1}}(\mathbf{z} - \eta_{k+1}(\mathbf{x})) \cdot \varepsilon && (3.8)\\
&= f_{k+1}(\mathbf{z}|\mathbf{x}) \cdot \varepsilon && (3.9)
\end{aligned}$$

where (3.7) follows from the fact that $|E_{\mathbf{z}-\eta_{k+1}(\mathbf{x})}| = \varepsilon$ is very small. If we use $f_{k+1}(\Theta_{\mathbf{z}}|\mathbf{x})$ as a likelihood function in Bayes' rule we get

$$\lim_{\varepsilon \searrow 0} f_{k+1|k+1}(\mathbf{x}|Z^k, \Theta_{\mathbf{z}_{k+1}}) \tag{3.10}$$

$$\triangleq \lim_{\varepsilon \searrow 0} \frac{f_{k+1}(\Theta_{\mathbf{z}_{k+1}}|\mathbf{x}) \cdot f_{k+1|k}(\mathbf{x}|Z^k)}{\int f_{k+1}(\Theta_{\mathbf{z}_{k+1}}|\mathbf{y}) \cdot f_{k+1|k}(\mathbf{y}|Z^k)d\mathbf{y}} \tag{3.11}$$

$$= \frac{f_{k+1}(\mathbf{z}_{k+1}|\mathbf{x}) \cdot f_{k+1|k}(\mathbf{x}|Z^k)}{\int f_{k+1}(\mathbf{z}_{k+1}|\mathbf{y}) \cdot f_{k+1|k}(\mathbf{y}|Z^k)d\mathbf{y}} \tag{3.12}$$

$$= f_{k+1|k+1}(\mathbf{x}|Z^k, \mathbf{z}_{k+1}). \tag{3.13}$$

Thus the conventional likelihood $f_{k+1}(\mathbf{z}|\mathbf{x})$ can be regarded as a limiting case of the unconventional likelihood $f_{k+1}(\Theta_{\mathbf{z}}|\mathbf{x})$, based on a different kind of measurement model. This is because:

- In the limit it produces the same posterior distribution as the conventional likelihood $f_{k+1}(\mathbf{z}|\mathbf{x})$.

My measurement modeling approach, described in more detail in subsequent chapters, is the obvious generalization of this scheme. *Arbitrary* random closed subsets $\Theta \subseteq \mathfrak{Z}_0$ model measurement ambiguities far more complex than those modeled by $\Theta_{\mathbf{z}}$. They are in fact structurally rich enough to incorporate fuzzy, Dempster-Shafer, and rule-based representations of measurement uncertainty.

I shall call Θ a *generalized measurement* and define its *generalized likelihood* to be

$$f_{k+1}(\Theta|\mathbf{x}) \triangleq \Pr(\eta_{k+1}(\mathbf{x}) \in \Theta). \tag{3.14}$$

Joint generalized likelihoods can be defined similarly:

$$f_{k+1}(\Theta_1, ..., \Theta_m|\mathbf{x}) \triangleq \Pr(\eta_{k+1}(\mathbf{x}) \in \Theta_1, ..., \eta_{k+1}(\mathbf{x}) \in \Theta_m). \tag{3.15}$$

These likelihoods can then be employed in the usual manner in the Bayes filter corrector step, (2.82):

$$f_{k+1|k+1}(\mathbf{x}|Z^{k+1}) = \frac{f_{k+1}(\Theta_{k+1}|\mathbf{x}) \cdot f_{k+1|k}(\mathbf{x}|Z^k)}{f_{k+1}(\Theta_{k+1}|Z^k)} \tag{3.16}$$

where

$$f_{k+1}(\Theta_{k+1}|Z^k) \triangleq \int f_{k+1}(\Theta_{k+1}|\mathbf{x}) \cdot f_{k+1|k}(\mathbf{x}|Z^k)d\mathbf{x}. \tag{3.17}$$

As we shall see in subsequent chapters, generalized likelihoods are usually highly non-Gaussian and highly nonlinear. Consequently,

- *Processing of nonconventional measurements will usually require a sophisticated real-time nonlinear filtering technique, such as sequential Monte Carlo approximation* (Section 2.5.3).

3.4.2 Imprecise, Slightly Random Measurements

If a measurement $\mathbf{z}$ is imprecise but not random, then it is known only to containment in some region $B_{\mathbf{z}}$ centered at $\mathbf{z}$. Then (3.14) would become

$$f_{k+1}(B_{\mathbf{z}}|\mathbf{x}) = \Pr(\eta_{k+1}(\mathbf{x}) \in B_{\mathbf{z}}) = \mathbf{1}_{B_{\mathbf{z}}}(\eta_{k+1}(\mathbf{x})). \tag{3.18}$$

Now reconsider the example of the previous section with the roles of randomness and imprecision reversed. That is, $\Theta_{\mathbf{z}} \triangleq B_{\mathbf{z}} - \mathbf{W}_{k+1}$, where $B - \mathbf{w} \triangleq \{\mathbf{z} - \mathbf{w} |\ \mathbf{z} \in B\}$. Here the probability density of $\mathbf{W}_{k+1}$ is $f_{\mathbf{W}_{k+1}}(\mathbf{w}) = \varepsilon^{-1} \cdot \mathbf{1}_E(\mathbf{w})$, where E is a very small neighborhood of the origin $\mathbf{0}$ that has (hyper) volume $\varepsilon = |E|$. In this case

$$\begin{aligned}
f_{k+1}(\Theta_{\mathbf{z}}|\mathbf{x}) &= \Pr(\eta_{k+1}(\mathbf{x}) \in B_{\mathbf{z}} - \mathbf{W}_{k+1}) && (3.19)\\
&= \Pr(\mathbf{W}_{k+1} \in B_{\mathbf{z}} - \eta_{k+1}(\mathbf{x})) && (3.20)\\
&= \int_{B_{\mathbf{z}} - \eta_{k+1}(\mathbf{x})} f_{\mathbf{W}_{k+1}}(\mathbf{w}) d\mathbf{w} && (3.21)\\
&= \int_E \mathbf{1}_{B_{\mathbf{z}} - \eta_{k+1}(\mathbf{x})}(\mathbf{w}) d\mathbf{w} \cong \mathbf{1}_{B_{\mathbf{z}} - \eta_{k+1}(\mathbf{x})}(\mathbf{0}) \cdot \varepsilon && (3.22)\\
&= \mathbf{1}_{B_{\mathbf{z}}}(\eta_{k+1}(\mathbf{x})) \cdot \varepsilon. && (3.23)
\end{aligned}$$

More accurately,

$$f_{k+1}(\Theta_{\mathbf{z}}|\mathbf{x}) = \int \mathbf{1}_{(B_{\mathbf{z}} - \eta_{k+1}(\mathbf{x})) \cap E}(\mathbf{w}) d\mathbf{w} \tag{3.24}$$

and so

$$\varepsilon^{-1} \cdot f_{k+1}(\Theta_{\mathbf{z}}|\mathbf{x}) = \begin{cases} 1 & \text{if} \quad (B_{\mathbf{z}} - \eta_{k+1}(\mathbf{x})) \cap E = E \\ 0 & \text{if} \quad (B_{\mathbf{z}} - \eta_{k+1}(\mathbf{x})) \cap E = \emptyset \\ \leq 1 & \text{if} \quad \text{otherwise} \end{cases}. \tag{3.25}$$

Thus, the introduction of a small degree of randomness to imprecision has the effect of slightly increasing the extent of the imprecision by "fuzzifying" the boundary of $B_{\mathbf{z}}$.

If we use $f_{k+1}(\Theta_{\mathbf{z}}|\mathbf{x})$ as a likelihood function in Bayes' rule we get

$$\lim_{\varepsilon \searrow 0} f_{k+1|k+1}(\mathbf{x}|Z^k, \Theta_{\mathbf{z}_{k+1}}) \tag{3.26}$$

$$\triangleq \lim_{\varepsilon \searrow 0} \frac{f_{k+1}(\Theta_{\mathbf{z}_{k+1}}|\mathbf{x}) \cdot f_{k+1|k}(\mathbf{x}|Z^k)}{\int f_{k+1}(\Theta_{\mathbf{z}_{k+1}}|\mathbf{y}) \cdot f_{k+1|k}(\mathbf{y}|Z^k) d\mathbf{y}} \tag{3.27}$$

$$= \frac{f_{k+1}(B_{\mathbf{z}_{k+1}}|\mathbf{x}) \cdot f_{k+1|k}(\mathbf{x}|Z^k)}{\int f_{k+1}(B_{\mathbf{z}_{k+1}}|\mathbf{y}) \cdot f_{k+1|k}(\mathbf{y}|Z^k) d\mathbf{y}} \tag{3.28}$$

$$= f_{k+1|k+1}(\mathbf{x}|Z^k, B_{\mathbf{z}_{k+1}}). \tag{3.29}$$

Thus the generalized likelihood $f_{k+1}(B_{\mathbf{z}}|\mathbf{x})$ of an imprecise measurement can be regarded as a limiting case of the generalized likelihood $f_{k+1}(\Theta_{\mathbf{z}}|\mathbf{x})$.

3.4.3 Nonrandom Vague Measurements

Assume that we are in the x-y plane and that objects have states of the form $\mathbf{x} = (x, y, v_x, v_y, c)$, where x, y are position coordinates, v_x, v_y velocity coordinates, and c an identity label. Consider the natural-language statement delivered to us by an observer:

$$\mathcal{S} = \text{'Gustav is NEAR the tower.'} \tag{3.30}$$

Let $\eta(\mathbf{x}) \triangleq (x, y)$. As a first attempt to capture the information in this sentence, we could draw a closed circular disk D on the ground centered at the tower. This is pictured in Figure 3.3. Points inside D are interpreted as "near" the tower. Expressed as a likelihood, this model might have the form

$$L_{\mathcal{S}}(\mathbf{x}) \triangleq \mathbf{1}_D(\eta(\mathbf{x})). \tag{3.31}$$

That is, $L_{\mathcal{S}}(\mathbf{x}) = 1$ if $(x, y) \in D$ and $L_{\mathcal{S}}(\mathbf{x}) = 0$ otherwise. If $\mathbf{x}_G$ is Gustav's state, then $L_{\mathcal{S}}(\mathbf{x}_G) = 1$ mathematically captures the information contained in $\mathcal{S}$. This information is *imprecise* in that Gustav's position can be specified only to within containment in D.

However, the concept "near" is not merely imprecise, it is *fuzzy* or *vague*. It is not possible to precisely specify D_0 for two reasons. First, "near" is a context-dependent concept that cannot be defined with precision in and of itself. Second, we are ignorant of its meaning to the observer who originated the report.

Consequently, we hedge by specifying a sequence $D_0 \subset ... \subset D_e$ of concentric disks of increasingly greater radius, all centered at the tower. The disk D_i is one interpretation of what "near" means, and we assign to it a belief-weight $w_i > 0$ that it is the correct one, with $w_1 + ... + w_e = 1$. If $D_e = \mathbb{R}^2$ is the entire x-y plane, then we are stating that, with belief w_e, there is a possibility that we know nothing whatsoever about Gustav's actual position. The measurement consists of the subsets $D_0, D_1, ..., D_e$ and their weights $w_0, w_1, ..., w_e$.

Now, interpret w_i as a probability: the probability that a random variable Θ_S takes D_i as an instantiation: $\Pr(\Theta_S = D_i) = w_i$. Then Θ_S is a *randomly varying closed disk*. It is one example of a *random (closed) subset* of the measurement space $\mathfrak{Z}_0 = \mathbb{R}^2$. The total probability that any point $\mathbf{z} = (x, y)$ lies within this random disk is

$$g_S(\mathbf{z}) \triangleq \Pr(\mathbf{z} \in \Theta_S) = \sum_{i=1}^{e} \Pr(\Theta = D_i, \mathbf{z} \in D_i) = \sum_{i=1}^{e} w_i \cdot \mathbf{1}_{D_i}(\mathbf{z}). \tag{3.32}$$

This defines a fuzzy membership function on $\mathfrak{Z}_0$. Letting $D_{-1} \triangleq \emptyset$, note that $g_S(\mathbf{z}) = \sum_{j=i}^{e} w_j$ for all $\mathbf{z} \in D_i - D_{i-1}$, for $i = 0, ..., e$. In particular, $g_S(\mathbf{z}) = 1$ for all $\mathbf{z} \in D_0$, so that Gustav is unequivocally regarded as being near the tower if he is inside D_0. On the other hand, $g_S(\mathbf{z}) = w_e$ for all $\mathbf{z} \in D_e - D_{e-1}$, so Gustav is not very near the tower if he is in $D_e - D_{e-1}$.

The fuzzy membership function $g_S(\mathbf{z})$ is what I will henceforth call a *vague measurement*. (I will elaborate on this example in Section 4.3.3.) The generalized likelihood for this measurement is

$$L_S(\mathbf{x}) \triangleq \Pr(\eta(\mathbf{x}) \in \Theta_S) = g_S(\eta(\mathbf{x})). \tag{3.33}$$

Once again, the equation $L_S(\mathbf{x}_G) = 1$ mathematically encapsulates the information contained in S.

The generalized measurement Θ_S is a *random set model* of the vague information S (or, more correctly, of its fuzzy model g_S). The generalized likelihood $L_S(\mathbf{x})$ translates the information in Θ_S into a form that can be used in the Bayes filter.

3.4.4 Nonrandom Uncertain Measurements

Consider now the natural-language statement

S = 'Gustav is probably near the tower, but it could be the smokestack; it's so foggy I can't say for sure.' (3.34)

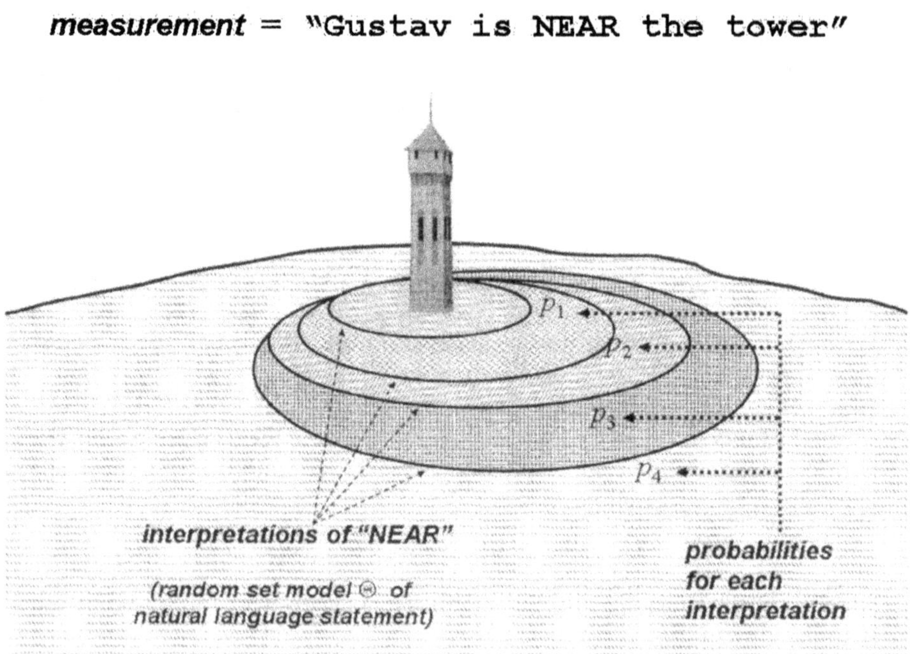

Figure 3.3 Simple illustration of the modeling of a vague concept—that of one thing being "near" another—as a randomly varying disk in the $x - y$ plane. (© 2004 IEEE. Reprinted from [148] with permission.)

In this case the observer confronts us not only with vagueness but also with uncertainty, in the form of three hypotheses. These alternatives are listed as follows:

- 'Gustav is near the tower';
- 'Gustav is near the smokestack';
- 'I'm not sure what I'm seeing.'

As in the previous example we begin by modeling the first statement as a closed circular disk D_1 centered at the tower and the second as a closed circular disk D_2 centered at the smokestack. The third statement we model as the "null hypothesis" $D_0 = \mathbb{R}^2$—that is, there is no constraint on Gustav's position.

Suppose that we also know that there are several other landmarks in the vicinity that could be mistaken for the tower or the smokestack. We place disks $D_3, ..., D_d$ about them as well. We model the observer's uncertainty with respect to these hypotheses by assigning weights $v_d \geq ... \geq v_2 > v_1 > v_0$ to these three hypotheses, with $\sum_{j=0}^{d} v_j = 1$.

As in the previous example, this model does not sufficiently capture the situation because of the vagueness of the concept "near." Consequently, we replace each D_i with a nested sequence $D_{j,0} \subset ... \subset D_{j,e(j)}$ of subsets with respective weights $w_{j,0}, ..., w_{j,e(j)}$ and $\sum_{i=1}^{e(j)} w_{j,i} = 1$ for $j = 1, ..., d$. Each sequence corresponds to a fuzzy membership function

$$g_j(\mathbf{z}) = \sum_{i=1}^{e(j)} w_{j,i} \mathbf{1}_{D_{j,i}}(\mathbf{z}) \tag{3.35}$$

on $\mathfrak{Z}_0$.

Define the random subset Θ_S of $\mathfrak{Z}_0 = \mathbb{R}^2$ by

$$\Pr(\Theta_S = D_{j,i}) = w_{j,i} \cdot v_j. \tag{3.36}$$

Then

$$g_S(\mathbf{z}) \triangleq \Pr(\mathbf{z} \in \Theta_S) = \sum_{j=1}^{d} \sum_{i=0}^{e(j)} w_{j,i} \cdot v_j \cdot \mathbf{1}_{D_{j,i}}(\mathbf{z}) = \sum_{j=1}^{d} v_j \cdot g_j(\mathbf{z}) \tag{3.37}$$

defines a fuzzy membership function on $\mathfrak{Z}_0$ (i.e., a vague measurement). The generalized likelihood function for this measurement is, therefore,

$$L_S(\mathbf{x}) \triangleq \Pr(\eta(\mathbf{x}) \in \Theta_S) = \sum_{j=1}^{d} v_j \cdot g_j(\eta(\mathbf{x})). \tag{3.38}$$

(Equivalently, this could have been reformulated from the outset as a specification of fuzzy membership functions $g_1(\mathbf{z}), ..., g_d(\mathbf{z})$ with weights $v_1, ..., v_d$—that is, as what I will shortly call a *fuzzy basic mass assignment* (fuzzy b.m.a.); see Section 4.6.)

Once again, we say that the generalized measurement Θ_S is a *random set model* of the uncertain information S (or, more correctly, its fuzzy model g_S). The generalized likelihood $L_S(\mathbf{x})$ translates the information in Θ_S into a form that can be used in the Bayes filter.

3.4.5 Ambiguity Versus Randomness

One special case is worth examining further. Suppose that the hypotheses in the previous example are precise: $D_j = \{\mathbf{z} - \mathbf{w}_j\}$ for $\mathbf{w}_1, ..., \mathbf{w}_d \in \mathbb{R}^2$ and $\mathbf{w}_j = (x_j, y_j)$. Then superficially,

$$\Pr(\Theta_S = \{\mathbf{z} - \mathbf{w}_j\}) = v_j \tag{3.39}$$

looks like a version of the example of Section 3.4.1 with a discrete random variable $\tilde{\mathbf{W}}_{k+1} \in \{\mathbf{w}_1, ..., \mathbf{w}_d\}$, such that

$$\Pr(\tilde{\mathbf{W}}_{k+1} = \mathbf{w}_j) = v_j. \tag{3.40}$$

However, *the semantics of the two models are fundamentally different*. This is because Section 3.4.1 deals with uncertainty due to randomness whereas this example deals with uncertainty due to ignorance.

In the last section, the observer attempts to describe what is being observed. Each $\mathbf{z} - \mathbf{w}_j$ is offered as an interpretation or hypothesis about what has been observed, with v_j being the degree of belief in that hypothesis. That is:

- *The observation Θ_S is deterministic (i.e., it is one instantiation of some random observation) even though it is modeled as a random set.*

In Section 3.4.1, however, the observer has only a single hypothesis about what has been observed—the measurement $\mathbf{z}$. From previous statistical analysis, however, he or she knows that his or her vision is subject to random error effects (say, because of poor eyeglasses). The observer models that error as the random set $E_{\mathbf{z}-\mathbf{W}_{k+1}}$ where the statistics of $\mathbf{W}_{k+1}$ are described by a probability density function $f_{\mathbf{W}_{k+1}}(\mathbf{w})$. This fact remains true even if $\mathbf{W}_{k+1}$ is discrete in the sense that $f_{\mathbf{W}_{k+1}}(\mathbf{w}) = \sum_{j=1}^{d} v_j \cdot \delta_{\mathbf{w}_j}(\mathbf{w})$. For in this case $\mathbf{W}_{k+1}$ is a random precise measurement drawn from $\mathbb{R}^2$ whereas $\tilde{\mathbf{W}}_{k+1}$ is a random variable on $\{\mathbf{w}_1, ..., \mathbf{w}_d\}$ that is used to model a deterministic but uncertain measurement.

As a consequence, the form of the generalized likelihood for the generalized measurement Θ_S is quite different from the likelihood for a conventional measurement $\mathbf{z}$:

$$L_S(\mathbf{x}) \triangleq \Pr(\eta(\mathbf{x}) \in \Theta_S) = \sum_{j=1}^{d} v_j \cdot \mathbf{1}_{\{\mathbf{z}-\mathbf{w}_j\}}(\eta(\mathbf{x})) \tag{3.41}$$

$$= \sum_{j=1}^{d} v_j \cdot \mathbf{1}_{\{\mathbf{z}-\mathbf{w}_j\}}(x, y) \tag{3.42}$$

$$= \begin{cases} v_j & \text{if} \quad \mathbf{x} = (z_x - x_j, z_y - y_j, v_x, v_y), \text{ some } v_x, v_y \\ 0 & \text{if} \quad \text{otherwise.} \end{cases} \tag{3.43}$$

To gain further insight into the distinction between uncertainty and randomness, suppose that $\mathbf{w}_j$ is one instantiation of a zero-mean random vector $\mathbf{W}_j$ where $\mathbf{W}_1, ..., \mathbf{W}_d$ are independent. Further assume that I is a random integer in $\{1, ..., d\}$ with distribution $\Pr(I = i) = v_i$. Assume that $I, \mathbf{W}_i$ are pairwise independent for all $i = 1, .., d$. Define the random set $\Theta_{I,\mathbf{W}_1, \quad ,\mathbf{W}_d}$ as

$$\Theta_{I,\mathbf{W}_1,...,\mathbf{W}_d} \triangleq E_{\mathbf{z}-\mathbf{W}_I} \tag{3.44}$$

where $E_{\mathbf{z}}$ is as in (3.3). Then

$$f(\Theta_{I,\mathbf{W}_1,...,\mathbf{W}_d}|\mathbf{x}) = \Pr(\eta(\mathbf{x}) \in \Theta_{I,\mathbf{W}_1,...,\mathbf{W}_d}) \tag{3.45}$$

$$= \sum_{i=1}^{d} \Pr(\eta(\mathbf{x}) \in E_{\mathbf{z}-\mathbf{W}_i}, I = i) \tag{3.46}$$

$$= \sum_{i=1}^{d} v_i \cdot \Pr(\eta(\mathbf{x}) \in E_{\mathbf{z}-\mathbf{W}_i}) \tag{3.47}$$

$$= \sum_{i=1}^{d} v_i \cdot \Pr(\mathbf{W}_i \in E_{\mathbf{z}-\eta(\mathbf{x})}) \tag{3.48}$$

$$\cong \varepsilon \cdot \sum_{i=1}^{d} v_i \cdot f_{\mathbf{W}_i}(\mathbf{z} - \eta(\mathbf{x})). \tag{3.49}$$

If $\mathbf{W}_i = \mathbf{w}_i$ is constant then this reduces to (3.41):

$$f(\Theta_{I,\mathbf{w}_1,...,\mathbf{w}_d}|\mathbf{x}) \cong \varepsilon \cdot \sum_{i=1}^{d} v_i \cdot \mathbf{1}_{\mathbf{w}_i}(\mathbf{z} - \eta(\mathbf{x})) \tag{3.50}$$

$$= \varepsilon \cdot \sum_{i=1}^{d} v_i \cdot \mathbf{1}_{\mathbf{z}-\mathbf{w}_i}(\eta(\mathbf{x})) \tag{3.51}$$

$$= \varepsilon \cdot \sum_{i=1}^{d} v_i \cdot \mathbf{1}_{\mathbf{z}-\mathbf{w}_i}(x, y). \tag{3.52}$$

Intuitively, then, $\Theta_{I,\mathbf{W}_1,...,\mathbf{W}_d}$ generalizes both $E_{\mathbf{z}-\mathbf{W}_{k+1}}$ of Section 3.4.1 and Θ_S of this section. On the one hand, it generalizes Θ_S by adding a small amount of imprecision to the singleton focal sets $\{\mathbf{z} - \mathbf{w}_1\}, ..., \{\mathbf{z} - \mathbf{w}_d\}$ to create new focal sets $E_{\mathbf{z}-\mathbf{w}_1}, ..., E_{\mathbf{z}-\mathbf{w}_d}$. These focal sets, together with their weights $v_1, ..., v_d$ constitute, however, a deterministic measurement (i.e., a single instantiation of a "random b.m.a."). The random variable $\Theta_{I,\mathbf{W}_1,..,\mathbf{W}_d}$ is the random set representation of this particular random b.m.a. in the same sense that Θ_S is the random set model of the deterministic b.m.a. $\{\mathbf{z} - \mathbf{w}_1\}, ..., \{\mathbf{z} - \mathbf{w}_d\}$, $v_1, ..., v_d$; and in the sense that $E_{\mathbf{z}-\mathbf{W}_{k+1}}$ is the (approximate) random set model of the conventional random vector $\mathbf{W}_{k+1}$. In this b.m.a., both the number d of focal sets and their weights $v_1, ..., v_d$ are constant. The focal sets themselves, however, are random.

Remark 2 *The random variable* $\mathbf{W}_I$ *is called a finite-mixture process, since the random integer* I *selects among a list* $\mathbf{W}_1, ..., \mathbf{W}_d$ *of random vectors. The likelihood function of* $\mathbf{W}_I$ *is the mixture distribution*

$$f_{\mathbf{W}_I}(\mathbf{z}|\mathbf{x}) = \sum_{i=1}^{d} v_i \cdot f_{\mathbf{W}_i}(\mathbf{z} - \eta(\mathbf{x})). \tag{3.53}$$

Thus a mixture process can be regarded as a special kind of random b.m.a. A general random b.m.a. can be regarded as a generalized mixture process in which the number d *of components and the distribution* $f_I(n)$ *of* I *are both random.*

3.5 THE CORE BAYESIAN APPROACH

In this section, I describe the application of formal Bayes modeling and the Bayes filter to general physical systems in Sections 3.5.1 and 3.5.2, respectively. In Section 3.5.3, I introduce the concept of a Bayes combination operator. I conclude, in Section 3.5.4, with the concept of Bayes-invariant conversion of measurements.

3.5.1 Formal Bayes Modeling in General

Formal Bayes filtering is based on the *dynamic state space model* [108, p. 2592]. The following eight stages are what I call *formal Bayes modeling* [126, 132, 144, 148].

- *State space*: Carefully define a state space $\mathfrak{X}$ that uniquely and exhaustively specifies the possible states ξ of the physical system—meaning those aspects of physical objects that are of interest to us.
- *Measurement space*: For each sensor carefully define a measurement space $\mathfrak{Z}$ that uniquely and exhaustively specifies the individual pieces of information ζ that the sensor can observe.
- *Integration*: Define integrals $\int_S f(\xi)d\xi$ and $\int_T g(\zeta)d\zeta$ of real-valued functions $f(\xi)$ and $g(\zeta)$ of state and measurement variables, respectively.
- *State-transition model*: Construct a state-transition model (typically, a motion model) $\xi \leftarrow \xi'$ describing how the object-state may change from ξ' at measurement collection time step k to ξ at measurement collection time step $k+1$.
- *State-transition density*: From this model construct a state-transition density $f_{k+1}(\xi|\xi', Z^k)$ that is normalized (i.e., $\int f_{k+1}(\xi|\xi', Z^k)d\xi = 1$) and that describes the likelihood that the object will have state ξ at time step $k+1$, if it had state ξ' at time step k and observations $Z^k : \zeta_1, ..., \zeta_k$ have been collected at previous time steps $i = 1, ..., k$.
- *Measurement model*: Construct a measurement model $\zeta \leftarrow \xi$ that describes how measurements ζ are generated by objects with specified states ξ.
- *Likelihood function*: From this model construct a likelihood function $L_\zeta(\xi) = f_{k+1}(\zeta|\xi, Z^k)$ that is normalized—$\int f_{k+1}(\zeta|\xi, Z^k)d\zeta = 1$—and that describes the likelihood of observing ζ if an object with state ξ is present and observations Z^k have been previously collected.

The following are some of the specific instances of formal Bayes modeling considered in this book:

- *Single-target multisource integration*: The state and measurement spaces are usually Euclidean spaces $\mathfrak{X} = \mathbb{R}^N$ and $\mathfrak{Z} = \mathbb{R}^M$ of vectors $\mathbf{x}$ and $\mathbf{z}$, respectively. Motion and measurement models, and likelihood functions and Markov transition densities, typically have the forms described in Section 2.4.4.
- *Multitarget-multisource integration*: Finite-set statistics (FISST) [70, 134, 144, 145, 148] must be used to extend formal Bayes modeling to many multitarget filtering problems [126, 132, 134]. State space is $\mathfrak{X}$ is the hyperspace of all finite subsets of single-target state space $\mathfrak{X}_0$. Measurement space is the hyperspace of all finite subsets of the measurement spaces of the individual sensors. This is one of the subjects of Part II of this book.
- *Unconventional data*: In Chapters 5 and 6, state space is $\mathfrak{X} = \mathfrak{X}_0 = \mathbb{R}^N \times C$ where C is finite. Measurement space $\mathfrak{Z}$ is a "hyperhyperspace." That is, it consists of all random closed subsets Θ of a "base" measurement space $\mathfrak{Z}_0 = \mathbb{R}^M \times D$, where D is finite.

3.5.2 The Bayes Filter in General

This filter [70, p. 238] temporally propagates the Bayes posterior $f_{k|k}(\xi|Z^k)$ through time steps $k = 0, 1, ...$ and consists of the following additional stages.

- *Initialization*: Select a prior density function $f_{0|0}(\xi|Z^0) = f_{0|0}(\xi)$;
- *Predictor*: Account for the increased ambiguity due to interim state-transition via the predictor equation, of which (2.121) and (2.81) are special cases:

$$f_{k+1|k}(\xi|Z^k) = \int f_{k+1}(\xi|\xi', Z^k) f_{k+1|k}(\xi'|Z^k) d\xi' \quad (3.54)$$

$$f_{k+1|k}(\xi|Z^k) = \int f_{k+1}(\xi|\xi') f_{k+1|k}(\xi'|Z^k) d\xi'. \quad (3.55)$$

As in Section 2.4.6, (3.54) is a statement of the total probability theorem. (3.55) results from the assumption $f_{k+1}(\xi|\xi', Z^k) = f_{k+1}(\xi|\xi')$ (i.e., the predicted future state of the object depends only on its current state);

- *Corrector*: Fuse current information ζ_{k+1} with previous information Z^k by using the corrector equation, of which (2.128) and (2.82) are special cases:

$$f_{k+1|k+1}(\xi|Z^{k+1}) \propto f_{k+1}(\zeta_{k+1}|\xi, Z^k) f_{k+1|k}(\xi|Z^k) \quad (3.56)$$

$$f_{k+1|k+1}(\xi|Z^{k+1}) \propto f_{k+1}(\zeta_{k+1}|\xi) f_{k+1|k}(\xi|Z^k). \quad (3.57)$$

  Equation (3.56) is just Bayes' rule, with the Bayes normalization factor being

$$f_{k+1}(\zeta_{k+1}|Z^{k+1}) = \int f_{k+1}(\zeta_{k+1}|\xi, Z^k) \cdot f_{k+1|k}(\xi|Z^k) d\xi. \quad (3.58)$$

  Equation (3.57) results from the assumption $f_{k+1}(\zeta|\xi, Z^k) = f_{k+1}(\zeta|\xi)$ (i.e., current observations depend only on the current state);

- *Fusion*: Multisensor data can be fused in the same manner as in Section 2.4.10. For example, measurements $\overset{1}{\zeta}_{k+1}$ and $\overset{2}{\zeta}_{k+1}$ from two sensors can be fused using their joint likelihood $\overset{12}{f}_{k+1}(\overset{1}{\zeta}_{k+1}, \overset{2}{\zeta}_{k+1}|\mathbf{x})$ in Bayes' rule (provided that it can be determined). If the sensors are independent then

$$\overset{12}{f}_{k+1}(\overset{1}{\zeta}_{k+1}, \overset{2}{\zeta}_{k+1}|\mathbf{x}) = \overset{1}{f}_{k+1}(\overset{1}{\zeta}_{k+1}|\mathbf{x}) \cdot \overset{2}{f}_{k+1}(\overset{2}{\zeta}_{k+1}|\mathbf{x}); \quad (3.59)$$

- *State estimation*: Estimate the current state of the object using a Bayes-optimal state estimator such as the maximum a posteriori (MAP) estimator $\hat{\xi}_{k+1|k+1} = \arg\sup_\xi f_{k+1|k+1}(\xi|Z^{k+1})$. Some care must be exercised, since standard state estimators may not be defined in certain applications;

- *Error estimation*: Characterize the uncertainty in the state using some measure of the statistical dispersion of $f_{k+1|k+1}(\xi|Z^{k+1})$, such as covariance and entropy.

3.5.3 Bayes Combination Operators

This section introduces a concept that will be fundamental to fusion of UGA measurements as developed in Chapter 5. Let $\zeta_1 \otimes \zeta_2$ be a commutative and associative binary operator on the elements ζ_1, ζ_2 of the measurement space $\mathfrak{Z}$ and let $f_0(\xi)$ be a prior distribution on state space $\mathfrak{X}$. We say that "$\otimes$" is a (conjunctive) *Bayes combination operator* [134, p. 64] if it generates *sufficient statistics*, that is,

$$f(\xi|\zeta_1 \otimes ... \otimes \zeta_m) = f(\xi|\zeta_1, ..., \zeta_m) \quad (3.60)$$

for all m. Here $f(\xi|\zeta_1 \otimes ... \otimes \zeta_m) \propto f(\zeta_1 \otimes ... \otimes \zeta_m|\xi) \cdot f_0(\xi)$ is the posterior distribution conditioned on $\zeta_1 \otimes ... \otimes \zeta_m$ and $f(\xi|\zeta_1, ..., \zeta_m) \propto f(\zeta_1, ..., \zeta_m|\xi) \cdot f_0(\xi)$ is the posterior distribution conditioned on $\zeta_1, ..., \zeta_m$.

That is, fusing evidence using the operator '$\otimes$' produces the same posterior distribution as would arise from fusing the same evidence using Bayes' rule alone. In this sense $\zeta_1 \otimes ... \otimes \zeta_m$ can be regarded as equivalent to Bayes' rule because it yields the same result as Bayes' rule.

Example 7 *Let state space be $\mathfrak{X}_0 = \mathbb{R}^N$ and let us be given a base measurement space $\mathfrak{Z}_0 = \mathbb{R}^M$. Let the measurement space $\mathfrak{Z}$ consist of all pairs $(C, \mathbf{c})$ where $\mathbf{c} \in \mathfrak{Z}_0$ and C is an $M \times M$ symmetric, positive-definite matrix. That is, $\mathfrak{Z}$ consists of the outputs of all possible Gaussian sensors that draw observations from $\mathfrak{Z}_0$. The outputs of two sensors can be combined using the rule*

$$(C_1, \mathbf{c}_1) \otimes (C_2, \mathbf{c}_2) = (C, \mathbf{c}) \tag{3.61}$$

where

$$C \triangleq (C_1^{-1} + C_2^{-1})^{-1}, \qquad \mathbf{c} \triangleq (C_1^{-1} + C_2^{-1})^{-1}(C_1^{-1}\mathbf{c}_1 + C_2^{-1}\mathbf{c}_2). \tag{3.62}$$

The operator $\otimes$ is easily shown to be commutative and associative (see Exercise 5). Define the likelihood of $(C, \mathbf{c})$ as

$$f(C, \mathbf{c}|\mathbf{x}) \triangleq N_C(\mathbf{x} - \mathbf{c}). \tag{3.63}$$

This is a generalized likelihood, since $\int \int N_C(\mathbf{x} - \mathbf{c}) d\mathbf{c} dC = \int 1 \cdot dC = \infty$.[3] *Assume that sensors are conditionally independent on the state, so that the joint likelihood is*

$$f((C_1, \mathbf{c}_1), (C_2, \mathbf{c}_2)|\mathbf{x}) = f(C_1, \mathbf{c}_1|\mathbf{x}) \cdot f(C_2, \mathbf{c}_2|\mathbf{x}). \tag{3.64}$$

From (2.36) it follows that

$$\begin{aligned} f((C_1, \mathbf{c}_1), (C_2, \mathbf{c}_2)|\mathbf{x}) &= N_{C_1}(\mathbf{x} - \mathbf{c}_1) \cdot N_{C_2}(\mathbf{x} - \mathbf{c}_2) && (3.65) \\ &= N_{C_1+C_2}(\mathbf{c}_1 - \mathbf{c}_2) \cdot N_C(\mathbf{x} - \mathbf{c}) && (3.66) \\ &\propto f((C_1, \mathbf{c}_1) \otimes (C_2, \mathbf{c}_2)|\mathbf{x}). && (3.67) \end{aligned}$$

Thus $\otimes$ is a Bayes combination operator.

3 Note, however, that it could be turned into a conventional one, if we restrict it to a bounded region of $\mathfrak{Z}_0$.

3.5.4 Bayes-Invariant Measurement Conversion

In Section 5.4.6 we will be concerned with the problem of converting one kind of measurement to another—for example, fuzzy to probabilistic and Dempster-Shafer to fuzzy. Thus suppose that we have two measurement spaces $\mathfrak{Z}$ and $\mathfrak{Z}'$. Suppose further that we wish to convert elements $\zeta \in \mathfrak{Z}$ to elements $\phi_\zeta \in \mathfrak{Z}'$ via some transformation

$$\phi : \mathfrak{Z} \to \mathfrak{Z}'. \tag{3.68}$$

Such conversion may, however, result in loss of information (a lossy data compression) from ζ into ϕ_ζ.

The path out of this quandary is to realize that, in a Bayesian formulation, all relevant information about the unknown state ξ is encapsulated in posterior distributions. The transformation of measurements ζ into posteriors $f(\xi|\zeta)$ may itself constitute a lossy form of data compression. Consequently, information loss due to conversion will cause no harm provided that it is consistent with the information loss already inherent to the construction of posteriors. The problem can, therefore, be restated thusly:

- *What conversions* $\zeta \mapsto \phi_\zeta$ *are Bayes-invariant—that is, leave posterior distributions unchanged in the sense that, for all* $m \geq 1$,

  $$f(\xi|\zeta_1, ..., \zeta_m) = f(\xi|\phi_{\zeta_1}, ..., \phi_{\zeta_m})? \tag{3.69}$$

  Stated differently, for every m the converted measurements $\phi_{\zeta_1}, ..., \phi_{\zeta_m}$ must be sufficient statistics for the original measurements $\zeta_1, ..., \zeta_m$.

Suppose further that $\mathfrak{Z}$ and $\mathfrak{Z}'$ are equipped with Bayes combination operators $\otimes$ and $\otimes'$, respectively. Then the following two procedures may result in very different outcomes:

1. Combining ζ_1 and ζ_2 into $\zeta_1 \otimes \zeta_2$ and then converting this to $\phi_{\zeta_1 \otimes \zeta_2}$;
2. Converting ζ_1 and ζ_2 to ϕ_{ζ_1} and ϕ_{ζ_2} and then combining these into $\phi_{\zeta_1} \otimes' \phi_{\zeta_2}$.

Conversion should be consistent with combination. Consistency in the strongest sense would mean that, for every m,[4]

$$\phi_{\zeta_1 \otimes ... \otimes \zeta_m} = \phi_{\zeta_1} \otimes' ... \otimes' \phi_{\zeta_m}. \tag{3.70}$$

4 Mathematically speaking, in this case the transformation $\zeta \mapsto \phi_\zeta$ is an *algebraic homomorphism* That is, it preserves fusion operations by transforming $\otimes$ into $\otimes'$.

If this is not true, however, then it should at least be true up to posterior conditioning,

$$f(\xi|\phi_{\zeta_1\otimes...\otimes\zeta_m}) = f(\xi|\phi_{\zeta_1}\otimes' ... \otimes' \phi_{\zeta_m}) \tag{3.71}$$

or, equivalently, that it is true at the likelihood function level,

$$f(\phi_{\zeta_1\otimes...\otimes\zeta_m}|\xi) = K \cdot f(\phi_{\zeta_1}\otimes' ... \otimes' \phi_{\zeta_m}|\xi) \tag{3.72}$$

for some K independent of ξ. If (3.69) and (3.71) are true, then we will say that the conversion $\zeta \mapsto \phi_\zeta$ is *Bayes-invariant.*

Equation (3.69) actually follows from (3.71), from the fact that $\otimes$ and $\otimes'$ are Bayes combination operators, and from (3.69) for $m = 1$. To see this, note that

$$\begin{aligned} f(\xi|\zeta_1, ..., \zeta_m) &= f(\xi|\zeta_1\otimes ... \otimes\zeta_m) = f(\xi|\phi_{\zeta_1\otimes...\otimes\zeta_m}) && (3.73)\\ &= f(\xi|\phi_{\zeta_1}\otimes' ... \otimes' \phi_{\zeta_m}) = f(\xi|\phi_{\zeta_1}, ..., \phi_{\zeta_m}). && (3.74)\end{aligned}$$

3.6 FORMAL MODELING OF GENERALIZED DATA

As already noted, one of the major goals of Part I is to extend the formal modeling process described in Sections 2.4.4 and 3.5.1 to generalized forms of data. In summary, such modeling will consist of the following steps:

- Determine what type of data D is being considered: measurements or state-estimates?
- Chapter 8: If a state-estimate, then what methodology?
- If a measurement, then what type?

By this last question we mean deciding among one of the following possibilities:

- *Chapter 2: Unambiguously generated unambiguous (UGU) measurements.* For completeness, this refers to conventional precise measurements with conventional likelihoods.
- *Chapter 5: Unambiguously generated ambiguous (UGA) measurements.* The four examples described in Section 3.4 all have one thing in common: the

existence of a *sensor transform model* $\eta(\mathbf{x})$ that precisely specifies how measurements are related to states. In this case the process by which states generate generalized measurements is also precisely specified. This covers situations when the relationship between measurements and states is clearly understood (e.g., measurements are positions), but in which human beings have been involved in interpreting what has been measured. Examples include natural-language statements and certain kinds of attributes.

- *Chapter 6: Ambiguously generated ambiguous (AGA) measurements.* The value of $\eta(\mathbf{x})$ itself may be known only imprecisely, or even only vaguely (in the sense of Section 3.4.3). This may occur when human interpretation processes are involved in the specification of model bases as well as in the interpretation of measurements. In this case, the generation of ambiguous measurements is itself ambiguous. Examples include natural-language statements, and attributes extracted from images, especially when models are based on human interpretation processes.

- *Chapter 7: Ambiguously generated unambiguous (AGU) measurements.* This type occurs when the data $\mathbf{z}$ itself is precise but its likelihood function $L_{\mathbf{z}}(\mathbf{x})$ cannot be specified precisely. Examples include Synthetic Aperture Radar (SAR) intensity images, in which statistically uncharacterizable real-world variability (dents, wet mud, and irregularly placed surface equipment.) make the precise specification of likelihoods essentially impossible.

Having determined the type of data, we then do the following:

- *Chapter 4:* Choose an appropriate uncertainty modeling formalism: Fuzzy logic? Dempster-Shafer? Fuzzy Dempster-Shafer? Rule-based?

- *Chapter 4:* Convert this representation into a *random set representation* (a generalized measurement) Θ_D;

- *Chapters 5-7:* Use this representation to construct generalized likelihood functions $f_{k+1}(D|\mathbf{x}) = f_{k+1}(\Theta_D|\mathbf{x})$;

- Fuse the data by employing joint generalized likelihood functions in the corrector step of the Bayes filter.

This process is illustrated schematically in Figure 3.4.

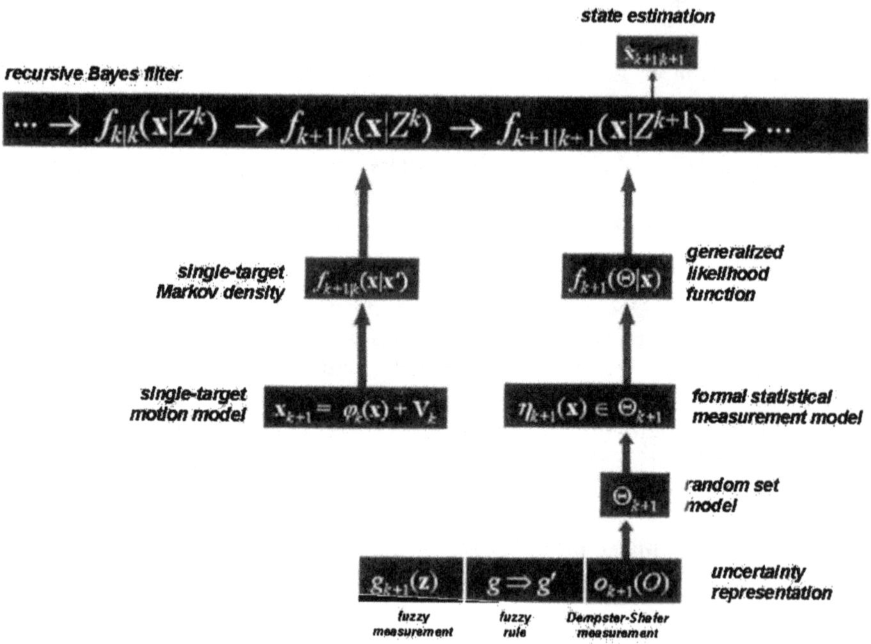

Figure 3.4 A schematic description of the Bayesian processing of generalized measurements Measurements are represented using the random set techniques of Chapters 4-8. Once these models have been converted to generalized likelihood functions, generalized measurements can be processed using the single-target Bayes filter of Section 2.4.

3.7 CHAPTER EXERCISE

Exercise 5 *Prove that the operator* $\otimes$ *of Example 7 in Section 3.5.3 is commutative and associative.*

Chapter 4

Random Set Uncertainty Representations

4.1 INTRODUCTION TO THE CHAPTER

As was noted in Chapter 1, beginning in the late 1970s, a number of researchers began uncovering close relationships between Mathéron's theory of random sets [75, 160] and many aspects of expert systems theory. This chapter describes random set formulations of the major expert system approaches in enough detail to facilitate understanding of the rest of Part I. As part of this discussion, I briefly review the elements of each of the expert systems approaches considered.

4.1.1 Summary of Major Lessons Learned

The following are the basic concepts, results, and formulas that the reader will discover in this chapter:

- The concept of an "event," as used in expert systems theory (Section 4.2);
- The concept of a "copula" and its centrality for modeling statistical dependence; see (4.37);
- Copula fuzzy logics, and their interpretation as models of statistical dependence between interpretations of fuzzy sets (Section 4.3.4);
- The concept of a fuzzy event, which models vagueness as a random interpretation; see (4.54);
- Normalized and unnormalized versions of Dempster's combination; see (4.87) and (4.95);

- Modified Dempster's combination; see (4.96);
- Voorbraak and pignistic probability distributions; see (4.115) and (4.119);
- Fuzzy version of Dempster' combination operator; see (4.129);
- The conventional wisdom regarding "Zadeh's paradox" results from misinterpretation based on inadequate modeling (Section 4.5.2);
- Contrary to conventional wisdom, Dempster-Shafer theory is not more general than Bayesian theory (Section 4.8);
- Conditional event algebra (CEA) method of combining first-order rules; see (4.145) and (4.148);
- Random sets can be used to represent uncertainty models based on fuzzy logic (section 4.3.2), Dempster-Shafer theory (Section 4.5.4), fuzzy Dempster-Shafer theory (Section 4.6.1), first-order rules (Section 4.7.3), composite rules (Section 4.7.4), and second-order rules (Section 4.7.5).

4.1.2 Organization of the Chapter

I begin in Section 4.2 by describing the concept of an "event" in a "universe of discourse." Then I will describe random set representations of fuzzy set theory (Section 4.3), Li's generalized fuzzy set theory (Section 4.4), the conventional and generalized Dempster-Shafer theories (Section 4.5), fuzzy Dempster-Shafer theory (Section 4.6), and rule-based inference (Section 4.7). I conclude with a discussion of the relationship between Bayesian probability and other expert systems formalisms in Section 4.8. Exercises for the chapter are in Section 4.9.

4.2 UNIVERSES, EVENTS, AND THE LOGIC OF EVENTS

Let $\mathfrak{U}$ be a discrete or a continuously infinite set: a "universe of discourse" or "frame of discernment." The elements of $\mathfrak{U}$ are assumed to be endowed with various *features* that can be used to distinguish them from one another. These features are assumed to be "crisp" in the sense that they can be distinguished from each other without uncertainty. Any given feature can be uniquely specified by forming the subset of all elements of $\mathfrak{U}$ that have that feature.

Features are therefore in one-to-one correspondence with the subsets of $\mathfrak{U}$, which are called "events." Features may be combined using the usual set-theoretic operations of intersection '$S \cap T$, union $S \cup T$, and complement S^c. The class of all events endowed with these operations is a *Boolean algebra*. This means that '$\cap$' and '$\cup$' are commutative,

$$S_1 \cap S_2 = S_2 \cap S_1, \qquad S_1 \cup S_2 = S_2 \cup S_1 \tag{4.1}$$

associative,

$$(S_1 \cap S_2) \cap S_3 = S_1 \cap (S_2 \cap S_3), \quad (S_1 \cup S_2) \cup S_3 = S_1 \cup (S_2 \cup S_3) \tag{4.2}$$

mutually distributive,

$$S_1 \cap (S_2 \cup S_3) = (S_1 \cap S_2) \cup (S_1 \cap S_3) \tag{4.3}$$

$$S_1 \cup (S_2 \cap S_3) = (S_1 \cup S_2) \cap (S_1 \cup S_3) \tag{4.4}$$

satisfy De Morgan's laws,

$$(S_1 \cap S_2)^c = S_1^c \cup S_2^c, \qquad (S_1 \cup S_2)^c = S_1^c \cap S_2^c \tag{4.5}$$

and have $\emptyset$ and $\mathfrak{U}$ as units:

$$\emptyset \cap S = \emptyset, \qquad \emptyset \cup S = S, \quad \mathfrak{U} \cap S = S, \quad \mathfrak{U} \cup S = \mathfrak{U}. \tag{4.6}$$

Example 8 *Let $\mathfrak{U}$ be a deck of 52 playing cards. Then the event KING is*

$$\mathit{KING} \triangleq \{\mathit{KS, KH, KC, KD}\}. \tag{4.7}$$

Likewise, the event HEART is

$$\mathit{HEART} \triangleq \{\mathit{KH, QH, JH, 10H, 9H, 8H, 7H, 6H, 5H, 4H, 3H, 2H, AH}\}. \tag{4.8}$$

It follows that $\mathit{KING} \cap \mathit{HEART} = \{\mathit{KH}\}$.

4.3 FUZZY SET THEORY

The reader may first wish to review the example presented in Section 3.4.3. A *fuzzy membership function* (or *fuzzy event*) on $\mathfrak{U}$ is a function $f(u)$ with values in $[0, 1]$. The value $f(u)$ is the *degree of membership* of u in the fuzzy set represented by

f. If $f(u) = 1$ for some $u \in \mathfrak{U}$ then f is "normal." Conventional or "crisp" subsets $S \subseteq \mathfrak{U}$ are represented by their indicator functions $\mathbf{1}_S(u)$ defined by $\mathbf{1}_S(u) = 1$ if $u \in S$ and $\mathbf{1}_S(u) = 0$ otherwise; see (5.38).

In what follows I discuss fuzzy logics (Section 4.3.1), random set representations of fuzzy sets (Section 4.3.2), finite-level fuzzy sets and their "synchronous" random set representations (Section 4.3.3), and copula fuzzy logics (Section 4.3.4). More general random set representations of fuzzy sets are discussed in Section 4.3.5.

4.3.1 Fuzzy Logics

A *fuzzy logic* consists of three operators defined on fuzzy membership functions: *conjunction* $f_1 \wedge f_2$, *disjunction* $f_1 \vee f_2$, and *complementation* f^c, with the last defined by $f^c(u) \triangleq 1 - f(u)$. The conjunction and disjunction operators must be commutative,

$$f_1 \wedge f_2 = f_2 \wedge f_1, \qquad f_1 \vee f_2 = f_2 \vee f_1, \tag{4.9}$$

associative,

$$(f_1 \wedge f_2) \wedge f_3 = f_1 \wedge (f_2 \wedge f_3), \quad (f_1 \vee f_2) \vee f_3 = f_1 \vee (f_2 \vee f_3), \tag{4.10}$$

and satisfy the De Morgan equalities,

$$(f_1 \wedge f_2)^c = f_1^c \vee f_2^c, \quad (f_1 \vee f_2)^c = f_1^c \wedge f_2^c. \tag{4.11}$$

Additionally, they must satisfy $0 \wedge f = 0$, $1 \wedge f = f$, $1 \vee f = 1$, and $0 \vee f = f$, where 1 and 0 denote the fuzzy membership functions that are identically equal to one and zero, respectively.

The Zadeh logic, by far the most familiar fuzzy logic, is defined by:

$$(f_1 \wedge f_2)(u) \triangleq \min\{f_1(u), f_2(u)\} \tag{4.12}$$

$$(f_1 \vee f_2)(u) \triangleq \max\{f_1(u), f_2(u)\}. \tag{4.13}$$

There is actually an infinitude of alternative fuzzy logics—see [80, pp. 225-227] for a compendium that includes infinite families of fuzzy logics. Two such alternatives are the *prodsum logic*

$$(f_1 \wedge f_2)(u) \triangleq f_1(u) f_2(u) \tag{4.14}$$

$$(f_1 \vee f_2)(u) \triangleq f_1(u) + f_2(u) - f_1(u) f_2(u) \tag{4.15}$$

and the *nilpotent logic*

$$(f_1 \wedge f_2)(u) \triangleq \max\{0, f_1(u) + f_2(u) - 1\} \quad (4.16)$$
$$(f_1 \vee f_2)(u) \triangleq \min\{1, f_1(u) + f_2(u)\}. \quad (4.17)$$

The Zadeh logic is distinguished among all others in that it is apparently the only one that also satisfies distributivity:

$$f_1 \wedge (f_2 \vee f_3) = (f_1 \wedge f_2) \vee (f_1 \wedge f_3) \quad (4.18)$$
$$f_1 \vee (f_2 \wedge f_3) = (f_1 \vee f_2) \wedge (f_1 \vee f_3). \quad (4.19)$$

4.3.2 Random Set Representation of Fuzzy Events

Fuzzy set theory can be formulated in random set form as follows [68]. Let Σ be a random subset of $\mathfrak{U}$. Goodman's *one point covering function* μ_Σ is the fuzzy membership function defined by

$$\mu_\Sigma(u) \triangleq \Pr(u \in \Sigma). \quad (4.20)$$

This is pictured in Figure 4.1.

Conversely, let $f(u)$ be a fuzzy membership function and let A be a uniformly distributed random number on $[0, 1]$. The *synchronous random set representation* of f is

$$\Sigma_A(f) \triangleq \{u|\ A \leq f(u)\}. \quad (4.21)$$

The random set $\Sigma_A(f)$ is pictured in Figure 4.2. For each $a \in [0, 1]$, the subset $\Sigma_a(f) \subseteq \mathfrak{U}$ is the level set of the function f where f is cut by the (hyper)plane defined by a.

Because A is uniform,

$$\mu_{\Sigma_A(f)}(u) = \Pr(u \in \Sigma_A(f)) = \Pr(A \leq f(u)) = f(u). \quad (4.22)$$

That is, the random set $\Sigma_A(f)$ faithfully represents the information contained in the fuzzy membership function f while attaching random content via A.

The instantiations of the random subset $\Sigma_A(f)$ are linearly ordered under set inclusion: for any $\omega, \omega' \in \Omega$, either $\Sigma_{A(\omega)}(f) \subseteq \Sigma_{A(\omega')}(f)$ or $\Sigma_{A(\omega')}(f) \subseteq \Sigma_{A(\omega)}(f)$.

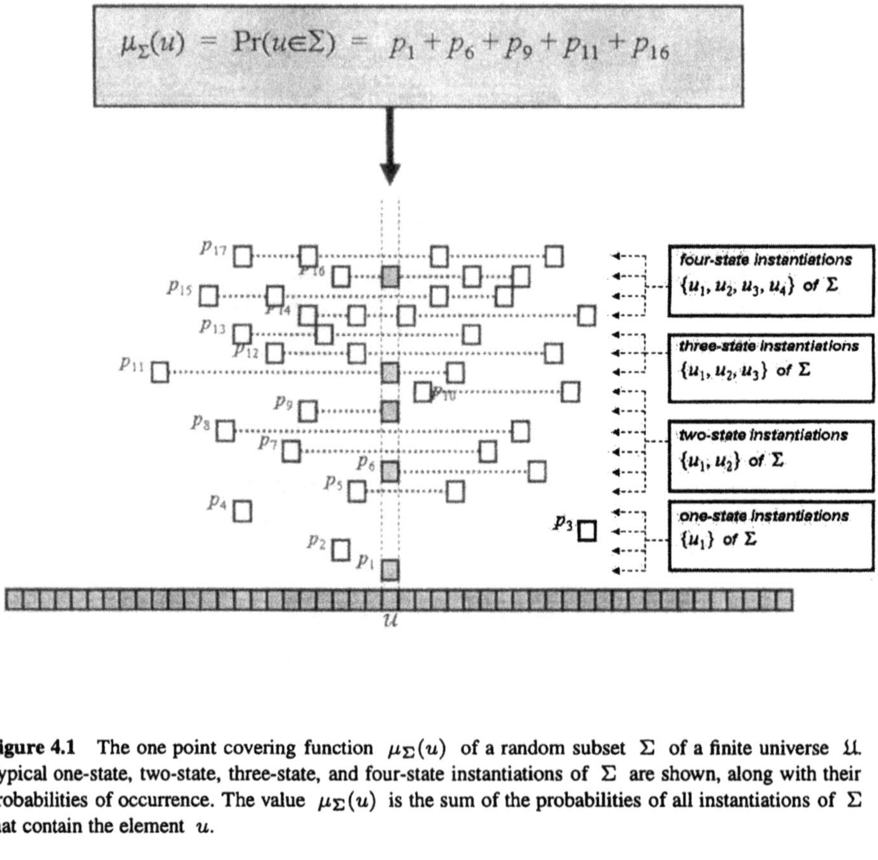

Figure 4.1 The one point covering function $\mu_\Sigma(u)$ of a random subset Σ of a finite universe $\mathfrak{U}$. Typical one-state, two-state, three-state, and four-state instantiations of Σ are shown, along with their probabilities of occurrence. The value $\mu_\Sigma(u)$ is the sum of the probabilities of all instantiations of Σ that contain the element u.

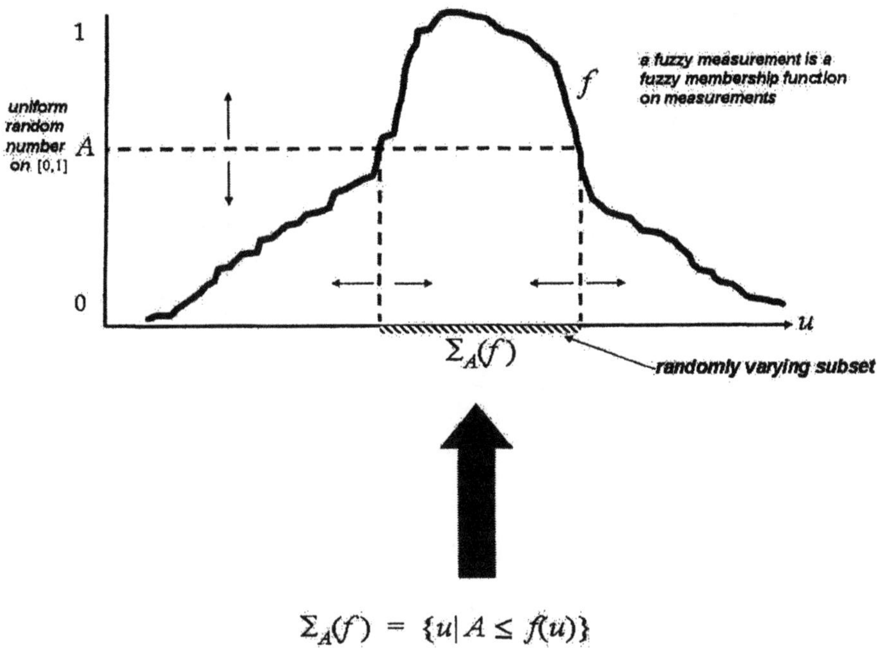

Figure 4.2 The synchronous random set representation $\Sigma_A(f)$ of a fuzzy subset $f(u)$ on $\mathfrak{U}$ is pictured. The instantiations of $\Sigma_A(f)$ are the level subsets of f selected by the uniform random number A on $[0, 1]$.

As with the "Gustav is near the tower" example of Section 3.4.3, the following are true.

- Each instantiation is a different interpretation of the meaning of a fuzzy concept;
- The random number A selects among the possible interpretations.

I must emphasize that $\Sigma_A(f)$ models a *deterministic* measurement. That is, even though it is a random set, it is in turn itself just one possible instantiation of some random variable.

To gain a better understanding of the distinction, suppose that ϕ is a *random fuzzy membership function* [113]. Then

$$\Sigma_A(\phi) \tag{4.23}$$

would be a random set model of ϕ. For any $A = a$ and $\phi = f$, $\Sigma_a(f)$ is one possible instantiation of $\Sigma_A(\phi)$.

It is left to the reader as Exercise 6 to show that the following are true for the Zadeh logic:

$$\Sigma_A(f) \cap \Sigma_A(g) = \Sigma_A(f \wedge g) \tag{4.24}$$

$$\Sigma_A(f) \cup \Sigma_A(g) = \Sigma_A(f \vee g). \tag{4.25}$$

However, $\Sigma_A(f)^c \neq \Sigma_A(f^c)$ since

$$\Sigma_A(f)^c = \{u|\ A > f(u)\} = \{u|\ 1 - A < 1 - f(u)\} \tag{4.26}$$

$$= \{u|\ 1 - A < f^c(u)\} \tag{4.27}$$

$$= \Sigma_{1-A}(f^c) - \{u|\ A = f(u)\}. \tag{4.28}$$

That is, from a random set viewpoint conventional fuzzy logic complementation can be regarded as the result of forcing algebraic closure on the random set representation.

4.3.3 Finite-Level Fuzzy Sets

An informative special case occurs when the value of $f(u)$ can take only a finite number $0 \leq l_1 < ... < l_e \leq 1$ of possibilities. In this case it is possible to establish a one-to-one correspondence between finite-level fuzzy membership functions, on the one hand, and discrete consonant random sets, on the other.

To see this consider Figure 4.3, from which it is clear that $\Sigma_A(f) = \{u|\ A \leq f(u)\}$ has only a finite number of instantiations. Let

$$S_i \triangleq \{u|\ l_i \leq f(u)\} \tag{4.29}$$

for $i = 1, ..., e$. For $0 \leq a \leq l_1$ we have $\Sigma_a(f) = S_1$, with $S_1 = \mathfrak{U}$ if $l_1 > 0$. For $l_1 < a \leq l_2$ we have $\Sigma_a(f) = S_2$. For $l_{e-1} < a \leq l_e$ we have $\Sigma_a(f) = S_e$. Finally, if $l_e < 1$ then $\Sigma_a(f) = \emptyset$ for $l_e < a \leq 1$. From Figure 4.3 it is clear that

$$\Pr(\Sigma_A(f) = S_i) = l_i - l_{i-1} \tag{4.30}$$

for $i = 0, ..., e+1$ where by convention $l_0 = 0$, $l_{e+1} = 1$, and $S_{e+1} = \emptyset$. Also, $f(u) = l_i$ if and only if $u \in S_i - S_{i+1}$.

The random set $\Sigma_A(f)$ is said to be *discrete* since it has a finite number of instantiations. It is *consonant* since its instantiations are linearly ordered under set inclusion: $S_1 \supset ... \supset S_e$.

Conversely, let Σ be a discrete consonant random subset of $\mathfrak{U}$ with $\Pr(\Sigma = T_i) = p_i$ for $i = 1, ..., e+1$ and $T_1 \supset ... \supset T_e \supset T_{e+1} = \emptyset$. Then one can construct a finite-level fuzzy membership function that is uniquely associated with it. Define $w_i = p_1 + ... + p_i$ for $i = 1, ..., e+1$. Then

$$\mu_\Sigma(u) = \Pr(u \in \Sigma) = \sum_{i=1}^{e+1} \Pr(\Sigma = T_i, u \in T_i) = \sum_{i=1}^{e+1} p_i \cdot \mathbf{1}_{T_i}(u). \tag{4.31}$$

So, $u \in T_i - T_{i+1}$ if and only if $\mu_\Sigma(u) = w_i$. Consequently, for $a \in (w_i, w_{i+1}]$

$$\Sigma_a(\mu_\Sigma) = \{u|\ a \leq \mu_\Sigma(u)\} = \bigcup_{j=i}^{e} \{u|\ \mu_\Sigma(u) = w_j\} \tag{4.32}$$

$$= (T_i - T_{i+1}) \cup ... \cup (T_{e-1} - T_e) \cup (T_e - T_{e+1}) \tag{4.33}$$

$$= T_i - T_{e+1} = T_i. \tag{4.34}$$

Likewise, for any uniformly distributed random number A on $[0, 1]$,

$$\Pr(\Sigma_A(\mu_\Sigma) = T_i) = w_i - w_{i-1} = p_i \tag{4.35}$$

Let A be chosen so that $A(\omega) \in (w_i, w_{i+1}]$ if and only if $\Sigma(\omega) = T_i$. Then

$$\Sigma_{A(\omega)}(\mu_\Sigma) = \Sigma(\omega) \tag{4.36}$$

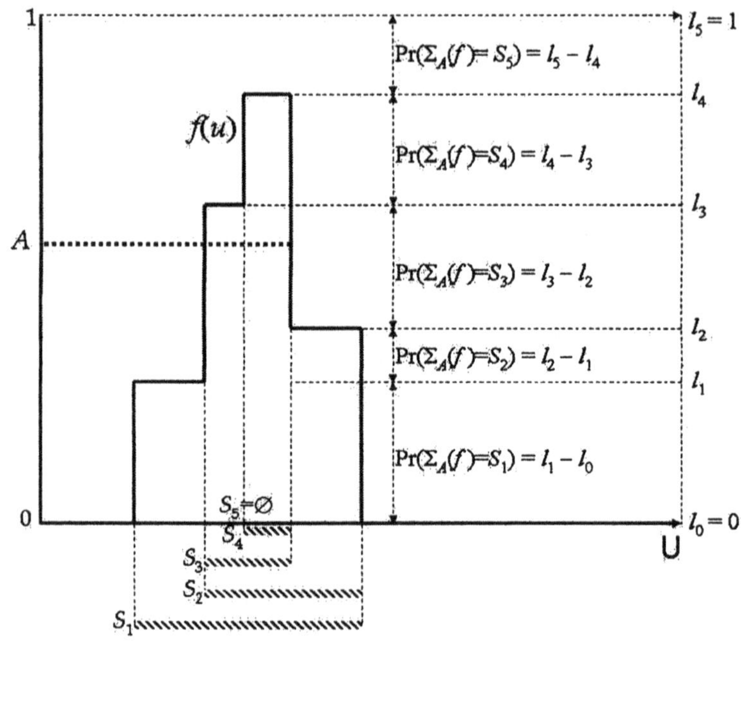

Figure 4.3 Relationship between a finite-level fuzzy membership function $f(u)$ and its random set representation $\Sigma_A(f)$. The function $f(u)$ has only five values, $l_0 \leq l_1 \leq l_2 \leq l_3 \leq l_4$ with $l_0 = 0$. Likewise, $\Sigma_A(f)$ has only five instantiations, $S_1 \supset S_2 \supset S_3 \supset S_4 \supset S_5$ with $S_5 = \emptyset$. The probability that $\Sigma_A(f)$ takes S_i as an instantiation is $l_i - l_{i-1}$

for all ω. In this sense, the transformations $\Sigma \mapsto \mu_\Sigma$ and $f \mapsto \Sigma_A(f)$ define a one-to-one correspondence between finite-level fuzzy membership functions on $\mathfrak{U}$, on the one hand, and discrete consonant random subsets of $\mathfrak{U}$, on the other.[1]

4.3.4 Copula Fuzzy Logics

Let A, A' be two uniformly distributed random numbers on $[0,1]$. Their "copula" [70, p. 457], [177, 178] is defined as

$$a \wedge_{A,A'} a' \triangleq \Pr(A \leq a, A' \leq a') \tag{4.37}$$

for $0 \leq a, a' \leq 1$.

Copulas completely characterize statistical dependencies between random numbers. Let X and X' be random numbers with cumulative distributions

$$\begin{aligned} P(x,x') &= \Pr(X \leq x, X' \leq x') && (4.38)\\ P(x) &= P(x,1) && (4.39)\\ P'(x') &= P(1,x'). && (4.40) \end{aligned}$$

Sklar's theorem [178, p. 12] shows that there are uniformly distributed random numbers A, A' on $[0,1]$ such that

$$P(x,x') = P(x) \wedge_{A,A'} P'(x') \tag{4.41}$$

for all x, x'. That is, $P(x,x')$ can be decomposed into two parts: the characterizations $P(x)$, $P'(x')$ of the statistics of X, X' individually; and the characterization $a \wedge_{A,A'} a'$ of their joint statistics. The converse is also true: If $P(x)$, $P'(x')$ are cumulative distributions, then $P(x,x') = P(x) \wedge_{A,A'} P'(x')$ is a joint cumulative distribution.

Note that

$$\begin{aligned} (f \wedge_{A,A'} f')(u) &\triangleq \Pr(u \in \Sigma_A(f) \cap \Sigma_{A'}(f')) && (4.42)\\ &= \Pr(A \leq f(u), A' \leq f'(u)) && (4.43)\\ &= f(u) \wedge_{A,A'} f'(u). && (4.44) \end{aligned}$$

Also note that $0 \wedge_{A,A'} a' = 0$ and $a \wedge_{A,A'} 0 = 0$, and that since A, A' are uniform, $1 \wedge_{A,A'} a' = a'$, $a \wedge_{A,A'} 1 = a$. If in addition '$\wedge_{A,A'}$' is commutative and

1 This fact is not necessarily true without the finiteness assumption; see [163]

associative, then (4.42) defines a fuzzy conjunction with corresponding disjunction

$$a \vee_{A,A'} a' \triangleq 1 - (1-a) \wedge_{A,A'} (1-a'). \quad (4.45)$$

We say that such fuzzy logics are *copula logics*.

The following inequality is satisfied if and only if a fuzzy conjunction '$\wedge$' is also a copula [72, p. 39]:

$$a \wedge b + a' \wedge b' \geq a \wedge b' + a' \wedge b \quad (4.46)$$

for all $a, a', b, b' \in [0,1]$ such that $a \leq a', b \leq b'$.

In summary, please note the following.

- Different copula logics can be regarded as *models of different statistical correlations between the interpretations modeled by different fuzzy sets.*

For each draw $A(\omega)$ of A the set $\Sigma_{A(\omega)}(f)$ is an interpretation of the fuzzy concept f. Likewise, for each draw $A'(\omega)$ of A' the set $\Sigma_{A'(\omega)}(f')$ is an interpretation of f'. Consequently, the interpretation $\Sigma_{A(\omega)}(f)$ of f corresponds to the interpretation $\Sigma_{A'(\omega)}(f')$ of f' and vice versa.

When $A = A' = a$ it is being stated that the interpretations $\Sigma_a(f)$ and $\Sigma_a(f')$ modeled by f and f' are in lock-step. In this case

$$a \wedge_{A,A} a' = \Pr(A \leq a, A' \leq a) = \Pr(A \leq \min\{a, a'\}) = \min\{a, a'\}. \quad (4.47)$$

Consequently the Zadeh '$\wedge$' corresponds to fusion of perfectly correlated interpretations.

The other extreme occurs when A, A' are independent. In this case the interpretations of f are completely independent of the interpretations of f. Thus

$$a \wedge_{A,A} a' = \Pr(A \leq a, A' \leq a') = \Pr(A \leq a) \cdot \Pr(A' \leq a') = a \cdot a' \quad (4.48)$$

which is the prodsum conjunction.

Finally, suppose that $A' = 1 - A$, so that the interpretations of f are anticorrelated with the interpretations of f'. Then

$$\begin{aligned} a \wedge_{A,A} a' &= \Pr(A \leq a, 1 - A \leq a') = \Pr(A \leq a, 1 - a' \leq A) && (4.49) \\ &= \Pr(A \leq a) - \Pr(A \leq a, 1 - a' > A) && (4.50) \\ &= a - \Pr(A < \min\{a, 1 - a'\}) && (4.51) \\ &= a - \min\{a, 1 - a'\} && (4.52) \\ &= \max\{0, a + a' - 1\} && (4.53) \end{aligned}$$

which is the nilpotent conjunction.

Thus in my formalization, a *fuzzy event* is a pair

$$(f, A). \tag{4.54}$$

When context is clear, as in the notation $f \wedge_{A,A'} f'$, the nonrandom content f of a fuzzy event (f, A) will represent that event, with its random content A left implicit.

Example 9 (Hamacher Fuzzy Logic) *The Hamacher fuzzy logic is defined by (see [58, p. 86] or [80, p. 225]):*

$$a \wedge a' \triangleq \frac{aa'}{a + a' - aa'} \tag{4.55}$$

$$a \vee a' \triangleq \frac{a + a' - 2aa'}{1 - aa'}. \tag{4.56}$$

It models the correlation between two random variables that are nearly statistically independent. It is left to the reader as Exercise 7 to show that $0 \leq a \wedge a' \leq 1$ *for all* a, a'*; and that '*$\wedge$*' is associative and is a copula.*

Example 10 (FGM Fuzzy Logics) *The infinite family of Farlie-Gumbel-Morgenstern (FGM) logics is defined, for* $-1 \leq \theta \leq 1$*, by*

$$a \wedge_\theta a' \triangleq aa' + \theta aa'(1 - a)(1 - a') \tag{4.57}$$

$$a \vee_\theta a' \triangleq a + a' - aa' - \theta aa'(1 - a)(1 - a'). \tag{4.58}$$

It is left to the reader as Exercise 8 to show that $0 \leq a \wedge_\theta a' \leq 1$ *for all* a, a', θ *and that '*$\wedge_\theta$*' is associative. The FGM conjunctions are the only copula conjunctions (in fact, the only copulas) that are expressible as quadratic polynomials in* a, a' *[178, p. 14]. The prodsum logic results from the choice* $\theta = 0$.

4.3.5 General Random Set Representations of Fuzzy Sets

The synchronous random set representation $\Sigma_A(f)$ of a fuzzy membership function $f(u)$, (4.21), is useful for most purposes. However, and as Goodman has repeatedly emphasized, it is by no means the only such representation.[2] In this

2 Goodman has mathematically characterized all random sets that have a given $f(u)$ as their one point covering function [68].

section I briefly introduce a more general random set representation that will prove useful in later chapters when I introduce the concept of a probability-generating functional (p.g.fl.).

In the synchronous representation, we assume that we have a single uniformly distributed random number A in $[0,1]$. More generally, let $\alpha(u)$ be a *uniform random scalar field*. That is, for each $u \in \mathfrak{U}$, $\alpha(u)$ is a uniformly distributed random number in $[0,1]$. Then the following defines a random subset of $\mathfrak{U}$:

$$\Sigma_\alpha(f) \triangleq \{u|\ \alpha(u) \leq f(u)\}. \tag{4.59}$$

It follows that $\Sigma_\alpha(f)$ is also a faithful random set representation of $f(u)$ since

$$\mu_{\Sigma_\alpha(f)}(u) = \Pr(u \in \Sigma_\alpha(f)) = \Pr(\alpha(u) \leq f(u)) = f(u). \tag{4.60}$$

Moreover, it is left to the reader as Exercise 9 to show that (4.24)-(4.25) remain valid:

$$\Sigma_\alpha(f) \cap \Sigma_\alpha(g) = \Sigma_\alpha(f \wedge g) \tag{4.61}$$
$$\Sigma_\alpha(f) \cup \Sigma_\alpha(g) = \Sigma_\alpha(f \vee g). \tag{4.62}$$

The synchronous random set representation of $f(u)$ results if we assume that $\alpha(u) = A$ for all $u \in \mathfrak{U}$. The opposite kind of representation results if we assume instead the following:

- For any integer $1 \leq n < \infty$ and any finite list $u_1, ..., u_n \in \mathfrak{U}$ of distinct elements of $\mathfrak{U}$, the random numbers $\alpha(u_1), ..., \alpha(u_n)$ are statistically independent.

In this case we call $\Sigma_\alpha(f)$ the *asynchronous* random set representation of $f(u)$.

How do the two representations differ? Let $U = \{u_1, ..., u_n\}$ be a finite subset of $\mathfrak{U}$ with $|U| = n$. For the synchronous representation,

$$\begin{aligned}
\Pr(U \subseteq \Sigma_A(f)) &= \Pr(A \leq f(u_1), ..., A \leq f(u_n)) && (4.63)\\
&= \Pr(A \leq \min\{f(u_1), ..., f(u_n)\}) && (4.64)\\
&= \min\{f(u_1), ..., f(u_n)\}. && (4.65)
\end{aligned}$$

For the asynchronous representation,

$$\begin{aligned}
\Pr(U \subseteq \Sigma_\alpha(f)) &= \Pr(\alpha(u_1) \leq f(u_1), ..., \alpha(u_n) \leq f(u_n)) && (4.66)\\
&= \Pr(\alpha(u_1) \leq f(u_1)) \cdots \Pr(\alpha(u_n) \leq f(u_n)) && (4.67)\\
&= f(u_1) \cdots f(u_n) && (4.68)
\end{aligned}$$

where the final equation results from the fact that $\alpha(u_1), ..., \alpha(u_n)$ are all uniformly distributed on $[0, 1]$. In other words, in the synchronous representation the degree of membership of different elements is determined, as the name indicates, synchronously. In the asynchronous representation, the degree of membership of different elements is determined completely independently.

4.4 GENERALIZED FUZZY SET THEORY

"Generalized fuzzy logic" was devised by Y. Li [119] to provide a probabilistic basis for fuzzy set theory. It allows us to extend (4.21) in such a manner as to easily construct random subsets with desired properties, especially in continuously infinite universes. Let $I \triangleq [0, 1]$ and define

$$\mathfrak{U}^* \triangleq \mathfrak{U} \times I. \tag{4.69}$$

Any subset of $\mathfrak{U}^*$ is called a "generalized fuzzy set" and the class of generalized fuzzy sets therefore defines a Boolean algebra.

If $f(u)$ is a fuzzy membership function on $\mathfrak{U}$ then

$$W_f \triangleq \{(u, a) | \; a \leq f(u)\} \tag{4.70}$$

is one example of a generalized fuzzy set. It is left to the reader as Exercise 10 to show that

$$W_f \cap W_g = W_{f \wedge g} \tag{4.71}$$

$$W_f \cup W_g = W_{f \vee g} \tag{4.72}$$

for Zadeh's fuzzy logic; but that, on the other hand, $W_f^c \neq W_{f^c}$.

Generalized fuzzy logic extends ordinary Zadeh fuzzy logic by embedding it in a Boolean algebra. This means that the law of the excluded middle, $W \cap W^c = \emptyset$ and $W \cup W^c = \mathfrak{U}$, is true for generalized fuzzy logic even though it is not true for ordinary fuzzy logic.

Every generalized fuzzy set gives rise to a fuzzy membership function as follows:

$$\mu_W(u) \triangleq \int_0^1 \mathbf{1}_W(u, a) da \tag{4.73}$$

(provided, of course, that the integral exists). Note that

$$\mu_{W_f}(u) = \int \mathbf{1}_{a \leq f(u)}(u, a) da = \int_0^{f(u)} da = f(u). \tag{4.74}$$

4.4.1 Random Set Representation of Generalized Fuzzy Events

Random set models of generalized fuzzy sets are constructed as follows. If A is a uniform random number on I and $W \subseteq \mathfrak{U}^*$ is a generalized fuzzy set, define the random subset $\Sigma_A(W)$ of $\mathfrak{U}$ as

$$\Sigma_A(W) \triangleq \{u|\ (u, A) \in W\}. \tag{4.75}$$

Figure 4.4 pictures $\Sigma_A(W)$. It generalizes Figure 4.2, in that $\Sigma_A(W)$ traces out the level sets of a many-valued function.

It is left to the reader as Exercise 11 to show that the following relationships are true:

$$\Sigma_A(V \cap W) = \Sigma_A(V) \cap \Sigma_A(W) \tag{4.76}$$

$$\Sigma_A(V \cup W) = \Sigma_A(V) \cup \Sigma_A(W) \tag{4.77}$$

$$\Sigma_A(W^c) = \Sigma_A(W)^c \tag{4.78}$$

$$\Sigma_A(\emptyset) = \emptyset, \quad \Sigma_A(\mathfrak{U}^*) = \mathfrak{U}. \tag{4.79}$$

The random set representation of a fuzzy membership function f, (4.21), is a special case of (4.75):

$$\Sigma_A(W_f) = \Sigma_A(f) \tag{4.80}$$

since

$$\Sigma_A(W_f) = \{u|\ (u, A) \in W_f\} = \{u|\ A \leq f(u)\} = \Sigma_A(f). \tag{4.81}$$

Remark 3 *One can define more general random set representations in the same manner as in Section 4.3.5:*

$$\Sigma_\alpha(W) \triangleq \{u|\ (u, \alpha(u)) \in W\}. \tag{4.82}$$

4.5 DEMPSTER-SHAFER THEORY

The reader may first wish to review the example presented in Section 3.4.4. A *basic mass assignment* (b.m.a.) on $\mathfrak{U}$ is a nonnegative function $m(U)$ defined on subsets $U \subseteq \mathfrak{U}$ such that: (1) $m(U) = 0$ for all but a finite number of U (called the *focal*

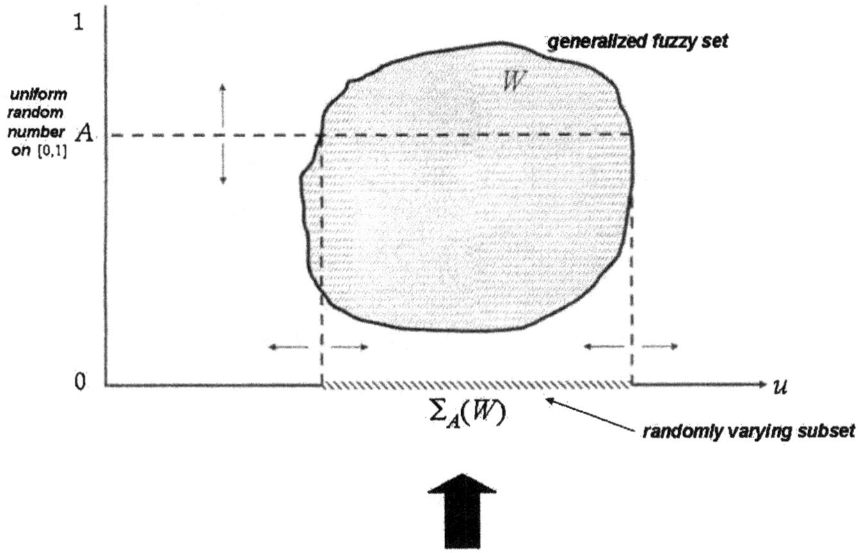

$$\Sigma_A(W) = \{u|\ (u,A) \in W\}$$

Figure 4.4 The synchronous random set representation $\Sigma_A(W)$ of a generalized fuzzy set W of $\mathfrak{U}$ is pictured. Its instantiations are the level subsets of W selected by the uniform random number A on $[0, 1]$.

sets of m); (2) $\sum_U m(U) = 1$; and (3) $m(\emptyset) = 0$. The quantity $m(U)$ is interpreted as the degree of belief in the hypothesis U that does not accrue to any more constraining hypothesis $V \subseteq U$. The choice $U = \mathfrak{U}$ indicates complete uncertainty and is called the "null hypothesis."

Expressed in different notation, a b.m.a. consists of a finite number $U_1, ..., U_d$ of focal subsets of $\mathfrak{U}$, that are assigned respective nonzero masses $m_1, ..., m_d$ with $m_1 + ... + m_d = 1$. Thus $m(U) = 0$ unless $U = U_i$ for some i, in which case $m(U_i) = m_i$.

From a b.m.a. m one can construct the belief, plausibility, and commonality functions

$$\mathrm{Bel}_m(U) \triangleq \sum_{V \subseteq U} m(V) \tag{4.83}$$

$$\mathrm{Pl}_m(U) \triangleq 1 - \mathrm{Bel}_m(U^c) = \sum_{V \cap U \neq \emptyset} m(V) \tag{4.84}$$

$$Q_m(U) \triangleq \sum_{V \supseteq U} m(V). \tag{4.85}$$

A b.m.a. m can be recovered from $\mathrm{Bel}_m(U)$ via the Möbius transform

$$m(U) = \sum_{V \subseteq U} (-1)^{|U-V|} \cdot \mathrm{Bel}_m(U). \tag{4.86}$$

This summation is defined, however, only if $\mathfrak{U}$ is finite.[3]

In what follows I discuss Dempster's combination (Section 4.5.1), Zadeh's paradox and its misinterpretation (Section 4.5.2), converting b.m.a.s into probability distributions (Section 4.5.3) and random set representation of b.m.a.s (Section 4.5.4).

4.5.1 Dempster's Combination

Assuming that they are "independent" in some sense, b.m.a.s are fused using *Dempster's combination*

$$(m_1 * m_2)(U) \triangleq \alpha^{-1} \sum_{U_1 \cap U_2 = U} m_1(U_1) \cdot m_2(U_2) \tag{4.87}$$

3 The set derivative, which will be introduced in Chapter 11, is a continuous-space analog of the Möbius transform. See (11.240).

if $U \neq \emptyset$ and $(m_1 * m_2)(\emptyset) \triangleq 0$ otherwise. The combination $m_1 * m_2$ is defined only if the *agreement* (absence of *conflict*) is nonzero:

$$0 \neq \alpha \overset{\text{abbr}}{=} \alpha_{\text{DS}}(m_1, m_2) \triangleq \sum_{U_1 \cap U_2 \neq \emptyset} m_1(U_1) \cdot m_2(U_2). \tag{4.88}$$

If the only focal subsets of m_2 are singletons, then the same is true of $m_1 * m_2$ and we get a relationship which resembles Bayes' rule:

$$(m_1 * m_2)(\{u\}) \propto L_{m_1}(u) \cdot m_2(\{u\}) \tag{4.89}$$

where $L_{m_1}(u) \triangleq \sum_{U \ni u} m_1(u)$. If in addition the only focal subsets of m_1 are singletons then

$$(m_1 * m_2)(u) \propto m_1(u) \cdot m_2(u). \tag{4.90}$$

If m_1 and m_2 are interpreted as posterior probability distributions, this is a special case of Bayes' rule known as *Bayes parallel combination* (see [60, p. 100] or [128, p. 30]):

$$(p_1 *_q p_2)(u) \propto p_1(u) \cdot p_2(u) \cdot q(u)^{-1} \tag{4.91}$$

of two posterior distributions $p_1(u)$ and $p_2(u)$ that are independent in that they are conditioned on independent information, and with the common prior distribution $q(u)$ assumed uniform.

To see why (4.91) is a special case of Bayes' rule, note that $p_1(u) \propto L_{z_1}(u) \cdot q(u)$ and $p_2(u) \propto L_{z_2}(u) \cdot q(u)$ for some observations z_1 and z_2. The posterior conditioned on both z_1 and z_2 is

$$p_{12}(u) \propto L_{z_1,z_2}(u) \cdot q(u) = L_{z_1}(u) \cdot L_{z_2}(u) \cdot q(u) \tag{4.92}$$

if z_1 and z_2 are assumed conditionally independent. However,

$$L_{z_1}(u) \cdot L_{z_2}(u) \cdot q(u) = L_{z_1}(u)q(u) \cdot L_{z_2}(u)q(u) \cdot q(u)^{-1} \tag{4.93}$$

$$\propto p_1(u) \cdot p_2(u) \cdot q(u)^{-1}. \tag{4.94}$$

Thus $p_{12}(u) \propto p_1(u) \cdot p_2(u) \cdot q(u)^{-1}$, which establishes (4.91).

4.5.1.1 Unnormalized Dempster's Combination

Some variants of Dempster-Shafer theory permit $m(\emptyset) \neq 0$ and use *unnormalized combination* [212]

$$(m_1 \cap m_2)(U) \triangleq \sum_{U_1 \cap U_2 = U} m_1(U_1) \cdot m_2(U_2). \tag{4.95}$$

In this case, I will refer to m as a *generalized b.m.a.* or *g.b.m.a.*, for short.

4.5.1.2 Modified Dempster's Combination

Fixsen and Mahler [60] proposed a probabilistic "modified Dempster's combination" to address the failure of Dempster's rule to account for nonuniform priors.[4].

$$(m_1 *_q m_2)(U) \triangleq \alpha^{-1} \cdot \sum_{U_1 \cap U_2 = U} m_1(U_1) \cdot m_2(U_2) \cdot \alpha_q(U_1, U_2) \qquad (4.96)$$

provided that the "modified agreement"

$$\alpha \overset{\text{abbr.}}{=} \alpha_q(m_1, m_2) \triangleq \sum_{U_1, U_2} m_1(U_1) \cdot m_2(U_2) \cdot \frac{q(U_1 \cap U_2)}{q(U_1) \cdot q(U_2)} \qquad (4.97)$$

is nonzero, where $q(U) \triangleq \sum_{u \in U} q(u)$. If m_1 and m_2 are b.m.a.s whose only focal sets are singletons, then

$$(m_1 *_q m_2)(u) \propto m_1(u) \cdot m_2(u) \cdot q(u)^{-1}. \qquad (4.98)$$

Thus modified combination reduces to (4.91) if b.m.a.s are posterior probability distributions on some state space.

4.5.2 "Zadeh's Paradox" and Its Misinterpretation

As noted in Section 3.1, Dempster's combination has been criticized for its supposed deficiencies in regard to resolution of conflicting evidence. Much of this criticism has arisen because of what has come to be called "Zadeh's paradox." The purpose of this section is to (1) describe this paradox; (2) show why it is actually an unknowing criticism of Bayes' rule; and (3) show why, once one realizes this, the apparent paradox can be resolved by more careful modeling.

Zadeh's paradox is as follows. Two doctors, Dr. A and Dr. B, independently examine a patient to determine if his symptoms indicate one of the following conditions: μ = meningitis, τ = brain tumor, or χ = concussion. They deliver their diagnoses as respective b.m.a.s m_A and m_B with $m_A(S) = 0$ and $m_B(S) = 0$ for all S such that $|S| \neq 1$ and, otherwise,

4 This combination operator had been proposed earlier by J. Yen, using a different underlying formalism [252]

$$m_A(\mu) = 0.99, \quad m_A(\tau) = 0.01, \quad m_A(\chi) = 0.0 \tag{4.99}$$

$$m_B(\mu) = 0.0, \quad m_B(\tau) = 0.01, \quad m_B(\chi) = 0.99. \tag{4.100}$$

That is, Dr. A believes that the patient almost certainly has meningitis, possibly a brain tumor, but definitely not a concussion. Dr. B believes essentially the opposite: The patient almost certainly has a concussion, possibly a brain tumor, but definitely not meningitis.

When these opinions are combined using Dempster's combination $m_{A,B} \overset{\text{abbr}}{=} m_A * m_B$ we get

$$m_{A,B}(\mu) = \frac{m_A(\mu) \cdot m_B(\mu)}{m_A(\mu)m_B(\mu) + m_A(\tau)m_B(\tau) + m_A(\chi)m_B(\chi)} \tag{4.101}$$

$$m_{A,B}(\tau) = \frac{m_A(\tau) \cdot m_B(\tau)}{m_A(\mu)m_B(\mu) + m_A(\tau)m_B(\tau) + m_A(\chi)m_B(\chi)} \tag{4.102}$$

$$m_{A,B}(\chi) = \frac{m_A(\chi) \cdot m_B(\chi)}{m_A(\mu)m_B(\mu) + m_A(\tau)m_B(\tau) + m_A(\chi)m_B(\chi)} \tag{4.103}$$

and thus

$$m_{A,B}(\mu) = 0.0, \qquad m_{A,B}(\tau) = 1.0, \qquad m_{A,B}(\chi) = 0.0. \tag{4.104}$$

In other words, the patient is unequivocally diagnosed as having a brain tumor, even though both doctors had concluded that this possibility was unlikely. Zadeh concludes, "This example and other easily constructed examples call into question the validity of Dempster's rule of combination when it involves a normalization of belief and plausibility" [253, p. 82].

This counterexample has been widely accepted as proof that Dempster's combination is deficient because it does not correctly resolve severe conflict. Numerous alternative combination operators have been generated in response. For example, the unnormalized rule of (4.95) has been described as an improvement since it "stores conflict in the empty set" as follows:

$$m_{A,B}(\mu) = 0.0, \quad m_{A,B}(\tau) = 0.0001, \quad m_{A,B}(\chi) = 0.0 \tag{4.105}$$

$$m_{A,B}(\emptyset) = 0.9999. \tag{4.106}$$

Here I have abbreviated $m_{A,B} \overset{\text{abbr.}}{=} m_A \cap m_B$. The hypothesis τ is slightly supported, but the conflict $m_{A,B}(\emptyset)$ is so large that it cannot be accepted as valid.

There is a hidden complication, however:

- *Zadeh's paradox is actually an unknowing criticism of Bayes' rule.*

Let us see why this is the case. The doctors observe symptoms z_A and z_B drawn from some measurement space but do not pass these on to us. Instead they pass on posterior probability distributions

$$m_A(x) = p(x|z_A) \propto L_{z_A}(x) \cdot p_A(x) \tag{4.107}$$
$$m_B(x) = p(x|z_B) \propto L_{z_B}(x) \cdot p_B(x) \tag{4.108}$$

on the state space $\mathfrak{X}_0 = \{\mu, \tau, \chi\}$. Here $L_{z_A}(x)$ and $L_{z_B}(x)$ are subjectively constructed likelihood functions and $p_A(x)$ and $p_B(x)$ are the doctors' respective priors. It is reasonable to assume that $p_A(x)$ and $p_A(x)$ are both uniform distributions, since otherwise the doctors would be prejudging the symptoms. Given this, (4.91) tells us that

$$m_{A,B}(x) = p(x|z_A, z_B) \propto L_{z_A,z_B}(x) = L_{z_A}(x) \cdot L_{z_B}(x) \tag{4.109}$$
$$= m_A(x) \cdot m_B(x) \tag{4.110}$$

is the Bayes posterior distribution conditioned on all observed symptoms, given that evidence is conditionally independent. *Thus Bayes' rule yields the same result as Dempster's combination, since in this case the two are identical.*

Are we to therefore conclude that the validity of Bayes' rule has been called into question and that it must be replaced by something better? The answer is no.

One of the basic lessons of practical probability theory is the following:

- *Assigning a zero probability to any state is very unwise.*

If the prior distribution $p_0(x)$ on $\mathfrak{X}_0$ vanishes on a given state x_0, then that state is always forbidden regardless of what evidence is collected. This is because

$$p(x|z_1, ..., z_m) \propto L_{z_1}(x) \cdots L_{z_m}(x) \cdot p_0(x) \tag{4.111}$$

vanishes for $x = x_0$. The underlying reason is that one is specifying at the outset that x_0 is not an allowable state of the system.

Likewise, if any subsequent likelihood function $L_{z_1}(x)$,...,$L_{z_m}(x)$ vanishes on x_0 then x_0 is similarly forbidden thereafter. One can never recover from the evidence $L_z(x_0) = 0$, even if all subsequent information strongly favors x_0.

The two doctors are consequently doing something much stronger than simply reporting diagnoses. Knowingly or otherwise, they are insisting that μ and χ are, respectively, *not actually admissible states of the system*. Meningitis cannot occur in Dr. A's conception of the state space, just as concussion cannot exist in Dr. B's.

Consequently, one should never allow zero likelihoods to occur. The supposed paradox is resolved if one applies the obvious corrective strategy. That is, assign arbitrarily small but nonzero values to $m_A(\chi)$ and $m_B(\mu)$. For example, choose

$$m_A(\mu) = 0.99, \quad m_A(\tau) = 0.009, \quad m_A(\chi) = 0.001 \quad (4.112)$$

$$m_B(\mu) = 0.001, \quad m_B(\tau) = 0.009, \quad m_B(\chi) = 0.99. \quad (4.113)$$

Then the composed b.m.a. is, according to Bayes' rule (4.91),

$$m_{A,B}(\mu) = 0.48, \quad m_{A,B}(\tau) = 0.04, \quad m_{A,B}(\chi) = 0.48. \quad (4.114)$$

Bayes' rule therefore resolves the conflict between the evidence by making the composed distribution uniform on τ and χ with equivocating values 0.48 and 0.48. This indicates a high state of uncertainty in the combined evidence in regard to τ and χ. Bayes' rule accounts for the lack of support for τ by assigning a much smaller probability to it in the composed distribution.

This returns us to my discussion of Section 3.2:

- *The fact that two information sources are greatly in conflict indicates that there is something wrong or incomplete about our understanding (and therefore modeling) of the problem at hand.*

4.5.3 Converting b.m.a.s to Probability Distributions

In this section we assume that $\mathfrak{U}$ is finite. Several methods for transforming a b.m.a. m into a probability distribution $p_m(u)$ have been proposed in the literature. Given $p_m(u)$, one can estimate what element $\hat{u}$ of $\mathfrak{U}$ is most supported by the evidence m using a standard state estimator such as the MAP estimator: $\hat{u} = \arg\max_u p_m(u)$. Because of their significance for my expert systems unification results in later chapters,[5] I consider only two transformation methods: *Voorbraak probability* and *pignistic probability*.

5 See (5.72), (5.158), and (8.44).

4.5.3.1 Voorbraak Probability

Voorbraak's "Bayesian approximation" of a b.m.a. m is [244]:

$$\nu_m(u) \triangleq \frac{\sum_{U \ni u} m(U)}{\sum_V m(V) \cdot |V|}. \tag{4.115}$$

It is left to the reader to show, as Exercise 12, that Voorbraak probability is consistent with Dempster's combination in the following sense:

$$\nu_{m_1 * m_2} = \nu_{m_1} * \nu_{m_2}. \tag{4.116}$$

Here $\nu_{m_1} * \nu_{m_2}$ denotes the special case of Bayes' rule known as Bayes parallel combination, (4.91).

The collapse $m \mapsto p_m$ of a b.m.a. into any probability distribution is a very lossy form of data compression. Equation (4.116) shows that one can:

- *Collapse all b.m.a.s into their Voorbraak distributions and then combine them using Bayes' rule without sacrificing any of the information gain that can be attributed to fusion using Dempster's combination.*

This is because the result will be the same as if one first fused the b.m.a.s using Dempster's combination and then collapsed the fused b.m.a. into its Voorbraak distribution.

Example 11 *Suppose x is a fixed element of $\mathfrak{U}$. Define $m(U) = 0$ for all $U \subseteq \mathfrak{U}$, with the exception of $m(\{x\}) = 1 - \varepsilon$ and $m(\mathfrak{U}) = \varepsilon$. Then*

$$\nu_m(x) = \frac{m(\{x\}) + m(\mathfrak{U})}{m(\{x\}) \cdot 1 + m(\mathfrak{U}) \cdot N} = \frac{1}{1 + \varepsilon(N-1)} \tag{4.117}$$

and, if $u \neq x$,

$$\nu_m(u) = \frac{m(\mathfrak{U})}{m(\{x\}) \cdot 1 + m(\mathfrak{U}) \cdot N} = \frac{\varepsilon}{1 + \varepsilon(N-1)}. \tag{4.118}$$

4.5.3.2 Pignistic Probability

Let $q(u)$ be a prior probability distribution on $\mathfrak{U}$, and let $q(U) \triangleq \sum_{u \in U} q(u)$. The "pignistic probability" of Smets, Kennes, and Wilson [213, 248] is, on the "betting frame" of all singletons,

$$\pi_m(u) \triangleq q(u) \cdot \sum_{U \ni u} \frac{m(U)}{q(U)}. \tag{4.119}$$

Fixsen and Mahler [60] showed that $\pi_m(u)$ can be regarded as a posterior probability distribution $\pi_m(u) \propto L_m(u) \cdot q(u)$ if the likelihood $L_m(u)$ is defined as

$$L_m(u) \triangleq \left(\sum_{U} \frac{m(U)}{q(U)} \right)^{-1} \left(\sum_{U \ni u} \frac{m(U)}{q(U)} \right). \tag{4.120}$$

It is left to the reader, as Exercise 13, to show that pignistic probability is consistent with the modified Dempster's combination of (4.96):

$$\pi_{m_1 *_q m_2} = \pi_{m_1} *_q \pi_{m_2}. \tag{4.121}$$

Once again, the parallel combination $\pi_{m_1} *_q \pi_{m_2}$ of probability distributions is a special case of Bayes' rule, (4.91). That is, one can:

- *Convert all b.m.a.s into their pignistic distributions and then combine them using Bayes' rule, without sacrificing any information gain attributable to fusion using modified Dempster's combination as defined in (4.96).*

Example 12 *Let $q(U) = |U|/N$ be the uniform distribution on $\mathfrak{U}$. As in Example 11, let x be a fixed element of $\mathfrak{U}$, and let $m(U) = 0$ for all $U \subseteq \mathfrak{U}$ except for $m(\{x\}) = 1 - \varepsilon$ and $m(\mathfrak{U}) = \varepsilon$. Then*

$$\pi_m(u) = \sum_{U \ni u} \frac{m(U)}{|U|} = \frac{m(\mathfrak{U})}{N} = \frac{\varepsilon}{N} \tag{4.122}$$

if $u \neq x$ and, otherwise,

$$\pi_m(x) = \frac{m(\{x\})}{1} + \frac{m(\mathfrak{U})}{N} = 1 - \varepsilon + \frac{\varepsilon}{N} = 1 - \varepsilon \cdot \frac{N-1}{N}. \tag{4.123}$$

4.5.4 Random Set Representation of Uncertain Events

Dempster-Shafer theory can be reformulated in terms of random sets [71, 84, 114, 179]. Define the random subset Σ_m of $\mathfrak{U}$ by

$$\Pr(\Sigma_m = U) = m(U). \tag{4.124}$$

Then Σ_m is a random set representation of m.

Plausibility, belief, and commonality as defined in (4.83)-(4.85) can then be redefined probabilistically:

$$
\begin{aligned}
Pl_m(U) &= \Pr(\Sigma_m \cap U \neq \emptyset) && (4.125)\\
Bel_m(U) &= \Pr(\Sigma_m \subseteq U) && (4.126)\\
Q_m(U) &= \Pr(\Sigma_m \supseteq U). && (4.127)
\end{aligned}
$$

Dempster's rule has the probabilistic interpretation

$$(m_1 * m_2)(U) = \Pr(\Sigma_{m_1} \cap \Sigma_{m_2} = U | \Sigma_{m_1} \cap \Sigma_{m_2} \neq \emptyset) \tag{4.128}$$

if $\Sigma_{m_1}, \Sigma_{m_2}$ are assumed statistically independent and if $\Pr(\Sigma_{m_1} \cap \Sigma_{m_2} \neq \emptyset) \neq 0$ [205]. The proof of (4.128) is left to the reader as Exercise 14.

4.6 FUZZY DEMPSTER-SHAFER THEORY

The reader may first wish to review the example presented in Section 3.4.4. Equation (5.60) can be generalized to include fuzzy Dempster-Shafer evidence [129, 252]. A fuzzy b.m.a. m generalizes a b.m.a. in that it assigns masses to fuzzy set membership functions that are not necessarily crisp. In other words, it is a nonnegative function $m(f)$ defined on fuzzy membership functions f of $\mathfrak{U}$ such that: (1) $m(f) = 0$ for all but a finite number of f; (2) $\sum_f m(f) = 1$; and (3) $m(\emptyset) = 0$.

Expressed in different notation, a fuzzy b.m.a. consists of a finite number $f_1, ..., f_d$ of focal fuzzy membership functions on $\mathfrak{U}$, that are assigned respective nonzero masses $m_1, ..., m_d$ with $m_1 + ... + m_d = 1$. Thus $m(f) = 0$ unless $f = f_i$ for some i, in which case $m(f_i) = m_i$.

For the purposes of this book, Dempster's combination as defined in (4.87) can be generalized to fuzzy b.m.a.s as follows:

$$(m_1 * m_2)(f) \triangleq \alpha^{-1} \sum_{f_1 \cdot f_2 = f} m_1(f_1) \cdot m_2(f_2) \tag{4.129}$$

provided that $f \neq 0$ and $(m_1 * m_2)(0) \triangleq 0$ otherwise and provided that the agreement is nonzero:

$$0 \neq \alpha \overset{\text{abbr.}}{=} \alpha_{\text{FDS}}(m_1, m_2) \triangleq \sum_{f_1 \cdot f_2 \neq 0} m_1(f_1) \cdot m_2(f_2). \tag{4.130}$$

Here, $(f_1 \cdot f_2)(u) \triangleq f_1(u) \cdot f_2(u)$; and $f \neq 0$ means $f(u) \neq 0$ for at least one u. It is left to the reader as Exercise 15 to verify that (4.129) reduces to ordinary Dempster's combination, (4.87), when all focal sets are crisp: $f(u) = \mathbf{1}_S(u)$, $f_1(u) = \mathbf{1}_{S_1}(u)$, $f_2(u) = \mathbf{1}_{S_2}(u)$.

The fuzzy version of Dempster's combination can be generalized in the obvious way to the unnormalized and modified Dempster's combinations. In the case of unnormalized combination,

$$(m_1 \cap m_2)(f) \triangleq \sum_{f_1 \cdot f_2 = f} m_1(f_1) \cdot m_2(f_2). \tag{4.131}$$

Here $m_1(0)$ and $m_2(0)$ can be nonzero, and unnormalized combination is defined by setting $\alpha = 1$ in (4.129).

In the case of modified Dempster's combination, for any fuzzy membership function $f(u)$ and prior distribution $q(u)$ on $\mathfrak{U}$, define $q(f) \triangleq \sum_u f(u) \cdot q(u)$. Then (4.96) becomes

$$(m_1 *_q m_2)(f) \triangleq \alpha^{-1} \cdot \sum_{f_1 \cdot f_2 = f} m_1(f_1) \cdot m_2(f_2) \cdot \alpha_q(f_1, f_2) \tag{4.132}$$

provided that the following is nonzero:

$$\alpha \overset{\text{abbr.}}{=} \alpha_q(m_1, m_2) \triangleq \sum_{f_1, f_2} m_1(f_1) \cdot m_2(f_2) \cdot \frac{q(f_1 \cap f_2)}{q(f_1) \cdot q(f_2)}. \tag{4.133}$$

4.6.1 Random Set Representation of Fuzzy DS Evidence

I use the generalized fuzzy set formalism of Section 4.4 to construct random set representations of fuzzy b.m.a.s. The basic idea is illustrated in Figure 4.5.

Let $f_1, ..., f_e$ be the fuzzy membership functions such that $o(f_i) > 0$ and let $o_i = o(f_i)$ be their respective weights. In (4.80) I noted that a fuzzy membership function f can be represented as the generalized fuzzy set $W_f = \{(u,a) | \ a \leq f(u)\}$. We divide $I = [0,1]$ up into intervals $I_1, ..., I_e$ of respective lengths $o_1, ..., o_e$. We vertically "shrink" W_{f_i} until it fits exactly within $\mathfrak{U} \times I_i$, thus creating a generalized fuzzy subset W_i that represents f_i, but such that the W_i are mutually disjoint. We then apply (4.75) to $W_o \triangleq W_1 \cup ... \cup W_e$ to get a random set representation.

In more detail, define $o_0^+ \triangleq 0$ and $o_i^+ \triangleq o_1 + ... + o_i$ for all $i = 1, .., e$. Thus $o_i^+ - o_{i-1}^+ = o_i$ for all $i = 1, ..., e$. Define W_i as

$$W_i \triangleq \{(u,a) | \ o_{i-1}^+ < a \leq o_{i-1}^+ + o_i f_i(u)\} \tag{4.134}$$

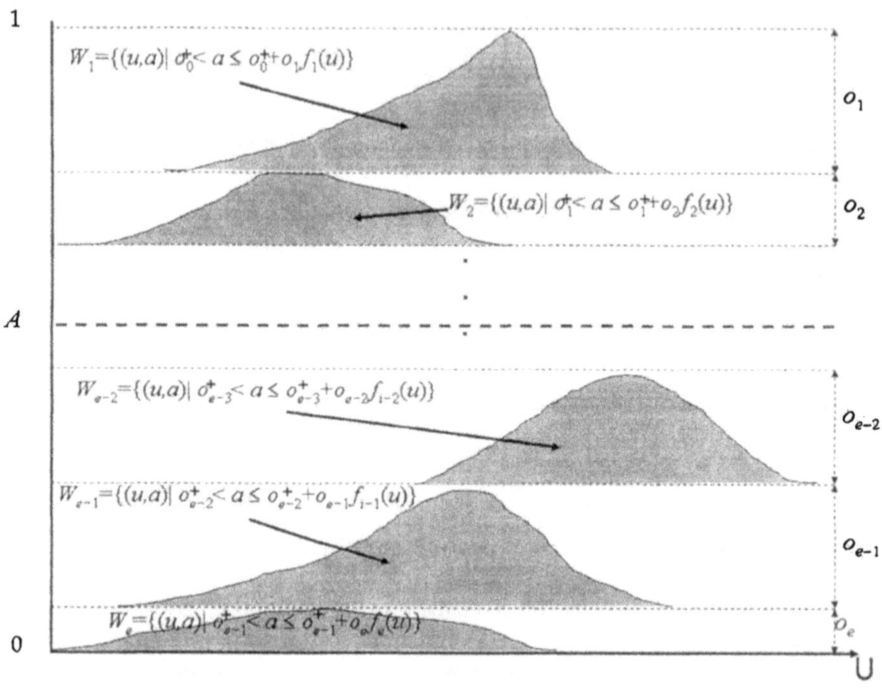

Figure 4.5 The basic idea behind the random set representation of a fuzzy basic mass assignment o. Here $f_1(u),...,f_e(u)$ are the focal fuzzy membership functions of o, and $o_1,...,o_e$ are their respective weights. The fraction of instances that the uniform random number A spends sweeping out level sets for the representation W_i of $f_i(u)$ is o_i.

and set

$$W_o \triangleq W_1 \cup \ldots \cup W_e. \tag{4.135}$$

Then $W_i \cap W_j \neq \emptyset$ for $i \neq j$. The random set representation of o is

$$\Sigma_A(W_o). \tag{4.136}$$

As can be seen in Figure 4.5, $\Sigma_A(W_o)$ generalizes the representation W_f of the fuzzy membership function f. The random number A sweeps out the level sets of the fuzzy membership function f_i for all instantiations $A = a$ such that $a \in I_i \triangleq (o_{i-1}^+, o_i^+]$. The probability that it does so is o_i.

4.7 INFERENCE RULES

In this section, I describe the random set representation of rule-based evidence. After describing the concept of a rule in Section 4.7.1, in Section 4.7.2, I introduce a methodology for combining rules called *conditional event algebra* (CEA). Random set representations of first-order, composite, and second-order rules are described in Sections 4.7.3-4.7.5.

4.7.1 What Are Rules?

Rule-based inference is based on the logical principle of *modus ponens*. This refers to the process of retrieving a suitable rule $X \Rightarrow S$ from a knowledge base when an antecedent event X has been collected, thereby allowing us to "fire" the rule and deduce the consequent event S and thus also $S \cap X$. I will consider the following kinds of rules:

- *First-order rules* have the form

$$X \Rightarrow S = \text{‘if } X \text{ then } S\text{’} \tag{4.137}$$

  where S, X are subsets of $\mathfrak{U}$;

- *Composite rules* allow for the possibility of choosing among alternative rules, as with

$$(X_1 \Rightarrow S_1) \text{ or } (X_2 \Rightarrow S_2); \tag{4.138}$$

- *Second-order rules* have the form

$$(X_1 \Rightarrow S_1) \Rightarrow (X_2 \Rightarrow S_2). \tag{4.139}$$

That is: if the rule $X_1 \Rightarrow S_1$ is found to apply then so does the rule $X_2 \Rightarrow S_2$;

- *Fuzzy rules* have the same form as crisp rules, except that the consequents and antecedents are fuzzy membership functions. Thus fuzzy first-order, composite, and second-order rules have the respective forms

$$\chi \Rightarrow \sigma \tag{4.140}$$
$$(\chi_1 \Rightarrow \sigma_1) \text{ or } (\chi_2 \Rightarrow \sigma_2) \tag{4.141}$$
$$(\chi_1 \Rightarrow \sigma_1) \Rightarrow (\chi_2 \Rightarrow \sigma_2) \tag{4.142}$$

where χ, σ, χ_1, σ_1, χ_2, and σ_2 are fuzzy membership functions on $\mathfrak{U}$.

4.7.2 Combining Rules Using Conditional Event Algebra

The probability theory of crisp events $S \subseteq \mathfrak{U}$ can be rigorously extended to a probability theory of crisp rules of all orders, based on a theory known as the Goodman-Nguyen-van Fraassen *product-space conditional event algebra (PS-CEA)* [69, 70, 73, 230].

This theory allows any prior distribution q defined on events $U \subseteq \mathfrak{U}$ to be naturally extended to a prior distribution $\hat{q}$ defined on arbitrary conditional events so that $\hat{q}(X \Rightarrow S) = q(S|X)$. That is, the probability of the rule is the conditional probability of its consequent given its antecedent. Rules ρ, ρ_1, and ρ_2 can be manipulated using Boolean logical operators $\rho_1 \wedge \rho_2$, $\rho_1 \vee \rho_2$, and ρ^c. In this algebra (as in any CEA), modus ponens becomes the algebraic equation

$$X \wedge (X \Rightarrow S) = S \wedge X = S \cap X. \tag{4.143}$$

The PS-CEA has a non-Boolean "first-order approximation" known as the *Goodman-Nguyen-Walker CEA* (GNW-CEA) (see [69] or [72, p. 93]). The author has shown how to extend this CEA to rules of any order [147]. Under GNW, rules $X_1 \Rightarrow S_1$ and $X_2 \Rightarrow S_2$ can be combined using the GNW conjunction and disjunction operators $\wedge$ and $\vee$ to get new first-order rules. Complementation is defined as

$$(X \Rightarrow S)^c \triangleq (X \Rightarrow S^c). \tag{4.144}$$

The GNW conjunction operator is defined as

$$(X_1 \Rightarrow S_1) \wedge (X_2 \Rightarrow S_2) \triangleq (Y \Rightarrow T) \tag{4.145}$$

where

$$Y \triangleq (S_1^c \cap X_1) \cup (S_2^c \cap X_2) \cup (X_1 \cap X_2) \tag{4.146}$$
$$T \triangleq S_1 \cap S_2. \tag{4.147}$$

The GNW disjunction operator is defined as

$$(X_1 \Rightarrow S_1) \vee (X_2 \Rightarrow S_2) \triangleq (Y \Rightarrow T) \tag{4.148}$$

where

$$Y \triangleq (S_1 \cap X_1) \cup (S_2 \cap X_2) \cup (X_1 \cap X_2) \tag{4.149}$$
$$T \triangleq S_1 \cup S_2. \tag{4.150}$$

It is left to the reader as Exercise 16 to show that $T \cap Y = (S_1 \cap X_1) \cup (S_2 \cap X_2)$.

A few basic properties are worth pointing out [69, p. 1687]:

- $X \Rightarrow S \cap X$ and $X \Rightarrow S$ are two different ways of expressing the same rule:
$$(X \Rightarrow S \cap X) = (X \Rightarrow S); \tag{4.151}$$
- The rule $\mathfrak{U} \Rightarrow S$ is the same as the ordinary event S:
$$(\mathfrak{U} \Rightarrow S) = S; \tag{4.152}$$
- *Generalized modus ponens*: The partial firing of a rule $X \Rightarrow S$ by a partial antecedent Y is
$$Y \wedge (X \Rightarrow S) = (X' \Rightarrow S') \tag{4.153}$$
where $X' = (X \cap Y) \cup Y^c$ and $S' = Y \cap X \cap S$.

Example 13 *Let* $X_1 = X_2 = X$. *Then*

$$(X \Rightarrow S_1) \vee (X \Rightarrow S_2) = (X \Rightarrow S_1 \cup S_2). \tag{4.154}$$

For from (4.149)-(4.150),

$$Y = (S_1 \cap X) \cup (S_2 \cap X_2) \cup X = X \tag{4.155}$$
$$T = (S_1 \cap X) \cup (S_2 \cap X) = (S_1 \cup S_2) \cap X. \tag{4.156}$$

The result follows from (4.151).

Example 14 *Referring to Example 8 of Section 4.2, let* $X_1 = KING$, $X_2 = HEART$, *and* $S_1 = S_2 = \{KH\}$. *Then the conjunction of the rules* $KING \Rightarrow KH$ *and* $HEART \Rightarrow KH$ *is*

$$(X_1 \Rightarrow S_1) \wedge (X_2 \Rightarrow S_2) = (Y \Rightarrow T) \tag{4.157}$$

where $T = S_1 \cap S_2 = \{KH\}$ *and* $Y = X_1 \cup X_2 \cup (S_1 \cap S_2)$. *So by (4.151) the conjoined rule is* $(KING \text{ or } HEART) \Rightarrow KH$. *The disjunction is*

$$(X_1 \Rightarrow S_1) \vee (X_2 \Rightarrow S_2) = (Y \Rightarrow T) \tag{4.158}$$

where $T = S_1 \cup S_2 = \{KH\}$ *and* $Y = S_1 \cup S_2 \cup (X_1 \cap X_2) = \{KH\}$. *So the disjoined rule is the tautology* $KH \Rightarrow KH$.

4.7.3 Random Set Representation of First-Order Rules

Let Φ be a *uniformly distributed* random subset of a *finite* universe $\mathfrak{U}$. That is, its probability distribution is $\Pr(\Phi = U) = 2^{-N}$ for all $U \subseteq \mathfrak{U}$, where N is the number of elements in $\mathfrak{U}$. It can be shown that a random set representation of the rule $X \Rightarrow S$ is [146, p. 54]:

$$\Sigma_\Phi(X \Rightarrow S) \triangleq (S \cap X) \cup (X^c \cap \Phi). \tag{4.159}$$

This representation is algebraically consistent with the GNW conditional event algebra [146, p. 54]. That is, if $\rho_1 = (X_1 \Rightarrow S_1)$ and $\rho_2 = (X_2 \Rightarrow S_2)$, then it can be shown [146, p. 54, (12-13)] that

$$\Sigma_\Phi(\rho_1 \wedge \rho_2) = \Sigma_\Phi(\rho_1) \cap \Sigma_\Phi(\rho_2) \tag{4.160}$$

$$\Sigma_\Phi(\rho_1 \vee \rho_2) = \Sigma_\Phi(\rho_1) \cup \Sigma_\Phi(\rho_2). \tag{4.161}$$

It is not true, however, that $\Sigma_\Phi(\rho)^c = \Sigma_\Phi(\rho^c)$.

Equation (4.159) is easily extended to fuzzy rules $\xi \Rightarrow \sigma$. Let A, A' be two uniformly distributed random numbers on $[0, 1]$. Let Ω be the underlying probability space. The fuzzy event ξ consists of instantiations $\Sigma_{A(\omega)}(\xi)$ over all $\omega \in \Omega$ and likewise for $\Sigma_{A'(\omega)}(\sigma)$, where the notation was defined in (4.21). The fuzzy rule $\xi \Rightarrow \sigma$ is to be interpreted as follows:

- *If the antecedent interpretation* $\Sigma_{A(\omega)}(\xi)$ *is observed then the consequent interpretation* $\Sigma_{A'(\omega)}(\sigma)$ *must be inferred.*

Given this, define $\Sigma_{\Phi,A,A'}(\xi \Rightarrow \sigma)$ by simple substitution $X \rightarrow \Sigma_{A'}(\xi)$ and $S \rightarrow \Sigma_A(\sigma)$ in (4.159):

$$\Sigma_{\Phi,A,A'}(\xi \Rightarrow \sigma) \triangleq (\Sigma_A(\sigma) \cap \Sigma_{A'}(\xi)) \cup (\Sigma_{A'}(\xi)^c \cap \Phi). \quad (4.162)$$

Caution: Equations (4.160) and (4.161) do not generalize to fuzzy rules.

4.7.4 Random Set Representation of Composite Rules

Let $\rho_1 = (X_1 \Rightarrow S_1)$ and $\rho_2 = (X_2 \Rightarrow S_2)$ be crisp rules. The random set representation of the composite rule 'ρ_1 or ρ_2' is just $\Sigma_\Phi(\rho_1 \vee \rho_2)$, where $\rho_1 \vee \rho_2$ denotes the disjunction of rules, (4.148)-(4.150):

$$\Sigma_\Phi(\rho_1 \vee \rho_2) = (T \cap Y) \cup (Y^c \cap \Phi) \quad (4.163)$$

where

$$T = S_1 \cup S_2 \quad (4.164)$$
$$Y = (S_1 \cap X_1) \cup (S_2 \cap X_2) \cup (X_1 \cap X_2). \quad (4.165)$$

Let $\rho_1 = (\xi_1 \Rightarrow \sigma_1)$ and $\rho_2 = (\xi_2 \Rightarrow \sigma_2)$ be fuzzy rules. A random set representation $\Sigma_{\Phi,A_1,A_1',A_2,A_2'}(\rho_1 \vee \rho_2)$ of the composite fuzzy rule $\rho_1 \vee \rho_2$ is constructed by substituting $S_1 \rightarrow \Sigma_{A_1}(\sigma_1)$, $X_1 \rightarrow \Sigma_{A_1'}(\xi_1)$, $S_2 \rightarrow \Sigma_{A_2}(\sigma_2)$, $X_2 \rightarrow \Sigma_{A_2'}(\xi_2)$ in (4.163):

$$\Sigma_{\Phi,A_1,A_1',A_2,A_2'}(\rho_1 \vee \rho_2) \triangleq (T \cap Y) \cup (Y^c \cap \Phi) \quad (4.166)$$

where

$$T = \Sigma_{A_1}(\sigma_1) \cup \Sigma_{A_2}(\sigma_2) \quad (4.167)$$
$$Y = (\Sigma_{A_1}(\sigma_1) \cap \Sigma_{A_1'}(\xi_1)) \cup (\Sigma_{A_2}(\sigma_2) \cap \Sigma_{A_2'}(\xi_2)) \cup(\Sigma_{A_1'}(\xi_1) \cap \Sigma_{A_2'}(\xi_2)). \quad (4.168)$$

Thus, by Exercise 16,

$$T \cap Y = (\Sigma_{A_1}(\sigma_1) \cap \Sigma_{A_2}(\xi)) \cup (\Sigma_{A_2}(\sigma_2) \cap \Sigma_{A_4}(\xi')). \quad (4.169)$$

4.7.5 Random Set Representation of Second-Order Rules

It can be shown [147, p. 591] that a random set representation of the second-order rule $(X \Rightarrow S) \Rightarrow (Y \Rightarrow T)$ is

$$\Sigma_{\Phi_1,\Phi_2}((X \Rightarrow S) \Rightarrow (Y \Rightarrow T)) \tag{4.170}$$
$$= (S \cap X \cap T \cap Y) \cup (S \cap X \cap Y^c \cap \Phi_1) \tag{4.171}$$
$$\cup((X - S) \cap \Phi_1^c) \cup (X^c \cap T \cap Y \cap (\Phi_1 \cup \Phi_2)) \tag{4.172}$$
$$\cup(X^c \cap Y^c \cap (\Phi_1 \cup \Phi_2)) \cup (X^c \cap (Y - T) \cap \Phi_1^c \cap \Phi_2). \tag{4.173}$$

Here Φ_1 and Φ_2 are uniformly distributed, independent random subsets. When $X = \mathfrak{U}$ this reduces to

$$\Sigma_\Phi(S \Rightarrow (Y \Rightarrow T)) = (S \cap T \cap Y) \cup (S \cap Y^c \cap \Phi) \cup (S^c \cap \Phi^c). \tag{4.174}$$

Equation (4.162) is easily extended to fuzzy second-order rules. Let A_1, A_2, A_3, A_4 be uniformly distributed random numbers on $[0, 1]$. Then a random set representation

$$\Sigma_{\Phi_1,\Phi_2,A_1,A_2,A_3,A_4}((\xi \Rightarrow \sigma) \Rightarrow (\theta \Rightarrow \tau)) \tag{4.175}$$

of a fuzzy second-order rule is obtained by substituting $X \rightarrow \Sigma_{A_1}(\xi)$, $S \rightarrow \Sigma_{A_2}(\sigma)$, $Y \rightarrow \Sigma_{A_3}(\theta)$, $T \rightarrow \Sigma_{A_4}(\tau)$ in (4.170).

4.8 IS BAYES SUBSUMED BY OTHER THEORIES?

In Chapter 5, I will show that certain aspects of the DS theory, fuzzy logic, and rule-based inference can, in a certain carefully specified sense, be seamlessly integrated within the Bayesian framework. Such an assertion may appear to be inherently contradictory to many. This is because conventional wisdom holds that Dempster-Shafer theory subsumes probability theory in general and Bayesian probability in particular. The basis for the conventional wisdom consists of the following assertions:

- *Assertion*: A probability distribution is just a special kind of Dempster-Shafer basic mass assignment in which the focal sets are all singletons.
- *Assertion*: Stated in different terms, expert systems theory is the study of general nonadditive measures (e.g., belief, plausibility, and possibility measures). Probability theory, on the other hand, is restricted to the study of additive measures—specifically, probability measures.

- *Assertion*: Stated in still different terms, random set theory is more general than Bayesian theory, since the former addresses probability distributions on the *subsets* of a universe, whereas the latter addresses only probability distributions on the *elements (singleton subsets)* of the universe.

  These assertions are based on fundamental misconceptions about the Bayesian approach. As should be clear from my discussion in Section 3.5.1, the general Bayesian approach is:

- *A theory of random variables on general measurement spaces and general state spaces.*

Random sets are just a particular kind of random variable and, as such, are subsumed by Bayesian theory.

For example, suppose that observations are *imprecise measurements*. That is, they are closed subsets of some space $\mathfrak{Z}_0$ of conventional observations. Then the actual measurement space $\mathfrak{Z}$ is the hyperspace of all closed subsets of $\mathfrak{Z}_0$, and a random observation is a random closed set of $\mathfrak{Z}_0$.

More generally, the observation space could be a space whose elements are fuzzy sets, or basic mass assignments, or rules. If one models fuzzy sets or basic mass assignments or rules as random sets, then the observation space is one whose elements are themselves random sets. Such an observation space can then be employed as part of a Bayesian analysis. This is precisely what I shall do in this book.

Similar comments apply to state spaces. In principle, one could have *imprecise states* (i.e., closed subsets of a conventional state space $\mathfrak{X}_0$). In this case the actual state space $\mathfrak{X}$ is the hyperspace of all closed subsets of $\mathfrak{X}_0$, and the Bayesian random state is a random closed subset of $\mathfrak{X}_0$.[6] One could similarly model more ambiguous kinds of states as b.m.a.s on $\mathfrak{X}_0$. Thus in this sense also, the Bayesian approach subsumes the other expert system approaches.

Moreover, the first two assertions above are obviously false for continuous universes. This can be seen from the example presented in Section 3.4.5. This example explains the semantics of a b.m.a., the focal sets of which are singleton sets of a continuous measurement space, and contrasts it with the semantics of a random variable on a continuous measurement space. Such a b.m.a. is not actually a probability distribution, in the sense that it is the distribution of a random *precise measurement*. Rather, it is an *individual instantiation of some random uncertain observation*, that has been modeled as a list of *ranked hypotheses* whose ranks sum

6 In fact, in Part II, I will model the states of a multitarget system as finite subsets $X = \{\mathbf{x}_1, ..., \mathbf{x}_n\}$ of ordinary states, so that the random state in a Bayesian analysis is a *random finite set*.

to one. The random observation in this case is, therefore, a random variable that takes such individual observations as instantiations.

The following is sometimes cited as another reason why probability theory is subsumed by Dempster-Shafer theory:

- *Assertion*: Equation (4.89),

$$(m_1 * m_2)(\{u\}) \propto L_{m_1}(u) \cdot m_2(\{u\}), \tag{4.176}$$

  shows that Bayes' rule can be regarded as a special case of Dempster's combination, when one of the b.m.a.s is a prior probability distribution, and the other is regarded as a likelihood function.

As we shall see in Section 5.3.4.2, there is some justification for this claim. Equation (4.89) indeed turns out to be a version of Bayes' rule—see (5.71).

However, this fact is valid only for universes that are *finite state spaces*. Equation (5.71) implicitly assumes that the universe $\mathfrak{U}$ is some state space $\mathfrak{U} = \mathfrak{X}_0$, since one of the b.m.a.s is a prior distribution. However, if $\mathfrak{X}_0$ is continuously infinite—say $\mathfrak{X}_0 = \mathbb{R}$—this b.m.a. cannot be a prior *probability distribution* on $\mathfrak{X}_0$. A b.m.a. has only a finite number of focal sets, and so the alleged "prior" would be a probability distribution on some finite subset of $\mathbb{R}$.

A true probability distribution on $\mathbb{R}$ must be a *density function defined on all elements* of $\mathbb{R}$. We already encountered the distinction between the two in the example in Section 3.4.5.

In summary, then, the following are true.

- It is indeed true that probability distributions in the everyday engineering sense are not mathematically rich enough to model general forms of data ambiguity. They are in this sense subsumed by more general uncertainty modeling formalisms;

- However, once these more general models have been constructed, they can be incorporated into a Bayesian analysis as summarized in Sections 3.5.1 and 3.5.2. They are, in this sense, subsumed within and by that analysis.

4.9 CHAPTER EXERCISES

Exercise 6 *Prove (4.24)-(4.25):* $\Sigma_A(f \wedge f') = \Sigma_A(f) \cap \Sigma_A(f')$ *and* $\Sigma_A(f \vee f') = \Sigma_A(f) \cup \Sigma_A(f')$.

Exercise 7 *Prove that the Hamacher fuzzy conjunction of Example 9 is (a) such that* $0 \leq a \wedge a' \leq 1$ *for all* a, a'*; (b) associative; and (c) a copula.*

Exercise 8 *Prove that the FGM conjunction of (4.57) is associative.*

Exercise 9 *Verify that (4.61) and (4.62) are true.*

Exercise 10 *Prove (4.71)-(4.72),* $W_f \cap W_g = W_{f \wedge g}$ *and* $W_f \cup W_g = W_{f \vee g}$*, but that* $W_f^c \neq W_{f^c}$*.*

Exercise 11 *Prove (4.76)-(4.79).*

Exercise 12 *Prove (4.116): Voorbraak probability is consistent with Dempster's combination.*

Exercise 13 *Prove (4.121): Pignistic probability is consistent with modified Dempster's combination.*

Exercise 14 *Prove (4.128): Dempster's combination can be expressed as a conditional probability defined in terms of random subsets.*

Exercise 15 *Prove that fuzzy Dempster combination, (4.129), reduces to ordinary Dempster's combination, (4.87), when all focal sets are crisp.*

Exercise 16 *Show that* $T \cap Y = (S_1 \cap X_1) \cup (S_2 \cap X_2)$*, where* T *and* Y *are defined in (4.149) and (4.150).*

Chapter 5

UGA Measurements

5.1 INTRODUCTION TO THE CHAPTER

In this chapter, I begin my development of formal Bayes modeling for generalized measurements, as introduced in the five examples of Section 3.4. I introduce the simplest kind of generalized measurement, the *unambiguously generated ambiguous (UGA) measurement*. In brief, a UGA measurement is characterized by two things:

- Modeling the measurement itself involves ambiguity;
- However, the relationship between measurements and target states can be described by a precise sensor-transform model of the familiar form $\mathbf{z} = \eta(\mathbf{x})$.

The following example illustrates the meaning of these statements.

Example 15 *Typical applications that require UGA measurement modeling are those that involve human-mediated feature extraction, but in which the features are associated in a clear-cut fashion with target types. Consider, for example, a feature extracted by a human operator from a synthetic aperture radar (SAR) image: the number* n *of hubs/tires. In this case it is known a priori that a target of type* v *will have a given number* $n = \eta(v)$ *of hubs (if a treaded vehicle) or* $n = \eta(v)$ *of tires (if otherwise). Suppose that the possible values of* n *are* $n = 1, ..., 8$. *The generalized observation* Θ *is a random subset of* $\mathfrak{Z}_0 = \{1, ..., 8\}$. *The generalized measurement model* $\eta(\mathbf{x}) \in \Theta$, *introduced in (5.17), models the matching of the data-model* Θ *with the known feature* $\eta(v)$ *associated with a target of type* v. *The* *generalized likelihood* $f(\Theta|v) = \Pr(v \in \Theta)$*, introduced as (5.18), is a measure of the degree of this matching.*

After establishing some notational conventions in Section 5.1.1, I summarize the primary lessons learned for the chapter in Section 5.1.2. I conclude, in Section 5.1.3, with a summary of the organization of the chapter.

5.1.1 Notation

In what follows I will presume the notation and concepts of the Bayesian "recipe" of Sections 3.5.1 and 3.5.2. In this section, I establish notational conventions that will also be assumed in Chapters 6-8.

We wish our results to be applicable to both continuous and finite spaces. We therefore assume that the target state space $\mathfrak{X} = \mathfrak{X}_0$ is a Cartesian product $\mathbb{R}^N \times C$ of a Euclidean space $\mathbb{R}^N$ and a finite set C, so that states can have continuous and discrete components. Integration is defined as

$$\int_S f(\mathbf{x})d\mathbf{x} \triangleq \sum_c \int \mathbf{1}_S(\mathbf{u}, c) \cdot f(\mathbf{u}, c)d\mathbf{u}. \tag{5.1}$$

The "base" measurement space $\mathfrak{Z}_0$ is a Cartesian product $\mathbb{R}^M \times D$ of a Euclidean space $\mathbb{R}^M$ and a finite set D, so that conventional measurements can have continuous and discrete components. Integration is defined by

$$\int_T g(\mathbf{z})d\mathbf{z} \triangleq \sum_e \int \mathbf{1}_T(\mathbf{v}, e) \cdot g(\mathbf{v}, e)d\mathbf{v}. \tag{5.2}$$

We abbreviate the prior distribution at the current time step, and its corresponding probability-mass function, as

$$f_0(\mathbf{x}) \triangleq f_{k+1|k}(\mathbf{x}|Z^k) \tag{5.3}$$

$$p_0(S) \triangleq \int_S f_0(\mathbf{x})d\mathbf{x}. \tag{5.4}$$

Let a new measurement ζ_{k+1} be collected. This measurement will not be an element of $\mathfrak{Z}_0$. Rather, it will be a random closed subset $\zeta_{k+1} = \Theta$ of $\mathfrak{Z}_0$. Thus the actual measurement space $\mathfrak{Z}$ is the class of all random closed subsets of $\mathfrak{Z}_0$. I abbreviate the generalized likelihood and the posterior distribution at the current time step, conditioned on the new measurement $\zeta_{k+1} = \Theta$, respectively, as

$$f(\Theta|\mathbf{x}) \triangleq f_{k+1}(\Theta|\mathbf{x}) \quad (5.5)$$

$$f(\mathbf{x}|\Theta) \triangleq f_{k+1|k+1}(\mathbf{x}|Z^{k+1}) = \frac{f_{k+1}(\Theta|\mathbf{x}) \cdot f_{k+1|k}(\mathbf{x}|Z^k)}{\int f_{k+1}(\Theta|\mathbf{y}) \cdot f_{k+1|k}(\mathbf{y}|Z^k)d\mathbf{y}}. \quad (5.6)$$

The fundamental problem we are to address is this (see Figure 3.4):

- *Given accumulated information through time step* $k+1$, *estimate the value of* $\mathbf{x}$ *as well as the error in this estimate.*

State-estimates can be obtained using some Bayes-optimal state estimator, as in Section 2.4.8. Estimates of error can be obtained using covariance or some other measure of dispersion, as in Section 2.4.9. So, our main task is to determine explicit formulas for the likelihood functions $f(\Theta|\mathbf{x})$.

5.1.2 Summary of Major Lessons Learned

The following are the major concepts, results, and formulas that the reader will encounter in this chapter:

- Definition of a UGA measurement (Section 5.2);
- Definition of the generalized likelihood of a UGA measurement, (5.18):

$$f(\Theta|\mathbf{x}) \triangleq \Pr(\eta(\mathbf{x}) \in \Theta); \quad (5.7)$$

- Formula for the generalized likelihood function for statistical measurements $\mathbf{z}$, see (5.24):

$$f(\Theta_{\mathbf{z}}|\mathbf{x}) \cong f_{\mathbf{W}}(\tau^{-1}_{\eta(\mathbf{x})}(\mathbf{z})) \cdot J_{\tau^{-1}_{\eta(\mathbf{x})}}(\mathbf{z}) \cdot \varepsilon; \quad (5.8)$$

- Formula for the generalized likelihood function for fuzzy measurements $g(\mathbf{z})$, see (5.29):

$$f(g|\mathbf{x}) = g(\eta(\mathbf{x})); \quad (5.9)$$

- Formula for the generalized likelihood function for Dempster-Shafer measurements $o(O)$, see (5.58):

$$f(o|\mathbf{x}) = \sum_O o(O) \cdot \mathbf{1}_O(\eta(\mathbf{x})); \quad (5.10)$$

- Formula for the generalized likelihood function for fuzzy Dempster-Shafer measurements $o(g)$, see (5.73):

$$f(o|\mathbf{x}) = \sum_{g} o(g) \cdot g(\eta(\mathbf{x})); \tag{5.11}$$

- Formula for the generalized likelihood function for first-order fuzzy rules $\xi \Rightarrow \sigma$ on measurements, see (5.80):

$$f(\xi \Rightarrow \sigma|\mathbf{x}) = (\xi \wedge \sigma)(\eta(\mathbf{x})) + \frac{1}{2}(1 - \xi(\eta(\mathbf{x}))); \tag{5.12}$$

- Formula for the generalized likelihood for a composite rule on measurements, see (5.102);

- Formula for the generalized likelihood function for second-order rules on measurements, see (5.111);

- These formulas for generalized likelihoods permit very disparate forms of information and uncertainty to be represented in a common probabilistic framework, and to be processed using a common fusion technique: the recursive Bayes filter of Chapter 2 (see Figure 3.4);

- From a Bayesian point of view, measurement fusion using normalized or unnormalized Dempster's combination is equivalent to measurement fusion using Bayes' rule (Section 5.4.1):

$$f(\mathbf{x}|o * o') = f(\mathbf{x}|o, o'), \qquad f(\mathbf{x}|o \cap o') = f(\mathbf{x}|o, o'); \tag{5.13}$$

- From a Bayesian point of view, measurement fusion using copula-type fuzzy conjunctions is equivalent to measurement fusion using Bayes' rule (Section 5.4.3):

$$f(\mathbf{x}|g \wedge_{A,A'} g') = f(\mathbf{x}|g, g'); \tag{5.14}$$

- From a Bayesian point of view, total or partial firing of fuzzy rules is equivalent to Bayes' rule (Section 5.4.4):

$$f(\mathbf{x}|\xi, \xi \Rightarrow \sigma) = f(\mathbf{x}|\xi \wedge \sigma); \tag{5.15}$$

- Generalized likelihood functions are strict likelihoods when measurement space is finite—meaning that in this case, we are strictly adhering to the Bayesian "recipe" of Section 3.5 (Section 5.4.5);

- It is possible to transform various uncertainty representations (probabilistic, fuzzy, Dempster-Shafer) into other uncertainty representations (probabilistic, fuzzy, Dempster-Shafer) without loss of *estimation-relevant* information (Section 5.4.6);
- A method for modeling imperfectly known statistical correlations in joint generalized likelihoods, see (5.173);
- A method for modeling unmodeled target types, see (5.176);
- Voorbraak probability distributions are special cases of posterior distributions conditioned on Dempster-Shafer measurements (Section 5.3.4.2);
- Predictor, corrector, and estimator equations for the *Kalman evidential filter* (KEF), a generalization of the Kalman filter that can process both conventional and fuzzy Dempster-Shafer measurements; see (5.223)-(5.247);
- From a Bayesian point of view, data update of the target state using Dempster's combination is equivalent to data update of the state using Bayes' rule, (5.225).

5.1.3 Organization of the Chapter

I introduce the concept of a UGA measurement in Section 5.2 and generalized likelihood functions for such measurements in Section 5.3. Specifically, I derive concrete formulas for the generalized likelihood functions for the following special cases of UGA measurements: fuzzy (Section 5.3.2), Li's generalized fuzzy (Section 5.3.3), Dempster-Shafer and generalized Dempster-Shafer (Section 5.3.4), fuzzy Dempster-Shafer (Section 5.3.5), first-order fuzzy rules (Section 5.3.6), composite rules (Section 5.3.7), and second-order fuzzy rules (Section 5.3.8).

I present my primary unification results in Section 5.4. Specifically, I show that the following measurement fusion techniques are equivalent to measurement fusion using Bayes' rule: copula-type fuzzy conjunctions (Section 5.4.3); normalized and unnormalized Dempster's combination (Section 5.4.1); and total and partial firing of fuzzy rules (Section 5.4.4). In this section, I also show that it is possible to transform various uncertainty representations into other uncertainty representations without loss of relevant information.

Section 5.5 addresses two other types of uncertainty: unknown statistical correlations, and unmodeled target types. I apply my results by deriving closed-form formulas for a generalization of the Kalman filter, the *Kalman evidential filter*

(KEF), that is capable of processing both conventional and fuzzy Dempster-Shafer measurements (Section 5.6). Exercises for the chapter are in Section 5.7.

5.2 WHAT IS A UGA MEASUREMENT?

The reader will have already gained some appreciation of the concept of a generalized measurement from the examples in Section 3.4. Here I codify the concepts introduced there with greater precision.

A data collection source observes a scene. It does not attempt to arrive at an a posteriori determination about the meaning (i.e., the state) of what has been observed. Rather, it attempts only to construct an interpretation of, or opinion about, what it has or has not observed. *Any uncertainties due to ignorance are therefore associated with the data collection process alone and not with the state.* As was explained in Section 3.3, this process has two basic parts: modeling of what has been observed (Section 5.2.1) and modeling of how the observables are generated by a target (Section 5.2.2).

5.2.1 Modeling UGA Measurements

Figure 3.2 portrayed a menagerie of measurements that mix different kinds and degrees of randomness and ignorance. The simplest kind of measurement is a precise and deterministic measurement vector, $\mathbf{z}$. If one randomizes this, one gets the most familiar instance of an ambiguous measurement: a random measurement vector $\mathbf{Z} \in \mathfrak{Z}_0$.

The simplest instance of nonstatistical ambiguity is an *imprecise measurement*. That is, the data-collection source cannot determine the value of $\mathbf{z}$ precisely but, rather, only to within containment in some measurement set $O \subseteq \mathfrak{Z}_0$. Thus O is the actual measurement. If one randomizes O, including randomization of position, size, and so on, one gets a random imprecise measurement. The random set $\Theta_{\mathbf{z}} = E_{\mathbf{z}-\mathbf{W}}$ described in Section 3.4 is a simple example of such a measurement.

Another kind of deterministic measurement is said to be *fuzzy* or *vague*. Because any single constraint O could be erroneous, the data-collection source specifies a *nested sequence* $O_0 \subset O_1 \subset ... \subset O_e$ of alternative constraints, with the constraint O_i assigned a belief $o_i \geq 0$ that it is the correct one, with $o_0 + ... + o_e = 1$. If $O_e = \mathfrak{Z}_0$ then the data-collection source is stipulating that there is some possibility that it may know nothing whatsoever about the value of $\mathbf{z}$. The nested constraint $O_0 \subset O_1 \subset ... \subset O_e$, taken together with its associated weights, is the actual "measurement." It can be represented as a random subset Θ of $\mathfrak{Z}_0$ by defining $\Pr(\Theta = O_i) = o_i$. In this case Θ is said to be "consonant" because its instantiations are linearly ordered under set inclusion and "discrete" because it has only a finite number of instantiations. (As noted in Section 4.3.3, nested constraints are properly called "fuzzy.")

If one randomizes all parameters of a vague measurement (centroid, size, shape, and number of its nested component subsets) one gets a random vague measurement. A (deterministic) vague measurement is just an instantiation of a random vague measurement [113]. An example of a random fuzzy measurement was described in (4.23).

Uncertainty, in the Dempster-Shafer sense, generalizes vagueness in that the component hypotheses no longer need be nested. In this case Θ is discrete but otherwise arbitrary: $\Pr(\Theta = O) = o(O)$ for all but a finite number of $O \subseteq \mathfrak{Z}_0$. As was stipulated in Section 4.5, $o(O)$ is the belief in O that does not accrue to any more constraining hypothesis $O' \subseteq O$. Once again, by randomizing all parameters one gets a random uncertain measurement.

All of the measurement types in this menagerie have one thing in common: they can be represented mathematically by a single probabilistic concept: a *random subset of measurement space* $\mathfrak{Z}_0$. Thus expressed with the greatest mathematical generality, a generalized measurement can be represented as an arbitrary random closed subset Θ of $\mathfrak{Z}_0$.[1] We conclude the following:

- *The observation space* $\mathfrak{Z}$ *of all generalized measurements is the set of all random closed subsets* Θ *of* $\mathfrak{Z}_0$.

We allow the possibility $\Pr(\Theta = \emptyset) \neq 0$. This means that:

- *Whenever* $\Theta = \emptyset$, *the data-collection source is electing to not propose any constraint on the measurement (i.e., it is indicating absence of opinion).*

One example of a generalized measurement is the random subset $\Sigma_A(g)$ of (4.21). It is consonant and therefore can be regarded as (in general) an infinite sequence of nested constraints.[2] The random subset $\Theta_W = \Sigma_A(W)$ of (4.75) provides another example.

1 Here randomness of (closed) subsets is defined in terms of the Mathéron "hit-or-miss" topology. See [160, 165, 70] or Appendix F.

2 Caution: nondiscrete consonant random sets do not necessarily have the simple form $\Sigma_A(g)$ [163].

5.2.2 Modeling the Generation of UGA Measurements

Once one has modeled an observable as a random subset $\Theta \subseteq \mathfrak{Z}_0$ as in the previous section, one must construct an a priori model of how the observables are generated. The fundamental assumption underlying the concept of a UGA measurement is that

- $\mathfrak{Z}_0$ and $\mathfrak{X}_0$ are related by the precise sensor state-to-measurement transform model

$$\mathbf{z} = \eta(\mathbf{x}) \overset{\text{abbr.}}{=} \eta_{k+1}(\mathbf{x}). \tag{5.16}$$

In other words, it is precisely known which observation will be generated by a target with a given state $\mathbf{x}$, if sensor noise is ignored. Given this model, and applying the reasoning introduced in the example of Section 3.4, we say that the ambiguous observation Θ is in agreement with the data-generation model $\eta(\mathbf{x})$ if

$$\eta(\mathbf{x}) \in \Theta. \tag{5.17}$$

5.3 LIKELIHOODS FOR UGA MEASUREMENTS

I introduced the generalized likelihood function of a generalized measurement Θ in (3.14):

$$f(\Theta|\mathbf{x}) \triangleq \Pr(\eta(\mathbf{x}) \in \Theta). \tag{5.18}$$

The likelihood $f(\Theta|\mathbf{x})$ is said to be "generalized" since $\int f(\Theta|\mathbf{x})d\Theta$ could be infinite. This deviation from the Bayesian "recipe" of Sections 3.5.1 and 3.5.2 does not stem from the lack of integrals $\int f(\Theta)d\Theta$ (integration theory is very general). Rather, it arises from the practical difficulty of constructing $f(\Theta|\mathbf{x})$ so that it meaningfully reflects real-world sensing conditions but also satisfies $\int f(\Theta|\mathbf{x})d\Theta = 1$. (In Section 5.4.5 I show that my approach strictly conforms to the "recipe" if we assume, as is common in expert systems theory, that state and measurement spaces are finite.)

Joint generalized likelihoods are defined as:

$$\begin{aligned} f(\Theta_1, ..., \Theta_m|\mathbf{x}) &\triangleq \Pr(\eta(\mathbf{x}) \in \Theta_1, ..., \eta(\mathbf{x}) \in \Theta_m) & (5.19)\\ &= \Pr(\eta(\mathbf{x}) \in \Theta_1 \cap ... \cap \in \Theta_m). & (5.20)\end{aligned}$$

We will sometimes assume that $\Theta_1, ..., \Theta_m$ are statistically independent, in which case

$$f(\Theta_1, .., \Theta_m|\mathbf{x}) = f(\Theta_1|\mathbf{x}) \cdots f(\Theta_m|\mathbf{x}). \tag{5.21}$$

Remark 4 *Equation (5.18) could be conditioned on* $\mathbf{X} = \mathbf{x}$*:* $f(\Theta|\mathbf{x}) \triangleq \Pr(\eta(\mathbf{x}) \in \Theta|\mathbf{x})$*. However, the uncertainty modeling in* Θ *is assumed to be a property of the data-collection process alone, in the same way that* $\Theta_{\mathbf{z}} = E_{\mathbf{z}-\mathbf{W}}$ *of Section 3.4 depends upon the sensor alone. So,* $\Pr(\eta(\mathbf{x}) \in \Theta|\mathbf{x}) = \Pr(\eta(\mathbf{x}) \in \Theta)$.

In the remainder of this section formulas are derived for generalized likelihoods for several special cases: statistical (Section 5.3.1), fuzzy (Section 5.3.2), generalized fuzzy (Section 5.3.3), Dempster-Shafer (Section 5.3.4), fuzzy Dempster-Shafer (Section 5.3.5), first-order fuzzy rules (Section 5.3.6), composite rules (Section 5.3.7), and second-order fuzzy rules (Section 5.3.8).

For each representation type τ we insert its random set model Θ_τ from Chapter 4 into (5.18) and derive the generalized likelihood for that type:

$$f(\tau|\mathbf{x}) \triangleq f(\Theta_\tau|\mathbf{x}) = \Pr(\eta(\mathbf{x}) \in \Theta_\tau). \tag{5.22}$$

5.3.1 Special Case: Θ Is Statistical

In Section 3.4, I introduced the generalized measurements $\Theta_{\mathbf{z}} \triangleq E_{\mathbf{z}-\mathbf{W}}$ corresponding to the additive measurement model $\mathbf{Z} = \eta(\mathbf{x}) + \mathbf{W}$. With modification, this reasoning can be extended to nonadditive models similar to those described in 2.4.4. We must assume more restrictive models of the form

$$\mathbf{Z} = \rho(\eta(\mathbf{x}), \mathbf{W}) \triangleq \rho_{\mathbf{W}}(\eta(\mathbf{x})) \triangleq \tau_{\eta(\mathbf{x})}(\mathbf{W}) \tag{5.23}$$

where $\rho_{\mathbf{z}}$ and $\tau_{\mathbf{z}}$ are families of nonsingular transformations of $\mathfrak{Z}_0$ indexed by $\mathbf{z}$. Then $\Theta_{\mathbf{z}} = \rho_{\mathbf{W}}^{-1} E_{\mathbf{z}}$ and

$$f(\Theta_{\mathbf{z}}|\mathbf{x}) \cong f_{\mathbf{W}}(\tau_{\eta(\mathbf{x})}^{-1}(\mathbf{z})) \cdot J_{\tau_{\eta(\mathbf{x})}^{-1}}(\mathbf{z}) \cdot \varepsilon. \tag{5.24}$$

To see this, note that

$$\begin{aligned} f(\Theta_{\mathbf{z}}|\mathbf{x}) &= \Pr(\eta(\mathbf{x}) \in \Theta_{\mathbf{z}}) \cong \Pr(\mathbf{W} \in \tau_{\eta(\mathbf{x})}^{-1} E_{\mathbf{z}}) && (5.25) \\ &= \int_{\tau_{\eta(\mathbf{x})}^{-1} E_{\mathbf{z}}} f_{\mathbf{W}}(\mathbf{w}) d\mathbf{w} && (5.26) \\ &= \int_{E_{\mathbf{z}}} f_{\mathbf{W}}(\tau_{\eta(\mathbf{x})}^{-1}(\mathbf{v})) \cdot J_{\tau_{\eta(\mathbf{x})}^{-1}}(\mathbf{v}) d\mathbf{v} && (5.27) \\ &\cong f_{\mathbf{W}}(\tau_{\eta(\mathbf{x})}^{-1}(\mathbf{z})) \cdot J_{\tau_{\eta(\mathbf{x})}^{-1}}(\mathbf{z}) \cdot \varepsilon && (5.28) \end{aligned}$$

where $J_{\tau^{-1}_{\eta(\mathbf{x})}}(\mathbf{z})$ is the Jacobian determinant of $\tau^{-1}_{\eta(\mathbf{x})}$. The third equation results from the integral change of variables formula, (2.111). Thus $f(\mathbf{z}|\mathbf{x})$ is a limiting case of $f(\Theta_{\mathbf{z}}|\mathbf{x})$ in the sense of (3.13) of Section 3.4.1.

5.3.2 Special Case: Θ Is Fuzzy

The random set representation of a fuzzy set was defined in Section 4.3. Let $g(\mathbf{z})$ be a fuzzy membership function on $\mathfrak{Z}_0$, and let $\Theta_g \triangleq \Sigma_A(g)$ be its random set model (4.21). Its generalized likelihood is $f(g|\mathbf{x}) \triangleq f(\Theta_g|\mathbf{x})$, in which case

$$f(g|\mathbf{x}) = g(\eta(\mathbf{x})). \tag{5.29}$$

To see this, note that

$$\begin{aligned} f(g|\mathbf{x}) &= \Pr(\eta(\mathbf{x}) \in \Sigma_A(g)) && (5.30) \\ &= \Pr(A \leq g(\eta(\mathbf{x}))) = g(\eta(\mathbf{x})) && (5.31) \end{aligned}$$

where the last equation is true because A is uniform.[3]

To gain a better understanding of (5.29), let the fuzzy measurement be Gaussian in form:

$$g_{C,\mathbf{c}}(\mathbf{z}) = \hat{N}_C(\mathbf{z} - \mathbf{c}) \tag{5.32}$$

where the normalized Gaussian is defined as

$$\begin{aligned} \hat{N}_C(\mathbf{z} - \mathbf{c}) &\triangleq N_C(\mathbf{0})^{-1} \cdot N_C(\mathbf{z} - \mathbf{c}) && (5.33) \\ &= \exp\left(-\frac{1}{2}(\mathbf{z} - \mathbf{c})^T C^{-1} (\mathbf{z} - \mathbf{c})\right). && (5.34) \end{aligned}$$

Then by (5.29) the likelihood is

$$f(g_{C,\mathbf{c}}|\mathbf{x}) = \exp\left(-\frac{1}{2}(\mathbf{c} - \eta(\mathbf{x}))^T C^{-1} (\mathbf{c} - \eta(\mathbf{x}))\right). \tag{5.35}$$

This differs by only a constant factor from the conventional likelihood $f(\mathbf{c}|\mathbf{x}) = N_C(\mathbf{c} - \eta(\mathbf{x}))$ corresponding to collection of an observation $\mathbf{c}$ by a Gaussian sensor with measurement covariance C. The corresponding posterior densities will

3 The same equation is true for the general random set models of fuzzy sets, as discussed in Section 4.3.5.

therefore be identical:

$$f(\mathbf{x}|\mathbf{c}) = f(\mathbf{x}|g_{C,\mathbf{c}}). \tag{5.36}$$

Despite this fact, and as in the example of Section 3.4.5,

- *The underlying semantics of the two cases are fundamentally different.*

On the one hand, $g_{C,\mathbf{c}}(\mathbf{z})$ is an observer's interpretation of what is being observed. Its value is $g_{C,\mathbf{c}}(\mathbf{z}) = 1$ for $\mathbf{z} = \mathbf{c}$, indicating that $\mathbf{c}$ is probably the best interpretation. However, the observer believes that $\mathbf{c}$ may only be known to containment within various possible constraint-sets $\Sigma_a(g_{C,\mathbf{c}}) = \{\mathbf{z}|\ a \leq g_{C,\mathbf{c}}(\mathbf{z})\}$. The value $f(g_{C,\mathbf{c}}|\mathbf{x})$ describes how likely it is that a target with state $\mathbf{x}$ could have generated the generalized observation $g_{C,\mathbf{c}}$.

On the other hand, the conventional likelihood $f(\mathbf{c}|\mathbf{x})$ indicates that the observer is certain about having observed $\mathbf{c}$, but from previous statistical analysis knows that $\mathbf{c}$ is a random perturbation of the true observation $\eta(\mathbf{x})$.

Example 16 *Consider the modeling of the hub/tire feature in Example 15. Random subsets of* $\{1,...,8\}$ *are difficult to visualize and to mathematically manipulate. Fuzzy sets are otherwise, and so it is often better to use them instead. We know from Section 4.3.2 that the random set representation* $\Sigma_A(g)$ *of a fuzzy membership function* $g(n)$ *faithfully preserves the information contained in* g. *So, we substitute* $\Theta = \Sigma_A(g_O)$, *where* $g_O(n)$ *is a fuzzy membership function on* $\{0,1,...,8\}$ *that models the human-mediated feature. For example, we may assign* $g_O(6) = 1$ *since the operator believes that* $n = 6$ *hubs/tires were observed; but assign* $g_O(5) = g_O(7) = 0.5$ *since there could be five or seven tires; and finally assign* $g_O(n) = 0.1$ *for* $n = 0,1,2,3,8$ *to hedge against the possibility that the operator was completely mistaken. The generalized likelihood function becomes, because of (5.29),*

$$f(g_O|v) = g_O(\eta(v)). \tag{5.37}$$

Those target types v *for which* $\eta(v) = 6$ *are most strongly supported, whereas those with* $\eta(v) = 5$ *or* $\eta(v) = 7$ *are less so. This is pictured in Figure 5.1.*

Example 17 *Let* $g(\mathbf{z}) = \mathbf{1}_O(\mathbf{z})$ *be the set indicator function of* $O \subseteq \mathfrak{Z}_0$:

$$\mathbf{1}_O(\mathbf{z}) = \left\{ \begin{array}{ccc} 1 & \text{if} & \mathbf{z} \in O \\ 0 & \text{if} & \text{otherwise} \end{array} \right. . \tag{5.38}$$

Then $\Theta_g = \Sigma_A(g)$ *has only one instantiation,* $\Theta_g = O$, *and is therefore an imprecise measurement. However, if* $g(\mathbf{z}) = (1-\varepsilon)\mathbf{1}_O(\mathbf{z}) + \varepsilon$ *for some small*

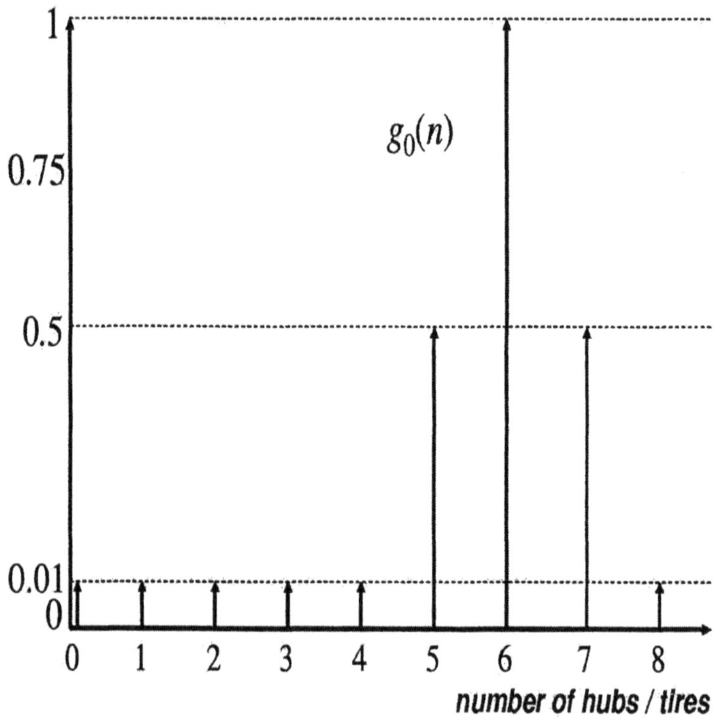

Figure 5.1 The fuzzy measurement $g_0(n)$ models a human operator declaring that a feature—the number of tires/hubs n extracted from an image of a target—is "about six." The measurement g_0 hedges against the uncertainties in this interpretation by assigning smaller but nonzero values to the other possible values of n.

positive number ε *then* $\Theta_g = O$ *or* $\Theta_g = \mathfrak{Z}_0$. *In this case* Θ_g *represents an opinion that the measurement is constrained by* O, *but that there is a small possibility that this belief is completely mistaken. (See Section 4.3.3 for generalized versions of this example.)*

Example 18 *Randomness and imprecision are combined by the model* $\Theta_O = O - \mathbf{W} \triangleq \{\mathbf{z} - \mathbf{W} |\ \mathbf{z} \in O\}$ *where* $\mathbf{W} \in \mathfrak{Z}_0$ *is a zero-mean random vector with probability density* $f_{\mathbf{W}}(\mathbf{z})$. *In this case the generalized likelihood is*

$$f(O|\mathbf{x}) \triangleq f(\Theta_O|\mathbf{x}) = \Pr(\eta(\mathbf{x}) \in O - \mathbf{W}) \tag{5.39}$$

$$= \Pr(\mathbf{W} \in O - \eta(\mathbf{x})) = \int_O f_{\mathbf{W}}(\mathbf{z} - \eta(\mathbf{x}))d\mathbf{z}. \tag{5.40}$$

5.3.3 Special Case: Θ Is Generalized Fuzzy

The random set representation of a generalized fuzzy set was defined in Section 4.4. The generalized fuzzy sets of $\mathfrak{Z}_0$ are the subsets of $\mathfrak{Z}_0^* \triangleq \mathfrak{Z}_0 \times I$ where $I \triangleq [0,1]$. Define $W_{\mathbf{z}} \subseteq I$ by $W_{\mathbf{z}} \triangleq \{a |\ (\mathbf{z}, a) \in W\}$. The generalized likelihood of a generalized fuzzy set is $f(W|\mathbf{x}) \triangleq \Pr(\eta(\mathbf{x}) \in \Sigma_A(W))$, in which case

$$f(W|\mathbf{x}) = \int_0^1 \mathbf{1}_W(\eta(\mathbf{x}), a)da = \mu_W(\eta(\mathbf{x})) \tag{5.41}$$

since

$$f(W|\mathbf{x}) = \Pr((\eta(\mathbf{x}), A) \in W) = \Pr(A \in W_{\eta(\mathbf{x})}) \tag{5.42}$$

$$= \int \mathbf{1}_{W_{\eta(\mathbf{x})}}(a) \cdot f_A(a)da = \int \mathbf{1}_{W_{\eta(\mathbf{x})}}(a)da \tag{5.43}$$

$$= \int_0^1 \mathbf{1}_W(\eta(\mathbf{x}), a)da \tag{5.44}$$

$$= \mu_W(\eta(\mathbf{x})) \tag{5.45}$$

where $f(a) = 1$ for all $a \in [0,1]$ is the probability density function for A; and where the notation $\mu_W(\mathbf{z})$ in the final equation was defined in (4.73).

Example 19 *Let* $g(\mathbf{z})$ *be a fuzzy membership function on* $\mathfrak{Z}_0$ *and let* $W_g = \{(\mathbf{z},a)|\ a \leq g(\mathbf{z})\}$. *Then*

$$
\begin{aligned}
f(W_g|\mathbf{x}) &= \int_0^1 \mathbf{1}_{\{(\mathbf{z},a)|\ a\leq g(\mathbf{z})\}}(\eta(\mathbf{x}),a)da &\quad (5.46)\\
&= \int_0^{g(\eta(\mathbf{x}))} da = g(\eta(\mathbf{x})) = f(g|\mathbf{x}) &\quad (5.47)
\end{aligned}
$$

which agrees with (5.30). On the other hand,

$$
\begin{aligned}
f(W_g^c|\mathbf{x}) &= \int_0^1 \mathbf{1}_{\{(\mathbf{z},a)|\ a> g(\mathbf{z})\}}(\eta(\mathbf{x}),a)da &\quad (5.48)\\
&= 1-\int_0^1 \mathbf{1}_{\{(\mathbf{z},a)|\ a\leq g(\mathbf{z})\}}(\eta(\mathbf{x}),a)da &\quad (5.49)\\
&= 1-g(\eta(\mathbf{x})) = g^c(\eta(\mathbf{x})) &\quad (5.50)\\
&= f(g^c|\mathbf{x}). &\quad (5.51)
\end{aligned}
$$

Thus $f(W_g^c|\mathbf{x}) = f(W_{g^c}|\mathbf{x})$ *even though* $W_g^c \neq W_{g^c}$.

Example 20 *More generally, given a generalized fuzzy set* W, *let* $\mu_W(\mathbf{z}) = \int_0^1 \mathbf{1}_W(\mathbf{z},a)da$ *be the induced fuzzy membership function defined in (4.73). Then note that*

$$
\begin{aligned}
f(\mu_W|\mathbf{x}) &= f(W_{\mu_W}|\mathbf{x}) = \mu_W(\eta(\mathbf{x})) &\quad (5.52)\\
&= \int_0^1 \mathbf{1}_W(\eta(\mathbf{x}),a)da &\quad (5.53)\\
&= f(W|\mathbf{x}). &\quad (5.54)
\end{aligned}
$$

That is, the likelihood of a generalized fuzzy measurement W *is identical to the likelihood of the fuzzy measurement* μ_W.

Example 21 *Let* $W_{O,J} = O \times J$ *where* $O \subseteq \mathfrak{Z}_0$ *and* $J = [a_1, a_2]$. *Then*

$$
\begin{aligned}
f(W_{O,J}|\mathbf{x}) &= \int_0^1 \mathbf{1}_{O\times J}(\eta(\mathbf{x}),a)da &\quad (5.55)\\
&= \mathbf{1}_O(\eta(\mathbf{x}))\cdot\int_0^1 \mathbf{1}_J(a)da &\quad (5.56)\\
&= (a_2-a_1)\cdot \mathbf{1}_{\eta^{-1}O}(\mathbf{x}). &\quad (5.57)
\end{aligned}
$$

Thus J *has no influence on the value of the posterior distribution* $f(\mathbf{x}|W_{O,J})$.

5.3.4 Special Case: Θ Is Discrete/Dempster-Shafer

The random set representations of basic mass assignments (b.m.a.s) and generalized b.m.a.s were defined in Section 4.6.1. If Θ is discrete, then $o(O) \triangleq \Pr(\Theta = O)$ is nonzero for only a finite number of $O \subseteq \mathfrak{Z}_0$. In this case it models a generalized b.m.a.—g.b.m.a., see (4.95)—and we write $\Theta = \Theta_o$. If in addition $o(\emptyset) = 0$ then Θ models an ordinary b.m.a. The generalized likelihood function is $f(o|\mathbf{x}) \triangleq \Pr(\eta(\mathbf{x}) \in \Theta_o)$ and

$$f(o|\mathbf{x}) = \sum_{O \ni \eta(\mathbf{x})} o(O). \tag{5.58}$$

To see this, note that

$$\begin{aligned} f(o|\mathbf{x}) &= \sum_{O} \Pr(\Theta_o = O, \eta(\mathbf{x}) \in O) & (5.59) \\ &= \sum_{O} o(O) \cdot \mathbf{1}_O(\eta(\mathbf{x})) & (5.60) \\ &= \sum_{O \ni \eta(\mathbf{x})} o(O). & (5.61) \end{aligned}$$

Thus $o(\emptyset) = \Pr(\Theta_o = \emptyset)$ has no effect on the value of $f(o|\mathbf{x})$. Section 3.41 described a concrete example of such a likelihood.

5.3.4.1 Dempster-Shafer Versus Generalized Dempster-Shafer Measurements

From the point of view of Bayesian measurement fusion, there is no essential distinction between a b.m.a. measurement and a g.b.m.a. measurement. Given a g.b.m.a. o, define the b.m.a. m_o by

$$m_o(O) \triangleq |o|^{-1} \cdot o(O) \tag{5.62}$$

for all $O \neq \emptyset$ and $m_o(\emptyset) \triangleq 0$ otherwise. Here

$$|o| \triangleq \sum_{Q \neq \emptyset} o(Q). \tag{5.63}$$

Then note that

$$f(o|\mathbf{x}) = \sum_{O \ni \eta(\mathbf{x})} o(O) = |o| \sum_{O \ni \eta(\mathbf{x})} |o|^{-1} o(O) \tag{5.64}$$

$$= |o| \sum_{O \ni \eta(\mathbf{x})} m_o(O) = |o| \cdot f(m_o|\mathbf{x}). \tag{5.65}$$

Consequently the posterior distributions conditioned on o and on m_o are identical since $|o|$ is constant: $f(\mathbf{x}|o) = f(\mathbf{x}|m_o)$.

5.3.4.2 Posterior Distributions and Voorbraak Probability

From (5.60) we know that the posterior distribution conditioned on o is

$$f(\mathbf{x}|o) = \frac{\sum_O o(O) \cdot \mathbf{1}_{\eta^{-1}O}(\mathbf{x}) \cdot f_0(\mathbf{x})}{\sum_Q o(Q) \cdot p_0(\eta^{-1}Q)}. \tag{5.66}$$

To see this, note that

$$f(\mathbf{x}|o) = \frac{f(o|\mathbf{x}) \cdot f_0(\mathbf{x})}{\int f(o|\mathbf{y}) \cdot f_0(\mathbf{y}) d\mathbf{y}} \tag{5.67}$$

$$= \frac{\sum_O o(O) \cdot \mathbf{1}_{\eta^{-1}O}(\mathbf{x}) \cdot f_0(\mathbf{x})}{\sum_Q o(Q) \cdot \int \mathbf{1}_{\eta^{-1}Q}(\mathbf{y}) \cdot f_0(\mathbf{y}) d\mathbf{y}} \tag{5.68}$$

$$= \frac{\sum_O o(O) \cdot \mathbf{1}_{\eta^{-1}O}(\mathbf{x}) \cdot f_0(\mathbf{x})}{\sum_Q o(Q) \cdot \int_{\eta^{-1}Q} f_0(\mathbf{y}) d\mathbf{y}} \tag{5.69}$$

$$= \frac{\sum_O o(O) \cdot \mathbf{1}_{\eta^{-1}O}(\mathbf{x}) \cdot f_0(\mathbf{x})}{\sum_Q o(Q) \cdot p_0(\eta^{-1}Q)} \tag{5.70}$$

where $\eta^{-1}O \triangleq \{\mathbf{x}|\ \eta(\mathbf{x}) \in O\}$ denotes the inverse image of O under η; and where $p_0(S) = \int_S f_0(\mathbf{x}) d\mathbf{x}$.

Suppose now that $\mathfrak{Z}_0 = \mathfrak{X}_0$, that $\mathfrak{Z}_0$ is finite with N elements, and that $\eta(x) = x$. Then (5.66) reduces to the special case of Dempster's combination described in (4.89):

$$f(x|o) \propto \left(\sum_{O \ni x} o(O) \right) \cdot f_0(x). \tag{5.71}$$

Further assume that the prior is the uniform distribution, $f_0(x) = N^{-1}$, so that $p_0(S) = N^{-1}|S|$. Then (5.71) further reduces to

$$f(x|o) = \frac{\sum_{O \ni x} o(O)}{\sum_Q o(Q) \cdot |Q|} = v_o(x) \tag{5.72}$$

where v_o is the Voorbraak transform as defined in (4.115). That is:

- Voorbraak probabilities are special cases of posterior probability distributions conditioned on Dempster-Shafer measurements.

5.3.5 Special Case: Θ Is Fuzzy Dempster-Shafer

The random set representation of a generalized fuzzy set was defined in Section 4.6. The generalized likelihood of a fuzzy b.m.a. is $f(o|\mathbf{x}) \triangleq f(W_o|\mathbf{x})$, where the generalized fuzzy set W_o was defined in (4.135). From (5.41), (5.60) generalizes to:

$$f(o|\mathbf{x}) = \sum_g o(g) \cdot g(\eta(\mathbf{x})). \tag{5.73}$$

To see this, let the focal sets of o be $g_1, ..., g_e$ and their weights be $o_1, ..., o_e$. Then

$$\begin{aligned} f(o|\mathbf{x}) &= \int_0^1 \mathbf{1}_{W_o}(\eta(\mathbf{x}), a)da &(5.74)\\ &= \sum_{i=1}^e \int_0^1 \mathbf{1}_{W_i}(\eta(\mathbf{x}), a)da &(5.75)\\ &= \sum_{i=1}^e \int_{o_{i-1}^+}^{o_{i-1}^+ + o_i g_i(\eta(\mathbf{x}))} da &(5.76)\\ &= \sum_{i=1}^e o_i \cdot g_i(\eta(\mathbf{x})), &(5.77)\end{aligned}$$

which agrees with (3.38) in Section 3.4.4.

Example 22 *Generalizing (5.35), one can choose*

$$g_i(\mathbf{z}) = \exp\left(-\frac{1}{2}(\mathbf{z} - \mathbf{c}_i)^T C_i^{-1}(\mathbf{z} - \mathbf{c}_i)\right) \tag{5.78}$$

in which case

$$f(o|\mathbf{x}) = \sum_{i=1}^{e} o_i \cdot \exp\left(-\frac{1}{2}(\mathbf{c}_i - \eta(\mathbf{x}))^T C_i^{-1}(\mathbf{c}_i - \eta(\mathbf{x}))\right). \tag{5.79}$$

5.3.6 Special Case: Θ Is a First-Order Fuzzy Rule

The random set representation $\Sigma_{\Phi,A,A'}(\rho)$ of a first-order fuzzy rule $\rho = (\xi \Rightarrow \sigma)$ was introduced in Section 4.7.3, where it was assumed that $\mathfrak{Z}_0$ is finite.[4] The likelihood function of ρ is $f(\rho|\mathbf{x}) \triangleq \Pr(\eta(\mathbf{x}) \in \Sigma_{\Phi,A,A'}(\rho)$. If $\mathfrak{Z}_0$ is finite it can be shown (see Appendix G.1) that

$$f(\rho|\mathbf{x}) \triangleq (\sigma \wedge_{A,A'} \xi)(\eta(\mathbf{x})) + \frac{1}{2}\xi^c(\eta(\mathbf{x})) \tag{5.80}$$

where

$$(\sigma \wedge_{A,A'} \xi)(\mathbf{z}) \triangleq \sigma(\mathbf{z}) \wedge_{A,A'} \xi(\mathbf{z}) \tag{5.81}$$

$$a \wedge_{A,A'} a' \triangleq \Pr(A \leq a,\ A' \leq a') \tag{5.82}$$

for $0 \leq a, a' \leq 1$ (see Section 4.3.4). When $\mathfrak{Z}_0$ is not finite we will adopt (5.80) as a *definition*.

Note that $f(\rho|\mathbf{x})$ is well defined, since

$$(\sigma \wedge_{A,A'} \xi)(\mathbf{z}) + \frac{1}{2} - \frac{1}{2}\xi(\mathbf{z}) \leq \xi(\mathbf{z}) + \frac{1}{2} - \frac{1}{2}\xi(\mathbf{z}) \leq \frac{1}{2} + \frac{1}{2} = 1. \tag{5.83}$$

If $A = A'$ (perfect correlation) then $a \wedge_{A,A} a' \triangleq \Pr(A \leq \min\{a, a'\}) = \min\{a, a'\} = a \wedge a'$ (Zadeh conjunction), and so

$$\rho(\xi \Rightarrow \sigma|\mathbf{x}) = (\sigma \wedge \xi)(\eta(\mathbf{x})) + \frac{1}{2}\xi^c(\eta(\mathbf{x})). \tag{5.84}$$

If on the other hand A, A' are independent, then $a \wedge_{A,A'} a' = aa'$, and so

$$\rho(\xi \Rightarrow \sigma|\mathbf{x}) = \sigma(\eta(\mathbf{x}))\xi(\eta(\mathbf{x})) + \frac{1}{2}\xi^c(\eta(\mathbf{x})). \tag{5.85}$$

4 This approach has been implemented and successfully tested in preliminary simulations [52].

Example 23 *Let* $\xi(\mathbf{z}) = \mathbf{1}_X(\mathbf{z})$ *and* $\sigma(\mathbf{z}) = \mathbf{1}_S(\mathbf{z})$ *for crisp subsets* $X, S \subseteq \mathfrak{Z}_0$. *Then*

$$f(X \Rightarrow S|\mathbf{x}) = \mathbf{1}_{S\cap X}(\eta(\mathbf{x})) + \frac{1}{2}\mathbf{1}_{X^c}(\eta(\mathbf{x})) \tag{5.86}$$

$$= \begin{cases} 1 & \textit{if} \quad \eta(\mathbf{x}) \in S \cap X \\ 0 & \textit{if} \quad \eta(\mathbf{x}) \in X - (S \cap X) \\ \frac{1}{2} & \textit{if} \quad \eta(\mathbf{x}) \in X^c \end{cases} . \tag{5.87}$$

This is in accordance with intuition. The rule says that if $\eta(\mathbf{x})$ *is in* X, *then it must also be in* S *and therefore in* $S \cap X$. *So, if* $\eta(\mathbf{x})$ *is in* X *it is impossible for it to be in* $X - (S \cap X)$, *but it is certain that it is in* $S \cap X$. *On the other hand, if* $\eta(\mathbf{x})$ *is outside of* X, *then it is not covered by the rule. There can be no constraint one way or the other on* $\mathbf{x}$, *so the likelihood assigns the equivocating value* $\frac{1}{2}$ *to such* $\mathbf{x}$*'s. The situation is pictured in Figure 5.2.*

Example 24 *Suppose that a second, imprecise, observation* T *is received, so that its likelihood is* $f(T|\mathbf{x}) = \mathbf{1}_T(\eta(\mathbf{x}))$. *Assuming conditional independence, Bayesian fusion of the rule and this observation is accomplished via the joint likelihood*

$$f(X \Rightarrow S, T|\mathbf{x}) = f(X \Rightarrow S|\mathbf{x}) \cdot f(T|\mathbf{x})$$

$$= \mathbf{1}_{S\cap X\cap T}(\eta(\mathbf{x})) + \frac{1}{2}\mathbf{1}_{X^c\cap T}(\eta(\mathbf{x})) \tag{5.88}$$

$$= \begin{cases} 1 & \textit{if} \quad \eta(\mathbf{x}) \in S \cap X \cap T \\ \frac{1}{2} & \textit{if} \quad \eta(\mathbf{x}) \in X^c \cap T \\ 0 & \textit{if} \quad \textit{otherwise} \end{cases} . \tag{5.89}$$

If $T = X$ *(i.e., the antecedent hypothesis is known to be exactly true), then* $f(X \Rightarrow S, T|\mathbf{x}) = \mathbf{1}_{S\cap X}(\eta(\mathbf{x}))$. *The rule has been completely "fired," and we infer that the consequent hypothesis* $S \cap X$ *is exactly true. If on the other hand* $T \neq X$ *with* $S \cap T \neq \emptyset$, *then the rule has been partially fired, to the extent consistent with supporting evidence. The situation is pictured in Figure 5.3.*

Example 25 *Let measurement and state spaces be* $\mathfrak{Z}_0 = \mathfrak{X}_0 = \mathbb{R}$, *and let* $\eta(x) = x$. *Define fuzzy measurements* ξ *and* σ *by*

$$\xi(x) = \exp\left(-x^2\right) \tag{5.90}$$

$$\sigma(x) = \exp\left(-x^2/8\right). \tag{5.91}$$

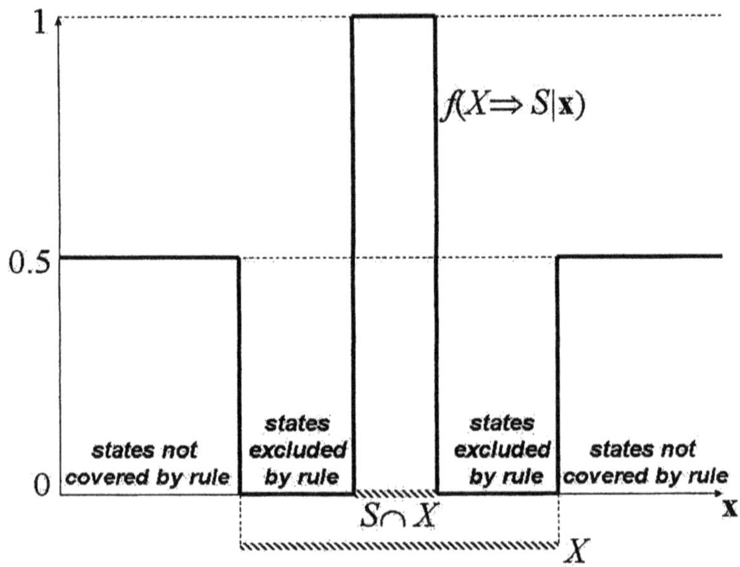

Figure 5.2 The likelihood function of the rule $X \Rightarrow S$ on crisp measurements X and S is pictured. If the antecedent X is observed, we conclude S and thus also $S \cap X$. So the state cannot be in $X - (S \cap X)$. The rule is equivocal about states in X^c.

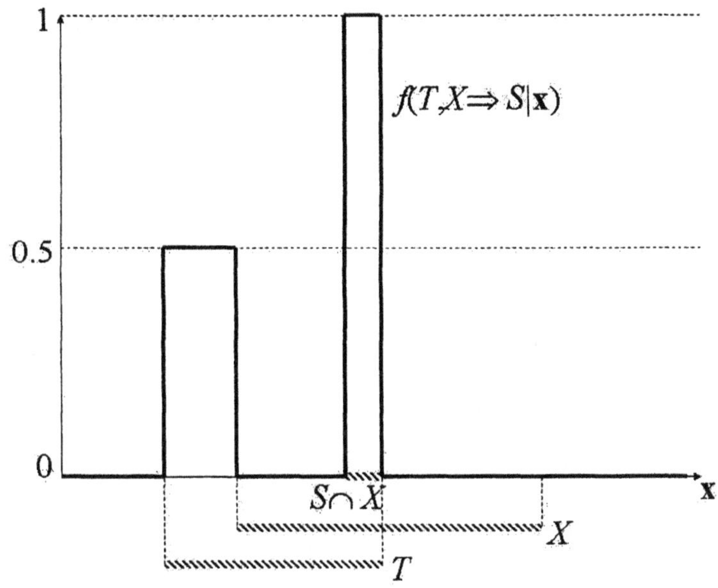

Figure 5.3 The likelihood function $f(T, X \Rightarrow S|\mathbf{x})$, expressing the partial firing of the crisp rule $X \Rightarrow S$ by a partial antecedent T, is pictured. Any state not in $S \cap X \cap T$ and $T \cap X^c$ is excluded. The states in $T \cap X^c$ are not covered by the rule.

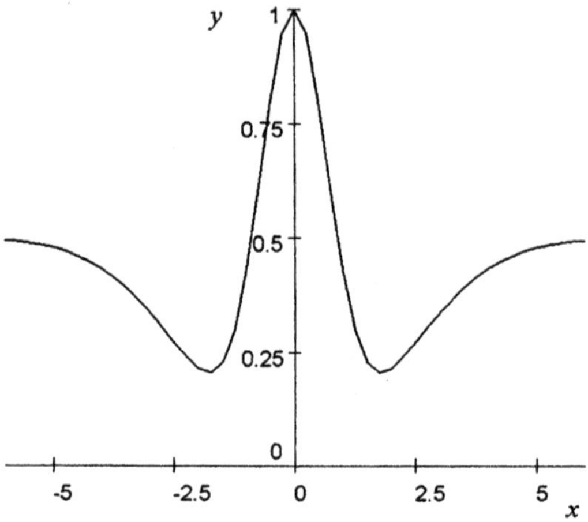

Figure 5.4 The likelihood function of the fuzzy rule $\rho = (\xi \Rightarrow \sigma)$ is depicted, where $\xi(x) = \exp(-x^2)$ and $\sigma(x) = \exp(-x^2/8)$.

Note that $\xi(x) \geq \sigma(x)$ *for all* x *and so from (5.80) the likelihood of the fuzzy rule* $\rho = (\xi \Rightarrow \sigma)$ *is (see Figure 5.4)*

$$f(\rho|x) = \exp(-x^2) + \frac{1}{2} - \frac{1}{2}\exp\left(-\frac{x^2}{8}\right). \tag{5.92}$$

Suppose now that the fuzzy observation ξ *is collected. The joint likelihood of* ξ *and* $\xi \Rightarrow \sigma$ *is*

$$f(\xi, \rho|x) = \Pr(x \in \Sigma_A(\xi) \cap \Sigma_{\Phi,A,A'}(\rho)). \tag{5.93}$$

Since

$$\Sigma_{\Phi,A,A'}(\rho) = (\Sigma_A(\sigma) \cap \Sigma_{A'}(\xi)) \cup (\Sigma_{A'}(\xi)^c \cap \Phi) \tag{5.94}$$

it follows that

$$\Sigma_A(\xi) \cap \Sigma_{\Phi,A,A'}(\rho) = \Sigma_A(\sigma) \cap \Sigma_{A'}(\xi) \tag{5.95}$$

and so

$$f(\xi, \rho|x) = \Pr(x \in \Sigma_A(\sigma) \cap \Sigma_{A'}(\xi)) = (\sigma \wedge_{A,A'} \xi)(x). \tag{5.96}$$

If $A = A'$, *then* $\wedge_{A,A'}$ *is the Zadeh conjunction and*

$$f(\xi, \rho|x) = \min\{\sigma(x), \xi(x)\} = \sigma(x). \tag{5.97}$$

If A *and* A' *are independent, then* $(\sigma \wedge_{A,A'} \xi)(x) = \sigma(x)\xi(x)$, *and so*

$$\begin{aligned} f(\xi, \rho|x) &= \exp(-\frac{x^2}{8}) \cdot \exp(-x^2) = \exp\left(-\frac{9}{8}x^2\right) & (5.98) \\ &\cong \exp(-x^2) = \sigma(x). & (5.99) \end{aligned}$$

5.3.7 Special Case: Θ Is a Composite Fuzzy Rule

The random set representation $\Sigma_{A_1,A_2,A_3,A_4,\Phi}(\rho \vee \rho')$ of the composite rule $\rho \vee \rho'$ of two fuzzy rules $\rho = (\xi \Rightarrow \sigma)$ and $\rho' = (\xi' \Rightarrow \sigma')$ was defined in Section 4.7.4. The likelihood function of $\rho \vee \rho'$ is

$$f(\rho \vee \rho'|\mathbf{x}) \triangleq \Pr(\eta(\mathbf{x}) \in \Sigma_{A_1,A_2,A_3,A_4,\Phi}(\rho \vee \rho')). \tag{5.100}$$

In what follows I will, for notational simplicity, suppress the A_1, A_2, A_3, A_4 by employing the following notational convention. For $0 \leq a_1, a_2, a_3, a_4 \leq 1$, I write

$$a_1 \barwedge a_2 \barwedge a_3 \barwedge a_4 \triangleq \Pr(A_1 \leq a_1, A_2 \leq a_2, A_3 \leq a_3, A_4 \leq a_4) \tag{5.101}$$

and then abbreviate $(\xi \barwedge \xi')(\mathbf{z}) = 1 \barwedge \xi(\mathbf{z}) \barwedge 1 \barwedge \xi'(\mathbf{z})$ and $(\sigma' \barwedge \xi')(\mathbf{z}) = 1 \barwedge 1 \barwedge \sigma'(\mathbf{z}) \barwedge \xi'(\mathbf{z})$, and so on. In Appendix G.2 it is shown that

$$f(\rho \vee \rho'|\mathbf{x}) = \tau(\eta(\mathbf{x})) + \frac{1}{2}\theta^c(\eta(\mathbf{x})) \tag{5.102}$$

where the fuzzy membership functions $\tau(\mathbf{z})$ and $\theta(\mathbf{z})$ on $\mathfrak{Z}_0$ are defined by

$$\begin{aligned} \tau &\triangleq \sigma \barwedge \xi + \sigma' \barwedge \xi' - \sigma \barwedge \xi \barwedge \sigma' \barwedge \xi' & (5.103) \\ \theta &\triangleq \sigma \barwedge \xi + \sigma' \barwedge \xi' + \xi \barwedge \xi' - \sigma \barwedge \xi \barwedge \xi' - \xi \barwedge \sigma' \barwedge \xi'. & (5.104) \end{aligned}$$

Example 26 *Let* $A_1 = A_2 = A_3 = A_4 = A$, *so that* $a_1 \barwedge a_2 \barwedge a_3 \barwedge a_4 = a_1 \wedge a_2 \wedge a_3 \wedge a_4$, *where '$\wedge$' is the Zadeh conjunction. If* $\xi = \xi'$, *then*

$$\begin{aligned} f(\rho \vee \rho'|\mathbf{x}) &= ((\sigma + \sigma' - \sigma \wedge \sigma') \wedge \xi)(\eta(\mathbf{x})) + \frac{1}{2}\xi^c(\eta(\mathbf{x})) & (5.105) \\ &= f(\xi \Rightarrow (\sigma + \sigma' - \sigma \wedge \sigma')|\mathbf{x}). & (5.106) \end{aligned}$$

Example 27 *Let* $A_1 = A_2 = A_3 = A_4 = A$ *as before. If* $\xi \wedge \xi' = 0$, *then*

$$f(\rho \vee \rho'|\mathbf{x}) = \frac{1}{2}\left(1 + \sigma \,\bar{\wedge}\, \xi + \sigma' \,\bar{\wedge}\, \xi'\right)(\eta(\mathbf{x})) \tag{5.107}$$

Thus

$$f(\rho \vee \rho'|\mathbf{x}) = \begin{cases} \frac{1}{2}\left(1 + \sigma \,\bar{\wedge}\, \xi\right)(\eta(\mathbf{x})) & \text{if} \quad \xi'(\eta(\mathbf{x})) = 0 \\ \frac{1}{2}\left(1 + \sigma' \,\bar{\wedge}\, \xi'\right)(\eta(\mathbf{x})) & \text{if} \quad \xi(\eta(\mathbf{x})) = 0 \end{cases}. \tag{5.108}$$

5.3.8 Special Case: Θ Is a Second-Order Fuzzy Rule

The random set representation of a second-order fuzzy rule was introduced in Section 4.7.5 where, once again, $\mathfrak{Z}_0$ was assumed finite. Given fuzzy rules $\rho = (\xi \Rightarrow \sigma)$ and $\rho' = (\xi' \Rightarrow \sigma')$, let $\rho \Rightarrow \rho'$ be a second-order rule. Its generalized likelihood is

$$f(\rho \Rightarrow \rho'|\mathbf{x}) \triangleq \Pr(\eta(\mathbf{x}) \in \Sigma_{\Phi_1,\Phi_2,A_1,A_2,A_3,A_4}(\rho \Rightarrow \rho')). \tag{5.109}$$

As we did for composite rules, write

$$a_1 \,\bar{\wedge}\, a_2 \,\bar{\wedge}\, a_3 \,\bar{\wedge}\, a_4 \triangleq \Pr(A_1 \leq a_1, A_2 \leq a_2, A_3 \leq a_3, A_4 \leq a_4) \tag{5.110}$$

and $(\xi \,\bar{\wedge}\, \xi')(\mathbf{z}) = 1 \,\bar{\wedge}\, \xi(\mathbf{z}) \,\bar{\wedge}\, 1 \,\bar{\wedge}\, \xi'(\mathbf{z})$ and $(\sigma' \,\bar{\wedge}\, \xi')(\mathbf{z}) = 1 \,\bar{\wedge}\, 1 \,\bar{\wedge}\, \sigma'(\mathbf{z}) \,\bar{\wedge}\, \xi'(\mathbf{z})$, and so on. In Appendix G.3 it is shown that

$$f(\rho \Rightarrow \rho'|\mathbf{x}) = (\sigma \,\bar{\wedge}\, \xi \,\bar{\wedge}\, \sigma' \,\bar{\wedge}\, \xi')(\eta(\mathbf{x})) + \frac{1}{2}(\xi \,\bar{\wedge}\, \xi')(\eta(\mathbf{x})) \tag{5.111}$$

$$-\frac{1}{2}(\sigma \,\bar{\wedge}\, \xi \,\bar{\wedge}\, \xi')(\eta(\mathbf{x})) + \frac{1}{2}(\sigma' \,\bar{\wedge}\, \xi')(\eta(\mathbf{x})) \tag{5.112}$$

$$-\frac{1}{2}(\xi \,\bar{\wedge}\, \sigma' \,\bar{\wedge}\, \xi')(\eta(\mathbf{x})) + \frac{1}{2}(\xi')^c(\eta(\mathbf{x})) + \frac{1}{4}\xi^c(\eta(\mathbf{x})). \tag{5.113}$$

If $\mathfrak{Z}_0$ is not finite, then we take this as a *definition* of $f(\rho \Rightarrow \rho'|\mathbf{x})$.

Example 28 *Suppose that* $\xi = 1$. *Then* $\rho \Rightarrow \rho'$ *becomes* $\sigma \Rightarrow (\xi' \Rightarrow \sigma')$ *and (5.111) is easily shown to reduce to*

$$f(\sigma \Rightarrow (\xi' \Rightarrow \sigma')|\mathbf{x}) = (\sigma \,\bar{\wedge}\, \sigma' \,\bar{\wedge}\, \xi')(\eta(\mathbf{x})) + \frac{1}{2}(\sigma \,\bar{\wedge}\, \xi')^c(\eta(\mathbf{x})). \tag{5.114}$$

Table 5.1
Likelihood Functions for UGA Measurements

Fuzzy	$f(g\|\mathbf{x}) = g(\eta(\mathbf{x}))$
Generalized fuzzy	$f(W\|\mathbf{x}) = \int_0^1 1_W(\eta(\mathbf{x}), a)da$
Dempster-Shafer	$f(o\|\mathbf{x}) = \sum_{O \ni \eta(\mathbf{x})} o(O)$
Fuzzy DS	$f(o\|\mathbf{x}) = \sum_g o(g) \cdot g(\eta(\mathbf{x}))$
First-order fuzzy rule	$\rho(\xi \Rightarrow \sigma\|\mathbf{x}) = (\sigma \wedge_{A,A'} \xi)(\eta(\mathbf{x})) + \frac{1}{2}\xi^c(\eta(\mathbf{x}))$

Suppose in addition that A_1, A_2, A_3, A_4 *are such that* $a_1 \bar{\wedge} a_2 \bar{\wedge} a_3 \bar{\wedge} a_4 = a_1 \wedge a_2 \wedge a_3 \wedge a_4$, *where '$\wedge$' is some copula-type fuzzy conjunction. (This is the case, for example, if* $A_1 = A_2 = A_3 = A_4$, *in which case '$\wedge$' is Zadeh conjunction.) Then*

$$f(\sigma \Rightarrow (\xi' \Rightarrow \sigma')|\mathbf{x}) \tag{5.115}$$

$$= (\sigma' \wedge (\sigma \wedge \xi'))(\eta(\mathbf{x})) + \frac{1}{2}(\sigma \wedge \xi')^c(\eta(\mathbf{x})) \tag{5.116}$$

$$= f((\sigma \wedge \xi') \Rightarrow \tau|\mathbf{x}) \tag{5.117}$$

where the second equality follows from (5.80).

The generalized likelihood functions for the various types of UGA measurements are collected in Table 5.1.

5.4 BAYES UNIFICATION OF UGA FUSION

In this section, I construct, assuming that measurements are UGA measurements, a seamless and conceptually parsimonious Bayesian unification of the measurement fusion techniques most commonly employed by practitioners. My unification presumes the following:

- *The primary aim of level 1 data fusion is to arrive at estimates of object states and estimates of the error in those state-estimates.*

In other words, regardless of the nature and complexity of the data that has been collected, in the end it must be reduced to summary information: a state-estimate $\hat{\mathbf{x}}$ and an error estimate such as an error covariance matrix $\hat{P}$. This process constitutes a very lossy form of data compression.

Because my approach is Bayesian, estimates of state and uncertainty are derived from posterior distributions. Because of this, compression consists of three steps:

1. Lossy compression of data D into a likelihood function $f(D|\mathbf{x})$;
2. Subsequent lossy compression of data D into a posterior distribution $f(\mathbf{x}|D) \propto f(D|\mathbf{x}) \cdot f_0(\mathbf{x})$;
3. Lossy compression of $f(\mathbf{x}|D)$ into $\hat{\mathbf{x}}$ and $\hat{P}$. (This process is illustrated graphically in Figure 5.5.)

Suppose, then, that D_1 and D_2 are measurements and that we have been given some conjunctive, commutative, associative operator '$\otimes$' that fuses D_1 and D_2 into a composite measurement $D_1 \otimes D_2$. Then there are two ways of estimating states and uncertainties:

- Constructing a posterior distribution $f(\mathbf{x}|D_1, D_2)$ conditioned on D_1, D_2;
- Constructing a posterior distribution $f(\mathbf{x}|D_1 \otimes D_2)$ conditioned on the fused measurement $D_1 \otimes D_2$.

If

$$f(\mathbf{x}|D_1, D_2) = f(\mathbf{x}|D_1 \otimes D_2) \tag{5.118}$$

identically, it follows that fusion using $\otimes$ produces the same results as fusion using Bayes' rule alone. It is, in this sense, equivalent to fusion using Bayes' rule. In particular, '$\otimes$' produces exactly the same estimates of state and uncertainty. In the terminology of Section 4.5.1, $\otimes$ is a *Bayes combination operator*. Stated differently, $\otimes$ has been *unified* (i.e., shown to be equivalent to Bayes' rule).

The primary purpose of this section is to show that, for UGA measurements, the following commonly employed fusion methods produce the same posterior distributions as Bayes' rule:

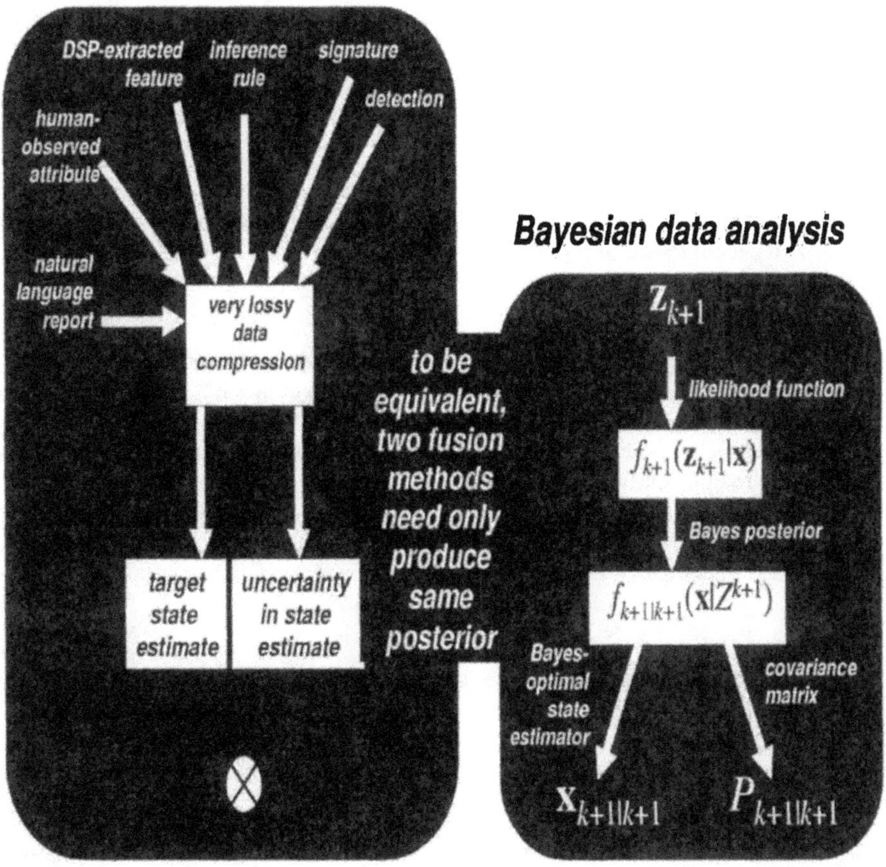

Figure 5.5 A schematic of the approach to unification of measurement fusion. *Left*: In the problems considered in this book, all information—regardless of how complicated it is—must ultimately be reduced to estimates of target state and error. *Middle*: Consequently, two fusion methods will be equivalent in practice so long as both result in the same "answer." *Right*: In a Bayesian approach, state and error estimates are derived from posterior distibutions. For two fusion methods to be equivalent, they must both produce the same posterior distribution.

- Normalized and unnormalized Dempster's combinations '$*$' and '$\cap$' (Section 5.4.1);
- The fuzzy versions of normalized and unnormalized Dempster's combination (Section 5.4.2);
- Any copula-type fuzzy conjunction '$\wedge$' (Section 5.4.3);
- The firing of a first-order fuzzy rule $\xi \Rightarrow \sigma$ by its antecedent ξ (Section 5.4.3);
- The partial firing of a fuzzy rule $\xi \Rightarrow \sigma$ by an event ξ' other than its antecedent (Section 5.4.3).

Since these results depend on *generalized* rather than strict likelihood functions, they are not strictly Bayesian. However, in Section 5.4.5, I demonstrate that generalized likelihoods are strict likelihoods, provided that the underlying measurement space $\mathfrak{Z}_0$ is finite. In this case it is possible to adhere to the Bayesian formal modeling "recipe" outlined in Section 3.5.1.

These results all relate to *fusion of measurements with measurements*. In Section 5.6.3, I will prove a unification result relating to *fusion of measurements with tracks*. Specifically, in (5.225) I show that, under certain assumptions,

- Updating data using Dempster's combination is equivalent to updating data using Bayes' rule.

It should also be pointed out here that, in Section 8.5, I will demonstrate an analogous result for modified Dempster's combination as previously introduced in (4.129).

A secondary purpose of the section is to show that:

- Various uncertainty representations (probabilistic, fuzzy, Dempster-Shafer) can be transformed into other uncertainty representations (probabilistic, fuzzy, Dempster-Shafer) without loss of any information relevant to the process of deriving target state-estimates and uncertainties in those estimates.

5.4.1 Bayes Unification of UGA Fusion Using Normalized and Unnormalized Dempster's Combinations

The Dempster-Shafer combination $o * o'$ of b.m.a.s o, o' was defined in (4.87). In Appendix G.4, I prove that

$$f(o * o'|\mathbf{x}) = \alpha^{-1} \cdot f(o|\mathbf{x}) \cdot f(o'|\mathbf{x}) \tag{5.119}$$

where $\alpha = \alpha_{\mathrm{DS}}(o, o')$. If o and o' are independent (i.e., their random set representations are independent), then from (5.21) $f(o, o'|\mathbf{x}) = f(o|\mathbf{x}) \cdot f(o'|\mathbf{x})$. So $f(o * o'|\mathbf{x}) = \alpha^{-1} \cdot f(o, o'|\mathbf{x})$, and so

$$f(\mathbf{x}|o * o') = f(\mathbf{x}|o, o'). \tag{5.120}$$

The unnormalized Dempster's combination $o \cap o'$ of g.b.m.a.s o, o' was defined in (4.95). In Appendix G.4 I prove that

$$f(o \cap o'|\mathbf{x}) = f(o|\mathbf{x}) \cdot f(o'|\mathbf{x}). \tag{5.121}$$

If o and o' are independent in the sense that their random set representations are independent, then $f(o, o'|\mathbf{x}) = f(o|\mathbf{x}) \cdot f(o'|\mathbf{x})$. So

$$f(\mathbf{x}|o, o') = f(\mathbf{x}|o \cap o'). \tag{5.122}$$

In summary, for every m, $o_1 * ... * o_m$ is a sufficient statistic for accumulated, independent b.m.a. measurements $o_1, ..., o_m$; and $o_1 \cap ... \cap o_m$ is a sufficient statistic for accumulated, independent g.b.m.a. measurements $o_1, ..., o_m$. Specifically,

- Both '$*$' and '$\cap$' are Bayes combination operators as defined in Section 3.5.3;
- Fusion of independent Dempster-Shafer measurements using Dempster's combination (normalized or unnormalized) is equivalent to fusing them using Bayes' rule alone.

Finally, let m_o be as in (5.64). In Appendix G.4, I prove that $m_{o \cap o'} = m_o * m_{o'}$. Thus from (5.64)

$$\begin{aligned} f(\mathbf{x}|o, o') &= f(\mathbf{x}|o \cap o') = f(\mathbf{x}|m_{o \cap o'}) \\ &= f(\mathbf{x}|m_o * m_{o'}) = f(\mathbf{x}|m_o, m_{o'}). \end{aligned} \tag{5.123}$$

Equation (5.123) shows that

- Normalized and unnormalized combinations are equivalent from the point of view of Bayesian measurement fusion.

One final thing should be pointed out. From (5.119) we see that

$$\begin{aligned} f(\mathbf{x}|m * m') &\propto f(m|\mathbf{x}) \cdot f(m'|\mathbf{x}) \cdot f_0(\mathbf{x}) && (5.124) \\ &= f(m|\mathbf{x}) f_0(\mathbf{x}) \cdot f(m'|\mathbf{x}) f_0(\mathbf{x}) \cdot f_0(\mathbf{x})^{-1} && (5.125) \\ &\propto f(\mathbf{x}|m) \cdot f(\mathbf{x}|m') \cdot f_0(\mathbf{x})^{-1} && (5.126) \end{aligned}$$

or in other words,

$$f(\cdot|m * m') = f(\cdot|m) *_{f_0} f(\cdot|m') \quad (5.127)$$

where '$*_{f_0}$' denotes Bayesian parallel combination as defined in (4.91).

Equation (5.127) generalizes the consistency relationship of the Voorbraak probability, see (4.116). Collapse of a b.m.a. m into the posterior distribution $f(\cdot|m)$ results, in general, in the loss of information. Equation (5.127) shows, however, that from a Bayesian point of view no *relevant* information has been lost.

By this, I mean information that is relevant to *estimating target states and estimating the error associated with those estimates.* One can collapse all b.m.a.s into their posteriors and then combine the posteriors using a special case of Bayes' rule (Bayesian parallel combination) without sacrificing any of the information gain that can be attributed to measurement fusion using Dempster's combination. This is because the result will be the same as if one first fused the b.m.a.s using Dempster's combination and then collapsed the fused b.m.a. into its posterior.

5.4.2 Bayes Unification of UGA Fusion Using Normalized and Unnormalized Fuzzy Dempster's Combinations

Equations (5.119)-(5.123) are easily generalized to the case when the focal sets of b.m.a.s and generalized b.m.a.s are fuzzy. This fact is left to the reader to prove as Exercise 18.

5.4.3 Bayes Unification of UGA Fusion Using Copula Fuzzy Conjunctions

Suppose that '$\wedge_{A,A'}$' is a copula fuzzy conjunction as in (4.37)-(4.46). Then

$$f(\mathbf{x}|g \wedge_{A,A'} g') = f(\mathbf{x}|g, g'). \quad (5.128)$$

To see this, note that from (4.42)

$$
\begin{aligned}
f(g,g'|\mathbf{x}) &\triangleq \Pr(\eta(\mathbf{x}) \in \Sigma_A(g),\ \eta(\mathbf{x}) \in \Sigma_{A'}(g')) && (5.129)\\
&= \Pr(A \leq g(\eta(\mathbf{x})),\ A' \leq g'(\eta(\mathbf{x}))) && (5.130)\\
&= (g \wedge_{A,A'} g')(\eta(\mathbf{x})) && (5.131)\\
&= \Pr(\eta(\mathbf{x}) \in \Sigma_{A''}(g \wedge_{A,A'} g')) && (5.132)\\
&= f(g \wedge_{A,A'} g'|\mathbf{x}). && (5.133)
\end{aligned}
$$

Thus '$\wedge_{A,A'}$' is a Bayes combination operator as defined in Section 3.5.3, since $g_1 \wedge_{A,A'} ... \wedge_{A,A'} g_m$ is a sufficient statistic for $g_1, ..., g_m$ for any m.

If $A = A'$ we get Zadeh conjunction, and $g_1 \wedge ... \wedge g_m$ is a sufficient statistic for fuzzy measurements with perfectly correlated interpretations. If A, A' are independent, $g \wedge_{A,A'} g' = gg'$ and $g_1 \cdots g_m$ is a sufficient statistic for fuzzy measurements with completely independent interpretations. In the first case the interpretations defined by g are in lock-step with the interpretations defined by g'. In the second case these constraints are independent of one another.

5.4.4 Bayes Unification of UGA Rule-Firing

Let (g, A), (g', A'), and (g'', A'') be fuzzy events on $\mathfrak{Z}_0$ in the sense of (4.54). In Appendix G.5, I show that

$$f(g'' \wedge (g \Rightarrow g')|\mathbf{x}) = f(g'', g \Rightarrow g'|\mathbf{x}) \tag{5.134}$$

where '$\wedge$' denotes the conjunction of the GNW conditional event algebra (Section 4.7.2). In particular, if $(g'', A'') = (g, A)$, then by (4.143) (i.e., modus ponens) (5.134) becomes

$$f(g, g \Rightarrow g'|\mathbf{x}) = f(g \wedge (g \Rightarrow g')|\mathbf{x}) = f(g \wedge g'|\mathbf{x}). \tag{5.135}$$

Consequently,

$$
\begin{aligned}
f(\mathbf{x}|g'' \wedge (g \Rightarrow g')) &= f(\mathbf{x}|g'', g \Rightarrow g') && (5.136)\\
f(\mathbf{x}|g \wedge (g \Rightarrow g')) &= f(\mathbf{x}|g, g \Rightarrow g'). && (5.137)
\end{aligned}
$$

Equation (5.136) states that partial firing of $g \Rightarrow g'$ by g'' is equivalent to Bayes' rule. The left-hand side of (5.137) is the posterior distribution resulting from using Bayes' rule to fuse g with the rule $g \Rightarrow g'$. The right-hand side is the result

$g \wedge_{A,A'} g'$ expected from firing $g \Rightarrow g'$. Consequently, total firing of the rule is equivalent to Bayes' rule.

For these concepts to make sense, I must provide some background. The "partial firing" of a crisp rule $O \Rightarrow O'$ by a crisp event O'' is defined by the GNW conjunction operation $O'' \wedge (O \Rightarrow O')$ defined in (4.145)-(4.147). The random set representation of $O \Rightarrow O'$ was given in (4.159):

$$\Sigma_\Phi(O \Rightarrow O') \triangleq (O' \cap O) \cup (O^c \cap \Phi). \tag{5.138}$$

So, the partial firing $g'' \wedge (g \Rightarrow g')$ of fuzzy rule $g \Rightarrow g'$ by fuzzy event g'' can be defined via a simple substitution $O \to \Sigma_A(g)$, $O' \to \Sigma_{A'}(g')$, and $O'' \to \Sigma_{A''}(g'')$:

$$\Sigma_{\Phi,A'',A,A'}(g'' \wedge (g \Rightarrow g')) \triangleq \Sigma_\Phi(\Sigma_{A''}(g'') \wedge (\Sigma_A(g) \Rightarrow \Sigma_{A'}(g'))). \tag{5.139}$$

The right-hand side of this can be simplified as follows

$$= \Sigma_\Phi(\Sigma_{A''}(g'')) \cap \Sigma_\Phi((\Sigma_A(g) \Rightarrow \Sigma_{A'}(g'))) \tag{5.140}$$
$$= \Sigma_{A''}(g'') \cap \Sigma_\Phi((\Sigma_A(g) \Rightarrow \Sigma_{A'}(g'))) \tag{5.141}$$
$$= \Sigma_{A''}(g'') \cap \Sigma_{\Phi,A,A'}(g \Rightarrow g'). \tag{5.142}$$

Equation (5.141) results from (4.160), and (5.142) results from (4.159). In summary, if $\rho = (f \Rightarrow f')$ is a fuzzy rule, then

$$\Sigma_{\Phi,A'',A,A'}(f'' \wedge \rho) = \Sigma_{A''}(f'') \cap \Sigma_{\Phi,A,A'}(\rho). \tag{5.143}$$

5.4.5 If $\mathfrak{Z}_0$ Is Finite, Then Generalized Likelihoods Are Strict Likelihoods

Let $\mathfrak{Z}_0$ have M elements, and let $M' = M - 1$. Then the space of b.m.a.s can be identified with the simplex S consisting of all $(a_1, ..., a_{M'})$ such that $a_1, ..., a_{M'} \geq 0$ and $a_1 + ... + a_{M'} = 1$. Consequently, the integral of $f(o|\mathbf{x})$ with respect to o is just

$$\int f(o|\mathbf{x})do \triangleq \int_S f(a_1, ..., a_{M'}|\mathbf{x})da_1 \cdots da_{M'}. \tag{5.144}$$

In Appendix G.6, I show that this integral is finite, nonzero, and has no dependence on $\mathbf{x}$. Therefore we need not appeal to the concept of a generalized likelihood function, since we can define a conventional normalized likelihood $\hat{f}(o|\mathbf{x}) =$

Table 5.2

Bayes Unification of Measurement Fusion

Copula fuzzy conjunctions	$f(\mathbf{x}\|g_1 \wedge ... \wedge g_m) = f(\mathbf{x}\|g_1, ..., g_m)$
Generalized fuzzy conjunction	$f(\mathbf{x}\|W_1 \cap ... \cap W_m) = f(\mathbf{x}\|W_1, ..., W_m)$
Fuzzy Dempster's combination	$f(\mathbf{x}\|o_1 * ... * o_m) = f(\mathbf{x}\|o_1, ..., o_m)$
Unnormalized combination	$f(\mathbf{x}\|o_1 \cap ... \cap o_m) = f(\mathbf{x}\|o_1, ..., o_m)$
Firing of fuzzy rules	$f(\mathbf{x}\|\xi, \xi \Rightarrow \sigma) = f(\mathbf{x}\|\xi \wedge \sigma)$
Partial firing of fuzzy rules	$f(\mathbf{x}\|\xi', \xi \Rightarrow \sigma) = f(\mathbf{x}\|\xi' \wedge (\xi \Rightarrow \sigma))$

$K^{-1} \cdot f(o|\mathbf{x})$ where $K = \int f(o|\mathbf{x})do$. We can strictly adhere to the Bayesian "recipe" of Sections 3.5.1 and 3.5.2.

This is not necessarily true when $\mathfrak{Z}_0$ is not finite. However, in this case we can choose some bounded region of $\mathfrak{Z}_0$ and discretize it.

The unification results of this section are summarized in Table 5.2.

5.4.6 Bayes-Invariant Conversions Between UGA Measurements

Much effort has been expended in the expert-systems literature on devising "conversions" of one uncertainty representation scheme to another—fuzzy to probabilistic, Dempster-Shafer to probabilistic, to fuzzy, and so on. Such efforts have been hindered by the fact that uncertainty representation formalisms vary considerably in the degree of complexity of information that they encode.

Most obviously, any conversion of a fuzzy b.m.a. $o(g)$ to a single fuzzy membership function $\mu_o(\mathbf{z})$ will result in a huge loss of information. As another example, $2^M - 1$ numbers are required to specify a (crisp) basic mass assignment $o(O)$ on a finite measurement space $\mathfrak{Z}_0$ with M elements, whereas only $M - 1$ numbers are required to specify a probability distribution $\varphi(z)$ on $\mathfrak{Z}_0$.

Consequently, any conversion of b.m.a.s o to probability distributions φ_o will result in a huge loss of information.

A second issue is the fact that conversion from one uncertainty representation formalism to another should be consistent with the data fusion methodologies intrinsic to these formalisms. For example, fusion of fuzzy b.m.a.s o and o' is commonly accomplished using Dempster's combination $o * o'$. Fusion of fuzzy membership functions μ_o and $\mu_{o'}$, on the other hand, is usually accomplished using fuzzy conjunction $\mu_o \wedge \mu_{o'}$. For information to be consistently converted, one should have $\mu_{o*o'} = \mu_o \wedge \mu_{o'}$ in some sense.

This last equation exposes a more subtle issue. Conversion does not make sense if different uncertainty representation formalisms are semantically incompatible. For example, we cannot fuse two fuzzy b.m.a.s o and o' into a single b.m.a. $o * o'$ using Dempster's combination '$*$' unless we first assume that o and o' are "independent" in some sense. This, in turn, means that the fuzzy membership functions μ_o and $\mu_{o'}$ must be independent in some sense. However, what might this sense be, and how does one correctly combine fuzzy membership functions given that we know what it is?

The path out of such quandaries is, as before, to assume that *the primary goal of level 1 information fusion (multisource integration) is to determine state-estimates and estimates of the uncertainty in those estimates.* No matter how complex or how disparate various kinds of information might be, ultimately they must be reduced to summary information—state estimates and, most typically, covariances. This reduction itself constitutes a very lossy form of data compression. Consequently, the conversion problem can be restated as follows (see Figure 5.5):

- What conversions between uncertainty representations lose no *estimation-relevant* information?

I addressed this issue in Section 3.5.4 by introducing the concept of *Bayes-invariant* conversions. In a Bayesian formalism, the conversion problem can further be restated as:

- What conversions between uncertainty representations are *Bayes-invariant* (i.e., leave posterior distributions unchanged)?

For example, the posterior distribution conditioned on the fuzzy b.m.a.s $o_1, ..., o_m$ is $f(\mathbf{x}|o_1, ..., o_m)$, and the posterior distribution conditioned on the conversions $\mu_{o_1}, ..., \mu_{o_m}$ of $o_1, .., o_m$ to fuzzy membership functions is $f(\mathbf{x}|\mu_{o_1}, ..., \mu_{o_m})$.

The conversion $o \mapsto \mu_o$ loses no estimation-relevant information if

$$f(\mathbf{x}|o_1, ..., o_m) = f(\mathbf{x}|\mu_{o_1}, ..., \mu_{o_m}) \tag{5.145}$$

identically for all m.

In the remainder of this section, I describe Bayes-invariant conversions between various uncertainty representation systems. I present only the basic conversions. Many other conversions can be derived by composing the basic ones. I will be making extensive use of (3.73) and (3.74).[5]

5.4.6.1 Generalized Fuzzy Dempster-Shafer to Fuzzy Dempster-Shafer

In (5.62) I defined a conversion $o \mapsto m_o$ of generalized (crisp) b.m.a.s o to ordinary (crisp) b.m.a.s m_o. I generalize this conversion as follows. Let $o(g)$ be a generalized fuzzy b.m.a. (i.e., $o(0)$ is not necessarily zero). Define

$$m_o(g) = |o|^{-1} \cdot o(g) \tag{5.146}$$

for all $g \neq 0$ and $m_o(0) = 0$ otherwise, where $|o| = \sum_{g \neq 0} o(g)$. It is left to the reader as Exercise 19 to show that

$$f(o|\mathbf{x}) = |o| \cdot f(m_o|\mathbf{x}) \tag{5.147}$$

$$m_{o \cap o'} = m_o * m_{o'}. \tag{5.148}$$

The fact that $o \mapsto m_o$ is Bayes-invariant then follows from (3.73).

5.4.6.2 Fuzzy to Fuzzy Dempster-Shafer

Given a fuzzy membership function $g(\mathbf{z})$ on $\mathfrak{Z}_0$, define the fuzzy b.m.a. $o(g')$ on fuzzy membership functions g' of $\mathfrak{Z}_0$ by

$$o_g(g') \triangleq \begin{cases} 1 & \text{if} \quad g' = g \\ 0 & \text{if} \quad \text{otherwise} \end{cases}. \tag{5.149}$$

It is left to the reader as Exercise 20 to show that

$$f(o_g|\mathbf{x}) = f(g|\mathbf{x}) \tag{5.150}$$

$$o_{g \cdot g'} = o_g * o_{g'} \tag{5.151}$$

where the second equation is true if $g \cdot g' \neq 0$. From these equations also show that the conversion $g \mapsto o_g$ is Bayes-invariant.

5 Much of this material originally appeared in [124].

5.4.6.3 Fuzzy Dempster-Shafer to Fuzzy

Given a fuzzy b.m.a. $o(g)$ on fuzzy membership functions g of $\mathfrak{Z}_0$, define the fuzzy membership function $\mu_o(\mathbf{z})$ on $\mathfrak{Z}_0$ by

$$\mu_o(\mathbf{z}) \triangleq \sum_g o(g) \cdot g(\mathbf{z}). \tag{5.152}$$

It is left to the reader as Exercise 21 to show that

$$f(o|\mathbf{x}) = f(\mu_o|\mathbf{x}) \tag{5.153}$$

$$\mu_{o*o'} = \alpha^{-1} \cdot \mu_o \cdot \mu_{o'} \tag{5.154}$$

where $\alpha = \alpha_{\text{FDS}}(o, o')$ is the fuzzy Dempster-Shafer agreement of (4.130) and where $(\mu_o \cdot \mu_{o'})(\mathbf{z}) \triangleq \mu_o(\mathbf{z}) \cdot \mu_{o'}(\mathbf{z})$. From these equations, show that the conversion $o \mapsto \mu_o$ is Bayes-invariant.

5.4.6.4 Fuzzy to Generalized Fuzzy

Let $g(\mathbf{z})$ be a fuzzy membership function on $\mathfrak{Z}_0$. In (4.70) I defined the generalized fuzzy subset W_g of $\mathfrak{Z}_0$ by

$$W_g \triangleq \{(\mathbf{z}, a) | \; a \leq g(\mathbf{z})\}. \tag{5.155}$$

From (5.46) and (4.71) we know that, respectively,

$$f(W_g|\mathbf{x}) = f(g|\mathbf{x}) \tag{5.156}$$

$$W_{g \wedge g'} = W_g \cap W_{g'} \tag{5.157}$$

where '$\wedge$' is Zadeh conjunction. From these equations it follows that the conversion $g \mapsto W_g$ is Bayes-invariant.

5.4.6.5 Fuzzy Dempster-Shafer to Probabilistic

Let $o(g)$ be a b.m.a. on fuzzy membership functions g of $\mathfrak{Z}_0$ such that $\int g(\mathbf{z})d\mathbf{z} < \infty$ whenever $o(g) > 0$. Define the probability density function $\varphi_o(\mathbf{z})$ on $\mathfrak{Z}_0$ by

$$\varphi_o(\mathbf{z}) \triangleq \frac{\sum_g o(g) \cdot g(\mathbf{z})}{\sum_g o(g) \int g(\mathbf{w})d\mathbf{w}}. \tag{5.158}$$

This equation generalizes the Voorbraak transformation of (4.115) to fuzzy b.m.a's.

Further, let the parallel combination of two probability density functions φ and φ' (with respect to an improper uniform prior) be defined by

$$(\varphi * \varphi')(\mathbf{z}) \triangleq \frac{\varphi(\mathbf{z}) \cdot \varphi'(\mathbf{z})}{\int \varphi(\mathbf{w}) \cdot \varphi'(\mathbf{w}) d\mathbf{w}}. \tag{5.159}$$

Also define

$$f(\varphi|\mathbf{x}) \triangleq \frac{\varphi(\eta(\mathbf{x}))}{\sup_{\mathbf{z}} \varphi(\mathbf{z})}. \tag{5.160}$$

It is left to the reader as Exercise 22 to show that

$$f(\varphi_o|\mathbf{x}) = K \cdot f(o|\mathbf{x}) \tag{5.161}$$

$$\varphi_{o*o'}(\mathbf{z}) = \frac{\varphi_o(\mathbf{z}) \cdot \varphi_{o'}(\mathbf{z})}{\int \varphi_o(\mathbf{w}) \cdot \varphi_{o'}(\mathbf{w}) d\mathbf{w}} = (\varphi_o * \varphi_{o'})(\mathbf{z}) \tag{5.162}$$

for some K independent of $\mathbf{x}$; and from these equations, to show that the conversion $o \mapsto \varphi_o$ is Bayes-invariant.

5.4.6.6 Probabilistic to Fuzzy

Given the probability density function $\varphi(\mathbf{z})$ on $\mathfrak{Z}_0$, define the fuzzy membership function $\mu_\varphi(\mathbf{z})$ by

$$\mu_\varphi(\mathbf{z}) = \frac{\varphi(\mathbf{z})}{\sup_{\mathbf{w}} \varphi(\mathbf{w})}. \tag{5.163}$$

It is left to the reader as Exercise 23 to show that

$$f(\mu_\varphi|\mathbf{x}) = f(\varphi|\mathbf{x}) \tag{5.164}$$

$$\mu_{\varphi * \varphi'} = K \cdot \mu_\varphi \cdot \mu_{\varphi'} \tag{5.165}$$

for some K independent of $\mathbf{x}$ and, from this, to show that the conversion $\varphi \mapsto \mu_\varphi$ is Bayes-invariant.

5.4.6.7 Fuzzy to Fuzzy Rule

Given a fuzzy membership function $g(\mathbf{z})$ on $\mathfrak{Z}_0$, define the fuzzy rule ρ_g by

$$\rho_g \triangleq (\mathbf{1} \Rightarrow g) \tag{5.166}$$

Table 5.3
Bayes-Invariant Uncertainty Model Conversions

Fuzzy $\mapsto$ fuzzy DS	$g \mapsto o_g(g') = \begin{cases} 1 & \text{if} \quad g' = g \\ 0 & \text{if} \quad \text{otherwise} \end{cases}$.
Fuzzy DS $\mapsto$ fuzzy	$o \mapsto \mu_o(\mathbf{z}) = \sum_g o(g) \cdot g(\mathbf{z})$
Fuzzy $\mapsto$ generalized fuzzy	$g \mapsto W_g = \{(\mathbf{z}, a) \mid a \leq g(\mathbf{z})\}$
Fuzzy DS $\mapsto$ probabilistic	$o \mapsto \varphi_o(\mathbf{z}) = \frac{\sum_g o(g) \cdot g(\mathbf{z})}{\sum_g o(g) \int g(\mathbf{w}) d\mathbf{w}}$
Probabilistic $\mapsto$ fuzzy	$\varphi \mapsto \mu_\varphi(\mathbf{z}) = \frac{\varphi(\mathbf{z})}{\sup_{\mathbf{w}} \varphi(\mathbf{w})}$
Fuzzy $\mapsto$ fuzzy rule	$g \mapsto \rho_g = (\mathbf{1} \Rightarrow g)$

where $\mathbf{1}(\mathbf{z})$ is the fuzzy membership function that is identically equal to one. Then

$$f(\rho_g|\mathbf{x}) = f(g|\mathbf{x}). \tag{5.167}$$

To see this, note that from (5.80) we have

$$f(\rho_g|\mathbf{x}) = f(\mathbf{1} \Rightarrow g|\mathbf{x}) = (g \wedge_{A,A'} \mathbf{1})(\eta(\mathbf{x})) + \frac{1}{2}\mathbf{1}^c(\eta(\mathbf{x})) \tag{5.168}$$

$$= g(\eta(\mathbf{x})) + 0 = f(g|\mathbf{x}). \tag{5.169}$$

The more important measurement conversions are listed in Table 5.3.

5.5 MODELING OTHER KINDS OF UNCERTAINTY

In this section, I consider two additional types of uncertainty caused by ignorance:

- Poorly characterized statistical dependencies between measurements;
- Target types that are unmodeled (because, for example, they were previously unknown).

I address these two types in turn.

5.5.1 Modeling Unknown Statistical Dependencies

If simultaneously collected generalized measurements Θ, Θ' can be assumed to be independent, then they are easily fused using Bayes' rule (Section 2.4.10). This is because we know that their joint generalized likelihood function is just the product of the generalized likelihood functions of the individual measurements:

$$f(\Theta, \Theta'|\mathbf{x}) = f(\Theta|\mathbf{x}) \cdot f(\Theta'|\mathbf{x}). \tag{5.170}$$

In general, however, independence cannot be assumed. For example, different features extracted from the same image or signature will typically be statistically correlated. Or, measurements from different sensors can be statistically dependent because the sensors share the same underlying noise phenomenology.

In some cases it may be possible to construct the joint likelihood function through statistical analysis. In many other cases, however, this is not possible. The most extreme example is that of fusing features extracted from the same image by an operator. These will be correlated, but the correlations are unknowable.

The fact that generalized likelihood functions always take values between zero and one provides us with one approach for dealing with uncharacterizable statistical dependency. By definition, see (5.19), the joint generalized likelihood is

$$f(\Theta, \Theta'|\mathbf{x}) = \Pr(\eta(\mathbf{x}) \in \Theta, \eta(\mathbf{x}) \in \Theta') = \Pr(\eta(\mathbf{x}) \in \Theta \cap \Theta'). \tag{5.171}$$

In (4.42) we noted that, in the particular case when $\Theta = \Sigma_A(g)$, $\Theta' = \Sigma_{A'}(g')$, one gets $\Pr(\mathbf{z} \in \Theta \cap \Theta') = g(\mathbf{z}) \wedge_{A,A'} g'(\mathbf{x})$. From this it follows that

$$f(\Theta, \Theta'|\mathbf{x}) = \Pr(\mathbf{z} \in \Theta \cap \Theta') = f(\Theta|\mathbf{x}) \wedge_{A,A'} f(\Theta'|\mathbf{x}) \tag{5.172}$$

where '$\wedge_{A,A'}$' is a copula-type fuzzy conjunction.

Copula conjunctions can be used to heuristically model statistical dependencies even when Θ and Θ' do not have the form $\Theta = \Sigma_A(g)$, $\Theta' = \Sigma_{A'}(g')$. That is, suppose that $\Theta_1, ..., \Theta_m$ are generalized measurements that we believe are not independent. Then one can define the heuristic joint generalized likelihood function by

$$f(\Theta_1, ..., \Theta_m|\mathbf{x}) \triangleq f(\Theta_1|\mathbf{x}) \wedge ... \wedge f(\Theta_m|\mathbf{x}). \tag{5.173}$$

For example, the Hamacher conjunction of (4.55) is

$$x \wedge y \triangleq \frac{xy}{x + y - xy} \tag{5.174}$$

for all $0 \leq x, y \leq 1$. It models a state of statistical dependence that is close to independence.

Remark 5 *This technique can also be used with the likelihoods for AGA and AGU measurements to be introduced in Chapters 6 and 7.*

5.5.2 Modeling Unknown Target Types

Model-based approaches, such as the ones proposed in this book, share major advantages over data-based approaches. They are more robust to overtraining on data, and they permit exploitation of a priori knowledge about targets.

In the realm of target identification, however, model-based methodologies share a well known Achilles heel: *unmodeled target types*. A given target type may be unmodeled because it has never been encountered before. Or, it may be unmodeled because real-time operation would be impossible if all possible target types in the theater of interest were to be modeled. In this case one has no choice but to model only the most crucial target types.

Regardless of the reason, a target identification algorithm that relies on an incomplete model-base will often perform poorly when confronted with a target of a type that is not in the model-base. Specifically, it will tend to confuse the unknown type with the modeled type or types that it most closely resembles.

Practical target identification algorithms must therefore have the capability of assigning unmodeled types to a generic *none of the above* (NOTA) target type. The purpose of this section is to propose a methodology for doing so, given that collected measurements are UGA measurements.[6]

The basic idea is as follows. Let $v_1, ..., v_n$ be the modeled target types. Suppose that a generalized measurement Θ has been collected from a target of an unmodeled type. The UGA measurement model $\eta(v_i) \in \Theta$ states that the observation matches the type v_i, and the generalized likelihood $f(\Theta|v_i) =$

6 I will extend the approach to AGA measurements in Section 6.6. In reduced-complexity simulations, this AGA approach has been shown to exhibit good behavior—see [215] and Section 6.5.

$\Pr(\eta(v_i) \in \Theta)$ is the probability that it does so. Likewise,

$$\Pr(\eta(v_1) \notin \Theta, ..., \eta(v_n) \notin \Theta) = \Pr(\{v_1, ..., v_n\} \subseteq \Theta^c) \quad (5.175)$$

is the probability that the observation matches none of the types. We append a new "none of the above" target type v_0 to the state space and define its generalized likelihood to be

$$f(\Theta|v_0) \triangleq \Pr(\eta(v_1) \notin \Theta, ..., \eta(v_n) \notin \Theta). \quad (5.176)$$

The following are specific formulas for the different UGA measurement types.

Example 29 *Suppose that* $\Theta_g = \Sigma_A(g)$ *is the random set model of a fuzzy measurement* g*. Then* $f(g|v_0) \triangleq f(\Theta_g|v_0)$ *and*

$$f(g|v_0) = \min\{g^c(\eta(v_1)), ..., g^c(\eta(v_n))\}. \quad (5.177)$$

To see this, note that

$$\begin{aligned} f(g|v_0) &= \Pr(A > g(\eta(v_1)), ..., A > g(\eta(v_n))) & (5.178) \\ &= \Pr(A > \max\{g(\eta(v_1)), ..., g(\eta(v_n))\}) & (5.179) \\ &= \Pr(1 - A < 1 - \max\{g(\eta(v_1)), ..., g(\eta(v_n))\}). & (5.180) \end{aligned}$$

Since $1 - A$ *is uniformly distributed on* $[0, 1]$*, we get*

$$\begin{aligned} f(g|v_0) &= 1 - \max\{g(\eta(v_1)), ..., g(\eta(v_n))\} & (5.181) \\ &= \min\{1 - g(\eta(v_1)), ..., 1 - g(\eta(v_n))\} & (5.182) \\ &= \min\{g^c(\eta(v_1), ..., g^c(\eta(v_n))\}. & (5.183) \end{aligned}$$

Example 30 *Suppose that* $\Theta_W = \Sigma_A(W)$ *is the random set model of the generalized fuzzy measurement* W*. Then* $f(W|v_0) \triangleq f(\Theta_W|v_0)$ *and*

$$f(W|v_0) = \int_0^1 \mathbf{1}_{W^c}(\eta(v_1), a) \cdots \mathbf{1}_{W^c}(\eta(v_n), a) da \quad (5.184)$$

since

$$
\begin{aligned}
f(W|v_0) &= \Pr(\eta(v_1) \in \Sigma_A(W)^c, ..., \eta(v_n) \in \Sigma_A(W)^c) && (5.185) \\
&= \Pr(\eta(v_1) \in \Sigma_A(W^c), ..., \eta(v_n) \in \Sigma_A(W^c)) && (5.186) \\
&= \Pr((\eta(v_1), A) \in W^c, ..., (\eta(v_n), A) \in W^c) && (5.187) \\
&= \int_0^1 \mathbf{1}_{W^c}(\eta(v_1), a) \cdots \mathbf{1}_{W^c}(\eta(v_n), a) da. && (5.188)
\end{aligned}
$$

Example 31 *Suppose that* Θ_o *is a random set model of a Dempster-Shafer measurement* $o(O)$*. Then* $f(o|v_0) \triangleq \Pr(\Theta_o|v_0)$ *and*

$$
f(o|v_0) = \sum_O o(O) \cdot \mathbf{1}_{O^c}(\eta(v_1)) \cdots \mathbf{1}_{O^c}(\eta(v_n)) \qquad (5.189)
$$

since

$$
\begin{aligned}
f(o|v_0) &= \sum_O \Pr(\Theta = O, \{\eta(v_1), ..., \eta(v_n)\} \subseteq O^c) && (5.190) \\
&= \sum_O o(O) \cdot (1 - \mathbf{1}_O(\eta(v_1))) \cdots (1 - \mathbf{1}_O(\eta(v_n))) && (5.191) \\
&= \sum_O o(O) \cdot \mathbf{1}_{O^c}(\eta(v_1)) \cdots \mathbf{1}_{O^c}(\eta(v_n)). && (5.192)
\end{aligned}
$$

Example 32 *Suppose that* Θ_o *is a random set model of a fuzzy b.m.a.* $o(g)$*. Then it is shown in Appendix G.7 that*

$$
f(o|v_0) = \sum_g o(g) \cdot \min\{g^c(\eta(v_1)), ..., g^c(\eta(v_n))\}. \qquad (5.193)
$$

Example 33 *In Example 29, assume that* $\mathfrak{Z}_0 = \mathbb{R}$, $\mathfrak{X}_0 = \{v_1, v_2\}$, $\eta(v_1) = -a$ *and* $\eta(v_2) = a$*. Also assume that we have collected the fuzzy measurement*

$$
g_{z'}(z) = \exp\left(-\frac{(z - z')^2}{2\sigma^2}\right). \qquad (5.194)
$$

Then from (5.29) the likelihoods of the states v_1 *and* v_2 *are*

$$
\begin{aligned}
f(g_{z'}|v_1) &= g_{z'}(\eta(v_1)) = \exp\left(-\frac{(a + z')^2}{2\sigma^2}\right) && (5.195) \\
f(g_{z'}|v_2) &= g_{z'}(\eta(v_2)) = \exp\left(-\frac{(a - z')^2}{2\sigma^2}\right). && (5.196)
\end{aligned}
$$

From (5.177) the likelihood of the NOTA state v_0 *is*

$$f(g_{z'}|v_0) = \min\left\{1 - \exp\left(-\frac{(a+z')^2}{2\sigma^2}\right), 1 - \exp\left(-\frac{(a-z')^2}{2\sigma^2}\right)\right\}. \quad (5.197)$$

If we assume a uniform prior on the augmented state space $\mathfrak{X}_0^+ = \{v_0, v_1, v_2\}$*, then the posterior distribution on* $\mathfrak{X}_0^+$ *conditioned on* $g_{z'}$ *is given by*

$$f(v_1|g_{z'}) = \frac{f(g_{z'}|v_1)}{f(g_{z'})} \quad (5.198)$$

$$f(v_2|g_{z'}) = \frac{f(g_{z'}|v_2)}{f(g_{z'})} \quad (5.199)$$

$$f(v_0|g_{z'}) = \frac{f(g_{z'}|v_0)}{f(g_{z'})} \quad (5.200)$$

where $f(g_{z'}) = f(g_{z'}|v_1) + f(g_{z'}|v_2) + f(g_{z'}|v_0)$ *is the Bayes normalization factor. Let* $\sigma = 1.0$ *and* $a = 1.0$*. The posterior values are plotted as a function of* z' *in Figure 5.6. The NOTA value* v_0 *will be chosen as the maximum a posteriori (MAP) estimate of target type only for sufficiently large positive or negative values of* z'*. Now choose* $\sigma = 1.0$ *and* $a = 2.0$*, so that there is greater separation between the models of* v_1 *and* v_2*. The posterior values are plotted as a function of* z' *in Figure 5.7. In this case* v_0 *will be chosen as the MAP estimate for sufficiently small positive and negative values of* z' *as well as for sufficiently large ones.*

5.6 THE KALMAN EVIDENTIAL FILTER (KEF)

Thus far I have proposed a systematic methodology for modeling UGA measurements and constructing generalized likelihood functions from them. In general, however, such measurements and likelihoods are *inherently very nonlinear and non-Gaussian*. Generally speaking, the single-target Bayes filter is required to process them, as illustrated in Figure 3.4. In turn, some computational approximation technique, such as the SMC approach of Section 2.5.3, must be used to implement it.

Other less computationally demanding techniques are desirable. In this section, I introduce a generalization of the Kalman filter that can accept not only conventional linear-Gaussian sensor measurements $\mathbf{z}_{k+1}$, R_{k+1} but also fuzzy DS measurements o_{k+1}.[7] The KEF filter is essentially a generalization of the Gaussian-mixture (GM) filter described in Section 2.5.2.

7 The material in this section first appeared in [125, 127].

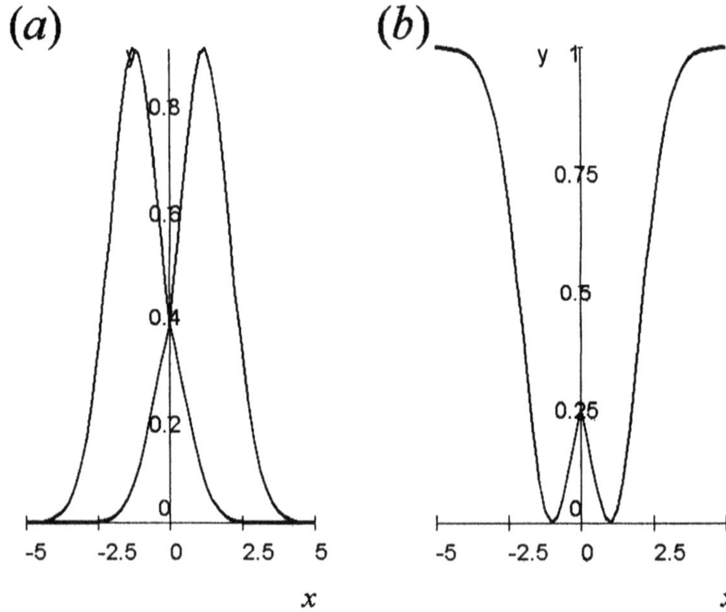

Figure 5.6 For $\sigma = 1$ and $a = 1$ (a) shows plots of the posterior probabilities for target types v_1 (left) and v_2 (right) and (b) shows the plot of the posterior probability for the none-of-the-above (NOTA) type v_0.

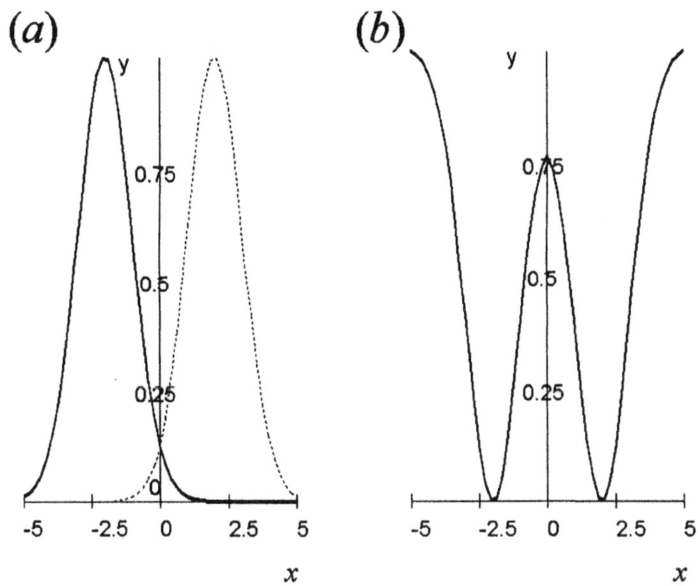

Figure 5.7 For $\sigma = 1$ and $a = 2$ (a) shows plots of the posterior probabilities for target types v_1 (left) and v_2 (right) and (b) shows the plot of the posterior probability for the NOTA type v_0.

We mimic the reasoning of Section 2.3, in which it was shown that the Kalman filter is a special case of the Bayes filter. Recall that I began by assuming a sensor that admits linear-Gaussian statistics: $f_{k+1}(\mathbf{z}|\mathbf{x}) = N_{R_{k+1}}(\mathbf{z} - H_{k+1}\mathbf{x})$. Given this, I asked: What simplified form must the Markov density $f_{k+1|k}(\mathbf{x}|\mathbf{x}')$ and the posteriors $f_{k|k}(\mathbf{x}|Z^k)$ and $f_{k+1|k}(\mathbf{x}|Z^k)$ have to permit the Bayes filter equations—see (2.81) and (2.82)—to be propagated in closed form? The answer was to assume the following, where $f(\mathbf{x}|\mathbf{c}, C) \triangleq N_C(\mathbf{x} - \mathbf{c})$:

1. Interim target motion is linear-Gaussian, $f_{k+1|k}(\mathbf{x}|\mathbf{x}') = N_{Q_k}(\mathbf{x} - F_k\mathbf{x}')$;
2. The posteriors admit linear-Gaussian sufficient statistics:

$$f_{k|k}(\mathbf{x}|Z^k) = f(\mathbf{x}|\mathbf{x}_{k|k}, P_{k|k}) \tag{5.201}$$

$$f_{k+1|k}(\mathbf{x}|Z^k) = f(\mathbf{x}|\mathbf{x}_{k+1|k}, P_{k+1|k}). \tag{5.202}$$

Then the Bayes filter equations propagate a time sequence

$$(\mathbf{x}_{0|0}, P_{0|0}) \to (\mathbf{x}_{1|0}, P_{1|0}) \to (\mathbf{x}_{1|1}, P_{1|1}) \to ... \tag{5.203}$$

$$\to (\mathbf{x}_{k|k}, P_{k|k}) \to (\mathbf{x}_{k+1|k}, P_{k+1|k}) \to (\mathbf{x}_{k+1|k+1}, P_{k+1|k+1}) \to ... \tag{5.204}$$

of sufficient statistics, as specified by the Kalman filter predictor and corrector equations.

Emulating this line of reasoning, I begin by assuming an information source that collects *fuzzy Dempster-Shafer UGA measurements* (Section 5.3.5) and that is therefore, by (5.73), described by the likelihood function

$$f_{k+1}(o|\mathbf{x}) = \sum_g o(g) \cdot g(\eta_{k+1}(\mathbf{x})). \tag{5.205}$$

Further assume that $\eta_{k+1}(\mathbf{x}) = H_{k+1}\mathbf{x}$ is linear and that all but one of the focal sets of o have the Gaussian form

$$g(\mathbf{z}) = \hat{N}_D(\mathbf{z} - \mathbf{z}_0) \tag{5.206}$$

where the normalized Gaussian distribution is

$$\hat{N}_D(\mathbf{z} - \mathbf{z}_0) \triangleq \exp\left(-\frac{1}{2}(\mathbf{z} - \mathbf{z}_0)^T D^{-1}(\mathbf{z} - \mathbf{z}_0)\right). \tag{5.207}$$

The single non-Gaussian focal set is assumed to be unity, $g(\mathbf{z}) \equiv 1$. This last assumption permits us to incorporate the null hypothesis—the possibility of complete ignorance—into fuzzy DS measurements. Note that $g(\mathbf{z}) \equiv 1$ can be treated as a limiting case of $g(\mathbf{z}) = \hat{N}_D(\mathbf{z} - \mathbf{z}_0)$ for which the matrix norm $\|D\|$ approaches infinity (i.e., the uncertainty in $\hat{N}_D(\mathbf{z} - \mathbf{z}_0)$ is arbitrarily large).

Given this I ask:

- *What simplified form must the Markov density and the posteriors have to permit the Bayes filter equations to be propagated in closed form?*

The answer is to assume that:

1. Interim target motion is linear-Gaussian, $f_{k+1|k}(\mathbf{x}|\mathbf{x}') = N_{Q_k}(\mathbf{x} - F_k\mathbf{x}')$;
2. The posteriors admit *fuzzy DS sufficient statistics*: $f_{k|k}(\mathbf{x}|Z^k) = f(\mathbf{x}|\mu_{k|k})$ and $f_{k+1|k}(\mathbf{x}|Z^k) = f(\mathbf{x}|\mu_{k+1|k})$ for all $k = 0, 1, ...$

Here $\mu_{k|k}$ and $\mu_{k+1|k}$ are "fuzzy DS states." These, along with the density function $f(\mathbf{x}|\mu)$ induced by a fuzzy DS state μ, will be defined in Section 5.6.1.

Given a sequence Z^k of fuzzy Dempster-Shafer UGA measurements, I will show that there is a time sequence

$$\mu_{0|0} \rightarrow \mu_{1|0} \rightarrow \mu_{1|1} \rightarrow \mu_{2|1} \rightarrow \mu_{2|2} \rightarrow \cdots \quad (5.208)$$

$$\rightarrow \mu_{k|k} \rightarrow \mu_{k+1|k} \rightarrow \mu_{k+1|k+1} \rightarrow \cdots \quad (5.209)$$

of fuzzy DS states such that, for all $k = 0, 1, ...,$ $f_{k|k}(\mathbf{x}|Z^k) = f(\mathbf{x}|\mu_{k|k})$ and $f_{k+1|k}(\mathbf{x}|Z^k) = f(\mathbf{x}|\mu_{k+1|k})$. Moreover I show that, for all $k = 0, 1, ...$

$$\mu_{k+1|k} = F_k\mu_{k|k} \quad (5.210)$$

$$\mu_{k+1|k+1} = H_{k+1}^{-1}o_{k+1} * \mu_{k+1|k} \quad (5.211)$$

where '$*$' is Dempster's combination, and where the notations $F\mu$ and $H^{-1}o$ will be defined momentarily in (5.222) and (5.215), respectively. In particular, (5.211) shows that:

- *Data update using Dempster's combination is equivalent to data update using Bayes' rule, in the same sense that data update using the Kalman corrector is equivalent to data update using Bayes' rule.*

The section is organized as follows. After beginning with some definitions in Section 5.6.1, I describe the KEF predictor step in Section 5.6.2 and the corrector steps for fuzzy DS measurements and for conventional measurements in Sections 5.6.3 and 5.6.4, respectively. Section 5.6.5 describes state estimation, and Section 5.6.6 contrasts the KEF with the Gaussian sum filter and the Kalman filter.

5.6.1 Definitions

5.6.1.1 Fuzzy DS States

A *fuzzy DS state* is a certain kind of fuzzy b.m.a. on state space $\mathfrak{X}_0$. Specifically, it is a nonnegative function $\mu(f)$ defined on fuzzy membership functions $f(\mathbf{x})$ on $\mathfrak{X}_0$ such that the following are true:

1. $\mu(f) = 0$ for all but a finite number of f (the focal sets);
2. If $\mu(f) > 0$, then for some $\mathbf{x}_0, C,$ $f(\mathbf{x})$ has the form

$$f(\mathbf{x}) = \hat{N}_C(\mathbf{x} - \mathbf{x}_0) \triangleq \exp\left(-\frac{1}{2}(\mathbf{x} - \mathbf{x}_0)^T C^{-1}(\mathbf{x} - \mathbf{x}_0)\right) \tag{5.212}$$

3. $\sum_f \mu(f) = 1$.

In particular, note from (5.212) that $\int f(\mathbf{x})d\mathbf{x} = N_C(\mathbf{0}) < \infty$ whenever $\mu(f) > 0$, and thus we can define

$$|\mu| \triangleq \sum_f \mu(f) \cdot \int f(\mathbf{x})d\mathbf{x} < \infty \tag{5.213}$$

and $f(\mathbf{x}|\mu)$ by

$$f(\mathbf{x}|\mu) \triangleq \frac{1}{|\mu|} \sum_f \mu(f) \cdot f(\mathbf{x}) . \tag{5.214}$$

5.6.1.2 The Fuzzy b.m.a. Induced by a Fuzzy DS Measurement

Any fuzzy DS measurement $o(g)$ induces a fuzzy b.m.a. $H^{-1}o$ on $\mathfrak{X}_0$ defined by

$$(H^{-1}o)(f) \triangleq \sum_{gH=f} o(g) \tag{5.215}$$

where the fuzzy membership function gH on $\mathfrak{X}_0$ is defined by

$$(gH)(\mathbf{x}) \triangleq g(H\mathbf{x}) \tag{5.216}$$

and where the summation is taken over all g such that $gH = f$.

Remark 6 (Caution) *In general, it is not true that* $\int f(\mathbf{x})d\mathbf{x} < \infty$ *if* f *is a focal set of* $H^{-1}o$. *So,* $H^{-1}o$ *is not necessarily a fuzzy DS state.*

5.6.2 KEF Predictor

Let us be given the fuzzy DS state $\mu_{k|k}$ from the previous time step. Enumerate the focal sets of $\mu_{k|k}$ as

$$f_i(\mathbf{x}) = \hat{N}_{C_i}(\mathbf{x} - \mathbf{x}_i) \tag{5.217}$$

for $i = 1, ..., e$, and abbreviate their weights as $\mu_i = \mu_{k|k}(f_i)$. In Appendix G.8 it is shown that, for $i = 1, ..., e$, the focal sets of the predicted fuzzy DS state $\mu_{k+1|k}$ are

$$f'_i(\mathbf{x}) = \hat{N}_{Q_k + F_k C_i F_k^T}(\mathbf{x} - F_k \mathbf{x}_i). \tag{5.218}$$

Their weights are $\mu_{k+1|k}(f_i) = \mu'_i$ where

$$\mu'_i \triangleq \frac{\omega_i \cdot \mu_i}{\sum_{i'=1}^{e} \omega_{i'} \cdot \mu_{i'}} \tag{5.219}$$

and where

$$\omega_i \triangleq \sqrt{\frac{\det C_i}{\det(Q_k + F_k C_i F_k^T)}}. \tag{5.220}$$

The predicted posterior distribution is

$$f_{k+1|k}(\mathbf{x}|Z^k) = f(\mathbf{x}|\mu_{k+1|k}). \tag{5.221}$$

We abbreviate the transformation $\mu_{k|k} \mapsto \mu_{k+1|k}$ as

$$\mu_{k+1|k} \overset{\text{abbr.}}{=} F_k \mu_{k|k}. \tag{5.222}$$

5.6.3 KEF Corrector (Fuzzy DS Measurements)

Let us be given the predicted fuzzy DS state $\mu_{k+1|k}$ and a new fuzzy DS measurement o_{k+1}. Enumerate the focal sets of $\mu_{k+1|k}$ as

$$f_i(\mathbf{x}) = \hat{N}_{D_i}(\mathbf{x} - \mathbf{x}_i) \tag{5.223}$$

for $i = 1, ..., e$, and abbreviate their weights as $\mu_i = \mu_{k+1|k}(f_i)$. Enumerate the focal sets of o_{k+1} as

$$g_j(\mathbf{z}) = \hat{N}_{C_j}(\mathbf{z} - \mathbf{z}_j) \tag{5.224}$$

for $j = 1, ..., d$, and abbreviate their weights as $o_j = o_{k+1}(g_j)$. In Appendix G.9 it is shown that the data-updated fuzzy DS state $\mu_{k+1|k+1}$ is given by the formula

$$\mu_{k+1|k+1} = H_{k+1}^{-1} o_{k+1} * \mu_{k+1|k} \tag{5.225}$$

where '$*$' is Dempster's combination and H_{k+1}^{-1} is defined in (5.215). It is further demonstrated there that, for $i = 1, ..., e$ and $j = 1, ..., d$, the focal sets of $\mu_{k+1|k+1}$ are

$$f_{i,j}(\mathbf{x}) \triangleq \hat{N}_{E_{i,j}}(\mathbf{x} - \mathbf{e}_{i,j}) \tag{5.226}$$

where

$$E_{i,j}^{-1} \triangleq D_i^{-1} + H_{k+1}^T C_j^{-1} H_{k+1} \tag{5.227}$$

$$E_{i,j}^{-1}\mathbf{e}_{i,j} \triangleq D_i^{-1}\mathbf{x}_i + H_{k+1}^T C_j^{-1}\mathbf{z}_j. \tag{5.228}$$

Or, rewritten in standard Kalman filter form,

$$E_{i,j} \triangleq (I - K_{i,j}H_{k+1})\,D_i \tag{5.229}$$

$$\mathbf{e}_{i,j} \triangleq \mathbf{x}_i + K_{i,j}\,(\mathbf{z}_j - H_{k+1}\mathbf{x}_i) \tag{5.230}$$

where

$$K_{i,j} \triangleq D_i H_{k+1}^T \left(H_{k+1} D_i H_{k+1}^T + C_j\right)^{-1}. \tag{5.231}$$

The weights of the $f_{i,j}$ are $\mu_{k+1|k+1}(f_{i,j}) = \mu_{i,j}$, where

$$\mu_{i,j} \triangleq \frac{\mu_i \cdot o_j \cdot \omega_{i,j} \cdot \hat{N}_{C_i + H_{k+1}^T D_j H_{k+1}}(\mathbf{z}_j - H_{k+1}\mathbf{x}_i)}{\sum_{i'=1}^{e} \sum_{j'=1}^{d} \mu_{i'} \cdot o_{j'} \cdot \omega_{i',j'} \cdot \hat{N}_{C_{i'} + H_{k+1}^T D_{j'} H_{k+1}}(\mathbf{z}_{j'} - H_{k+1}\mathbf{x}_{i'})}. \tag{5.232}$$

and where

$$\omega_{i,j} \triangleq \sqrt{\frac{\det C_i \cdot \det D_j}{\det E_{i,j} \cdot \det(C_i + H_{k+1}^T D_j H_{k+1})}}. \tag{5.233}$$

The data-updated posterior distribution is

$$f_{k+1|k+1}(\mathbf{x}|Z^{k+1}) = f(\mathbf{x}|\mu_{k+1|k+1}). \tag{5.234}$$

5.6.4 KEF Corrector (Conventional Measurements)

Let us be given the predicted fuzzy DS state $\mu_{k+1|k}$ and a new conventional measurement $\mathbf{z}_{k+1}$. Enumerate the focal sets of $\mu_{k+1|k}$ as

$$f_i(\mathbf{x}) = \hat{N}_{D_i}(\mathbf{x} - \mathbf{x}_i) \tag{5.235}$$

for $i = 1, ..., e$, and abbreviate their weights as $\mu_i = \mu_{k+1|k}(f_i)$. Let the likelihood function for a linear-Gaussian sensor be

$$f_{k+1}(\mathbf{z}|\mathbf{x}) = N_{R_{k+1}}(\mathbf{z} - H_{k+1}\mathbf{z}_j). \tag{5.236}$$

Then, for $i = 1, ..., e$, the focal sets of the data-updated fuzzy DS state $\mu_{k+1|k+1}$ are

$$f'_i(\mathbf{x}) \triangleq \hat{N}_{E_i}(\mathbf{x} - \mathbf{e}_i) \tag{5.237}$$

where

$$E_i^{-1} \triangleq D_i^{-1} + H_{k+1}^T R_{k+1}^{-1} H_{k+1} \tag{5.238}$$

$$E_i^{-1}\mathbf{e}_i \triangleq D_i^{-1}\mathbf{x}_i + H_{k+1}^T R_{k+1}^{-1} \mathbf{z}_{k+1}. \tag{5.239}$$

Or, rewritten in standard Kalman filter form,

$$E_i \triangleq (I - K_i H_{k+1}) D_i \tag{5.240}$$

$$\mathbf{e}_i \triangleq \mathbf{x}_i + K_i (\mathbf{z}_{k+1} - H_{k+1}\mathbf{x}_i) \tag{5.241}$$

where

$$K_i \triangleq D_i H_{k+1}^T \left(H_{k+1} D_i H_{k+1}^T + R_{k+1}\right)^{-1}. \tag{5.242}$$

The weights of the f_i are $\mu_{k+1|k}(f_i) = \mu'_i$, where

$$\mu'_i \triangleq \frac{\mu_i \cdot \omega_i \cdot \hat{N}_{C_i + H_{k+1}^T R_{k+1} H_{k+1}}(\mathbf{z}_j - H_{k+1}\mathbf{x}_i)}{\sum_{i'=1}^{e} \mu_{i'} \cdot \omega_{i'} \cdot \hat{N}_{C_{i'} + H_{k+1}^T R_{k+1} H_{k+1}}(\mathbf{z}_{j'} - H_{k+1}\mathbf{x}_{i'})} \tag{5.243}$$

and where

$$\omega_i \triangleq \sqrt{\frac{\det C_i \cdot \det R_{k+1}}{\det E_{i,j} \cdot \det(C_i + H_{k+1}^T R_{k+1} H_{k+1})}}. \tag{5.244}$$

The data-updated posterior distribution is

$$f_{k+1|k+1}(\mathbf{x}|Z^{k+1}) = f(\mathbf{x}|\mu_{k+1|k+1}). \tag{5.245}$$

Equations (5.237)-(5.244) follow immediately from (5.226)-(5.233). This is because, formally, we can regard the likelihood

$$f_{k+1}(\mathbf{z}|\mathbf{x}) = N_{R_k}(\mathbf{z} - H_{k+1}\mathbf{z}_j) = N_{R_k}(\mathbf{0}) \cdot \hat{N}_{R_k}(\mathbf{z} - H_{k+1}\mathbf{z}_j) \tag{5.246}$$

as a fuzzy DS measurement with $d = 1$, single focal set $g_1(\mathbf{z}) = \hat{N}_{R_k}(\mathbf{z}-H_{k+1}\mathbf{z}_j)$ and single weight $o_1 = N_{R_k}(\mathbf{0})$. Since o_1 is constant, it cancels out in (5.232), thus resulting in (5.243).

5.6.5 KEF State Estimation

The EAP estimator of (2.133) can be evaluated in closed form. Assume that the focal sets and masses of $\mu_{k+1|k+1}$ are $f_{i,j}(\mathbf{x}) = \hat{N}_{E_{i,j}}(\mathbf{x} - \mathbf{e}_{i,j})$ and $\mu_{i,j}$. Then

$$\mathbf{x}^{\text{EAP}}_{k+1|k+1} = \frac{\sum_{i,j} \mu_{i,j} \cdot \sqrt{\det E_{i,j}} \cdot \mathbf{e}_{i,j}}{\sum_{i,j} \mu_{i,j} \cdot \sqrt{\det E_{i,j}}}. \tag{5.247}$$

To see this, note that

$$\mathbf{x}^{\text{EAP}}_{k+1|k+1} = \frac{\sum_{i,j} \mu_{i,j} \cdot \int \mathbf{x} \cdot f_{i,j}(\mathbf{x})d\mathbf{x}}{\sum_{i,j} \mu_{i,j} \cdot \int f_{i,j}(\mathbf{x})d\mathbf{x}} \tag{5.248}$$

$$= \frac{\sum_{i,j} \mu_{i,j} \cdot N_{E_{i,j}}(\mathbf{0})^{-1} \cdot \int \mathbf{x} \cdot N(\mathbf{x} - \mathbf{e}_{i,j})d\mathbf{x}}{\sum_{i,j} \mu_{i,j} \cdot N_{E_{i,j}}(\mathbf{0})^{-1} \cdot \int N(\mathbf{x} - \mathbf{e}_{i,j})d\mathbf{x}} \tag{5.249}$$

$$= \frac{\sum_{i,j} \mu_{i,j} \cdot N_{E_{i,j}}(\mathbf{0})^{-1} \cdot \mathbf{e}_{i,j}}{\sum_{i,j} \mu_{i,j} \cdot N_{E_{i,j}}(\mathbf{0})^{-1}} \tag{5.250}$$

$$= \frac{\sum_{i,j} \mu_{i,j} \cdot \sqrt{\det E_{i,j}} \cdot \mathbf{e}_{i,j}}{\sum_{i,j} \mu_{i,j} \cdot \sqrt{\det E_{i,j}}}. \tag{5.251}$$

5.6.6 KEF Compared to Gaussian-Mixture and Kalman Filters

We illustrate the KEF by examining it in a special case. Suppose that $\mu_{k|k}$ has a single focal set $f(\mathbf{x}) = \hat{N}_{C_{k+1|k}}(\mathbf{x} - \mathbf{x}_{k+1|k})$. Then (G.129)-(G.133) reduce to a fuzzy DS state $\mu_{k+1|k}$ with a single focal set

$$f(\mathbf{x}) = \hat{N}_{Q_k + F_k C_{k+1|k} F_k^T}(\mathbf{x} - F_k \mathbf{x}_{k+1|k}). \tag{5.252}$$

Similarly assume that o_{k+1} has a single focal set $g(\mathbf{z}) = \hat{N}_{D_{k+1}}(\mathbf{z} - \mathbf{z}_{k+1})$ and that $\mu_{k+1|k}$ has a single focal set $f(\mathbf{x}) = \hat{N}_{C_{k+1|k}}(\mathbf{x} - \mathbf{x}_{k+1|k})$. Then (5.226)-(5.232) reduce to a fuzzy DS state $\mu_{k+1|k+1}$ with single focal set

$$f_{k+1|k+1}(\mathbf{x}) = \hat{N}_{E_{k+1|k+1}}(\mathbf{x} - \mathbf{x}_{k+1|k+1}) \tag{5.253}$$

where

$$\begin{aligned} E^{-1}_{k+1|k+1} &\triangleq C^{-1}_{k+1} + H^T_{k+1} D^{-1}_{k+1} H_{k+1} \\ E^{-1}_{k+1|k+1} \mathbf{e}_{k+1|k+1} &\triangleq C^{-1}_{k+1|k} \mathbf{x}_{k+1|k} + H^T_{k+1} D^{-1}_{k+1} \mathbf{z}_{k+1}. \end{aligned}$$

In other words, the KEF strongly resembles the Kalman filter in this special case. However, it is important to understand that the semantics of the two are not the same. For the Kalman filter, observational uncertainty is entirely due to randomness. For this special case of the KEF, it is due entirely to ignorance.

More generally, the KEF resembles the conventional single-target Gaussian-mixture filter (GMF) described in Section 2.5.2, since in both cases posterior distributions are weighted sums of Gaussian distributions.

Once again, however, the underlying semantics of the two filters are different. For the Gaussian sum filter, observational uncertainty is entirely due to randomness. For the KEF, it is due to ignorance.

5.7 CHAPTER EXERCISES

Exercise 17 *Show that*

$$f(W_1, ..., W_m|\mathbf{x}) = f(W_1 \cap ... \cap W_m|\mathbf{x}). \tag{5.254}$$

That is, '$\cap$' is a Bayes combination operator in the sense of Section 3.5.3. Stated differently: Measurement fusion of generalized fuzzy sets $W_1, ..., W_m$ using set-theoretic intersection produces the same results as measurement fusion of $W_1, ..., W_m$ using Bayes' rule alone.

Exercise 18 *Generalize (5.119)-(5.123) to the case when the focal sets of b.m.a.s and generalized b.ma.'s are fuzzy.*

Exercise 19 *Prove (5.147) and (5.148), from them, that the conversion* $o \mapsto m_o$ *of generalized fuzzy b.m.a.s to fuzzy b.m.a.s is Bayes-invariant.*

Exercise 20 *Prove (5.150) and (5.151) and, from them, that the conversion* $g \mapsto o_g$ *of fuzzy membership functions to fuzzy b.m.a.s is Bayes-invariant.*

Exercise 21 *Prove (5.153) and (5.154) and, from them, that the conversion* $o \mapsto \mu_o$ *from fuzzy b.m.a.s to fuzzy membership functions is Bayes-invariant.*

Exercise 22 *Prove (5.161) and (5.162) and, from them, show that the conversion* $o \mapsto \varphi_o$ *of fuzzy b.m.a.s to probability density functions is Bayes-invariant.*

Exercise 23 *Prove (5.164) and (5.165) and, from them, show that the conversion* $\varphi \mapsto \mu_\varphi$ *of probability density functions to fuzzy membership functions is Bayes-invariant.*

Chapter 6

AGA Measurements

6.1 INTRODUCTION TO THE CHAPTER

In this chapter, I continue my development of formal Bayes modeling of generalized measurements, by extending my approach to the more complicated case of *ambiguously generated ambiguous (AGA) measurements.* A UGA measurement is characterized by two things: (1) modeling of the measurement itself involves ambiguity (because, for example, of human operator interpretation processes); but (2) the association of measurements with target states can be precisely specified by a measurement equation of the form $\mathbf{z} = \eta(\mathbf{x})$. AGA measurements, on the other hand, involve even greater ambiguity because

- $\eta(\mathbf{x})$ itself cannot be unambiguously specified.

Example 34 *Consider the hub/tire feature in Examples 15 and 16. Suppose that it is not possible to define a function that assigns a hub/tire number* $\eta(v)$ *to every target type* v. *Rather, our understanding of at least some targets may be incomplete, so that we cannot say for sure that type* v *has a specific number of hubs/tires. In this case it is not only the human-mediated observation that must be modeled as a fuzzy measurement* g_0 *as in Example 16. One must also employ a fuzzy model* $\eta_v(n)$. *For example, suppose that target type* $T1$ *is believed to have* $n = 5$ *hubs/tires, but that we are not entirely sure about this fact. So we assign* $\eta_{T1}(5) = 1$, $\eta_{T1}(4) = \eta_{T1}(6) = 0.5$, *and* $\eta_{T1}(n) = 0.1$ *for* $n = 1, 2, 3, 7, 8$.

Example 35 *Typical applications that require AGA measurement modeling are those that involve not only human-mediated observations, as with UGA measurements, but also the construction of human-mediated model bases. Consider, for example, a passive-acoustic feature* ϕ *(a frequency line, for example) of a particular* *type extracted by a human operator from a sonogram. The ambiguity associated with ϕ is encapsulated in the generalized observation Θ_ϕ that models ϕ. Different features are associated in varying degrees with different target types, a fact that makes target identification possible. Ideally, one would have a precise model of what value $\eta(v)$ of the feature is associated with a given target of type v. This is often not possible with passive-acoustic data, since frequency attributes exhibit unstable bandwidths, center frequencies, and so on, and there can be literally thousands of them. Also, different features exhibit varying degrees of identity-resolving power. So, one must rely on the expertise of acoustic intelligence analysts to construct a model of ϕ and other features. This introduces a human interpretation process, and the ambiguities associated with it necessitate more complicated models in place of $\eta(v)$.*

We presume the same notational conventions introduced in Section 5.1.1. After summarizing the primary lessons learned for the chapter in Section 6.1.1, I describe the organization of the chapter in Section 6.1.2.

6.1.1 Summary of Major Lessons Learned

The following are the major concepts, results, and formulas that the reader will encounter while working through this chapter:

- Definition of an AGA measurement (Section 6.2);
- Definition of the generalized likelihood of an AGA measurement (Section 6.2):
$$f(\Theta|\mathbf{x}) = \Pr(\Theta \cap \Sigma_\mathbf{x} \neq \emptyset); \tag{6.1}$$
- Formula for the generalized likelihood function for AGA fuzzy measurements; see (6.11):
$$f(g|\mathbf{x}) = \sup_\mathbf{z} \min\{g(\mathbf{z}),\ \eta_\mathbf{x}(\mathbf{z})\}; \tag{6.2}$$
- Formula for the generalized likelihood function for AGA Dempster-Shafer measurements; see (6.19):
$$f(o|\mathbf{x}) = \alpha_{\mathrm{DS}}(o, \sigma_\mathbf{x}); \tag{6.3}$$
- Formula for the generalized likelihood function for AGA fuzzy Dempster-Shafer measurements; see (6.24):
$$f(o|\mathbf{x}) = \alpha_{\mathrm{FDS}}(o, \sigma_\mathbf{x}); \tag{6.4}$$

- As was the case with UGA measurements, formulas for generalized likelihoods permit even more disparate forms of information and uncertainty to be represented in a common probabilistic framework, and permit them to be processed using the recursive Bayes filter of Chapter 2 (see Figure 3.4);
- Application of Bayes filtering to robust data fusion and target identification using datalink INT features, Section 6.7, including robustness against unmodeled target types, Sections 6.6 and 6.8.

6.1.2 Organization of the Chapter

I introduce the concepts of generalized sensor-transform models and AGA measurements in Section 6.2. I define generalized likelihood functions for such measurements in Section 6.3. I examine the Bayes filter for the special case of fuzzy AGA measurements in Section 6.4.

In Section 6.5, I illustrate my concepts with an extended example: target identification when measurements are imprecise and target models are fuzzy. In Section 6.6, I address the case when a target identification algorithm must robustly deal with unmodeled target types.

In Section 6.7, I describe a concrete application: robust data fusion and target identification based on intelligence-information (INT) features extracted from signatures by human operators and then transmitted in alphanumeric form over datalinks. This is extended to unmodeled target types in Section 6.8. Exercises for the chapter are in Section 6.9.

6.2 AGA MEASUREMENTS DEFINED

The sensor-transform model $\eta(\mathbf{x})$ of (5.17) is *unambiguous* in the sense that the value of $\eta(\mathbf{x})$ is known *precisely*—there is no ambiguity in associating the measurement value $\eta(\mathbf{x})$ with the state value $\mathbf{x}$. However, in some applications this assumption may not be valid. This includes human-mediated information such as observed target features or natural-language reports [134, pp. 69-72]—see Examples 36 and 37 in Section 6.3.

Mathematically speaking, $\eta(\mathbf{x})$ may be known only up to containment within some constraint $\eta(\mathbf{x}) \in H_{0,\mathbf{x}} \subseteq \mathfrak{Z}_0$. In this case $\eta(\mathbf{x})$ is set-valued: $\eta(\mathbf{x}) = H_{0,\mathbf{x}}$.

Or, we may need to specify a nested sequence of constraints $H_{0,\mathbf{x}} \subset H_{1,\mathbf{x}} \subset \ldots \subset H_{e,\mathbf{x}}$ with associated probabilities $\eta_{i,\mathbf{x}}$ that $H_{i,\mathbf{x}}$ is the correct constraint.

If we define the random subset $\Sigma_{\mathbf{x}}$ of $\mathfrak{Z}_0$ by $\Pr(\Sigma_{\mathbf{x}} = H_{i,\mathbf{x}}) = \eta_{i,\mathbf{x}}$ then $\eta(\mathbf{x})$ is random-set valued: $\eta(\mathbf{x}) = \Sigma_{\mathbf{x}}$. In general, $\Sigma_{\mathbf{x}}$ can be any random closed subset of $\mathfrak{Z}_0$ such that $\Pr(\Sigma_{\mathbf{x}} = \emptyset) = 0$.

An *ambiguously generated ambiguous (AGA) measurement* is, then, a generalized measurement Θ as in Section 5.2.2, whose relationship with the state variable $\mathbf{x}$ is also ambiguous.

6.3 LIKELIHOODS FOR AGA MEASUREMENTS

Given that models are ambiguous, how do we generalize the UGA measurement model $\eta(\mathbf{x}) \in \Theta$? That is, how do we state that the measurement Θ matches the generalized model $\eta(\mathbf{x}) = \Sigma_{\mathbf{x}}$? One could choose

$$f(\Theta|\mathbf{x}) \stackrel{?}{=} \Pr(\Sigma_{\mathbf{x}} \subseteq \Theta) \tag{6.5}$$

since this mimics the basic set-theoretic structure of (5.17), namely $\Pr(\eta(\mathbf{x}) \in \Theta)$. However, it has been observed that, in at least some practical applications, Bayes filters based on this definition do not behave robustly if the degree of ambiguity of the $\Sigma_{\mathbf{x}}$ is high.

I have instead employed the following less restrictive model [135, 134, 145]:

$$f(\Theta|\mathbf{x}) \triangleq \Pr(\Theta \cap \Sigma_{\mathbf{x}} \neq \emptyset). \tag{6.6}$$

That is:

- *The generalized measurement* Θ *matches the generalized model* $\Sigma_{\mathbf{x}}$ *if it does not flatly contradict it.*

This is a weak specification of the concept of matching a model, whereas $\Theta \supseteq \Sigma_{\mathbf{x}}$ is a restrictive one. This fact accounts for the greater robustness observed with the choice of (6.6).

Unfortunately, the unification results for UGA measurements (see Section 5.4) cannot be extended to AGA measurements if such a definition is assumed.

Joint generalized likelihood functions for AGA measurements are defined as follows:

$$f(\Theta_1, ..., \Theta_m|\mathbf{x}) \triangleq \Pr(\Theta_1 \cap \Sigma^1_{\mathbf{x}} \neq \emptyset, ..., \Theta_m \cap \Sigma^m_{\mathbf{x}} \neq \emptyset) \tag{6.7}$$

where $\Sigma^1_{\mathbf{x}}, ..., \Sigma^m_{\mathbf{x}}$ are different realizations of $\Sigma_{\mathbf{x}}$. Thus, for example, if $\Theta_1, \Sigma^1_{\mathbf{x}}$ are independent of $\Theta_2, \Sigma^2_{\mathbf{x}}$ then

$$
\begin{aligned}
f(\Theta_1, \Theta_2|\mathbf{x}) &\triangleq \Pr(\Theta_1 \cap \Sigma^1_{\mathbf{x}} \neq \emptyset, \Theta_2 \cap \Sigma^2_{\mathbf{x}} \neq \emptyset) && (6.8)\\
&= \Pr(\Theta_1 \cap \Sigma^1_{\mathbf{x}} \neq \emptyset) \cdot \Pr(\Theta_2 \cap \Sigma^2_{\mathbf{x}} \neq \emptyset) && (6.9)\\
&= f(\Theta_1|\mathbf{x}) \cdot f(\Theta_2|\mathbf{x}). && (6.10)
\end{aligned}
$$

Example 36 *Suppose that an observed target feature is a frequency* $\phi \in \mathfrak{Z}_0 = \mathbb{R} = (-\infty, \infty)$ *extracted from a signature by a human operator. The ambiguity in the interpretation process might be modeled as a random interval of the form*

$$\Theta = [f_0 + F - b_0 - B,\ f_0 + F + b_0 + B]$$

where f_0 *is the nominal center frequency and* b_0 *the nominal bandwidth of the feature; and where the zero-mean random numbers* F *and* B *model variation in* f_0 *and* b_0, *respectively. Suppose that our understanding of how target types* v *are associated with the feature is similarly ambiguous. Then a model base must be constructed in which similar intervals*

$$\Sigma_v = [f_v + F_v - b_v - B_v,\ f_v + F_v + b_v + B_v]$$

model this association. The matching of Θ *with models* Σ_{T1} *and* Σ_{T2} *of two target types* $T1$ *and* $T2$ *is pictured in Figure 6.1. The random interval* Σ_{T1} *tends to be concentrated around one region of the frequency line, whereas the other random interval* Σ_{T2} *tends to be concentrated around a different region. The random interval* Θ *tends to be more concentrated in that part of the line associated with* Σ_{T1} *than with* Σ_{T2}. *Thus, intuitively speaking,* Θ *will have nonempty intersection with* Σ_{T1} *more frequently than with* Σ_{T2}. *Consequently, the generalized likelihood* $f(\Theta|T1) = \Pr(\Theta \cap \Sigma_{T1} \neq \emptyset)$ *for* $T1$ *will be larger than that for* $T2$. *We conclude that the generalized observation* Θ *supports the existence of a target of type* $T1$ *rather than of type* $T2$.

In the remainder of this section, I derive concrete formulas for the generalized likelihoods for the following types of AGA measurements: fuzzy (Section 6.3.1), generalized fuzzy (Section 6.3.2), Dempster-Shafer (Section 6.3.3), and fuzzy Dempster-Shafer (Section 6.3.4).

6.3.1 Special Case: Θ and $\Sigma_{\mathbf{x}}$ Are Fuzzy

Suppose that the sensor transform models and observations are all fuzzy and that the interpretations of all fuzzy events are statistically completely correlated. That

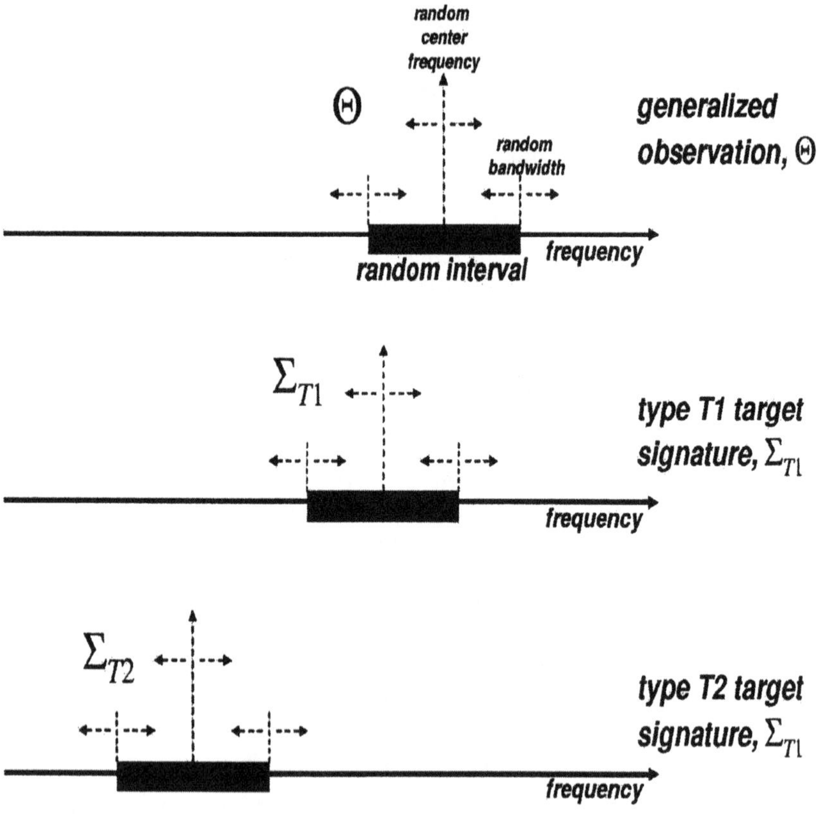

Figure 6.1 A generalized measurement Θ on the frequency line is to be compared with the generalized signatures Σ_{T1} and Σ_{T2} of two target types $T1$ and $T2$. The random interval Θ tends to intersect the random interval Σ_{T2} much more frequently than the random interval Σ_{T1}. This indicates that the generalized measurement was generated by a target of type $T2$.

is, $\Sigma_{\mathbf{x}} = \Sigma_A(\eta_{\mathbf{x}})$ and $\Theta_g = \Sigma_A(g)$, where as usual A is a uniform random number in $[0,1]$ and where $\Sigma_A(g)$ was defined in (4.21). Then the generalized likelihood $f(g|\mathbf{x})$ of g is $f(g|\mathbf{x}) \triangleq \Pr(\Theta_g \cap \Sigma_{\mathbf{x}} \neq \emptyset)$, and it is shown in Appendix G.10 that it is

$$f(g|\mathbf{x}) = \sup_{\mathbf{z}} \min\{g(\mathbf{z}),\ \eta_{\mathbf{x}}(\mathbf{z})\}. \tag{6.11}$$

Example 37 *Consider the modeling of a frequency feature using random intervals, as described in Example 36. Since random sets are more difficult to understand and to mathematically manipulate than fuzzy sets, it will often be more useful to replace the former with the latter. We know from Section 4.3.2 that the random set representation $\Sigma_A(g)$ of a fuzzy membership function $g(\phi)$ faithfully preserves the information contained in g. So, we substitute $\Theta = \Sigma_A(g)$ and $\Sigma_v = \Sigma_A(\eta_v)$. In this case the generalized likelihood function becomes, because of (6.11),*

$$f(g|v) = \Pr(\Sigma_A(g) \cap \Sigma_A(\eta_v) \neq \emptyset) = \sup_{\phi} \min\{g(\phi),\ \eta_v(\phi)\}$$

where ϕ denotes values on the frequency line. This is pictured in Figure 6.2. The fuzzy model η_{T1} for target $T1$ tends to be concentrated around one region of the frequency line, whereas the model η_{T2} for target $T2$ tends to be concentrated around a different region. The fuzzy measurement g tends to be more concentrated in that part of the line associated with η_{T1} than with η_{T2}. Thus $g \wedge \eta_{T1}$ will tend to have larger values than $g \wedge \eta_{T2}$. Consequently, the largest value $\sup_\phi \min\{g(\phi),\ \eta_{T1}(\phi)\}$ of $g \wedge \eta_{T1}$ will be greater than the largest value $\sup_\phi \min\{g(\phi),\ \eta_{T2}(\phi)\}$ of $g \wedge \eta_{T2}$. Thus the fuzzy observation g supports the existence of target type $T1$.

Example 38 *Suppose that $\eta_{\mathbf{x}}(\mathbf{z}) = \delta_{\mathbf{z},\eta(\mathbf{x})}$, where $\delta_{\mathbf{z},\mathbf{z}'}$ denotes the Kronecker delta on elements of $\mathfrak{Z}_0$. Then it is left to the reader as Exercise 24 to verify that (6.11) reduces to (5.29): $f(g|\mathbf{x}) = g(\eta(\mathbf{x}))$. That is, UGA fuzzy measurements are a special case of AGA fuzzy measurements.*

Example 39 *Suppose that $\mathfrak{Z}_0 = \mathbb{R}$ and that*

$$g(z) \triangleq \exp\left(-\frac{(z-z_0)^2}{2\sigma_0^2}\right) \tag{6.12}$$

$$\eta_x(z) \triangleq \exp\left(-\frac{(z-z_x)^2}{2\sigma_x^2}\right). \tag{6.13}$$

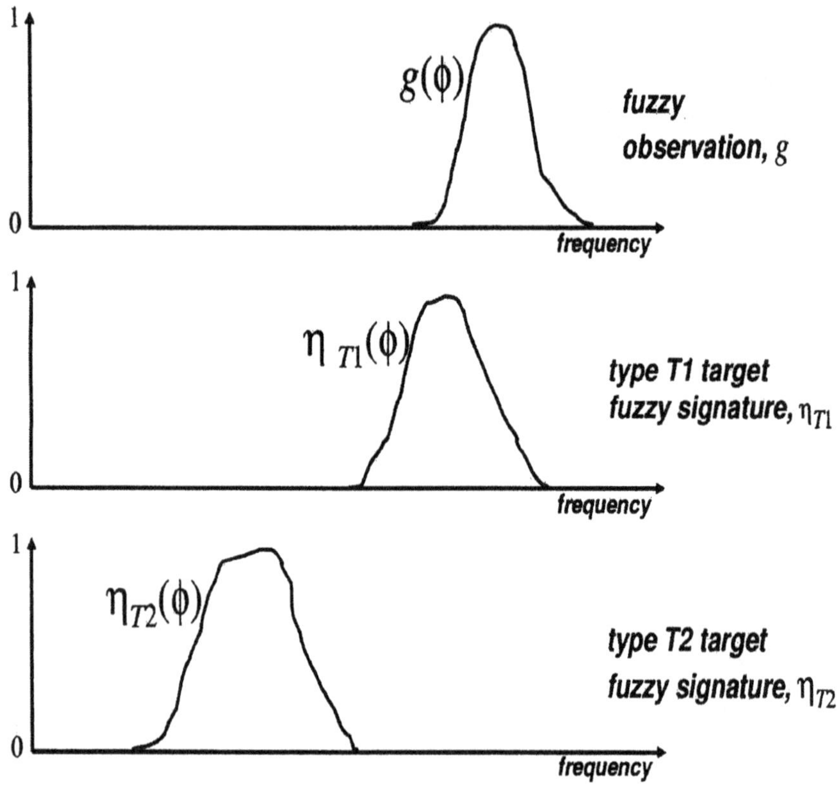

Figure 6.2 A fuzzy measurement g on the frequency line is to be compared with the fuzzy signatures η_{T1} and η_{T2} of two target types $T1$ and $T2$. Since g tends to intersect η_{T2} more strongly than η_{T1}, this indicates that the generalized measurement was generated by a target of type $T2$.

Then it is left to the reader as Exercise 25 to show that

$$f(g|x) = \exp\left(-\frac{(z_0 - z_x)^2}{2(\sigma_0 + \sigma_x)^2}\right). \tag{6.14}$$

6.3.2 Special Case: Θ and $\Sigma_{\mathbf{x}}$ Are Generalized Fuzzy

Suppose that the sensor transform models and observations are all generalized-fuzzy and statistically completely correlated:

$$\Sigma_{\mathbf{x}} = \Sigma_A(W_{\mathbf{x}}), \qquad \Theta_W = \Sigma_A(W), \tag{6.15}$$

where as usual A is a uniform random number in $[0,1]$ and where $\Sigma_A(W)$ was defined in (4.21). Then the generalized likelihood $f(g|\mathbf{x})$ of g is

$$f(W|\mathbf{x}) \triangleq \Pr(\Theta_W \cap \Sigma_{\mathbf{x}} \neq \emptyset). \tag{6.16}$$

In Appendix G.11, it is shown to be

$$f(W|\mathbf{x}) = \sup_{\mathbf{z}} \int_0^1 \mathbf{1}_W(\mathbf{z}, a) \cdot \mathbf{1}_{W_{\mathbf{x}}}(\mathbf{z}, a) da. \tag{6.17}$$

Example 40 *Suppose that* $W = W_g$ *and* $W_{\mathbf{x}} = W_{\eta_{\mathbf{x}}}$ *where* $g(\mathbf{z})$ *and* $\eta_{\mathbf{x}}(\mathbf{z})$ *are fuzzy membership functions on* $\mathfrak{Z}_0$ *and where the notation* W_g *was defined in (4.70). It is left to the reader as Exercise 26 to verify that*

$$f(W_g|\mathbf{x}) = f(g|\mathbf{x}). \tag{6.18}$$

6.3.3 Special Case: Θ and $\Sigma_{\mathbf{x}}$ Are Dempster-Shafer

Let $o(O)$ be a DS measurement (i.e., o is a b.m.a. on $\mathfrak{Z}_0$). Likewise, let $\sigma_{\mathbf{x}}(O)$ be a b.m.a. on $\mathfrak{Z}_0$ that models the ambiguity associated with generation of measurements by targets with state $\mathbf{x}$. Let Θ_o and $\Sigma_{\mathbf{x}}$ be random set representations of o and $o_{\mathbf{x}}$, respectively: $\Pr(\Theta = O) = o(O)$ and $\Pr(\Sigma_{\mathbf{x}} = O) = \sigma_{\mathbf{x}}(O)$. Assume that Θ_o and $\Sigma_{\mathbf{x}}$ are statistically independent. The generalized likelihood of o is $f(o|\mathbf{x}) \triangleq \Pr(\Theta_o \cap \Sigma_{\mathbf{x}} \neq \emptyset)$ and is

$$f(o|\mathbf{x}) = \alpha_{\mathrm{DS}}(o, \sigma_{\mathbf{x}}) \tag{6.19}$$

where the Dempster-Shafer agreement $\alpha_{\text{DS}}(o, o')$ of two b.m.a.s o, o' was defined in (4.88). To see this, note that

$$
\begin{aligned}
f(o|\mathbf{x}) &= \sum_{O,O'} \Pr(\Theta_o = O, \Sigma_{\mathbf{x}} = O', O \cap O' \neq \emptyset) && (6.20)\\
&= \sum_{O \cap O' \neq \emptyset} \Pr(\Theta_o = O) \cdot \Pr(\Sigma_{\mathbf{x}} = O') && (6.21)\\
&= \sum_{O \cap O' \neq \emptyset} o(O) \cdot \sigma_{\mathbf{x}}(O') && (6.22)\\
&= \alpha_{\text{DS}}(o, \sigma_{\mathbf{x}}). && (6.23)
\end{aligned}
$$

Example 41 *Suppose that $O_{\mathbf{x}}$ is the unique focal set of $\sigma_{\mathbf{x}}$. Then it is left to the reader as Exercise 27 to verify that $f(o|\mathbf{x}) = Pl_o(O_{\mathbf{x}})$ where 'Pl_o' is the plausibility function of o as defined in (4.84). In particular suppose that $O_{\mathbf{x}} = \{\eta(\mathbf{x})\}$. Then it is left to the reader in the same exercise to verify that $f(o|\mathbf{x})$ reduces to (5.58): $f(o|\mathbf{x}) = \sum_{O \ni \eta(\mathbf{x})} o(O)$. That is, a UGA Dempster-Shafer measurement is a special case of an AGA Dempster-Shafer measurement.*

6.3.4 Special Case: Θ and $\Sigma_{\mathbf{x}}$ Are Fuzzy DS

Let $o(g)$ be a fuzzy DS measurement (i.e., o is a fuzzy b.m.a. on $\mathfrak{Z}_0$). Let $\sigma_{\mathbf{x}}(g)$ be a fuzzy b.m.a. that models the ambiguity associated with generation of measurements by targets with state $\mathbf{x}$. Let $\Theta_o = \Sigma_A(W_o)$ and $\Sigma_{\mathbf{x}} = \Sigma_{A'}(W_{\sigma_{\mathbf{x}}})$ be random set models of o and $\sigma_{\mathbf{x}}$, respectively, as defined in Section 4.6.1. Assume that Θ_o and $\Sigma_{\mathbf{x}}$ are statistically independent (i.e., A, A' are statistically independent). Then the generalized likelihood of o is $f(o|\mathbf{x}) \triangleq \Pr(\Theta_o \cap \Sigma_{\mathbf{x}} \neq \emptyset)$. This is shown in Appendix G.12 to be

$$f(o|\mathbf{x}) = \alpha_{\text{FDS}}(o, \sigma_{\mathbf{x}}) \tag{6.24}$$

where the agreement $\alpha_{\text{FDS}}(o, o')$ of two fuzzy b.m.a.s was defined in (4.130).

6.4 FILTERING WITH FUZZY AGA MEASUREMENTS

Let states x and measurements z be real numbers. Also assume that fuzzy data $g_1, ..., g_k$ and fuzzy signature models η_x have the following forms:

$$g_k(z) = \exp\left(-\frac{(z-z_k)^2}{2\sigma_k^2}\right) \tag{6.25}$$

$$\eta_x(z) = \exp\left(-\frac{(z-z_x)^2}{2\sigma_x^2}\right) \tag{6.26}$$

for all z, z_0, x. From (6.14) the corresponding generalized likelihood is

$$f(g_{k+1}|x) = \exp\left(-\frac{(z_{k+1}-z_x)^2}{2(\sigma_{k+1}+\sigma_x)^2}\right). \tag{6.27}$$

Assume that $\sigma_x = \sigma$ is independent of x. Then

$$f(g_{k+1}|x) \propto N_{(\sigma_{k+1}+\sigma)^2}(z_{k+1}-z_x) = f_{k+1}(z_{k+1}|x) \tag{6.28}$$

where $f_{k+1}(z_{k+1}|x)$ is the (conventional) likelihood function for the nonlinear measurement model $Z = z_x + V$, where V is zero-mean Gaussian noise with variance $(\sigma_{k+1}+\sigma)^2$. Assume further that target state does not change between measurements. Then

$$f_{k+1|k+1}(x|g_1, ..., g_{k+1}) \propto f(g_{k+1}|x) \cdots f(g_1|x) \cdot f_0(x) \tag{6.29}$$

$$\propto f_{k+1}(z_{k+1}|x) \cdots f_1(z_1|x) \cdot f_0(x) \tag{6.30}$$

$$\propto f_{k+1|k+1}(x|z_1, ..., z_{k+1}) \tag{6.31}$$

and so

$$f_{k+1|k+1}(x|g_1, ..., g_{k+1}) = f_{k+1|k+1}(x|z_1, ..., z_{k+1}) \tag{6.32}$$

for all $k \geq 1$.

That is:

- If the fuzzy models η_x have identical shapes then a nonlinear filter drawing upon measurements $g_1, ..., g_k$ from a single fuzzy source behaves essentially the same way as a nonlinear filter drawing upon conventional data $z_1, ..., z_k$ collected by a nonfuzzy sensor with certain noise statistics.

Despite this fact, it should be emphasized that the underlying semantics are not the same. In the case of the conventional measurement z_k, there is no uncertainty about what has been observed: It is z_k. The uncertainty associated with z_k lies in the fact that, though precise, it is the result of a draw from a random variable.

In the case of a fuzzy measurement g_k, on the other hand, it is not clear what has been observed, and g_k models uncertainty in the observation itself. In this case, the observation g_k is entirely deterministic—it is not the result of a random draw.

6.5 EXAMPLE: FILTERING WITH POOR DATA

Consider the following simple problem.[1] A sensor collects instantaneous radio frequency center frequencies from different frequency-agile radars carried by aircraft of five different types, $u = T1,...,T5$. We are to identify an aircraft on the basis of information collected from it by the sensor. The only information we have about these radars is summarized in the following table:

	$T1$	$T2$	$T3$	$T4$	$T5$
Nominal center fr (MHz), $f_u =$	80	90	100	110	115
Hopping bandwidth (MHz), $\Delta f_u =$	5	5	10	5	5

Note that there is a great deal of overlap between the signatures, reflecting the fact that the features typically generated by the different radar types do not differ greatly from each other.

The only other thing we know is what measurements the sensor collects. Any measurement that is collected from a target of type u is some interval J on the frequency line. The actual sensor measurement model, which is unknown to us, combines randomness with imprecision:

$$J = [f_u + Z - \frac{1}{2}\Delta f_0,\ f_u + Z + \frac{1}{2}\Delta f_0] \tag{6.33}$$

where Z is a zero-mean, Gaussian-distributed random number with variance σ_0^2. That is, the value of the actual frequency is known only to within containment in an interval J of constant width Δf_0. Moreover, the center of J varies randomly about the nominal center frequency f_u for the target type.

1 This material first appeared in [155, pp. 66-69].

In what follows I describe a robust-Bayes classifier (Section 6.5.1) and describe its performance in two simulations: high imprecision with high randomness (Section 6.5.2) and greater precision with lesser randomness (Section 6.5.3). The apparently paradoxical results are interpreted in Section 6.5.4 as due to an unanticipated interaction between randomness and imprecision in the measurements.

6.5.1 A Robust-Bayes Classifier

Since we do not have detailed information about the transmitters, we must specify *robust models* for the five target types:

$$\eta_u(z) = \exp\left(-\frac{1}{2}\left(\frac{z-f_u}{\frac{1}{2}\Delta f_u}\right)^2\right) = \exp\left(-\frac{2(z-f_u)^2}{\Delta^2 f_u}\right). \tag{6.34}$$

These are pictured in Figure 6.3.

These models encapsulate what we do know about the transmitters, while hedging against any uncertainties in this information.

Suppose that we collect an interval measurement J. Since the measurements and the models are fuzzy membership functions, we use the likelihood of (6.11):

$$\begin{aligned} f(J|u) &\triangleq \sup_z \min\{\mathbf{1}_J(z), \eta_u(z)\} = \sup_{z\in J} \eta_u(z) && (6.35)\\ &= \sup_{z\in J} \exp\left(-\frac{(z-f_u)^2}{2(\frac{1}{2}\Delta f_u)^2}\right) && (6.36)\\ &= \exp\left(-\frac{2\cdot\inf_{z\in J}(z-f_u)^2}{\Delta^2 f_u}\right). && (6.37)\end{aligned}$$

The joint likelihood of a series $J_1, ..., J_m$ of conditionally independent measurements is:

$$\begin{aligned} f(J_1,...,J_m|u) &= f(J_1|u)\cdots f(J_m|u) && (6.38)\\ &= \exp\left(-\frac{2}{\Delta^2 f_u}\sum_{i=1}^m \inf_{z\in J_i}(z-f_u)^2\right). && (6.39)\end{aligned}$$

The posterior probability is, assuming a uniform prior,

$$f(u|J_1,..,J_m) = \frac{f(J_1|u)\cdots f(J_m|u)}{\sum_v f(J_1|v)\cdots f(J_m|v)}.$$

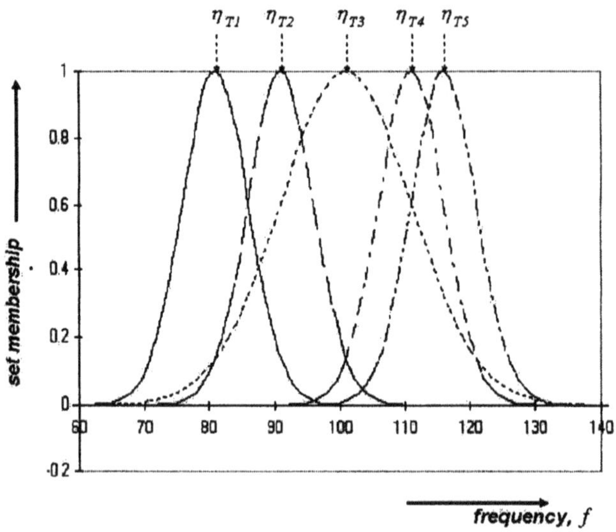

Figure 6.3 Ambiguous signature models for the RF sources. The models $\eta_{T1}(z)$, $\eta_{T2}(z)$, $\eta_{T3}(z)$, $\eta_{T4}(z)$, $\eta_{T5}(z)$ for five target types $T1, T2, T3, T4, T5$ are plotted from left to right. The signatures for $T1$ and $T2$ are easily confused, and those for $T4$ and $T5$ even more so. The signature for $T3$ can be potentially confused with all others, but especially with $T2$ and $T4$.

The MAP estimate of target type coincides with the MLE:

$$\hat{u}_m = \arg\min_u \frac{1}{\Delta^2 f_u} \sum_{i=1}^{m} \inf_{z \in J_i} (z - f_u)^2. \tag{6.40}$$

6.5.2 Simulation 1: More Imprecise, More Random

The following table summarizes the results of applying the Bayes filter with $\Delta f_0 =$ 30 MHz and $\sigma_0 = 10$ MHz, as illustrated in Figure 6.4.

Figures 6.5 and 6.6 display typical results if the actual target types are $T1$ and $T5$, respectively.

Actual target type, $u =$	$T1$	$T2$	$T3$	$T4$	$T5$
Observations for $\geq 90\%$ correct. I.D.	16	40	8	60	63

In other words, the Bayes filter correctly determines target I.D. with high confidence, despite the following:

1. Extreme ambiguity of the features;
2. The poor resolving power of the features;
3. The fact that our measurement model $f(J|u)$ only crudely approximates the actual feature generation process.

6.5.3 Simulation 2: Less Imprecise, Less Random

The simulation was repeated with what would appear to be less stressing conditions, namely lesser imprecision $\Delta f_0 = 20$ MHz and less intense randomness $\sigma_0 = 5$ MHz, as illustrated in Figure 6.7. Two typical cases are shown in Figures 6.8 and 6.9.

Complete results are listed in the following table.

Actual target type, $u =$	$T1$	$T2$	$T3$	$T4$	$T5$
I.D. after 800 measurements =	$T1$	$T2$	$T3$	$T4, T5$	$T4, T5$
Confidence in I.D. estimate =	73%	40%	100%	70%	100%

Paradoxically,

- *Filter performance deteriorates even though precision has increased and randomness has decreased.*

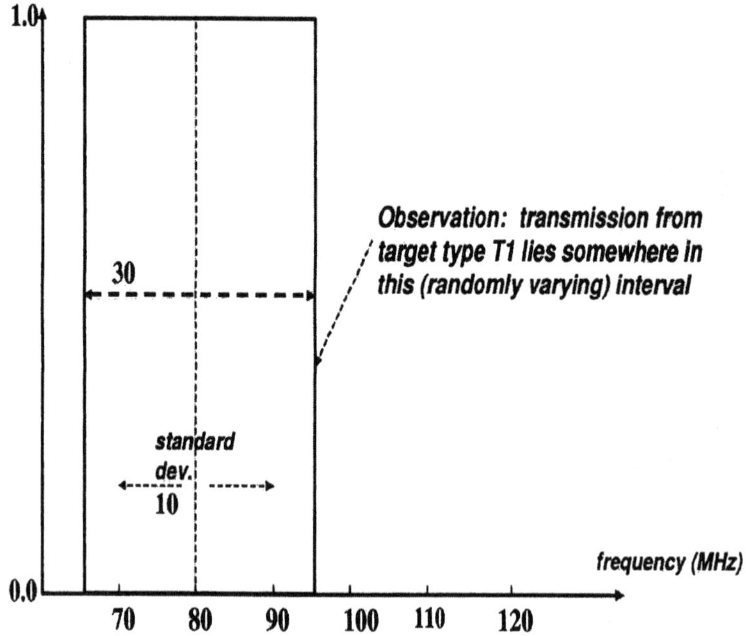

Figure 6.4 A typical measurement for the first simulation. Measurements are intervals of width 30 MHz and their centers vary randomly with standard deviation 10 MHz. The bandwidth of 30 MHz is so large that it simultaneously overlaps the model signatures of adjacent target types.

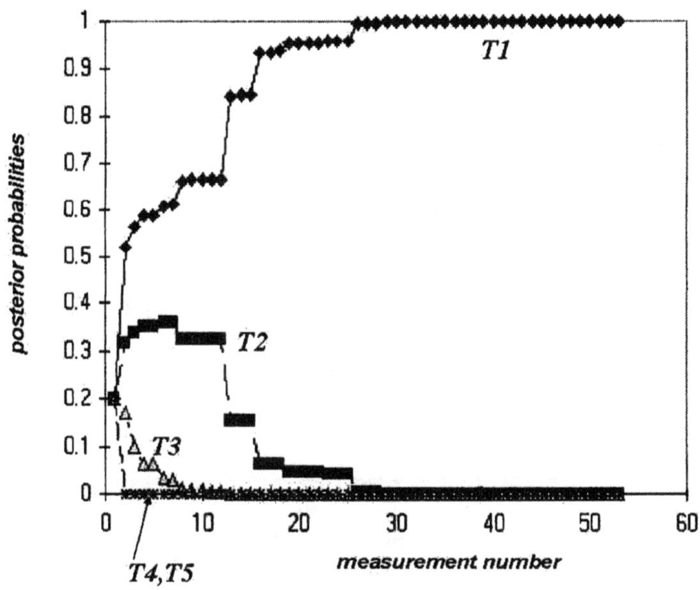

Figure 6.5 The posterior values for the five target types are plotted as functions of measurement number for the first simulation. The posterior for the correct type, $T1$, rapidly dominates the others.

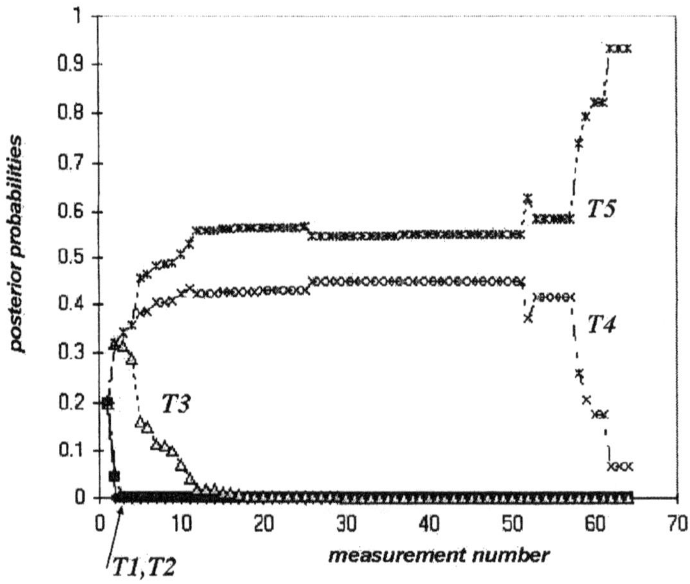

Figure 6.6 In this case $T5$ is the correct target type. Because of the greater similarity between the models for $T4$ and $T5$, the latter is clearly separated from the former only until well after the 50th measurement has been processed.

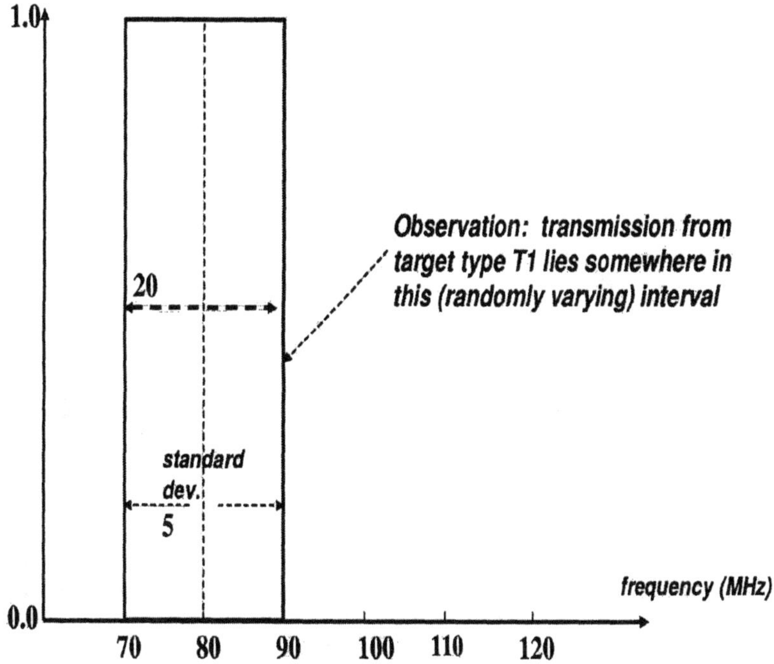

Figure 6.7 A typical measurement for the second simulation. Both imprecision and randomness are less severe than in the first simulation: Measurements are intervals of width 20 MHz, and their centers vary randomly with standard deviation 5 MHz.

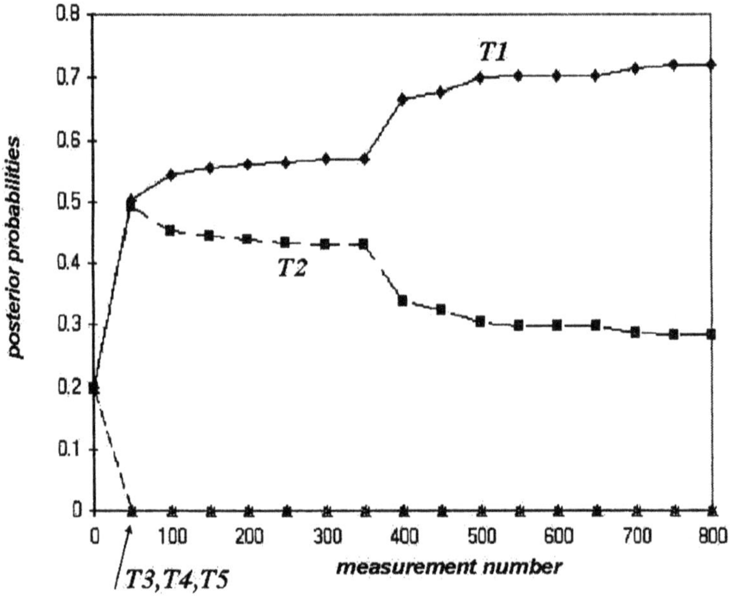

Figure 6.8 The posterior values for the five target types are plotted as functions of measurement number for the second simulation. The posterior probability for the correct type, $T1$, dominates. However, far more measurements are required to achieve this decision than was the case in the first simulation.

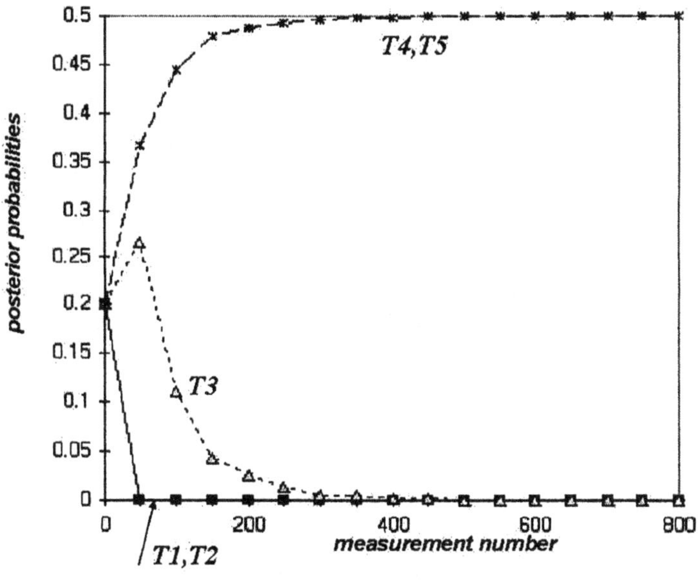

Figure 6.9 Even after 800 measurements have been collected, the classifier is unable to determine whether or not the target is type $T4$ or type $T5$. This is paradoxical behavior since both imprecision and randomness have decreased in severity compared to the first simulation. The paradox is resolvable, since it can be shown to be due to an unexpected interaction between imprecision and randomness.

6.5.4 Interpretation of the Results

The apparent paradox is due to *the interaction between the imprecision and randomness of the measurements.*

In the first simulation, any observation J is a fixed interval, the width of which is 30 MHz. However, a 30-MHz interval spans a huge portion of the graphs of any of the five models $\eta_{T1},...,\eta_{T5}$. Because there is so much overlap between the graphs of these functions, the quantity $f(J|u)$ tends to be approximately constant as J varies—*unless the center frequency of J undergoes a major excursion from the mean.* Consequently,

- *Only outlier measurements will add useful information.*

Given a data stream consisting entirely of nonoutliers, the Bayes filter will be unable to resolve target pairs, such as $T4, T5$, whose signatures exhibit considerable overlap. After the collection of an outlier, the filter may suddenly decide in favor of one target type because the likelihood of the outlier now differs enough on the tails of the models η_{T4}, η_{T5} to resolve the two types.

This behavior is more pronounced in the second simulation than the first. Even though precision has increased, there will be fewer outlier observations because the degree of randomness has decreased. Outliers are not generated frequently enough to sufficiently resolve types $T4$ and $T5$.

6.6 UNMODELED TARGET TYPES

In Section 5.5.2 we introduced an approach for robustly dealing with target types that, for whatever reason, are not among those target types $v_1, ..., v_n$ that have models in our model base. Recall that, given a UGA generalized measurement Θ, we defined the likelihood of the "none of the above" or NOTA state v_0 to be

$$f(\Theta|v_0) = \Pr(\{\eta(v_1), ..., \eta(v_n)\} \cap \Theta = \emptyset). \tag{6.41}$$

This section extends the approach to AGA generalized measurements. Just as there are potentially many ways to define the generalized likelihood function of an AGA measurement, so there are many potential ways to define $f(\Theta|v_0)$.

The direct generalization of (6.41) is the following "strong" definition:

$$\begin{aligned} f(\Theta|v_0) &\triangleq \Pr(\Theta \cap (\Sigma_{v_1} \cup ... \cup \Sigma_{v_n}) = \emptyset) & (6.42) \\ &= \Pr(\Theta \subseteq (\Sigma_{v_1} \cup ... \cup \Sigma_{v_n})^c). & (6.43) \end{aligned}$$

Note that

$$\Sigma_{v_0} \triangleq (\Sigma_{v_1} \cup ... \cup \Sigma_{v_n})^c = \Sigma_{v_1}^c \cap ... \cap \Sigma_{v_n}^c \tag{6.44}$$

can be interpreted as a *random set model of the NOTA state*, in that it embodies the logical expression '*not* v_1 *and not* v_2 *and ... not* v_n' in mathematical form. Thus (6.44) defines an AGA measurement originating with an unmodeled target as one that is in complete agreement with the NOTA model.

Alternatively, one could employ Σ_{v_0} as an explicit model in the usual AGA generalized likelihood function of (6.6):

$$f(\Theta|v_0) \triangleq \Pr(\Theta \cap \Sigma_{v_0} \neq \emptyset) = \Pr(\Theta \cap \Sigma_{v_1}^c \cap ... \cap \Sigma_{v_n}^c \neq \emptyset). \tag{6.45}$$

This is a "weak" definition of the NOTA state. It defines an AGA measurement originating with an unmodeled target as one that does not flatly contradict the NOTA model Σ_{v_0}.

As an example of the strong NOTA definition, let $\Theta = \Sigma_A(g)$ be a fuzzy measurement and let the models be fuzzy: $\Sigma_{v_i} = \Sigma_A(\eta_{v_i})$ for $i = 1, ..., n$. Then

$$f(g|v_0) = 1 - \sup_{\mathbf{z}} \min\{g(\mathbf{z}), \max\{\eta_{v_1}(\mathbf{z}), ..., \eta_{v_n}(\mathbf{z})\}\}. \tag{6.46}$$

To see this, note that

$$\begin{aligned} f(g|v_0) &= \Pr(\Sigma_A(g) \cap (\Sigma_A(\eta_{v_1}) \cup ... \cup \Sigma_A(\eta_{v_n})) = \emptyset) & (6.47) \\ &= \Pr(\Sigma_A(g \wedge (\eta_{v_1} \vee ... \vee \eta_{v_n}))) = \emptyset) & (6.48) \\ &= 1 - \sup_{\mathbf{z}} g \wedge (\eta_{v_1} \vee ... \vee \eta_{v_n}) & (6.49) \\ &= 1 - \sup_{\mathbf{z}} \min\{g(\mathbf{z}), \max\{\eta_{v_1}(\mathbf{z}), ..., \eta_{v_n}(\mathbf{z})\}\} & (6.50) \end{aligned}$$

where (6.48) follows from (4.24-4.25) and (6.49) from (G.176).

With the same assumptions we get an example of the weak NOTA definition:

$$\begin{aligned} f(\Theta|v_0) &= \Pr(\Sigma_A(g) \cap (\Sigma_A(\eta_{v_1}) \cap ... \cap \Sigma_A(\eta_{v_n}))^c \neq \emptyset) & (6.51) \\ &= \Pr(\Sigma_A(g) \cap \Sigma_A(\eta_{v_1} \vee ... \vee \eta_{v_n})^c \neq \emptyset) & (6.52) \\ &= \Pr(\Sigma_A(g) \cap \Sigma_{1-A}((\eta_{v_1} \vee ... \vee \eta_{v_n})^c) \neq \emptyset) & (6.53) \\ &= \Pr(\Sigma_A(g) \cap \Sigma_{1-A}(\eta_{v_1}^c \wedge ... \wedge \eta_{v_n}^c) \neq \emptyset) & (6.54) \end{aligned}$$

where (6.52) results from (4.25) and (6.53) from (4.28). It is not evident what a closed-form formula for the final expression might be. However, one can "cheat" and use the approximation

$$f(\Theta|v_0) = \sup_{\mathbf{z}} \min\{g(\mathbf{z}), \eta^c_{v_1}(\mathbf{z}), ..., \eta^c_{v_n}(\mathbf{z})\} \tag{6.55}$$

that results from replacing $1 - A$ with A:

$$\begin{aligned} f(\Theta|v_0) &= \Pr(\Sigma_A(g) \cap \Sigma_A(\eta^c_{v_1} \wedge ... \wedge \eta^c_{v_n}) \neq \emptyset) & (6.56)\\ &= \Pr(\Sigma_A(g \wedge \eta^c_{v_1} \wedge ... \wedge \eta^c_{v_n}) \neq \emptyset) & (6.57)\\ &= \sup_{\mathbf{z}} \min\{g(\mathbf{z}), \eta^c_{v_1}(\mathbf{z}), ..., \eta^c_{v_n}(\mathbf{z})\}. & (6.58)\end{aligned}$$

Remark 7 *The strong NOTA likelihood is often much less computationally demanding than the weak one. Suppose, for example, that the fuzzy measurements are Gaussian in form:* $g(\mathbf{z}) = \hat{N}_C(\mathbf{z} - \hat{\mathbf{z}})$ *and* $\eta_{v_i}(\mathbf{z}) = \hat{N}_{C_i}(\mathbf{z} - \mathbf{z}_i)$ *where, as usual,* $\hat{N}_D(\mathbf{z}) = N_D(\mathbf{0})^{-1} N_D(\mathbf{z})$ *denotes the normalized Gaussian distribution. Then (6.42) requires the computation of*

$$\sup_{\mathbf{z}} \min\{g(\mathbf{z}), \max\{\eta_{v_1}(\mathbf{z}), ..., \eta_{v_n}(\mathbf{z})\}\}. \tag{6.59}$$

This is easily achieved provided that the intercepts of the g, $\eta_{v_1}, ..., \eta_{v_n}$ *with each other are known. Because* g, $\eta_{v_1}, ..., \eta_{v_n}$ *are normalized Gaussians, however, their intercepts can be computed in closed form by solving the equations* $\log g(\mathbf{z}) = \log \eta_{v_i}(\mathbf{z})$ *and* $\log \eta_{v_j}(\mathbf{z}) = \log \eta_{v_i}(\mathbf{z})$*. Equation (6.55) cannot, however, be computed in closed form—the intercepts must be computed numerically, using steepest descent or some other iterative technique.*

Example 42 *In Example 33 of Section 5.5.2, I illustrated my NOTA approach for UGA measurements. Here I extend this example to AGA measurements to the case when the target models* $\eta(v_1)$ *and* $\eta(v_2)$ *are no longer precise but fuzzy:*

$$\begin{aligned} \eta_{v_1}(z) &= \exp\left(-\frac{(a+z)^2}{2\sigma^2}\right) & (6.60)\\ \eta_{v_2}(z) &= \exp\left(-\frac{(a-z)^2}{2\sigma^2}\right). & (6.61)\end{aligned}$$

Assume that we have collected the fuzzy measurement

$$g_{z'}(z) = \exp\left(-\frac{(z-z')^2}{2\sigma^2}\right) \tag{6.62}$$

By (6.14) the likelihoods for target types v_1 *and* v_2 *are*

$$f(g_{z'}|v_1) = \exp\left(-\frac{(a+z')^2}{8\sigma^2}\right) \tag{6.63}$$

$$f(g_{z'}|v_2) = \exp\left(-\frac{(a-z')^2}{8\sigma^2}\right). \tag{6.64}$$

From (6.42), the "strong" likelihood function for the NOTA state v_0 *is*

$$f(g_{z'}|v_0) = 1 - \sup_{\mathbf{z}} \min\{g_{z'}(z), \max\{\eta_{v_1}(z), \eta_{v_2}(z)\}\}. \tag{6.65}$$

By symmetry, $\eta_{v_1}(z)$ *and* $\eta_{v_2}(z)$ *intercept at* $z' = 0$*, and there* $\eta_{v_1}(0) = \eta_{v_2}(0) = e^{-a^2/2\sigma^2}$*. So,* $\max\{\eta_{v_1}(z), \eta_{v_2}(z)\} = \eta_{v_1}(z)$ *for* $z \leq 0$ *and* $\max\{\eta_{v_1}(z), \eta_{v_2}(z)\} = \eta_{v_2}(z)$ *for* $z \geq 0$*. Thus* $f(g_{z'}|v_0) = 1 - f(g_{z'}|v_1)$ *for* $z' \leq 0$ *and* $f(g_{z'}|v_0) = 1 - f(g_{z'}|v_2)$ *for* $z' \geq 0$*. Consequently, the likelihood for* v_0 *is*

$$f(g_{z'}|v_0) = \begin{cases} 1 - \exp\left(-\frac{(a+z')^2}{8\sigma^2}\right) & if \quad z' < 0 \\ 1 - \exp\left(-\frac{(a-z')^2}{8\sigma^2}\right) & if \quad z' \geq 0 \end{cases}. \tag{6.66}$$

If we assume a uniform prior on the augmented state space $\mathfrak{X}_0^+ = \{v_0, v_1, v_2\}$ *then the posterior distribution on* $\mathfrak{X}_0^+$ *conditioned on* $g_{z'}$ *is*

$$f(v_1|g_{z'}) = \frac{f(g_{z'}|v_1)}{f(g_{z'})} \tag{6.67}$$

$$f(v_2|g_{z'}) = \frac{f(g_{z'}|v_2)}{f(g_{z'})} \tag{6.68}$$

$$f(v_0|g_{z'}) = \frac{f(g_{z'}|v_0)}{f(g_{z'})} \tag{6.69}$$

where $f(g_{z'}) = f(g_{z'}|v_1) + f(g_{z'}|v_2) + f(g_{z'}|v_0)$*. Let* $\sigma = 1.0$ *and* $a = 2.0$*. The posterior values are plotted as a function of* z' *in Figure 6.10. Compare this to Figure 5.7. This time* v_0 *will not be chosen as the MAP estimate for sufficiently small values of* z'*. This is because the models for* v_1 *and* v_2 *are uncertain enough that it is more difficult to distinguish them from each other. Now repeat this for* $\sigma = 1.0$ *and* $a = 3.0$*, so that the models for* v_1 *and* v_2 *are more distinguishable. The posterior values are plotted in Figure 6.11. The NOTA state will now be chosen as the MAP estimate for sufficiently small as well as sufficiently large values of* z'*.*

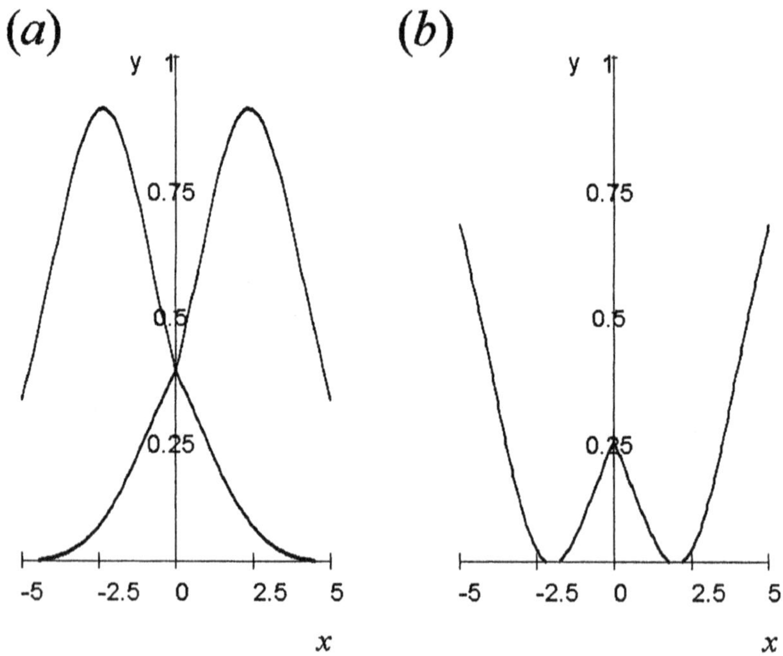

Figure 6.10 For $\sigma = 1$ and $a = 2$ (a) shows plots of the posterior probabilities for target types v_1 (left) and v_2 (right), and (b) shows a plot of the posterior probability for the "strong" NOTA type v_0.

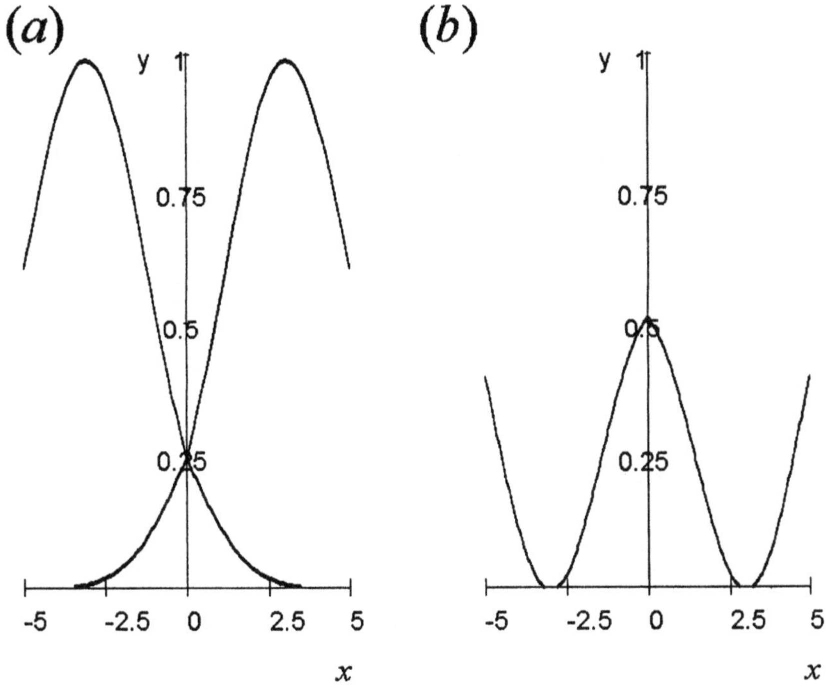

Figure 6.11 For $\sigma = 1$ and $a = 3$ (a) shows plots of the posterior probabilities for target types v_1 (left) and v_2 (right), and (b) shows a plot of the posterior probability for the "strong" NOTA type v_0 (solid curve).

Example 43 *I consider the same situation as in Example 42, but using the weak NOTA approach of (6.55).*[2] *We assume* $\sigma = 1.0$ *and* $a = 2.0$. *As noted in Remark 7, computation of the weak likelihood functions require knowledge of the up to four possible intercepts between normalized Gaussians and inverted normalized Gaussians. These are found using Newton-Raphson search. Figure 6.12(a) lists five fuzzy measurements. Figure 6.12(b) and Figure 6.12(c) display the posterior values conditioned on these measurements without and with weak NOTA, respectively. The most dramatic difference between the two tables occurs for* g_4 *and* g_5. *Without NOTA, the MAP estimate for both measurements is* v_2 *and the values of* $p(v_2|g)$ *are large, indicating very high confidence in this estimate. Stated differently, in both cases the measurement most likely originated with an unmodeled target, but it has been confused with that modeled type with which it has the greatest resemblance, however minimal. With NOTA, however, the MAP estimate for both measurements is, with high confidence,* v_0. *A less dramatic but significant difference between the tables occurs for* g_1. *In this case the measurement is exactly midway between the models. Because it has the same degree of agreement with both models* v_1 *and* v_2, *the classifier equivocates by assigning probabilities of* 0.5 *to both classes. The classifier equipped with weak NOTA, however, notes that the agreement of* g_1 *with either model is not strong. The MAP estimate is, therefore,* v_0.

6.7 EXAMPLE: TARGET ID USING LINK INT DATA

In many surveillance aircraft, workstation operators examine various forms of data such as images, extract "features" from this data (e.g., number of hubs or wheels or presence or absence of a gun), and then transmit these features on a datalink in a standard alphanumeric format.[3] Features of this kind are corrupted by certain *irreducible uncertainties*, such as: "fat-fingering" (i.e., the operator accidentally hit the wrong key); degradations in operator effectiveness due to fatigue and data overload; differences in training; or differences in individual ability.

In principal, any individual operator could be "calibrated" to statistically characterize these effects. This is not possible because "calibration data" cannot be collected in practice. Thus one can never know any given operator's likelihood function with any certainty. It is necessary, therefore, to construct a *robust-Bayes*

2 This example originally appeared in [215, p. 284].

3 The work reported in this section was completed for the Air Force Research Laboratory by Scientific Systems Company, Inc., in cooperation with Lockheed Martin MS2 Tactical Systems and Summit Research Corporation [50].

a

report	center frequency	behavior
g_1	$z' = 0$	*no significant overlap with other model*
g_2	$z' = -1$	*biased toward* v_1
g_3	$z' = -2$	*identical to* v_1
g_4	$z' = 5$	*moving out of model* v_1 *range*
g_5	$z' = 10$	*well out of range of both models*

b

report	$p(v_1 \mid g)$	$p(v_2 \mid g)$
$g = g_1$	0.5	0.5
$g = g_2$	0.731	0.269
$g = g_3$	0.881	0.190
$g = g_4$	0.006	0.994
$g = g_5$	0.0	1.0

c

report	$p(v_1 \mid g)$	$p(v_2 \mid g)$	$p(v_0 \mid g)$
$g = g_1$	0.292	0.292	0.416
$g = g_2$	0.446	0.164	0.390
$g = g_3$	0.612	0.083	0.305
$g = g_4$	0.002	0.246	0.752
$g = g_5$	$\cong 0.0$	0.003	$\cong 1.0$

Figure 6.12 Results of none-of-the-above (NOTA) simulations: (a) summarizes five fuzzy reports fed to the robust-Bayes classifier; (b) shows that without NOTA, the classifier misidentifies the unmodeled target as type v_2; and (c) shows that with NOTA, the classifier correctly identifies the target as of unknown type.

classifier algorithm that hedges against the uncharacterizable uncertainties involved in a human interpretation process.

In what follows I describe the classifier algorithm (Section 6.7.1) and tests of it against two kinds of simulated data: poor-quality "pseudodata" (Section 6.7.2) and higher-quality but "as-real-as-possible-but-unclassified data" drawn from targets in a Persian Gulf scenario (Section 6.7.3).

6.7.1 Robust-Bayes Classifier

We assume that measurements consist of features extracted from datalink message lines derived from the following sources:

- Synthetic aperture radar (SAR);
- Long range optical (LRO);
- Moving target indicator (MTI) radar;
- Electronic intelligence (ELINT);
- Signals intelligence (SIGINT).

A typical message line is assumed to look like this one:

$$z_{\mathrm{SAR}} = \mathrm{IMA/SAR/26.0F/32.0F/12.0F/Y/L/Y/S/L/T/7/N//.} \tag{6.70}$$

It is a conceptual/notional message line that describes "ideal" SAR features of a T-72 tank extracted by an operator from a SAR image. That is, this message line states that a T-72 tank has nominal hull length 26.0 feet; nominal forward length 32.0 feet; nominal width 12.0 feet; has a gun (Y); the gun is large (L) rather than small (S); has a turret (Y); the turret is of square type (S) rather than rounded (R) and is large (L) rather than small (S); is tracked (T) rather than wheeled (W); has 7 hubs; and has no radar antenna (N).

We construct a generalized likelihood function $f_{\mathrm{SAR}}(z_{\mathrm{SAR}}|v)$ that models the features in the message line. The message line contains eleven features, each drawn from a different continuous or finite measurement space. The uncertainties in these features are encoded as fuzzy membership functions $g_1, ..., g_{11}$ on those respective measurement spaces. For continuous features, such as target length, we choose a Gaussian-type representation as in (6.14):

$$g_1(l) = \exp\left(-\frac{(l-l_0)^2}{2\sigma_1^2}\right) \tag{6.71}$$

where l_0 is the operator's estimate of target length and where σ_1 quantifies our assessment of the operator's degree of trustworthiness. For a discrete feature, such as number of hubs/wheels, we assign values $g_{10}(n) = n_i$, where the largest n_i corresponds to the operator's estimate of n, and the other values are set to some small but nonzero value.

The corresponding features in the model base are likewise encoded as fuzzy membership functions $\eta_1^v, ..., \eta_{11}^v$ for each target type v.

If one assumed that the features are independent, the generalized likelihood function for the entire SAR message line would be

$$f_{\text{SAR}}(z_{\text{SAR}}|v) = f_1(g_1|v) \cdots f_{11}(g_{11}|v) \tag{6.72}$$

where

$$f_i(g_i|v) \triangleq \Pr(\Sigma_A(g_i) \cap \Sigma_A(g_i^v) \neq \emptyset). \tag{6.73}$$

However, and as noted in Section 6.6, features extracted from the same sensor source are typically correlated to some degree. So as was proposed in that section, we instead define

$$f_{\text{SAR}}(z_{\text{SAR}}|v) = f_1(g_1|v) \wedge ... \wedge f_{11}(g_{11}|v) \tag{6.74}$$

where $x \wedge y \triangleq xy/(x + y - xy)$ is the Hamacher fuzzy conjunction introduced in Example 9 of Section 4.3.4. This process is pictured in Figure 6.13. A similar process is used to construct likelihood functions $f_{\text{LRO}}(z_{\text{LRO}}|v)$, $f_{\text{MTI}}(z_{\text{MTI}}|v)$, $f_{\text{ELINT}}(z_{\text{ELINT}}|v)$, $f_{\text{SIGINT}}(z_{\text{SIGINT}}|v)$ for the other possible message lines.

The classifier algorithm consists of recursive application of Bayes' rule followed by MAP estimation. If at time step $k+1$ a message set contains a single attribute-containing message line, say $z_{k+1} = z_{\text{LRO}}$, then one updates the posterior distribution using the corresponding likelihood:

$$f_{k+1|k+1}(v|Z^{k+1}) \propto f_{\text{LRO}}(z_{\text{LRO}}|v) \cdot f_{k|k}(v|Z^k). \tag{6.75}$$

If a message set contains more than one attribute-containing message line, say $z_{k+1} = (z_{\text{LRO}}, z_{\text{ELINT}})$, then one updates the posterior distribution using the corresponding joint likelihood:

$$f_{k+1|k+1}(v|Z^{k+1}) \propto f_{\text{LRO}}(z_{\text{LRO}}|v) \cdot f_{\text{ELINT}}(z_{\text{ELINT}}|v) \cdot f_{k|k}(v|Z^k). \tag{6.76}$$

Once the posterior is updated, a map estimate of target type is extracted:

$$\hat{v}_{k+1|k+1} = \arg\max_v \; f_{k+1|k+1}(v|Z^{k+1}). \tag{6.77}$$

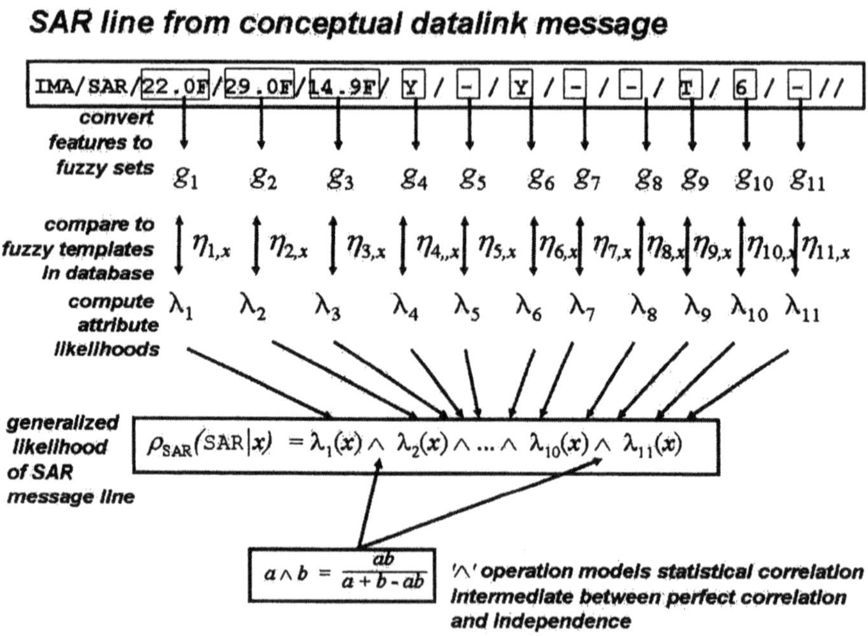

Figure 6.13 Schematic illustration of the construction of the generalized likelihood function for a notional datalink message line containing attributes extracted from a synthetic aperture radar (SAR) image by a human operator.

The corresponding probability $p_{k+1|k+1} = f_{k+1|k+1}(\hat{v}_{k+1|k+1}|Z^{k+1})$ serves as a measure of the confidence in $\hat{v}_{k+1|k+1}$: A value near unity indicates great confidence, and smaller values indicate lesser degrees of confidence.

6.7.2 "Pseudodata" Simulation Results

The robust Bayes classifier was tested against nine measurement streams consisting of low-quality synthetic "pseudodata," highly corrupted by erroneous clutter features. The possible target types v in the scenario are tanks, self-propelled guns, missile launchers, and personnel carriers. These are assigned integer labels as follows: (1) T-72 tank; (2) M1A1 tank; (3) BMP-2 amphibious fighting vehicle; (4) M-2 Bradley light armored vehicle; (5) BTR-60PB armored personnel carrier; (6) M110 self-propelled Howitzer; (7) M109A2 self-propelled Howitzer; (8) SA-8 GECKO surface-to-air missile launcher; and (9) GIAT Industries 155 mm self-propelled gun.

The robust-Bayes classifier correctly identified target type with high confidence (greater than 98%) in all nine instances. The results are summarized in the following table, where the third row indicates the number of reports required for the posterior distribution to attain or exceed the indicated probability:

Target type:	1	2	3	4	5	6	7	8	9
Certainty:	.98	.98	.98	.98	.98	.98	.98	.98	.98
Reports needed:	6	7	14	8	19	19	14	3	13

6.7.3 "LONEWOLF-98" Simulation Results

The robust-Bayes classifier was also tested against 16 targets of unknown type in an "as-real-as-possible-but-unclassified" simulated scenario called "*LONEWOLF-98*" [50]. The target types were the same as those assumed before. Taken together, these measurements consisted of nearly 100 datalink messages containing a total of 145 feature-containing message lines. The robust-Bayes classifier identified the correct target in all instances with very high confidence (98% or greater in 10 cases and 93% or greater in the remaining seven). The results are summarized in the following tables:

Target type:	1	1	5	5	8	3	3	7
Certainty:	.98	.98	.98	.98	.95	.93	.93	.98
Reports needed:	6	6	6	4	6	6	7	6

Target type:	7	2	2	4	4	6	6	8
Certainty:	.98	.98	.95	.98	.95	.98	.98	.98
Reports needed:	6	6	6	6	6	4	4	4

6.8 EXAMPLE: UNMODELED TARGET TYPES

The problem addressed is that of robustly identifying the type of ground or air targets using relatively realistic operator-extracted INT datalink attributes.[4] Seventy-five target types are modeled, including eight classes of tanks, 12 classes of armored personnel carriers (APCs), and four classes of MIG fighter jets.

The test scenario consists of 32 targets from friendly, neutral, and hostile forces. Over the duration of the scenario, the classifier algorithm receives from three to 12 measurements from each target in the scenario. Report-to-track association (i.e., target tracking) is assumed to be perfect. Thus we know a priori the measurement stream $Z^k : g_1, ..., g_k$ generated by any given target in the scenario.

The attributes are extracted from the following sensors:

1. Moving target indicator (MTI) radar: speed, force structure attributes;
2. Electronic intelligence (ELINT): source type, transmitter name, transmitter bandwidth;
3. Communications/signals intelligence (COMINT/SIGINT): speed, country of origin, call sign;
4. Images of ground targets: size, weapons, engine type, radar type;
5. Images of air targets: size, shape, engine type.

The basic classifier methodology is the "weak" NOTA approach described in Example 43 of Section 6.6. The target types are modeled as fuzzy membership functions. We use normalized Gaussian membership functions to model continuous attributes for both the measurements and the models. Each time a new measurement g_{k+1} is collected from a given target, use (6.55) to compute its generalized likelihood $f(g_{k+1}|v_i)$ for $i = 0, 1, ..., 75$. The current posterior distribution is recursively computed from the likelihoods, given the previous posterior:

$$p_{k+1|k+1}(v_i|Z^{k+1}) \propto f(g_{k+1}|v_i) \cdot p_{k|k}(v_i|Z^k). \tag{6.78}$$

4 The material in this section originally appeared in [215, pp. 285-287].

Target type at time step $k+1$ is computed as a MAP estimate:

$$\hat{v}_{k+1|k+1} = \arg\sup_{v} p_{k+1|k+1}(v|Z^{k+1}). \tag{6.79}$$

Figure 6.14 shows a typical classification result when all measurements originate with one of the modeled target types, in this case a MIG-21. The figure shows the time evolution of the posteriors

$$p_{k|k}(v_0|Z^k),\ p_{k|k}(v_1|Z^k), ..., p_{k|k}(v_{75}|Z^k) \tag{6.80}$$

for all 75 types and the NOTA type, as a total of seven measurements are collected. The correct type quickly dominates, though the NOTA class momentarily collects some mass because the sixth measurement is bad.

Figure 6.15 shows the time evolution of the posteriors when seven measurements are collected from an unmodeled target type. This sequence of reports was generated by randomly varying entries in an existing measurement sequence for a modeled target. For the first two measurements, a "plateau" of probability builds up over the airborne types. By the time of collection of the third report, the probability for the NOTA state dominates and remains dominant thereafter. If the NOTA class had not been used, the plateau would have extended out through the time of the seventh measurement.

6.9 CHAPTER EXERCISES

Exercise 24 *Verify that (6.11) reduces to (5.29):* $f(g|\mathbf{x}) = g(\eta(\mathbf{x}))$.

Exercise 25 *Prove (6.14).*

Exercise 26 *Prove (6.18):* $f(W_g|\mathbf{x}) = f(g|\mathbf{x})$.

Exercise 27 *Show that (6.19) reduces to* $f(o|\mathbf{x}) = Pl_o(O_\mathbf{x})$ *(the plausibility of* $O_\mathbf{x}$ *as defined in (4.84)) when the model* $\sigma_\mathbf{x}(O)$ *has* $O_\mathbf{x}$ *as its unique focal set. Show that (6.19) further reduces to (5.58) when* $O_\mathbf{x} = \{\eta(\mathbf{x})\}$.

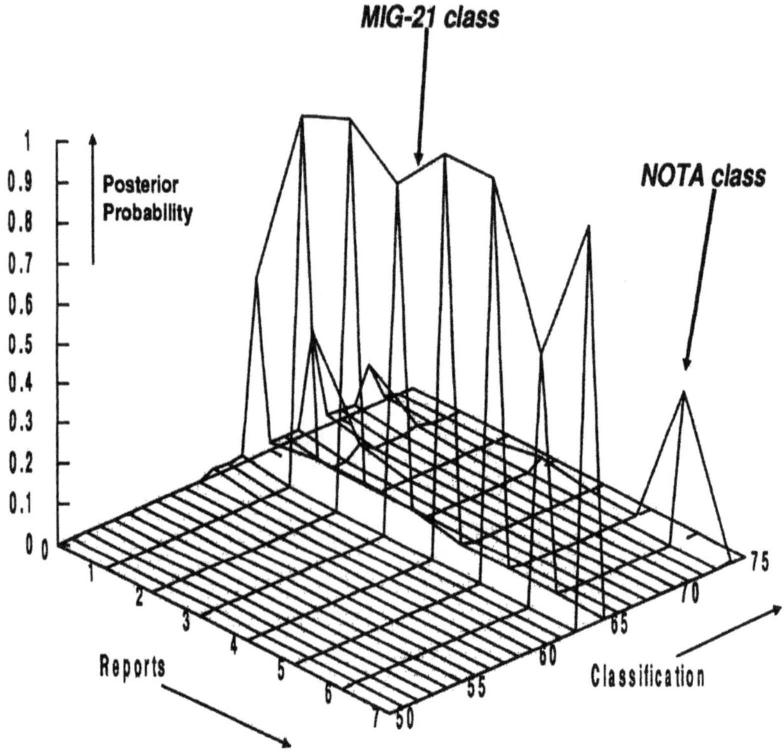

Figure 6.14 The recursive time evolution of the posterior probabilities as seven measurements are collected from a MIG-21. The correct type rapidly dominates. The sixth measurement is bad, resulting in momentary assignment of some probability mass to the NOTA state. (© 2001 SPIE. Reprinted from [215] with permission.)

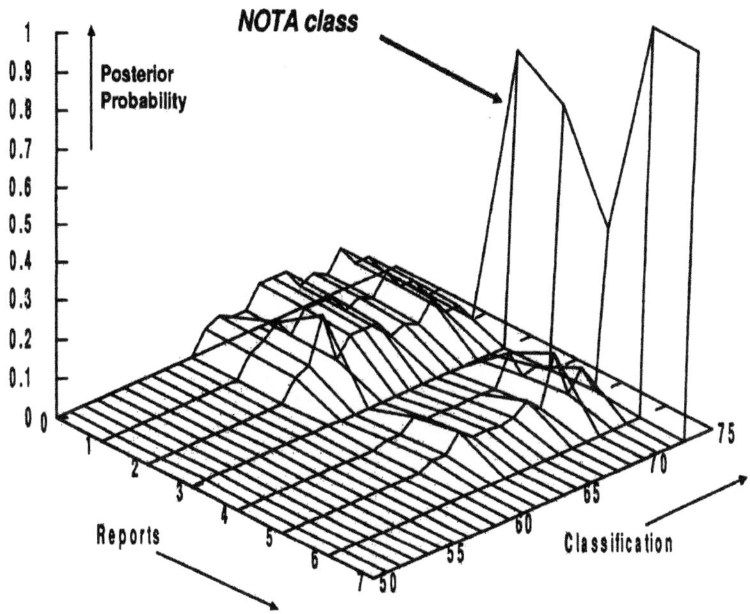

Figure 6.15 Recursive time evolution of the posterior probabilities as seven measurements are collected from an unmodeled target type. The NOTA type rapidly dominates. (© 2001 SPIE. Reprinted from [215] with permission.)

Chapter 7

AGU Measurements

7.1 INTRODUCTION TO THE CHAPTER

In this chapter, I complete my development of formal Bayes modeling of generalized measurements by extending my methods to the still more complicated case of *ambiguously generated unambiguous (AGU) measurements*. Thus far I have considered UGA measurements: measurements that are ambiguous but have precise target models $\mathbf{z} = \eta(\mathbf{x})$. I have also considered AGA measurements: measurements that do not have precise target models. AGU measurements differ from UGA and AGA measurements as follows:

- The measurement $\mathbf{z}$ itself can be precisely specified.
- Its corresponding likelihood function $f(\mathbf{z}|\mathbf{x})$ cannot.

If the uncertainty in $f(\mathbf{z}|\mathbf{x})$ is due to uncertainty in the sensor state-to-measurement transform model $\eta(\mathbf{x})$, then AGA techniques could be directly applied. In this case $\eta(\mathbf{x})$ is replaced by a random set model $\Sigma_{\mathbf{x}}$ and we just set $\Theta = \{\mathbf{z}\}$. Applying (6.6), the definition of an AGA generalized likelihood, we get:

$$f(\mathbf{z}|\mathbf{x}) = \Pr(\{\mathbf{z}\} \cap \Sigma_{\mathbf{x}} \neq 0) = \Pr(\mathbf{z} \in \Sigma_{\mathbf{x}}). \tag{7.1}$$

In the special case $\Sigma_{\mathbf{x}} = \Sigma_A(\eta_{\mathbf{x}})$, for example, we get

$$f(\mathbf{z}|\mathbf{x}) = \Pr(\mathbf{z} \in \Sigma_A(\eta_{\mathbf{x}})) = \Pr(A \leq \eta_{\mathbf{x}}(\mathbf{z})) = \eta_{\mathbf{x}}(\mathbf{z}). \tag{7.2}$$

Unfortunately, it is often difficult to devise meaningful, practical models of the form $\Sigma_{\mathbf{x}}$. In such cases a different random set technique must be used: the *random error bar*.

Example 44 *Applications in which AGU techniques are required are those in which statistically uncharacterizable real-world variations make the specification of conventional likelihoods* $f(\mathbf{z}|\mathbf{x})$ *difficult or impossible. Typical applications include ground-target identification using synthetic aperture radar (SAR) intensity-pixel images or high range-resolution radar (HRRR) intensity range-bin signatures. SAR images of ground targets, for example, can vary greatly because of the following phenomenologies: dents, wet mud, irregular placement of standard equipment (e.g., grenade launchers) on the surface of a target, placement of nonstandard equipment (e.g., a hibachi welded above the motor of a tank), turret articulation for a tank, and so on. If one could obtain the cooperation of an adversary and periodically obtain typical pieces of field equipment for RF turntable tests, one could in principle construct reasonably precise likelihood functions* $f(\mathbf{z}|\mathbf{x})$. *In actuality of course, this is not possible. So, one has no choice but to develop techniques that allow one to hedge against unknowable real-world uncertainties.*

In what follows we as usual presume the notational conventions introduced in Section 5.1.1. After summarizing the primary lessons learned for the chapter in Section 7.1.1, in Section 7.1.2, I address an existing approach for dealing with AGU data, *robust statistics*. I describe the organization of the chapter in Section 7.1.3.

7.1.1 Summary of Major Lessons Learned

The following are the major concepts, results, and formulas of the chapter:

- The concept of modeling AGA measurements using random error bars, Section 7.2;
- The concept of modeling fuzzy AGA measurements using fuzzy error bars, Section 7.4;
- Formula for generalized likelihood functions for AGA measurements, (7.27);
- Application of the approach to robust automatic target recognition (ATR) of motionless ground targets using synthetic aperture radar (SAR), Section 7.5.

7.1.2 Why Not Robust Statistics?

Robust statistics [91, 105] is a classical theory that one might use to address problems of the sort considered in this chapter. One would assume that the imperfectly known likelihood function $\ell_{\mathbf{z}}(\mathbf{x}) = f(\mathbf{z}|\mathbf{x})$ belongs to some suitably well behaved, prespecified class $\mathfrak{F}$ of functions $L(\mathbf{x})$. This family models the likelihood functions that are plausible because of uncharacterizable statistical variations.

For example, $\mathfrak{F}$ could consist of all functions $L(\mathbf{x})$ that are "near" a nominal likelihood function $\mathring{\ell}_{\mathbf{z}}$ in the sense that L is in $\mathfrak{F}$ if and only if $d(L, \mathring{\ell}_{\mathbf{z}}) < \varepsilon$ for some small $\varepsilon > 0$ and some metric $d(L, L')$ defined on functions $L(\mathbf{x}), L'(\mathbf{x})$. Alternatively, one could assume that L is in $\mathfrak{F}$ if it has the form $L(\mathbf{x}) = \varepsilon \mathring{\ell}_{\mathbf{z}}(\mathbf{x}) + (1 - \varepsilon) l(\mathbf{x})$ where the $l(\mathbf{x})$ are drawn from some other family of functions. (This approach is often called an "epsilon-contamination" model of uncertainty.) Given that $\mathfrak{F}$ is suitably chosen, it is possible to mathematically prove that $\mathbf{x}$ can be optimally estimated in a robust manner, given the degree of uncertainty modeled by the family.

Such an approach is useful in certain circumstances. For example, epsilon-contamination may accurately model uncertainty in situations where signatures consist of a dominant random process that is contaminated by a sporadic (and therefore poorly understood) secondary random process.

However I believe that, in general, conventional robust statistics approaches are self-contradictory because

- *Their claimed optimality presumes accurate models of ignorance.*

The small handful of function families $\mathfrak{F}$ normally employed in robust statistics are chosen for their *mathematical tractability*—that is, the fact that it is possible to prove optimality results if one assumes them—rather than for their fidelity to the actual structure of uncertainty in any given problem. Under conditions of uncertainty due to ignorance, an optimizable uncertainty structure $\mathfrak{F}$ may bear so little resemblance to the actual but unknowable structure of uncertainty that any claim of optimality based on it is meaningless.

The approach taken in this chapter differs in that it takes the following stance:

- The best one can do when the structure of uncertainty is unknowable, is to construct likelihood functions that *hedge* against that uncertainty.

7.1.3 Organization of the Chapter

I begin in Section 7.2 by describing AGA measurements—that is, measurements that are themselves precise, but whose likelihood functions exhibit varying degrees of uncertainty. I show how to construct the generalized likelihood function of an AGA measurement in Section 7.3. A more practical version of the approach, using fuzzy set representations, is formulated in Section 7.4.

The chapter concludes in Section 7.5 with a summary of the application of the approach to ATR of motionless ground targets using SAR intensity images.

7.2 RANDOM SET MODELS OF UGA MEASUREMENTS

Suppose that we are given an ambiguous likelihood $L_{\mathbf{z}}(\mathbf{x}) = f(\mathbf{z}|\mathbf{x})$. It should be regarded as a *random function* $\mathbf{L}_{\mathbf{z}}(\mathbf{x})$, the probability law of which is largely unknowable. If it were otherwise we could just set $\bar{L}_{\mathbf{z}}(\mathbf{x}) = \mathrm{E}[\mathbf{L}_{\mathbf{z}}^{i}(\mathbf{x})]$, that is, the expected value.

7.2.1 Random Error Bars

Since the probability law of $\mathbf{L}_{\mathbf{z}}(\mathbf{x})$ is unknown, the best that one can do is specify, for any state $\mathbf{x}$ and any measurement $\mathbf{z}$, some *error bar constraint* on the value of $L_{\mathbf{z}}(\mathbf{x})$:

$$L_{\mathbf{z}}(\mathbf{x}) \in J_{\mathbf{z}}(\mathbf{x}) \tag{7.3}$$

where $J_{\mathbf{z}}(\mathbf{x})$ is some closed interval of real numbers. However, the constraint $J_{\mathbf{z}}(\mathbf{x})$ may be only an educated guess. To hedge against this fact we specify a nested family $J_{\mathbf{z}}^{1}(\mathbf{x}) \subset ... \subset J_{\mathbf{z}}^{e}(\mathbf{x})$ of successively less constraining bounds, each $J_{\mathbf{z}}^{j}(\mathbf{x})$ having attached to it a guess $0 \leq p_{\mathbf{z}}^{j}(\mathbf{x}) \leq 1$ about how plausible this bound is, with $\sum_{j=1}^{e} p_{\mathbf{z}}^{j}(\mathbf{x}) = 1$ for each $\mathbf{z}, \mathbf{x}$. The tightest constraint $J_{\mathbf{z}}^{1}(\mathbf{x})$ describes what we believe we know about the value of $L_{\mathbf{z}}(\mathbf{x})$. We hedge against our uncertainty in this belief by allowing the possibility of alternative, successively looser constraints $J_{\mathbf{z}}^{2}(\mathbf{x}), ..., J_{\mathbf{z}}^{e}(\mathbf{x})$.

Stated differently, we are modeling our uncertainty as a *random error bar*. A random error bar is a random interval $\Sigma_{\mathbf{z}}(\mathbf{x})$ such that the probability of the instantiation $\Sigma_{\mathbf{z}}(\mathbf{x}) = J_{\mathbf{z}}^{j}(\mathbf{x})$ is $p_{\mathbf{z}}^{j}(\mathbf{x})$.

Still more generally, we can allow $\Sigma_{\mathbf{z}}(\mathbf{x})$ to have a continuously infinite number of instantiations. The random interval $\Sigma_{\mathbf{z}}(\mathbf{x})$ can be regarded as a *random interval-valued likelihood function* that models the uncertainties associated with the actual likelihood function. This is illustrated in Figure 7.1.

7.2.2 Random Error Bars: Joint Likelihoods

If $L_{\mathbf{z}}(\mathbf{x}) = L_{\mathbf{z}_1}(\mathbf{x}) \cdots L_{\mathbf{z}_m}(\mathbf{x})$ is a joint likelihood for conditionally independent observations $\mathbf{z} = (\mathbf{z}_1^T, ..., \mathbf{z}_m^T)^T$, we can still apply the same reasoning using *interval arithmetic* [167]. Suppose that $I = [a, b]$ and $I' = [a', b']$ are intervals

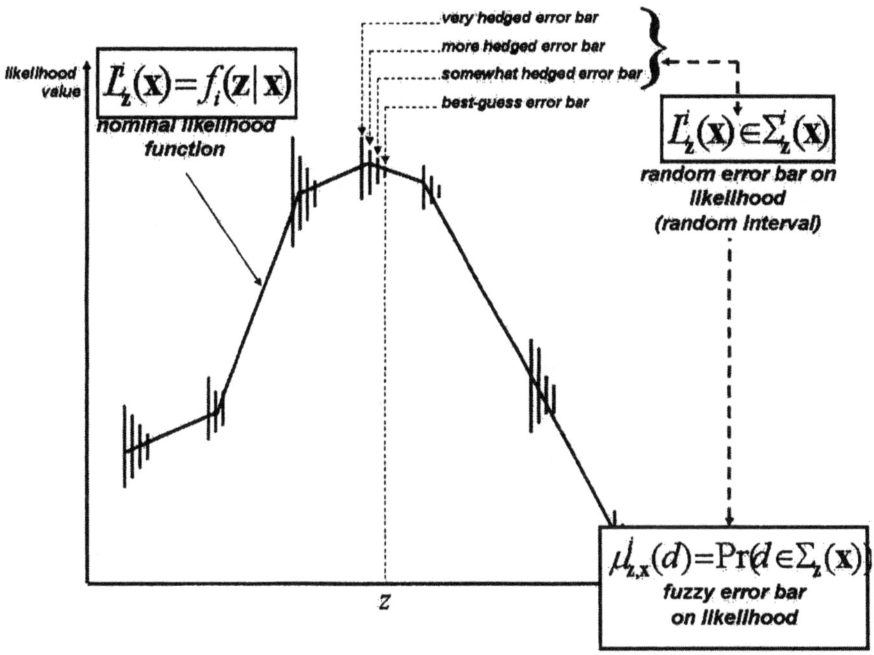

Figure 7.1 For each measurement $\mathbf{z}$ and each state $\mathbf{x}$, the uncertainty in the value $L_{\mathbf{z}}(\mathbf{x})$ of the likelihood function is modeled as a random error bar $\Sigma_{\mathbf{z}}(\mathbf{x})$. Each instantiation of $\Sigma_{\mathbf{z}}(\mathbf{x})$ is a guess about which is the correct error bar.

and define

$$\overline{I} = b, \qquad \underline{I} = a, \qquad I \cdot I' = [aa', bb'] \tag{7.4}$$

The product interval $I \cdot I'$ is just the set of all possible products $x \cdot x'$ with $x \in I$ and $x' \in I$; and $\overline{I}, \underline{I}$ denote the upper and lower bounds, respectively, of the interval I. If the intervals $J_{\mathbf{z}_1}(\mathbf{x}), ..., J_{\mathbf{z}_m}(\mathbf{x})$ impose bounds on the respective possible values of $L_{\mathbf{z}_1}(\mathbf{x}), ..., L_{\mathbf{z}_m}(\mathbf{x})$, then the product interval

$$J_{\mathbf{z}}(\mathbf{x}) = J_{\mathbf{z}_1}(\mathbf{x}) \cdots J_{\mathbf{z}_m}(\mathbf{x}) \tag{7.5}$$

describes the consequent bounds on the joint likelihood

$$L_{\mathbf{z}}(\mathbf{x}) = L_{\mathbf{z}_1}(\mathbf{x}) \cdots L_{\mathbf{z}_m}(\mathbf{x}). \tag{7.6}$$

More generally, if the random intervals $\Sigma_{\mathbf{z}_1}(\mathbf{x}), ..., \Sigma_{\mathbf{z}_m}(\mathbf{x})$ impose variable bounds on the respective values $L_{\mathbf{z}_1}(\mathbf{x}), ..., L_{\mathbf{z}_m}(\mathbf{x})$, then the *random interval-valued likelihood function*

$$\Sigma_{\mathbf{z}}(\mathbf{x}) = \Sigma_{\mathbf{z}_1}(\mathbf{x}) \cdots \Sigma_{\mathbf{z}_m}(\mathbf{x}) \tag{7.7}$$

describes the consequent (random) constraints on $L_{\mathbf{z}}(\mathbf{x})$.

7.3 LIKELIHOODS FOR AGU MEASUREMENTS

Suppose for the moment that we have a single bound $J_{\mathbf{z}}(\mathbf{x})$ on the possible values of $L_{\mathbf{z}}(\mathbf{x})$. Then each function $L(\mathbf{x})$ such that $L(\mathbf{x}) \in J_{\mathbf{z}}(\mathbf{x})$ for all $\mathbf{x}$ is a plausible likelihood function, given this particular structure of uncertainty. Its corresponding maximum likelihood estimate (MLE) $\hat{\mathbf{x}}_L = \arg\sup_{\mathbf{x}} L(\mathbf{x})$ is a plausible estimate of target state. (In general, $\hat{\mathbf{x}}_L$ could be a subset of target states.)

Let $\hat{S}_{\mathbf{z}}$ denote the set of MLE's $\hat{\mathbf{x}}_L$ for all plausible likelihoods L. Then this is the subset of all target states that could have plausibly produced the measurement $\mathbf{z}$, given our presumed structure of uncertainty. We call this the *interval MLE* and write

$$\hat{S}_{\mathbf{z}} = \text{int}\arg\sup_{\mathbf{x}} \ J_{\mathbf{z}}(\mathbf{x}) \tag{7.8}$$

where the quantity on the right is called the *interval argsup*.

In Appendix G.13, I show that the interval argsup can be computed using the following formula:

$$\hat{S}_{\mathbf{z}} = \{\mathbf{x} | \ \overline{J_{\mathbf{z}}(\mathbf{x})} \geq \sup_{\mathbf{y}} \underline{J_{\mathbf{z}}(\mathbf{y})}\}. \tag{7.9}$$

More generally,

$$\hat{\Gamma}_{\mathbf{z}} \triangleq \{\mathbf{x} | \overline{\Sigma_{\mathbf{z}}(\mathbf{x})} \geq \sup_{\mathbf{y}} \underline{\Sigma_{\mathbf{z}}(\mathbf{y})}\} \tag{7.10}$$

is the *random subset of plausible target states*, given the structure of uncertainty modeled by the random intervals $\Sigma_{\mathbf{z}}(\mathbf{x})$.

The *generalized likelihood* of the AGU measurement $\mathbf{z}$ is the probability that a target with state $\mathbf{x}$ could have plausibly generated the observation $\mathbf{z}$:

$$\lambda_{\mathbf{z}}(\mathbf{x}) \triangleq f(\mathbf{z}|\mathbf{x}) \triangleq \Pr(\mathbf{x} \in \hat{\Gamma}_{\mathbf{z}}) \tag{7.11}$$

$$= \Pr(\overline{\Sigma_{\mathbf{z}}(\mathbf{x})} \geq \sup_{\mathbf{y}} \underline{\Sigma_{\mathbf{z}}(\mathbf{y})}). \tag{7.12}$$

Example 45 *Suppose that our knowledge about the likelihood is perfect, so that* $\Sigma_{\mathbf{z}}(\mathbf{x}) = \{L_{\mathbf{z}}(\mathbf{x})\}$ *for all* $\mathbf{z}$ *and* $\mathbf{x}$. *Then (7.11) becomes*

$$\lambda_{\mathbf{z}}(\mathbf{x}) = \Pr(L_{\mathbf{z}}(\mathbf{x}) \geq \sup_{\mathbf{y}} L_{\mathbf{z}}(\mathbf{y})) \tag{7.13}$$

$$= \begin{cases} 1 & \text{if } \mathbf{x} \text{ supremizes } L_{\mathbf{z}} \\ 0 & \text{if otherwise} \end{cases}. \tag{7.14}$$

This is in accord with the construction of $\lambda_{\mathbf{z}}(\mathbf{x})$. *If the likelihood is certain then the random set* $\hat{\Gamma}_{\mathbf{z}}$ *of plausible MLE's reduces to the nonrandom set of MLE values of* $L_{\mathbf{z}}$.

Example 46 *Suppose that* $\Sigma_{\mathbf{z}}(\mathbf{x}) = [L_{\mathbf{z}}(\mathbf{x}) - \varepsilon/2, L_{\mathbf{z}}(\mathbf{x}) + \varepsilon/2]$ *for all* $\mathbf{z}$ *and* $\mathbf{x}$. *Then*

$$\lambda_{\mathbf{z}}(\mathbf{x}) = \Pr(L_{\mathbf{z}}(\mathbf{x}) + \varepsilon/2 \geq \sup_{\mathbf{y}} L_{\mathbf{z}}(\mathbf{y}) - \varepsilon/2) \tag{7.15}$$

$$= \mathbf{1}_{\varepsilon \geq \sup_{\mathbf{y}} L_{\mathbf{z}}(\mathbf{y}) - L_{\mathbf{z}}(\mathbf{x})}(\mathbf{x}). \tag{7.16}$$

7.4 FUZZY MODELS OF AGU MEASUREMENTS

Since random sets are neither as easy to understand nor as easy to mathematically manipulate as fuzzy sets, it is useful to reformulate Sections 7.2 and 7.3 in terms of the latter. Suppose that for each $\mathbf{z}$ and $\mathbf{x}$ we choose

$$\Sigma_{\mathbf{z}}(\mathbf{x}) = \Sigma_A(\theta_{\mathbf{z},\mathbf{x}}) \tag{7.17}$$

where $\theta_{\mathbf{z},\mathbf{x}}(y)$ is a fuzzy membership function on $[0,\infty)$ and where the notation $\Sigma_A(f)$ was defined in (4.21). The function $\theta_{\mathbf{z},\mathbf{x}}$ is interpreted as a *fuzzy error bar* on the value of the likelihood $L_{\mathbf{z}}(\mathbf{x})$, for each choice of $\mathbf{z}$ and $\mathbf{x}$.

If we write

$$\theta_{\mathbf{x}}(\mathbf{z},d) \triangleq \theta_{\mathbf{z},\mathbf{x}}(d) \tag{7.18}$$

then $\theta_{\mathbf{x}}$ is a fuzzy membership function on $\mathfrak{Z}_0 \times [0,\infty)$. This is pictured in Figure 7.2. When the likelihood function is precisely known—that is, when

$$\theta_{\mathbf{z},\mathbf{x}}(d) = \delta_{f(\mathbf{z}|\mathbf{x}),d} \tag{7.19}$$

where $\delta_{e,d}$ is the Kronecker delta—then the fuzzy error bar is pictured in Figure 7.3.

We assume that $\theta_{\mathbf{z},\mathbf{x}}$ is normal: $\theta_{\mathbf{z},\mathbf{x}}(y) = 1$ for some y. (In practice, it is also useful to assume that $\theta_{\mathbf{z},\mathbf{x}}$ is unimodal and that $\theta_{\mathbf{z},\mathbf{x}}(L_{\mathbf{z}}(\mathbf{x})) = 1$ where $L_{\mathbf{z}}(\mathbf{x})$ is the nominal likelihood of $\mathbf{z}$ given $\mathbf{x}$.)

The random subsets $\Sigma_A(\theta_{\mathbf{z},\mathbf{x}})$ of $\mathfrak{Z}_0$ are *consonant*. That is, the instantiations of $\Sigma_A(\theta_{\mathbf{z},\mathbf{x}})$ are linearly ordered by set inclusion. In Appendix G.14 I show that the same is true of the random subset

$$\hat{\Gamma}_{\mathbf{z}} = \{\mathbf{x}|\ \overline{\Sigma_A(\theta_{\mathbf{z},\mathbf{x}})} \geq \sup_{\mathbf{y}} \underline{\Sigma_A(\theta_{\mathbf{z},\mathbf{y}})}\} \tag{7.20}$$

of $\mathfrak{X}_0$. So, it too behaves like the random set representation of some fuzzy membership function. Note that

$$\begin{aligned}
\overline{\Sigma_a(\theta_{\mathbf{z},\mathbf{x}})} &= \sup\{y|\ a \leq \theta_{\mathbf{z},\mathbf{x}}(y)\} && (7.21)\\
&= \sup\{y|\ \theta_{\mathbf{z},\mathbf{x}}(y) = a\} && (7.22)\\
&= \sup \theta_{\mathbf{z},\mathbf{x}}^{-1}\{a\} && (7.23)
\end{aligned}$$

and

$$\begin{aligned}
\underline{\Sigma_a(\theta_{\mathbf{z},\mathbf{y}})} &= \inf\{y|\ a \leq \theta_{\mathbf{z},\mathbf{y}}(y)\} && (7.24)\\
&= \inf\{y|\ \theta_{\mathbf{z},\mathbf{y}}(y) = a\} && (7.25)\\
&= \inf\ \theta_{\mathbf{z},\mathbf{y}}^{-1}\{a\} && (7.26)
\end{aligned}$$

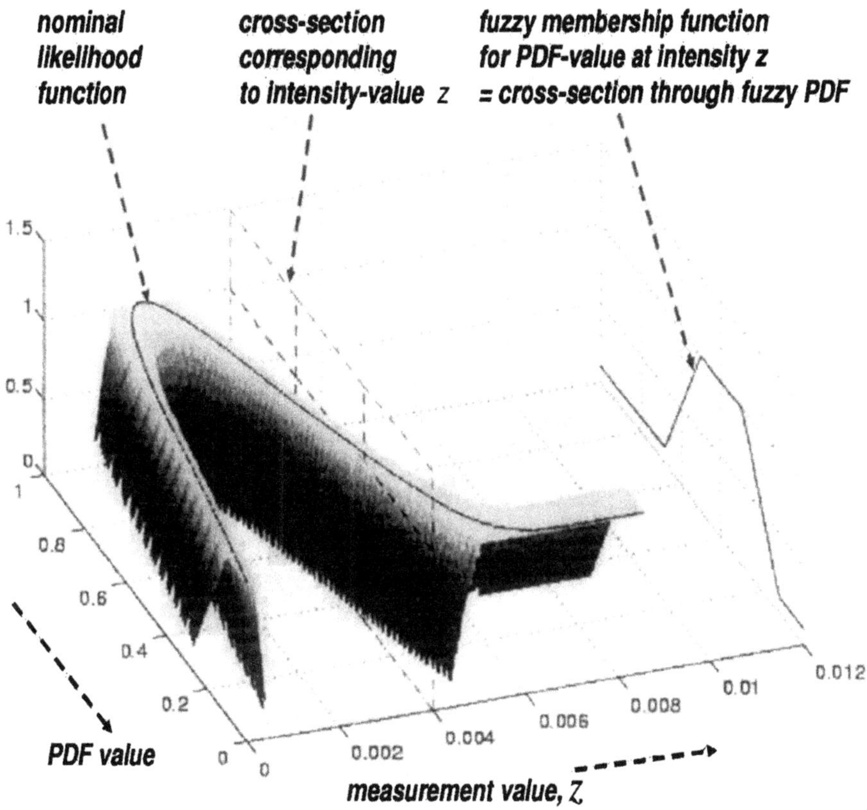

Figure 7.2 The fuzzy error bar for a generalized likelihood function for AGU measurements. It is a fuzzy membership function on the space $\mathfrak{Z}_0 \times [0, \infty)$. It is unity on the nominal likelihood function and smaller than unity elsewhere. (© 1998 SPIE. Reprinted from [157] with permission.)

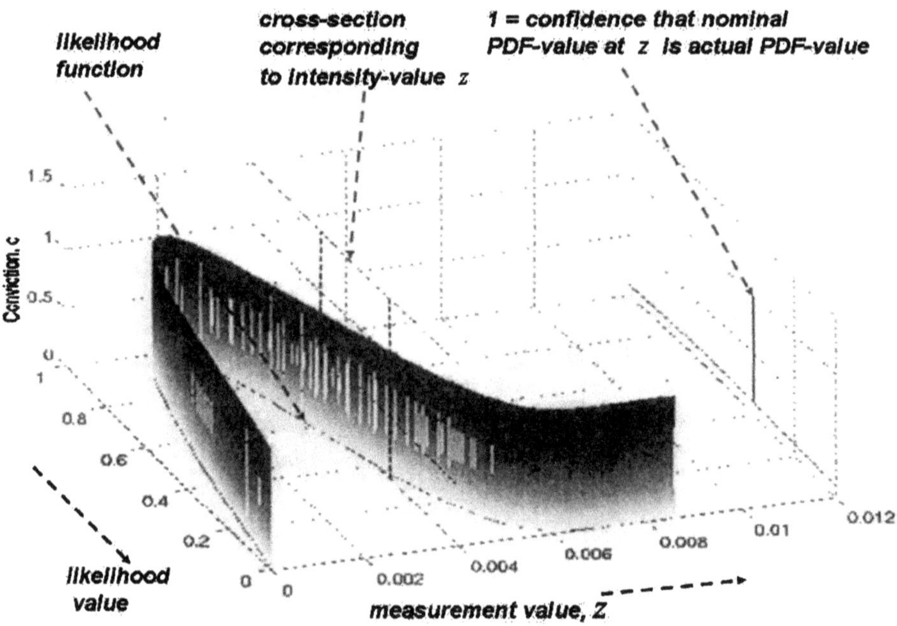

Figure 7.3 The fuzzy error bar when the generalized likelihood function is precise. (© 1998 SPIE. Reprinted from [157] with permission.)

so that we can write

$$\lambda_{\mathbf{z}}(\mathbf{x}) = \Pr\left(\sup_{\mathbf{y}} \inf\ \theta^{-1}_{\mathbf{z},\mathbf{y}}\{A\} \leq \sup \theta^{-1}_{\mathbf{z},\mathbf{x}}\{A\}\right). \tag{7.27}$$

Example 47 *We are to determine the state x of a target based on collection of a single measurement z. A nominal likelihood function $0 < L_z(x) = f(z|x) \leq 1$ is known, but its values may be erroneous. For each z and x we model the possible error in $\ell_{z,x} \triangleq L_z(x)$ using the triangular fuzzy error bars defined by*

$$\theta_{z,x}(y) \triangleq \max\{0,\ 1 - 2\alpha^{-1}_{z,x}|y - \ell_{z,x}|\} + \varepsilon \tag{7.28}$$

where $\varepsilon > 0$ is a small number. That is, the maximum of $\theta_{z,x}(y)$ is at the nominal likelihood $y = \ell_{z,x}$, and the triangle has y-intercepts at $y = \ell_{z,x} - \frac{1}{2}\alpha_{z,x}$, and at $y = \ell_{z,x} + \frac{1}{2}\alpha_{z,x}$. The number ε creates a "tail" on the fuzzy membership function $\theta_{z,x}(y)$ to prevent it from taking a zero value. Because of it, no value of y—and hence of $\ell_{z,x}$—is irrevocably excluded. Thus it hedges against the possibility that we know absolutely nothing about the value of $\ell_{z,x}$. Solving $1 - 2\alpha^{-1}_{z,x}|y - \ell_{z,x}| = a$ for y yields $y = \ell_{z,x} \pm \frac{1}{2}\alpha_{z,x}(1-a)$. Consequently,

$$\sup \theta^{-1}_{z,x}\{a\} = \ell_{z,x} + \frac{1}{2}\alpha_{z,x}(1-a) \tag{7.29}$$

$$\inf\ \theta^{-1}_{z,x}\{a\} = \ell_{z,x} - \frac{1}{2}\alpha_{z,x}(1-a) \tag{7.30}$$

when $a > \varepsilon$; and $\sup \theta^{-1}_{z,x}\{a\} = \infty$ and $\inf\ \theta^{-1}_{z,x}\{a\} = -\infty$ otherwise. From (7.27) the generalized likelihood function is

$$\begin{aligned}\lambda_z(x) &= \Pr\left(\sup_{x} \inf \theta^{-1}_{z,x}\{A\} \leq \sup \theta^{-1}_{z,x}\{A\}\right) &(7.31)\\ &= \varepsilon + \Pr\left(A > \varepsilon, \sup_{w} \inf \theta^{-1}_{z,w}\{A\} \leq \sup \theta^{-1}_{z,x}\{A\}\right) &(7.32)\end{aligned}$$

$$= \varepsilon + \Pr\left(\begin{array}{c} A > \varepsilon, \\ \sup_w \ell_{z,w} - \frac{1}{2}\alpha_{z,w}(1-A)) \leq \ell_{z,x} + \frac{1}{2}\alpha_{z,x}(1-A) \end{array}\right). \tag{7.33}$$

Now suppose that the fuzzy error bars have identical shapes regardless of target state—that is, $\alpha_{z,x} = \alpha_z$. *Some algebra shows that*

$$\lambda_z(x) \tag{7.34}$$

$$= \varepsilon + \Pr\left(\begin{array}{c} A > \varepsilon, \\ \sup_w \ell_{z,w} - \frac{1}{2}\alpha_z(1-A) \leq \ell_{z,x} + \frac{1}{2}\alpha_z(1-A) \end{array}\right) \tag{7.35}$$

$$= \varepsilon + \Pr\left(\varepsilon < A \leq 1 - \frac{1}{\alpha_z}(\sup_w \ell_{z,w} - \ell_{z,x})\right) \tag{7.36}$$

$$= \varepsilon + \max\left\{0,\ 1 - \frac{1}{\alpha_z}(\sup_w \ell_{z,w} - \ell_{z,x}) - \varepsilon\right\} \tag{7.37}$$

$$= \max\left\{\varepsilon,\ 1 - \frac{1}{\alpha_z}(\sup_w L_z(w) - L_z(x))\right\}. \tag{7.38}$$

As a specific instance, let the nominal likelihood be

$$L_z(x) = \frac{1}{\sqrt{2\pi}\sigma} \exp\left(-\frac{(x-z)^2}{2\sigma^2}\right) \tag{7.39}$$

in which case $\sup_w L_z(w) = 1/\sqrt{2\pi}\sigma$. *Thus*

$$\lambda_z(x) = \max\left\{\varepsilon,\ 1 - \frac{1}{\sqrt{2\pi}\sigma\alpha_z} + \frac{1}{\sqrt{2\pi}\sigma\alpha_z} \exp\left(-\frac{(x-z)^2}{2\sigma^2}\right)\right\}. \tag{7.40}$$

Choose $\varepsilon = 0.05$ *and* $\sigma = 1$. *If* $\alpha_z = 0.01$ *and thus there is little uncertainty in the values of* $L_z(x)$), *then the situation is as pictured in Figure 7.4. On the other hand, suppose* $\alpha_z = 0.5$ *and thus that there is significant uncertainty in the values of* $L_z(x)$. *Then the situation is as pictured in Figure 7.5. Finally, suppose that* $\alpha_z = 1.0$ *and thus there is considerable uncertainty in the values of* $L_z(x)$. *Then the situation is as pictured in Figure 7.6.*

7.5 ROBUST ATR USING SAR DATA

As was remarked in Example 44, synthetic aperture radar (SAR) images [21] of ground targets can vary greatly.[1] This is because target signatures can be

1 The work reported in this section was completed for the Air Force Research Laboratory by Scientific Systems Company, Inc., in cooperation with Lockheed Martin MS2 Tactical Systems [88].

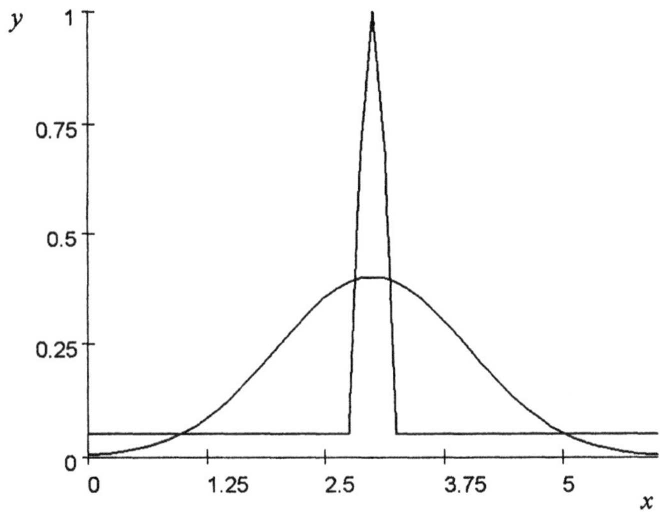

Figure 7.4 The generalized likelihood function $\lambda_z(x)$ (peaky curve) is compared to the nominal likelihood $L_z(x)$ (flatter curve) when $z = 3.0$, $\varepsilon = 0.05$, $\sigma = 1.0$, and $\alpha_z = 0.01$. Since there is little uncertainty in $L_z(x)$, $\lambda_z(x)$ is concentrated at the MLE value $x = 3$ of $L_z(x)$.

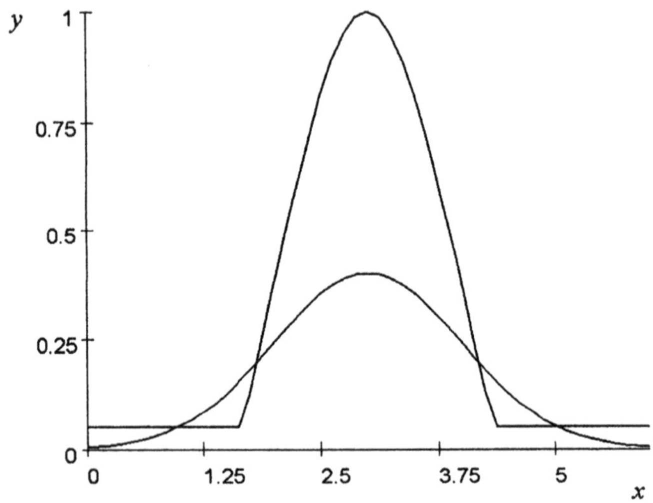

Figure 7.5 The generalized likelihood function $\lambda_z(x)$ (peaky curve) is compared to the nominal likelihood $L_z(x)$ (flatter curve) when $z = 3.0$, $\varepsilon = 0.05$, $\sigma = 1.0$, and $\alpha_z = 0.5$. Because of the significant uncertainty in $L_z(x)$, $\lambda_z(x)$ is much less concentrated at the MLE value.

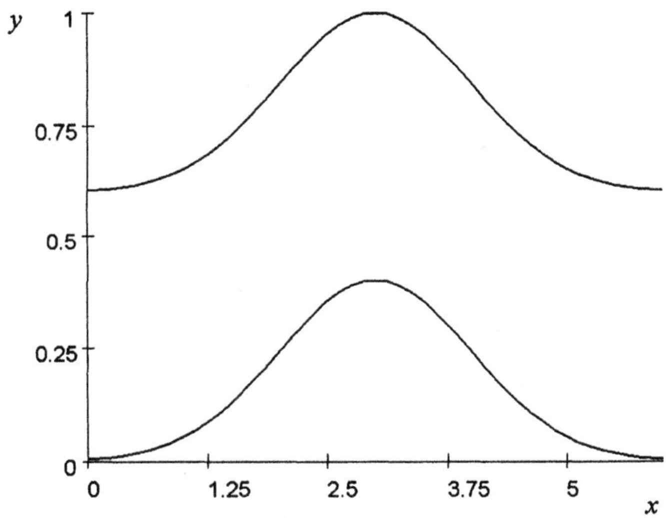

Figure 7.6 The generalized likelihood function $\lambda_z(x)$ (top curve) is compared to the nominal likelihood $L_z(x)$ (bottom curve) when $z = 3.0$, $\varepsilon = 0.05$, $\sigma = 1.0$, and $\alpha_z = 1.0$. Because of the great uncertainty in $L_z(x)$, $\lambda_z(x)$ is even less concentrated at the MLE value.

greatly altered by the presence of dents, wet mud, irregular placement of standard equipment, placement of nonstandard equipment, turret articulation, and so on. We have applied the AGU measurement techniques described in the previous sections to the problem of *automated target recognition* (ATR) of ground targets using SAR data. Because of space limitations I can only summarize the major elements of this work. Greater detail can be found in [88].

Under the ground rules of the project, we inherited a legacy ATR paradigm from the DARPA MSTAR program.[2] A pixelated SAR image is mathematically represented as an image vector $\mathbf{z} = (z_1, ..., z_N)^T$ where z_i is the SAR intensity in the ith pixel, out of a total of N pixels. Under favorable situations the generation of intensity values in the ith pixel can be described with high fidelity by a pixel likelihood function $\ell^i_z(c) = f_i(z|c)$. This describes the likelihood of observing the intensity z in the ith pixel, given that a target of class c is present. To keep things simple, we assumed that targets are correctly registered in the image and that interpixel likelihoods are statistically independent. In this case the generation of the entire SAR image can be described by an image likelihood function

$$\ell_{\mathbf{z}}(c) = f(\mathbf{z}|c) = f_1(z_1|c) \cdots f_N(z_N|c). \tag{7.41}$$

The ℓ^i_z cannot be known precisely, however, because of the uncharacterizable statistical variations just mentioned.

7.5.1 Summary of Methodology

Two target identification algorithms were designed in accordance with the theory described in earlier sections, called quantile (QUANT) and standard deviation (STDDEV). For both algorithms, the fuzzy error bars $\mu^i_{z,c}(x)$ are assumed to be trapezoidal in shape. The level set of the trapezoid at $0 \leq a \leq 1$ has lower and upper bounds

$$x^{i,a}_{z,c} = x^{i,z,c}_1(1-a) + x^{i,z,c}_2 a, \qquad y^{i,a}_{z,c} = x^{i,z,c}_3 a + x^{i,z,c}_4(1-a). \tag{7.42}$$

Here,

$$x^{i,z,c}_1 \leq x^{i,z,c}_2 \leq x^{i,z,c}_3 \leq x^{i,z,c}_4 \tag{7.43}$$

are the abscissas of the vertices

$$(x^{i,z,c}_1, 0),\ (x^{i,z,c}_2, 1), (x^{i,z,c}_3, 1), (x^{i,z,c}_4, 0) \tag{7.44}$$

2 http://www.sdms.afrl.af.mil/datasets/mstar/.

of the trapezoid. Our goal is to use training data to select the shapes of these trapezoids. That is, we are to estimate the proper values of the abscissas $x_1^{i,z,c}$, $x_2^{i,z,c}$, $x_3^{i,z,c}$, $x_4^{i,z,c}$ for all i, z, c. Both algorithms are designed to be trained on a set of data. This data is assumed to consist of a list $\ell_z^{i,1}(c),...,\ell_z^{i,E}(c)$ of likelihoods, where $\ell_z^{i,e}(c)$ is the likelihood value for the intensity value z in the ith pixel in the eth training exemplar, given the presence of a target of type c.

7.5.1.1 Quantile (QUANT) Classifier

If $0 \leq \alpha \leq 1$, the αth quantile of a random number Y is the value q_α such that $\Pr(Y < q) = \alpha$. For fixed i, z, c we treat the values $\ell_z^{i,1}(c), ..., \ell_z^{i,E}(c)$ as the instantiations of some random number Y. Choose $0 \leq \alpha_1 \leq \alpha_2 \leq \alpha_3 \leq \alpha_4 \leq 1$ and let $q_1^{i,z,c}, q_2^{i,z,c}, q_3^{i,z,c}, q_4^{i,z,c}$, respectively, denote the α_1'th, α_2'th, α_3'th, and α_4'th quantiles of Y, which we compute according to the formula

$$q_k^{i,z,c} = \ell_z^{i,j}(c) \tag{7.45}$$

where j is such that

$$\frac{\left|\{k \leq E \,|\, \ell_z^{i,k}(c) < \ell_z^{i,j}(c)\}\right|}{E} \cong \alpha_k. \tag{7.46}$$

Then, we choose trapezoidal fuzzy error bars according to the formulas

$$x_{z,c}^{i,a} = q_1^{i,z,c}(1-a) + q_2^{i,z,c}a, \qquad y_{z,c}^{i,a} = q_3^{i,z,c}a + q_4^{i,z,c}(1-a) \tag{7.47}$$

Then, from (7.11) the generalized likelihood function is

$$\lambda_{\mathbf{z}}^{\mathrm{QUANT}}(c) = \sup_{0 \leq a \leq 1} \left\{ a \,\middle|\, y_{z_1,c}^{1,a} \cdots y_{z_N,c}^{N,a} \geq \max_d x_{z_1,d}^{1,a} \cdots x_{z_N,d}^{N,a} \right\}. \tag{7.48}$$

We chose $\alpha_1 = 0.4$, $\alpha_2 = 0.5$, $\alpha_3 = 0.5$, and $\alpha_4 = 0.6$, in which case the trapezoidal fuzzy error bars are actually triangular in shape.

7.5.1.2 Standard Deviation (STDDEV) Classifier

The more uncertain we are about the mean, the larger the size of the random interval around the mean. Thus we define the vertices of the trapezoid to be functions of the mean and standard deviation over the training set. Once again, for fixed i, z, c

regard $\ell_z^{i,1}(c), ..., \ell_z^{i,E}(c)$ to be instantiations of some random number Y. Let $\bar{\ell}^i_{z,c}$ and $\sigma^i_{z,c}$ be the mean and standard deviations of this random number. Then define the trapezoidal fuzzy error bars according to the formulas

$$x^{i,a}_{z,c} = (\bar{\ell}^i_{z,c} - \alpha_1 \sigma^i_{z,c})(1-a) + (\bar{\ell}^i_{z,c} - \alpha_0 \sigma^i_{z,c})a \tag{7.49}$$

$$y^{i,a}_{z,c} = (\bar{\ell}^i_{z,c} + \alpha_3 \sigma^i_{z,c})(1-a) + (\bar{\ell}^i_{z,c} + \alpha_2 \sigma^i_{z,c})a \tag{7.50}$$

for some choice of $\alpha_0 \leq \alpha_1 \leq \alpha_2 \leq \alpha_3$ (which should not be confused with the quantile α's used for the QUANT classifier). The generalized likelihood function is

$$\lambda^{\text{STDDEV}}_{\mathbf{z}}(c) = \sup_{0 \leq a \leq 1} \left\{ a \,\middle|\, y^{1,a}_{z_1,c} \cdots y^{N,a}_{z_N,c} \geq \max_d \, x^{1,a}_{z_1,d} \cdots x^{N,a}_{z_N,d} \right\}. \tag{7.51}$$

We chose $\alpha_0 = -1.0$, $\alpha_1 = 0.0$, $\alpha_2 = 0.0$, and $\alpha_3 = 1.0$, which define triangular-shaped fuzzy error bars.

7.5.2 Experimental Ground Rules

We obtained training data for the following six target classes: (1) BMP2 infantry combat vehicle, (2) BRDM2 reconnaissance vehicle, (3) M60 tank, (4) MTU20 armored combat support vehicle, (5) T62 tank, and (6) T72 tank.

We examined five types of statistically uncharacterizable phenomena, called "extended operating conditions" (EOCs):

- Part removal (the fuel barrels of the T62 and T72 can be absent or present);
- Roll (target roll angle can vary within the range of -5 degrees to +5 degrees at a zero degree pose angle);
- Pitch (target pitch angle can vary within the range of -5 degrees to +5 degrees at a zero degree pose angle);
- Turret articulation (the articulation angle of a tank turret can vary within a range of -10 degrees and +10 degrees);
- Pose (the pose angle can vary within a range of -10 degrees to +10 degrees about the zero degree target rotation angle).

These EOCs were chosen because they are more easily simulated than other EOCs such as mud and dents.

Each exemplar used for training is an example of a particular EOC for a particular target type,[3] and consists of the following:

- A SAR intensity image (a "chip");
- A family of nominal likelihood functions.

For each of the five possible EOCs, 100 exemplars were generated, resulting in a total of 500 exemplars numbered 101 through 600. These EOCs were generated for each of the six target types, resulting in a total of 3,000 exemplars.

For our experiments, we utilized these exemplars as follows. We are given $E' = 100$ exemplars $e = 1, ..., E, ..., E'$ of a particular kind of EOC. We used the likelihood functions for the first $E = 50$ of these exemplars to train our algorithms. We then tested the algorithms on the list $\mathbf{z}_{E+1}, ..., \mathbf{z}_{E'}$ of 50 chips from the remaining exemplars. We preregistered the targets within all frames to ensure that classifier algorithms would not be inadvertently trained to classify targets on the basis of position within a frame.

Each chip in an exemplar consists of a 100-by-100 array of pixels, so that the dimension of the image vector $\mathbf{z}$ is $N = 10,000$. Each chip is the simulated output of an X-band spotlight mode, one-foot resolution SAR at a fixed depression angle. This applies only to target-generated pixels since both target shadow and clutter background were not simulated—the intensity of any nontarget pixel in any exemplar chip is set to zero. (The lack of realistic backgrounds in these images induced us to create data with more realistic backgrounds.)

The family of nominal likelihoods in an exemplar has the form $\ell_z^{i,e}(c) = f_i^e(z|c)$ for each pixel i in the image, each target class c, and each pixel intensity-value z. These likelihoods are generated by a "predictor"algorithm that determines all of the contributors to each pixel from physics-based CAD-model scatterers at the specified azimuth, elevation, and squint parameters for a particular target type. For each pixel, the predictor makes use of the contributor list to generate statistical moments for that pixel. No interpixel correlation statistics are developed. The moments are used to compute the types and parameters of the likelihood function for that pixel. The likelihood functions $\ell_z^{i,e}(c)$ are specified as normalized 100-bin histograms.

3 The exemplars used for training and testing were generated by Veridian-ERIM Corporation. [88].

7.5.3 Summary of Experimental Results

7.5.3.1 Experiments with Target Occlusion

In this experiment the STDDEV and QUANT classifiers were trained on only the Pose EOC, using exemplars 101-120 for each of the six target types. Then the image was progressively occluded by successively zeroing pixels. Four different occlusion tests were performed: left to right, right to left, top to bottom, and bottom to top. The performance of the classifiers proved to be directly related to the percentage of on-target pixels that are not occluded. Thus, to get a performance of 0.3 in the fourth graph, we require roughly 70% of pixels to be on-target. When all target pixels are occluded, both classifiers chose the BRDM2 as their estimate. This is because the BRDM2 has the fewest number of pixels on target.

7.5.3.2 Experiments with Background Clutter

All EOC exemplars had zero-energy backgrounds. To determine how well our classifier algorithms might perform against real SAR chips, we modified our exemplars by inserting natural backgrounds extracted from MSTAR clutter data. Since these backgrounds did not contain the SAR shadows of the targets and since such shadows contain much target-related information, the resulting data could be expected to have, in some ways, less information than real data. Even so, our classifiers performed well against these modified exemplars: classifier performance degraded but not dramatically so.

The MSTAR clutter data we used was generated by an X-band, strip map mode, one foot resolution SAR having roughly the same depression angle as that assumed in the generation of our exemplars. We were unable to determine if the pixel intensities for the exemplars and the MSTAR data were comparable by direct examination of the data file headers. So, we examined the statistics of the pixels themselves and concluded that they were indeed comparable. Then we overlaid the two images by simply replacing those pixels in the clutter image that corresponded to target images in the exemplar image.

Next, we modified our training exemplars by assuming that the clutter statistics in each background pixel could be modeled by a Rayleigh distribution with mean 0.0437, and that clutter noise was independent pixel to pixel. Statistical tests verified that this was a reasonable assumption. Our results are summarized in Table 7.1.

Table 7.1

Performance of Classifiers Against Background Clutter

	Articulation	Part Removal	Pitch	Roll	Pose
STDDEV	0.3008	**0.1983**	**0.0900**	**0.0822**	**0.0311**
QUANT	**0.1541**	**0.1983**	0.1361	0.1930	0.6413

It displays the values of the performance metric for each of the three classifier algorithms when trained on each of the five EOCs. The performance values are averaged over testing images for all target types and all EOCs.

Note that performance has degraded relative to our previous experiments. It is not clear if this degradation is occurring due to the nonindependent nature of the clutter or to some other phenomenon. The biggest surprise is the unexpected collapse of the QUANT classifier when trained on the Pose EOC. In particular, the QUANT classifier confuses everything with the M60, for every EOC.

7.5.3.3 Experiments with Image Coarsening

The purpose of this experiment was to stress our classifier algorithms by reducing the amount of information contained in the images fed to them. We accomplished this by coarsening the images. That is, we subdivided any given 100-by-100 array of pixels into as many disjoint n-by-n blocks as could be inscribed within them, and then collapsed each block into a single pixel by summing the intensities of the pixels within the block. Surprisingly, we found that the performance of our classifiers continued to be good. For example, we could correctly classify with up to 90% accuracy on 20-by-20 images. However, performance degraded rapidly as images decreased further in size. In general, we found that classifier performance roughly corresponded to the percentage of pixels available for the target. For "reasonable" performance, our classifiers require roughly 70% of the original pixels to be on-target.

Chapter 8

Generalized State-Estimates

8.1 INTRODUCTION TO THE CHAPTER

In the most general sense, measurements are opinions about, or interpretations of, what is being observed. Some data sources do not supply measurements, however. Instead they supply:

- A posteriori opinions about *target state*, based on measurements that they do not choose to pass on to us.

This kind of information is far more complex and difficult to process than measurements.

The most familiar example is a radar that feeds its measurements into a Kalman tracker and then passes on only the time-evolving track data $\mathbf{x}_{k|k}$, $P_{k|k}$—or, equivalently, the time-evolving posterior distribution $f_{k|k}(\mathbf{x}|Z^k) = N_{P_{k|k}}(\mathbf{x} - \mathbf{x}_{k|k})$. Successive data in the data-stream supplied by such a source will be temporally correlated, and special techniques (e.g., inverse Kalman filters or other "tracklet" techniques) must be used to decorrelate it [44].

As another example, suppose that we are to deduce the state of a single target based on state-estimates supplied by two sources. Even if individual measurements are independent, these state-estimates will be spatially correlated ("double counted") if at least some of the sources' measurements have been obtained from the same sensors. If it is known which measurements are shared then the sources can, at least in principle, be fused using a form of Bayes' rule, see (8.36) of [26]. If otherwise, and if the correlation is believed to be large, then a density-level generalization of the Julier-Uhlmann covariance intersection technique [101] must be used to address the unknown correlation [140, 95].

In this chapter I concentrate on the simplest kind of estimate-source fusion problem—which is to say, ones like that addressed in Section 4.5.2 during my discussion of Zadeh's paradox. There, two doctors supplied their *independent* diagnoses to us in the form of posterior probability distributions on a state space $\mathfrak{X}_0 = \{\mu, \tau, \chi\}$ of disease types.

General two-expert problems of this kind can be mathematically formulated in the same manner as in Section 4.5.2. The sources send us posterior distributions $f_1(\mathbf{x})$ and $f_2(\mathbf{x})$, encapsulating their respective assessments regarding the state $\mathbf{x}$ of a target. From Bayes' rule we know that

$$f_1(\mathbf{x}) = f(\mathbf{x}|\mathbf{z}^1_{k+1}) \propto f^1_{k+1}(\mathbf{z}^1_{k+1}|\mathbf{x}) \cdot f_{0,1}(\mathbf{x}) \tag{8.1}$$

$$f_2(\mathbf{x}) = f(\mathbf{x}|\mathbf{z}^2_{k+1}) \propto f^2_{k+1}(\mathbf{z}^2_{k+1}|\mathbf{x}) \cdot f_{0,2}(\mathbf{x}) \tag{8.2}$$

where the data $\mathbf{z}^1_{k+1}$, $\mathbf{z}^2_{k+1}$ and the likelihoods $f^1_{k+1}(\mathbf{z}^1|\mathbf{x})$, $f^2_{k+1}(\mathbf{z}^2|\mathbf{x})$ are unknown to us. Assume that

1. The sources are conditionally independent of target state, that is

$$f^{12}_{k+1}(\mathbf{z}^1, \mathbf{z}^2|\mathbf{x}) = f^1_{k+1}(\mathbf{z}^1|\mathbf{x}) \cdot f^2_{k+1}(\mathbf{z}^2|\mathbf{x}). \tag{8.3}$$

2. The priors $f_{0,1}(\mathbf{x})$ and $f_{0,2}(\mathbf{x})$ are known and equal,

$$f_0(\mathbf{x}) \triangleq f_{0,1}(\mathbf{x}) = f_{0,2}(\mathbf{x}). \tag{8.4}$$

Then we can fuse the two posteriors into a single joint posterior using the special case of Bayes' rule cited in (4.91)—that is, Bayes parallel combination:

$$f_{12}(\mathbf{x}) = f(\mathbf{x}|\mathbf{z}^1_{k+1}, \mathbf{z}^2_{k+1}) \propto f_1(\mathbf{x}) \cdot f_2(\mathbf{x}) \cdot f_0(\mathbf{x})^{-1}. \tag{8.5}$$

The purpose of this chapter is to extend (8.5) to posterior information that exhibits nonstatistical forms of ambiguity.

For example, state-estimates can be imprecise in that the two sources can specify the target state $\mathbf{x}$ only within respective constraints defined by subsets S_1 and S_2 of $\mathfrak{X}_0$. State-estimates could be vague, so that expert information comes in the form of fuzzy membership functions $\sigma_1(\mathbf{x})$ and $\sigma_2(\mathbf{x})$ on $\mathfrak{X}_0$. State-estimates could be uncertain, being represented as crisp or fuzzy b.m.a.s $o(S)$ or $o(\sigma)$ on $\mathfrak{X}_0$. How can one fuse state-estimates that take these disparate forms?

Most of the material in this section has been rigorously treated elsewhere using the theory of random conditional events [60, 128]. My development here is simpler and intended to show that modified Dempster's combination arises out of the UGA measurement theory of Chapter 5; see (8.28) and (8.53).

In what follows we employ the same notational conventions introduced in Section 5.1.1. Thus

$$f_0(\mathbf{x}) = f_{k+1|k}(\mathbf{x}|Z^k) \tag{8.6}$$

is the current prior distribution and

$$p_0(S) = \int_S f_0(\mathbf{x}) d\mathbf{x} \tag{8.7}$$

is its corresponding probability-mass function. If Θ is a generalized measurement, then

$$f(\mathbf{x}|\Theta) = f_{k+1|k+1}(\mathbf{x}|Z^{k+1}) \propto f(\Theta|\mathbf{x}) \cdot f_0(\mathbf{x}) \tag{8.8}$$

denotes the posterior distribution conditioned on Θ.

8.1.1 Summary of Major Lessons Learned

The following are the major concepts, results, and formulas that the reader will encounter in this chapter:

- Characterization of a UGA Dempster-Shafer state-estimate s as a transformation $s = \eta^{\sim 1} o$ of some UGA Dempster-Shafer measurement o; see (8.13);
- The likelihood of a UGA Dempster-Shafer state-estimate s; see (8.29):

$$f(s|\mathbf{x}) \triangleq \left(\sum_T \frac{s(T)}{p_0(T)} \right)^{-1} \left(\sum_T \frac{s(T)}{p_0(T)} \cdot \mathbf{1}_T(\mathbf{x}) \right); \tag{8.9}$$

- The posterior distribution conditioned on a UGA Dempster-Shafer state-estimate s; see (8.30):

$$f(\mathbf{x}|s) = \sum_S \frac{s(S)}{p_0(S)} \cdot \mathbf{1}_S(\mathbf{x}) \cdot f_0(\mathbf{x}); \tag{8.10}$$

- Pignistic probability distributions are special cases of posterior distributions conditioned on UGA Dempster-Shafer state-estimates; see (8.44);

- The primary distinction between Voorbraak and pignistic probabilities is that the former is associated with measurement fusion whereas the latter is associated with state-estimate fusion (Section 8.4.3);
- The transformation $o \mapsto \eta^{\sim 1} o$ preserves all Bayes-relevant information contained in the UGA Dempster-Shafer measurements $o_1, ..., o_m$; see (8.52):

$$f(\mathbf{x}|o_1, ..., o_m) = f(\mathbf{x}|\eta^{\sim 1} o_1 , ..., \eta^{\sim 1} o_m); \tag{8.11}$$

- Definition of modified Dempster's combination for fuzzy Dempster-Shafer state-estimates; see (8.62);
- Fusing independent UGA fuzzy Dempster-Shafer state-estimates $s_1, ..., s_m$ using modified Dempster's combination '$*_0$' is equivalent to fusing them using Bayes' rule alone; see (8.77):

$$f(\mathbf{x}|s_1 *_0 ... *_0 s_m) = f(\mathbf{x}|s_1, ..., s_m). \tag{8.12}$$

8.1.2 Organization of the Chapter

In Section 8.2 I discuss the concept of a generalized state-estimate and, in Section 8.3, define its special case: the UGA Dempster-Shafer state-estimate. I describe likelihoods and posterior distributions for such state-estimates in Section 8.4, and also show that pignistic probability is a special kind of posterior distribution. The fact that fusion of state-estimates using modified Dempster's combination is equivalent to Bayes' rule is demonstrated in Section 8.5. I show that the transformation $o \mapsto \eta^{\sim 1} o$ of an UGA Dempster-Shafer measurement into its corresponding state-estimate is Bayes invariant in Section 8.6. I briefly show how to extend these results to fuzzy Dempster-Shafer state-estimates in Section 8.7, leaving most of the work as Exercises for the reader in Section 8.8.

8.2 WHAT IS A GENERALIZED STATE-ESTIMATE?

The most familiar generalized state-estimate is the random state vector $\mathbf{X}_{k|k} \in \mathfrak{X}_0$ corresponding to the posterior distribution $f_{k|k}(\mathbf{x}|Z^k)$ transmitted by, say, a remote Kalman tracking algorithm.

The simplest instance of a nonrandom generalized state-estimate is an *imprecise state*, that is, a nonempty subset $S_0 \subseteq \mathfrak{X}_0$ that constrains the possible state $\mathbf{x}$.

If we are uncertain about this constraint we specify a sequence of nested constraints $S_0 \subset S_1 .. \subset S_e$, each assigned a probability s_i that it is the correct one. A nested list of constraints (a *fuzzy state*) can be modeled as a random closed subset Γ with $\Pr(\Gamma = S_i) = s_i$.

In general, a generalized state-estimate is an arbitrary random closed subset $\Gamma \subseteq \mathfrak{X}_0$. Intuitively, the situation is the same as in Figure 3.2, except that all subsets now reside in the state space $\mathfrak{X}_0$ rather than in some measurement space $\mathfrak{Z}_0$.

However, this is only a minimal characterization. A generalized state-estimate Γ differs fundamentally from a generalized measurement Θ in that it depends, in some as yet undetermined fashion, on the prior information contained in $f_0(\mathbf{x})$. If we are to fuse Γ with some other generalized state-estimate Γ', then this dependence on the prior must somehow be characterized. That is, we must first determine what form Γ should have, based on the measurement theory developed in earlier chapters.

The most obvious model for Γ would be as follows. If $\eta(\mathbf{x}) \overset{\text{abbr.}}{=} \eta_{k+1}(\mathbf{x})$ is the state-to-measurement transform model and if $\eta^{-1}O = \{\mathbf{x}|\ \eta(\mathbf{x}) \in O\}$ denotes its inverse image on some subset $O \subseteq \mathfrak{Z}_0$, then $\Gamma = \eta^{-1}\Theta_{k+1}$ is a random subset of state space that encapsulates the information contained in the current generalized measurement Θ_{k+1}. However, $\eta^{-1}\Theta_{k+1}$ does not capture the concept of a generalized state-estimate since it has no dependence upon the prior. How do we characterize this dependence?

I do not know the answer to this question in general. To arrive at a partial answer I must restrict to a special case as in the next section.

8.3 WHAT IS A UGA DS STATE-ESTIMATE?

Assume that Γ is discrete, so that $s(S) \triangleq \Pr(\Gamma = S)$ is zero for all but a finite number of nonempty $S \subseteq \mathfrak{X}_0$. In this section I show that, to be a generalized state-estimate, a b.m.a. $s(S)$ on $\mathfrak{X}_0$ must have the general form

$$s(S) = (\eta^{\sim 1} o)(S) \tag{8.13}$$

$$\triangleq \frac{p_0(S) \cdot \sum_{\eta^{-1}O=S} o(O)}{\sum_T o(T) \cdot p_0(\eta^{-1}T)} \tag{8.14}$$

where $o(O)$ is a UGA Dempster-Shafer measurement on $\mathfrak{Z}_0$. Note that $s(S) = 0$ whenever $p_0(S) = 0$ so that, in particular, $s(\emptyset) = 0$. (The summation $\sum_{\eta^{-1}O=S} o(O)$ is taken over all focal subsets O such that $\eta^{-1}O = S$.)

How do we arrive at (8.13)? From (5.66) we know that the posterior distribution conditioned on a UGA Dempster-Shafer measurement o is

$$f(\mathbf{x}|o) = \frac{\sum_O o(O) \cdot \mathbf{1}_{\eta^{-1}O}(\mathbf{x})}{\sum_Q o(Q) \cdot p_0(\eta^{-1}Q)} \cdot f_0(\mathbf{x}). \tag{8.15}$$

Thus for any region $S \subseteq \mathfrak{X}_0$, the value of the corresponding probability mass function $p(S|o) \triangleq \int_S f(\mathbf{x}|o)d\mathbf{x}$ evaluated on S is

$$\begin{aligned} p(S|o) &= \frac{\sum_O o(O) \cdot \int_S \mathbf{1}_{\eta^{-1}O}(\mathbf{x}) f_0(\mathbf{x}) d\mathbf{x}}{\sum_Q o(Q) \cdot p_0(\eta^{-1}Q)} && (8.16) \\ &= \frac{\sum_O o(O) \cdot \int_{S \cap \eta^{-1}O} f_0(\mathbf{x}) d\mathbf{x}}{\sum_Q o(Q) \cdot p_0(\eta^{-1}Q)} && (8.17) \\ &= \frac{\sum_O o(O) \cdot p_0(S \cap \eta^{-1}O)}{\sum_Q o(Q) \cdot p_0(\eta^{-1}Q)} && (8.18) \\ &= \frac{\sum_T (\eta^{-1}o)(T) \cdot p_0(S \cap T)}{\sum_R (\eta^{-1}o)(R) \cdot p_0(R)} && (8.19) \end{aligned}$$

where

$$(\eta^{-1}o)(T) \triangleq \sum_{\eta^{-1}O = T} o(O). \tag{8.20}$$

Examining the right-hand side of (8.19), can we discern a "natural" b.m.a. on $\mathfrak{X}_0$ that uniquely encapsulates the information contained in $p(S|o)$? We immediately see that

$$\eta^{\sim 1}(T) \triangleq \frac{(\eta^{-1}o)(T) \cdot p_0(T)}{\sum_R (\eta^{-1}o)(R) \cdot p_0(R)} \tag{8.21}$$

is such a b.m.a. Thus we can rewrite the right-hand side of (8.19) as

$$\begin{aligned} p(S|o) &= \sum_T (\eta^{\sim 1}o)(T) \cdot \frac{p_0(S \cap T)}{p_0(T)} && (8.22) \\ &= p_0(S) \cdot \sum_T (\eta^{\sim 1}o)(T) \cdot \frac{p_0(S \cap T)}{p_0(S) \cdot p_0(T)}. && (8.23) \end{aligned}$$

The factor furthest to the right on the right-hand side of this equation should be familiar. Recall that if $s(S)$ and $s'(S')$ are b.m.a.s on $\mathfrak{X}_0$, then the modified agreement with respect to the prior distribution $q(S) = p_0(S)$, (4.97), is

$$\alpha_0(s, s') = \sum_{S,S'} s(S) \cdot s'(S') \cdot \alpha_0(S, S') \tag{8.24}$$

$$= \sum_{S,S'} s(S) \cdot s'(S') \cdot \frac{p_0(S \cap S')}{p_0(S) \cdot p_0(S')} \tag{8.25}$$

Thus we can rewrite (8.19) as

$$p(S|o) = p_0(S) \cdot \sum_T (\eta^{\sim 1} o)(T) \cdot \alpha_0(S, T) \tag{8.26}$$

$$= p_0(S) \cdot \alpha_0(S, \eta^{\sim 1} o). \tag{8.27}$$

Equivalently,[1]

$$\frac{p(S|o)}{p_0(S)} = \alpha_0(S, \eta^{\sim 1} o). \tag{8.28}$$

In other words, the degree to which posterior probability mass $p(S|o)$ increases or decreases inside S, compared to the corresponding prior mass $p_0(S)$, is equal to the degree to which S agrees, in the Fixsen-Mahler sense, with $\eta^{\sim 1} o$. Thus $\eta^{\sim 1} o$ *characterizes* the posterior information supplied by o, compared to already-existing prior information.

We conclude:

- To be a UGA Dempster-Shafer state-estimate, a b.m.a. $s(S)$ on $\mathfrak{X}_0$ must have the form $s = \eta^{\sim 1} o$ of (8.13) where o is some UGA Dempster-Shafer measurement.

8.4 POSTERIOR DISTRIBUTIONS AND STATE-ESTIMATES

In this section, I define the concept of a posterior distribution conditioned on an UGA Dempster-Shafer state-estimate, and show that it is consistent with the concept of a posterior distribution conditioned on an UGA Dempster-Shafer measurement.

1 Equation (8.26) does not determine $\eta^{\sim 1} o$ uniquely. This is because $\alpha_0(o_1, o') = \alpha_0(o_2, o')$ for all o' does not necessarily imply that $o_1 = o_2$. In particular, $\alpha_0(o, o') = \alpha_0(v_o, o')$ for all o, o' where v_o is the b.m.a. on singleton subsets induced by the Voorbraak probability distribution of o.

8.4.1 The Likelihood of a DS State-Estimate

Let us be given a generalized state-estimate $s(S)$ such that $p_0(S) > 0$ whenever $s(S) > 0$. Define the "likelihood" of s in analogy with (4.120):

$$f(s|\mathbf{x}) \triangleq \left(\sum_T \frac{s(T)}{p_0(T)}\right)^{-1} \left(\sum_T \frac{s(T)}{p_0(T)} \cdot \mathbf{1}_T(\mathbf{x})\right). \tag{8.29}$$

8.4.2 Posterior Distribution Conditioned on a DS State-Estimate

The posterior distribution with respect to this likelihood is

$$f(\mathbf{x}|s) = \sum_S \frac{s(S)}{p_0(S)} \cdot \mathbf{1}_S(\mathbf{x}) \cdot f_0(\mathbf{x}). \tag{8.30}$$

To see this, note that

$$f(\mathbf{x}|s) = \frac{f(s|\mathbf{x}) \cdot f_0(\mathbf{x})}{\int f(s|\mathbf{y}) \cdot f_0(\mathbf{y}) d\mathbf{y}} \tag{8.31}$$

$$= \frac{\sum_S \frac{s(S)}{p_0(S)} \cdot \mathbf{1}_S(\mathbf{x}) \cdot f_0(\mathbf{x})}{\sum_T \frac{s(T)}{p_0(T)} \cdot \int \mathbf{1}_T(\mathbf{y}) f_0(\mathbf{y}) d\mathbf{y}} \tag{8.32}$$

$$= \frac{\sum_S \frac{s(S)}{p_0(S)} \cdot \mathbf{1}_S(\mathbf{x}) \cdot f_0(\mathbf{x})}{\sum_T \frac{s(T)}{p_0(T)} \cdot p_0(T)} \tag{8.33}$$

$$= \frac{\sum_S \frac{s(S)}{p_0(S)} \cdot \mathbf{1}_S(\mathbf{x}) \cdot f_0(\mathbf{x})}{\sum_T s(T)} \tag{8.34}$$

$$= \sum_S \frac{s(S)}{p_0(S)} \cdot \mathbf{1}_S(\mathbf{x}) \cdot f_0(\mathbf{x}). \tag{8.35}$$

Further note that

$$\int_S f(\mathbf{x}|s) d\mathbf{x} = \sum_T \frac{s(T)}{p_0(T)} \cdot \int_S \mathbf{1}_T(\mathbf{x}) f_0(\mathbf{x}) d\mathbf{x} \tag{8.36}$$

$$= \sum_T s(T) \cdot \frac{p_0(S \cap T)}{p_0(T)} \tag{8.37}$$

$$= p_0(S) \cdot \alpha_0(S, s). \tag{8.38}$$

Thus, from (8.26),

$$\int_S f(\mathbf{x}|o)d\mathbf{x} = p_0(S) \cdot \alpha_0(S, \eta^{\sim 1} o) = \int_S f(\mathbf{x}|\eta^{\sim 1} o)d\mathbf{x}. \tag{8.39}$$

Since this is for all $S \subseteq \mathfrak{X}_0$,

$$f(\mathbf{x}|o) = f(\mathbf{x}|\eta^{\sim 1} o) \tag{8.40}$$

almost everywhere. We conclude:

- *The state-estimate $\eta^{\sim 1} o$ contains exactly the same information as the UGA measurement o insofar as a Bayesian analysis is concerned.*

Example 48 *Let $\mathfrak{X}_0 = \{x_1, ..., x_N\}$ be finite and let $\{x_1\}, ..., \{x_N\}$ be the only focal sets of s, with respective weights $p_i = s(\{x_i\})$. Then (8.30) becomes*

$$\begin{aligned} f(x_j|s) &= \sum_{i=1}^{N} \frac{s(\{x_i\})}{p_0(x_i)} \cdot \mathbf{1}_{\{x_i\}}(x_j) \cdot p_0(x) & (8.41) \\ &= \sum_{i=1}^{N} \frac{p_i}{p_0(x_i)} \cdot \delta_{i,j} \cdot p_0(x_j) & (8.42) \\ &= p_j. & (8.43) \end{aligned}$$

8.4.3 Posterior Distributions and Pignistic Probability

In Section 5.3.4.2 I showed that Voorbraak probability distributions as defined in (4.115) are a special case of posterior distributions conditioned on UGA Dempster-Shafer measurements. Similarly, pignistic probability distributions as defined in (4.119) are a special case of posterior distributions conditioned on UGA Dempster-Shafer state-estimates.

To see this, let $\mathfrak{X}_0$ be finite and $q = p_0$. Then (8.30) becomes

$$f(x|s) = \sum_S \frac{s(S)}{q(S)} \cdot \mathbf{1}_S(x) \cdot q(x) = q(x) \cdot \sum_{S \ni x} \frac{s(S)}{q(S)} \tag{8.44}$$

which is just (4.119). That is:

- Pignistic probability distributions are special cases of posterior distributions conditioned on fuzzy DS state-estimates.

This also leads us to infer the following distinction between Voorbraak and pignistic probabilities:

- *Voorbraak probability is an inherent consequence of Bayesian processing of measurement information, whereas pignistic probability is an inherent consequence of Bayesian processing of state-estimate information.*

8.5 UNIFICATION OF STATE-ESTIMATE FUSION USING MODIFIED DEMPSTER'S COMBINATION

In Section 5.4.1, I showed that fusing UGA Dempster-Shafer *measurements* o and o' using Dempster's combination is equivalent to fusing them using Bayes' rule alone. In this section I show that modified Dempster's combination of (4.96) plays an analogous role for UGA Dempster-Shafer *state-estimates*. Specifically, let '$*_0$' denote modified combination with respect to the prior distribution $f_0(\mathbf{x})$. Then in Appendix G.15 I show that

$$f(s *_0 s'|\mathbf{x}) = f(s|\mathbf{x}) \cdot f(s'|\mathbf{x}) \tag{8.45}$$

From this follows

$$f(\mathbf{x}|s *_0 s') = f(\mathbf{x}|s, s') \tag{8.46}$$

since

$$\begin{aligned} f(\mathbf{x}|s *_0 s') &\propto f(s *_0 s'|\mathbf{x}) \cdot f_0(\mathbf{x}) && (8.47)\\ &= f(s|\mathbf{x}) \cdot f(s'|\mathbf{x}) \cdot f_0(\mathbf{x}) && (8.48)\\ &\propto f(\mathbf{x}|s, s'). && (8.49) \end{aligned}$$

More generally,

$$f(\mathbf{x}|s_1, ..., s_m) = f(\mathbf{x}|s_1 \ *_0 \ldots *_0 \ s_m). \tag{8.50}$$

Thus fusing Dempster-Shafer state-estimates using modified Dempster's combination is equivalent to fusing them using Bayes' rule alone.

8.6 BAYES-INVARIANT TRANSFORMATION

In Section 5.4.6, I showed that various uncertainty representations for generalized measurements can be converted into each other without loss of relevant information.

In this section, I show that the transformation

$$o \mapsto \eta^{\sim 1} o \tag{8.51}$$

of UGA Dempster-Shafer measurements o into UGA Dempster-Shafer state-estimates $\eta^{\sim 1} o$ is similarly Bayes-invariant in that this transformation leaves posterior information unchanged:

$$f(\mathbf{x}|o_1, ..., o_m) = f(\mathbf{x}|\eta^{\sim 1} o_1 \,, ..., \; \eta^{\sim 1} o_m). \tag{8.52}$$

This provides further assurance that $\eta^{\sim 1} o$ is the "natural" form that a Dempster-Shafer state-estimate should have.

Equation (8.52) is a consequence of the following facts. First, let o, o' be UGA Dempster-Shafer measurements. In Appendix G.16, I show that, provided that everything is defined,

$$\eta^{\sim 1}(o \cap o') = \eta^{\sim 1}(o * o') = \eta^{\sim 1} o \; *_0 \; \eta^{\sim 1} o' \tag{8.53}$$

where '$*_0$' denotes modified Dempster's combination with respect to the prior distribution $q = p_0$. That is, fusion of UGA measurements using Dempster's rule is the same thing as fusion of their corresponding state-estimates $\eta^{\sim 1} o$ and $\eta^{\sim 1} o'$ using modified Dempster's combination. Consequently,

$$f(\mathbf{x}|o, o') = f(\mathbf{x}|\eta^{\sim 1} o, \; \eta^{\sim 1} o'). \tag{8.54}$$

This is because

$$f(\mathbf{x}|o, o') = f(\mathbf{x}|o * o') \tag{8.55}$$

$$= f(\mathbf{x}|\eta^{\sim 1}(o * o')) \tag{8.56}$$

$$= f(\mathbf{x}|\eta^{\sim 1} o \; *_0 \; \eta^{\sim 1} o') \tag{8.57}$$

$$= f(\mathbf{x}|\eta^{\sim 1} o, \; \eta^{\sim 1} o'). \tag{8.58}$$

Here, (8.55) follows from (5.120); (8.56) from (8.40); (8.57) from (8.53); and (8.58) from (8.50). Equation (8.52) then follows from (3.73)-(3.74) and (8.55) and (8.58).

8.7 EXTENSION TO FUZZY DS STATE-ESTIMATES

The results of the previous sections can be extended to generalized state-estimates that take the form of fuzzy b.m.a.s. I sketch only the outlines of this generalization, leaving most work to the reader as exercises.

First of all, extend the probability mass function $p_O(S) = \int_S f_O(\mathbf{x})d\mathbf{x}$ from crisp sets S to fuzzy sets f by defining

$$p_O(f) \triangleq \int f(\mathbf{x}) \cdot f_O(\mathbf{x})d\mathbf{x}. \tag{8.59}$$

Note that $0 \leq p_O(f) \leq \int f_O(\mathbf{x})d\mathbf{x} = 1$. Likewise, extend modified agreement from crisp to fuzzy b.m.a.s by defining

$$\alpha_O(f, f') \triangleq \frac{p_O(f \cdot f')}{p_O(f) \cdot p_O(f')} \tag{8.60}$$

$$\alpha_O(o, o') \triangleq \sum_{f,f'} o(f) \cdot o'(f') \cdot \alpha_O(f, f') \tag{8.61}$$

where $(f \cdot f')(\mathbf{x}) \triangleq f(\mathbf{x}) \cdot f'(\mathbf{x})$. Extend modified Dempster's combination from crisp to fuzzy b.m.a.s as follows:

$$(s *_O s')(f'') \triangleq \alpha_O(o, o')^{-1} \sum_{f \cdot f' = f''} s(f) \cdot s'(f') \cdot \alpha_O(f, f'). \tag{8.62}$$

Finally, given a fuzzy b.m.a. o define

$$(\eta^{\sim 1} o)(f) \triangleq \frac{\left(\sum_{\eta^{-1} g = f} o(g)\right) \cdot p_O(f)}{\sum_{g'} o(g') \cdot p_O(\eta^{-1} g')} \tag{8.63}$$

where

$$(\eta^{-1} g)(\mathbf{x}) \triangleq g(\eta(\mathbf{x})). \tag{8.64}$$

Given this, from (5.77) we know that the posterior probability-mass function conditioned on a fuzzy b.m.a. o is

$$\int_S f(\mathbf{x}|o)d\mathbf{x} = \frac{\int_S f(o|\mathbf{x}) \cdot f_O(\mathbf{x})d\mathbf{x}}{\int f(o|\mathbf{y}) \cdot f_O(\mathbf{y})d\mathbf{y}} \tag{8.65}$$

$$= \frac{\sum_g o(g) \cdot \int_S g(\eta(\mathbf{x})) \cdot f_O(\mathbf{x})d\mathbf{x}}{\sum_{g'} o(g') \cdot \int g'(\eta(\mathbf{y})) \cdot f_O(\mathbf{y})d\mathbf{y}} \tag{8.66}$$

$$= \frac{\sum_g o(g) \cdot p_O(\mathbf{1}_S \cdot \eta^{-1} g)}{\sum_{g'} o(g') \cdot p_O(\eta^{-1} g')} \tag{8.67}$$

and thus

$$\int_S f(\mathbf{x}|o)d\mathbf{x} = p_0(\mathbf{1}_S) \cdot \sum_f (\eta^{\sim 1} o)(f) \cdot \frac{p_0(\mathbf{1}_S \cdot f)}{p_0(\mathbf{1}_S) \cdot p_0(f)} \tag{8.68}$$

$$= p_0(\mathbf{1}_S) \cdot \sum_f (\eta^{\sim 1} o)(f) \cdot \alpha_0(\mathbf{1}_S, f) \tag{8.69}$$

$$= p_0(\mathbf{1}_S) \cdot \alpha_0(S, \eta^{\sim 1} o) \tag{8.70}$$

$$= \int_S f(\mathbf{x}|\eta^{\sim 1} o) d\mathbf{x}. \tag{8.71}$$

Consequently, define the likelihood of a fuzzy b.m.a. s on $\mathfrak{X}_0$ as

$$f(s|\mathbf{x}) \triangleq \left(\sum_f \frac{s(f)}{p_0(f)} \right)^{-1} \left(\sum_f \frac{s(f)}{p_0(f)} \cdot f(\mathbf{x}) \right). \tag{8.72}$$

It is left to the reader as Exercise 28 to show that the posterior distribution conditioned on s is

$$f(\mathbf{x}|s) = \sum_f \frac{s(f)}{p_0(f)} \cdot f(\mathbf{x}) \cdot p_0(\mathbf{x}) \tag{8.73}$$

and consequently that

$$f(\mathbf{x}|o) = f(\mathbf{x}|\eta^{\sim 1} o). \tag{8.74}$$

It is further left to the reader as Exercise 29 to show that if $o * o'$ denotes Dempster's combination of fuzzy b.m.a.s as defined in (4.129), then

$$\eta^{\sim 1}(o * o') = \eta^{\sim 1} o_1 *_0 \eta^{\sim 1} o' \tag{8.75}$$

and, consequently, that the transformation $o \mapsto \eta^{\sim 1} o$ is Bayes-invariant:

$$f(\mathbf{x}|o_1, ..., o_m) = f(\mathbf{x}|\eta^{\sim 1} o_1, ..., \eta^{\sim 1} o_m). \tag{8.76}$$

It is further left to the reader as Exercise 30 to show that fusing independent UGA fuzzy state-estimates using modified Dempster's combination is equivalent to fusing them using Bayes' rule alone:

$$f(\mathbf{x}|s_1 *_0 ... *_0 s_m) = f(\mathbf{x}|s_1, ..., s_m) \tag{8.77}$$

Example 49 *Two independent expert sources provide Dempster-Shafer state-estimates* s_A *and* s_B *regarding the state of a target, as follows. The focal fuzzy sets for the first source are* $f_1(\mathbf{x})$ *and* $\mathbf{1}_{\mathfrak{X}_0}(\mathbf{x}) = 1(\mathbf{x}) \equiv 1$ *with respective weights* $1-\varepsilon$ *and* ε*. The focal fuzzy sets for the second source are* $f_2(\mathbf{x})$ *and* $\mathbf{1}_{\mathfrak{X}_0}(\mathbf{x}) = 1(\mathbf{x}) \equiv 1$ *with respective weights* $1-\varepsilon$ *and* ε*. We may assume that both sources are using the same prior information. Assume also that neither* f_1 *nor* f_2 *are identically equal to one. The focal sets of the combined information* $s_{A,B} = s_A *_0 s_B$ *are* $f_1 \cdot f_2$, $f_1 \cdot 1 = f_1$, $1 \cdot f_2 = f_2$, *and* $1 \cdot 1 = 1$*. The modified agreement is*

$$\alpha_0(s_A, s_B) = (1-\varepsilon)^2\alpha_0(f_1 \cdot f_2) + \varepsilon(1-\varepsilon)\alpha_0(1, f_2) \tag{8.78}$$

$$+\varepsilon(1-\varepsilon)\alpha_0(f_2, 1) + \varepsilon^2\alpha_0(1,1) \tag{8.79}$$

$$= (1-\varepsilon)^2\alpha_0(f_1 \cdot f_2) + \varepsilon(1-\varepsilon) + \varepsilon(1-\varepsilon) + \varepsilon^2 \tag{8.80}$$

$$= (1-\varepsilon)^2\alpha_0(f_1 \cdot f_2) + 2\varepsilon - \varepsilon^2 \tag{8.81}$$

where

$$\alpha_0(f_1 \cdot f_2) = \frac{p_0(f_1 \cdot f_2)}{p_0(f_1) \cdot p_0(f_2)} \tag{8.82}$$

$$= \frac{\int f_1(\mathbf{x}) \cdot f_2(\mathbf{x}) \cdot f_0(\mathbf{x}) d\mathbf{x}}{\left(\int f_1(\mathbf{x}) f_0(\mathbf{x}) d\mathbf{x}\right)\left(\int f_2(\mathbf{x}) f_0(\mathbf{x}) d\mathbf{x}\right)}. \tag{8.83}$$

The respective weights of the four focal fuzzy sets are

$$s_{A,B}(f_1 \cdot f_2) = \frac{(1-\varepsilon)^2 \cdot \alpha_0(f_1, f_2)}{\alpha_0(s_A, s_B)} \tag{8.84}$$

$$= \frac{(1-\varepsilon)^2 \cdot \alpha_0(f_1, f_2)}{(1-\varepsilon)^2\alpha_0(f_1 \cdot f_2) + 2\varepsilon - \varepsilon^2} \tag{8.85}$$

$$s_{A,B}(f_1) = \frac{\varepsilon(1-\varepsilon)}{\alpha_0(s_A, s_B)} \tag{8.86}$$

$$= \frac{\varepsilon(1-\varepsilon)}{(1-\varepsilon)^2\alpha_0(f_1 \cdot f_2) + 2\varepsilon - \varepsilon^2} \tag{8.87}$$

and

$$s_{A,B}(f_2) = \frac{\varepsilon(1-\varepsilon)}{\alpha_0(s_A, s_B)} \tag{8.88}$$

$$= \frac{\varepsilon(1-\varepsilon)}{(1-\varepsilon)^2\alpha_0(f_1 \cdot f_2) + 2\varepsilon - \varepsilon^2} \tag{8.89}$$

$$s_{A,B}(\mathbf{1}) = \frac{\varepsilon^2}{\alpha_0(s_A, s_B)} \tag{8.90}$$

$$= \frac{\varepsilon^2}{(1-\varepsilon)^2\alpha_0(f_1 \cdot f_2) + 2\varepsilon - \varepsilon^2}. \tag{8.91}$$

Note that if $\varepsilon = 0$*, so that we are fusing only the focal sets* f_1 *and* f_2*, the combined fuzzy b.m.a. has only a single focal fuzzy membership function* $f_1 \cdot f_2$*, the weight of which is* 1*. We conclude: Prior information has no effect on the fusion of purely fuzzy state-estimates.*

8.8 CHAPTER EXERCISES

Exercise 28 *Prove (8.73) and (8.74).*

Exercise 29 *Prove (8.75) and (8.76).*

Exercise 30 *Prove (8.77).*

Chapter 9

Finite-Set Measurements

9.1 INTRODUCTION TO THE CHAPTER

In previous chapters I have implicitly assumed that measurements collected from single targets and their backgrounds can be adequately represented as vectors in some Euclidean space $\mathbb{R}^M$ or at least in some continuous-finite hybrid space $\mathbb{R}^M \times C$. Many data types that occur in application cannot be accurately represented in this fashion. The most prominent examples are *detection-type* measurements and their special case, *extended-target* measurements. These are by no means the only instances.

The purpose of this chapter is twofold: first, to introduce the concept of a *finite-set measurement,* and second, to argue that finite-set measurements constitute a major theoretical and practical challenge. This is because:

- The vector-based Bayesian measurement formalism we relied on in Chapter 2 cannot be directly brought to bear on such measurements. In its place, a *finite set-based* formalism must be devised and elaborated.

The details of a measurement theory for finite-set measurements must be deferred to Chapter 12. My purpose here is only to describe how finite-set measurements arise and to cast a spotlight on the gaps that must be filled to achieve such a measurement theory. In this sense, this chapter is a continuation of Chapter 3.

9.1.1 Summary of Major Lessons Learned

The following are the major ideas and concepts to be introduced in this chapter:

- Finite-set measurements are encountered in many practical applications.
- Such applications do not necessarily involve detection-type data.
- Formal Bayes modeling can be generalized to finite-set measurements.
- This requires a generalized integro-differential calculus based on the *set integral* and its inverse operation, the *set derivative*.

9.1.2 Organization of the Chapter

In Section 9.2, I describe several real-world sensing scenarios in which finite-set measurements are routinely encountered. In Section 9.3, I sketch the manner by which, in subsequent chapters, I will construct measurement models and true likelihood functions for finite-set measurements.

9.2 EXAMPLES OF FINITE-SET MEASUREMENTS

In this section I examine several of the more familiar applications in which finite-set measurements arise. These include *signal detections* for ground-to-air surveillance radars (Section 9.2.1), *clutter detections* for ground moving target indicator (GMTI) radars (Section 9.2.2), *extended-target detections* for millimeter-wave seeker radars (Section 9.2.3), *feature extraction-sets* for imaging sensors (Section 9.2.4), collections of *generalized measurements* (Section 9.2.5), and *multisensor measurements* (Section 9.2.6).

9.2.1 Ground-to-Air Radar Detection Measurements

Detection-type measurements are commonly encountered with radar sensors [211]. Consider, for example, a radar with a mechanically rotating antenna. This radar periodically projects radio frequency (RF) pulses into space within a major lobe aligned in the forward direction of the antenna.[1] As a pulse spreads, it propagates into space as a shell. When this shell impinges on an aircraft surface, some of the RF energy of the pulse is back-reflected and collected by the radar antenna. Range to the aircraft can be estimated using the time delay elapsed between transmission of the pulse and collection of its reflection.

Since the pulse is generally very brief, the range resolution of the radar is typically good. The azimuthal resolution is typically not nearly as good, since there

1 There are typically also much smaller sidelobes that can generate clutter.

is a limit to how narrow the major lobe can be made. Resolution in elevation for this type of radar is typically poor to nonexistent. For convenience I will assume that the radar has good resolution in both azimuth α and range r, and none at all in elevation.

As the radar antenna rotates, the back reflections it collects create an RF-intensity signature of the form $s(r, \alpha)$—that is, an intensity-valued function of range r and azimuth angle α. The function $s_r(\alpha) \triangleq s(r, \alpha)$ is pictured in Figure 9.1(a) for a fixed range r. If the target has significant size and reflectivity in the RF band of interest, target-generated reflection-intensities will be significantly stronger than the surrounding noise floor of the radar. Thus, as is also pictured in Figure 9.1(a), the target is easily extracted from the signature by setting a threshold value τ that is high enough to avoid the noise floor but small enough to detect the target-generated intensity. The event $s_r(\alpha) > \tau$ is called a *detection*. Since the detection indicates the presence of a target with range r and azimuth α, it can be represented as a vector $\mathbf{z} = (r, \alpha)$ or perhaps as $\mathbf{z} = (r, \alpha, I)$, where $I = s(r, \alpha)$.

On occasion, the target-generated signal may be too small to exceed the threshold, in which case the target is not detected by the test $s(r, \alpha) > \tau$. In this case we say that a *missed detection* has occurred. The probability of such an event is called the *probability of missed detection* and is denoted $1-p_D \overset{\text{abbr.}}{=} 1-p_D(r, \alpha; \tau)$. If the target is just within the maximum detection range of the radar, target-generated intensities are not so obviously separated from the noise floor. As depicted in Figure 9.1(b), the test $s(r, \alpha) > \tau$ will single out not only a target detection but also a large number of false detections, called *false alarms*. If τ is set too large, it may be impossible to detect the target at all. If τ is set small enough to ensure target detection, there is a penalty: a large number of false alarms. The smaller the value of τ the larger their average number, denoted $\lambda \overset{\text{abbr.}}{=} \lambda(r, \alpha; \tau)$.

This trade-off between increased probability of detection versus increased false alarms is what is commonly described by a receiver operating characteristic (ROC) curve [231, pp. 36-46].

When multiple targets are present in a scene, one can get multiple target detections. In such situations it is possible that a single signature peak is due to a superposition of the signals of two or more targets that are very close together (relative to the resolution of the sensor). Such targets are said to be *unresolved.*[2]

Generally speaking, the result of a detection process is *a measurement that is not a vector*. Let $\mathbf{z}_1 = (r_1, \alpha_1), \ldots, \mathbf{z}_m = (r_m, \alpha_m)$ be the collected detections, false or otherwise. The vector $(\mathbf{z}_1, \ldots, \mathbf{z}_m)$ has fixed dimension m, but the

2 Another common usage is to describe the *joint measurement* generated by closely spaced targets as being unresolved [16, p. 15].

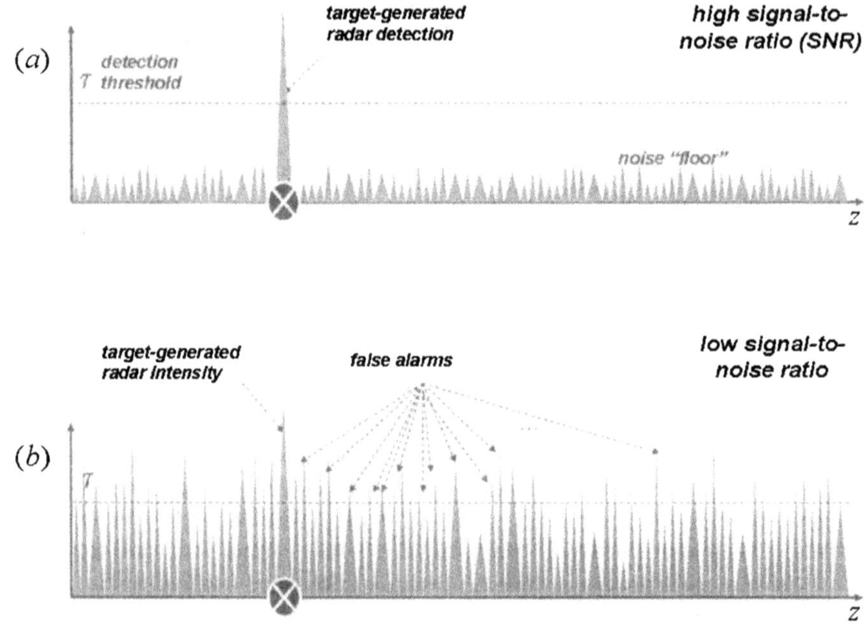

Figure 9.1 The intensity signature from a mechanically rotating radar is depicted, for a constant range-bin. If signal-to-noise ratio (SNR) is large, the signal return from a target greatly exceeds the underlying sensor noise "floor." The target range and azimuth can be deduced by applying a suitable theshold τ to the signature, as in (a). If the noise floor is higher, however, any value of τ small enough to detect the target will also detect a large number of false detections (false alarms), as shown in (b).

number of detections m can vary from zero to some arbitrarily large number. Also, the components $\mathbf{z}_1, ..., \mathbf{z}_m$ of $(\mathbf{z}_1, ..., \mathbf{z}_m)$ have fixed order, whereas the detections $\mathbf{z}_1, ..., \mathbf{z}_m$ have no inherent physical order. Thus the actual measurement is more accurately represented as a *finite observation set*, the elements of which are individual detections:

$$Z = \{\mathbf{z}_1, ..., \mathbf{z}_m\} = \{(r_1, \alpha_1), ..., (r_m, \alpha_m)\}. \tag{9.1}$$

Stated differently, for detection-type measurements

- *The measurement space* $\mathfrak{Z}$ *is the hyperspace of all finite subsets of an underlying measurement space* $\mathfrak{Z}_0$.

Not only the individual detections (r_i, α_i) will be randomly varying, but also their number m. Thus the random measurement is a *randomly varying finite subset* Σ of $\mathfrak{Z}_0$. The various possibilities are:

$$Z = \emptyset \quad \text{(no detections collected)} \tag{9.2}$$
$$Z = \{\mathbf{z}_1\} \quad \text{(a single detection } \mathbf{z}_1 \text{ is collected)} \tag{9.3}$$
$$Z = \{\mathbf{z}_1, \mathbf{z}_2\} \quad \text{(two detections } \mathbf{z}_1 \neq \mathbf{z}_2 \text{ are collected)} \tag{9.4}$$
$$\vdots$$
$$Z = \{\mathbf{z}_1, ..., \mathbf{z}_m\} \quad (m \text{ detections } \mathbf{z}_1 \neq ... \neq \mathbf{z}_m \text{ are collected}) \tag{9.5}$$
$$\vdots$$

This type of measurement will be addressed in generality in Section 12.3.

9.2.2 Air-to-Ground Doppler Detection Measurements

Ground moving target indicator (GMTI) radars [204] provide a second but somewhat different example of detection-type measurements. A GMTI radar is a Doppler radar. The measurement collected from a target is a measure of the relative speed of the target along the line segment connecting sensor to target. Target returns normally appear as intensities significantly larger than the general noise background, and thus can be extracted from the GMTI signature as detections.

One can also collect *clutter detections*. Anything on the ground with significant motion relative to the sensor can produce a GMTI detection—rotating windmills, helicopter rotors, and so on. So in addition to the false alarm process, which is an artifact of the detection technique, one can also have a *clutter process*, that consists of detections generated by "confuser" targets [16, p. 8].

9.2.3 Extended-Target Detection Measurements

The measurements generated by *extended targets* provide another example of measurements that are finite sets of conventional measurements. Suppose that a millimeter-wave seeker radar radiates a pulse at a ground target that is relatively near the radar. If the wavelength of the pulse is sufficiently small, it will tend to be back-reflected much more strongly by those areas on the target surface that resemble corner reflectors. When the back-reflections are collected, the result is an image of the target consisting of a large number of point intensities—detections, essentially—which markedly stand out from the rest of the back-reflection.

Thus any observation generated by an extended target will, as with conventional surveillance radar detections, also be a finite set Z of vectors. The difference, of course, lies in the differing sensor phenomenologies. In the surveillance radar case most of the elements of a measurement set Z will not provide any information about the target state. In the millimeter-wave radar case, most of them will. This type of measurement process is addressed in detail in Section 12.7. See also Figure 12.2.

9.2.4 Features Extracted from Images

The type of detection measurements we have considered thus far involve a particular type of *feature*—the exceeding of a threshold value—extracted from a signature [118]. This is not the only kind of detection measurement that can be extracted from a signature by a digital signal processing (DSP) algorithm.

Consider real-time imaging sensors such as electro-optical (EO) or infrared (IR) cameras. An edge detector can be applied to an image, extracting the outlines of objects in it. The centroids of these outlines can then be treated as "detections." What results is a randomly varying finite set of point observations.

As another example, one can apply a corner detector to an image, isolating significant points on the surfaces of objects. The resulting random set of points resemble the observations of extended targets as described in Section 9.2.3.

9.2.5 Human-Mediated Features

Many feature types are not produced by automated algorithms but are, rather, generated by one or more human operators. Chapters 5-7 were devoted to the modeling of such "generalized measurements" as random closed subsets Θ of one or more underlying measurement spaces $\mathfrak{Z}_0$. Human-mediated information-collection processes can result in finite measurement sets of the form $Z = \{\Theta_1, ..., \Theta_m\}$ where now the individual measurements are random closed subsets Θ_j rather than conventional measurement vectors.

9.2.6 General Finite-Set Measurements

Detection-type measurements, however defined, do not exhaust the situations in which collected data will consist of finite sets of conventional measurements. Suppose, for example, that multiple sensors collect observations from some scenario, multitarget or otherwise. If the number of sensors were of known constant number m, the joint measurement collected by all sensors would be a finite set of the form $\{\overset{1}{\mathbf{z}}, ..., \overset{s}{\mathbf{z}}\}$ where $\overset{i}{\mathbf{z}}$ originates with the ith sensor. Since sensors may enter and leave a scenario or may fail to deliver their observations (e.g., because of transmission drop-outs), the number s of sensors is actually randomly varying. Once again, such observations are not accurately modeled as vectors.

9.3 MODELING FINITE-SET MEASUREMENTS?

In this section, I sketch the formal Bayes measurement modeling process for finite-set measurements to be described in more detail in subsequent chapters. After reviewing conventional formal modeling in Section 9.3.1, I introduce the concept of a multiobject *set integral* in Section 9.3.2. In Section 9.3.3 I present a simple example: measurement modeling of detection-type finite-set measurements. The corresponding true likelihood function for this model is described, without derivation, in Section 9.3.4. I conclude with a short discussion of the *set derivative* and its use in systematic construction of likelihood functions from measurement models (Section 9.3.5).

9.3.1 Formal Modeling of Finite-Set Measurements

The basic concepts and procedures of formal Bayes modeling and filtering for general measurement and state spaces were summarized in Section 3.5.1. Their application to vector-based observations was described in Section 2.4.4. For the sake of clarity, I review what we learned in the latter section.

9.3.1.1 Conventional Formal Modeling (A Review)

Given measurements $\mathbf{z}$ in some measurement space $\mathfrak{Z}_0$, we must construct a likelihood function $L_{\mathbf{z}}(\mathbf{x}) = f(\mathbf{z}|\mathbf{x})$ that describes the likelihood that a single sensor will collect a vector observation $\mathbf{z}$ from a target with vector state $\mathbf{x}$. This likelihood function must be "true" in the sense that it faithfully describes the sensor's statistical behavior. That is, it faithfully incorporates the information in the measurement model, without inadvertently introducing information extraneous to the model.

True likelihoods are constructed by first formulating a formal statistical measurement model. For example, $\mathbf{Z} = \eta(\mathbf{x}) + \mathbf{W}$ where $\mathbf{x} \mapsto \eta(\mathbf{x})$ models the deterministic generation of measurements from states and where the random vector $\mathbf{W}$ models the random variation in these measurements caused by sensor noise.

In Section 2.4.4 I demonstrated that the true likelihood function corresponding to $\mathbf{Z} = \eta(\mathbf{x}) + \mathbf{W}_{k+1}$ is $f(\mathbf{z}|\mathbf{x}) = f_{\mathbf{W}}(\mathbf{z} - \eta(\mathbf{x}))$. I did this as follows. I began with the fundamental probability integral

$$p(S|\mathbf{x}) = \int_T f(\mathbf{z}|\mathbf{x})d\mathbf{z} \tag{9.6}$$

where

$$p(S|\mathbf{x}) \triangleq \Pr(\mathbf{Z} \in T|\mathbf{x})$$

is the conditional probability-mass function of $\mathbf{Z}$ given $\mathbf{x}$. Using elementary algebra and basic properties of integrals, I showed that

$$\int_T f(\mathbf{z}|\mathbf{x})d\mathbf{z} = \int_T f(\mathbf{z} - \eta(\mathbf{x}))d\mathbf{z}. \tag{9.7}$$

Finally, I noted that if this is true for all T then $f(\mathbf{z}|\mathbf{x}) = f_{\mathbf{W}}(\mathbf{z} - \eta(\mathbf{x}))$ almost everywhere.

9.3.1.2 Formal Modeling for Finite-Set Measurements

Now consider finite-set measurements. Continue to assume that a single sensor observes a single target with unknown state $\mathbf{x}$, but that at any instant it collects a finite set $Z = \{\mathbf{z}_1, ..., \mathbf{z}_m\}$ of conventional measurements $\mathbf{z}_1, ..., \mathbf{z}_m$ from a conventional measurement space $\mathfrak{Z}_0$:

$$
\begin{array}{lcl}
Z = & \emptyset & \text{(no measurements collected)} \\
Z = & \{\mathbf{z}_1\} & \text{(one measurement } \mathbf{z}_1 \text{ collected)} \\
Z = & \{\mathbf{z}_1, \mathbf{z}_2\} & \text{(two measurements } \mathbf{z}_1, \mathbf{z}_2 \text{ collected)} \\
\vdots & & \vdots \\
Z = & \{\mathbf{z}_1, ..., \mathbf{z}_m\} & (m \text{ measurements } \mathbf{z}_1, ..., \mathbf{z}_m \text{ collected)} \\
\vdots & \vdots & \vdots
\end{array}
\tag{9.8}
$$

We must construct a true likelihood function $f(Z|\mathbf{x})$ that describes the likelihood that any given measurement set Z will be collected if a target with state $\mathbf{x}$ is present. What mathematical machinery will allow us to accomplish this?

I begin by summarizing how a pure mathematician would address this question.[3] I will then summarize how, in subsequent chapters, we will reformulate the approach in a more practitioner friendly form.

Since the finite set Z is just one instantiation of some random variable Σ, it follows that Σ must be a *random finite set (RFS)*. This RFS draws its instantiations from the *actual sensor measurement space* $\mathfrak{Z}$—that is, the hyperspace whose elements are the *finite subsets* of $\mathfrak{Z}_0$. The conditional probability-mass function of Σ must have the form $\Pr(\Sigma \in O|\mathbf{x})$ where O is any (measurable) subset of $\mathfrak{Z}$. We must define integrals of the form

$$\int_O f(Z|\mathbf{x}) d\mu(Z) \tag{9.9}$$

with respect to some measure μ on $\mathfrak{Z}$ such that

$$\Pr(\Sigma \in O|\mathbf{x}) = \int_O f(Z|\mathbf{x}) d\mu(Z) \tag{9.10}$$

and such that, in particular, $\int f(Z|\mathbf{x}) d\mu(Z) = 1$ when $O = \mathfrak{Z}$. We must devise some systematic method for constructing measurement models of the form $\Sigma = \eta(\mathbf{x})$ where $\eta(\mathbf{x})$ is a RFS-valued function of $\mathbf{x}$ that relates measurement sets with target state vectors $\mathbf{x}$. Finally, we must devise some systematic method for explicitly constructing specific and concrete formulas for $f(Z|\mathbf{x})$.

The mathematician's formulation is inconvenient from a practical point of view. The concept of a *hyperspace*—a space $\mathfrak{Z}$ whose "points" are subsets of some

3 Greater detail can be found in Appendix F.

other space $\mathfrak{Z}_0$ [175]—is confusingly abstract. Most signal processing practitioners already find the measure theoretic theory of integration intimidating. Measure theoretic integrals $\int_O f(Z|\mathbf{x})d\mu(Z)$ defined on the measurable subsets O of a hyperspace $\mathfrak{Z}$ of finite subsets are even more so. It would be desirable to largely dispense with the hyperspace $\mathfrak{Z}$ and work entirely within the ordinary space $\mathfrak{Z}_0$ of conventional sensor measurements. FISST exploits the fact that it is possible to do this.

9.3.1.3 Finite-Set Statistics (FISST)

In summary, I will define a *set integral* $\int_T f(Z|\mathbf{x})\delta Z$ defined on subsets $T \subseteq \mathfrak{Z}_0$ of conventional measurement space, such that

$$\beta(T|\mathbf{x}) = \int_T f(Z|\mathbf{x})\delta Z \tag{9.11}$$

where

$$\beta(T|\mathbf{x}) \triangleq \Pr(\Sigma \subseteq T|\mathbf{x}) \tag{9.12}$$

is the (conditional) *belief-mass function* of Σ. In particular,

$$\int f(Z|\mathbf{x})\delta Z \triangleq \int_T f(Z|\mathbf{x})\delta Z = 1 \tag{9.13}$$

when $T = \mathfrak{Z}_0$. I will devise a systematic method for constructing measurement models $\Sigma_\mathbf{x}$ and, from them, specific and concrete formulas for $f(Z|\mathbf{x})$.

This will be accomplished as follows. For any $Z, T \subseteq \mathfrak{Z}_0$ I will show how to construct the *set derivative*

$$\frac{\delta\beta}{\delta Z}(T|\mathbf{x}) \tag{9.14}$$

of a belief-mass function $\beta(S)$. From this it will follow that the true likelihood function can be constructed by setting $T = \emptyset$:

$$f(Z|\mathbf{x}) = \left[\frac{\delta\beta}{\delta Z}(T|\mathbf{x})\right]_{T=\emptyset} = \frac{\delta\beta}{\delta Z}(\emptyset|\mathbf{x}). \tag{9.15}$$

It is the true likelihood function because it can be shown that

$$\int_T \frac{\delta\beta}{\delta Z}(\emptyset|\mathbf{x})\delta Z = \Pr(\Sigma \subseteq T|\mathbf{x}) = \int_T f(Z|\mathbf{x})\delta Z \tag{9.16}$$

for all T and hence

$$\frac{\delta\beta}{\delta Z}(\emptyset|\mathbf{x}) = f(Z|\mathbf{x}) \tag{9.17}$$

almost everywhere. That is:

- The belief-mass function $\beta(T|\mathbf{x})$ and the likelihood function $f(Z|\mathbf{x})$ can be constructed each from the other without loss of information, or without introducing extraneous information.

In the following sections, I summarize the approach in greater detail.

9.3.2 Multiobject Integrals

Formal modeling requires a more general definition of an integral, one that takes the random variability of the number of measurements into account. I will provide a more careful definition in (11.96)-(11.99). For my current purposes, I define the set integral of the function $f(Z|\mathbf{x})$ with respect to the finite-set variable Z, concentrated in a region T of $\mathfrak{Z}_0$, to be:

$$\int_T f(Z|\mathbf{x})\delta Z \triangleq f(\emptyset|\mathbf{x}) + \int_T f(\{\mathbf{z}\}|\mathbf{x})d\mathbf{z} \tag{9.18}$$

$$+\frac{1}{2!}\int_{T\times T} f(\{\mathbf{z}_1, \mathbf{z}_2\}|\mathbf{x})d\mathbf{z}_1 d\mathbf{z}_2 \tag{9.19}$$

$$+\frac{1}{3!}\int_{T\times T\times T} f(\{\mathbf{z}_1, \mathbf{z}_2, \mathbf{z}_3\}|\mathbf{x})d\mathbf{z}_1 d\mathbf{z}_2 d\mathbf{z}_3 + ... \tag{9.20}$$

The mth term in this sum

$$\frac{1}{m!}\int_{\underbrace{T\times ... \times T}_{m}} f(\{\mathbf{z}_1, ..., \mathbf{z}_m\}|\mathbf{x})d\mathbf{z}_1 \cdots d\mathbf{z}_m \tag{9.21}$$

is the total probability that m measurements will be collected and that they are all contained in T. If $T = \mathfrak{Z}_0$ then we write

$$\int f(Z|\mathbf{x})\delta Z \triangleq \int_{\mathfrak{Z}_0} f(Z|\mathbf{x})\delta Z \tag{9.22}$$

$$= f(\emptyset|\mathbf{x}) + \int f(\{\mathbf{z}\}|\mathbf{x})d\mathbf{z} \tag{9.23}$$

$$+\frac{1}{2!}\int f(\{\mathbf{z}_1,\mathbf{z}_2\}|\mathbf{x})d\mathbf{z}_1 d\mathbf{z}_2 \tag{9.24}$$

$$+\frac{1}{3!}\int f(\{\mathbf{z}_1,\mathbf{z}_2,\mathbf{z}_3\}|\mathbf{x})d\mathbf{z}_1 d\mathbf{z}_2 d\mathbf{z}_3 + ... \tag{9.25}$$

Therefore

$$p(m|\mathbf{x}) \triangleq \frac{1}{m!}\int f(\{\mathbf{z}_1, ..., \mathbf{z}_m\}|\mathbf{x})d\mathbf{z}_1 \cdots d\mathbf{z}_m \tag{9.26}$$

is the total probability that m conventional measurements will be collected. Note that it has no units of measurement. (The significance of this fact will become clear shortly.)

Thus the following are true.

- $f(\emptyset|\mathbf{x})$ is the likelihood that no measurements will be collected;
- $f(\{\mathbf{z}\}|\mathbf{x})$ is the likelihood that a single measurement $\mathbf{z}$ will be collected;
- $f(\{\mathbf{z}_1, ..., \mathbf{z}_m\}|\mathbf{x})$ is the likelihood that m measurements will be collected and that they are $\mathbf{z}_1, ..., \mathbf{z}_m$; and so on.

Several things should be pointed out here. First, since the number of measurements can be arbitrarily large, the sum in (9.22) is, in general, an infinite sum.

Second, for purposes of integration, each term in (9.22) must be regarded as a conventional function

$$f_m(\mathbf{z}_1, ..., \mathbf{z}_m|\mathbf{x}) \triangleq \frac{1}{m!} \cdot f(\{\mathbf{z}_1, ..., \mathbf{z}_m\}|\mathbf{x}) \tag{9.27}$$

in m vector variables. The factor $m!$ occurs since the probability density of $\{\mathbf{z}_1, ..., \mathbf{z}_m\}$ must be equally distributed among the $m!$ possible vectors $(\mathbf{z}_{\sigma 1}, ..., \mathbf{z}_{\sigma m})$ for all permutations σ of $1, ..., m$.

Also, since the cardinality of $\{\mathbf{z}_1, ..., \mathbf{z}_m\}$ will be less than m if $\mathbf{z}_i = \mathbf{z}_j$ for any $i \neq j$, it must be the case that $f_m(\mathbf{z}_1, ..., \mathbf{z}_m|\mathbf{x}) = 0$ if $\mathbf{z}_i = \mathbf{z}_j$ for any $i \neq j$.[4]

Third, the likelihood $f(Z|\mathbf{x})$ differs from conventional likelihoods in that

- *Its units of measurement can vary as a function of* $|Z|$, *the number of elements in* Z.

Suppose, for example, that the measurement space $\mathfrak{Z}_0 = \mathbb{R}$ is the real line, and that the units of measurement are kilometers. The likelihood $f(\emptyset|\mathbf{x})$ that no measurement will be collected is a unitless probability. The likelihood $f(\{\mathbf{z}_1\}|\mathbf{x})$ that a single measurement $\mathbf{z}_1$ will be collected has units of km^{-1}. The likelihood $f(\{\mathbf{z}_1, \mathbf{z}_2\}|\mathbf{x})$ that two measurements $\mathbf{z}_1, \mathbf{z}_2$ will be collected has units of km^{-2}. In general, the likelihood $f(\{\mathbf{z}_1, ..., \mathbf{z}_m\}|\mathbf{x})$ that m measurements $\mathbf{z}_1, ..., \mathbf{z}_m$ will be collected has units of km^{-m}.

9.3.3 Finite-Set Measurement Models

At this stage of analysis $f(Z|\mathbf{x})$ is *just a mathematical abstraction.* That is, it *has no obvious connection to real-world sensor phenomenology*. As in the single-sensor case, the bare fact that $f(Z|\mathbf{x})$ is a conditional probability is unhelpful "mathemagic." It does not tell us how to interpret $f(Z|\mathbf{x})$ except as a mathematical abstraction. If we are to connect this likelihood to the real world we must follow the formal modeling paradigm.

9.3.3.1 Case 1: No Missed Detections

Formal modeling requires that we show how to construct a careful statistical measurement model and then how to construct $f(Z|\mathbf{x})$ from it. Consider the detections collected by a surveillance sensor observing a single target, as described in Section 9.2.1. Since target-generated measurements are always collected, the measurement model must have the general form

$$\underbrace{\Sigma_{k+1}}_{\text{measurement set}} = \underbrace{\{\mathbf{Z}\}}_{\text{target measurement}} \cup \underbrace{C}_{\text{false measurements}}. \tag{9.28}$$

Here $\mathbf{Z} = \eta(\mathbf{x}, \mathbf{W})$ is the random observation-vector generated by the target at time step $k+1$; and C is the random observation set produced by the false alarm

4 If $\mathfrak{Z}_0$ is continuously infinite, this last requirement is moot. This is because any diagonal set $\mathbf{z}_i = \mathbf{z}_j$ $(i \neq j)$ will contribute no mass to the integral in (9.26) if the integrand is finite-valued.

process. The total random observation set Σ is the union of these two observation sets.

How do we concretely model C? We will address this issue in depth in Chapter 12. For the time being, we sketch the basic ideas.

The usual approach is to assume that C is independent of $\mathbf{Z}$ and that it is a *multidimensional Poisson process*; see (11.122). That is, the number of elements $|C|$ in C is a Poisson-distributed random nonnegative integer. In other words, $f_C(m) \triangleq \Pr(|C| = m) = e^{-\mu}\mu^m/m!$, where μ is the expected value of $|C|$. Given that $|C| = m$, the elements of C are assumed to be statistically independent and each element is distributed spatially according to a probability density $c(\mathbf{z})$. That is,

$$f_m(\mathbf{z}_1, ..., \mathbf{z}_m|m) \triangleq \frac{f_m(\mathbf{z}_1, ..., \mathbf{z}_m)}{f_C(m)} = c(\mathbf{z}_1) \cdots c(\mathbf{z}_m). \tag{9.29}$$

Most typically, $c(\mathbf{z})$ is assumed to be uniform in some region.

9.3.3.2 Poisson False Alarm Model

This false alarm model may appear arbitrary, but it is not [11, pp. 102,103]. To see why, partition the region of interest in measurement space $\mathfrak{Z}_0$ into M disjoint cells. Assume that any measurement in a cell is a nonnegative real number in a intensity-signal signature. Let sensor noise in the cells be modeled by respective random numbers $A_1, ..., A_M$. Assume that $A_1, ..., A_M$ are statistically independent and identically distributed with common density function $f_A(a)$. The probability that a noise-generated detection is declared in any given cell is[5]

$$p = \Pr(A_i > \tau) = \int_\tau^\infty f_A(a)da. \tag{9.30}$$

We then ask, What is the probability that m false detections will be declared?

This problem is conceptually identical to the problem of tossing a weighted coin M times, with p being the probability of drawing "heads." The probability of drawing m "heads" (= m detections) in M trials (= M cells) is $B_{M,p}(m)$ where $B_{M,p}(m) = C_{M,m}p^m (1-p)^{M-m}$ is the binomial distribution with mean value $\mu = Mp$. Since the probability of declaring a detection is identical in each cell, the distribution of noise-generated detections is uniformly distributed over the region of interest.

5 See [207] for a discussion of the Swerling and other models for $P(A_i > \tau)$ in radar applications.

Now let the size of the cells go to zero, so that their number M increases without bound. Hold the expected number Mp of false detections in the scene constant and equal to μ. Then $B_M(m)$ converges to the Poisson distribution.

Thus:

- False alarms in a continuous measurement space can be modeled as distributed uniformly in space, with Poisson time-arrivals.

False alarms are not necessarily spatially uniform. For some sensors, false alarms tend to be clustered around the true target, for example. This means that they are not only not uniformly distributed, but that their distribution depends on target state. Their distribution can have the form $c(\mathbf{z}|\mathbf{x})$ in general. In such cases we say that the false alarm process is *state-dependent*. For the time being, I will assume that $c(\mathbf{z}|\mathbf{x})$ is independent of $\mathbf{x}$ and write $c(\mathbf{z}) = c(\mathbf{z}|\mathbf{x})$.

As we shall see in Chapter 11, the probability distribution of the Poisson false alarm model C is

$$f_C(Z) = \begin{cases} e^{-\lambda} & \text{if} \quad Z = \emptyset \\ e^{-\lambda} \prod_{\mathbf{z} \in Z} \lambda c(\mathbf{z}) & \text{if} \quad \text{otherwise} \end{cases} .$$

It is left to the reader as Exercise 31 to verify that

$$\Pr(C \subseteq T) = \int_T f_C(Z) \delta Z = e^{\lambda p_c(T) - \lambda} \tag{9.31}$$

where $p_c(T) \triangleq \int_T c(\mathbf{z}) d\mathbf{z}$.

9.3.3.3 Case 2: Missed Detections

In this case $\{\mathbf{Z}\}$ is an oversimplified model of the observation set generated by the target. The model should have the more accurate form

$$\overset{\text{measurement set}}{\Sigma_{k+1}} = \overset{\text{target measurement set}}{\Upsilon} \cup \overset{\text{false measurements}}{C} . \tag{9.32}$$

Here, either $\Upsilon = \{\mathbf{z}\}$ (target is detected and the detection is $\mathbf{z}$) or $\Upsilon = \emptyset$ (target is not detected and so there is no target-generated observation). The probability that $\Upsilon = \emptyset$ is $1 - p_D(\mathbf{x})$. If the target detection and sensor noise processes are independent, then the probability density that $\Upsilon = \{\mathbf{z}\}$ (conditioned on $\mathbf{x}$) is $p_D(\mathbf{x}) \cdot f(\mathbf{z}|\mathbf{x})$ where $f(\mathbf{z}|\mathbf{x})$ is the distribution of $\mathbf{Z}$.

9.3.4 True Likelihoods for Finite-Set Measurements

It will be shown in (12.68) that the formula for the true likelihood is $f(Z|\mathbf{x}) \overset{\text{abbr.}}{=} f_\Sigma(Z|\mathbf{x})$ where

$$f(Z|\mathbf{x}) = f_C(Z) \cdot \left(1 - p_D(\mathbf{x}) + p_D(\mathbf{x}) \sum_{\mathbf{z} \in Z} \frac{f(\mathbf{z}|\mathbf{x})}{\lambda \cdot c(\mathbf{z})}\right) \tag{9.33}$$

and where, by convention, the summation vanishes if $Z = \emptyset$:

$$f(\emptyset|\mathbf{x}) = f_C(\emptyset) \cdot (1 - p_D(\mathbf{x})) = e^{-\lambda} (1 - p_D(\mathbf{x})) . \tag{9.34}$$

How do we know that (9.33) is not a heuristic contrivance or that it has not been erroneously constructed? In other words, how do we know that it is the *true* likelihood for the sensor? That is, how do we know that

$$\int_T f(Z|\mathbf{x}) \delta Z = \Pr(\Sigma \subseteq T) \tag{9.35}$$

and therefore that $f(Z|\mathbf{x})$ faithfully reflects the information in the model $\Sigma = \Upsilon \cup C$?

9.3.5 Constructive Likelihood Functions

While (9.33) turns out to be the true likelihood function for the model, how could we have possibly known that this was the case? It is hardly self-evident, after all, that this rather complicated formula is the correct answer. We will not be in a position to answer this question in any detail until Section 12.3.3.

For now, I refer back to (9.11) and note that the belief-mass function corresponding to the model $\Sigma = \Upsilon \cup C$ is

$$\beta(T|\mathbf{x}) = (1 - p_D(\mathbf{x}) + p_D(\mathbf{x}) \cdot p(T|\mathbf{x})) \cdot e^{\lambda p_c(T) - \lambda} \tag{9.36}$$

where $p_c(T) = \int_T c(\mathbf{z})d\mathbf{z}$. For,

$$\beta(T|\mathbf{x}) \tag{9.37}$$

$$= \Pr(\Sigma \subseteq T|\mathbf{x}) = \Pr(\Upsilon \cup C \subseteq T|\mathbf{x}) \tag{9.38}$$

$$= \Pr(\Upsilon \subseteq T, C \subseteq T|\mathbf{x}) \tag{9.39}$$

$$= \Pr(\Upsilon \subseteq T|\mathbf{x}) \cdot \Pr(C \subseteq T) \tag{9.40}$$

$$= [\Pr(\Upsilon = \emptyset|\mathbf{x}) + \Pr(\Upsilon \subseteq T, \Upsilon \neq \emptyset|\mathbf{x})] \cdot e^{\lambda p_c(T) - \lambda} \tag{9.41}$$

$$= [1 - p_D(\mathbf{x}) + p_D(\mathbf{x}) \cdot \Pr(\Upsilon \subseteq T|\Upsilon \neq \emptyset, \mathbf{x})] \cdot e^{\lambda p_c(T) - \lambda} \tag{9.42}$$

$$= [1 - p_D(\mathbf{x}) + p_D(\mathbf{x}) \cdot \Pr(\mathbf{Z} \in T|\mathbf{x})] \cdot e^{\lambda p_c(T) - \lambda} \tag{9.43}$$

$$= [1 - p_D(\mathbf{x}) + p_D(\mathbf{x}) \cdot p(T|\mathbf{x})] \cdot e^{\lambda p_c(T) - \lambda}. \tag{9.44}$$

Thus $\beta(T|\mathbf{x})$ can be entirely expressed as an algebraic combination of the probability-mass functions $p(T|\mathbf{x})$ and $p_c(T)$.

I have claimed that (9.15) can be used to derive the true likelihood function:

$$f(Z|\mathbf{x}) = \left[\frac{\delta\beta}{\delta Z}(T|\mathbf{x})\right]_{T=\emptyset} \tag{9.45}$$

Thus it would be possible to derive a concrete formula for $f(Z|\mathbf{x})$ if we had at our disposal basic rules for computing the set derivatives of the following:

- Sums, products, transcendental functions, and so on;
- Probability-mass functions such as $p(T|\mathbf{x})$ and $p_C(T)$.

Specifying such rules is one of the primary purposes of Chapter 11.

9.4 CHAPTER EXERCISES

Exercise 31 *Prove (9.31):* $\int_T f_C(Z)\delta Z = e^{\lambda p_c(T) - \lambda}$.

Part II

Unified Multitarget-Multisource Integration

Chapter 10

Conventional Multitarget Filtering

10.1 INTRODUCTION TO THE CHAPTER

The multitarget tracking and data fusion literature is so vast that it would be beyond the scope of this book to offer even a limited survey. For this purpose the reader is directed to tutorial articles such as Blackman [16], Uhlmann [228, 229, 227], and Poore, Lu, and Suchomel [187]; surveys such as Pulford [188]; or to canonical reference texts such as Bar-Shalom and Li [11], Blackman [17], and Blackman and Popoli [18].

The purpose of this chapter is much more limited. In Chapter 2, I summarized the basic concepts of conventional single-sensor, single-target detection, tracking, localization, and identification. The purpose of this summary was to provide the reader with an entry point into Chapters 3-9. This chapter serves an analogous purpose. By sketching the basic concepts of conventional multitarget detection and tracking, I will create an entry point into the chapters that follow.

Conventional multitarget filtering approaches typically employ bottom-up, divide-and-conquer strategies that partition a multitarget problem into a family of parallel single-target problems. To accomplish this, they commonly begin with the following assumptions:

- Observations of targets are *signal detections* of the kind described in Section 9.2.1. That is, they are generated by applying a threshold to some signal-intensity signature;
- Targets are neither extended nor unresolved. That is, any detection measurement is generated by no more than a single target and no target generates more than a single detection;

- At any given time, we have at hand a *track table*—that is, a list of state-estimates and error covariance estimates of what we believe are the currently possible targets;
- Target motions are statistically independent of each other.

Given these assumptions, one can employ one of the following three broad divide-and-conquer strategies. At each measurement collection time, try to identify any given report as having been one of the following.

- *Optimally generated* by some prior track; or, otherwise, by some new track; or, otherwise, by a false track (false alarm). In this case, each track is uniquely associated with a single time sequence of detections. Once initiated, any given track is propagated using a single-target filter (usually an extended Kalman filter). At any given time, the current track table itself is accepted as the best model of multitarget reality;
- *Optimally or suboptimally generated* by some prior track; or, otherwise, by some new track; or, otherwise, by a false track (false alarm). In this case, one has a sequence of ranked optimal and suboptimal *hypotheses*, each being some subset of the track table and each being a different model of multitarget reality;
- *Proportionately generated* by all or some prior tracks. Having assigned a degree to which each measurement contributed to each track and using this to construct a composite state vector or innovation for each track, use a single-target filter to propagate each track.

The most familiar instance of the first approach is *single-hypothesis tracking (SHT)*; of the second, *multihypothesis tracking (MHT)* and its variants; and of the third, *joint probabilistic data association (JPDA)* and its variants. In the following sections I will sketch the basic concepts of each of these in turn.

10.1.1 Summary of Major Lessons Learned

The following are the major ideas and concepts to be encountered in this chapter:

- The "standard"—that is, most commonly encountered single-sensor, multi-target measurement model (Section 10.2.1);
- The "standard"—that is, most commonly encountered multitarget motion model (Section 10.2.2);

- *Association distance* $d(\mathbf{z};\mathbf{x}')$ between a measurement $(\mathbf{z},R)$ and a predicted track estimate $(\mathbf{x}',P')$; see (10.29):

$$d(\mathbf{z};\mathbf{x}')^2 \triangleq (\mathbf{z}-H\mathbf{x}')^T\,(HP'H^T+R)^{-1}\,(\mathbf{z}-H\mathbf{x}')\,; \tag{10.1}$$

- *Global measurement-to-track association* (also known as *association hypothesis*, Section 10.5.1) between m measurements $(\mathbf{z}_1,R)$,...,$(\mathbf{z}_m,R)$ and n predicted track estimates $(\mathbf{x}'_1,P'_1)$,....,$(\mathbf{x}'_n,P'_n)$. I consider this for three respective cases: no missed detections or false alarms (Section 10.5.2),

$$\sigma:\{1,...,n\}\to\{1,...,n\}, \tag{10.2}$$

false alarms but no missed detections (Section 10.5.3),

$$\chi:\{1,...,n\}\to\{1,...,m\}, \tag{10.3}$$

and false alarms and missed detections (Section 10.5.4):

$$\theta:\{1,...,n\}\to\{0,1,...,m\}; \tag{10.4}$$

- *Global association distance* d_χ for a given association $\chi:\{1,...,n\}\to\{1,...,m\}$; see (10.33):

$$d_\chi^2=\sum_{i=1}^{n}\left(\mathbf{z}_{\chi(i)}-H\mathbf{x}'_i\right)^T\,(HP'_iH^T+R)^{-1}\left(\mathbf{z}_{\chi(i)}-H\mathbf{x}'_i\right); \tag{10.5}$$

- *Global association likelihood* ℓ_σ, ℓ_χ, ℓ_θ for cases σ, χ, θ (Sections 10.5.2-10.5.4);

- The likelihood function of a general association hypothesis θ; see (10.53):

$$f_{k+1}(\mathbf{z}_1,...,\mathbf{z}_m|\mathbf{x}_1,...,\mathbf{x}_n,\theta); \tag{10.6}$$

- The posterior distribution on tracks $\mathbf{x}_1,...,\mathbf{x}_n$ and associations θ; see (10.56):

$$f_{k+1|k+1}(\mathbf{x}_1,...,\mathbf{x}_n,\theta|\mathbf{z}_1,...,\mathbf{z}_m); \tag{10.7}$$

- The *hypothesis probability* of an association hypothesis for three respective cases: no missed detections or false alarms; see (10.68),

$$p_\sigma=\frac{\exp\left(-\frac{1}{2}d_\sigma^2\right)}{\sum_{\sigma'}\exp\left(-\frac{1}{2}d_{\sigma'}^2\right)} \tag{10.8}$$

false alarms but no missed detections; see (10.90),

$$p_X = \frac{\left(\prod_{\mathbf{z}\in Z-W_X} c(\mathbf{z})\right)\cdot \exp\left(-\frac{1}{2}d_X^2\right)}{\sum_{X'}\left(\prod_{\mathbf{z}\in Z-W_{X'}} c(\mathbf{z})\right)\cdot \exp\left(-\frac{1}{2}d_{X'}^2\right)} \tag{10.9}$$

and false alarms and missed detections; see (10.115) and (10.116):

$$p_0 = \frac{1}{1+\sum_{\theta'\neq 0} K^{n_{\theta'}}\cdot \exp\left(-\frac{1}{2}d_{\theta'}^2\right)\cdot \prod_{\mathbf{z}\in W_{\theta'}} c(\mathbf{z})^{-1}} \tag{10.10}$$

$$p_\theta = \frac{K^{n_\theta}\cdot \exp\left(-\frac{1}{2}d_{\theta}^2\right)\cdot \prod_{\mathbf{z}\in W_{\theta}} c(\mathbf{z})^{-1}}{1+\sum_{\theta'\neq 0} K^{n_{\theta'}}\cdot \exp\left(-\frac{1}{2}d_{\theta'}^2\right)\cdot \prod_{\mathbf{z}\in W_{\theta'}} c(\mathbf{z})^{-1}}; \tag{10.11}$$

- Single-hypothesis correlation (SMC) filters, which at each time step select the globally optimal report-to-track assignment (Section 10.4);
- SMC filters perform well when targets are not too closely spaced (relative to the spatial resolution of the observing sensors) and signal-to-noise ratio is high (Section 10.7.1);
- Ideal multihypothesis correlation (MHC) filters, which at each time step propagate the full set of possible association hypotheses (Section 10.5);
- With appropriate approximations, MHC filters perform tractably and well under moderately low signal-to-noise ratios and for moderately closely spaced targets (Section 10.7.1);
- Composite-hypothesis correlation (CHC) filters, the most well known of which are the joint probabilistic data association (JPDA) filters (Section 10.6);
- Conventional multitarget correlation approaches will fail when targets are too closely spaced, signal-to-noise ratio is too low, target-motion linearities are too large, sensor nonlinearities are too large, and so on (Section 10.7.1);
- The concept of an association hypothesis may not be consistent with the requirements of a strict multitarget Bayesian analysis (Section 10.7.2);
- Conventional multitarget correlation techniques can be extended to allow SMC, MHC, and CHC approaches to process fuzzy and Dempster-Shafer measurements (Section 10.8).

10.1.2 Organization of the Chapter

The chapter is organized as follows. Section 10.2 describes the multitarget measurement and motion models most commonly assumed in multitarget tracking theory. The concepts of local association distance and likelihood, global association distance and likelihood, and global nearest-neighbor measurement-to-track association are introduced in Section 10.3.

I then turn to an analysis of the three most common conventional multitarget correlation and tracking methods. The simplest such approach, single-hypothesis correlation (SHC) filtering, is described in Section 10.4. The concepts of multiple hypotheses, hypothesis probabilities, and multihypothesis correlation (MHC) filtering are introduced in Section 10.5. The elements of CHC filtering, in the form of joint probabilistic data association (JPDA), are described in Section 10.6. Section 10.7 describes the strengths and weaknesses of conventional multitarget correlation and tracking approaches.

The chapter concludes with a generalization of conventional multitarget correlation and tracking to unconventional evidence. In Section 10.8, I sketch the basics of integrating the Kalman Evidential Filter of Section 5.6 into conventional multitarget tracking schemes.

10.2 STANDARD MULTITARGET MODELS

In this section I introduce the multitarget measurement and motion models that are typically assumed in conventional multitarget tracking theory [62, 193]. I refer to them as the "standard" models.

10.2.1 Standard Multitarget Measurement Model

The most familiar and widely used multitarget tracking (MTT) algorithms presume some version of the "standard" multitarget measurement model described in (9.28) and (9.32). This model is based on the following assumptions [62, 193]:

- A single sensor observes a scene involving an unknown number of unknown targets.
- No target-generated measurement is generated by more than a single target.

- A single target generates either a single measurement (a detection) with probability p_D or no measurement (a missed detection) with probability $1 - p_D$.
- The false alarm process is Poisson-distributed in time and uniformly distributed in space.
- Target-generated measurements are conditionally independent of state.
- The false alarm process and target-measurement processes are statistically independent.

We generalize (9.28) and (9.32) to a multitarget measurement model in three increasingly more complex stages. We assume that the scene is being observed by a single sensor with single-target Kalman measurement model $\mathbf{Z} = H\mathbf{x} + \mathbf{W}$.

10.2.1.1 Standard Measurement Model: No Missed Detections or False Alarms

Let n targets with state vectors $\mathbf{x}_1, ..., \mathbf{x}_n$ be present and let $X = \{\mathbf{x}_1, ..., \mathbf{x}_n\}$. The random observation generated by the ith target is, therefore, $\mathbf{Z}_i = H\mathbf{x}_i + \mathbf{W}_i$ for $i = 1, ..., n$ where $\mathbf{W}_1, ..., \mathbf{W}_n$ are independent, identically distributed zero-mean Gaussian random vectors with covariance matrix R. Consequently, all targets generate a *measurement set* of the following form

$$\overset{\text{total measurement set}}{Z} = \overset{\text{target-generated measurements}}{\{\mathbf{z}_1, ..., \mathbf{z}_n\}} \tag{10.12}$$

where the total number of measurements is $m = n$.

10.2.1.2 Standard Measurement Model: False Alarms, No Missed Detections

Measurements will consist not only of target-generated observations, but also of observations generated by the false alarm process. In this case the finite measurement set has the form

$$\overset{\text{total measurement set}}{Z} = \overset{\text{target measurement set}}{\{\mathbf{z}_1, ..., \mathbf{z}_n\}} \cup \overset{\text{false measurements}}{\{\mathbf{c}_1, ..., \mathbf{c}_{m'}\}}. \tag{10.13}$$

Here, the false measurements $\mathbf{c}_i$ are assumed to be uniformly distributed and their number m' is Poisson-distributed. The average number of measurements is $m = n + \lambda$ where λ is the average number of false alarms.

10.2.1.3 Standard Measurement Model: False Alarms, Missed Detections

The measurement set is

$$\overset{\text{measurement set}}{Z} = \overset{\text{target-generated measurements}}{Z(\mathbf{x}_1) \cup \ldots \cup Z(\mathbf{x}_n)} \cup \overset{\text{false measurements}}{\{\mathbf{c}_1, \ldots, \mathbf{c}_{m'}\}}. \tag{10.14}$$

Here, each $\{\mathbf{z}_i\}$ in the previous case has been replaced by a set $Z(\mathbf{x}_i)$ that can have one of the following two forms:

- $Z(\mathbf{x}_i) = \emptyset$ (the ith target is not detected);
- $Z(\mathbf{x}_i) = \{\mathbf{z}_i\}$ (the ith target is detected and generated measurement $\mathbf{z}_i$).

If p_D is the probability of detection then the probability that $Z(\mathbf{x}_i) = \emptyset$ is $1 - p_D$ and the probability density that $Z(\mathbf{x}_i) = \{\mathbf{z}_i\}$ is $p_D \cdot f(\mathbf{z}_i|\mathbf{x}_i)$. On average, there will be $m = p_D \cdot n + \lambda$ measurements, where λ is the average number of false alarms.

10.2.2 Standard Multitarget Motion Model

Conventional multitarget tracking algorithms are typically based on the following assumptions. These assumptions are directly analogous to the measurement model assumptions:

- The motion of any individual target is governed by a Kalman motion model $\mathbf{x} = F\mathbf{x}' + \mathbf{V}$;
- The motions of the targets are statistically independent;
- New targets appear in the scene uniformly distributed and with Poisson time-arrivals [193, p. 847].

We will proceed through three increasingly more complex cases.

Remark 8 (Target "Birth" and Target "Death") *In the chapters that follow, the terms "target birth" and "target death" are to be understood as shorthand synonyms for "target appearance" and "target disappearance." No additional connotations should be read into them. Targets "die" if they disappear from the scene (e.g., if they pass outside of the sensor field of view). Targets are "born" if they appear in the scene (e.g., if they enter the field of view or if they are spawned by other targets).*

10.2.2.1 Standard Motion Model: No Target Birth or Death

Suppose that no targets appear or disappear from the scene. Let n targets with state vectors $\mathbf{x}'_1, ..., \mathbf{x}'_{n'}$ be present at time step k and let $X' = \{\mathbf{x}'_1, ..., \mathbf{x}'_n\}$. These are the "prior targets." The motion model at time step $k+1$ generated by the ith target is, therefore, $\mathbf{x}_i = F\mathbf{x}'_i + \mathbf{V}_i$ for $i = 1, ..., n$ where $\mathbf{V}_1, ..., \mathbf{V}_n$ are independent, identically distributed zero-mean Gaussian random vectors with covariance matrix Q. Consequently, all targets taken together generate the following finite state set

$$\overbrace{X}^{\text{total predicted state set}} = \overbrace{\{\mathbf{x}_1, ..., \mathbf{x}_n\}}^{\text{predicted targets}}. \tag{10.15}$$

Here the total number of predicted tracks is n.

10.2.2.2 Standard Motion Model: Target Birth, No Target Death

If targets may appear but not disappear then

$$\overbrace{X}^{\text{total predicted state set}} = \overbrace{\{\mathbf{x}_1, ..., \mathbf{x}_{n'}\}}^{\text{predicted targets}} \cup \overbrace{\{\mathbf{b}_1, ..., \mathbf{b}_\nu\}}^{\text{new targets}} \tag{10.16}$$

where the new targets $\mathbf{b}_i$ are assumed to be uniformly distributed and their number ν is Poisson-distributed with expected value μ. The average number of predicted targets is, therefore, $n = n' + \mu$.

10.2.2.3 Standard Motion Model: Target Birth and Death

If targets may appear as well as disappear then

$$\overbrace{X}^{\text{total predicted state set}} = \overbrace{X(\mathbf{x}'_1) \cup ... \cup X(\mathbf{x}'_{n'})}^{\text{persisting targets}} \cup \overbrace{\{\mathbf{b}_1, ..., \mathbf{b}_\nu\}}^{\text{new targets}}. \tag{10.17}$$

Here $X(\mathbf{x}'_i)$ can have one of the following two forms.

- $X(\mathbf{x}'_i) = \emptyset$ (the ith target vanishes) or
- $X(\mathbf{x}'_i) = \{\mathbf{x}_i\}$ (the ith target persists and attains state $\mathbf{x}_i$).

If p_S is the probability that a target will persist from time step k to time step $k+1$, then the probability that $X(\mathbf{x}'_i) = \emptyset$ is $1 - p_S$ and the probability density that $X(\mathbf{x}'_i) = \{\mathbf{x}_i\}$ is $p_S \cdot f_{k+1|k}(\mathbf{x}_i|\mathbf{x}'_i)$. On average, the number of predicted tracks is $n = p_S \cdot n' + \mu$ where μ is the average number of newly appearing targets.

10.3 MEASUREMENT-TO-TRACK ASSOCIATION

The purpose of this section is to introduce the concept of measurement-to-track association. For an analysis of the conceptual and computational complexities underlying this concept, see the tutorial papers by Uhlmann [227, 228, 229] and the more technical overview by Poore, Lu, and Suchomel [187].

In what follows, $\mathbf{x}'_1, ..., \mathbf{x}'_n$ is the list of predicted tracks at time step $k+1$; $P'_1, ..., P'_n$ are their respective predicted error covariance matrices. Also, $\mathbf{Z} = H\mathbf{x}' + \mathbf{W}$ is the sensor measurement model at time step $k+1$ where $\mathbf{W}$ has covariance matrix R, and so the likelihood function is $f(\mathbf{z}|\mathbf{x}') = N_R(\mathbf{z} - H\mathbf{x}')$.

10.3.1 Distance Between Measurements and Tracks

Before we can begin, we must define a concept of "distance" $d(\mathbf{z}|\mathbf{x}'_i)$ between a predicted track $\mathbf{x}'_i$, P'_i and a measurement $\mathbf{z}$. The obvious choice—the Euclidean distance $d(\mathbf{x}_i|\mathbf{z}) = \|\mathbf{z} - H\mathbf{x}'_i\|$ between $\mathbf{z}$ and the predicted measurement $H\mathbf{x}'_i$—will not serve for this purpose.

To see why, assume that the uncertainties R in $\mathbf{z}$ and P_i in $\mathbf{x}'_i$ are both small compared to $\|\mathbf{z} - H\mathbf{x}'_i\|$. Then it is unlikely that $\mathbf{z}$ could have been generated by a target with state $\mathbf{x}'_i$. On the other hand, assume that both uncertainties are large compared to $\|\mathbf{z} - H\mathbf{x}'_i\|$. Then it is likely that $\mathbf{z}$ was generated by such a target. In summary:

- $d(\mathbf{x}_i|\mathbf{z})$ must take the uncertainties R, P_i into account.

To arrive at a defensible concept of association distance, we must take a Bayesian approach.

10.3.1.1 Likelihood of Association

Suppose that we collect a measurement $\mathbf{z}$. We are to estimate which of the predicted tracks $\mathbf{x}'_1, ..., \mathbf{x}'_n$ most likely generated it. Let

$$f(\mathbf{x}'|i) = N_{P_i}(\mathbf{x}' - \mathbf{x}'_i) \tag{10.18}$$

be the track density for $\mathbf{x}'_i$. The total likelihood that $\mathbf{z}$ was generated by $\mathbf{x}'_i$ is

$$\ell(\mathbf{z}|i) \triangleq \int f(\mathbf{z}|\mathbf{x}) \cdot f(\mathbf{x}|i) d\mathbf{x}. \tag{10.19}$$

Let p_i be the prior probability of the ith track. Since there is no a priori reason to prefer one track over another, p_i is a uniform distribution. Thus the posterior probability that the ith track generated $\mathbf{z}_j$ is

$$p(i|\mathbf{z}) = \frac{\ell(\mathbf{z}|i) \cdot p_i}{\sum_{e=1}^{n} \ell(\mathbf{z}|e) \cdot p_e} = \frac{\ell(\mathbf{z}|i)}{\sum_{e=1}^{n} \ell(\mathbf{z}|e)}. \tag{10.20}$$

The track that most probably generated $\mathbf{z}$ is the MAP estimate (which, in this case, is also a maximum likelihood estimate):

$$\hat{\imath} = \arg\max_i \ p(i|\mathbf{z}) = \arg\max_i \ \ell(\mathbf{z}|i). \tag{10.21}$$

We could then define $d(\mathbf{x}'_i|\mathbf{z}) \triangleq -\log \ p(i|\mathbf{z})$ to be the (unitless) association distance between $\mathbf{z}$ and the ith track.

10.3.1.2 Association Distance

This analysis is not what is usually presumed in conventional application. Under our linear-Gaussian assumptions,

$$\begin{aligned} \ell(\mathbf{z}|i) &= \int N_R(\mathbf{z} - H\mathbf{x}) \cdot N_{P_i}(\mathbf{x} - \mathbf{x}'_i) d\mathbf{x} && (10.22) \\ &= \int N_{HP_iH^T+R}(\mathbf{z} - H\mathbf{x}'_i) \cdot N_C(\mathbf{x} - \mathbf{c}) d\mathbf{x} && (10.23) \\ &= N_{HP_iH^T+R}(\mathbf{z} - H\mathbf{x}'_i) && (10.24) \\ &= \frac{1}{\sqrt{\det 2\pi (HP_iH^T + R)}} && (10.25) \\ &\quad \cdot \exp\left(-\frac{1}{2} (\mathbf{z} - H\mathbf{x}'_i)^T (HP_iH^T + R)^{-1} (\mathbf{z} - H\mathbf{x}'_i)\right) && (10.26) \end{aligned}$$

where (10.23) results from the fundamental identity for Gaussian densities, (D.1). Therefore,

$$\begin{aligned} -\log \ell(\mathbf{z}|i) &= \frac{1}{2} \log \det 2\pi \left((HP_iH^T + R)\right) && (10.27) \\ &\quad + \frac{1}{2} (\mathbf{z} - H\mathbf{x}'_i)^T (HP_iH^T + R)^{-1} (\mathbf{z} - H\mathbf{x}'_i). && (10.28) \end{aligned}$$

In practice the first term in this sum is ignored. The second term, a Mahalanobis distance, is the definition of association distance between $\mathbf{z}$ and $\mathbf{x}_i$ [11, p. 95]:

$$d(\mathbf{x}'_i|\mathbf{z})^2 \triangleq (\mathbf{z} - H\mathbf{x}'_i)^T (HP_iH^T + R)^{-1} (\mathbf{z} - H\mathbf{x}'_i). \tag{10.29}$$

10.3.1.3 Global Association Distance

Let a measurement set Z be collected. Begin by assuming that the number of measurements equals the number n of tracks: $Z = \{\mathbf{z}_1, ..., \mathbf{z}_n\}$ with $|Z| = n$. The simplest association approach, known as *nearest neighbor* (NN), associates to predicted track $\mathbf{x}'_i$ that measurement $\mathbf{z}_j$ that minimizes $d(\mathbf{x}'_i|\mathbf{z}_j)$, for each fixed choice of i.

This approach will perform well if the targets are well separated compared to sensor resolution. The more that this assumption is violated, however, the more that the performance of the NN approach deteriorates.

The more common approach, known as *global nearest neighbor* (GNN), is based on the fact that all possible associations between measurements and predicted tracks have the form

$$\mathbf{z}_{\sigma 1} \leftrightarrow H\mathbf{x}'_1 \quad ,..., \quad \mathbf{z}_{\sigma n} \leftrightarrow H\mathbf{x}'_n \tag{10.30}$$

where σ is any one of the $n!$ possible permutations of the numbers $1, ..., n$. Define the *global association distance* d_σ of an association σ between the measurements and the predicted tracks as follows:

$$\begin{aligned} d^2_\sigma &\triangleq \sum_{i=1}^{n} d(\mathbf{z}_{\sigma i}, H\mathbf{x}'_i)^2 \qquad (10.31) \\ &= \sum_{i=1}^{n} (\mathbf{z}_{\sigma i} - H\mathbf{x}'_i)^T (HP_iH^T + R)^{-1} (\mathbf{z}_{\sigma i} - H\mathbf{x}'_i). \end{aligned}$$

The "optimal association" between measurements and predicted tracks is that permutation $\sigma = \hat{\sigma}$ that minimizes d^2_σ. Various algorithms (Munkres, JVC, auction, and so on) have been devised to efficiently solve this minimization problem (see [16] or [18, p. 342]).

More generally, suppose that the number of tracks does not exceed the number of measurements, $m \geq n$. Then all associations between measurements and predicted tracks have the form

$$\mathbf{z}_{\chi(1)} \leftrightarrow H\mathbf{x}'_1 \quad ,..., \quad \mathbf{z}_{\chi(m)} \leftrightarrow H\mathbf{x}'_m \tag{10.32}$$

where $\chi : \{1, ..., n\} \to \{1, ..., m\}$ is any of the possible one-to-one functions.[1] The optimal association problem can be solved by minimizing

$$d_\chi^2 \triangleq \sum_{i=1}^n d(\mathbf{z}_{\chi(i)}, H\mathbf{x}'_i)^2 \tag{10.33}$$

$$= \sum_{i=1}^n \left(\mathbf{z}_{\chi(i)} - H\mathbf{x}'_i\right)^T (HP_iH^T + R)^{-1} \left(\mathbf{z}_{\chi(i)} - H\mathbf{x}'_i\right) \tag{10.34}$$

over all such functions χ.

10.3.1.4 Global Association Likelihood

For our purposes we require a more general concept than global association distance. Begin with the case when numbers of measurements and state vectors are both n. The *global association likelihood* of an association σ is

$$\ell_\sigma \triangleq \ell(\mathbf{z}_{\sigma 1}|1) \cdots \ell(\mathbf{z}_{\sigma n}|n) \tag{10.35}$$

(assuming that measurements are conditionally independent of target states). That is, the total likelihood of σ is the product of the likelihoods that $\mathbf{z}_{\sigma 1}$ is associated with track $\mathbf{x}_1$, and that $\mathbf{z}_{\sigma 2}$ is associated with track $\mathbf{x}_2$, and so on.

Under our current assumptions it is easily shown that

$$\ell_\sigma = \frac{1}{\prod_{i=1}^n \det 2\pi(HP_iH^T + R)} \cdot \exp\left(-\frac{1}{2}d_\sigma^2\right). \tag{10.36}$$

More generally, suppose that $n \leq m$. If

$$\ell_\chi \triangleq \ell(\mathbf{z}_{\chi(1)}|1) \cdots \ell(\mathbf{z}_{\chi(n)}|n) \tag{10.37}$$

is the association likelihood of an association χ then

$$\ell_\chi = \frac{1}{\prod_{i=1}^n \det 2\pi(HP_iH^T + R)} \cdot \exp\left(-\frac{1}{2}d_\chi^2\right). \tag{10.38}$$

1 A one-to-one function $\chi : S \to S'$ between two sets S, S' is one for which $\chi(s_1) = \chi(s_2)$ implies $s_1 = s_2$. If $S = \{s_1, ..., s_e\}$ is finite with $|S| = e$ elements, then the image set $\chi(S) = \{\chi(s_1), ..., \chi(s_e)\}$ is also a finite subset of S' with e elements.

10.4 SINGLE-HYPOTHESIS CORRELATION (SHC)

In this section, I outline the simplest type of measurement-to-track correlation filter. In what follows, I assume a single sensor with single-target likelihood function $f_{k+1}(\mathbf{z}|\mathbf{x})$, and constant probability of detection p_D.

10.4.1 SHC: No Missed Detections, No False Alarms

Assume for the moment that n targets are known to be present in the scene. Through time step $k+1$ we collect a sequence $Z_1, ..., Z_{k+1}$ of observation sets $Z_i = \{\mathbf{z}_i^1, ..., \mathbf{z}_i^n\}$. Suppose that at time step k we have a *track table*

$$\mathcal{T}_{k|k} = \{(l_{k|k}^1, \mathbf{x}_{k|k}^1, P_{k|k}^1), ..., (l_{k|k}^n, \mathbf{x}_{k|k}^n, P_{k|k}^n)\} \tag{10.39}$$

of statistically independent tracks. That is, each 3-tuple $(l_{k|k}^i, \mathbf{x}_{k|k}^i, P_{k|k}^i)$ consists of the following:

- An estimate $\mathbf{x}_{k|k}^i$ of the state of the track;
- An estimate $P_{k|k}^i$ of the error in $\mathbf{x}_{k|k}^i$;
- A *track label* or *track tag* $l_{k|k}^i$.

The track label uniquely identifies the 3-tuple $(l_{k|k}^i, \mathbf{x}_{k|k}^i, P_{k|k}^i)$ as the incarnation, at time step k, of a continuously evolving sequence of such 3-tuples. That is, $(l_{k|k}^i, \mathbf{x}_{k|k}^i, P_{k|k}^i)$ is part of a continuously evolving "target track" in the temporal rather than the instantaneous sense.

Based on $\mathcal{T}_{k|k}$ and the new observation set Z_{k+1} we are to construct a new track table

$$\mathcal{T}_{k+1|k+1} = \{(l_{k+1|k+1}^1, \mathbf{x}_{k+1|k+1}^1, P_{k+1|k+1}^1), ..., (l_{k+1|k+1}^n, \mathbf{x}_{k+1|k+1}^n, P_{k+1|k+1}^n)\}. \tag{10.40}$$

First, extrapolate the tracks in $\mathcal{T}_{k|k}$ to time step $k+1$ using the Kalman predictor equations (2.6) and (2.7):

$$\mathbf{x}_{k+1|k}^i = F_k \mathbf{x}_{k|k}^i \tag{10.41}$$

$$P_{k+1|k}^i = F_k \mathbf{x}_{k|k}^i F_k^T + Q_k. \tag{10.42}$$

The predicted tracks keep the same labels as their predecessors: $l^i_{k+1|k} = l^i_{k|k}$ for $i = 1, ..., n$. This produces the predicted track table

$$\mathcal{T}_{k+1|k} = \{(l^1_{k+1|k}, \mathbf{x}^1_{k+1|k}, P^1_{k+1|k}), ..., (l^n_{k+1|k}, \mathbf{x}^n_{k+1|k}, P^n_{k+1|k})\}. \tag{10.43}$$

Second, use (10.33) to choose the measurement-to-track association $\hat{\sigma}$ that has minimal global association distance. Reorder the measurements accordingly:

$$\tilde{\mathbf{z}}^i_{k+1} \triangleq \mathbf{z}^{\hat{\sigma}i}_{k+1}. \tag{10.44}$$

Third, use the Kalman corrector equations (2.9) and (2.10), to update the predicted tracks:

$$\mathbf{x}^i_{k+1|k+1} = \mathbf{x}^i_{k+1|k} + K^i_{k+1}(\tilde{\mathbf{z}}^i_{k+1} - H_k\mathbf{x}^i_{k+1|k}) \tag{10.45}$$

$$P^i_{k+1|k+1} = \left(I - K^i_{k+1}H_{k+1}\right) P^i_{k+1|k}. \tag{10.46}$$

The data-updated tracks keep the same labels as their predecessors: $l^i_{k+1|k+1} = l^i_{k+1|k}$ for $i = 1, ..., n$. This results in the data-updated track table $\mathcal{T}_{k+1|k+1}$. Since the measurements in Z_{k+1} are conditionally independent upon state, the tracks in $\mathcal{T}_{k+1|k+1}$ are also statistically independent.

This table constitutes a unique *hypothesis* about what multitarget reality actually looks like. The finite state set

$$X_{k+1|k+1} = \{\mathbf{x}^1_{k+1|k+1}, ..., \mathbf{x}^n_{k+1|k+1}\} \tag{10.47}$$

is a *multitarget state-estimate*—that is, the SHC filter's best estimate of the states of the n targets.

The SHC filter is essentially a bank of n Kalman filters operating in parallel, the ith Kalman filter being applied to the time sequence of measurements

$$\tilde{Z}^{k+1}_i : \tilde{\mathbf{z}}^i_1, ..., \tilde{\mathbf{z}}^i_{k+1}. \tag{10.48}$$

Note that if $i \neq j$ then $\tilde{Z}^{k+1}_i$ and $\tilde{Z}^{k+1}_j$ have no measurements in common. Thus the n Kalman filters produce statistically independent track sequences, each with its own uniquely identifying label.

10.4.2 SHC: Missed Detections and False Alarms

In what follows we assume that new tracks can enter the scene. Let $Z_{k+1} = \{\mathbf{z}^1_{k+1}, ..., \mathbf{z}^{m(k+1)}_{k+1}\}$ be the measurement set collected at time step $k+1$ and let $\mathbf{z}^i_{k+1} \in Z_{k+1}$. Then there are three possibilities:

- $\mathbf{z}_{k+1}^i$ is associated with some existing predicted track in $X_{k+1|k}$, in which case the data-updated track inherits the predicted tracks' label;
- $\mathbf{z}_{k+1}^i$ is associated with a new track, to which a new label is assigned;
- $\mathbf{z}_{k+1}^i$ is not associated with an existing or a new track (i.e., it is a false alarm).

In more naïve approaches to SMC, (10.33) is used to determine the best global association of measurements with predicted tracks. More sophisticated approaches use the MHC techniques to be shortly discussed in Section 10.5. Any measurements not thus associated must be either false alarms or new tracks.

The logic used to determine whether or not a measurement is a new track or a false alarm depends on the application at hand. In applications such as surveillance, in which the false alarm rate is relatively high and immediate action is not crucial, the principle is:

- *A measurement that cannot be associated with an existing track is a false alarm until proven to be a new track.* In this case, measurements consistent with $\mathbf{z}_{k+1}^i$ must be collected during the next few scans before a new track can be declared.

In applications in which false alarm rate is low and early target detection is crucial (e.g., ballistic missile boost-phase tracking), the opposite principle applies:

- *A measurement that cannot be associated with an existing track is a new track until proven to be a false alarm.* In this case, if no measurements are consistent with $\mathbf{z}_{k+1}^i$ during the next few scans, the new track is discarded.

SHC filters include internal logic that allows them to discard an existing track if no measurements have been collected from it over some period of time.

10.5 MULTIHYPOTHESIS CORRELATION (MHC)

The performance of SHC filters deteriorates when targets are closely spaced and SNR is small. A major reason is that, at each time step, an SHC filter is forced to declare that a single measurement-to-track association is the correct one. However, the optimal association will often be the wrong one. Moreover, because of the high relative density of measurements, it will be essentially impossible to determine which association is actually correct.

An "ideal" MHC filter sidesteps this difficulty by propagating not only a track table $\mathcal{T}_{k|k}$, but also a *hypothesis table* $\mathcal{H}_{k|k}$. Because $\mathcal{H}_{k|k}$ contains tracks created from the time-accumulation of *all possible time sequences of feasible measurement-to-track associations*, it also contains the correct associations.

In this section, I outline the basic ideas behind the type of MHC filtering known as "measurement oriented."[2] For greater detail, see [16], [18, pp. 360-369], [11, pp. 334-354] and Reid's founding paper [193]. For the sake of greater clarity, we will assume that the number of targets is known to be constant and equal to $n > 0$. Also, for the sake of clarity, I will describe the MHC process only for time steps $k = 0$ and $k = 1$. (Processing in subsequent steps is essentially the same.)

At time step $k = 0$ we begin with an initial track table

$$\mathcal{T}_{0|0} = \{(l''_1, \mathbf{x}''_1, P''_1), ..., (l''_{n''}, \mathbf{x}''_{n''}, P''_{n''})\} \tag{10.49}$$

and an initial hypothesis table $\mathcal{H}_{0|0} = \{(1, \mathcal{T}_{0|0})\}$. That is, initially there is only the a priori model of reality—the track table $\mathcal{T}_{0|0}$. The probability that it is correct is, by definition, unity. The tracks are time-updated to the first time step using the Kalman predictor equations, resulting in a predicted track table

$$\mathcal{T}_{1|0} = \{(l'_1, \mathbf{x}'_1, P'_1), ..., (l'_{n'}, \mathbf{x}'_{n'}, P'_{n'})\} \tag{10.50}$$

and a predicted hypothesis-set $\mathcal{H}_{1|0} = \{(1, \mathcal{T}_{1|0})\}$.

We then collect the first observation set $Z = \{\mathbf{z}_1, ..., \mathbf{z}_m\}$, which contains m measurements. Using Z we are to construct a data-updated track table

$$\mathcal{T}_{1|1} = \{(l_1, \mathbf{x}_1, P_1), ..., (l_n, \mathbf{x}_n, P_n)\} \tag{10.51}$$

and a data-updated hypothesis table

$$\mathcal{H}_{1|1} = \{(p_1, \mathcal{S}_1), ..., (p_e, \mathcal{S}_e)\}. \tag{10.52}$$

Here, each $\mathcal{S}_i$ is a subset of $\mathcal{T}_{1|1}$ that serves as a hypothesized model of multitarget reality, and p_i is the probability that it is the correct model. Each $\mathcal{S}_i$ is called a *hypothesis* and each p_i is its *hypothesis probability*. We assume that, taken together, all hypotheses exhaustively model reality: $p_1 + ... + p_e = 1$.

2 As opposed to "track-oriented" MHC, which does not propagate the hypothesis set at each stage. Rather, it discards the current hypothesis set and reconstitutes a new one from the updated track table. See, for example, [17, 187].

10.5.1 Elements of MHC

We will proceed through three stages of increasing complexity: no missed detections or false alarms (Section 10.5.2), false alarms but no missed detections (Section 10.5.3), and both missed detections and false alarms (Section 10.5.4).

In each case we will employ the following methodology. We will define an "association" to be a certain kind of function θ from track indexes $1, ..., n$ to measurement indexes $0, 1, ..., m$ where '0' will represent the index of the null or empty measurement. Typical examples of associations are illustrated in Figure 10.1. Three kinds of associations are represented in Figure 10.2.

Given an association θ, we will construct a likelihood of the form[3]

$$f(\mathbf{z}_1, ..., \mathbf{z}_m | \mathbf{x}_1, ..., \mathbf{x}_n, \theta). \tag{10.53}$$

That is, it is the likelihood that measurements $Z = \{\mathbf{z}_1, ..., \mathbf{z}_m\}$ were generated by a scene containing the targets $X = \{\mathbf{x}_1, ..., \mathbf{x}_n\}$, given the assignment hypothesized by θ. Given this, the joint posterior probability distribution for the unknown states $\mathbf{x}_1, ..., \mathbf{x}_n$ and the unknown assignment θ, conditioned on all measurements $Z = \{\mathbf{z}_1, ..., \mathbf{z}_n\}$, is

$$f_{1|1}(\mathbf{x}_1, ..., \mathbf{x}_n, \theta) \propto f(Z | \mathbf{x}_1, ..., \mathbf{x}_n, \theta) \cdot f_{1|0}(\mathbf{x}_1, ..., \mathbf{x}_n) \cdot p'_\theta \tag{10.54}$$

where $f_{1|0}(\mathbf{x}_1, ..., \mathbf{x}_n)$ is the predicted joint probability density for all targets; where p'_σ is the prior probability distribution on σ; and where θ has been assumed independent of $\mathbf{x}_1, ..., \mathbf{x}_n$. By assumption, the predicted tracks are independent and linear-Gaussian:

$$f_{1|0}(\mathbf{x}_1, ..., \mathbf{x}_n) = f_1(\mathbf{x}_1) \cdots f_n(\mathbf{x}_n) = N_{P'_1}(\mathbf{x}_1 - \mathbf{x}'_1) \cdots N_{P'_n}(\mathbf{x}_n - \mathbf{x}'_n) \tag{10.55}$$

where $f_i(\mathbf{x}) \triangleq N_{P'_i}(\mathbf{x} - \mathbf{x}'_i)$ is the predicted distribution of the ith track. There is no a priori reason for preferring one association over another, so p'_θ is uniform. Thus the data-updated posterior distribution has the form

$$f_{1|1}(\mathbf{x}_1, ..., \mathbf{x}_n, \theta) \propto f(Z | \mathbf{x}_1, ..., \mathbf{x}_n, \theta) \cdot f_1(\mathbf{x}_1) \cdots f_n(\mathbf{x}_n). \tag{10.56}$$

3 Actually, this function will differ by a constant multiple from the actual likelihood, since it may not be the case that $\int f(\mathbf{z}_1, ..., \mathbf{z}_n | \mathbf{x}_1, ..., \mathbf{x}_n, \alpha) d\mathbf{z}_1 \cdots d\mathbf{z}_n = 1$.

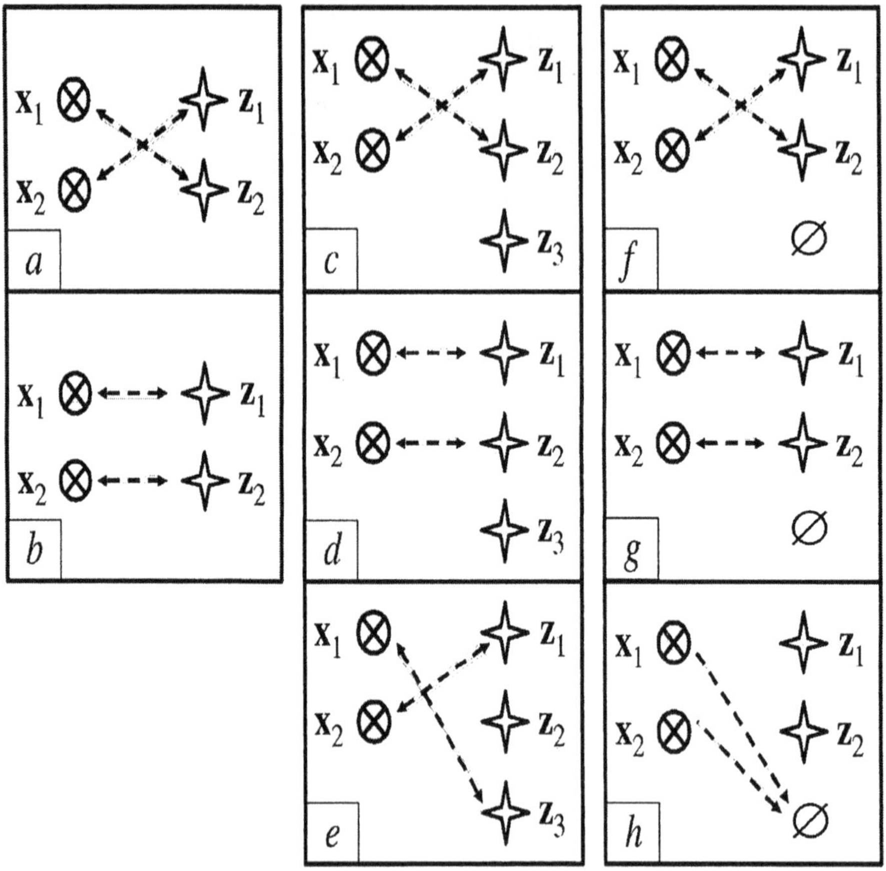

Figure 10.1 Examples of associations between two tracks $\mathbf{x}_1, \mathbf{x}_2$ and measurements $\mathbf{z}_1, \mathbf{z}_2, \mathbf{z}_3, \emptyset$, where '$\emptyset$' is the null measurement. The first column presumes no missed or false detections; and (a) and (b) depict the only two possible associations. The second column presumes false but no missed detections; and (c), (d), and (e) depict three of the six possible associations between two targets and three measurements. The third column presumes missed but no false detections; and (f)-(h) show three of the seven possible associations. In particular, (h) depicts a double missed detection.

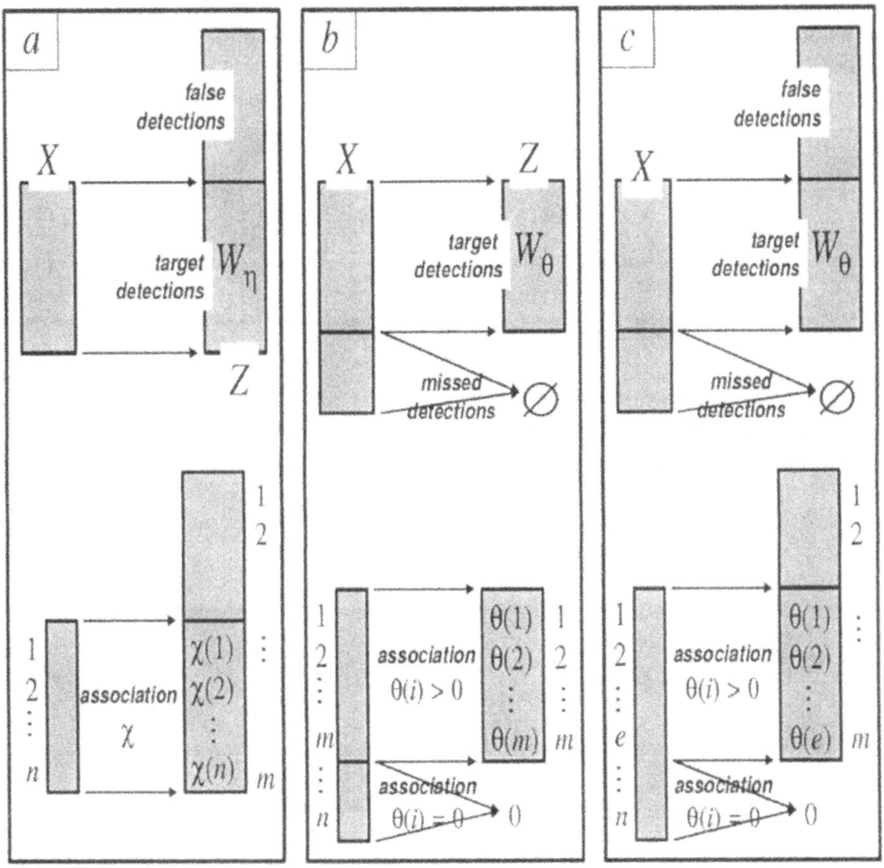

Figure 10.2 Three increasingly complicated kinds of association θ are illustrated. (a) *False alarms but no missed detections.* An association maps all targets in X to detections $W_\chi \subseteq Z$. Thus $Z - W_\chi$ consists of false alarms. If arbitrary a priori orderings are chosen for X and Z then the association can be represented as a one-to-one function χ from $\{1, ..., n\}$ to $\{1, ..., m\}$. That is, state $\mathbf{x}_i$ generates measurement $\mathbf{z}_{\chi(i)}$. (b) *Missed detections but no false alarms.* Some or all states in X fail to produce detections $W_\theta \subseteq Z$ while others do not. This association is represented as a function θ from $\{1, ..., n\}$ to $\{0, 1, ..., m\}$ with the property that $\theta(i) = \theta(i') > 0$ implies $i = i'$. Those $\mathbf{x}_i$ with $\theta(i) = 0$ generate no detections, while $\mathbf{x}_i$ with $\theta(i) > 0$ generates $\mathbf{z}_{\theta(i)}$. (c) *Missed detections and false alarms.* The association is represented by the same type of function, except that now $Z - W_\theta$ consists of false alarms.

10.5.1.1 Hypothesis Probabilities

Integrating out the state variables $\mathbf{x}_1, ..., \mathbf{x}_n$ in (10.56), we get the marginal posterior probability distribution on assignments:

$$p_\theta = \int f_{1|1}(\mathbf{x}_1, ..., \mathbf{x}_n, \theta|Z) d\mathbf{x}_1 \cdots d\mathbf{x}_n \tag{10.57}$$

$$\propto \int f(\mathbf{z}_1, ..., \mathbf{z}_n|\mathbf{x}_1, ..., \mathbf{x}_n, \theta) \cdot f_1(\mathbf{x}) \cdots f_n(\mathbf{x}) \tag{10.58}$$

$$\cdot d\mathbf{x}_1 \cdots d\mathbf{x}_n. \tag{10.59}$$

10.5.1.2 Track Probabilities and Firm Tracks

Since it is not feasible or desirable to display all hypotheses to an operator, an estimate of current multitarget reality is typically displayed instead. The two most common methods are the following:

- Displaying the tracks in the hypothesis with largest p_θ;
- Displaying the firm tracks.

The latter method tends to produce the most stable display of information.

Firm tracks are defined as follows. The *track probability* $p_{(\mathbf{x},P)}$ of the track $(\mathbf{x}, P)$ in $\mathcal{T}_{1|1}$ is the sum of the probabilities p_i of all of the hypotheses $\mathcal{S}_i$ in $\mathcal{H}_{1|1}$ which contain $(\mathbf{x}, P)$. A track is said to be *firm* if its track probability exceeds some sufficiently high threshold, say, 0.9.

10.5.2 MHC: No Missed Detections or False Alarms

Under these assumptions every target gives rise to a measurement, and no measurements arise in any other way. Thus the number of measurements is $m = n$. An assignment in this case is a permutation σ on the numbers $1, ..., n$. That is, it is a function

$$\sigma : \{1, ..., n\} \to \{1, ..., n\} \tag{10.60}$$

that is a one-to-one correspondence. It represents the following hypothesis:

- For every $i = 1, ..., n$, the observation $\mathbf{z}_{\sigma i}$ is uniquely associated with (i.e., uniquely originated with) the track $\mathbf{x}_i$.

10.5.2.1 Update of Track and Hypothesis Tables

Given a permutation σ, the Kalman corrector equations are used to update the predicted track table

$$\mathcal{T}_{1|0} = \{(l'_1, \mathbf{x}'_1, P'_1), ..., (l'_n, \mathbf{x}'_n, P'_n)\} \tag{10.61}$$

to a corresponding track table

$$\mathcal{T}^{\sigma}_{1|1} = \{(l_1, \mathbf{x}^{\sigma}_1, P_1), ..., (l_n, \mathbf{x}^{\sigma}_n, P_n)\} \tag{10.62}$$

where

$$\mathbf{x}^{\sigma}_1 \triangleq \mathbf{x}'_1 + K_1(\mathbf{z}_{\sigma 1} - H\mathbf{x}'_1) \;,...,\; \mathbf{x}^{\sigma}_n \triangleq \mathbf{x}'_n + K(\mathbf{z}_{\sigma n} - H\mathbf{x}'_n) \tag{10.63}$$

$$P_1 = (I - K_1 H) P'_1 \;,....,\; P_n = (I - K_n H) P'_n \tag{10.64}$$

and where

$$K_1 = P'_1 H^T \left(H P'_1 H^T + R\right)^{-1} ,..., K_n = P'_n H^T \left(H P'_n H^T + R\right)^{-1}. \tag{10.65}$$

Each data-updated track inherits the label of the predicted track from which it came. This results in a branching of each original track into several possible alternative predicted futures. If $e = n!$ and if $\sigma_1, ..., \sigma_e$ denote all of the permutations, then the union of these e new track tables results in the updated track table

$$\mathcal{T}_{1|1} = \mathcal{T}^{\sigma_1}_{1|1} \cup ... \cup \mathcal{T}^{\sigma_e}_{1|1}. \tag{10.66}$$

This table can have as many as n^2 distinct tracks, corresponding to the assignment of each measurement $\mathbf{z}_j$ to each track $\mathbf{x}_i$. Each $\mathcal{T}^{\sigma}_{1|1}$ is an alternative hypothesis about multitarget reality. Thus the data-updated hypothesis table $\mathcal{H}_{1|1}$ must have the form

$$\mathcal{H}_{1|1} = \{(p_{\sigma_1}, \mathcal{T}^{\sigma_1}_{1|1}), ..., (p_{\sigma_e}, \mathcal{T}^{\sigma_e}_{1|1})\}. \tag{10.67}$$

10.5.2.2 Computation of Hypothesis Probabilities

The explicit formula for p_σ is

$$p_\sigma = \frac{\ell_\sigma}{\sum_{\sigma'} \ell_{\sigma'}} = \frac{\exp\left(-\frac{1}{2} d^2_\sigma\right)}{\sum_{\sigma'} \exp\left(-\frac{1}{2} d^2_{\sigma'}\right)} \tag{10.68}$$

where

$$d_\sigma^2 \triangleq \sum_{i=1}^{n} (\mathbf{z}_{\sigma i} - H\mathbf{x}_i')^T (R + HP_i'H^T)^{-1} (\mathbf{z}_{\sigma i} - H\mathbf{x}_i') \tag{10.69}$$

$$\ell_\sigma \triangleq \frac{1}{\prod_{i=1}^{n} \sqrt{\det 2\pi(R + HP_i'H^T)}} \cdot \exp\left(-\frac{1}{2}d_\sigma^2\right) \tag{10.70}$$

are the global association distance and global association likelihood as defined in (10.31) and (10.36), respectively.

To see why (10.68) is true, note that under current assumptions the likelihood function of (10.53) must have the form

$$f(\mathbf{z}_1, ..., \mathbf{z}_n | \mathbf{x}_1, ..., \mathbf{x}_n, \sigma) = \overbrace{f(\mathbf{z}_{\sigma 1}|\mathbf{x}_1) \cdots f(\mathbf{z}_{\sigma n}|\mathbf{x}_n)}^{\text{target measurements}}. \tag{10.71}$$

Given this, (10.57) reduces to:

$$p_\sigma \propto \int f(\mathbf{z}_{\sigma 1}|\mathbf{x}_1) \cdots f(\mathbf{z}_{\sigma n1}|\mathbf{x}_n) \cdot f_1(\mathbf{x}_1) \cdots f_n(\mathbf{x}_n) \tag{10.72}$$

$$\cdot d\mathbf{x}_1 \cdots d\mathbf{x}_n \tag{10.73}$$

$$= \left(\int f(\mathbf{z}_{\sigma 1}|\mathbf{x}) \cdot f_1(\mathbf{x}) d\mathbf{x}\right) \cdots \left(\int f(\mathbf{z}_{\sigma n}|\mathbf{x}) \cdot f_n(\mathbf{x}) d\mathbf{x}\right) \tag{10.74}$$

$$= \left(\int N_R(\mathbf{z}_{\sigma 1} - H\mathbf{x}) \cdot N_{P_1'}(\mathbf{x} - \mathbf{x}_1') d\mathbf{x}\right) \tag{10.75}$$

$$\cdots \left(\int N_R(\mathbf{z}_{\sigma n} - H\mathbf{x}) \cdot N_{P_n'}(\mathbf{x} - \mathbf{x}_n') d\mathbf{x}\right). \tag{10.76}$$

From the fundamental identity for Gaussian distributions, see (D.1),

$$p_\sigma \propto N_{R+HP_1'H^T}(\mathbf{z}_{\sigma 1} - H\mathbf{x}_1') \cdots N_{R+HP_n'H^T}(\mathbf{z}_{\sigma n} - H\mathbf{x}_n') \tag{10.77}$$

$$= \frac{1}{\prod_{i=1}^{n} \det 2\pi(R + HP_1'H^T)} \cdot \exp\left(-\frac{1}{2}d_\sigma^2\right) \tag{10.78}$$

$$= \ell_\sigma. \tag{10.79}$$

Thus, as claimed, the posterior probability of σ is

$$p_\sigma = \frac{\ell_\sigma}{\sum_{\sigma'} \ell_{\sigma'}} = \frac{\exp\left(-\frac{1}{2}d_\sigma^2\right)}{\sum_{\sigma'} \exp\left(-\frac{1}{2}d_{\sigma'}^2\right)}. \tag{10.80}$$

10.5.3 MHC: False Alarms, No Missed Detections

Under these assumptions every target gives rise to a measurement. Thus the number of measurements is $m = n + m'$ where m' is the number of false alarms. An assignment is a one-to-one function

$$\chi : \{1, ..., n\} \to \{1, ..., m\}. \tag{10.81}$$

It represents the following hypothesis—see Figure 10.2(a):

- For every $i = 1, ..., n$, the observation $\mathbf{z}_{\chi(i)}$ is uniquely associated with (uniquely originated with) the track $\mathbf{x}_i$.

If

$$W_\chi \triangleq \{\mathbf{z}_{\chi(1)}, ..., \mathbf{z}_{\chi(n)}\} \tag{10.82}$$

then W_χ can be any subset of Z that has exactly n elements. The measurements in the complement $Z - W_\chi$ must be false alarms. By assumption, the false alarm process has the Poisson form

$$\kappa(\mathbf{z}_1, ..., \mathbf{z}_m) = \frac{e^{-\lambda}\lambda^m}{m!} \cdot c(\mathbf{z}_1) \cdots c(\mathbf{z}_m) \tag{10.83}$$

where λ is the average number of false alarms and where $c(\mathbf{z})$ is the spatial distribution of any given false alarm.

10.5.3.1 Update of Track and Hypothesis Tables

Given an association χ, just as in Section 10.5.2 the Kalman corrector equations are used to update the predicted track table

$$\mathcal{T}_{1|0} = \{(l'_1, \mathbf{x}'_1, P'_1), ..., (l'_n, \mathbf{x}'_n, P'_n)\} \tag{10.84}$$

to a corresponding track table

$$\mathcal{T}^\chi_{1|1} = \{(l_1, \mathbf{x}^\chi_1, P_1), ..., (l_1, \mathbf{x}^\chi_n, P_n)\} \tag{10.85}$$

where

$$\mathbf{x}^\chi_1 \triangleq \mathbf{x}'_1 + K_1(\mathbf{z}_{\chi(1)} - H\mathbf{x}'_1) \;,...,\; \mathbf{x}^\chi_n \triangleq \mathbf{x}'_n + K(\mathbf{z}_{\chi(n)} - H\mathbf{x}'_n) \tag{10.86}$$

$$P_1 = (I - K_1 H)\, P'_1 \;,....,\; P_n = (I - K_n H)\, P'_n. \tag{10.87}$$

Once again, each data-updated track inherits the label of the predicted track from which it came.

Let e be the total number of associations and $\chi_1, ..., \chi_e$ a listing of all of them. Then the union of these e new track tables results in the updated track and hypotheses tables:

$$\mathcal{T}_{1|1} = \mathcal{T}_{1|1}^{\chi_1} \cup ... \cup \mathcal{T}_{1|1}^{\chi_e} \quad (10.88)$$

$$\mathcal{H}_{1|1} = \{(p_{\chi_1}, \mathcal{T}_{1|1}^{\chi_1}), ..., (p_{\chi_e}, \mathcal{T}_{1|1}^{\chi_e})\}. \quad (10.89)$$

10.5.3.2 Computation of Hypothesis Probabilities

I show that the explicit formula for p_χ is

$$p_\chi = \frac{\left(\prod_{\mathbf{z}\in Z-W_\chi} c(\mathbf{z})\right) \cdot \ell_\chi}{\sum_{\chi'} \left(\prod_{\mathbf{z}\in Z-W_{\chi'}} c(\mathbf{z})\right) \cdot \ell_{\chi'}} \quad (10.90)$$

$$= \frac{\left(\prod_{\mathbf{z}\in Z-W_\chi} c(\mathbf{z})\right) \cdot \exp\left(-\frac{1}{2}d_\chi^2\right)}{\sum_{\chi'} \left(\prod_{\mathbf{z}\in Z-W_{\chi'}} c(\mathbf{z})\right) \cdot \exp\left(-\frac{1}{2}d_{\chi'}^2\right)} \quad (10.91)$$

where

$$d_\chi^2 \triangleq \sum_{i=1}^{n} (\mathbf{z}_{\chi(i)} - H\mathbf{x}'_i)^T (R + HP'_iH^T)^{-1} (\mathbf{z}_{\chi(i)} - H\mathbf{x}'_i) \quad (10.92)$$

$$\ell_\chi \triangleq \frac{1}{\prod_{i=1}^n \sqrt{\det 2\pi(R + HP'_iH^T)}} \cdot \exp\left(-\frac{1}{2}d_\chi^2\right) \quad (10.93)$$

are the global association distance and global association likelihood as defined in (10.33) and (10.38), respectively.

To see why (10.90) is true, note that under current assumptions the likelihood function of (10.53) must have the form

$$f(\mathbf{z}_1, ..., \mathbf{z}_m | \mathbf{x}_1, ..., \mathbf{x}_n, \chi) = \overbrace{f(\mathbf{z}_{\chi(1)}|\mathbf{x}_1) \cdots f(\mathbf{z}_{\chi(n)}|\mathbf{x}_n)}^{\text{target measurements}} \quad (10.94)$$

$$\cdot \overbrace{e^{-\lambda} \prod_{\mathbf{z}\in Z-W_\chi} \lambda c(\mathbf{z})}^{\text{false alarms}}. \quad (10.95)$$

Abbreviate

$$S_i \stackrel{\text{abbr.}}{=} R + HP'_iH^T , \qquad \mathbf{s}_i \stackrel{\text{abbr.}}{=} \mathbf{z}_{\chi(i)} - H\mathbf{x}'_i \tag{10.96}$$

for $i = 1, .., n$. Then (10.57) reduces to

$$p_\chi \propto \left(e^{-\lambda} \prod_{\mathbf{z} \in Z - W_\chi} \lambda c(\mathbf{z}) \right) \tag{10.97}$$

$$\cdot \int f(\mathbf{z}_{\chi(1)}|\mathbf{x}_1) \cdots f(\mathbf{z}_{\chi(n)}|\mathbf{x}_n) \cdot f_1(\mathbf{x}_1) \cdots f_n(\mathbf{x}_n) \tag{10.98}$$

$$d\mathbf{x}_1 \cdots d\mathbf{x}_n \tag{10.99}$$

and so

$$= e^{-\lambda} \lambda^{m-n} \left(\prod_{\mathbf{z} \in Z - W_\chi} c(\mathbf{z}) \right) \cdot N_{S_1}(\mathbf{s}_1) \cdots N_{S_n}(\mathbf{s}_n) \tag{10.100}$$

$$= e^{-\lambda} \lambda^{m-n} \left(\prod_{\mathbf{z} \in Z - W_\chi} c(\mathbf{z}) \right) \cdot \left(\prod_{i=1}^{n} \frac{1}{\sqrt{2\pi \det S_i}} \right) \tag{10.101}$$

and so

$$\cdot \exp \left(-\frac{1}{2} \sum_{i=1}^{n} \mathbf{s}_i^T S_i^{-1} \mathbf{s}_i \right) \tag{10.102}$$

$$= e^{-\lambda} \lambda^{m-n} \cdot \left(\prod_{\mathbf{z} \in Z - W_\chi} c(\mathbf{z}) \right) \cdot \ell_\chi. \tag{10.103}$$

Thus, as claimed,

$$p_\chi = \frac{\left(\prod_{\mathbf{z} \in Z - W_\chi} c(\mathbf{z}) \right) \cdot \ell_\chi}{\sum_{\chi'} \left(\prod_{\mathbf{z} \in Z - W_{\chi'}} c(\mathbf{z}) \right) \cdot \ell_{\chi'}} \tag{10.104}$$

$$= \frac{\left(\prod_{\mathbf{z} \in Z - W_\chi} c(\mathbf{z}) \right) \cdot \exp\left(-\frac{1}{2} d_\chi^2\right)}{\sum_{\chi'} \left(\prod_{\mathbf{z} \in Z - W_{\chi'}} c(\mathbf{z}) \right) \cdot \exp\left(-\frac{1}{2} d_{\chi'}^2\right)}. \tag{10.105}$$

10.5.4 MHC: Missed Detections and False Alarms

Under these assumptions it is not necessarily the case that any given target will generate a measurement. The number of measurements is $m = n - n' + m'$ where m' is the number of false alarms and n' is the number of missed detections. In this case an assignment is no longer necessarily a one-to-one function. The number '0', which symbolizes a null observation (missed detection), must be appended to $\{1, ..., m\}$, resulting in an augmented set $\{0, 1, ..., m\}$ of measurement indices. In this case an association is a function

$$\theta : \{1, ..., n\} \to \{0, 1, ..., m\} \tag{10.106}$$

that has the following property: $\theta(i) = \theta(i') > 0$ implies $i = i'$. It represents the following hypothesis—see Figures 10.2(b) and 10.2(c):

- For every $i = 1, ..., n$, if $\theta(i) > 0$ then the observation $\mathbf{z}_{\theta(i)}$ is uniquely associated with the track $\mathbf{x}_i$; but if $\theta(i) = 0$ then no observation is associated with $\mathbf{x}_i$ (the target $\mathbf{x}_i$ was not detected).

At one extreme, if $\theta(i) = 0$ for all i then every measurement in Z must be a false alarm. In this case we abbreviate $\theta \overset{\text{abbr.}}{=} 0$ and adopt the convention $d_\theta = 0$. At the other extreme, if $\theta(i) > 0$ for all i then every target gave rise to a measurement.

If $\theta = 0$ define $W_\theta = W_0 = \emptyset$. Otherwise,

$$W_\theta \triangleq \bigcup_{i:\theta(i)>0} \{\mathbf{z}_{\theta(i)}\} \tag{10.107}$$

is the set of target-generated measurements, given the hypothesis θ. Taken over all θ, W_θ can be any subset of Z with no more than n elements, the null set included. The number of target-generated measurements is therefore the number

$$n_\theta \triangleq |W_\theta| \tag{10.108}$$

of those $i \in \{1, ..., n\}$ such that $\theta(i) > 0$. In particular, $n_0 = 0$.

10.5.4.1 Update of Track and Hypothesis Tables

Given an association θ, the Kalman corrector equations are once again used to update the predicted track table

$$\mathcal{T}_{1|0} = \{(l'_1, \mathbf{x}'_1, P'_1), ..., (l'_n, \mathbf{x}'_n, P'_n)\} \tag{10.109}$$

to a track table

$$\mathcal{T}^{\theta}_{1|1} = \{(l_1, \mathbf{x}^{\theta}_1, P_1), ..., (l_1, \mathbf{x}^{\theta}_n, P_n)\}. \tag{10.110}$$

Here, $\mathbf{x}^{\theta}_i = \mathbf{x}_i$ if $\theta(i) = 0$. Otherwise,

$$\mathbf{x}^{\theta}_i \triangleq \mathbf{x}'_i + K_i(\mathbf{z}_{\theta(i)} - H\mathbf{x}'_1) \tag{10.111}$$

$$P_i = (I - K_i H)\, P'_i. \tag{10.112}$$

As usual, each track inherits its predecessor's label.[4]

If e is the number of possible associations, this results in updated track and hypotheses tables:

$$\mathcal{T}_{1|1} = \mathcal{T}^{\theta_1}_{1|1} \cup ... \cup \mathcal{T}^{\theta_e}_{1|1} \tag{10.113}$$

$$\mathcal{H}_{1|1} = \{(p_{\theta_1}, \mathcal{T}^{\theta_1}_{1|1}), ..., (p_{\theta_e}, \mathcal{T}^{\theta_e}_{1|1})\}. \tag{10.114}$$

10.5.4.2 Computation of Hypothesis Probabilities

I show that the formulas for p_0 and for p_θ with $\theta \neq 0$ are

$$p_0 = \frac{1}{1 + \sum_{\theta' \neq 0} K^{n_{\theta'}} \cdot \ell_{\theta'} \cdot \left(\prod_{\mathbf{z} \in W_{\theta'}} c(\mathbf{z})^{-1}\right)} \tag{10.115}$$

$$p_\theta = \frac{K^{n_\theta} \cdot \ell_\theta \cdot \left(\prod_{\mathbf{z} \in W_\theta} c(\mathbf{z})^{-1}\right)}{1 + \sum_{\theta' \neq 0} K^{n_{\theta'}} \cdot \ell_{\theta'} \cdot \left(\prod_{\mathbf{z} \in W_{\theta'}} c(\mathbf{z})^{-1}\right)} \tag{10.116}$$

where

$$K \triangleq \frac{p_D}{\lambda\,(1 - p_D)}$$

and where

$$d^2_\theta \triangleq \sum_{i:\theta(i)>0} (\mathbf{z}_{\theta(i)} - H\mathbf{x}'_i)^T (R + HP'_i H^T)^{-1} (\mathbf{z}_{\theta(i)} - H\mathbf{x}'_i)$$

$$\ell_\theta \triangleq \frac{1}{\prod_{\theta(i)>0} \sqrt{\det 2\pi S_i}} \cdot \exp\left(-\frac{1}{2} d^2_\theta\right) \tag{10.117}$$

4 In these examples we have been assuming that target number is constant and known. If new targets were to appear, they would be assigned new labels.

are essentially the global association distance and global association likelihood as defined in (10.33) and (10.38), respectively.

The derivation of this result is more complicated than in the previous sections. It can be found in Appendix G.17.

Example 50 *Suppose that only a single target is in the scene:* $n = 1$. *Then any non-null association has the form* $\theta_j(1) = j$ *for* $j = 1, ..., m$, *in which case we write* $p_j = p_{\theta_j}$, $d_j = d_{\theta_j}$, $n_j = n_{\theta_j} = 1$, $\ell_j = \ell_{\theta_j}$, *and* $W_j = W_{\theta_j} = \{\mathbf{z}_j\}$. *Also, assume that* $c(\mathbf{z}) = d_0^{-1}$ *is uniform. Then (10.115) and (10.116) reduce to:*

$$p_0 = \frac{b}{b + \sum_{e=1}^{m} \exp\left(-\frac{1}{2}d_e^2\right)} \tag{10.118}$$

$$p_j = \frac{\exp\left(-\frac{1}{2}d_j^2\right)}{b + \sum_{e=1}^{m} \exp\left(-\frac{1}{2}d_e^2\right)} \tag{10.119}$$

where[5]

$$b \triangleq \frac{\lambda(1 - p_D)\sqrt{\det 2\pi(R + HP'H^T)}}{d_0 p_D}. \tag{10.120}$$

To see this, abbreviate $S \overset{\text{abbr.}}{=} R + HP'H^T$. *On the one hand, for* p_θ *with* $\theta = 0$ *we get from (10.115)*

$$p_0 = \frac{1}{1 + \sum_{\theta' \neq 0} K^{n_{\theta'}} \cdot \ell_{\theta'} \cdot \prod_{\mathbf{z} \in W_{\theta'}} c(\mathbf{z})^{-1}} \tag{10.121}$$

$$= \frac{1}{1 + \sum_{e=1}^{m} K \cdot \ell_e \cdot c(\mathbf{z}_e)^{-1}} \tag{10.122}$$

$$= \frac{1}{1 + \sum_{e=1}^{m} \frac{p_D}{\lambda(1-p_D)} \cdot \frac{1}{\sqrt{\det 2\pi S}} \cdot \exp\left(-\frac{1}{2}d_e^2\right) \cdot d_0} \tag{10.123}$$

$$= \frac{b}{b + \sum_{e=1}^{m} \exp\left(-\frac{1}{2}d_e^2\right)}. \tag{10.124}$$

5 Equations (10.118)-(10.120) should be familiar from the theory of the probabilistic data association (PDA) filter [11, p. 136]. See Section 10.6.3.

On the other hand, for p_θ *with* $\theta \neq 0$ *we get from (10.116)*

$$p_j = \frac{K \cdot \ell_j \cdot c(\mathbf{z}_j)^{-1}}{1 + \sum_{e=1}^{m} K \cdot \ell_e \cdot c(\mathbf{z}_e)^{-1}} \quad (10.125)$$

$$= \frac{K \cdot \frac{1}{\sqrt{\det 2\pi S}} \cdot \exp\left(-\frac{1}{2}d_j^2\right) \cdot d_0}{1 + \sum_{e=1}^{m} K \cdot \frac{1}{\sqrt{\det 2\pi S}} \cdot \exp\left(-\frac{1}{2}d_e^2\right) \cdot d_0} \quad (10.126)$$

$$= \frac{\exp\left(-\frac{1}{2}d_j^2\right)}{b + \sum_{e=1}^{m} \exp\left(-\frac{1}{2}d_e^2\right)}. \quad (10.127)$$

10.6 COMPOSITE-HYPOTHESIS CORRELATION (CHC)

MHC filters avoid the performance limitations of SHC filters, but at the price of increased computational complexity. Other approaches attempt to steer a middle road between SHC and MHC by, so to speak, creating a single "composite" hypothesis. The *Joint Probabilistic Data Association* (JPDA) filter, due to Fortmann, Bar-Shalom, and Scheffe [62], is the most well known such approach. In this section, I can only sketch the basic ideas of JPDA. For more details, see [107], [62, pp. 174-177], [11, pp. 310-319], [198], or [18, pp. 350-360].

10.6.1 Elements of CHC

As usual we assume that the number of targets is constant and known. At time step $k+1$ let $Z = \{\mathbf{z}_1, ..., \mathbf{z}_m\}$ be the measurement set and $X' = \{\mathbf{x}'_1, ..., \mathbf{x}'_n\}$ the predicted target state set. We assume that $m \geq n$. Abbreviate $R \overset{\text{abbr.}}{=} R_{k+1}$ and $H \overset{\text{abbr.}}{=} H_{k+1}$. For the sake of clarity, ignore the issue of validation-gating.

A SHC filter determines the optimal measurement-to-track association $\mathbf{z}_{\hat{\sigma}i} \leftrightarrow \mathbf{x}_i$ and a MHC filter preserves all $n!$ associations $\mathbf{z}_{\sigma i} \leftrightarrow \mathbf{x}_i$. A JPDA filter, by way of contrast, assumes that all measurements are associated to some extent with every track. The point is to determine the degree to which each measurement contributed to each track, and then create composite tracks that reflect the correct degree of influence of the original measurements.

In more detail, let $(\mathbf{x}'_1, P'_1), ..., (\mathbf{x}'_n, P'_n)$ be the predicted tracks at time step $k+1$. We are to construct data-updated tracks $(\mathbf{x}_1, P_1), ..., (\mathbf{x}_n, P_n)$.

Let the following be true:

- β_j^1 is the probability that measurement $\mathbf{z}_j$ originated with track $\mathbf{x}'_1$.

- β_j^2 is the probability that $\mathbf{z}_j$ originated with $\mathbf{x}'_2$, and so on.
- β_0^1 is the probability that no measurement was generated by $\mathbf{x}'_1$.
- β_0^2 is the probability that no measurement was generated by $\mathbf{x}'_2$ and so on.

We assume that $\sum_{j=0}^{n} \beta_j^i = 1$ for each $i = 1, ..., n$. Let $\mathbf{s}_j^i \triangleq \mathbf{z}_j - H\mathbf{x}'_i$ for $j = 1, ..., m$ be the innovation corresponding to $\mathbf{z}_j$ and $\mathbf{x}_i$. Since β_j^i is the proportion to which $\mathbf{z}_j$ contributed to $\mathbf{x}'_i$, then the weighted innovations

$$\mathbf{s}_1 \triangleq \sum_{j=1}^{n} \beta_j^1 \cdot \mathbf{s}_j^1 \quad ,....., \quad \mathbf{s}_n \triangleq \sum_{j=1}^{n} \beta_j^n \cdot \mathbf{s}_j^n \tag{10.128}$$

are composite innovations. They reflect the degree to which each measurement is associated with each track $\mathbf{x}'_i$. If we data-update $\mathbf{x}'_i$ using $\mathbf{s}_i$ we therefore get

$$\mathbf{x}_1 = \mathbf{x}'_1 + K_1\mathbf{s}_1 \quad ,....., \quad \mathbf{x}_n = \mathbf{x}'_n + K_n\mathbf{s}_n \tag{10.129}$$

where

$$K_1 = P'_1 H^T \left(HP'_1 H^T + R\right)^{-1} \;,...., K_n = P'_n H^T \left(HP'_n H^T + R\right)^{-1}. \tag{10.130}$$

(Determining the corresponding covariance matrices is considerably more complex and will not be discussed here.)

The probabilities β_j^i are constructed as follows. Consider all associations θ such that $\theta(i) = j$ (that is, those associations that associate $\mathbf{z}_j$ with $\mathbf{x}'_i$). Then the total degree to which $\mathbf{z}_j$ is associated with $\mathbf{x}'_i$ is

$$d_j^i \triangleq \sum_{\theta:\theta(i)=j} p_\theta \tag{10.131}$$

where p_θ is the probability of the association hypothesis θ. Normalizing, we get

$$\beta_j^i \triangleq \frac{d_j^i}{\sum_{l=1} d_l^i}. \tag{10.132}$$

The critical step is determining the values of p_θ. We have already accomplished this chore in Section 10.5.

In what follows I sketch the JPDA procedure in three increasingly more complicated steps: (1) no missed detections or false alarms, (2) a single target in missed detections and false alarms, and (3) a known number n of targets in missed detections and false alarms.

10.6.2 CHC: No Missed Detections or False Alarms

In this case $m = n$ and an association hypothesis θ is just a permutation σ on the numbers $1, ..., n$. From (10.68)-(10.69) we know that the probability of the association σ is

$$p_\sigma = \frac{\exp\left(-\frac{1}{2}d_\sigma^2\right)}{\sum_{\sigma'} \exp\left(-\frac{1}{2}d_{\sigma'}^2\right)} \tag{10.133}$$

where

$$d_\sigma^2 = \sum_{i=1}^{n} (\mathbf{z}_{\sigma i} - H\mathbf{x}'_i)^T (R + HP'_i H^T)^{-1} (\mathbf{z}_{\sigma i} - H\mathbf{x}'_i). \tag{10.134}$$

Thus

$$d_j^i = \sum_{\sigma:\sigma i = j} p_\sigma \tag{10.135}$$

and

$$\beta_j^i \triangleq \frac{d_j^i}{\sum_{l=1} d_l^i}. \tag{10.136}$$

10.6.3 CHC: Probabilistic Data Association (PDA)

Consider the case described in Example 50: missed detections and false alarms, but only a single target: $n = 1$. As noted in that example, all non-null associations have the form $\theta_j(1) = j$ for $j = 1, ..., m$. The probabilities $\beta_j \triangleq \beta_j^1$ are

$$\beta_j \triangleq \frac{d_j}{\sum_{l=0} d_l} \tag{10.137}$$

where, by (10.118) and (10.119),

$$d_0 = \sum_{\theta:\theta(1)=0} p_\theta = p_0 = \frac{b}{b + \sum_{e=1}^{m} \exp\left(-\frac{1}{2}d_e^2\right)} \tag{10.138}$$

$$d_j = \sum_{\theta:\theta(1)=j} p_\theta = p_j = \frac{\exp\left(-\frac{1}{2}d_j^2\right)}{b + \sum_{e=1}^{m} \exp\left(-\frac{1}{2}d_e^2\right)} \tag{10.139}$$

and where

$$b \triangleq \frac{\lambda(1 - p_D)\sqrt{\det 2\pi (R + HP'_i H^T)}}{d_0 p_D}. \tag{10.140}$$

Consequently,

$$\sum_{j=0}^{m} d_j = \sum_{j=0}^{m} p_j = 1 \tag{10.141}$$

and so

$$\beta_j = p_j \tag{10.142}$$

for $j = 0, 1, ..., m$. This special case of JPDA is known as *probabilistic data association* (PDA).

10.6.4 CHC: Missed Detections, False Alarms

In this case n is fixed and known but otherwise arbitrary. The probabilities β_j^i are

$$\beta_j^i \triangleq \frac{d_j^i}{\sum_{l=1} d_l^i} \tag{10.143}$$

where $d_j^i = \sum_{\theta:\theta(i)=j} p_\theta$ for all $i = 1, ..., n$ and $j = 0, 1, ..., m$ and where p_θ is defined as in (10.115) and (10.116).

10.7 CONVENTIONAL FILTERING: LIMITATIONS

In this section, I briefly assess the strengths and weaknesses of the most well known conventional approaches to multitarget filtering.

The question of real-time performance is addressed in Section 10.7.1. A more subtle theoretical issue—*is the MHC approach actually Bayesian?*—is addressed in Section 10.7.2.

10.7.1 Real-Time Performance

The ideal MHC approach is computationally intractable. For example, assume the situation in Section 10.5.2: no missed detections or false alarms. The process outlined there is repeated recursively, with each hypothesis in $\mathcal{H}_{k|k}$ leading to $n!$ new hypotheses in $\mathcal{H}_{k+1|k+1}$. The updated track table $\mathcal{T}_{k|k}$ will have as many as n^k distinct tracks, and the updated hypothesis table $\mathcal{H}_{k|k}$ will have as many as $(n!)^k$ hypotheses.

Various techniques must be used to reduce the numbers of tracks and hypotheses. These techniques include: track clustering, merging of similar tracks, merging of similar hypotheses, pruning of improbable tracks, and pruning of improbable hypotheses [16, 18]. Alternatively, track-oriented versions of MHC are used instead of the measurement-oriented type described here [16, 18].

When such techniques are applied, MHC filters are both computationally tractable and more effective than SHC filters. As Blackman notes [16, p. 13],

> A number of studies...have indicated that an MHT tracker will provide performance that is comparable to the conventional single hypothesis (GNN) method at 10 to 100 times the false alarm density of the GNN method.

MHC-based algorithms have become the "workhorses" for real-time multitarget tracking. MHC trackers are beginning to achieve dominance in fielded systems. Techniques such as covariance-gating of observations allow them to detect and track in moderately low SNRs. They can be rendered robust to strong target maneuvers via such techniques as large plant noise or interacting multiple model (IMM) techniques. Very high track densities can be maintained using a priori terrain information and other domain-specific "tricks."

The reach of MHC trackers can be further extended using adaptive filter selection. That is, an MHC tracker can be modified to adaptively select filter type according to individual target motion: simple alpha-beta filters for nonmaneuvering targets, EKFs for moderately maneuvering targets, IMM-EKFs for strongly maneuvering targets, and nonlinear filters for targets whose motions are highly nonlinear (e.g., restricted by sharp terrain constraints).

This all having been acknowledged, MHC trackers have inherent limitations. When signal-to-noise ratio (SNR) decreases to the point that false alarms or clutter returns cannot be effectively rejected through validation gating, MHC tracks will prove ineffective in detecting or tracking targets. When target returns are sporadic in comparison to even moderately noisy backgrounds, or blocked by terrain obscurations, an MHC tracker's track tables and hypothesis lists will combinatorially explode in size. If during a crossing two (or more) targets are closely spaced relative to sensor resolution and if their velocities are different but not greatly so, an MHC-based algorithm may be unable to successfully track the targets through the crossing.

It is in applications such as these that the multitarget recursive Bayes filter, to be formally introduced in Chapter 14, becomes potentially important.

10.7.2 Is a Hypothesis Actually a State Variable?

A more subtle issue involves the theoretical basis of MHC algorithms from a Bayesian point of view: *Is the MHC approach Bayesian?* We ask the following:

- What role does the concept of an association hypothesis play in a strict Bayesian analysis?

On the one hand, an association is clearly not an observable. There is no way that one could, in general, observe the correct data association. On the other hand, the MHC methodology, and especially the notation for likelihoods

$$f_{k+1}(\mathbf{z}_1, ..., \mathbf{z}_m|\mathbf{x}_1, ..., \mathbf{x}_n, \theta) \tag{10.144}$$

and posterior densities

$$f_{k+1|k+1}(\mathbf{x}_1, ..., \mathbf{x}_n, \theta|\mathbf{z}_1, ..., \mathbf{z}_m) \tag{10.145}$$

as introduced in (10.53) and (10.56), imply that an association θ is a state variable.

Given that this is the case, a complete representation of a multitarget state for an MHC filter would consist of the following:

- The unknown number n of targets;
- The set $\{\mathbf{x}_1, ..., \mathbf{x}_n\}$ of unknown state vectors;
- Some choice of an integer $m \geq n$;
- Some choice of a function $\theta : \{1, ..., n\} \rightarrow \{0, 1, ..., m\}$ such that $\theta(i) = \theta(i') > 0$ implies $i = i'$.

The third requirement implies that m is a state variable, the value of which is to be estimated. This is inherently suspicious. A carefully specified state space should not include observation parameters as state parameters for the following reasons.

First, we cannot declare m to be the cardinality of a set of future measurements to which $\mathbf{x}_1, ..., \mathbf{x}_n$ are to be associated via the function θ. Otherwise, m would clearly be an observation variable that has been improperly introduced as a state variable. If so, what meaning does m have as a *state variable*?

Second, it is not possible to estimate θ as a state parameter without implicitly introducing knowledge of the observation process into the definition of a multitarget state. Suppose that a measurement set $Z = \{\mathbf{z}_1, ..., \mathbf{z}_m\}$ has been collected. Then as an association, θ implicitly depends on the existence of some specific ordering $(\mathbf{z}_1, ..., \mathbf{z}_m)$ of the measurements of Z.

However, measurement scans have no a priori order in general. Any permutation σ on $1, ..., m$ determines a different possible ordering $(\mathbf{z}_{\sigma 1}, ..., \mathbf{z}_{\sigma m})$. So, the function θ could just as easily imply the association $\mathbf{x}_i \leftrightarrow \mathbf{z}_{\theta(\sigma i)}$ as the association $\mathbf{x}_i \leftrightarrow \mathbf{z}_{\theta(i)}$. In choosing an a priori order, are we introducing extraneous information—and thus also a statistical bias?

That is:

- *It does not appear that the correct association hypothesis θ can be estimated as a state variable, without introducing into the state representation an a priori ordering of measurements that have not yet been collected.*

Thus I pose the following question:

- Since in a Bayesian analysis the unknown association θ must be a random state variable, is the MHC procedure consistent with the Bayesian "recipe" described in Section 3.5?

The conundrum is resolved if we replace the individual association likelihood functions

$$f_{k+1}(\mathbf{z}_1, ..., \mathbf{z}_m | \mathbf{x}_1, ..., \mathbf{x}_n, \theta) \tag{10.146}$$

of (10.144) with an average:

$$f_{k+1}(\mathbf{z}_1, ..., \mathbf{z}_m | \mathbf{x}_1, ..., \mathbf{x}_n) \triangleq K^{-1} \sum_{\theta} f_{k+1}(\mathbf{z}_1, ..., \mathbf{z}_m | \mathbf{x}_1, ..., \mathbf{x}_n, \theta) \tag{10.147}$$

where K is a normalization constant. To see this, note that $f_{k+1}(\mathbf{z}_1, ..., \mathbf{z}_m | \mathbf{x}_1, ..., \mathbf{x}_n)$ thus defined would be completely independent of the $\mathbf{x}_i \leftrightarrow \mathbf{z}_{\theta(\sigma i)}$. If $(\theta\sigma)(i) \triangleq \theta(\sigma i)$ then

$$\sum_{\theta} f_{k+1}(\mathbf{z}_1, ..., \mathbf{z}_m | \mathbf{x}_1, ..., \mathbf{x}_n, \theta\sigma) = \sum_{\theta} f_{k+1}(\mathbf{z}_1, ..., \mathbf{z}_m | \mathbf{x}_1, ..., \mathbf{x}_n, \theta). \tag{10.148}$$

We will return to this issue in Section 12.4.

10.8 MHC WITH FUZZY DS MEASUREMENTS

In Section 5.6, I introduced a single-target filter, the *Kalman evidential filter* (KEF), that processes both conventional and fuzzy Dempster-Shafer (fuzzy DS) measurements. In this section I show how conventional multitarget correlation algorithms can be modified to include unconventional measurements of this type.[6]

It should be clear from previous sections of this chapter that the KEF can be substituted for the EKF in SHC, MHC, and CHC approximations, provided that we have a suitable means of determining associations between fuzzy DS measurements o and fuzzy DS states μ. This, in turn, means that the concept of local and global association likelihood must be extended to this situation.

Recall that we applied (10.19) to define the association likelihood between a measurement $\mathbf{z}$ and a predicted track described by a track density $f(\mathbf{x}'|i)$:

$$\ell(\mathbf{z}|i) \triangleq \int f(\mathbf{z}|\mathbf{x}) \cdot f(\mathbf{x}|i) d\mathbf{x} \tag{10.149}$$

where $f(\mathbf{z}|\mathbf{x})$ is the likelihood function for the source. Then, given a non-null measurement-to-track association $\theta : \{1, ..., n\} \to \{0, 1, ..., m\}$, in (10.117) we defined its global association likelihood to be

$$\ell_\theta \triangleq \prod_{i:\theta(i)>0} \ell(\mathbf{z}_{\theta(i)}|i). \tag{10.150}$$

Assuming linear-Gaussian measurement statistics, we derived specific formulas for ℓ_θ. These in turn permitted computation of the association probabilities p_θ required for the SHC, MHC, and CHC approaches.

We utilize the same definition of association likelihood. In this case it takes the form $\ell(o|\mu)$ of a likelihood between a fuzzy DS measurement o and a predicted fuzzy DS state-estimate μ. We derive the specific formula

$$\ell(o|\mu) = \sum_{i=1}^{e} \sum_{j=1}^{d} \omega_{i,j} \cdot o_j \cdot \mu_i \cdot \hat{N}_{C_j + HD_iH^T}(H\mathbf{x}'_i - \mathbf{z}_j) \tag{10.151}$$

where

$$\omega_{i,j} \triangleq \sqrt{\frac{\det 2\pi C_j}{\det 2\pi \left(C_j + HD_iH^T\right)}} \cdot \frac{\sqrt{\det 2\pi D_i}}{\sum_{i'=1}^{e} \mu_{i'} \cdot \sqrt{\det 2\pi D_{i'}}}. \tag{10.152}$$

Given these preliminaries, one can compute the association probabilities using (10.115) and (10.116). Then one can apply the SHC, MHC, and CHC approaches to targets that generate fuzzy DS measurements.

Equations (10.151) and (10.152) are established in Appendix G.22.

6 Some of this material first appeared in [125].

Chapter 11

Multitarget Calculus

11.1 INTRODUCTION TO THE CHAPTER

Scientists and engineers are introduced at an early stage in their educations to the analytical power of *integral transform* methods. Let $f(t)$ be a real-valued, time-varying signal with $-\infty < t < \infty$. The *Fourier transform* $\mathcal{F}_f(\omega)$ and its inverse are [192, 218]

$$\mathcal{F}_f(\omega) \triangleq \int_{-\infty}^{\infty} e^{-\iota\omega t} \cdot f(t)\, dt, \quad \mathcal{F}_F^{-1}(t) \triangleq \frac{1}{2\pi} \int_{-\infty}^{\infty} e^{\iota\omega t} \cdot F(\omega)\, d\omega \tag{11.1}$$

where $-\infty < \omega < \infty$ is radian frequency and where $\iota = \sqrt{-1}$ is the imaginary unit. The *Laplace transform* is fundamental to the analysis of continuous-time signals [192, 218]:

$$\mathcal{L}_f(s) \triangleq \int_0^{\infty} e^{-st} \cdot f(t)\, dt \tag{11.2}$$

where s is a complex-number variable. The *z-transform* is fundamental to the analysis of sampled discrete-time signals $a(n) = f(n \cdot \Delta t)$. It is the function of the complex variable z defined by [192, 218]

$$\mathcal{Z}_a(z) \triangleq \sum_{n=0}^{\infty} \frac{a(n)}{z^n}. \tag{11.3}$$

The power of integral transform methods arises in part from the fact that many operations that are mathematically challenging in the time domain are more simply addressed in the ω, s, or z domains. Consider, for example, the convolution

$$(f_1 * f_2)(t) = \int_{-\infty}^{\infty} f_1(\tau) \cdot f_2(t-\tau) d\tau \tag{11.4}$$

of two signals $f_1(t)$ and $f_2(t)$. The Laplace transform converts convolutions into products:

$$\mathcal{L}_{f_1 * f_2}(s) = \mathcal{L}_{f_1}(s) \cdot \mathcal{L}_{f_2}(s). \tag{11.5}$$

It also converts derivatives $\dot{f}(t)$ and integrals $F(t) = \int_0^t f(\tau) d\tau$ into simple algebraic expressions:

$$\mathcal{L}_{\dot{f}}(s) = s \cdot \mathcal{L}_f(s), \qquad \mathcal{L}_F(s) = s^{-1} \cdot \mathcal{L}_f(s). \tag{11.6}$$

11.1.1 Transform Methods in Conventional Statistics

Integral transforms are fundamental to probability and statistics as well. Let A be a random real number with density function $f_A(a)$. Its *characteristic function* $\chi_A(x)$ and *moment generating function* are, respectively,

$$\chi_A(x) \triangleq \int_{-\infty}^{\infty} e^{\iota x a} \cdot f_A(a) da = \mathrm{E}[e^{\iota x A}] \tag{11.7}$$

$$M_A(x) \triangleq \int_{-\infty}^{\infty} e^{x \cdot a} \cdot f_A(a) da = \mathrm{E}[e^{A \cdot x}]. \tag{11.8}$$

where x is a real variable. The nth statistical moments $m_{A,n} = \int x^n \cdot f_A(a) da$ of A can be recovered from $\chi_A(x)$ and $M_A(x)$:

$$\mu_{A,n} = \frac{d^n M_A}{dx^n}(0) = \frac{1}{\iota^n} \frac{d^n \chi_A}{dx^n}(0). \tag{11.9}$$

Let $J \geq 0$ be a random nonnegative integer with $p_J(n) \triangleq \Pr(J = n)$. Its *probability-generating function (p.g.f.)* or *geometric transform* is

$$G_J(x) \triangleq \sum_{n=0}^{\infty} p_J(n) \cdot x^n = \mathrm{E}[x^J]. \tag{11.10}$$

The terminology "probability-generating function" arises because the probability distribution of J can be recovered from $G_J(x)$ via the following inverse operation:

$$p_J(n) = \frac{1}{n!} \frac{d^n G_J}{dx^n}(0). \tag{11.11}$$

Typical examples are: $G_J(x) = e^{\lambda x - \lambda}$ and $G_J(x) = (1 - q + qx)^\nu$ for Poisson- and binomial-distributed random integers, respectively. The p.g.f. can be expressed in terms of z-transforms: $G_J(x) = \mathcal{Z}_{p_J}(x^{-1})$.

As J. E. Gray [78] has usefully reminded us, the utility of transform methods in probability and statistics arises, as in signal processing, in part from the fact that they can transform mathematically complicated concepts and problems into simpler ones.

For example, what is the probability density function of a sum $A + A'$ of two independent random numbers A, A'? The characteristic function of $A + A'$ is

$$\begin{aligned} \chi_{A+A'}(x) &= \int_{-\infty}^{\infty} \int_{-\infty}^{\infty} e^{\iota x(a+a')} \cdot f_A(a) \, f_{A'}(a') \, da da' \\ &= \left(\int_{-\infty}^{\infty} e^{\iota x a} \cdot f_A(a) \, da \right) \left(\int_{-\infty}^{\infty} e^{\iota x a'} \cdot f_{A'}(a') \, da' \right) \\ &= \chi_A(x) \cdot \chi_{A'}(x). \end{aligned}$$

Applying a table of Fourier transform pairs [192, p. 137] shows that the convolution is the inverse transform of the product $\chi_A \cdot \chi_{A'}$ and thus that:

$$f_{A+A'}(x) = (f_A * f_{A'})(x). \tag{11.12}$$

Similar simplified derivations apply to other situations, such as finding the probability density function of a product AA'. Without transform methods, such results would require involved and lengthy derivations using direct brute-force methods. With transform methods, the derivations consist of some algebraic manipulations and a few simple rules from undergraduate calculus.

11.1.2 Transform Methods in Multitarget Statistics

Integral transforms are no less important in *multitarget statistics*. The purpose of this chapter is to introduce the elements of multitarget derivative, multitarget integral, and the multitarget integral transform known as the *probability-generating functional* (p.g.fl.).

There are a number of *direct mathematical parallels* between the world of single-sensor, single-target statistics and the world of multisensor-multitarget statistics. This parallelism is so close that general statistical methodologies can, with a bit of prudence, be directly "translated" from the single-sensor, single-target case to the multisensor-multitarget case.

That is, the parallelism can be thought of as a dictionary that establishes a direct correspondence between the words and grammar in the random-vector language and cognate words and grammar of the random-set language. Consequently, any "sentence" (any concept or algorithm) phrased in the random-vector language can, in principle, be directly "translated" into a corresponding sentence (corresponding concept or algorithm) in the random-set language.

As with any translation process, the correspondence between dictionaries is not precisely one-to-one. As just one example, vectors can be added and subtracted whereas finite sets cannot. Nevertheless, the parallelism is complete enough that, provided one exercises some care, 100 years of accumulated knowledge about single-sensor, single-target statistics can be *directly* brought to bear on multisensor-multitarget problems.

This process can be thought of as a general strategy for attacking multisource-multitarget information fusion problems. It will directly inform much of the remainder of the book. For example, the concepts of true multitarget likelihood functions (Chapter 12) and true multitarget Markov densities (Chapter 13) are direct analogs of likelihood functions and Markov densities in the single-target case. An analogy with the single-target, constant-gain Kalman filter inspires the PHD and CPHD approximation techniques of Chapter 16. A more general analogy with the Kalman filter underlies the multi-Bernoulli approximation approach of Chapter 17, and so on.

In summary, a major goal of this book is to establish Table 11.1.

11.1.3 Summary of Major Lessons Learned

The following are the major ideas and concepts to be encountered in this chapter:

- The random finite set (RFS) is a natural extension of the concept of random vector (Section 11.2).
- RFSs provide a natural way of modeling certain sensing phenomenologies (Section 11.2).
- The belief-mass function $\beta_\Psi(S)$ of an RFS Ψ is a natural generalization of the probability-mass function of a random vector (Section 11.3.2).
- The probability density function $f_\Psi(Y)$ of an RFS Ψ is a natural generalization of the concept of probability density function of a random vector (Section 11.3.3).

Table 11.1

Single-Target Versus Multitarget Statistics

Random Vector $\mathbf{Y}$	**Random Finite Set** Ψ	**Random Finite Set** Ψ
observation-vector $\mathbf{Z}$	observation-set Z	
sensor model $\mathbf{Z} = \eta(\mathbf{x}, \mathbf{W})$	multitarget sensor model $\Xi = \Gamma(X) \cup C(X)$	
motion model $\mathbf{X}_{k+1} = f_k(\mathbf{x}, \mathbf{V})$	multitarget motion model $X_{k+1} = \Upsilon_k(X) \cup B_k(X)$	
probability-mass func. $p_{\mathbf{Y}}(S) = \Pr(\mathbf{Y} \in S)$	belief-mass func. $\beta_\Psi(S) = \Pr(\Psi \subseteq S)$	probability generating functional (p.g.fl.) $G_\Psi[h]$
Radon-Nikodým derivative $\frac{dp_{\mathbf{Y}}}{d\mathbf{y}}$	set derivative $\frac{\delta \beta_\Psi}{\delta Y}(S)$	functional derivative $\frac{\delta G_\Psi}{\delta Y}[h]$
density function $f_{\mathbf{Y}}(\mathbf{y}) = \frac{dp_{\mathbf{Y}}}{d\mathbf{y}}$	multiobject density $f_\Psi(Y) = \frac{\delta \beta_\Psi}{\delta Y}(\emptyset)$	multiobject density $f_\Psi(Y) = \frac{\delta G_\Psi}{\delta Y}[0]$
Lebesgue integral $\int_S f_{\mathbf{Y}}(\mathbf{y}) d\mathbf{y} = p_{\mathbf{Y}}(S)$	set integral $\int_S f_\Psi(Y) \delta Y = \beta_\Psi(S)$	set integral $\int_S h^Y f_\Psi(Y) \delta Y = G_\Psi[h]$
expected value $\bar{\mathbf{Y}} = \int \mathbf{y} \cdot f_{\mathbf{Y}}(\mathbf{y}) d\mathbf{y}$	probability hypothesis density (PHD) $D_\Psi(\mathbf{y}) = \int f_\Psi(\{\mathbf{y}\} \cup Y) \delta Y$	PHD $D_\Psi(\mathbf{y}) = \frac{\delta G_\Psi}{\delta \mathbf{y}}[1]$
likelihood function $f_k(\mathbf{z}\|\mathbf{x})$	multitarget likelihood $f_k(Z\|X)$	
Markov density $f_{k+1\|k}(\mathbf{x}\|\mathbf{x}')$	multitarget Markov density $f_{k+1\|k}(X\|X')$	
posterior density $f_{k\|k}(\mathbf{x}\|Z^k)$	multitarget posterior $f_{k\|k}(X\|Z^{(k)})$	

- The probability-generating functional $G_\Psi[h]$ of an RFS Ψ is a natural generalization of belief-mass function from crisp to fuzzy sets (Remark 13).
- A multitarget analog of differential calculus can be based upon a concept borrowed from quantum physics: the *functional derivative* (Section 11.4.1).
- Set derivatives of belief-mass functions are special cases of functional derivatives of p.g.fl.s (Section 11.4.2).
- Set derivatives are continuous-space generalizations of Möbius transforms; see (11.240).
- Set and functional derivatives can be computed using rules similar to those of undergraduate calculus (Section 11.6).
- The statistical descriptors $\beta_\Psi(S)$ and $f_\Psi(Y)$ and $G_\Psi[h]$ contain identical information about Ψ. Because of multitarget calculus, any one of them can be derived from any of the others (Section 11.5).

11.1.4 Organization of the Chapter

I formally introduce and illustrate the concepts of set integral $\int \delta Y$ and random finite set Ψ in Section 11.2.

Section 11.3 introduces and illustrates three fundamental statistical descriptors of a finite set: the belief-mass function $\beta_\Psi(S)$, the multiobject probability density function $f_\Psi(Y)$, and the probability-generating functional (p.g.fl.) $G_\Psi[h]$.

Section 11.4 introduces the basic elements of multitarget calculus: the functional derivative of a p.g.fl.; and its special case, the set derivative of a belief-mass function. Two central results for multitarget calculus, the fundamental theorem and the Radon-Nikodým theorem, are described in Section 11.5. A list of basic rules for the functional and set derivatives is the subject of Section 11.6.

Exercises for the chapter may be found in Section 11.7.

11.2 RANDOM FINITE SETS

The reader is already familiar with the following types of random variables:

- *Random integer,* J: A random variable that draws its instantiations $J = j$ from the set $\mathbb{Z}$ of all integers, the set $\mathbb{Z}^+$ of nonnegative integers, and so on;

- *Random number,* A: A random variable that draws its instantiations $A = a$ from the set $\mathbb{R}$ of real numbers, from the set $\mathbb{R}^+$ of nonnegative reals, from the set $[0, 1]$ of numbers $0 \leq a \leq 1$, and so on;
- *Random vector,* $\mathbf{Y}$: A random variable that draws its instantiations $\mathbf{Y} = \mathbf{y}$ from a Euclidean space $\mathbb{R}^N$ of vectors $\mathbf{y}$, or (somewhat carelessly speaking) from a continuous-discrete hybrid space $\mathbb{R}^N \times C$ where C is a finite set.

The reader is likely much less familiar with one of the fundamental statistical concepts of this book:

- *Random finite set,* Ψ: A random variable that draws its instantiations $\Psi = Y$ from the hyperspace $\mathfrak{Y}$ of all finite subsets Y (the null set $\emptyset$ included) of some underlying space $\mathfrak{Y}_0$.

For most of Part II we will assume that $\mathfrak{Y}_0 = \mathbb{R}^n$ is a Euclidean vector space and that, therefore, $\mathfrak{Y}$ consists of all finite sets of the form $Y = \emptyset$ or $Y = \{\mathbf{y}_1, ..., \mathbf{y}_n\}$ where $n \geq 1$ is an arbitrary positive integer and $\mathbf{y}_1, ..., \mathbf{y}_n$ are arbitrary vectors in $\mathbb{R}^n$. However, we also allow $\mathfrak{Y}_0 = \mathbb{R}^n \times C$ where C is a finite set.

The following five simple examples draw from a common everyday experience: observing stars in the nighttime sky. They illustrate the concept of a random finite set and of a random finite-set model.

Example 51 shows that a random vector is a special case of a random set. Example 52 shows how random sets add greater versatility to the random vector concept by allowing it to model a star that "twinkles." Examples 53 and 54 illustrate random finite sets with two elements that are independent or perfectly correlated, respectively. Example 55 considers the random set modeling of two twinkling stars.

In all five examples, we will do the following:

1. Construct a random set model Ψ of the sensing situation being observed;
2. Compute the probability $\Pr(\Psi \subseteq S)$ that the observations can be contained in some region S;
3. From $\Pr(\Psi \subseteq S)$ deduce a multiobject probability density function $f_\Psi(Y) =$ "$\Pr(\Psi = Y)$" that equivalently characterizes Ψ;
4. Show that $\Pr(\Psi \subseteq S)$ and $f_\Psi(Y)$ are related by the concept of a set integral.

Example 51 (Random Vector as a Random Set) *Consider a single relatively bright star in the night sky. Though always visible, its position appears to waver randomly* *due to intervening atmospheric turbulence. It can be mathematically modeled as a single randomly varying point set of the form* $\Psi = \{\mathbf{Y}\}$ *where* $\mathbf{Y}$ *is a two-dimensional random vector. The random finite set* Ψ *always contains exactly one element, and its instantiations are singleton sets* $\{\mathbf{y}\}$. *Let* S *be some region of the sky. The probability that* Ψ *will be in* S, *which we denote as* $\beta_\Psi(S)$, *is*

$$\beta_\Psi(S) = \Pr(\Psi \subseteq S) \tag{11.13}$$

$$= \Pr(\mathbf{Y} \in S) = p_\mathbf{Y}(S) = \int_S f_\mathbf{Y}(\mathbf{y})d\mathbf{y} \tag{11.14}$$

where $f_\mathbf{Y}(\mathbf{y})$ *is the probability density function of* $\mathbf{Y}$. *Thus* $\beta_\Psi(S)$ *is completely characterized by the function* $f(Y)$ *of a finite-set variable* Y *defined by*

$$f_\Psi(Y) \triangleq \begin{cases} 0 & \textit{if} \quad Y = \emptyset \\ f_\mathbf{Y}(\mathbf{y}) & \textit{if} \quad Y = \{\mathbf{y}\} \\ 0 & \textit{if} \quad Y = \{\mathbf{y}_1, \mathbf{y}_2\}, \ \mathbf{y}_1 \neq \mathbf{y}_2 \\ 0 & \textit{if} \quad \textit{otherwise} \end{cases}. \tag{11.15}$$

To see why this is true, recall that in (9.18) I introduced the concept of a "set integral." Then $\beta_\Psi(S)$ *can be expressed as a set integral of* $f_\Psi(Y)$:

$$\beta_\Psi(S) = \int_S f_\mathbf{Y}(\mathbf{y})d\mathbf{y} \tag{11.16}$$

$$= 0 + \int_S f_\mathbf{Y}(\mathbf{y})d\mathbf{y} + \frac{1}{2}\int_{S\times S} 0 \cdot d\mathbf{y}_1 d\mathbf{y}_2 + 0 + \ldots \tag{11.17}$$

$$= f_\Psi(\emptyset) + \int_S f_\Psi(\{\mathbf{y}\})d\mathbf{y} \tag{11.18}$$

$$+\frac{1}{2}\int_{S\times S} f_\Psi(\{\mathbf{y}_1, \mathbf{y}_2\})d\mathbf{y}_1 d\mathbf{y}_2 + \ldots \tag{11.19}$$

$$= \int_S f_\Psi(Y)\delta Y. \tag{11.20}$$

Example 52 (A "Twinkling" Random Vector) *Random sets allow us to increase the versatility of the random vector concept. Consider the case of a star that twinkles because it is relatively dim in comparison to intervening atmospheric turbulence. To the eye, it seems to appear and disappear randomly and its apparent position varies whenever it is visible. Let* $1 - p$ *be the probability that it disappears. The star can be modeled as a randomly varying finite set* Ψ *such that: (1)* $\Psi = \emptyset$

with probability $1-p$, *and (2)* $|\Psi| = 1$ *with probability* p. *In the latter case,* $\Psi = \{\mathbf{Y}\}$ *where* $\mathbf{Y}$ *is a random vector conditioned on the fact that* $\Psi \neq \emptyset$. *We can express* Ψ *in the form*

$$\Psi = \emptyset^p \cap \{\mathbf{Y}\} \tag{11.21}$$

where $\emptyset^p$ *is the discrete random set defined by*

$$\Pr(\emptyset^p = S) \triangleq \begin{cases} 1-p & \text{if} \quad S = \emptyset \\ p & \text{if} \quad S = \mathfrak{Y}_0 \\ 0 & \text{if} \quad \text{otherwise} \end{cases}. \tag{11.22}$$

The probability that Ψ *will be contained within some region* S *is*

$$\beta_\Psi(S) = 1 - p + p \cdot p_{\mathbf{Y}}(S) = 1 - p + p \int_S f_{\mathbf{Y}}(\mathbf{y}) d\mathbf{y}. \tag{11.23}$$

To see this, note that

$$\begin{aligned} \beta_\Psi(S) &= \Pr(\Psi \subseteq S) && (11.24) \\ &= \Pr(\Psi \subseteq S, \Psi = \emptyset) + \Pr(\Psi \subseteq S, \Psi \neq \emptyset) && (11.25) \\ &= \Pr(\Psi = \emptyset) + \Pr(\Psi \neq \emptyset) \cdot \Pr(\Psi \subseteq S | \Psi \neq \emptyset) && (11.26) \\ &= 1 - p + p \cdot \Pr(\mathbf{Y} \in S | \Psi \neq \emptyset) && (11.27) \\ &= 1 - p + p \cdot p_{\mathbf{Y}}(S) = 1 - p + p \int_S f_{\mathbf{Y}}(\mathbf{y}) d\mathbf{y}. && (11.28) \end{aligned}$$

Define

$$f_\Psi(Y) \triangleq \begin{cases} 1-p & \text{if} \quad Y = \emptyset \\ p \cdot f_{\mathbf{Y}}(\mathbf{y}) & \text{if} \quad Y = \{\mathbf{y}\} \\ 0 & \text{if} \quad Y = \{\mathbf{y}_1, \mathbf{y}_2\},\ \mathbf{y}_1 \neq \mathbf{y}_2 \\ 0 & \text{if} \quad \text{otherwise} \end{cases}. \tag{11.29}$$

Then $\beta_\Psi(S)$ can be expressed as a set integral of $f_\Psi(Y)$:

$$\beta_\Psi(S) = 1 - p + p\int_S f_{\mathbf{Y}}(\mathbf{y})d\mathbf{y} \tag{11.30}$$

$$= 1 - p + \int_S p \cdot f_{\mathbf{Y}}(\mathbf{y})d\mathbf{y} + \frac{1}{2}\int_{S\times S} 0 \cdot d\mathbf{y}_1 d\mathbf{y}_2 + 0 + ... \tag{11.31}$$

$$= f_\Psi(\emptyset) + \int_S f_\Psi(\{\mathbf{y}\})d\mathbf{y} \tag{11.32}$$

$$+\frac{1}{2}\int_{S\times S} f_\Psi(\{\mathbf{y}_1, \mathbf{y}_2\})d\mathbf{y}_1 d\mathbf{y}_2 + ... \tag{11.33}$$

$$= \int_S f_\Psi(Y)\delta Y. \tag{11.34}$$

Example 53 (Independent Random Vectors) *Consider two stars that, though always visible, are far enough apart that their randomness is independent of each other. The two stars can be modeled as a randomly varying two point set of the form $\Psi = \{\mathbf{Y}_1, \mathbf{Y}_2\}$ where $\mathbf{Y}_1, \mathbf{Y}_2$ are random vectors. The probability that Ψ will be contained within S is*

$$\beta_\Psi(S) = \Pr(\Psi \subseteq S) = \Pr(\{\mathbf{Y}_1, \mathbf{Y}_2\} \subseteq S) = \Pr(\mathbf{Y}_1 \in S,\ \mathbf{Y}_2 \in S). \tag{11.35}$$

Since $\mathbf{Y}_1, \mathbf{Y}_2$ are independent, this becomes

$$\beta_\Psi(S) = \Pr(\mathbf{Y}_1 \in S) \cdot \Pr(\mathbf{Y}_2 \in S) = p_{\mathbf{Y}_1}(S) \cdot p_{\mathbf{Y}_2}(S) \tag{11.36}$$

$$= \left(\int_S f_{\mathbf{Y}_1}(\mathbf{y})d\mathbf{y}\right) \cdot \left(\int_S f_{\mathbf{Y}_2}(\mathbf{y})d\mathbf{y}\right) \tag{11.37}$$

$$= \int_S \int_S f_{\mathbf{Y}_1}(\mathbf{y}_1) \cdot f_{\mathbf{Y}_2}(\mathbf{y}_2)d\mathbf{y}_1 d\mathbf{y}_2 \tag{11.38}$$

$$= \int_{S\times S} f_{\mathbf{Y}_1}(\mathbf{y}_1) \cdot f_{\mathbf{Y}_2}(\mathbf{y}_2)d\mathbf{y}_1 d\mathbf{y}_2 \tag{11.39}$$

$$= \frac{1}{2}\int_{S\times S} \left(f_{\mathbf{Y}_1}(\mathbf{y}_1)f_{\mathbf{Y}_2}(\mathbf{y}_2) + f_{\mathbf{Y}_1}(\mathbf{y}_2)f_{\mathbf{Y}_2}(\mathbf{y}_1)\right) d\mathbf{y}_1 d\mathbf{y}_2. \tag{11.40}$$

Thus $\beta_\Psi(S)$ *is purely quadratic in the set variable* S*. It is completely described by the function* $f(Y)$ *of a finite-set variable* Y *defined by*

$$f_\Psi(Y) \triangleq \begin{cases} 0 & \textit{if} \quad Y = \emptyset \\ 0 & \textit{if} \quad Y = \{\mathbf{y}\} \\ \begin{array}{c} f_{\mathbf{Y}_1}(\mathbf{y}_1) f_{\mathbf{Y}_2}(\mathbf{y}_2) \\ + f_{\mathbf{Y}_1}(\mathbf{y}_2) f_{\mathbf{Y}_2}(\mathbf{y}_1) \end{array} & \textit{if} \quad \left\{ \begin{array}{c} Y = \{\mathbf{y}_1, \mathbf{y}_2\} \\ \mathbf{y}_1 \neq \mathbf{y}_2 \end{array} \right. \\ 0 & \textit{if} \quad \textit{otherwise} \end{cases}. \tag{11.41}$$

Furthermore, $\beta_\Psi(S)$ *can be expressed as the set integral of* $f_\Psi(Y)$*:*

$$\beta_\Psi(S) \tag{11.42}$$

$$= \frac{1}{2} \int_{S \times S} (f_{\mathbf{Y}_1}(\mathbf{y}_1) \cdot f_{\mathbf{Y}_2}(\mathbf{y}_2) + f_{\mathbf{Y}_1}(\mathbf{y}_2) \cdot f_{\mathbf{Y}_2}(\mathbf{y}_1)) \, d\mathbf{y}_1 d\mathbf{y}_2 \tag{11.43}$$

and so

$$= 0 + \int_S 0 \cdot d\mathbf{y} \tag{11.44}$$

$$+\frac{1}{2} \int_{S \times S} (f_{\mathbf{Y}_1}(\mathbf{y}_1) \cdot f_{\mathbf{Y}_2}(\mathbf{y}_2) + f_{\mathbf{Y}_1}(\mathbf{y}_2) \cdot f_{\mathbf{Y}_2}(\mathbf{y}_1)) \tag{11.45}$$

$$\cdot d\mathbf{y}_1 d\mathbf{y}_2 \tag{11.46}$$

$$+0 + ... \tag{11.47}$$

$$= f_\Psi(\emptyset) + \int_S f_\Psi(\{\mathbf{y}\}) d\mathbf{y} \tag{11.48}$$

$$+\frac{1}{2} \int_{S \times S} f_\Psi(\{\mathbf{y}_1, \mathbf{y}_2\}) d\mathbf{y}_1 d\mathbf{y}_2 + ... \tag{11.49}$$

$$= \int_S f_\Psi(Y) \delta Y. \tag{11.50}$$

Example 54 (Two Perfectly Correlated Random Vectors) *Suppose that the two stars are so close together that their observability is affected by the same underlying turbulence. In particular, suppose that they are so close that they appear to waver in lockstep. In this case* $\Psi = \{\mathbf{Y}, \mathbf{y}_0 + \mathbf{Y}\}$ *for some* $\mathbf{y}_0$*. Abbreviate* $S - \mathbf{y}_0 \overset{\textit{abbr.}}{=}$

$\{\mathbf{y}-\mathbf{y}_0 |\ \mathbf{y} \in S\}$. *Then*

$$\beta_\Psi(S) = \Pr(\Psi \subseteq S) = \Pr(\mathbf{Y} \in S,\ \mathbf{y}_0 + \mathbf{Y} \in S) \quad (11.51)$$

$$= \Pr(\mathbf{Y} \in S,\ \mathbf{Y} \in S - \mathbf{y}_0) \quad (11.52)$$

$$= \Pr(\mathbf{Y} \in S \cap (S - \mathbf{y}_0)) = p_\mathbf{Y}(S \cap (S - \mathbf{y}_0)) \quad (11.53)$$

$$= \int \mathbf{1}_S(\mathbf{y}) \cdot \mathbf{1}_{S-\mathbf{y}_0}(\mathbf{y}) \cdot f_\mathbf{Y}(\mathbf{y}) d\mathbf{y} \quad (11.54)$$

$$= \int \mathbf{1}_S(\mathbf{y}) \cdot \mathbf{1}_S(\mathbf{y}+\mathbf{y}_0) \cdot f_\mathbf{Y}(\mathbf{y}) d\mathbf{y} \quad (11.55)$$

$$= \int \mathbf{1}_S(\mathbf{y}_1) \cdot \mathbf{1}_S(\mathbf{y}_2) \cdot \delta_{\mathbf{y}_2}(\mathbf{y}_1+\mathbf{y}_0) \cdot f_\mathbf{Y}(\mathbf{y}_1) d\mathbf{y}_1 d\mathbf{y}_2 \quad (11.56)$$

$$= \int_{S\times S} \delta_{\mathbf{y}_2}(\mathbf{y}_1+\mathbf{y}_0) \cdot f_\mathbf{Y}(\mathbf{y}_1) d\mathbf{y}_1 d\mathbf{y}_2 \quad (11.57)$$

$$= \frac{1}{2}\int_{S\times S} \left(\begin{array}{c} \delta_{\mathbf{y}_2}(\mathbf{y}_1+\mathbf{y}_0) \cdot f_\mathbf{Y}(\mathbf{y}_1) \\ +\delta_{\mathbf{y}_1}(\mathbf{y}_2+\mathbf{y}_0) \cdot f_\mathbf{Y}(\mathbf{y}_2) \end{array} \right) d\mathbf{y}_1 d\mathbf{y}_2 \quad (11.58)$$

where $\delta_{\mathbf{y}'}(\mathbf{y})$ *is the Dirac delta density concentrated at* $\mathbf{y}'$. *Thus* $\beta_\Psi(S)$ *is quadratic in form but this time is described by*

$$f_\Psi(Y) \triangleq \left\{ \begin{array}{ccc} 0 & \text{if} & Y = \emptyset \\ 0 & \text{if} & Y = \{\mathbf{y}\} \\ \begin{array}{c} \delta_{\mathbf{y}_2}(\mathbf{y}_1+\mathbf{y}_0) f_\mathbf{Y}(\mathbf{y}_1) \\ +\delta_{\mathbf{y}_1}(\mathbf{y}_2+\mathbf{y}_0) f_\mathbf{Y}(\mathbf{y}_2) \end{array} & \text{if} & \left\{ \begin{array}{c} Y = \{\mathbf{y}_1, \mathbf{y}_2\} \\ \mathbf{y}_1 \neq \mathbf{y}_2 \end{array} \right. \\ 0 & \text{if} & \text{otherwise} \end{array} \right. . \quad (11.59)$$

As usual, $\beta_\Psi(Y)$ *can be expressed as a set integral of* $f_\Psi(Y)$*:*

$$\beta_\Psi(S) \quad (11.60)$$

$$= \frac{1}{2}\left(\begin{array}{c} \int_{S\times S} \delta_{\mathbf{y}_2}(\mathbf{y}_1+\mathbf{y}_0) \cdot f_\mathbf{Y}(\mathbf{y}_1) \\ +\delta_{\mathbf{y}_1}(\mathbf{y}_2+\mathbf{y}_0) \cdot f_\mathbf{Y}(\mathbf{y}_2) \end{array} \right) d\mathbf{y}_1 d\mathbf{y}_2 \quad (11.61)$$

$$= 0 + \int_S 0 \cdot d\mathbf{y} \quad (11.62)$$

$$+\frac{1}{2}\left(\int_{S\times S} \delta_{\mathbf{y}_2}(\mathbf{y}_1+\mathbf{y}_0) \cdot f_\mathbf{Y}(\mathbf{y}_1) + \delta_{\mathbf{y}_1}(\mathbf{y}_2+\mathbf{y}_0) \cdot f_\mathbf{Y}(\mathbf{y}_2) \right) + 0 + \ldots \quad (11.63)$$

$$= f_\Psi(\emptyset) + \int_S f_\Psi(\{\mathbf{y}\}) d\mathbf{y} + \frac{1}{2}\int_{S\times S} f_\Psi(\{\mathbf{y}_1, \mathbf{y}_2\}) d\mathbf{y}_1 d\mathbf{y}_2 + \ldots \quad (11.64)$$

$$= \int_S f_\Psi(Y)\delta Y. \tag{11.65}$$

Example 55 (Two "Twinkling" Random Vectors) *Finally, consider two relatively dim but well-separated stars. They can be considered to be statistically independent and their rates of twinkling may be different. The two stars are mathematically modeled as the union* $\Psi = \Psi_1 \cup \Psi_2$ *where* Ψ_1, Ψ_2 *are statistically independent and where*

$$\Psi_1 = \emptyset^{p_1} \cap \{\mathbf{Y}_1\}, \qquad \Psi_2 = \emptyset^{p_2} \cap \{\mathbf{Y}_2\}. \tag{11.66}$$

Thus Ψ *can be empty, have one point, or have two points. So*

$$\begin{aligned}
\beta_\Psi(S) &= \Pr(\Psi \subseteq S) = \Pr(\Psi_1 \subseteq S, \Psi_2 \subseteq S) & (11.67)\\
&= \Pr(\Psi_1 \subseteq S) \cdot \Pr(\Psi_2 \subseteq S) & (11.68)\\
&= [1 - p_1 + p_1 p_{\mathbf{Y}_1}(S)] \cdot [1 - p_2 + p_2 p_{\mathbf{Y}_2}(S)] & (11.69)\\
&= (1-p_1)(1-p_2) + p_1(1-p_2)p_{\mathbf{Y}_1}(S) & (11.70)\\
&\quad +(1-p_1)p_2 \cdot p_{\mathbf{Y}_2}(S) + p_1 p_2 \cdot p_{\mathbf{Y}_1}(S) p_{\mathbf{Y}_2}(S). & (11.71)
\end{aligned}$$

Using the results of Examples 52 and 53 this can be rewritten as

$$\begin{aligned}
& \beta_\Psi(S) & (11.72)\\
= \; & (1-p_1)(1-p_2) & (11.73)\\
& + \int_S [p_1(1-p_2) f_{\mathbf{Y}_1}(\mathbf{y}) + (1-p_1)p_2 f_{\mathbf{Y}_2}(\mathbf{y})]\, d\mathbf{y} & (11.74)\\
& + \frac{p_1 p_2}{2} \int_{S\times S} (f_{\mathbf{Y}_1}(\mathbf{y}_1) f_{\mathbf{Y}_2}(\mathbf{y}_2) + f_{\mathbf{Y}_1}(\mathbf{y}_2) f_{\mathbf{Y}_2}(\mathbf{y}_1)) \,. & (11.75)
\end{aligned}$$

Thus $\beta_\Psi(S)$ *is completely described by*

$$f_\Psi(Y) \triangleq \begin{cases} (1-p_1)(1-p_2) & \text{if} \quad Y = \emptyset \\ \begin{array}{c} p_1(1-p_2) f_{\mathbf{Y}_1}(\mathbf{y}) \\ +(1-p_1)p_2 f_{\mathbf{Y}_2}(\mathbf{y}) \end{array} & \text{if} \quad Y = \{\mathbf{y}\} \\ p_1 p_2 \cdot \left(\begin{array}{c} f_{\mathbf{Y}_1}(\mathbf{y}_1) \cdot f_{\mathbf{Y}_2}(\mathbf{y}_2) \\ + f_{\mathbf{Y}_1}(\mathbf{y}_2) \cdot f_{\mathbf{Y}_2}(\mathbf{y}_1) \end{array} \right) & \text{if} \quad \left\{ \begin{array}{c} Y = \{\mathbf{y}_1, \mathbf{y}_2\} \\ \mathbf{y}_1 \neq \mathbf{y}_2 \end{array} \right. \\ 0 & \text{if} \quad \text{otherwise} \end{cases} . \tag{11.76}$$

Once again, $\beta_\Psi(S)$ *can be written as a set integral of* $f_\Psi(Y)$*:*

$$\begin{aligned}
\beta_\Psi(S) &= (1-p_1)(1-p_2) && (11.77)\\
&\quad+\int_S [p_1(1-p_2)f_{\mathbf{Y}_1}(\mathbf{y})+(1-p_1)p_2 f_{\mathbf{Y}_2}(\mathbf{y})]\,d\mathbf{y} && (11.78)\\
&\quad+\frac{1}{2}\int_{S\times S} p_1 p_2\,[f_{\mathbf{Y}_1}(\mathbf{y}_1)f_{\mathbf{Y}_2}(\mathbf{y}_2)+f_{\mathbf{Y}_1}(\mathbf{y}_2)f_{\mathbf{Y}_2}(\mathbf{y}_1)] && (11.79)\\
&= f_\Psi(\emptyset)+\int_S f_\Psi(\{\mathbf{y}\})d\mathbf{y} && (11.80)\\
&\quad+\frac{1}{2}\int_{S\times S} f_\Psi(\{\mathbf{y}_1,\mathbf{y}_2\})d\mathbf{y}_1 d\mathbf{y}_2+... && (11.81)\\
&= \int_S f_\Psi(Y)\delta Y. && (11.82)
\end{aligned}$$

11.3 FUNDAMENTAL STATISTICAL DESCRIPTORS

Examples 51-55 described the modeling of a simple everyday application: observing stars with the naked eye. Two points should be made about these examples.

First, they are instances of typical detection-type measurement models of the kind described in Section 9.2.1. The description of a star includes not only its apparent position but also its apparent intensity. That is, the eye collects a measurement that can be modeled as a vector $\mathbf{z}=(\alpha,\theta,a)$ where α is azimuth, θ is altitude, and a is intensity. When its intensity falls below a certain point, a star is no longer detectable by naked eye.

In these examples we ignored the intensity variable and declared a star sighting to be (α,θ) if it is visible and $\emptyset$ if not.

Second, in each of Examples 51-55 we adhered to the formal modeling paradigm:

- Construct a statistical model of the problem as a random finite set Ψ.
- Construct the *belief-mass function* $\beta_\Psi(S)$.
- From $\beta_\Psi(S)$, construct the *probability density function* $f_\Psi(Y)$.

In the following sections, I introduce these concepts more precisely, while also introducing a third fundamental multiobject statistical descriptor: the *probability-generating functional* (p.g.fl.) $G_\Psi[h]$ of a finite random set.

The section is organized as follows. I provide motivation for the concepts of belief-mass function, probability density function, and p.g.fl. in Section 11.3.1. I describe these concepts in more detail in Section 11.3.2 (belief-mass functions), Sections 11.3.3 and 11.3.4 (multiobject probability densities), and Section 11.3.5 (p.g.fl.s).

11.3.1 Multitarget Calculus—Why?

As we shall see in Section 11.5, $f_\Psi(Y)$ and $\beta_\Psi(S)$ and $G_\Psi[h]$ are *equivalent* in the sense that they contain exactly the same information about Ψ. Why do we need three statistical descriptors rather than just one? The reason is that each addresses a different aspect of multitarget problems.

11.3.1.1 Why Are Multiobject Density Functions Important?

Multiobject probability density functions $f_\Psi(Y)$ are central to the theoretical foundation of multitarget detection, tracking, localization, and classification: the *multitarget Bayes filter* (to be discussed in Chapter 14). This filter propagates multitarget probability density functions through time:

$$\cdots \to f_{k|k}(X|Z^{(k)}) \to f_{k+1|k}(X|Z^{(k)}) \to f_{k+1|k+1}(X|Z^{(k+1)}) \to \cdots \quad (11.83)$$

The multitarget probability distribution $f_{k|k}(X|Z^{(k)})$ contains all relevant information about the numbers and states of the targets at time steps k and $k+1$, respectively. Likewise, the distribution $f_{k+1|k}(X|Z^{(k)})$ contains all relevant information about the predicted numbers and predicted states of the targets at time step $k+1$. Multiobject state-estimation techniques, to be discussed in Section 14.5, can then be used to extract useful information from the multitarget posterior distributions.

11.3.1.2 Why Are Belief-Mass Functions Important?

Belief-mass functions $\beta_\Psi(S)$ are central to multisource-multitarget formal Bayes modeling. Specifically, they are central to the process of constructing true multitarget likelihood functions from multitarget measurement models (Chapter 12) and multitarget Markov density functions from multitarget motion models (Chapter 13).

For example, given a multitarget measurement model $\Sigma_{k+1} = \Upsilon_{k+1}(X)$ we can construct explicit formulas for $\beta_{k+1}(T|X) = \Pr(\Sigma_{k+1} \subseteq T|X)$ and, from this, the true multitarget likelihood function

$$f_{k+1}(Z|X) = \frac{\delta \beta_{k+1}}{\delta Z}(\emptyset|X) \tag{11.84}$$

where "$\delta/\delta Z$" denotes a *set derivative* as defined in Section 11.4.2.

Likewise, given a multitarget motion model $\Xi_{k+1|k} = \Gamma_{k+1|k}(X')$ we can construct explicit formulas for $\beta_{k+1|k}(S|X') = \Pr(\Xi_{k+1|k} \subseteq S|X')$ and, from this, the true multitarget Markov density function

$$f_{k+1|k}(X|X') = \frac{\delta \beta_{k+1|k}}{\delta X}(\emptyset|X'). \tag{11.85}$$

11.3.1.3 Why Are Probability-Generating Functionals (p.g.fl.s) Important?

Probability-generating functionals can often transform difficult mathematical problems into simpler ones. For example, the predictor step

$$f_{k|k}(X|Z^{(k)}) \rightarrow f_{k+1|k}(X|Z^{(k)}) \tag{11.86}$$

of (11.83) can be transformed into an equation of the form

$$G_{k+1|k}[h] = G_{k|k}[T[h]] \tag{11.87}$$

(see Section 14.8). Here $G_{k|k}[h]$ is the p.g.fl. of $f_{k|k}(X|Z^{(k)})$; $G_{k+1|k}[h]$ is the p.g.fl. of $f_{k+1|k}(X|Z^{(k)})$; and $T[h]$ is a functional transformation that encapsulates the information contained in the multitarget motion model. Consequently, the problem of mathematically analyzing the multitarget filter predictor step reduces to a technique of multitarget calculus—*chain rules for functional derivatives* (see Section 11.6).

As another example, suppose that we are to restrict the focus of multitarget filtering to *targets of interest* (ToIs)—that is, targets that for whatever reason are more tactically interesting or important than other targets.

Tactical importance can be modeled in terms of a relative-importance ranking function $0 \leq \rho(\mathbf{x}) \leq 1$. Here $\mathbf{x}$ is the state of a target and $\rho(\mathbf{x}) = 1$ indicates that a target with state $\mathbf{x}$ has maximal tactical importance and $\rho(\mathbf{x}) = 0$ indicates that a target with state $\mathbf{x}$ is of no tactical interest at all.

Let $G_{k|k}[h]$ be the p.g.fl. for all targets (irrespective of tactical interest) at time step k. Then as I shall show in Section 14.9,

$$G^{\rho}_{k|k}[h] = G_{k|k}[1 - \rho + h\rho] \tag{11.88}$$

is the p.f.gl. for the current or potential tactically significant targets at time step k. Thus the problem of target prioritization is converted into a purely mathematical problem—using a chain rule for functional derivatives.

11.3.1.4 Why Is Multiobject Calculus Important?

By now the answer to this question should be evident.

- Set derivatives and set integrals show that multiobject density probability distributions, belief-mass functions, and p.g.fl.s are all equivalent statistical representations of the same multiobject probability law.
- Set derivatives and set integrals are central to the process of constructing new, principled approximation techniques as discussed in Part III.
- Set derivatives are necessary to derive true multitarget Markov densities and true multisource-multitarget likelihood functions from, respectively, multi-target motion models and multisource-multitarget measurement models (see Chapters 12 and 13).

Some readers may ask:

- Since I can just copy these formulas from this book, why do I need to bother with the calculus used to derive them?

Those readers whose problems are directly addressed in this book indeed need not bother with the derivations. However, the bulk of Chapters 12 and 13 are devoted to so-called standard multitarget motion and measurement models. Anyone who wishes to apply finite-set statistics to real-world problems not modeled in this book may require the multiobject calculus to derive likelihoods and Markov densities appropriate to their problem.

11.3.2 Belief-Mass Functions

A *set function* on $\mathfrak{Y}_0$ is a real-valued function $\phi(S)$ of a closed-set variable S. The following are examples of set functions that occur regularly in the information fusion literature:

- *Probability-mass functions*: If $\mathbf{Y}$ is a random vector on $\mathfrak{Y}_0$, then

$$p_{\mathbf{Y}}(S) \triangleq \Pr(\mathbf{Y} \in S); \tag{11.89}$$

- *Possibility measures*: If $\mu(\mathbf{y})$ is a fuzzy membership function on $\mathfrak{Y}_0$,

$$\pi_\mu(S) \triangleq \sup_{\mathbf{y}\in S} \mu(\mathbf{y}); \tag{11.90}$$

- *Belief-mass functions*: If Ψ is a finite random subset of $\mathfrak{Y}_0$,

$$\beta_\Psi(S) \triangleq \Pr(\Psi \subseteq S). \tag{11.91}$$

  If $\Psi \neq \emptyset$, then $\beta_\Psi(S)$ is called a *belief function* or *belief measure* in the Dempster-Shafer literature.

- *Plausibility-mass functions*:

$$Pl_\Psi(S) = 1 - \beta_\Psi(S^c) = \Pr(\Psi \cap S \neq \emptyset). \tag{11.92}$$

  If $\Psi \neq \emptyset$, then $Pl_\Psi(S)$ is called a *plausibility function* or *plausibility measure* in the Dempster-Shafer literature.

- *Sugeno (also known as uncertainty also known as fuzzy) measures* [76, p. 110]: These are set functions $\sigma(S)$ such that: (1) $\phi(S) \geq 0$ for all S; (2) $\phi(\emptyset) = 0$; $\phi(\mathfrak{Y}_0) = 1$; (3) $\phi(S_1) \leq \phi(S_2)$ whenever $S_1 \subseteq S_2$; and (4) $\lim_{n\to\infty} \phi(S_n) = \phi(\cup_{n\geq 0} S)$ for any infinite nested sequence $S_1 \subseteq S_2 \subseteq \ldots$.

In this book we will be primarily interested in set functions $\phi(S)$ that are algebraic and transcendental combinations of one or more belief-mass functions; for example,

$$\phi(S) = \beta_\Psi(S)^n \tag{11.93}$$

$$\phi(S) = e^{a\cdot\beta_\Psi(S)} \tag{11.94}$$

$$\phi(S) = \beta_{\Psi_1}(S)\cdots\beta_{\Psi_n}(S). \tag{11.95}$$

11.3.3 Multiobject Density Functions and Set Integrals

In (9.18), I introduced the concept of a *set integral*. I begin by defining this concept with greater precision.

11.3.3.1 Set Integrals

Let $f(Y)$ be a real-valued function of a finite-set variable Y. Then its set integral concentrated in a region S of $\mathfrak{Y}_0$ is[1]

$$\int_S f(Y)\delta Y \triangleq \sum_{n=0}^{\infty} \frac{1}{n!} \int_{\underbrace{S \times \ldots \times S}_{n}} f(\{\mathbf{y}_1, \ldots, \mathbf{y}_n\}) d\mathbf{y}_1 \cdots d\mathbf{y}_n \quad (11.96)$$

$$= f(\emptyset) + \int_S f(\{\mathbf{y}\}) d\mathbf{y} \quad (11.97)$$

$$+ \frac{1}{2} \int_{S \times S} f(\{\mathbf{y}_1, \mathbf{y}_2\}) d\mathbf{y}_1 d\mathbf{y}_2 + \ldots \quad (11.98)$$

The integrals in (11.96) are not well defined without further explanation. For each $n \geq 2$, define the function $f_n(\mathbf{y}_1, \ldots, \mathbf{y}_n)$ in n vector variables $\mathbf{y}_1, \ldots, \mathbf{y}_n$ by

$$f_n(\mathbf{y}_1, \ldots, \mathbf{y}_n) \triangleq \begin{cases} \frac{1}{n!} f(\{\mathbf{y}_1, \ldots, \mathbf{y}_n\}) & \text{if} \quad \mathbf{y}_1, \ldots, \mathbf{y}_n \text{ are distinct} \\ 0 & \text{if} \quad \text{otherwise} \end{cases} . \quad (11.99)$$

Then each term in (11.96) is defined to mean[2]

$$\int_{\underbrace{S \times \ldots \times S}_{n}} f(\{\mathbf{y}_1, \ldots, \mathbf{y}_n\}) d\mathbf{y}_1 \cdots d\mathbf{y}_n \quad (11.100)$$

$$\triangleq n! \int_{\underbrace{S \times \ldots \times S}_{n}} f_n(\mathbf{y}_1, \ldots, \mathbf{y}_n) d\mathbf{y}_1 \cdots d\mathbf{y}_n. \quad (11.101)$$

Furthermore, note that the summation in (11.96) is not defined unless the function $f(Y)$ has a specific form. Since $f(\emptyset)$ must be a unitless probability, every integral in (11.96) must also have units of measurement. Let u be the unit of measurement in $\mathfrak{Y}_0$. Then $d\mathbf{y}_1 \cdots d\mathbf{y}_n$ has units of measurement u^n. If (11.99) is

1 Set integrals appear routinely in the statistical theory of gases and liquids [85, pp. 234, 266], though not explicitly identified as being multiobject integrals.

2 If $\mathfrak{Y}_0$ is infinite and $f(Y) < \infty$ for all Y, then this definition is redundant since "diagonal" events of the form $\mathbf{y}_i = \mathbf{y}_j$ with $i \neq j$ are zero-probability. If $\mathfrak{Y}_0$ is a finite set of state cells, however, then two distinct targets could be in the same state cell. This problem is avoided by using a careful definition of target state: see Remark 19 in Chapter 12.

to be unitless, the units of $f_n(\mathbf{y}_1, ..., \mathbf{y}_n)$ must be u^{-n}. Consequently, set integrals $\int f(Y)\delta Y$ will be well defined only if $f(Y)$ is a specific type of function as defined in the next section.

11.3.3.2 Multiobject Density Functions

A *multiobject density function* on $\mathfrak{Z}_0$ is a real-valued function $f(Y)$ of a finite-subset variable $Y \subseteq \mathfrak{Z}_0$ such that:[3]

- If $\mathfrak{Z}_0$ has a unit of measurement u, then the unit of measurement of $f(Y)$ is $u^{-|Y|}$ where $|Y|$ is the number of elements in Y.

It is clear that set integrals of multiobject density functions f are linear in f:

$$\int_S (a_1 f_1(Y) + a_2 f_2(Y))\, \delta Y = a_1 \int_S f_1(Y)\delta Y + a_2 \int_S f_2(Y)\delta Y. \quad (11.102)$$

It is equally clear, however, that they cannot be additive in S even when $S_1 \cap S_2 = \emptyset$:[4]

$$\int_{S_1 \cup S_2} f(Y)\delta Y \neq \int_{S_1} f(Y)\delta Y + \int_{S_2} f(Y)\delta Y. \quad (11.103)$$

A multiobject density function $f(Y)$ is a *multiobject probability density function* if $f(Y) \geq 0$ for all Y and if

$$\int f(Y)\delta Y = 1. \quad (11.104)$$

The functions $f_\Psi(Y)$ constructed in Examples 51-55 are all probability density functions. To see this, note that in each of those examples I showed that

$$\int_S f_\Psi(Y)\delta Y = \Pr(\Psi \subseteq S). \quad (11.105)$$

Setting $S = \mathfrak{Z}_0$ we get $\int f_\Psi(Y)\delta Y = \Pr(\Psi \subseteq \mathfrak{Z}_0) = 1$.

3 Multiobject density functions are the same thing as what, in early publications such as [70], I called "global densities," "global probability density functions," "global posterior distributions," etc.

4 More theoretically attuned readers may object that a set integral cannot be called an "integral" because of this. There are two reasons why such an objection would be misplaced. First, the concept of nonadditive measure and integral is common in the literature, especially in expert systems theory. See, for example, the book by Denneberg [39]; Sugeno (a.k.a. fuzzy) measures and integrals [76, pp. 128-133]; and the Graf integral [77]. Second, the set integral is closely related (though not identical) to the usual concept of integral as understood in measure theory—see Appendix F.

11.3.3.3 Multiobject Probability Distributions

The *probability density function of a random finite set* Ψ is, if it exists, that function $f_\Psi(Y)$, such that

$$\int_S f_\Psi(Y)\delta Y = \Pr(\Psi \subseteq S) \tag{11.106}$$

for all S.[5]

In the sequel we will use the following notational conventions:

$$\begin{aligned}
\int_{\Psi\subseteq S} f_\Psi(Y)\delta Y &\triangleq \Pr(\Psi \subseteq S) && (11.107)\\
&= \int_S f_\Psi(Y)\delta Y && (11.108)\\
\int_{|\Psi|=n} f_\Psi(Y)\delta Y &\triangleq \Pr(|\Psi| = n) && (11.109)\\
&= \frac{1}{n!}\int f_\Psi(\{\mathbf{y}_1,...,\mathbf{y}_n\})d\mathbf{y}_1\cdots d\mathbf{y}_n && (11.110)\\
\int_{\Psi\ni\mathbf{y}} f_\Psi(Y)\delta Y &\triangleq \Pr(\mathbf{y}\in\Psi) && (11.111)\\
&= \int f_\Psi(\{\mathbf{y}\}\cup W)\delta W && (11.112)
\end{aligned}$$

and so on.

11.3.3.4 Cardinality Distributions

I call the probability distribution

$$\begin{aligned}
p_\Psi(n) &\triangleq p_{|\Psi|}(n) && (11.113)\\
&\triangleq \Pr(|\Psi| = n) && (11.114)\\
&= \frac{1}{n!}\int f_\Psi(\{\mathbf{y}_1,...,\mathbf{y}_n\})d\mathbf{y}_1\cdots d\mathbf{y}_n && (11.115)
\end{aligned}$$

of (11.109) the *cardinality distribution* of the finite random set Ψ.

5 In general $f_\Psi(Y)$ may involve complicated expressions involving Dirac delta functions.

Remark 9 (Set Notation Versus Vector Notation) *Multitarget distributions can also be expressed in vector notation. For example,* $f(\{\mathbf{y}_1, \mathbf{y}_2\}) = 2 \cdot f(\mathbf{y}_1, \mathbf{y}_2)$ *since the probability (density) assigned to the two element set* $\{\mathbf{y}_1, \mathbf{y}_2\} = \{\mathbf{y}_2, \mathbf{y}_1\}$ *must be distributed equally over the two possible vectors* $(\mathbf{y}_1, \mathbf{y}_2)$ *and* $(\mathbf{y}_2, \mathbf{y}_1)$*. In general, the two notations are related by the relationship [70, p. 229]*

$$f(\{\mathbf{y}_1, ..., \mathbf{y}_n\}) = n!\, f(\mathbf{y}_1, ..., \mathbf{y}_n). \tag{11.116}$$

The reason is that the probability assigned to the finite set $\{\mathbf{y}_1, ..., \mathbf{y}_n\}$ *must be equally distributed among the* $n!$ *possible vectors* $(\mathbf{y}_{\sigma 1}, ..., \mathbf{y}_{\sigma n})$ *with the same elements. (Here* σ *is a permutation on the numbers* $1, ..., n$.*)*

Example 56 (Two Targets in One Dimension) *Suppose that targets lie on the one-dimensional real line, with distance measured in meters. Suppose that there are two statistically independent targets located at* $x = a$ *and* $x = b$ *with* $a \neq b$*, both of which have track variance* σ^2*. Then the multitarget distribution has the form* $f(X) = 0$ *if* $|X| \neq 2$ *and, for all* $x \neq y$*,*

$$f(\{x, y\}) \tag{11.117}$$

$$= N_{\sigma^2}(x-a) \cdot N_{\sigma^2}(y-b) + N_{\sigma^2}(x-b) \cdot N_{\sigma^2}(y-a) \tag{11.118}$$

$$= \frac{1}{2\pi\sigma^2} \left[\begin{array}{c} \exp\left(-\frac{(x-a)^2+(y-b)^2}{2\sigma^2}\right) \\ +\exp\left(-\frac{(x-b)^2+(y-a)^2}{2\sigma^2}\right) \end{array} \right]. \tag{11.119}$$

Expressed in vector notation the posterior is, therefore,

$$f(x, y) = \frac{1}{4\pi\sigma^2} \left[\begin{array}{c} \exp\left(-\frac{(x-a)^2+(y-b)^2}{2\sigma^2}\right) \\ +\exp\left(-\frac{(x-b)^2+(y-a)^2}{2\sigma^2}\right) \end{array} \right]. \tag{11.120}$$

The function $4\pi\sigma^2 f(x, y)$ *is plotted in Figure 11.1 for* $\sigma = 1$*,* $a = 1$*, and* $b = 4$*. It is bimodal with two peaks near* $(x, y) = (1, 4)$ *and* $(x, y) = (4, 1)$*. The function* $f(X)$ *of a finite-set variable* $X = \{x, y\}$*, by way of contrast, is unimodal with a peak near* $X = \{1, 4\}$*.*

11.3.4 Important Multiobject Probability Distributions

In this section, I introduce a number of examples of multiobject probability density functions, namely those corresponding to (1) independent, identically distributed (i.i.d.) cluster processes, (2) multiobject Poisson processes, (3) multiobject

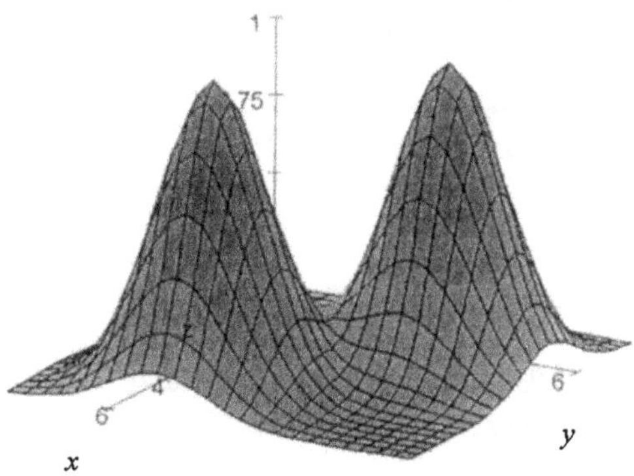

Figure 11.1 The graph of a two target multitarget probability distribution $f(X)$ as a function $f(x, y)$ of distances x, y along the real axes. Whereas $f(X)$ is unimodal with peak located near $X = \{a, b\}$ the function $f(x, y)$ is bimodal with peaks located near $(x, y) = (a, b)$ and $(x, y) = (b, a)$.

Dirac delta densities, (4) multiobject uniform processes, and (5) multiobject multi-Bernoulli processes.

11.3.4.1 Independent, Identically Distributed (i.i.d.) Cluster Processes

Let $p(n)$ be a probability distribution on the nonnegative integers and let $f(\mathbf{y})$ be a probability density function on $\mathfrak{Y}_0$. For any $Y = \{\mathbf{y}_1, ..., \mathbf{y}_n\}$ with $|Y| = n$, define

$$f(Y) \triangleq n! \cdot p(n) \cdot f(\mathbf{y}_1) \cdots f(\mathbf{y}_n). \tag{11.121}$$

It is left to the reader as Exercise 34 to show that $f(Y)$ is a probability density: $\int f(Y)\delta Y = 1$.

An *i.i.d. cluster process* [36, pp. 123, 124, 145] is any random finite set Ψ that has $f(Y)$ as its distribution, for some choice of $p(n)$ and $f(\mathbf{y})$.

11.3.4.2 Multiobject Poisson Processes

If in (11.121) we set $p(n) = e^{-\lambda}\lambda^n/n!$ (the Poisson distribution with Poisson parameter λ), then

$$f(Y) \triangleq e^{-\lambda} \cdot \lambda^n \cdot f(\mathbf{y}_1) \cdots f(\mathbf{y}_n). \tag{11.122}$$

This is called a *multidimensional Poisson distribution.* Any random finite set having $f(Y)$ as its distribution is a *multiobject Poisson process*. The function

$$I(\mathbf{y}) \triangleq \lambda \cdot f(\mathbf{y}) \tag{11.123}$$

is called the *intensity density* of the Poisson process.

11.3.4.3 Multiobject Dirac Delta Densities

The density $\delta_{Y'}(Y)$ is defined as follows:

$$\delta_{Y'}(Y) \triangleq \begin{cases} 0 & \text{if} \quad |Y| \neq |Y'| \\ 1 & \text{if} \quad Y = Y' = \emptyset \\ \sum_\sigma \delta_{\mathbf{y}'_{\sigma 1}}(\mathbf{y}_1) \cdots \delta_{\mathbf{y}'_{\sigma n}}(\mathbf{y}_n) & \text{if} \quad \left\{ \begin{array}{l} Y = \{\mathbf{y}_1, ..., \mathbf{y}_n\} \\ Y' = \{\mathbf{y}'_1, ..., \mathbf{y}'_n\} \end{array} \right. \end{cases} \tag{11.124}$$

where the summation is taken over all permutations on the numbers $1, ..., n$. It is left to the reader as Exercise 36 to show that $\delta_{Y'}(Y)$ satisfies the defining property of a Dirac delta density; see (B.2):

$$\int \delta_{Y'}(Y) \cdot f(Y)\delta Y = f(Y') \tag{11.125}$$

for any function $f(Y)$ of a finite-set variable Y.

11.3.4.4 Multiobject Uniform Processes

Let $\mathfrak{Y}_0 = \mathbb{R}^N$. Suppose that there can be no more than $\mathring{n}$ objects in a domain $D \subseteq \mathfrak{Y}_0$ of finite (hyper)volume $|D|$. If $Y = \{\mathbf{y}_1, ..., \mathbf{y}_n\}$ with $|Y| = n$ then the multitarget uniform distribution is defined as [70, p. 144]:

$$u_{D,\mathring{n}}(Y) = \frac{n!}{|D|^n \cdot (\mathring{n} + 1)} \tag{11.126}$$

if $Y \subseteq D$ and $u_{D,\mathring{n}}(Y) = 0$ otherwise. It is left to the reader as Exercise 35 to show that $u_{D,\mathring{n}}(Y)$ thus defined is a probability density. It is called a multiobject uniform density.

Remark 10 *Equation (11.126) does not reduce to a uniform distribution in the discrete case. Choose some bounded region* D *of the space* $\mathfrak{Y}_0$*. Partition* D *into* ν *equal-sized rectangular cells* $c_1, ..., c_\nu$ *of very small hypervolume* ε*, so that* $\nu\varepsilon = |D|$ *is the total (hyper)volume of* D*. Let* $\hat{\mathbf{y}}_i$ *denote the center of cell* c_i*. The uniform distribution on the discretized space is*

$$u(c_{i_1}, ..., c_{i_n}) = \frac{1}{\sum_{j=0}^{\mathring{n}} \nu^j}. \tag{11.127}$$

On the other hand, the discretized version of $u_{D,\mathring{n}}(Y)$ *is*

$$\tilde{u}_{D,\mathring{n}}(c_{i_1}, ..., c_{i_n}) \triangleq \int_{c_{i_1} \times ... \times c_{i_n}} u_{D,\mathring{n}}(\mathbf{y}_1, ..., \mathbf{y}_n) d\mathbf{y}_1 \cdots d\mathbf{y}_n \tag{11.128}$$

$$= \int_{c_{i_1} \times ... \times c_{i_n}} \frac{n!}{|D|^n \cdot (\mathring{n} + 1)} d\mathbf{y}_1 \cdots d\mathbf{y}_n \tag{11.129}$$

$$= \frac{\varepsilon^n \cdot n!}{|D|^n \cdot (\mathring{n} + 1)} = \frac{n!}{\nu^n \cdot (\mathring{n} + 1)} \tag{11.130}$$

for all $0 \leq n \leq \mathring{n}$ *and, for each* n*, all* $1 \leq i_1, ..., i_n \leq \nu$*. This does not look anything like a discrete uniform distribution. So why have I described* $u_{D,\mathring{n}}(Y)$ *as* *a multiobject uniform distribution? Because it behaves like a uniform distribution insofar as the marginal multitarget (MaM) estimator of (14.98) and (14.99) is concerned—see Example 82.*

11.3.4.5 Multiobject Multi-Bernoulli Processes

The processes discussed in this section are central to the "standard" multitarget measurement and motion models to be discussed in Sections 12.3 and 13.2, respectively. They are also central to Chapter 17.

Let $f_1(\mathbf{y}), ..., f_\nu(\mathbf{y})$ be probability density functions on $\mathfrak{Y}_0$ and let $0 \leq q_1, ..., q_\nu < 1$ be numbers. Then a *multiobject multi-Bernoulli process* defined by $b_1(\mathbf{y}), ..., b_\nu(\mathbf{y})$ and $q_1, ..., q_\nu$ is any random finite set that has a probability distribution of the form

$$f(Y) \triangleq Q_0 \triangleq (1-q_1)\cdots(1-q_\nu) \tag{11.131}$$

when $Y = \emptyset$; and when $Y = \{\mathbf{y}_1, ..., \mathbf{y}_n\}$ with $|Y| = n$, by either of the following two expressions:

$$f(\{\mathbf{y}_1, ..., \mathbf{y}_n\}) \tag{11.132}$$

$$\triangleq \sum_{1 \leq i_1 \neq ... \neq i_n \leq \nu} Q_{i_1,...,i_n} \cdot f_{i_1}(\mathbf{y}_1) \cdots f_{i_n}(\mathbf{y}_n) \tag{11.133}$$

$$= n! \sum_{1 \leq i_1 < ... < i_n \leq \nu} Q_{i_1,...,i_n} \cdot f_{i_1}(\mathbf{y}_1) \cdots f_{i_n}(\mathbf{y}_n). \tag{11.134}$$

The second summation is taken over all $i_1, ..., i_n$ such that $1 \leq i_1 < ... < i_n \leq \nu$. Also,

$$Q_{i_1,...,i_n} \triangleq (1-q_1)\cdots(1-q_\nu) \cdot \frac{q_{i_1}}{1-q_{i_1}} \cdots \frac{q_{i_n}}{1-q_{i_n}}. \tag{11.135}$$

The corresponding cardinality distribution, (11.113), is the multi-Bernoulli[6] distribution[7]

$$\begin{aligned} & B_{q_1,...,q_\nu}(n) && (11.136) \\ \triangleq \quad & \sum_{1\leq i_1<...<i_n\leq\nu} Q_{i_1,...,i_n} && (11.137) \\ = \quad & (1-q_1)\cdots(1-q_\nu) && (11.138) \\ & \cdot \sum_{1\leq i_1<...<i_n\leq\nu} \frac{q_{i_1}}{1-q_{i_1}}\cdots\frac{q_{i_n}}{1-q_{i_n}} && (11.139) \\ = \quad & \left(\prod_{i=1}^{\nu}(1-q_i)\right)\cdot\sigma_{\nu,n}\left(\frac{q_1}{1-q_1},...,\frac{q_\nu}{1-q_\nu}\right). && (11.140) \end{aligned}$$

Here

$$\sigma_{\nu,n}(y_1,...,y_\nu) \triangleq \sum_{1\leq i_1<...<i_n\leq\nu} y_{i_1}\cdots y_{i_n} \tag{11.141}$$

for $0 \leq n \leq \nu$ is called the *elementary symmetric function of degree* n *in* ν *variables.*[8]

We verify that (11.131) and (11.132) define a multiobject probability distribution. The elementary symmetric functions satisfy the following well known identity:[9]

$$\sum_{n=0}^{\nu}\sigma_{\nu,n}(y_1,...,y_\nu) = (1+y_1)\cdots(1+y_\nu). \tag{11.142}$$

6 The terminology "multi-Bernoulli," "multiple-Bernoulli," "multivariate Bernoulli," and so on, has become widespread in language processing to describe a certain kind of relevance model for text [161, (2)]. Strictly speaking, a multi-Bernoulli distribution has the form $p(i_1,...,i_n) = (n!)^{-1}\cdot Q_{i_1,...,i_n}$ for all $n \geq 0$ and all $i_1,...,i_n$ with $1 \leq i_1 \neq ... \neq i_n \leq \nu$. We are abusing terminology by referring to the marginal distribution $p(n) = \sum_{1\leq i_1\neq...\neq i_n\leq\nu} Q_{i_1,...,i_n}$ as "multi-Bernoulli."

7 The binomial distribution $B_{\nu,q}(n) = C_{\nu,n}q^n(1-q)^{\nu-n}$ gives the probability of drawing n heads from ν tosses (ν Bernoulli trials) of a biased coin, where q is the probability of drawing heads on a single toss. The multi-Bernoulli process generalizes this. Suppose that we have ν different biased coins, with q_i being the probability of drawing a head from the ith coin. If we toss each of the ν coins exactly once, then $B_{q_1,...,q_\nu}(n)$ is the probability of drawing n heads.

8 Thus, for example, $\sigma_{\nu,0}(y_1,...,y_\nu) = 1$ and $\sigma_{\nu,1}(y_1,...,y_\nu) = y_1 + ... + y_\nu$ and $\sigma_{\nu,\nu}(y_1,...,y_\nu) = y_1\cdots y_\nu$.

9 Equation (11.142) is a generalization of the binomial formula.

It follows that

$$\int f(Y)\delta Y = \sum_{n=0}^{\infty} \frac{1}{n!} \int f(\{\mathbf{x}_1, ..., \mathbf{x}_n\}) d\mathbf{x}_1 \cdots d\mathbf{x}_n \quad (11.143)$$

$$= \sum_{n=0}^{\nu} B_{q_1,...,q_\nu}(n) \quad (11.144)$$

$$= (1-q_1)\cdots(1-q_\nu) \quad (11.145)$$

$$\cdot \sum_{n=0}^{\nu} \sigma_{\nu,n}\left(\frac{q_1}{1-q_1}, ..., \frac{q_\nu}{1-q_\nu}\right). \quad (11.146)$$

Thus from (11.141) and (11.142) we get, as claimed,

$$\int f(Y)\delta Y = (1-q_1)\cdots(1-q_\nu) \quad (11.147)$$

$$\cdot \left(1+\frac{q_1}{1-q_1}\right)\cdots\left(1+\frac{q_\nu}{1-q_\nu}\right) \quad (11.148)$$

$$= (1-q_1+q_1)\cdots(1-q_\nu+q_\nu) = 1. \quad (11.149)$$

11.3.4.6 Binomial i.i.d. Cluster Processes

If $q_1 = ... = q_\nu = q$ and $f_1(\mathbf{y}) = ... = f_\nu(\mathbf{y}) = f(\mathbf{y})$ then (11.132) reduces to an i.i.d. cluster process with a binomial cardinality distribution $f(n) = B_{\nu,q}(n) = C_{\nu,n} q^n (1-q)^{\nu-n}$:

$$f(\{\mathbf{y}_1, ..., \mathbf{y}_n\}) \triangleq n! \cdot B_{\nu,q}(n) \cdot f(\mathbf{y}_1) \cdots f(\mathbf{y}_n). \quad (11.150)$$

11.3.5 Probability-Generating Functionals (p.g.fl.s)

In what follows let $h(\mathbf{y})$ be a nonnegative real-valued function of $\mathbf{y} \in \mathfrak{Y}_0$ that *has no units of measurement*. Hereafter I will call $h(\mathbf{y})$ a *test function*. Most frequently but not always, we will assume that $0 \leq h(\mathbf{y}) \leq 1$. The simplest example of a test function is the indicator function $\mathbf{1}_S(\mathbf{y})$ of a subset $S \subseteq \mathfrak{Y}_0$; see (5.38)

$$\mathbf{1}_S(\mathbf{y}) \triangleq \begin{cases} 1 & \text{if} \quad \mathbf{y} \in S \\ 0 & \text{if} \quad \text{otherwise} \end{cases}. \quad (11.151)$$

A *functional* is a real-valued function $F[h]$ of h that has no units of measurement.

The simplest nontrivial functionals are the *linear functionals* defined by

$$f[h] \triangleq \int h(\mathbf{y}) \cdot f(\mathbf{y}) d\mathbf{y} \tag{11.152}$$

where $f(\mathbf{y})$ is any density function on $\mathfrak{Y}_0$. Note that $f[h]$ has no units of measurement and that $f[\mathbf{1}_S] = \int_S f(\mathbf{y}) d\mathbf{y}$.

Remark 11 (Notation for Linear Functionals) *Linear functionals, and the notation in (11.152) used to express them, will play a central part in the rest of the book.*

If $Y \subseteq \mathfrak{Y}_0$ is any finite set and $h(\mathbf{y})$ a test function, define the *power* of h with respect to Y to be

$$h^Y \triangleq \begin{cases} 1 & \text{if} \quad Y = \emptyset \\ \prod_{\mathbf{y} \in Y} h(\mathbf{y}) & \text{if} \quad \text{otherwise} \end{cases} . \tag{11.153}$$

If, for example, $h(\mathbf{y}) = y$ is a constant function with value y, then $h^Y = y^Y = y^{|Y|}$.

Let $f_\Psi(Y)$ be the probability density function of a finite random set Ψ. Then its *probability-generating functional* is

$$G_\Psi[h] \triangleq \int h^Y \cdot f_\Psi(Y) \delta Y \tag{11.154}$$

$$= f_\Psi(\emptyset) + \int h(\mathbf{y}) \cdot f_\Psi(\{\mathbf{y}\}) d\mathbf{y} \tag{11.155}$$

$$+ \frac{1}{2} \int h(\mathbf{y}_1) \cdot h(\mathbf{y}_2) \cdot f_\Psi(\{\mathbf{y}_1, \mathbf{y}_2\}) d\mathbf{y}_1 d\mathbf{y}_2 + ... \tag{11.156}$$

11.3.5.1 Basic Properties of p.g.fl.s

Note that:

- $G_\Psi[h]$ has no units of measurement;
- $G_\Psi[0] = f_\Psi(\emptyset)$;
- $G_\Psi[1] = \int f_\Psi(Y) \delta Y = 1$;
- $G_\Psi[h_1] \leq G_\Psi[h_2]$ whenever $h_1 \leq h_2$.

If $h(\mathbf{y}) = y$ is a constant nonnegative real number for all $\mathbf{y}$, then

$$G_\Psi[y] = G_{|\Psi|}(y) \tag{11.157}$$

where $G_{|\Psi|}(y)$ is the probability-generating function (p.g.f.) of the random integral $|\Psi|$ as defined in (11.10). To see this, note that

$$\begin{aligned} G_\Psi[y] &= \int y^{|Y|} \cdot f_\Psi(Y)\delta Y && (11.158)\\ &= f_\Psi(\emptyset) + y \int f_\Psi(\{\mathbf{y}\})d\mathbf{y} && (11.159)\\ &\quad +\frac{1}{2}y^2 \int f_\Psi(\{\mathbf{y}_1, \mathbf{y}_2\})d\mathbf{y}_1 d\mathbf{y}_2 + ... && (11.160)\\ &= p_\Psi(0) + y \cdot p_\Psi(1) + y^2 \cdot p_\Psi(2) + ... && (11.161)\\ &= G_{|\Psi|}(y) && (11.162)\\ &\triangleq G_\Psi(y). && (11.163) \end{aligned}$$

Here $p_\Psi(n)$ is the cardinality distribution as defined in (11.113).

Two basic properties of p.g.f.s should be pointed out here. If N_Ψ is the expected value of $|\Psi|$ and if σ^2_Ξ is its variance, then

$$\begin{aligned} N_\Psi &= G'_\Psi(1) && (11.164)\\ \sigma^2_\Psi &= G''_\Psi(1) - N^2_\Psi + N_\Psi && (11.165) \end{aligned}$$

where $G'_\Psi(x)$ and $G''_\Psi(x)$ are the first and second derivatives of $G_\Psi(x)$. Verification of these facts are left to the reader as Exercise 37.

I leave the proof of the following fact to the reader as Exercise 40. Let $\Psi = \Psi_1 \cup ... \cup \Psi_s$ with $\Psi_1 \cup ... \cup \Psi_s$ statistically independent. Then

$$G_\Psi[h] = G_{\Psi_1}[h] \cdots G_{\Psi_s}[h]. \tag{11.166}$$

The following two remarks point out additional properties of p.g.fl.s.

Remark 12 (p.g.fl.'s Are Integral Transforms) *The p.g.fl.* $G_\Psi[h]$ *can be regarded as an integral transform of the probability density function* $f_\Psi(Y)$. *Following (11.10) we noted that the probability-generating function (p.g.f.) of a random nonnegative integer* J *can be regarded as a modified z-transform. Since p.g.fl.s generalize p.g.f.s to multiobject statistics, it follows that they can be regarded as generalized multiobject z-transforms.*

Remark 13 (p.g.fl.s Are Generalized Belief-Mass Functions) *On the one hand, belief-mass functions are special cases of p.g.fl.s:*

$$G_\Psi[\mathbf{1}_S] = \int \mathbf{1}_S^Y \cdot f_\Psi(Y)\delta Y = \int_S f_\Psi(Y)\delta Y = \beta_\Psi(S). \tag{11.167}$$

On the other hand, suppose that $0 \leq h(\mathbf{y}) \leq 1$ *is a fuzzy membership function on* $\mathfrak{Z}_0$. *Then the p.g.fl.* $0 \leq G_\Psi[h] \leq 1$ *can be interpreted as the probability that* Ψ *is contained in the fuzzy set represented by* $h(\mathbf{y})$. *To see this, recall that in (4.59) I introduced the "asynchronous" random set representation of a fuzzy membership function* $h(\mathbf{y})$,

$$\Sigma_\alpha(h) \triangleq \{\mathbf{y} | \, \alpha(\mathbf{y}) \leq h(\mathbf{y})\} \tag{11.168}$$

where for every $\mathbf{y} \in \mathfrak{Z}_0$, *the quantity* $\alpha(\mathbf{y})$ *is a uniformly distributed random number in* $[0, 1]$. *If* Ψ *and* α *are independent, then*

$$\Pr(\Psi \subseteq \Sigma_\alpha(h)) = \int \Pr(Y \subseteq \Sigma_\alpha(h)) \cdot f_\Psi(Y)\delta Y. \tag{11.169}$$

If $Y = \emptyset$ *then* $\Pr(Y \subseteq \Sigma_\alpha(h)) = 1 = h^\emptyset$. *Otherwise, we know from (4.68) that since* $\Sigma_\alpha(h)$ *is asynchronous,*

$$\Pr(Y \subseteq \Sigma_\alpha(h)) = \prod_{\mathbf{y} \in Y} h(\mathbf{y}) = h^Y \tag{11.170}$$

Thus

$$\Pr(\Psi \subseteq \Sigma_\alpha(h)) = \int h^Y \cdot f_\Psi(Y)\delta Y = G_\Psi[h]. \tag{11.171}$$

In the special case $h = \mathbf{1}_S$ *we have* $\Sigma_\alpha(\mathbf{1}_S) = S$. *In this sense,* $G_\Psi[h]$ *is the probability that* Ψ *is completely contained in the fuzzy set with membership function* $h(\mathbf{y})$.

11.3.5.2 Examples of p.g.fl.s

I present several examples of p.g.fl.s and random finite sets.

Example 57 (p.g.fl. of a Poisson Process) *The multiobject Poisson process was introduced in (11.122). It is left to the reader as Exercise 34 to show that its p.g.fl. is*

$$G_f[h] = e^{\lambda f[h] - \lambda} \tag{11.172}$$

where $f[h] \triangleq \int h(\mathbf{y}) \cdot f(\mathbf{y})d\mathbf{y}$.

Example 58 (p.g.fl. of an i.i.d. Cluster Process) *The i.i.d. cluster process was defined in (11.121). Using the same reasoning as in the previous example it is easily shown that its p.g.fl. is*

$$G[h] = G(f[h]) \tag{11.173}$$

where $G(y) = \sum_{n=0}^{\infty} p(n) y^n$ *is the p.g.f. of the probability distribution* $p(n)$ *and where* $f[h] = \int h(\mathbf{y}) \cdot f(\mathbf{y}) d\mathbf{y}$.

Example 59 (p.g.fl. of a Multiobject Dirac Delta Density) *The multiobject Dirac delta* $\delta_{Y'}(Y)$ *was defined in (11.124). It is left to the reader as Exercise 38 to show that its p.g.fl. is*

$$G_{Y'}[h] = h^{Y'} \tag{11.174}$$

for any Y'*, where the notation* $h^{Y'}$ *was defined in (11.154).*

Example 60 (p.g.fl. of a Multi-Bernoulli Process) *The multiobject multi-Bernoulli process was defined in (11.131)-(11.135). Its p.g.fl. is*

$$G[h] = (1 - q_1 + q_1 \cdot f_1[h]) \cdots (1 - q_\nu + q_\nu \cdot f_\nu[h]). \tag{11.175}$$

To see this, abbreviate $Q \overset{\text{abbr.}}{=} \prod_{i=1}^{\nu} (1 - q_i)$ *and* $f_i[h] \overset{\text{abbr.}}{=} \int h(\mathbf{y}) \cdot f_i(\mathbf{y}) d\mathbf{y}$*. Then from (11.131)-(11.135),*

$$G[h] = \int h^Y \cdot f(Y) \delta Y \tag{11.176}$$

$$= \sum_{n=0}^{\infty} \frac{1}{n!} \int h(\mathbf{y}_1) \cdots h(\mathbf{y}_n) \cdot f(\{\mathbf{y}_1, ..., \mathbf{y}_n\}) d\mathbf{y}_1 \cdots d\mathbf{y}_n \tag{11.177}$$

$$= Q \sum_{n=0}^{\infty} \frac{1}{n!} \int \left(n! \sum_{1 \leq i_1 < ... < i_n \leq \nu} \prod_{e=1}^{n} \frac{q_{i_e} f_{i_e}(\mathbf{y}_e) \cdot h(\mathbf{y}_e)}{1 - q_{i_e}} \right) \tag{11.178}$$

$$\cdot d\mathbf{y}_1 \cdots d\mathbf{y}_n \tag{11.179}$$

$$= Q \sum_{n=0}^{\infty} \sum_{1 \leq i_1 < ... < i_n \leq \nu} \prod_{e=1}^{n} \frac{q_{i_e} \int f_{i_e}(\mathbf{y}_e) \cdot h(\mathbf{y}_e) d\mathbf{y}_e}{1 - q_{i_e}} \tag{11.180}$$

$$= Q \sum_{n=0}^{\infty} \sum_{1 \leq i_1 < ... < i_n \leq \nu} \prod_{e=1}^{n} \frac{q_{i_e} f_{i_e}[h]}{1 - q_{i_e}} \tag{11.181}$$

From (11.141) and (11.142) it follows that

$$
\begin{aligned}
G[h] &= Q \cdot \sum_{n=0}^{\infty} \sigma_{\nu,n}\left(\frac{q_1 f_1[h]}{1-q_1}, ..., \frac{q_\nu f_\nu[h]}{1-q_\nu}\right) && (11.182)\\
&= Q \cdot \left(1+\frac{q_1 f_1[h]}{1-q_1}\right) \cdots \left(1+\frac{q_\nu f_\nu[h]}{1-q_\nu}\right) && (11.183)\\
&= (1-q_1+q_1 f_1[h]) \cdots (1-q_\nu+q_\nu f_\nu[h]). && (11.184)
\end{aligned}
$$

Example 61 (p.g.fl. of a Multiobject Uniform Distribution) *The multiobject uniform distribution was defined in (11.126). It is left to the reader as Exercise 39 to show that its p.g.fl. is*

$$
G[h] = \frac{1}{\mathring{n}+1} \sum_{n=0}^{\mathring{n}} \left(\frac{\mathbf{1}_D[h]}{|D|}\right)^n \tag{11.185}
$$

where as usual $\mathbf{1}_D[h] \triangleq \int h(\mathbf{y}) \cdot \mathbf{1}_D(\mathbf{y}) d\mathbf{y} = \int_D h(\mathbf{y}) d\mathbf{y}$.

11.4 FUNCTIONAL DERIVATIVES AND SET DERIVATIVES

In Appendix C the *gradient (also known as the Frechét or directional) derivative* of a function is reviewed. A *functional derivative* (Section 11.4.1) is a special kind of gradient derivative. A *set derivative* (Section 11.4.2) is a special kind of functional derivative.

11.4.1 Functional Derivatives

Because functionals $F[h]$ are themselves just functions defined on a variable h which is an ordinary function, one can speak of the gradient derivative $(\delta F/\delta g)[h]$ of a functional $F[h]$ in the direction of a function g. A *functional derivative* is the gradient derivative $(\delta F/\delta g)[h]$ of a functional $F[h]$ in the direction of a Dirac delta density $g = \delta_{\mathbf{y}}$.

Specifically, let $F[h]$ be a functional on $\mathfrak{Z}_0$ and let $\mathbf{y} \in \mathfrak{Z}_0$. The *functional derivative* of $F[h]$ at $\mathbf{y}$ is (see [199, pp. 173, 174], [137, pp. 140, 141], or [144, p. 12])

$$
\frac{\delta F}{\delta \mathbf{y}}[h] \triangleq \lim_{\varepsilon \searrow 0} \frac{F[h+\varepsilon\delta_{\mathbf{y}}] - F[h]}{\varepsilon} \tag{11.186}
$$

where as usual $\delta_{\mathbf{y}}(\mathbf{w})$ is the Dirac delta function concentrated at $\mathbf{y}$.[10] Note that because test functions h must always be nonnegative, ε must always be nonnegative and thus the limit must always be taken from above: $\varepsilon \searrow 0$ rather than $\varepsilon \to 0$.

General gradient derivatives can be expressed as infinite linear combinations of functional derivatives. That is, we can abbreviate the equation

$$g(\mathbf{y}) = \int g(\mathbf{w})\delta_{\mathbf{w}}(\mathbf{y})d\mathbf{w} \tag{11.187}$$

as

$$g = \int g(\mathbf{w})\delta_{\mathbf{w}}d\mathbf{w}. \tag{11.188}$$

So

$$\frac{\partial F}{\partial g}[h] = \int g(\mathbf{w}) \cdot \frac{\partial F}{\partial \delta_{\mathbf{w}}}[h]d\mathbf{w} = \int g(\mathbf{w}) \cdot \frac{\delta F}{\delta \mathbf{w}}[h]d\mathbf{w}. \tag{11.189}$$

The third equation results from the fact that, by definition, the functional defined by $g \mapsto (\partial F \partial g)[h]$ is linear and continuous for each choice of h.

Given $\mathbf{y}_1, ..., \mathbf{y}_n \in \mathfrak{Z}_0$, the *iterated* functional derivatives of $F[h]$ at $\mathbf{y}_1, ..., \mathbf{y}_n$ are defined recursively by

$$\frac{\delta^n F}{\delta \mathbf{y}_n \cdots \delta \mathbf{y}_1}[h] \triangleq \frac{\delta}{\delta \mathbf{y}_n} \frac{\delta^{n-1} F}{\delta \mathbf{y}_{n-1} \cdots \delta \mathbf{y}_1}[h]. \tag{11.190}$$

If Y is a finite subset of $\mathfrak{Z}_0$ then we define the functional derivative of $F[h]$ at Y to be

$$\frac{\delta F}{\delta Y}[h] \triangleq \begin{cases} F[h] & \text{if} \quad Y = \emptyset \\ \frac{\delta^n F}{\delta \mathbf{y}_1 \cdots \delta \mathbf{y}_n}[h] & \text{if} \quad \begin{cases} Y = \{\mathbf{y}_1, ..., \mathbf{y}_n\} \\ |Y| = n \end{cases} \end{cases}. \tag{11.191}$$

Remark 14 *The concept of a functional derivative is borrowed from quantum physics (see Remark 16). I employ it because I believe that, from a practitioner's point of view, it is more intuitive and easier to use than conventional gradient derivatives as employed by mathematicians. These advantages come with a potential cost, however. Dirac delta functions must be treated with care because they do not necessarily behave as one might intuitively expect. For example, if H is a*

10 I am abusing notation here. In my definition, the test function h in a functional $F[h]$ typically takes its values in $[0, 1]$. Here we are allowing it to include Dirac deltas. See [136, p. 1161] for a more precise and complete definition.

nonsingular matrix one might expect that $\delta_{\mathbf{y}}(H\mathbf{w}) = \delta_{H^{-1}\mathbf{y}}(\mathbf{w})$. *However, this is not true, as is easily seen in the one-dimensional case with* $Hw \triangleq a \cdot w$:

$$\int h(w) \cdot \delta_y(aw)dw = \int h(a^{-1}v) \cdot a^{-1} \cdot \delta_y(v)dv \quad (11.192)$$

$$= h(a^{-1}y) \cdot a^{-1} \quad (11.193)$$

$$= \int h(w) \cdot a^{-1} \cdot \delta_{a^{-1}y}(w)dw. \quad (11.194)$$

Since this is for all h *it follows that* $\delta_y(aw) = a^{-1} \cdot \delta_{a^{-1}y}(w)$. *To assist the reader, in Appendix B I have tabulated the major properties of delta functions.*

Remark 15 (Units of Measurement) *Functional derivatives behave like density functions. Suppose that the units of measurement of* $\mathfrak{Z}_0$ *are* u. *Then the units of measurement of the density* $\delta_{\mathbf{y}}(\mathbf{w})$ *are* u^{-1}. *Since* $h + \varepsilon\delta_{\mathbf{y}}$ *must be a unitless function, the units of* ε *must be* u. *Consequently, the units of* $(\delta F/\delta\mathbf{y})[h]$ *are those of a density function:* u^{-1} *since* ε *appears in the denominator of the defining limit. Thus the units of* $(\delta F/\delta Y)[h]$ *are* $u^{-|Y|}$ *since this is a first-order functional derivative iterated* $|Y|$ *times. Thus the set integrals*

$$\int_S \frac{\delta F}{\delta Y}[h]\delta Y \quad (11.195)$$

are always mathematically well defined.

The following three examples illustrate the direct computation of functional derivatives using (11.186).

Example 62 (Derivative of a Linear Functional) *[136, p. 1162] Let*

$$f[h] = \int h(\mathbf{y}) \cdot f(\mathbf{y})d\mathbf{y} \quad (11.196)$$

be the linear functional corresponding to a function $f(\mathbf{y})$, *as defined in (11.152). Then*

$$\frac{\delta f}{\delta Y}[h] = \begin{cases} f[h] & \text{if} \quad Y = \emptyset \\ f(\mathbf{y}) & \text{if} \quad Y = \{\mathbf{y}\} \\ 0 & \text{if} \quad |Y| \geq 2 \end{cases}. \quad (11.197)$$

To see this, note that

$$\frac{\delta f}{\delta \mathbf{y}}[h] = \lim_{\varepsilon \searrow 0} \frac{f[h+\varepsilon\delta_{\mathbf{y}}] - f[h]}{\varepsilon} \quad (11.198)$$

$$= \lim_{\varepsilon \searrow 0} \frac{f[h] + \varepsilon \cdot f[\delta_{\mathbf{y}}] - f[h]}{\varepsilon} \quad (11.199)$$

$$= f[\delta_{\mathbf{y}}] = f(\mathbf{y}). \quad (11.200)$$

Since $\frac{\delta f}{\delta \mathbf{y}}[h]$ *no longer depends on the variable* h, *it is constant with respect to functional differentiation. Thus* $\frac{\delta f}{\delta Y}[h] = 0$ *for any* Y *containing two or more elements.*

Example 63 (Derivative of a Quadratic Functional) *Let* $F[h] = f_1[h] \cdot f_2[h]$ *where* $f_1(\mathbf{y})$ *and* $f_2(\mathbf{y})$ *are density functions. Then*

$$\frac{\delta F}{\delta \mathbf{y}_1}[h] \quad (11.201)$$

$$= \frac{\delta}{\delta \mathbf{y}_1} f_1[h] \cdot f_2[h] \quad (11.202)$$

$$= \lim_{\varepsilon \searrow 0} \frac{F[h+\varepsilon\delta_{\mathbf{y}_1}] - F[h]}{\varepsilon} \quad (11.203)$$

$$= \lim_{\varepsilon \searrow 0} \frac{f_1[h+\varepsilon\delta_{\mathbf{y}_1}] \cdot f_2[h+\varepsilon\delta_{\mathbf{y}_1}] - f_1[h] \cdot f_2[h}{\varepsilon} \quad (11.204)$$

and so

$$= \lim_{\varepsilon \searrow 0} \frac{\begin{array}{c} f_1[h+\varepsilon\delta_{\mathbf{y}_1}] \cdot f_2[h+\varepsilon\delta_{\mathbf{y}_1}] - f_1[h+\varepsilon\delta_{\mathbf{y}_1}] \cdot f_2[h] \\ +f_1[h+\varepsilon\delta_{\mathbf{y}_1}] \cdot f_2[h] - f_1[h] \cdot f_2[h] \end{array}}{\varepsilon} \quad (11.205)$$

$$= \lim_{\varepsilon \searrow 0} f_1[h+\varepsilon\delta_{\mathbf{y}_1}] \cdot \lim_{\varepsilon \searrow 0} \frac{f_2[h+\varepsilon\delta_{\mathbf{y}_1}] - f_2[h]}{\varepsilon} \quad (11.206)$$

$$+ \lim_{\varepsilon \searrow 0} \frac{f_1[h+\varepsilon\delta_{\mathbf{y}_1}] - f_1[h]}{\varepsilon} \cdot \lim_{\varepsilon \searrow 0} f_2[h] \quad (11.207)$$

$$= f_1[h] \cdot f_2(\mathbf{y}_1) + f_1(\mathbf{y}_1) \cdot f_2[h]. \quad (11.208)$$

Thus

$$\frac{\delta^2 F}{\delta \mathbf{y}_2 \delta \mathbf{y}_1}[h] \tag{11.209}$$

$$= \lim_{\varepsilon \searrow 0} \frac{\begin{array}{c} f_1[h+\varepsilon\delta_{\mathbf{y}_2}] \cdot f_2(\mathbf{y}_1) + f_1(\mathbf{y}_1) \cdot f_2[h+\varepsilon\delta_{\mathbf{y}_2}] \\ -\left(f_1[h] \cdot f_2(\mathbf{y}_1) + f_1(\mathbf{y}_1) \cdot f_2[h]\right) \end{array}}{\varepsilon} \tag{11.210}$$

$$= \lim_{\varepsilon \searrow 0} \frac{\begin{array}{c} f_1[h] \cdot f_2(\mathbf{y}_1) + \varepsilon \cdot f_1[\delta_{\mathbf{y}_2}] \cdot f_2(\mathbf{y}_1) \\ +f_1(\mathbf{y}_1) \cdot f_2[h] + \varepsilon \cdot f_1(\mathbf{y}_1) \cdot f_2[\delta_{\mathbf{y}_2}] \\ -\left(f_1[h] \cdot f_2(\mathbf{y}_1) + f_1(\mathbf{y}_1) \cdot f_2[h]\right) \end{array}}{\varepsilon} \tag{11.211}$$

$$= \lim_{\varepsilon \searrow 0} \frac{\varepsilon \cdot f_1[\delta_{\mathbf{y}_2}] \cdot f_2(\mathbf{y}_1) + \varepsilon \cdot f_1(\mathbf{y}_1) \cdot f_2[\delta_{\mathbf{y}_2}]}{\varepsilon} \tag{11.212}$$

$$= f_1[\delta_{\mathbf{y}_2}] \cdot f_2(\mathbf{y}_1) + f_1(\mathbf{y}_1) \cdot f_2[\delta_{\mathbf{y}_2}] \tag{11.213}$$

$$= f_1(\mathbf{y}_2) \cdot f_2(\mathbf{y}_1) + f_1(\mathbf{y}_1) \cdot f_2(\mathbf{y}_2). \tag{11.214}$$

Since this no longer depends on the variable h, the functional derivatives of order three and greater vanish identically.

Example 64 (Chain Rule for Affine Functional Transformations) *[136, p. 1162] Let $h_1(\mathbf{y})$ and $h_2(\mathbf{y})$ be unitless functions and $F[h]$ a functional. Let $T[h] = h_1 h + h_2$ be an affine functional transformation. Then*

$$\frac{\delta}{\delta \mathbf{y}} F[h_1 h + h_2] = h_1(\mathbf{y}) \cdot \frac{\delta F}{\delta \mathbf{y}}[h_1 h + h_2]. \tag{11.215}$$

To see this, note that

$$\frac{\delta}{\delta \mathbf{y}} F[h_1 h + h_2] \tag{11.216}$$

$$= \lim_{\varepsilon \searrow 0} \frac{F[h_1 h + h_1 \varepsilon \delta_{\mathbf{y}} + h_2] - F[h_1 h + h_2]}{\varepsilon} \tag{11.217}$$

$$= \lim_{\varepsilon \searrow 0} \frac{F[h_1 h + h_2 + h_1(\mathbf{y}) \varepsilon \delta_{\mathbf{y}}] - F[h_1 h + h_2]}{\varepsilon}. \tag{11.218}$$

If $h_1(\mathbf{y}) = 0$ *then this is identically zero and we are finished. If on the other hand* $h_1(\mathbf{y}) \neq 0$,

$$\frac{\delta}{\delta \mathbf{y}} F[h_1 h + h_2] \tag{11.219}$$

$$= h_1(\mathbf{y}) \cdot \lim_{\varepsilon \searrow 0} \frac{F[h_1 h + h_2 + h_1(\mathbf{y})\varepsilon\delta_{\mathbf{y}}] - F[h_1 h + h_2]}{h_1(\mathbf{y})\varepsilon} \tag{11.220}$$

$$= h_1(\mathbf{y}) \cdot \frac{\delta F}{\delta \mathbf{y}}[h_1 h + h_2] \tag{11.221}$$

and we are finished.

11.4.2 Set Derivatives

Every functional $F[h]$ gives rise to a set function

$$\phi_F(S) \triangleq F[\mathbf{1}_S]. \tag{11.222}$$

When similarly restricted, functional derivatives reduce to a special case called a *set derivative* of the set function $\phi_F(S)$:

$$\frac{\delta \phi_F}{\delta Y}(S) \triangleq \frac{\delta F}{\delta Y}[\mathbf{1}_S]. \tag{11.223}$$

Let $E_{\mathbf{y}}$ be some very small neighborhood of $\mathbf{y}$ with (hyper)volume $|E_{\mathbf{y}}| = \varepsilon$. Then note that

$$\delta_{\mathbf{y}}(\mathbf{w}) \cong \frac{1}{|E_{\mathbf{y}}|} \cdot \mathbf{1}_{E_{\mathbf{y}}}(\mathbf{w}) \tag{11.224}$$

and so $\varepsilon\delta_{\mathbf{y}} \cong \mathbf{1}_{E_{\mathbf{y}}}$. In the limit,

$$\frac{\delta F}{\delta \mathbf{y}}[\mathbf{1}_S] = \lim_{\varepsilon \searrow 0} \frac{F[\mathbf{1}_S + \varepsilon\delta_{\mathbf{y}}] - F[\mathbf{1}_S]}{\varepsilon} \tag{11.225}$$

$$= \lim_{|E_{\mathbf{y}}| \searrow 0} \frac{F[\mathbf{1}_S + \mathbf{1}_{E_{\mathbf{y}}}] - F[\mathbf{1}_S]}{|E_{\mathbf{y}}|}. \tag{11.226}$$

Assume that S and $E_{\mathbf{y}}$ are disjoint. Then $\mathbf{1}_S + \mathbf{1}_{E_{\mathbf{y}}} = \mathbf{1}_{S \cup E_{\mathbf{y}}}$ and so

$$\frac{\delta F}{\delta \mathbf{y}}[\mathbf{1}_S] = \lim_{|E_{\mathbf{y}}| \searrow 0} \frac{F[\mathbf{1}_{S \cup E_{\mathbf{y}}}] - F[\mathbf{1}_S]}{|E_{\mathbf{y}}|} \tag{11.227}$$

$$= \lim_{|E_{\mathbf{y}}| \searrow 0} \frac{\phi_F(S \cup E_{\mathbf{y}}) - \phi_F(S)}{|E_{\mathbf{y}}|}. \tag{11.228}$$

That is, the restriction of a first functional derivative of a functional F to indicator functions of sets—see (5.38)— results in an expression defined entirely in terms of the associated set function $\phi_F(S)$.

Consequently, given a set function $\phi(S)$, I define its *set derivative* at $\mathbf{y}$ to be[11]

$$\frac{\delta \phi}{\delta \mathbf{y}}(S) \triangleq \lim_{|E_{\mathbf{y}}| \searrow 0} \frac{\phi_F(S \cup E_{\mathbf{y}}) - \phi_F(S)}{|E_{\mathbf{y}}|}. \tag{11.229}$$

Iterated set derivatives are defined in the obvious manner:

$$\frac{\delta^n \phi}{\delta \mathbf{y}_n \cdots \delta \mathbf{y}_1}(S) \triangleq \frac{\delta}{\mathbf{y}_n} \frac{\delta^{n-1} \phi}{\delta \mathbf{y}_{n-1} \cdots \delta \mathbf{y}_1}(S). \tag{11.230}$$

The general set derivative is

$$\frac{\delta \phi}{\delta Y}(S) \triangleq \left\{ \begin{array}{ccc} \phi(S) & \text{if} & Y = \emptyset \\ \frac{\delta^n \phi}{\delta \mathbf{y}_n \cdots \delta \mathbf{y}_1}(S) & \text{if} & \left\{ \begin{array}{c} Y = \{\mathbf{y}_1, ..., \mathbf{y}_n\} \\ |Y| = n \end{array} \right. \end{array} \right. . \tag{11.231}$$

General set derivatives and general functional derivatives are related as follows:

$$\frac{\delta \phi_F}{\delta Y}(S) = \frac{\delta F}{\delta Y}[\mathbf{1}_S]. \tag{11.232}$$

Example 65 (Set Derivative of a Probability-Mass Function) *Let*

$$p_{\mathbf{Y}}(S) = \int_S f_{\mathbf{Y}}(\mathbf{w}) d\mathbf{w} \tag{11.233}$$

11 In [70, pp. 150, 151] I defined set derivatives directly using a more rigorous form of this limit, that is known as a constructive Radon-Nikodým derivative. In later papers I showed that the set derivative is a special case of the functional derivative [136, 137].

be the probability-mass function of the random vector $\mathbf{Y}$. *Then from (11.152) we know that* $p_{\mathbf{Y}}(S) = f_{\mathbf{Y}}[\mathbf{1}_S]$. *From (11.232) and Example 62 we therefore get*

$$\frac{\delta p_{\mathbf{Y}}}{\delta Y}(S) = \frac{\delta f_{\mathbf{Y}}}{\delta Y}[\mathbf{1}_S] = \left\{ \begin{array}{ccc} p_{\mathbf{Y}}(S) & \text{if} & Y = \emptyset \\ f_{\mathbf{Y}}(\mathbf{w}) & \text{if} & Y = \{\mathbf{y}\} \\ 0 & \text{if} & |Y| \geq 2 \end{array} \right. . \tag{11.234}$$

Example 66 (Nonsingular Transformations of RFSs) *Let* $T : \mathfrak{Y}_0 \to \mathfrak{Y}_0$ *be a nonsingular vector transformation. If* Y *is a finite subset of* $\mathfrak{Y}_0$ *define* $TY \triangleq \{T\mathbf{y} | \mathbf{y} \in Y\}$ *if* $Y \neq \emptyset$ *and* $TY \triangleq \emptyset$ *otherwise. If* Ψ *is a random finite subset of* $\mathfrak{Y}_0$, *so is* $T\Psi$. *Its multiobject probability density is*

$$f_{T\Psi}(Y) = f_{\Psi}(T^{-1}Y) \cdot \frac{1}{J_T^Y} \tag{11.235}$$

where $J_T^Y \triangleq \prod_{\mathbf{y} \in Y} J_T(\mathbf{y})$ *and where* $J_T(\mathbf{y})$ *is the Jacobian determinant of* T. *This result follows from the second chain rule for functional derivatives, (11.282), by setting* $T[h](\mathbf{y}) \triangleq h(T^{-1}(\mathbf{y}))$:

$$f_{T\Psi}(Y) = \left[\frac{\delta}{\delta Y} F[T[h]]\right]_{h=0} = \frac{1}{J_T^Y} \cdot \frac{\delta F}{\delta T^{-1}Y}[T^{-1}0] \tag{11.236}$$

$$= \frac{1}{J_T^Y} \cdot \frac{\delta F}{\delta T^{-1}Y}[0] = \frac{1}{J_T^Y} \cdot f_{\Psi}(T^{-1}Y). \tag{11.237}$$

Remark 16 (Notation for Set and Functional Derivatives) *Our notation for functionals, set derivatives, and functional derivatives was not chosen arbitrarily. It is an abbreviation of a common notation employed in quantum field theory (see [199, pp. 173, 174] or [137, pp. 140, 141]). The following equation displays three different notations for iterated functional derivatives: mathematics (bottom), physics (right), and FISST (left):*

$$\overbrace{\frac{\delta^n F}{\delta \mathbf{y}_1 \cdots \delta \mathbf{y}_n}[h]}^{\text{FISST}} = \overbrace{\frac{\delta^n F}{\delta h(\mathbf{y}_1) \cdots \delta h(\mathbf{y}_n)}[h]}^{\text{physics}} \tag{11.238}$$

$$= \overbrace{\frac{\partial^n F}{\partial \delta_{\mathbf{y}_1} \cdots \partial \delta_{\mathbf{y}_n}}[h]}^{\text{mathematics}} . \tag{11.239}$$

Remark 17 (Set Derivative Is Continuous Möbius Transform) *Let* $\phi(S)$ *be a set function and let* $\mathbf{y}_1, ..., \mathbf{y}_k$ *be distinct elements of* $\mathfrak{Y}_0$*. Define the function* $\phi_i(Y)$ *of a finite-set variable* $Y = \{\mathbf{y}_1, ..., \mathbf{y}_k\}$ *by*

$$\phi_\varepsilon(\{\mathbf{y}_1, ..., \mathbf{y}_n\}) \triangleq \phi(E_{\mathbf{y}_1} \cup ... \cup E_{\mathbf{y}_n})$$

for all distinct $\mathbf{y}_1, ..., \mathbf{y}_n$*, where as usual* $E_\mathbf{y}$ *denotes a very small region of hypervolume* $\varepsilon = |E_\mathbf{y}|$ *surrounding* $\mathbf{y}$*. If* $Y = \{\mathbf{y}_1, ..., \mathbf{y}_n\}$ *with* $|Y| = n$ *then it can be shown that [70, p. 149, Prop. 9]:*

$$\frac{\delta \phi}{\delta Y}(\emptyset) = \lim_{\varepsilon \searrow 0} \frac{\sum_{W \subseteq Y} (-1)^{|Y-W|} \; \phi_\varepsilon(W)}{\varepsilon^{|Y|}} \tag{11.240}$$

where the numerator is a Möbius transform of the set function $\phi_\varepsilon(S)$ *as defined in (4.86). That is, the set derivative is a generalization of the Möbius transform from finite to continuously infinite spaces.*

11.5 KEY MULTIOBJECT-CALCULUS FORMULAS

The undergraduate fundamental theorem of calculus states that

$$\int_a^y \frac{df}{dy}(w)dw = f(y) - f(a) \tag{11.241}$$

$$\frac{d}{dy} \int_a^y f(w)dw = f(y). \tag{11.242}$$

That is, differentiation and integration are essentially inverse operations. Another basic result of integration theory, the Radon-Nikodým theorem, indicates that, for any random vector $\mathbf{Y}$, there is an essentially unique function $f_\mathbf{Y}(\mathbf{y})$ such that, for all S,

$$\int_S f_\mathbf{Y}(\mathbf{y})d\mathbf{y} = \Pr(\mathbf{Y} \in S). \tag{11.243}$$

The density function $f_\mathbf{Y}(\mathbf{y})$ is the probability density function of $\mathbf{Y}$.

My purpose in this section is to present the corresponding results for multi-object calculus:

- The fundamental theorem of multiobject calculus, (11.244)-(11.247);

- Radon-Nikodým theorems for multiobject calculus, (11.248) and (11.250).

Taken together, these equations show that the following are *equivalent statistical descriptors* of a random finite set Ψ:

- The belief-mass function $\beta_\Psi(S)$;
- The probability density function $f_\Psi(Y)$;
- The probability-generating functional $G_\Psi[h]$.

This fact will, in turn, allow us to carry out the systematic formal modeling strategy outlined in Section 9.3.5. That is, in Chapter 12 we will be able to:

- Construct a formal measurement model in terms of a random finite set Σ;
- Determine a formula for the corresponding belief-mass function $\beta_\Sigma(S)$;
- From it construct the true multitarget likelihood function $f_\Sigma(Z|X)$ corresponding to the model.

In Chapter 13 we will be able to do the same thing for multitarget motion models and multitarget Markov densities.

11.5.1 Fundamental Theorem of Multiobject Calculus

This is [70, pp. 159-161]

$$\phi(S) = \int_S \frac{\delta\phi}{\delta Y}(\emptyset)\delta Y \tag{11.244}$$

$$\left[\frac{\delta}{\delta Y}\int_S f(W)\delta W\right]_{S=\emptyset} = f(Y). \tag{11.245}$$

The set integral and set derivative are inverse operations. The left-hand side of (11.245) means that the set derivative $\delta/\delta Y$ is applied to the set function $\phi(S) = \int_S f(W)\delta W$, thus obtaining another set function $(\delta\phi/\delta Y)(S)$. Then we set $S = \emptyset$ to get $(\delta\phi/\delta Y)(\emptyset)$.

The fundamental theorem can also be expressed in terms of functionals:

$$F[h] = \int h^Y \cdot \frac{\delta F}{\delta Y}[0]\delta Y \tag{11.246}$$

$$\left[\frac{\delta}{\delta Y}\int h^W \cdot f(W)\delta W\right]_{h=0} = f(Y). \tag{11.247}$$

The left-hand side of (11.247) is interpreted as follows. The functional derivative $\delta/\delta Y$ is applied to the functional $F[h] = \int h^W \cdot f(W)\delta W$, thus obtaining another functional $(\delta F/\delta Y)[h]$. Then we set $h = 0$ to get $(\delta F/\delta Y)[0]$.

11.5.2 Radon-Nikodým Theorems

The first Radon-Nikodým theorem is [70, pp. 159-161]:

$$\int_S \frac{\delta \beta_\Psi}{\delta Y}(\emptyset)\delta Y = \Pr(\Psi \subseteq S) = \beta_\Psi(S). \tag{11.248}$$

That is, the probability density function for a random finite set Ψ can be constructed by taking set derivatives of the belief-mass function of Ψ:

$$\beta_\Psi(S) = \int_S f_\Psi(Y)\delta Y, \quad \text{where} \quad f_\Psi(Y) = \frac{\delta \beta_\Psi}{\delta Y}(\emptyset). \tag{11.249}$$

Equation (11.249) can be generalized to the set derivatives of $\beta_\Psi(S)$ as follows [139, p. 252]:

$$\frac{\delta \beta_\Psi}{\delta Y}(S) = \int_S f_\Psi(Y \cup W)\delta W \tag{11.250}$$

and further generalized to p.g.fl.s,

$$\frac{\delta G_\Psi}{\delta Y}[h] = \int h^W \cdot f_\Psi(Y \cup W)\delta W. \tag{11.251}$$

11.5.3 Fundamental Convolution Formula

Let $\Psi = \Psi_1 \cup ... \cup \Psi_n$ where $\Psi_1, ..., \Psi_n$ are statistically independent random finite subsets. Thus $\beta_\Psi(S) = \beta_{\Psi_1}(S) \cdots \beta_{\Psi_n}(S)$. The probability density of Ψ is related to the probability densities of $\Psi_1, ..., \Psi_n$ as follows:

$$f_\Psi(Y) = \sum_{W_1 \uplus ... \uplus W_n = Y} f_{\Psi_1}(W_1) \cdots f_{\Psi_n}(W_n) \tag{11.252}$$

where the summation is taken over all mutually disjoint subsets $W_1, ..., W_n$ of Y such that $W_1 \cup ... \cup W_n = Y$. If $n = 2$, this reduces to a formula that resembles convolution

$$f_\Psi(Y) = \sum_{W \subseteq Y} f_{\Psi_1}(W) \cdot f_{\Psi_2}(Y - W). \tag{11.253}$$

Equation (11.252) is a direct consequence of the general product rule for the set derivative, (11.273) of Section 11.6:

$$\frac{\delta \beta_\Psi}{\delta Y}(S) = \sum_{W_1 \uplus \ldots \uplus W_n = Y} \frac{\delta \beta_{\Psi_1}}{\delta W_1}(S) \cdots \frac{\delta \beta_{\Psi_n}}{\delta W_n}(S). \tag{11.254}$$

Substituting $S = \emptyset$, (11.249) tells us that, as claimed,

$$f_\Psi(Y) = \frac{\delta \beta_\Psi}{\delta Y}(\emptyset) \tag{11.255}$$

$$= \sum_{W_1 \uplus \ldots \uplus W_n = Y} \frac{\delta \beta_{\Psi_1}}{\delta W_1}(\emptyset) \cdots \frac{\delta \beta_{\Psi_n}}{\delta W_n}(\emptyset) \tag{11.256}$$

$$= \sum_{W_1 \uplus \ldots \uplus W_n = Y} f_{\Psi_1}(W_1) \cdots f_{\Psi_n}(W_n). \tag{11.257}$$

11.6 BASIC DIFFERENTIATION RULES

One of the features of undergraduate calculus that makes it useful for practitioners is the fact that much of it can be reduced to simple formulas such as the product rule and the chain rule. Because of these rules, there is usually no need to bother with the formal limit-based definitions of the derivative or integral.

The purpose of this section is to list similar procedural rules for multiobject derivatives. For the sake of clarity and completeness I will express these separately for set derivatives and for functional derivatives (even though the rules for the former are just special cases of the corresponding rules for the latter). These rules are:

- Constant rule;
- Linear rule;
- Monomial rule;
- Power rule;
- Sum rule;

- Product rule;
- General product rule (two factors);
- General product rule (n factors);
- First chain rule;
- Second chain rule;
- Third chain rule;
- Fourth chain rule.
- *Constant rule:* Let $\phi(S) = K$ be a constant set function and $F[h] = K$ a constant functional. Then

   ▷ *Set derivative version:*

$$\frac{\delta}{\delta Y} K = 0. \tag{11.258}$$

   ▷ *Functional derivative version:*

$$\frac{\delta}{\delta Y} K = 0. \tag{11.259}$$

- *Linear rule:* Let $f(\mathbf{y})$ be a real-valued function of $\mathbf{y}$. Define the additive set function $p_f(S) = \int_S f(\mathbf{y})d\mathbf{y}$ for all S and the linear functional $f[h] = \int h(\mathbf{y})f(\mathbf{y})d\mathbf{y}$ for all h. Then

   ▷ *Set derivative version:*

$$\frac{\delta}{\delta Y} p_f(S) = \begin{cases} p_f(S) & \text{if} \quad Y = \emptyset \\ f(\mathbf{y}) & \text{if} \quad Y = \{\mathbf{y}\} \\ 0 & \text{if} \quad |Y| \geq 2 \end{cases}. \tag{11.260}$$

   ▷ *Functional derivative version:*

$$\frac{\delta}{\delta Y} f[h] = \begin{cases} f[h] & \text{if} \quad Y = \emptyset \\ f(\mathbf{y}) & \text{if} \quad Y = \{\mathbf{y}\} \\ 0 & \text{if} \quad |Y| \geq 2 \end{cases}. \tag{11.261}$$

- *Monomial rule:* Let $p_f(S)$ and $f[h]$ be as in the linear rule. Then if $|Y| = n$ and N is a nonnegative integer,

▷ *Set derivative version:*

$$\frac{\delta}{\delta Y} p_f(S)^N = \begin{cases} p_f(S)^N & \text{if} \quad Y = \emptyset \\ n! C_{N,n} \cdot p_f(S)^{N-n} \cdot f(\mathbf{y}_1) \cdots f(\mathbf{y}_n) & \text{if} \quad \begin{cases} Y = \{\mathbf{y}_1, ..., \mathbf{y}_n\} \\ |Y| = n \leq N \end{cases} \\ 0 & \text{if} \quad |Y| > N \end{cases}. \tag{11.262}$$

▷ *Functional derivative version:*

$$\frac{\delta}{\delta Y} f[h]^N = \begin{cases} f[h]^N & \text{if} \quad Y = \emptyset \\ n! C_{N,n} \cdot f[h]^{N-n} \cdot f(\mathbf{y}_1) \cdots f(\mathbf{y}_n) & \text{if} \quad \begin{cases} Y = \{\mathbf{y}_1, ..., \mathbf{y}_n\} \\ |Y| = n \leq N \end{cases} \\ 0 & \text{if} \quad |Y| > N \end{cases}. \tag{11.263}$$

- *Power rule:* Let $\phi(S)$ be a set function and let $F[h]$ be a functional. Then

▷ *Set derivative version:*

$$\frac{\delta}{\delta \mathbf{y}} \phi(S)^N = N \cdot \phi(S)^{N-1} \cdot \frac{\delta \phi}{\delta \mathbf{y}}(S). \tag{11.264}$$

▷ *Functional derivative version:*

$$\frac{\delta}{\delta \mathbf{y}} F[h]^N = N \cdot F[h]^{N-1} \cdot \frac{\delta F}{\delta \mathbf{y}}[h]. \tag{11.265}$$

- *Sum rule* [70, p. 151]: Let $\phi_1(S), \phi_2(S)$ be set functions, let $F_1[h], F_2[h]$ be functionals, and let a_1, a_2 be real numbers. Then

▷ *Set derivative version:*

$$\frac{\delta}{\delta Y} (a_1 \phi_1(S) + a_2 \phi_2(S)) = a_1 \frac{\delta \phi_1}{\delta Y}(S) + a_2 \frac{\delta \phi_2}{\delta Y}(S). \tag{11.266}$$

▷ *Functional derivative version:*

$$\frac{\delta}{\delta Y} (a_1 F_1[h] + a_2 F_2[h]) = a_1 \frac{\delta F_1}{\delta Y}[h] + a_2 \frac{\delta F_2}{\delta Y}[h]. \tag{11.267}$$

- *Product rule:* Let $\phi_1(S), \phi_2(S)$ be set functions and let $F_1[h], F_2[h]$ be functionals. Then

▷ *Set derivative version:*

$$\frac{\delta}{\delta \mathbf{y}} (\phi_1(S) \cdot \phi_2(S)) = \frac{\delta \phi_1}{\delta \mathbf{y}}(S) \cdot \phi_2(S) + \phi_1(S) \cdot \frac{\delta \phi_2}{\delta \mathbf{y}}(S). \tag{11.268}$$

▷ *Functional derivative version:*

$$\frac{\delta}{\delta \mathbf{y}}\left(F_1[h]\cdot F_2[h]\right)=\frac{\delta F_1}{\delta \mathbf{y}}[h]\cdot F_2[h]+F_1[h]\cdot\frac{\delta F_2}{\delta \mathbf{y}}[h]. \tag{11.269}$$

- *General product rule, two factors* [70, p. 151]: Let $\phi_1(S), \phi_2(S)$ be set functions and let $F_1[h], F_2[h]$ be functionals. Then

▷ *Set derivative version:*

$$\frac{\delta}{\delta Y}\left(\phi_1(S)\cdot\phi_2(S)\right)=\sum_{W\subseteq Y}\frac{\delta\phi_1}{\delta W}(S)\cdot\frac{\delta\phi_2}{\delta(Y-W)}(S). \tag{11.270}$$

▷ *Functional derivative version:*

$$\frac{\delta}{\delta Y}\left(F_1[h]\cdot F_2[h]\right)=\sum_{W\subseteq Y}\frac{\delta F_1}{\delta W}[h]\cdot\frac{\delta F_2}{\delta(Y-W)}[h]. \tag{11.271}$$

Remark 18 *This generalizes the general product rule for iterated derivatives of ordinary functions* $f_1(y), f_2(y)$ *[192, p. 137],*

$$\frac{d^n}{dy^n}\left(f_1(y)\cdot f_2(y)\right)=\sum_{j=0}^{n}C_{n,j}\cdot\frac{d^j f_1}{dy^j}(y)\cdot\frac{d^{n-j}f_2}{dy^{n-j}}(y). \tag{11.272}$$

- *General product rule, n factors* [70, p. 151]: Let $\phi_1(S),...,\phi_n(S)$ be set functions and let $F_1[h],...,F_n[h]$ be functionals. Then

▷ *Set derivative version:*

$$\frac{\delta}{\delta Y}\left(\phi_1(S)\cdots\phi_n(S)\right)=\sum_{W_1\uplus\ldots\uplus W_n=Y}\frac{\delta\phi_1}{\delta W_1}(S)\cdots\frac{\delta\phi_n}{\delta W_n}(S). \tag{11.273}$$

▷ *Functional derivative version:*

$$\frac{\delta}{\delta Y}\left(F_1[h]\cdots F_n[h]\right)=\sum_{W_1\uplus\ldots\uplus W_n=Y}\frac{\delta F_1}{\delta W_1}[h]\cdots\frac{\delta F_n}{\delta W_n}[h]. \tag{11.274}$$

Here the summations are taken over all mutually disjoint subsets $W_1,...,W_n$ of Z whose union is Z.

- *First chain rule:* Let $f(y_1,...,y_n)$ and $f(y)$ be real-valued functions of the real variables $y_1,...,y_n,y$. Let $\phi_1(S),...,\phi_n(S),\phi(S)$ be set functions and let $F_1[h],...,F_n[h],F[h]$ be functionals. Then

  $\triangleright$ *Set derivative version:*

$$\frac{\delta}{\delta \mathbf{y}} f(\phi_1(S),...,\phi_n(S)) \tag{11.275}$$

$$= \sum_{j=1}^{n} \frac{\partial f}{\partial y_j}(\phi_1(S),...,\phi_n(S)) \cdot \frac{\delta \phi_j}{\delta \mathbf{y}}(S). \tag{11.276}$$

$$\frac{\delta}{\delta \mathbf{y}} f(\phi(S)) = \frac{df}{dy}(\phi(S)) \cdot \frac{\delta \phi}{\delta \mathbf{y}}(S). \tag{11.277}$$

  $\triangleright$ *Functional derivative version:*

$$\frac{\delta}{\delta \mathbf{y}} f(F_1[h],...,F_n[h]) \tag{11.278}$$

$$= \sum_{j=1}^{n} \frac{\partial f}{\partial y_j}(F_1[h],...,F_n[h]) \cdot \frac{\delta F_j}{\delta \mathbf{y}}[h]. \tag{11.279}$$

$$\frac{\delta}{\delta \mathbf{y}} f(F[h]) = \frac{df}{dy}(F[h]) \cdot \frac{\delta F}{\delta \mathbf{y}}[h]. \tag{11.280}$$

- *Second chain rule:* (This generalizes Example 66.) Let $T : \mathfrak{Y}_0 \to \mathfrak{Y}_0$ be a nonsingular vector transformation. For any test function $h(\mathbf{y})$, define $(T^{-1}h)(\mathbf{y}) \triangleq h(T(\mathbf{y}))$ for all $\mathbf{y}$. Also define $T^{-1}S \triangleq \{T^{-1}\mathbf{y} | \ \mathbf{y} \in S\}$ and let $\phi(S)$ be a set function and $F[h]$ a functional. Let $J_T^Y \triangleq \prod_{\mathbf{y} \in Y} J_T(\mathbf{y})$ where $J_T(\mathbf{y})$ is the Jacobian determinant of T. Then it is left to the reader as Exercise 47 to show that

  $\triangleright$ *Set derivative version:*

$$\frac{\delta}{\delta Y} \phi(T^{-1}S) = \frac{1}{J_T^Y} \cdot \frac{\delta \phi}{\delta T^{-1}Y}(T^{-1}S). \tag{11.281}$$

  $\triangleright$ *Functional derivative version:*

$$\frac{\delta}{\delta Y} F[T^{-1}h] = \frac{1}{J_T^Y} \cdot \frac{\delta F}{\delta T^{-1}Y}[T^{-1}h]. \tag{11.282}$$

- *Third chain rule:* Let $s(y)$ be a real-valued function of the real variable y and let $F[h]$ be a functional. Define $s_h(\mathbf{y}) \triangleq s(h(\mathbf{y}))$ for all $\mathbf{y}$. The proof of the following result is left to the reader as Exercise 41:

▷ *Functional derivative version only:*

$$\frac{\delta}{\delta \mathbf{y}} F[s_h] = \frac{\delta F}{\delta \mathbf{y}}[s_h] \cdot \frac{ds}{dy}(h(\mathbf{y})). \tag{11.283}$$

- *Fourth chain rule:* The second and third chain rules are actually special cases of this more general chain rule (see Exercises 47 and 48). Let $T[h]$ be a functional transformation (i.e., it transforms test functions h to test functions $T[h]$). Define the functional derivative $(\delta T/\delta \mathbf{y})[h]$ of $T[h]$ to be the pointwise functional derivative. That is, for all $\mathbf{y}$ and $\mathbf{w}$, define

$$\frac{\delta T}{\delta \mathbf{y}}[h](\mathbf{w}) \triangleq \frac{\delta}{\delta \mathbf{y}} T[h](\mathbf{w}) = \lim_{\varepsilon \searrow 0} \frac{T[h + \varepsilon \delta_{\mathbf{y}}](\mathbf{w}) - T[h](\mathbf{w})}{\varepsilon}. \tag{11.284}$$

Let $F[h]$ be a functional. Then it is left to the reader as Exercise 45 to show that:

▷ *Functional derivative version only:*

$$\frac{\delta}{\delta \mathbf{y}} F[T[h]] = \int \frac{\delta T}{\delta \mathbf{y}}[h](\mathbf{w}) \cdot \frac{\delta F}{\delta \mathbf{w}}[T[h]] d\mathbf{w}. \tag{11.285}$$

Example 67 (Two "Twinkling" Vectors, Revisited) *In Example 55 of Section 11.2 we considered* $\Psi = \Psi_1 \cup \Psi_2$ *where* Ψ_1, Ψ_2 *are statistically independent and where*

$$\Psi_1 = \emptyset^{p_1} \cap \{\mathbf{Y}_1\}, \qquad \Psi_2 = \emptyset^{p_2} \cap \{\mathbf{Y}_2\}. \tag{11.286}$$

Using indirect methods, we derived the probability density function $f_\Psi(Y)$ *of* Ψ *in (11.76). In this example, we derive* $f_\Psi(Y)$ *again using the constructive methods introduced in this chapter. The belief-mass-function of* Ψ *is*

$$\begin{aligned} \beta_\Psi(S) &= \Pr(\Psi \subseteq S) = \Pr(\Psi_1 \subseteq S, \Psi_2 \subseteq S) && (11.287) \\ &= \Pr(\Psi_1 \subseteq S) \cdot \Pr(\Psi_2 \subseteq S) && (11.288) \\ &= \beta_{\Psi_1}(S) \cdot \beta_{\Psi_2}(S) && (11.289) \end{aligned}$$

where, for $i = 1, 2$,

$$\Pr(\Psi_i \subseteq S) = \Pr(\Psi_i = \emptyset) + \Pr(\Psi_i \subseteq S, \Psi_i \neq \emptyset) \tag{11.290}$$

$$\begin{aligned} &= \Pr(\Psi_i = \emptyset) + \Pr(\Psi_i \neq \emptyset) \cdot \Pr(\Psi_i \subseteq S | \Psi_i \neq \emptyset) && (11.291) \\ &= 1 - p_i + p_i \cdot p_{\mathbf{Y}_i}(S). && (11.292) \end{aligned}$$

Thus

$$\beta_\Psi(S) = (1 - p_1 + p_1 p_{\mathbf{Y}_1}(S)) \cdot (1 - p_2 + p_2 p_{\mathbf{Y}_2}(S)). \tag{11.293}$$

By (11.249), we are to compute $f_\Psi(Y) = (\delta\beta_\Psi/\delta Y)(\emptyset)$. *First we compute the set derivatives* $(\delta\beta_\Psi/\delta Y)(S)$. *By (11.231),* $(\delta\beta_\Psi/\delta\emptyset)(S) = \beta_\Psi(S)$. *The first set derivative of* β_Ψ *is, using the product rule of (11.268),*

$$\frac{\delta\beta_\Psi}{\delta\mathbf{y}_1}(S) \tag{11.294}$$

$$= \frac{\delta}{\delta\mathbf{y}_1}((1 - p_1 + p_1 p_{\mathbf{Y}_1}(S)) \cdot (1 - p_2 + p_2 p_{\mathbf{Y}_2}(S))) \tag{11.295}$$

$$= \left(\frac{\delta}{\delta\mathbf{y}_1}(1 - p_1 + p_1 p_{\mathbf{Y}_1}(S))\right)(1 - p_2 + p_2 p_{\mathbf{Y}_2}(S)) \tag{11.296}$$

$$+ (1 - p_1 + p_1 p_{\mathbf{Y}_1}(S)) \tag{11.297}$$

$$\cdot \left(\frac{\delta}{\delta\mathbf{y}_1}(1 - p_2 + p_2 p_{\mathbf{Y}_2}(S))\right). \tag{11.298}$$

However, by the sum rule (11.266), constant rule (11.258), and linear rule (11.260),

$$\frac{\delta}{\delta\mathbf{y}_1}(1 - p_i + p_i p_{\mathbf{Y}_i}(S)) = p_i \cdot \frac{\delta}{\delta\mathbf{y}_1} p_{\mathbf{Y}_i}(S) = p_i f_{\mathbf{Y}_i}(\mathbf{y}_1) \tag{11.299}$$

for $i = 1, 2$. *So,*

$$\frac{\delta\beta_\Psi}{\delta\mathbf{y}_1}(S) = p_1 \cdot f_{\mathbf{Y}_1}(\mathbf{y}_1) \cdot (1 - p_2 + p_2 p_{\mathbf{Y}_2}(S)) \tag{11.300}$$

$$+ (1 - p_1 + p_1 \cdot p_{\mathbf{Y}_1}(S)) \cdot p_2 f_{\mathbf{Y}2}(\mathbf{y}_1). \tag{11.301}$$

The second functional derivative is, applying the sum, constant, and linear rules,

$$\frac{\delta^2 \beta_\Psi}{\delta \mathbf{y}_2 \delta \mathbf{y}_1}(S) = \frac{\delta}{\delta \mathbf{y}2} \frac{\delta \beta_\Psi}{\delta \mathbf{y}_1}(S) \tag{11.302}$$

$$= \frac{\delta}{\delta \mathbf{y}_2} \left(\begin{array}{c} p_1 \cdot f_{\mathbf{Y}_1}(\mathbf{y}_1) \cdot (1 - p_2 + p_2 p_{\mathbf{Y}_2}(S)) \\ + (1 - p_1 + p_1 \cdot p_{\mathbf{Y}_1}(S)) \cdot p_2 f_{\mathbf{Y}2}(\mathbf{y}_1) \end{array} \right) \tag{11.303}$$

$$= p_1 f_{\Psi_1}(\mathbf{y}_1) \cdot \frac{\delta}{\delta \mathbf{y}_2} (1 - p_2 + p_2 p_{\mathbf{Y}_2}(S)) \tag{11.304}$$

$$+ \left(\frac{\delta}{\delta \mathbf{y}_2} (1 - p_1 + p_1 \cdot p_{\mathbf{Y}_1}(S)) \right) \cdot p_2 f_{\mathbf{Y}_2}(\mathbf{y}_1) \tag{11.305}$$

$$= p_1 f_{\mathbf{Y}_1}(\mathbf{y}_1) \cdot p_2 \frac{\delta}{\delta \mathbf{y}_2} p_{\mathbf{Y}_2}(S) \tag{11.306}$$

$$+ p_1 \cdot \frac{\delta}{\delta \mathbf{y}_2} p_{\mathbf{Y}_1}(S) \cdot p_2 f_{\mathbf{Y}2}(\mathbf{y}_1) \tag{11.307}$$

$$= p_1 f_{\mathbf{Y}_1}(\mathbf{y}_1) \cdot p_2 f_{\mathbf{Y}_2}(\mathbf{y}_2) \tag{11.308}$$

$$+ p_1 \cdot f_{\mathbf{Y}_1}(\mathbf{y}_2) \cdot p_2 f_{\mathbf{Y}_2}(\mathbf{y}_1). \tag{11.309}$$

The final expression no longer has any functional dependence upon the free variable S. *Thus all higher set derivatives will vanish identically. To summarize,*

$$\frac{\delta \beta_\Psi}{\delta \emptyset}(S) = (1 - p_1 + p_1 p_{\mathbf{Y}_1}(S)) \tag{11.310}$$

$$\cdot (1 - p_2 + p_2 p_{\mathbf{Y}_2}(S)) \tag{11.311}$$

$$\frac{\delta \beta_\Psi}{\delta \mathbf{y}_1}(S) = p_1 \cdot f_{\Psi_1}(\mathbf{y}_1) \cdot (1 - p_2 + p_2 p_{\mathbf{Y}_2}(S)) \tag{11.312}$$

$$+ (1 - p_1 + p_1 \cdot p_{\mathbf{Y}_1}(S)) \cdot p_2 f_{\Psi 2}(\mathbf{y}_1) \tag{11.313}$$

$$\frac{\delta^2 \beta_\Psi}{\delta \mathbf{y}_2 \delta \mathbf{y}_1}(S) = p_1 f_{\mathbf{Y}_1}(\mathbf{y}_1) \cdot p_2 f_{\mathbf{Y}_2}(\mathbf{y}_2) \tag{11.314}$$

$$+ p_1 \cdot f_{\mathbf{Y}_1}(\mathbf{y}_2) \cdot p_2 f_{\mathbf{Y}_2}(\mathbf{y}_1) \tag{11.315}$$

$$\frac{\delta^n \beta_\Psi}{\delta \mathbf{y}_n \cdots \delta \mathbf{y}_1}(S) = 0 \quad \textit{for} \quad n \geq 3. \tag{11.316}$$

Therefore, substituting $S = \emptyset$ *and using the fact that* $p_{\mathbf{Y}_i}(\emptyset) = 0$,

$$\frac{\delta \beta_\Psi}{\delta \emptyset}(\emptyset) = (1 - p_1) \cdot (1 - p_2) \quad (11.317)$$

$$\frac{\delta \beta_\Psi}{\delta \mathbf{y}_1}(\emptyset) = p_1 (1 - p_2) f_{\mathbf{Y}_1}(\mathbf{y}_1) \quad (11.318)$$

$$+ (1 - p_1) p_2 f_{\mathbf{Y}_2}(\mathbf{y}_1) \quad (11.319)$$

$$\frac{\delta^2 \beta_\Psi}{\delta \mathbf{y}_2 \delta \mathbf{y}_1}(\emptyset) = p_1 p_2 f_{\mathbf{Y}_1}(\mathbf{y}_1) f_{\mathbf{Y}_2}(\mathbf{y}_2) \quad (11.320)$$

$$+ p_1 p_2 \cdot f_{\mathbf{Y}_1}(\mathbf{y}_2) f_{\mathbf{Y}2}(\mathbf{y}_1) \quad (11.321)$$

$$\frac{\delta^n \beta_\Psi}{\delta \mathbf{y}_n \cdots \delta \mathbf{y}_1}(\emptyset) = 0 \quad \text{for} \quad n \geq 3. \quad (11.322)$$

Comparing these four equations to (11.76) of Example 55 we verify that

$$f_\Psi(\emptyset) = \frac{\delta \beta_\Psi}{\delta \emptyset}(\emptyset) \quad (11.323)$$

$$f_\Psi(\{\mathbf{y}_1\}) = \frac{\delta \beta_\Psi}{\delta \mathbf{y}_1}(\emptyset) \quad (11.324)$$

$$f_\Psi(\{\mathbf{y}_1, \mathbf{y}_2\}) = \frac{\delta^2 \beta_\Psi}{\delta \mathbf{y}_2 \delta \mathbf{y}_1}(\emptyset) \quad (11.325)$$

$$f_\Psi(\{\mathbf{y}_1, ..., \mathbf{y}_n\}) = \frac{\delta^n \beta_\Psi}{\delta \mathbf{y}_n \cdots \delta \mathbf{y}_1}(\emptyset) \quad \text{for} \quad n \geq 3. \quad (11.326)$$

That is, $f_\Psi(Y)$ *computed using set derivatives leads to the same formulas for* $f_\Psi(Y)$ *as were previously computed using indirect methods.*

For the convenience of the reader, the basic rules for functional derivatives are listed in Table 11.2.

11.7 CHAPTER EXERCISES

Exercise 32 *Let* $\mathbf{Y}$ *be a random vector with distribution* $f_\mathbf{Y}(\mathbf{y}) = \delta_{\mathbf{y}_0}(\mathbf{y})$. *Show that the p.g.fl. of* $\Psi = \{\mathbf{Y}\}$ *is* $G_\Psi[h] = h(\mathbf{y}_0)$.

Exercise 33 *Let* $\mathbf{w}_1, \mathbf{w}_2 \in Y_0$ *be fixed and let* $\Psi = \{\mathbf{Y}_1, \mathbf{Y}_2\}$ *where* $\mathbf{Y}_1, \mathbf{Y}_2$ *are independent random vectors with respective distributions* $f_{\mathbf{Y}_1}(\mathbf{y}) = \delta_{\mathbf{w}_1}(\mathbf{y})$

Table 11.2

Basic Rules for Functional Derivatives

Constant	$\frac{\delta}{\delta Y} K = 0$
Linear	$\frac{\delta}{\delta Y} f[h] = \begin{cases} f[h] & \text{if} \quad Y = \emptyset \\ f(\mathbf{y}) & \text{if} \quad Y = \{\mathbf{y}\} \\ 0 & \text{if} \quad \lvert Y \rvert \geq 2 \end{cases}$
Monomial	$\frac{\delta}{\delta Y} F[h]^N = \begin{cases} F[h]^N & \text{if} \quad Y = \emptyset \\ n! C_{N,n} \cdot F[h]^{N-n} \cdot f(\mathbf{y}_1) \cdots f(\mathbf{y}_n) & \text{if} \quad Y = \{\mathbf{y}_1, ..., \mathbf{y}_n\} \\ 0 & \text{if} \quad \lvert Y \rvert > N \end{cases}$
Power	$\frac{\delta}{\delta \mathbf{y}} F[h]^N = N \cdot F[h]^{N-1} \cdot \frac{\delta F}{\delta \mathbf{y}}[h]$
Sum	$\frac{\delta}{\delta Y} (a_1 F_1[h] + a_2 F_2[h]) = a_1 \frac{\delta F_1}{\delta Y}[h] + a_2 \frac{\delta F_2}{\delta Y}[h]$
Product	$\frac{\delta}{\delta \mathbf{y}} (F_1[h] \cdot F_2[h]) = \frac{\delta F_1}{\delta \mathbf{y}}[h] \cdot F_2[h] + F_1[h] \cdot \frac{\delta F_2}{\delta \mathbf{y}}[h]$
Product	$\frac{\delta}{\delta Y} (F_1[h] \cdot F_2[h]) = \sum_{W \subseteq Y} \frac{\delta F_1}{\delta W}[h] \cdot \frac{\delta F_2}{\delta (Y-W)}[h]$
Product	$\frac{\delta}{\delta Y} (F_1[h] \cdots F_n[h]) = \sum_{W_1 \uplus ... \uplus W_n = Y} \frac{\delta F_1}{\delta W_1}[h] \cdots \frac{\delta F_n}{\delta W_n}[h]$
Chain 1	$\frac{\delta}{\delta \mathbf{y}} f(F_1[h], ..., F_n[h]) = \sum_{j=1}^{n} \frac{\partial f}{\partial y_j}(F_1[h], ..., F_n[h]) \cdot \frac{\delta F_j}{\delta \mathbf{y}}[h]$
Chain 1	$\frac{\delta}{\delta \mathbf{y}} f(F[h]) = \frac{df}{dy}(F[h]) \cdot \frac{\delta F}{\delta \mathbf{y}}[h]$
Chain 2	$\frac{\delta}{\delta Y} F[T^{-1}h] = \frac{1}{J_T^Y} \cdot \frac{\delta F}{\delta T^{-1} Y}[T^{-1}h], \quad (T^{-1}h)(\mathbf{y}) \triangleq h(T(\mathbf{y}))$
Chain 3	$\frac{\delta}{\delta \mathbf{y}} F[s_h] = \frac{\delta F}{\delta \mathbf{y}}[s_h] \cdot \frac{ds}{dy}(h(\mathbf{y})), \quad s_h(\mathbf{y}) \triangleq s(h(\mathbf{y}))$
Chain 4	$\frac{\delta}{\delta \mathbf{y}} F[T[h]] = \int \frac{\delta T}{\delta \mathbf{y}}[h](\mathbf{w}) \cdot \frac{\delta F}{\delta \mathbf{w}}[T[h]] d\mathbf{w}$

and $f_{\mathbf{Y}_2}(\mathbf{y}) = \delta_{\mathbf{w}_2}(\mathbf{y})$. *Show that the probability density function of* Ψ *is*

$$f_\Psi(Y) = \begin{cases} \delta_{\mathbf{y}_1}(\mathbf{w}_1)\cdot\delta_{\mathbf{y}_2}(\mathbf{w}_2) + \delta_{\mathbf{y}_2}(\mathbf{w}_1)\cdot\delta_{\mathbf{y}_1}(\mathbf{w}_2) & \text{if } \begin{cases} Y = \{\mathbf{y}_1, \mathbf{y}_2\} \\ |Y| = 2 \end{cases} \\ 0 & \text{if } \quad |Y| \neq 2 \end{cases} . \tag{11.327}$$

Hint: Show that $G_\Psi[h] = h(\mathbf{w}_1)\cdot h(\mathbf{w}_2)$.

Exercise 34 *Show that (11.121) defines a probability density function:*

$$\int f(Y)\delta Y = 1.$$

Further show that the corresponding belief-mass function and p.g.fl. of a Poisson process are, respectively, $\beta(S) = e^{\lambda p_f(S)-\lambda}$ *and* $F[h] = e^{\lambda f[h]-\lambda}$ *where* $p_f(S) \triangleq \int_S f(\mathbf{y})d\mathbf{y}$ *and* $f[h] \triangleq \int h(\mathbf{y})f(\mathbf{y})d\mathbf{y}$.

Exercise 35 *Show that the multiobject uniform density function of (11.126) is a probability density:* $\int f(Y)\delta Y = 1$.

Exercise 36 *Verify (11.125). That is, that* $\delta_{Y'}(Y)$ *satisfies the defining property of a Dirac delta density:*

$$\int \delta_{Y'}(Y)\cdot f(Y)\delta Y = f(Y'). \tag{11.328}$$

Exercise 37 *Verify (11.164) and (11.165):*

$$N_\Psi = G'_\Psi(1), \qquad \sigma^2_\Psi = G''_\Psi(1) - N^2_\Psi + N_\Psi. \tag{11.329}$$

Exercise 38 *Verify (11.174):* $G_{Y'}[h] = h^{Y'}$.

Exercise 39 *Verify (11.185):*

$$G[h] = \frac{1}{\mathring{n}+1}\sum_{n=0}^{\mathring{n}}\left(\frac{\mathbf{1}_D[h]}{|D|}\right)^n. \tag{11.330}$$

Exercise 40 *Prove (11.166):* $G_{\Psi_1\cup\ldots\cup\Psi_s}[h] = G_{\Psi_1}[h]\cdots G_{\Psi_s}[h]$ *if* $\Psi_1, ..., \Psi_s$ *are independent.*

Exercise 41 *Prove the third chain rule, (11.283). Hint: Use (11.186) and expand* $s_{h+\varepsilon\delta_{\mathbf{y}}} \cong s_h + \varepsilon\frac{ds}{dy}(h(\mathbf{y}))\delta_{\mathbf{y}}$ *in a Taylor series to first order about* $h(\mathbf{y})$.

Exercise 42 (Convolution Formula for a Randomly Translated RFS) *Let* $\mathbf{Y}$ *be a random vector with density* $f_{\mathbf{Y}}(\mathbf{y})$, *let* Ψ *be a random finite set with density* $f_\Psi(Y)$, *and assume that* $\Psi, \mathbf{Y}$ *are independent. Define* $\Psi' = \Psi + \mathbf{Y}$ *where, for any finite set* Y *and any vector* $\mathbf{y}$,

$$Y + \mathbf{y} \triangleq \begin{cases} \emptyset & \text{if } \; Y = \emptyset \\ \{\mathbf{w} + \mathbf{y} |\; \mathbf{w} \in Y\} & \text{if } \; Y \neq \emptyset \end{cases}.$$

Show that

$$f_{\Psi'}(Y) = \int f_\Psi(Y - \mathbf{y}) \cdot f_{\mathbf{Y}}(\mathbf{y}) d\mathbf{y}. \tag{11.331}$$

Also, note that $f_{\Psi'}(\emptyset) = f_\Psi(\emptyset)$.

Exercise 43 (Probability Density of a Censored RFS) *[70, pp. 164, 165] Let* Ψ *be a random finite set and let* $T \subseteq \mathfrak{Y}_0$ *be a closed set. Then* $\Psi \cap T$ *is the random finite subset of* $\mathfrak{Y}_0$ *that results from "censoring" anything in* Ψ *that is not in* T. *Show that the probability density function of* $\Psi \cap T$ *is*

$$f_{\Psi \cap T}(Y) = \begin{cases} \frac{\delta \beta_\Psi}{\delta Y}(T^c) & \text{if } \; Y \subseteq T \\ 0 & \text{if } \; \text{otherwise} \end{cases}. \tag{11.332}$$

Exercise 44 *[70, p. 165] Given a random finite set* Ψ, *define the geometric conditional belief-mass function* $\beta_{\Psi|T}(S)$ *by*

$$\beta_{\Psi|T}(S) \triangleq \frac{\beta_\Psi(S \cap T)}{\beta_\Psi(T)}. \tag{11.333}$$

Show that the probability density function of $\beta_{\Psi|T}(S)$ *is*

$$f_{\Psi|T}(Y) = \begin{cases} \frac{f_\Psi(Y)}{\beta_\Psi(T)} & \text{if } \; Y \subseteq T \\ 0 & \text{if } \; \text{otherwise} \end{cases}. \tag{11.334}$$

Exercise 45 *Prove the fourth chain rule, (11.285). Hint: Note that to first order,*

$$T[h + \varepsilon \delta_{\mathbf{y}}] \cong T[h] + \varepsilon \cdot \frac{\delta T}{\delta \mathbf{y}}[h] \tag{11.335}$$

and therefore that

$$\frac{\delta}{\delta \mathbf{y}} F[T[h]] = \frac{\partial F}{\partial g_{h,\mathbf{y}}}[h] \quad where \quad g_{h,\mathbf{y}}(\mathbf{w}) \triangleq \frac{\delta T}{\delta \mathbf{y}}[h](\mathbf{w}). \tag{11.336}$$

Then apply (11.189).

Exercise 46 *In Exercise 45, let*

$$T[h](\mathbf{y}) = \sum_{i=0}^{e} h_i(\mathbf{y}) \cdot h(\mathbf{y})^i \tag{11.337}$$

or, abbreviated,

$$T[h] = \sum_{i=0}^{e} h_i h^i. \tag{11.338}$$

Let $F[h]$ *be a functional. Show that*

$$\frac{\delta T}{\delta \mathbf{y}}[h] = \delta_{\mathbf{y}} \sum_{i=1}^{e} i \cdot h_i(\mathbf{y}) \cdot h(\mathbf{y})^{i-1} \tag{11.339}$$

and thus that

$$\frac{\delta}{\delta \mathbf{y}} F[T[h]] = \left(\sum_{i=1}^{e} i \cdot h_i(\mathbf{y}) \cdot h(\mathbf{y})^{i-1} \right) \cdot \frac{\delta F}{\delta \mathbf{y}}[T[h]]. \tag{11.340}$$

Exercise 47 *Show that the third chain rule (11.283) follows from the fourth chain rule (11.285).*

Exercise 48 *Show that the second chain rule (11.282) follows from the fourth chain rule, (11.285).*

Chapter 12

Multitarget Likelihood Functions

12.1 INTRODUCTION TO THE CHAPTER

Because of the material introduced in Chapter 11, we are in a position to develop a systematic formal modeling approach for multisource-multitarget measurements as summarized in Sections 9.3 and 11.3.

- Suppose that, at time step $k+1$, we have succeeded in constructing a model $\Sigma_{k+1} = \Upsilon_{k+1}(X)$ of the randomly varying measurement set likely to be collected by our sensors, given that targets with state set X are present;
- The statistics of Σ_{k+1} are completely characterized by the belief-mass function $\beta_{k+1}(T|X) = \Pr(\Sigma_{k+1} \subseteq T|X)$;
- From (11.249) we know that the probability density function of Σ_{k+1} can be constructed as a set derivative

$$f_{k+1}(Z|X) \triangleq \frac{\delta \beta_{k+1}}{\delta Z}(\emptyset|X) = \left[\frac{\delta \beta_{k+1}}{\delta Z}(T|X)\right]_{T=\emptyset}. \tag{12.1}$$

From Section 11.5 we also know that $f_{k+1}(Z|X)$ and $\beta_{k+1}(T|X)$ contain exactly the same information about Σ_{k+1}. This is because $f_{k+1}(Z|X)$ and $\beta_{k+1}(T|X)$ can be derived from each other using the fundamental formulas of multiobject calculus of Section 11.5.1:

$$\beta_{k+1}(T|X) = \int_T f_{k+1}(Z|X)\delta Z. \tag{12.2}$$

In other words:

- $f_{k+1}(Z|X)$ as defined in (12.1) is the *true multitarget likelihood function* for the sensor(s);
- That is, it faithfully encapsulates the information in the measurement model $\Sigma_{k+1} = \Upsilon_{k+1}(X)$, without introducing extraneous information;
- Equation (12.1) is the *central formula of Bayesian multisensor-multitarget measurement modeling.*

Multisensor-multitarget likelihoods $f_{k+1}(Z|X)$ are unlike conventional single-sensor, single-target likelihoods $f_{k+1}(\mathbf{z}|\mathbf{x})$ in that they *comprehensively describe the sensor(s).* They incorporate not just sensor noise models, but any other modeling necessary to describe the behavior of the sensor(s). This includes models of the following:

- Probabilities of detection;
- Sensor fields of view (FoVs);
- False alarms;
- Clutter;
- Data-transmission drop-outs due to occlusions, atmospheric losses, and so on;
- Data-transmission noise, bandwidth, and so on.

Example 68 (Simple Example) *Suppose that a single Gaussian sensor with likelihood $f(z|x) = N_{\sigma^2}(z - x)$ observes two targets in one dimension, with the targets located at $x = a$ and $x = b$ with $a \neq b$. There are no missed detections or false alarms. The multitarget likelihood $f(Z|X)$ for this problem is nonzero only if $|Z| = |X| = 2$, in which case for all $z \neq z'$ and $x \neq x'$,*

$$f(\{z, z'\}|\{x, x'\}) \tag{12.3}$$

$$= N_{\sigma^2}(z - x) \cdot N_{\sigma^2}(z' - x') + N_{\sigma^2}(z' - x) \cdot N_{\sigma^2}(z - x') \tag{12.4}$$

$$= \frac{1}{2\pi\sigma^2} \left[\begin{array}{c} \exp\left(-\frac{(z-x)^2 + (z'-x')^2}{2\sigma^2} \right) \\ + \exp\left(-\frac{(z'-x)^2 + (z-x')^2}{2\sigma^2} \right) \end{array} \right]. \tag{12.5}$$

12.1.1 Summary of Major Lessons Learned

The following are other major concepts to be described in this chapter:

- The most mathematically concise and accurate representation of a multitarget state is a finite-state set (Section 12.2.1).
- Sensors also have states, and these must be taken into account in measurement models (Section 12.2.2).
- A careful statistical specification of the "standard" single-sensor, multitarget measurement model (Section 12.3.1).
- Derivation of the true multitarget likelihood function for the standard measurement model; see (12.139):

$$f_{k+1}(Z|X) = e^{\lambda} f_C(Z) \cdot f_{k+1}(\emptyset|X) \tag{12.6}$$

$$\cdot \sum_{\theta} \prod_{i:\theta(i)>0} \frac{p_D(\mathbf{x}_i) \cdot f_{k+1}(\mathbf{z}_{\theta(i)}|\mathbf{x}_i)}{(1 - p_D(\mathbf{x}_i)) \cdot \lambda c(\mathbf{z}_{\theta(i)})}; \tag{12.7}$$

- True multitarget likelihoods result when we average over all possible association hypotheses (Section 12.4).
- Generalization of the standard model to state-dependent false alarms; see (12.186):

$$f_{k+1}(Z|X) = e^{\lambda(X)} \cdot f_{C(X)}(Z) \cdot f_{k+1}(\emptyset|X) \tag{12.8}$$

$$\cdot \sum_{\theta} \prod_{i:\theta(i)>0} \frac{p_D(\mathbf{x}_i) \cdot f_{k+1}(\mathbf{z}_{\theta(i)}|\mathbf{x}_i)}{(1 - p_D(\mathbf{x}_i)) \cdot \lambda(X) \cdot c(\mathbf{z}_{\theta(i)}|X)}; \tag{12.9}$$

- Generalization of the standard model to account for transmission drop-outs; see (12.193):

$$\mathring{f}_{k+1}(Z|X) = \left\{ \begin{array}{lll} 1 - \mathring{p}_T + \mathring{p}_T \cdot f_{k+1}(\emptyset|X) & \text{if} & Z = \emptyset \\ \mathring{p}_T \cdot f_{k+1}(Z|X) & \text{if} & \text{otherwise} \end{array} \right. ; \tag{12.10}$$

- Generalization of the standard model to extended targets; see (12.208):

$$f_{k+1}(Z|\mathbf{x}) = e^{\lambda} f_C(Z) f_{k+1}(\emptyset|\mathbf{x}) \sum_{\theta} \prod_{\theta(\ell)>0} \frac{p_D^{\ell}(\mathbf{x}) \cdot f_{k+1}^{\ell}(\mathbf{z}_{\theta(\ell)}|\mathbf{x})}{(1 - p_D^{\ell}(\mathbf{x})) \cdot \lambda c(\mathbf{z}_{\theta(\ell)})}; \tag{12.11}$$

- Generalization of the standard model to unresolved targets; see (12.285):

$$f_{k+1}(Z|\mathring{X}) = \sum_{W_1 \uplus ... \uplus W_n = Z} |W_1|! \cdots |W_n|! \tag{12.12}$$

$$\cdot B_{a_1, p_D(\mathbf{x}_1)}(|W_1|) \cdots B_{a_n, p_D(\mathbf{x}_n)}(|W_n|) \tag{12.13}$$

$$\cdot \left(\prod_{\mathbf{z} \in W_1} f_{k+1}(\mathbf{z}|\mathbf{x}_1) \right) \cdots \left(\prod_{\mathbf{z} \in W_n} f_{k+1}(\mathbf{z}|\mathbf{x}_n) \right). \tag{12.14}$$

- Multitarget likelihoods for the unresolved-targets model are *continuous in target number*; see (12.290)-(12.291).
- A general formula for multisensor-multitarget true likelihood functions, given conditional independence of state; see (12.294):

$$f_{k+1}(Z|X) = \overset{1}{f}_{k+1}(\overset{1}{Z}|X) \cdots \overset{s}{f}_{k+1}(\overset{s}{Z}|X); \tag{12.15}$$

- A likelihood function for multiple bearing-only sensors due to Vihola (Section 12.10):

$$f_{k+1}(\{a\}|X) = q \cdot p_{FA} \cdot \kappa_{k+1}(a) + \frac{q\,(1 - p_{FA})}{n} \sum_{i=1}^{n} f_{k+1}(a|\mathbf{x}_i); \tag{12.16}$$

- A likelihood function for extracting the number and shapes of "soft" data-clusters in dynamically evolving data; see (12.355):

$$f(Z|Q, \mathbf{p}) \tag{12.17}$$

$$= m! \cdot p_{\mathbf{p}}(m) \tag{12.18}$$

$$\cdot \frac{\begin{array}{c}(a_1 f(\mathbf{z}_1|\mathbf{u}_1) + ... + a_\gamma f(\mathbf{z}_1|\mathbf{u}_\gamma)) \\ \cdots (a_1 f(\mathbf{z}_m|\mathbf{u}_1) + ... + a_\gamma f(\mathbf{z}_m|\mathbf{u}_\gamma))\end{array}}{(a_1 + ... + a_\gamma)^m}. \tag{12.19}$$

12.1.2 Organization of the Chapter

I begin, in Section 12.2, by carefully defining the concepts of *multitarget state space*, *multisensor state space*, and *multisensor-multitarget measurement space*.

Then, in Section 12.3, we return to the "standard" single-sensor, multitarget measurement model first introduced in Sections 9.3.3 and 10.2.1, and derive its true multitarget likelihood function. In Section 12.4, I compare my method with the conventional MHC modeling methodology described in Section 10.5.

I then turn to various generalizations of the standard model and specification of their true likelihoods: state-dependent false alarms (Section 12.5), transmission drop-outs (Section 12.6), extended targets (Section 12.7), and unresolved targets (Section 12.8).

Multisensor-multitarget measurement models and their true likelihood functions are addressed in Section 12.9.

The chapter concludes with two "nonstandard" measurement models. Both are based on finite-mixture modeling in the sense of Remark 2 of Section 3.4.5. The first (Section 12.10) models bearings-only sensors in multitarget environments. The second is designed to extract unknown numbers of "soft" data clusters from dynamically evolving measurement sets (Section 12.11). Exercises for the chapter are in Section 12.12.

12.2 MULTITARGET STATE AND MEASUREMENT SPACES

I cannot discuss the construction of multitarget measurement models and their true likelihoods without first carefully defining multitarget measurement and state spaces. In what follows I describe the following:

- Multitarget state spaces (Section 12.2.1);
- Sensor state spaces (Section 12.2.2);
- Single-sensor, multitarget measurement spaces (Section 12.2.3);
- Multisensor-multitarget measurement spaces (Section 12.2.4).

See [139, pp. 245-248] for a still more systematic treatment.

12.2.1 Multitarget State Spaces

The complete description of a multitarget system requires a *unified state representation*. Suppose that we have n targets with respective state vectors $\mathbf{x}_1, ..., \mathbf{x}_n$, where in general $\mathbf{x}_i$ has the form $\mathbf{x}_i = (\mathbf{y}, c)$ where $\mathbf{y}$ contains all of the continuous state variables and c all of the discrete ones, especially target identity/label.

The vector $(\mathbf{x}_1, ..., \mathbf{x}_n)$ has fixed dimension n, but the number n of targets can vary from zero to some arbitrarily large number. Also, the components $\mathbf{x}_1, ..., \mathbf{x}_n$ of $(\mathbf{x}_1, ..., \mathbf{x}_n)$ have fixed order, whereas the target states $\mathbf{x}_1, ..., \mathbf{x}_n$

have no inherent order. Thus the actual *multitarget state* is more accurately represented as a *finite state set*, the elements of which are the individual target states:

$$X = \{\mathbf{x}_1, ..., \mathbf{x}_n\}. \quad (12.20)$$

This *unified multitarget state representation* accounts for the fact that n is variable and that targets have no physically inherent order. The various possibilities for X are:

$$X = \emptyset \quad \text{(no targets present)} \quad (12.21)$$

$$X = \{\mathbf{x}_1\} \quad \text{(single target with state } \mathbf{x}_1 \text{ present)} \quad (12.22)$$

$$X = \{\mathbf{x}_1, \mathbf{x}_2\} \quad \text{(two targets with states } \mathbf{x}_1 \neq \mathbf{x}_2 \text{ present)} \quad (12.23)$$

$$\vdots$$

$$X = \{\mathbf{x}_1, ..., \mathbf{x}_n\} \quad (n \text{ targets, states } \mathbf{x}_1 \neq ... \neq \mathbf{x}_m) \quad (12.24)$$

$$\vdots$$

Stated differently,

- *The multitarget state space $\mathfrak{X}$ is the hyperspace of all finite subsets of the single-target state space $\mathfrak{X}_0$.*

Thus $\{\mathbf{x}_1, \mathbf{x}_2\} = \{\mathbf{x}_2, \mathbf{x}_1\}$ is a single unified state model of two targets with state vectors $\mathbf{x}_1, \mathbf{x}_2$. On the other hand, vectors $(\mathbf{x}_1, \mathbf{x}_2) \neq (\mathbf{x}_2, \mathbf{x}_1)$ do not accurately represent the physical multitarget state. This is because they do so redundantly—for example, two distinct representations $(\mathbf{x}_1, \mathbf{x}_2)$, $(\mathbf{x}_2, \mathbf{x}_1)$ of the same two target state. It is also because they cannot model the inherent permutation symmetry of multitarget states.

Multitarget states are inherently more computationally challenging than single-target states. For example, if the single-target state vector has the form

$$x = (x, y, z, v_x, v_y, v_z, a_x, a_y, a_z, c) \quad (12.25)$$

then 10 parameters (nine real numbers $x, y, z, v_x, v_y, v_z, a_x, a_y, a_z$ and one discrete value c) are required to specify any state of the target. Thus the following are true:

- $1 + 10 = 11$ parameters are required to describe a system with no more than a single target.

- $1+10+20=31$ parameters are required to describe a system with no more than two targets.
- $1+10+20+30=61$ parameters are required for a system with no more than three targets; and so on.

The multitarget state space $\mathfrak{X}$ requires, in general, an infinite number of state parameters to describe it. Furthermore, in a careful Bayesian approach the unknown state must be a random variable. Consequently:

- *The unknown state set at time step k must be a randomly varying finite subset $\Xi_{k|k}$ of $\mathfrak{X}_0$.*[1]

Remark 19 (Careful Multitarget-State Representation) *A basic principle of state modeling is that, ideally, there should be a one-to-one correspondence between physical states and their mathematical representations. No two distinct physical states should have the same state representation; and conversely, no mathematical states should represent impossible physical states. Consequently, a careful, systematic definition of multitarget state space is more involved than has been implied here (see [70, pp. 196-199]). Such a definition requires that (1) every single-target state vector $\mathbf{x}=(\mathbf{w},c)$ must have a discrete identity state parameter c as well as geokinematic state parameters $\mathbf{w}$; and that (2) no single target can have two or more kinematic states at the same time. For example, $X=\{(\mathbf{w}_1,c),(\mathbf{w}_2,c)\}$ with $\mathbf{w}_1\neq\mathbf{w}_2$ is an impossible multitarget state. The simplified state representation $\mathbf{x}_1=\mathbf{w}_1$ and $\mathbf{x}_2=\mathbf{w}_2$ typically employed in conventional multitarget tracking theory reduces notational complexity. However, it does so at the expense of permitting degenerate multitarget states. If the trajectories of two distinct targets intersect simultaneously, for example, their kinematic states will be identical at the intersection point: $\mathbf{w}_1=\mathbf{w}_2=\mathbf{w}$. In a systematic representation, by way of contrast, the multitarget state at the crossing point would be $X=\{(\mathbf{w},c_1),(\mathbf{w},c_2)\}$ with $c_1\neq c_2$.*

1 One cannot define a random variable of any kind without, typically, first defining a topology on the space of objects to be randomized and then defining random elements of that space in terms of the Borel-measurable subsets O. The space of state sets is topologized using the Mathéron "hit-or-miss" topology. Once this is done, the probability law of a finite random state set $\Xi_{k|k}$ is its probability-mass function (also known as a probability measure) $p_\Xi(O)=\Pr(\Xi\in O)$. See Appendix F for greater detail.

12.2.2 Multisensor State Spaces

That sensors themselves have states is a central fact of control-theoretic tracking applications.[2] For example, the current state of a missile-tracking camera can be described as a vector $\overset{*}{\mathbf{x}} = (\alpha, \theta, \rho)$ where α is azimuth, θ is elevation, and ρ is focal length. Alternatively, it can be described by a vector $\overset{*}{\mathbf{x}} = (e_\alpha, e_\theta, e_\rho)$ where $e_\alpha, e_\theta, e_\rho$ are voltages that, when applied to the camera's servomotors, correspond to azimuth, elevation, and focal length α, θ, ρ.

More generally, sensors are often carried on physical platforms such as airplanes and unattended aerial vehicles (UAVs), the states of which must also be characterized.

As a more specific example, assume a two-dimensional problem in which the sensor is on a platform that executes coordinated turns. That is, the body frame axis of the platform is always tangent to the platform trajectory. Then we could have

$$\overset{*}{\mathbf{x}} = (\overset{*}{x}, \overset{*}{y}, \overset{*}{v}_x, \overset{*}{v}_y, \overset{*}{\omega}, \overset{*}{\ell}, \mu, \chi) \tag{12.26}$$

where $\overset{*}{x}, \overset{*}{y}$ are position parameters, $\overset{*}{v}_x, \overset{*}{v}_y$ are velocity parameters, $\overset{*}{\omega}$ is turn radius, $\overset{*}{\ell}$ is fuel level, μ is the sensor mode, and χ is the datalink transmission channel currently used by the sensor.

When $s > 1$ sensors are in use, a state variable must be introduced to distinguish them from each other. Thus we assume that each sensor has a unique integer identifier $j = 1, ..., s$ called a *sensor tag*. Once this tag is incorporated into the state vector as a (known) state variable, the sensors will have their own respective state spaces $\overset{*1}{\mathfrak{X}}_0, ..., \overset{*s}{\mathfrak{X}}_0$ with individual sensor states denoted as $\overset{*1}{\mathbf{x}} \in \overset{*1}{\mathfrak{X}}_0, ..., \overset{*s}{\mathbf{x}} \in \overset{*s}{\mathfrak{X}}_0$. The joint state space for all sensors will be the disjoint union of the individual state spaces:

$$\overset{*}{\mathfrak{X}}_0 = \overset{*1}{\mathfrak{X}}_0 \uplus ... \uplus \overset{*s}{\mathfrak{X}}_0. \tag{12.27}$$

2 In single-sensor, single-target control theory, the sensor and target are regarded as a jointly evolving stochastic system with joint state vector $(\mathbf{x}, \overset{*}{\mathbf{x}})$. In a multisensor-multitarget sensor management problem, the sensors and targets should likewise be regarded as a single jointly evolving multiobject stochastic system. In this case, a state of the joint system has the form $\breve{X} = X \uplus \overset{*}{X} = \{\mathbf{x}_1, ..., \mathbf{x}_n, \overset{*}{\mathbf{x}}_1, ..., \overset{*}{\mathbf{x}}_{\overset{*}{n}}\}$. This fact will not further concern us in this book. See [139, pp. 244-249] for more detail.

We will write the state of a sensor with unidentified sensor tag as $\mathring{\mathbf{x}}$, so that a multisensor system will have the state

$$\mathring{X} = \{\mathring{\mathbf{x}}_1, ..., \mathring{\mathbf{x}}_{\mathring{n}}\}. \tag{12.28}$$

The space of all such multisensor states is denoted by $\mathring{\mathfrak{X}}$.

12.2.3 Single-Sensor, Multitarget Measurement Spaces

The multitarget measurement space of a single sensor is essentially the same as described in Section 9.3.1. Suppose that the sensor has collected m measurement vectors $\mathbf{z}_1, ..., \mathbf{z}_m$. As with states, a *multitarget measurement* is more accurately represented as a *finite measurement set*, the elements of which are the individual measurements:

$$Z = \{\mathbf{z}_1, ..., \mathbf{z}_m\}. \tag{12.29}$$

The various possibilities are:

$$Z = \emptyset \quad \text{(no measurements collected)} \tag{12.30}$$

$$Z = \{\mathbf{z}_1\} \quad \text{(a single measurement } \mathbf{z}_1 \text{ is collected)} \tag{12.31}$$

$$Z = \{\mathbf{z}_1, \mathbf{z}_2\} \quad \text{(two measurements } \mathbf{z}_1 \neq \mathbf{z}_2 \text{ collected)} \tag{12.32}$$

$$\vdots$$

$$Z = \{\mathbf{z}_1, ..., \mathbf{z}_m\} \quad (m \text{ measurements } \mathbf{z}_1 \neq ... \neq \mathbf{z}_m) \tag{12.33}$$

$$\vdots$$

Thus:

- *A single-sensor multitarget measurement space $\mathfrak{Z}$ is the hyperspace of all finite subsets of the underlying single-sensor measurement space $\mathfrak{Z}_0$.*

Remark 20 (Unconventional Measurements) *We have been presuming a conventional information source that supplies conventional observations. If the source is unconventional then it will be delivering something else, such as unambiguously generated ambiguous (UGA) measurements. In this case each measurement consists of a random set model Θ. Thus, as noted in Section 9.2.5, a measurement set must have the form $Z = \{\Theta_1, ..., \Theta_m\}$ rather than $Z = \{\mathbf{z}_1, ..., \mathbf{z}_m\}$. We will return to such issues in Section 14.4.2.*

12.2.4 Multisensor-Multitarget Measurement Spaces

We assume that any observation collected by a given sensor has that sensor's tag attached to it as an observation parameter. Consequently, the s sensors will have their own respective and unique measurement spaces $\overset{1}{\mathfrak{Z}}_0,...,\overset{s}{\mathfrak{Z}}_0$ with individual measurements denoted as $\overset{1}{\mathbf{z}} \in \overset{1}{\mathfrak{Z}}_0,...,\overset{s}{\mathbf{z}} \in \overset{s}{\mathfrak{Z}}_0$. The total multisensor measurement space will be the disjoint union of the individual measurement spaces:

$$\mathfrak{Z}_0 = \overset{1}{\mathfrak{Z}}_0 \uplus ... \uplus \overset{s}{\mathfrak{Z}}_0. \tag{12.34}$$

As a consequence, an observation collected by whatever sensors might be present will be a finite subset of $\mathfrak{Z}_0$ of the form

$$Z = \overset{1}{Z} \cup ... \cup \overset{s}{Z} = \{\overset{1}{\mathbf{z}}_{1,1}, ..., \overset{1}{\mathbf{z}}_{1,\overset{1}{m}}, ..., \overset{s}{\mathbf{z}}_{s,1}, ..., \overset{s}{\mathbf{z}}_{s,\overset{s}{m}}\}. \tag{12.35}$$

This indicates that the first sensor has collected $\overset{1}{m} \geq 0$ observations $\overset{1}{Z} = \{\overset{1}{\mathbf{z}}_{1,1}, ..., \overset{1}{\mathbf{z}}_{1,\overset{1}{m}}\}$; the second sensor has collected $\overset{2}{m} \geq 0$ observations $\overset{2}{Z} = \{\overset{2}{\mathbf{z}}_{2,1}, ..., \overset{2}{\mathbf{z}}_{2,\overset{2}{m}}\}$; and so on. The set of all finite subsets of $\mathfrak{Z}_0$ will be denoted by $\mathfrak{Z}$. I will denote a measurement with unidentified sensor tag as $\mathbf{z} \in \mathfrak{Z}_0$ and a finite subset of such observations as

$$Z = \{\mathbf{z}_1, ..., \mathbf{z}_m\} \tag{12.36}$$

where $m = \overset{1}{m} + ... + \overset{s}{m}$.

12.3 THE STANDARD MEASUREMENT MODEL

The standard multitarget measurement model was summarized in Sections 9.3.3 and 10.2.1. It is most commonly used to model detection-type measurements collected by surveillance and tracking radars, as explained in Section 9.2.1. In this section I consider a somewhat more general version of this model based on the following assumptions (see Figure 12.1):

- A single sensor with state $\overset{*}{\mathbf{x}}$ and likelihood function

$$f_{k+1}(\mathbf{z}|\mathbf{x}) \overset{\text{abbr.}}{=} f_{k+1}(\mathbf{z}|\mathbf{x}, \overset{*}{\mathbf{x}}) = f_{\mathbf{W}_{k+1}}(\mathbf{z} - \eta_{k+1}(\mathbf{x}, \overset{*}{\mathbf{x}})) \tag{12.37}$$

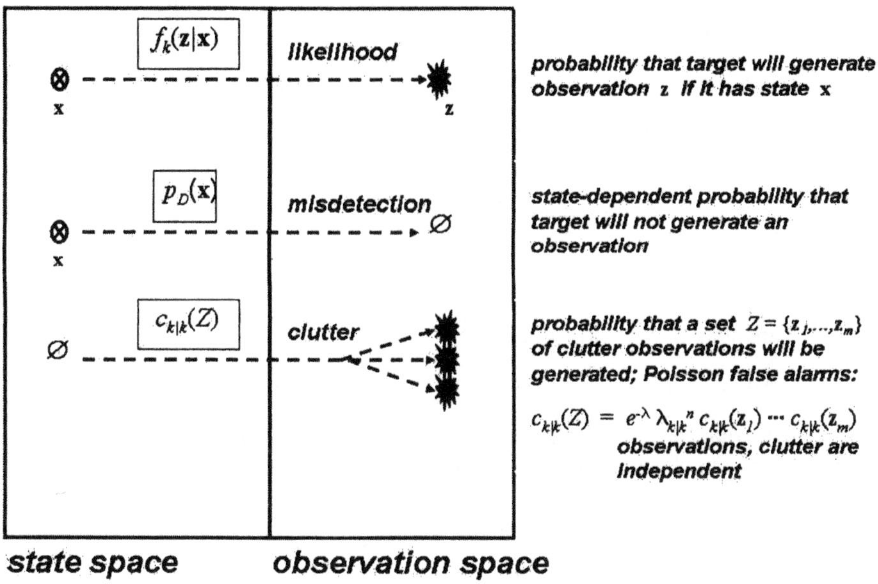

Figure 12.1 A summary of the major assumptions underlying the "standard" single-sensor, multitarget measurement model.

observes a scene involving an unknown number of unknown targets;[3]

- No measurement is generated by more than a single target.
- A single target with state $\mathbf{x}$ generates either a single measurement—a target detection, occurring with probability

$$p_D(\mathbf{x}) \overset{\text{abbr.}}{=} p_D(\mathbf{x}, \mathring{\mathbf{x}}) \tag{12.38}$$

  or a missed detection (no measurement at all), occurring with probability $1 - p_D(\mathbf{x})$.

- The false alarm process $C \overset{\text{abbr.}}{=} C(\mathring{\mathbf{x}})$ is Poisson-distributed in time with expected value $\lambda \overset{\text{abbr.}}{=} \lambda(\mathring{\mathbf{x}})$, and distributed in space according to an arbitrary density $c(\mathbf{z}) \overset{\text{abbr.}}{=} c(\mathbf{z}|\mathring{\mathbf{x}})$.
- The false alarm process C and target-measurement process

$$\Upsilon(X) \overset{\text{abbr.}}{=} \Upsilon_{k+1}(X, \mathring{\mathbf{x}}) \tag{12.39}$$

  are statistically independent.

- All measurements are conditionally independent of target state.

In what follows I derive the *true multitarget likelihood function* $f_{k+1}(Z|X)$ *for the standard model.* I show that if $Z = \emptyset$ then

$$f_{k+1}(\emptyset|X) = e^{-\lambda} \prod_{\mathbf{x} \in X} (1 - p_D(\mathbf{x})) \tag{12.40}$$

and, if $Z \neq \emptyset$, that

$$f_{k+1}(Z|X) = e^{\lambda} f_C(Z) \cdot f_{k+1}(\emptyset|X) \tag{12.41}$$

$$\cdot \sum_{\theta} \prod_{i:\theta(i)>0} \frac{p_D(\mathbf{x}_i) \cdot f_{k+1}(\mathbf{z}_{\theta(i)}|\mathbf{x}_i)}{(1 - p_D(\mathbf{x}_i)) \cdot \lambda c(\mathbf{z}_{\theta(i)})} \tag{12.42}$$

where the summation is taken over all association hypotheses θ as defined in Section 10.5.4; where $X = \{\mathbf{x}_1, ..., \mathbf{x}_n\}$ with $|X| = n$; and where $Z =$

3 We are assuming an additive measurement model for the sake of clarity. More general measurement models can also be used.

$\{\mathbf{z}_1, ..., \mathbf{z}_m\}$ with $|Z| = m$. Also,

$$f_{C(X)}(Z) = e^{-\lambda(X)} \prod_{\mathbf{z} \in Z} \lambda \cdot c(\mathbf{z}). \tag{12.43}$$

I begin in Section 12.3.1 by constructing the equations for the standard model. Then, to both build up experience with the multitarget formal modeling process and to break the problem into smaller, more intuitively accessible steps, I construct the true likelihood function for a series of increasingly more complex cases:

- Case I: No target is present (Section 12.3.2).
- Case II: No more than one target is present (Section 12.3.3).
- Case III: Arbitrary number of targets, but no missed detections or false alarms (Section 12.3.4).
- Case IV: Arbitrary number of targets with missed detections but no false alarms (Section 12.3.5).
- Case V: Arbitrary number of targets with missed detections and false alarms (Section 12.3.6).

Finally, in Section 12.3.7, I list explicit formulas for the probability-generating functional (p.g.fl.) $G_{k+1}[g|X]$ of the multitarget likelihood $f_{k+1}(Z|X)$ for each of these five cases.

12.3.1 Measurement Equation for the Standard Model

For a given predicted multitarget state $X = \{\mathbf{x}_1, ..., \mathbf{x}_n\}$, the random measurement set collected by the sensor will have the form

$$\overset{\text{measurement set}}{\Sigma_{k+1}} = \overset{\text{target detection set}}{\Upsilon(X)} \cup \overset{\text{false detections}}{C} \tag{12.44}$$

where $\Upsilon(X)$ has the form

$$\Upsilon(X) = \overset{\text{detection set for state } \mathbf{x}_1}{\Upsilon(\mathbf{x}_1)} \cup ... \cup \overset{\text{detection set for state } \mathbf{x}_n}{\Upsilon(\mathbf{x}_n)} \tag{12.45}$$

and where C is Poisson with mean value λ and spatial distribution $c(\mathbf{z})$. I assume that $\Upsilon(\mathbf{x}_1)$,...,$\Upsilon(\mathbf{x}_n)$,C are statistically independent.

Any given target can generate either a single observation or no observation at all. Consequently, $\Upsilon(\mathbf{x}_i)$ must have the form of (11.21):

$$\Upsilon(\mathbf{x}_i) = \emptyset^{p_D(\mathbf{x})} \cap \{\mathbf{Z}_i\}. \tag{12.46}$$

Here $\mathbf{Z}_i = \eta(\mathbf{x}_i, \mathbf{W}_i)$ is the sensor-noise model associated with the ith state $\mathbf{x}_i$; and $\emptyset^p$ is the discrete random subset of $\mathfrak{Z}_0$; see (11.22)

$$\Pr(\emptyset^p = T) = \begin{cases} 1-p & \text{if} \quad T = \emptyset \\ p & \text{if} \quad T = \mathfrak{Z}_0 \\ 0 & \text{if} \quad \text{otherwise} \end{cases}. \tag{12.47}$$

The fact that the false alarm process C is Poisson means that it must have the following form:

$$C = \{\mathbf{C}_1, ..., \mathbf{C}_M\} \tag{12.48}$$

where $M = |C|$ is a random nonnegative integer with probability distribution

$$p_M(m) = \frac{e^{-\lambda}\lambda^m}{m!}; \tag{12.49}$$

and where, conditioned on $M = m$, $\mathbf{C}_1, ..., \mathbf{C}_m$ are independent, identically distributed (i.i.d.) random measurement vectors with probability density $c(\mathbf{z}) = f_C(\mathbf{z})$.

12.3.2 Case I: No Target Is Present

That is, $X = \emptyset$ but Z can be arbitrary. In this case, (12.44) reduces to

$$\overset{\text{measurement set}}{\Sigma_{k+1}} = \overset{\text{false detections}}{C}. \tag{12.50}$$

I show that $f_{k+1}(\emptyset|\emptyset) = f_C(\emptyset) = e^{-\lambda}$ and that, otherwise, if $Z \neq \emptyset$,

$$f_{k+1}(Z|\emptyset) = f_C(Z) = e^{-\lambda} \prod_{\mathbf{z} \in Z} \lambda c(\mathbf{z}). \tag{12.51}$$

Begin by constructing the belief-mass function of the model. This is

$$\beta_{k+1}(T|\emptyset) = e^{\lambda p_C(T) - \lambda} \tag{12.52}$$

where $p_{\mathbf{C}}(T) = \int_T c(\mathbf{z})d\mathbf{z} \triangleq p_c(T)$. To see this, note that

$$\beta_{k+1}(T|\emptyset) = \Pr(C \subseteq T|\emptyset) \tag{12.53}$$

$$= \sum_{m=0}^{\infty} \Pr(C \subseteq T, |C| = m|\emptyset) \tag{12.54}$$

$$= \sum_{m=0}^{\infty} \Pr(|C| = m|\emptyset) \cdot \Pr(C \subseteq T|\emptyset,\ |C| = m) \tag{12.55}$$

$$= e^{-\lambda} \sum_{m=0}^{\infty} \frac{\lambda^m}{m!} \cdot \Pr(\{\mathbf{C}_1, ..., \mathbf{C}_\nu\} \subseteq T) \tag{12.56}$$

$$= e^{-\lambda} \sum_{m=0}^{\infty} \frac{\lambda^m}{m!} \cdot \Pr(\mathbf{C} \in T)^m \tag{12.57}$$

$$= e^{-\lambda} \sum_{m=0}^{\infty} \frac{\lambda^m}{m!} \cdot p_{\mathbf{C}}(T)^m \tag{12.58}$$

$$= e^{-\lambda} \cdot e^{\lambda p_{\mathbf{C}}(T)} = e^{\lambda p_{\mathbf{C}}(T) - \lambda}. \tag{12.59}$$

Therefore,

$$\frac{\delta \beta_{k+1}}{\delta Z}(T|\emptyset) = \frac{\delta}{\delta Z} e^{\lambda p_c(T) - \lambda} = \frac{\delta^m}{\delta \mathbf{z}_m \cdots \delta \mathbf{z}_1} e^{\lambda p_c(T) - \lambda} \tag{12.60}$$

$$= \frac{\delta^{m-1}}{\delta \mathbf{z}_m \cdots \delta \mathbf{z}_2} \frac{\delta}{\delta \mathbf{z}_1} e^{\lambda p_c(T) - \lambda} \tag{12.61}$$

$$= \frac{\delta^{m-1}}{\delta \mathbf{z}_m \cdots \delta \mathbf{z}_2} e^{\lambda p_c(T) - \lambda} \frac{\delta}{\delta \mathbf{z}_1} (\lambda p_c(T) - \lambda) \tag{12.62}$$

$$= \lambda \cdot c(\mathbf{z}_1) \cdot \frac{\delta^{m-2}}{\delta \mathbf{z}_m \cdots \delta \mathbf{z}_3} \frac{\delta}{\delta \mathbf{z}_2} e^{\lambda p_c(T) - \lambda} \tag{12.63}$$

$$= \lambda^2 \cdot c(\mathbf{z}_1) \cdot c(\mathbf{z}_2) \cdot \frac{\delta^{m-2}}{\delta \mathbf{z}_m \cdots \delta \mathbf{z}_3} e^{\lambda p_c(T) - \lambda} \tag{12.64}$$

$$= ... = \lambda^m \cdot c(\mathbf{z}_1) \cdots c(\mathbf{z}_m) \cdot e^{\lambda p_c(T) - \lambda}. \tag{12.65}$$

Here I have repeatedly applied the first chain rule (11.277) followed by the sum rule (11.266) and the linear rule (11.260). Setting $T = \emptyset$ and applying (12.1) I get, as claimed,

$$f_{k+1}(Z|\emptyset) = \frac{\delta \beta_{k+1}}{\delta Z}(\emptyset|\emptyset) = e^{-\lambda} \lambda^m c(\mathbf{z}_1) \cdots c(\mathbf{z}_m). \tag{12.66}$$

12.3.3 Case II: One Target Is Present

That is, $|X| = 1$ and so X must have the form $X = \{\mathbf{x}\}$ for some $\mathbf{x}$ in $\mathfrak{X}_0$. In (9.32), I listed a special case of (12.44)

$$\overset{\text{measurement set}}{\Sigma_{k+1}} = \overset{\text{target detection set}}{\Upsilon(\mathbf{x})} \cup \overset{\text{false detections}}{C}. \tag{12.67}$$

In (9.33) I stated that the true likelihood function corresponding to this model is

$$f_{k+1}(Z|\mathbf{x}) = f_C(Z) \cdot \left(1 - p_D(\mathbf{x}) + p_D(\mathbf{x}) \sum_{\mathbf{z} \in Z} \frac{f(\mathbf{z}|\mathbf{x})}{\lambda c(\mathbf{z})}\right). \tag{12.68}$$

We verify this fact. The belief-mass function for the measurement model is

$$\begin{aligned} \beta_{k+1}(T|\mathbf{x}) &= \Pr(\Sigma_{k+1} \subseteq T|\mathbf{x}) = \Pr(\Upsilon(\mathbf{x}) \subseteq T, C \subseteq T|\mathbf{x}) & (12.69) \\ &= \Pr(\Upsilon(\mathbf{x}) \subseteq T|\mathbf{x}) \cdot \Pr(C \subseteq T) & (12.70) \\ &= [1 - p_D(\mathbf{x}) + p_D(\mathbf{x}) \cdot p(T|\mathbf{x})] \cdot e^{\lambda p_c(T) - \lambda}. & (12.71) \end{aligned}$$

Setting $T = \emptyset$, we conclude on the basis of (12.1),

$$f_{k+1}(\emptyset|\mathbf{x}) = \frac{\delta \beta_{k+1}}{\delta \emptyset}(\emptyset|\mathbf{x}) = \beta_{k+1}(\emptyset|\mathbf{x}) = (1 - p_D(\mathbf{x})) \cdot e^{-\lambda}. \tag{12.72}$$

Assume now that $T \neq \emptyset$. From the general product rule, (11.270), we know that

$$\begin{aligned} \frac{\delta \beta_{k+1}}{\delta Z}(T|\mathbf{x}) &= \sum_{W \subseteq Z} \left(\frac{\delta}{\delta W}[1 - p_D(\mathbf{x}) + p_D(\mathbf{x}) \cdot p(T|\mathbf{x})]\right) & (12.73) \\ &\quad \cdot \left(\frac{\delta}{\delta(Z - W)} e^{\lambda p_c(T) - \lambda}\right). & (12.74) \end{aligned}$$

By the sum rule (11.266) and linear rule (11.260),

$$\begin{aligned} \frac{\delta}{\delta \mathbf{z}}[1 - p_D(\mathbf{x}) + p_D(\mathbf{x}) \cdot p(T|\mathbf{x})] &= p_D(\mathbf{x}) \cdot \frac{\delta}{\delta \mathbf{z}} p(T|\mathbf{x}) & (12.75) \\ &= p_D(\mathbf{x}) \cdot f_{k+1}(\mathbf{z}|\mathbf{x}). & (12.76) \end{aligned}$$

From the constant rule (11.258) we also see that the set derivatives of order two and greater vanish identically. So, the only surviving terms in the sum in (12.73) are those for which $W = \emptyset$ and $W = \{\mathbf{z}\}$. Thus

$$\frac{\delta \beta_{k+1}}{\delta Z}(T|\mathbf{x}) = \left(\frac{\delta}{\delta \emptyset}[1 - p_D(\mathbf{x}) + p_D(\mathbf{x}) \cdot p(T|\mathbf{x})]\right) \tag{12.77}$$

$$\cdot \left(\frac{\delta}{\delta Z} e^{\lambda p_c(T) - \lambda}\right) \tag{12.78}$$

$$+ \sum_{\mathbf{z} \in Z} \left(\frac{\delta}{\delta \mathbf{z}}[1 - p_D(\mathbf{x}) + p_D(\mathbf{x}) \cdot p(T|\mathbf{x})]\right) \tag{12.79}$$

$$\cdot \left(\frac{\delta}{\delta(Z - \{\mathbf{z}\})} e^{\lambda p_c(T) - \lambda}\right) \tag{12.80}$$

$$= [1 - p_D(\mathbf{x}) + p_D(\mathbf{x}) \cdot p(T|\mathbf{x})] \left(\frac{\delta}{\delta Z} e^{\lambda p_c(T) - \lambda}\right) \tag{12.81}$$

$$+ \sum_{\mathbf{z} \in Z} p_D(\mathbf{x}) \cdot f_{k+1}(\mathbf{z}|\mathbf{x}) \cdot \frac{\delta}{\delta(Z - \{\mathbf{z}\})} e^{\lambda p_c(T) - \lambda}. \tag{12.82}$$

Because of (12.51),

$$\frac{\delta \beta_{k+1}}{\delta Z}(T|\mathbf{x}) = [1 - p_D(\mathbf{x}) + p_D(\mathbf{x}) \cdot p(T|\mathbf{x})] \cdot e^{\lambda p_c(T) - \lambda} \tag{12.83}$$

$$\cdot \prod_{\mathbf{z} \in Z} \lambda c(\mathbf{z}) + \sum_{\mathbf{z} \in Z} p_D(\mathbf{x}) \cdot f_{k+1}(\mathbf{z}|\mathbf{x}) \cdot e^{\lambda p_c(T) - \lambda} \tag{12.84}$$

$$\cdot \prod_{\mathbf{z} \in Z - \{\mathbf{z}\}} \lambda c(\mathbf{z}). \tag{12.85}$$

From (12.1) we conclude, as claimed, that

$$f_{k+1}(Z|\mathbf{x}) = \frac{\delta \beta_{k+1}}{\delta Z}(\emptyset|\mathbf{x}) \tag{12.86}$$

$$= e^{-\lambda}(1 - p_D(\mathbf{x})) \cdot \prod_{\mathbf{z} \in Z} \lambda c(\mathbf{z}) \tag{12.87}$$

$$+ e^{-\lambda} p_D(\mathbf{x}) \sum_{\mathbf{z} \in Z} f_{k+1}(\mathbf{z}|\mathbf{x}) \prod_{\mathbf{z} \in Z - \{\mathbf{z}\}} \lambda c(\mathbf{z}) \tag{12.88}$$

$$= e^{-\lambda} \left(\prod_{\mathbf{z} \in Z} \lambda c(\mathbf{z})\right) \tag{12.89}$$

$$\cdot \left(1 - p_D(\mathbf{x}) + p_D(\mathbf{x}) \cdot \sum_{\mathbf{z} \in Z} \frac{f_{k+1}(\mathbf{z}|\mathbf{x})}{\lambda c(\mathbf{z})}\right). \tag{12.90}$$

12.3.4 Case III: No Missed Detections or False Alarms

In this case the number m of measurements equals the number n of targets. I show that $f_{k+1}(Z|X) = 0$ if $|Z| \neq n$ and that, if otherwise with $Z = \{\mathbf{z}_1, ..., \mathbf{z}_n\}$,

$$f_{k+1}(Z|X) = \sum_{\sigma} f_{k+1}(\mathbf{z}_1|\mathbf{x}_{\sigma 1}) \cdots f_{k+1}(\mathbf{z}_n|\mathbf{x}_{\sigma n}). \tag{12.91}$$

The summation is taken over all permutations σ of the numbers $1, ..., n$.[4]

Equation (12.44) becomes

$$\overset{\text{measurement set}}{\Sigma_{k+1}} = \overset{\text{target detection set}}{\{\mathbf{Z}_1(\mathbf{x}_1), ..., \mathbf{Z}_n(\mathbf{x}_n)\}} \tag{12.92}$$

The belief-mass function corresponding to this model is

$$\begin{aligned}
\beta_{k+1}(T|X) &= \Pr(\Sigma_{k+1} \subseteq T|X) && (12.93)\\
&= \Pr(\mathbf{Z}_1(\mathbf{x}_1) \in T, ..., \mathbf{Z}_n(\mathbf{x}_n) \in T|X) && (12.94)\\
&= \Pr(\mathbf{Z}_1(\mathbf{x}_1) \in T|\mathbf{x}_1) \cdots \Pr(\mathbf{Z}_n(\mathbf{x}_n) \in T|\mathbf{x}_n) && (12.95)\\
&= \Pr(\eta_{k+1}(\mathbf{x}_1) + \mathbf{W}_1 \in T|\mathbf{x}_1) && (12.96)\\
&\quad \cdots \Pr(\eta_{k+1}(\mathbf{x}_n) + \mathbf{W}_n \in T|\mathbf{x}_n) && (12.97)\\
&= \left(\int_T f_{k+1}(\mathbf{z}|\mathbf{x}_1)d\mathbf{z}\right) \cdots \left(\int_T f_{k+1}(\mathbf{z}|\mathbf{x}_n)d\mathbf{z}\right) && (12.98)\\
&= p_1(T) \cdots p_n(T) && (12.99)
\end{aligned}$$

where we have abbreviated $p_i(T) \overset{\text{abbr.}}{=} \int_T f_{k+1}(\mathbf{z}|\mathbf{x}_i)d\mathbf{z}$. The fundamental convolution formula, (11.252), yields

$$\frac{\delta \beta_{k+1}}{\delta Z}(T|X) = \sum_{W_1 \uplus ... \uplus W_n = Z} \frac{\delta p_1}{\delta W_1}(T) \cdots \frac{\delta p_n}{\delta W_n}(T) \tag{12.100}$$

4 Recall that, in the terminology of Chapter 10, σ is an *association hypothesis* under our current assumptions.

where the summation is taken over all mutually disjoint subsets $W_1, ..., W_n$ of Z such that $W_1 \cup ... \cup W_n = Z$. Setting $T = \emptyset$ we get

$$f_{k+1}(Z|X) = \frac{\delta\beta_{k+1}}{\delta Z}(\emptyset|X) \tag{12.101}$$

$$= \sum_{W_1 \uplus ... \uplus W_n = Z} \frac{\delta p_1}{\delta W_1}(\emptyset) \cdots \frac{\delta p_n}{\delta W_n}(\emptyset). \tag{12.102}$$

However,

$$\frac{\delta p_i}{\delta \emptyset}(\emptyset) = p_i(\emptyset) = 0 \tag{12.103}$$

$$\frac{\delta p_i}{\delta\{\mathbf{z}\}}(\emptyset) = f_{k+1}(\mathbf{z}|\mathbf{x}_i) \tag{12.104}$$

$$\frac{\delta p_i}{\delta Z}(\emptyset) = 0 \quad \text{if} \quad |Z| > 1. \tag{12.105}$$

The only surviving terms in (12.102) are those for which $|W_1| = ... = |W_n| = 1$. Thus all terms will vanish if $|Z| \neq n$. Otherwise, $Z = W_1 \uplus ... \uplus W_n$ where the $W_1, ..., W_n$ must have the form $W_1 = \{\mathbf{z}_{\sigma 1}\}, ..., W_n = \{\mathbf{z}_{\sigma n}\}$ and where σ is some permutation on $1, ..., n$. So, (12.101) becomes

$$f_{k+1}(Z|X) = \sum_\sigma \frac{\delta p_1}{\delta \mathbf{z}_{\sigma 1}}(\emptyset) \cdots \frac{\delta p_n}{\delta \mathbf{z}_{\sigma n}}(\emptyset) \tag{12.106}$$

$$= \sum_\sigma f_{k+1}(\mathbf{z}_{\sigma 1}|\mathbf{x}_1) \cdots f_{k+1}(\mathbf{z}_{\sigma n}|\mathbf{x}_n) \tag{12.107}$$

$$= \sum_\sigma f_{k+1}(\mathbf{z}_1|\mathbf{x}_{\sigma 1}) \cdots f_{k+1}(\mathbf{z}_n|\mathbf{x}_{\sigma n}). \tag{12.108}$$

We can verify that this yields the true likelihood function by taking the set integral:

$$\int_T f_{k+1}(Z|X)\delta Z \tag{12.109}$$

$$= \frac{1}{n!} \int_{T \times ... \times T} \left(\sum_\sigma f_{k+1}(\mathbf{z}_1|\mathbf{x}_{\sigma 1}) \cdots f_{k+1}(\mathbf{z}_n|\mathbf{x}_{\sigma n}) \right) \tag{12.110}$$

$$\cdot d\mathbf{z}_1 \cdots d\mathbf{z}_n, \tag{12.111}$$

and so

$$= \frac{1}{n!}\sum_{\sigma}\left(\int_T f_{k+1}(\mathbf{z}|\mathbf{x}_{\sigma 1})d\mathbf{z}\right)\cdots\left(\int_T f_{k+1}(\mathbf{z}|\mathbf{x}_{\sigma 1})d\mathbf{z}\right) \quad (12.112)$$

$$= \frac{1}{n!}\sum_{\sigma} p_{\sigma 1}(T)\cdots p_{\sigma n}(T), \quad (12.113)$$

and so

$$= p_1(T)\cdots p_n(T) = \Pr(\mathbf{Z}_1 \in T|\mathbf{x}_1)\cdots\Pr(\mathbf{Z}_n \in T|\mathbf{x}_1) \quad (12.114)$$

$$= \Pr(\{\mathbf{Z}_1,...,\mathbf{Z}_n\} \in T|X) = \beta_{k+1}(T|X). \quad (12.115)$$

Example 69 *Suppose that two targets moving along the real line are being interrogated by a single position-measuring sensor, without missed detections or false alarms. Measurements are assumed conditionally independent of target state. Target states have the form* $\mathbf{x} = (x, v)$ *and the measurement model is*

$$z = x + W \quad (12.116)$$

where W *is a zero-mean Gaussian random number with variance* σ^2. *For two targets* $X = \{(x_1, v_1), (x_2, v_2)\}$ *the corresponding single-target likelihood is*

$$f_{k+1}(z|x, v) = f_W(z - x) = N_{\sigma^2}(z - x). \quad (12.117)$$

Typical single-sensor measurement sets have the form $Z = \{z_1, z_2\}$. *Therefore the corresponding multitarget likelihood is, by (12.91),*

$$f_{k+1}(Z|X) = f_W(z_1 - x_1)\cdot f_W(z_2 - x_2) + f_W(z_2 - x_1)\cdot f_W(z_1 - x_2). \quad (12.118)$$

See [70, pp. 248-252] for a more extensive treatment of this example.

12.3.5 Case IV: Missed Detections, No False Alarms

I show that if $X = \{\mathbf{x}_1, ..., \mathbf{x}_n\}$ with $|X| = n$,

$$f_{k+1}(\emptyset|X) = (1 - p_D(\mathbf{x}_1))\cdots(1 - p_D(\mathbf{x}_n)) \quad (12.119)$$

and that if $Z \neq \emptyset$ where $Z = \{\mathbf{z}_1, ..., \mathbf{z}_m\}$ with $|Z| = m \leq n$,

$$f_{k+1}(Z|X) = f_{k+1}(\emptyset|X)\sum_{\theta}\prod_{i:\theta(i)>0}\frac{p_D(\mathbf{x}_i)\cdot f_{k+1}(\mathbf{z}_{\theta(i)}|\mathbf{x}_i)}{1 - p_D(\mathbf{x}_i)} \quad (12.120)$$

where the summation is taken over all one-to-one functions $\theta : \{1,...,n\} \rightarrow \{0,1,...,m\}$ as defined in Section 10.5.4 and Figure 10.2(b). That is, the θ have the property that $\theta(i) = \theta(i') > 0$ implies $i = i'$. Since $m \leq n$ this means that $W_\theta = Z$ where W_θ was defined in (10.106). For, if $\theta(i) = 0$ identically then one would have to have $m = 0$. If $\theta(i) = 0$ except for a single value of i, then $m = 1$. In general, m must equal the number of i such that $\theta(i) > 0$. That is, in the notation of Section 10.5.4 and Figure 10.2(b), $W_\theta = Z$.

In this case (12.44) has the form

$$\overset{\text{measurement set}}{\Sigma_{k+1}} = \overset{\text{detection set for state } \mathbf{x}_1}{\Upsilon(\mathbf{x}_1)} \cup ... \cup \overset{\text{detection set for state } \mathbf{x}_n}{\Upsilon(\mathbf{x}_1)} . \tag{12.121}$$

From (12.101) we have

$$f_{k+1}(Z|X) = \sum_{W_1 \uplus ... \uplus W_n = Z} \frac{\delta p_1}{\delta W_1}(\emptyset) \cdots \frac{\delta p_n}{\delta W_n}(\emptyset) \tag{12.122}$$

where $Z = \{\mathbf{z}_1, ..., \mathbf{z}_m\}$ with $m \leq n$ and

$$\frac{\delta p_i}{\delta \emptyset}(\emptyset) = p_i(\emptyset) = 1 - p_D(\mathbf{x}_i) \tag{12.123}$$

$$\frac{\delta p_i}{\delta \{\mathbf{z}\}}(\emptyset) = p_D(\mathbf{x}_i) \cdot f_{k+1}(\mathbf{z}|\mathbf{x}_i) \tag{12.124}$$

$$\frac{\delta p_i}{\delta Z}(\emptyset) = 0 \quad \text{if} \quad |Z| > 1. \tag{12.125}$$

The only surviving terms in the summation in (12.122) are those for which $|W_1| \leq 1,...,|W_n| \leq 1$. If $Z = \emptyset$ then

$$f_{k+1}(\emptyset|X) = \frac{\delta p_1}{\delta \emptyset}(\emptyset) \cdots \frac{\delta p_n}{\delta \emptyset}(\emptyset) = p_1(\emptyset) \cdots p_n(\emptyset) \tag{12.126}$$

$$= \prod_{i=1}^{n} (1 - p_D(\mathbf{x}_i)) . \tag{12.127}$$

If $Z \neq \emptyset$, then we get

$$
\begin{aligned}
& f_{k+1}(Z|X) && (12.128)\\
= \; & f_{k+1}(\emptyset|X) \sum_{W_1 \uplus ... \uplus W_n = Z} \frac{\frac{\delta p_1}{\delta W_1}(\emptyset) \cdots \frac{\delta p_n}{\delta W_n}(\emptyset)}{\prod_{i=1}^n (1 - p_D(\mathbf{x}_i))} && (12.129)\\
= \; & f_{k+1}(\emptyset|X) && (12.130)\\
& \cdot \sum_{1 \leq i_1 \neq ... \neq i_m \leq n} \frac{\frac{\delta p_{i_1}}{\delta \mathbf{z}_1}(\emptyset) \cdots \frac{\delta p_{i_m}}{\delta \mathbf{z}_m}(\emptyset)}{(1 - p_D(\mathbf{x}_{i_1})) \cdots (1 - p_D(\mathbf{x}_{i_m}))} && (12.131)
\end{aligned}
$$

and so

$$
\begin{aligned}
= \; & f_{k+1}(\emptyset|X) \sum_{1 \leq i_1 \neq ... \neq i_m \leq n} \prod_{j=1}^m \frac{p_D(\mathbf{x}_{i_j}) \cdot f_{k+1}(\mathbf{z}_j|\mathbf{x}_{i_j})}{1 - p_D(\mathbf{x}_{i_j})} && (12.132)\\
= \; & n! \cdot f_{k+1}(\emptyset|X) && (12.133)\\
& \cdot \sum_{1 \leq i_1 < ... < i_m \leq n} \prod_{j=1}^m \frac{p_D(\mathbf{x}_{i_j}) \cdot f_{k+1}(\mathbf{z}_j|\mathbf{x}_{i_j})}{1 - p_D(\mathbf{x}_{i_j})}. && (12.134)
\end{aligned}
$$

Each m-tuple $(i_1, ..., i_m)$ with $1 \leq i_1 \neq ... \neq i_m \leq n$ determines a one-to-one function $\tau : \{1, ..., m\} \to \{1, ..., n\}$ by $\tau(j) = i_j$. For each such m-tuple, define

$$\theta : \{1, ..., n\} \to \{0, 1, ..., m\} \tag{12.135}$$

by $\theta(i) = \tau^{-1}(i)$ if i is in the image of τ, and $\theta(i) = 0$ otherwise. Then the m-tuples are in one-to-one correspondence with the associations θ as defined in Section 10.5.4 or Figure 10.2(b) and we can write

$$f_{k+1}(Z|X) = f_{k+1}(\emptyset|X) \sum_{\theta} \prod_{i:\theta(i)>0} \frac{p_D(\mathbf{x}_i) \cdot f_{k+1}(\mathbf{z}_{\theta(i)}|\mathbf{x}_i)}{1 - p_D(\mathbf{x}_i)}. \tag{12.136}$$

12.3.6 Case V: Missed Detections and False Alarms

In this section I am finally addressing the standard model in its entirety. In this case the measurement model has already been defined in (12.44):

$$\overset{\text{measurement set}}{\Sigma_{k+1}} = \overset{\text{target detection set}}{\Upsilon(X)} \cup \overset{\text{false detections}}{C} \tag{12.137}$$

where

$$\Upsilon(X) = \Upsilon(\mathbf{x}_1) \cup ... \cup \Upsilon(\mathbf{x}_n). \tag{12.138}$$

I show that

$$f_{k+1}(Z|X) = e^{\lambda} f_C(Z) \cdot f_{k+1}(\emptyset|X) \tag{12.139}$$

$$\cdot \sum_{\theta} \prod_{i:\theta(i)>0} \frac{p_D(\mathbf{x}_i) \cdot f_{k+1}(\mathbf{z}_{\theta(i)}|\mathbf{x}_i)}{(1 - p_D(\mathbf{x}_i)) \cdot \lambda c(\mathbf{z}_{\theta(i)})} \tag{12.140}$$

where the summation is taken over all association hypotheses $\theta : \{1, ..., n\} \rightarrow \{0, 1, ..., m\}$ as defined in Section 10.5.4; and where

$$f_C(Z) = e^{-\lambda} \prod_{\mathbf{z} \in Z} \lambda c(\mathbf{z}) \tag{12.141}$$

$$f_{k+1}(\emptyset|X) = e^{-\lambda} \prod_{\mathbf{x} \in X} (1 - p_D(\mathbf{x})). \tag{12.142}$$

This result is established in Appendix G.18.

12.3.7 p.g.fl.s for the Standard Measurement Model

The purpose of this section is to derive a p.g.fl. representation for the standard measurement model. This representation will prove useful in Chapter 16.

The p.g.fl. of a multiobject probability density function was defined in (11.154). Given a multitarget likelihood function $f_{k+1}(Z|X)$, its p.g.fl. is

$$G_{k+1}[g|X] \triangleq \int g^Z \cdot f_{k+1}(Z|X) \delta Z \tag{12.143}$$

where, as defined in (11.153), $g^Z = 1$ if $Z = \emptyset$ and, otherwise, $g^Z \triangleq \prod_{\mathbf{z} \in Z} g(\mathbf{z})$. Define

$$c[g] \triangleq \int g(\mathbf{z}) \cdot c(\mathbf{z}) d\mathbf{z} \tag{12.144}$$

$$p_g(\mathbf{x}) \triangleq \int g(\mathbf{z}) \cdot f_{k+1}(\mathbf{z}|\mathbf{x}) d\mathbf{z} \tag{12.145}$$

$$p_g^X \triangleq \prod_{\mathbf{x} \in X} p_g(\mathbf{x}). \tag{12.146}$$

In Appendix G.19, I show that the p.g.fl.s for the five cases of the standard model can be written in terms of relatively simple formulas. These are, respectively:

- *Case I: No targets present*:

$$G_{k+1}[g|\emptyset] = e^{\lambda c[g]-\lambda}; \tag{12.147}$$

- *Case II: A single target is present*:

$$G_{k+1}[g|\mathbf{x}] = [1 - p_D(\mathbf{x}) + p_D(\mathbf{x}) \cdot p_g(\mathbf{x})] \cdot e^{\lambda c[g]-\lambda}; \tag{12.148}$$

- *Case III: No missed detections or false alarms*:

$$G_{k+1}[g|X] = p_g^X; \tag{12.149}$$

- *Case IV: Missed detections, no false alarms*:

$$G_{k+1}[g|X] = (1 - p_D + p_D p_g)^X; \tag{12.150}$$

- *Case V: Missed detections and Poisson false alarms*:

$$G_{k+1}[g|X] = (1 - p_D + p_D p_g)^X \cdot e^{\lambda c[g]-\lambda}. \tag{12.151}$$

12.4 RELATIONSHIP WITH MHC

In Section 10.7.2, I pointed out a possible theoretical difficulty with the multihypothesis correlation (MHC) approach from a Bayesian point of view. We are now in a position to better understand it. As a result, we will also better understand the relationship between the MHC filter and the multitarget Bayes recursive filter to be introduced in Chapter 14.

In Section 10.5.4, I derived formulas for the hypothesis probabilities for the MHC hypothesis probabilities, assuming the standard measurement model. This derivation was, in turn, based on (G.241) in Appendix G.17. There I showed that the likelihood function for a given association hypothesis θ is:

$$f(\mathbf{z}_1, ..., \mathbf{z}_m | \mathbf{x}_1, ..., \mathbf{x}_n, \theta) \tag{12.152}$$

$$= e^{-\lambda} \lambda^{m-n_\theta} p_D^{n_\theta} (1 - p_D)^{n-n_\theta} \tag{12.153}$$

$$\cdot \left(\prod_{i:\theta(i)>0} f(\mathbf{z}_{\theta(i)}|\mathbf{x}_i) \right) \cdot \left(\prod_{\mathbf{z} \in Z - W_\theta} c(\mathbf{z}) \right). \tag{12.154}$$

Let p'_θ be the prior probability of the association θ. Since there is no a priori reason to prefer one association over another, we can assume that p'_θ is uniform. Thus the likelihood of collecting a measurement set Z if targets with state set X are present is:

$$
\begin{aligned}
& f(\mathbf{z}_1, ..., \mathbf{z}_m | \mathbf{x}_1, ..., \mathbf{x}_n) && (12.155) \\
\propto\ & \sum_\theta f(\mathbf{z}_1, ..., \mathbf{z}_m | \mathbf{x}_1, ..., \mathbf{x}_n, \theta) \cdot p'_\theta && (12.156) \\
=\ & e^{-\lambda} \sum_\theta \lambda^{m-n_\theta} p_D^{n_\theta} (1-p_D)^{n-n_\theta} \cdot \left(\prod_{i:\theta(i)>0} f(\mathbf{z}_{\theta(i)} | \mathbf{x}_i) \right) && (12.157) \\
& \cdot \left(\prod_{\mathbf{z} \in Z - W_\theta} c(\mathbf{z}) \right) && (12.158) \\
=\ & f_C(Z) \cdot \left(\begin{array}{c} \sum_\theta (1-p_D)^{n-n_\theta} \\ \cdot \left(\prod_{i:\theta(i)>0} p_D f(\mathbf{z}_{\theta(i)} | \mathbf{x}_i) \right) \\ \cdot \left(\prod_{\mathbf{z} \in W_\theta} \frac{1}{\lambda c(\mathbf{z})} \right) \end{array} \right) && (12.159) \\
=\ & e^\lambda f_C(Z) f(Z|\emptyset) \left(\begin{array}{c} \sum_\theta \left(\prod_{i:\theta(i)>0} p_D f(\mathbf{z}_{\theta(i)} | \mathbf{x}_i) \right) \\ \cdot \left(\prod_{\mathbf{z} \in W_\theta} \frac{1}{(1-p_D) \lambda c(\mathbf{z})} \right) \end{array} \right) && (12.160) \\
=\ & e^\lambda f_C(Z) f(Z|\emptyset) \sum_\theta \prod_{i:\theta(i)>0} \frac{p_D f(\mathbf{z}_{\theta(i)} | \mathbf{x}_i)}{(1-p_D) \lambda c(\mathbf{z}_{\theta(i)})} && (12.161)
\end{aligned}
$$

where, recall,

$$
f_C(Z) = e^{-\lambda} \left(\prod_{\mathbf{z} \in Z} \lambda c(\mathbf{z}) \right), \quad f(Z|\emptyset) = e^{-\lambda} (1-p_D)^n. \qquad (12.162)
$$

Compare (12.161) with the formula for the likelihood function for the standard model, (12.41), setting p_D constant:

$$
\begin{aligned}
f_{k+1}(Z|X) \quad = \quad & e^\lambda f_C(Z) \cdot f_{k+1}(\emptyset|X) && (12.163) \\
& \cdot \sum_\theta \prod_{i:\theta(i)>0} \frac{p_D \cdot f_{k+1}(\mathbf{z}_{\theta(i)} | \mathbf{x}_i)}{(1-p_D) \cdot \lambda c(\mathbf{z}_{\theta(i)})}. && (12.164)
\end{aligned}
$$

We see that $f(\mathbf{z}_1, ..., \mathbf{z}_m | \mathbf{x}_1, ..., \mathbf{x}_n)$ differs from $f_{k+1}(Z|X)$ at most by a constant multiple. From this we conclude the following:

- *The true multitarget likelihood arises as a consequence of averaging over all association hypotheses, taking all association hypotheses into account equally;*
- *Each association hypothesis occurs as an additive term in the formula for the multitarget likelihood.*

12.5 STATE-DEPENDENT FALSE ALARMS

In Section 12.3 we implicitly assumed that the false alarm process is independent of the states of the targets. In the case of many sensors, however, false alarms can be more concentrated in target-containing regions than elsewhere. The purpose of this section is to show that the analysis of Section 12.3 is easily modified so as to extend the standard model to include state-dependent false alarms.

The measurement model now has the form

$$\overbrace{\Sigma_{k+1}}^{\text{measurement set}} = \overbrace{\Upsilon(X)}^{\text{target detections}} \cup \overbrace{C(X)}^{\text{false detections}} ; \tag{12.165}$$

where

$$\Upsilon(X) = \overbrace{\Upsilon(\mathbf{x}_1)}^{\text{detection set for } \mathbf{x}_1} \cup ... \cup \overbrace{\Upsilon(\mathbf{x}_n)}^{\text{detection set for } \mathbf{x}_n} ; \tag{12.166}$$

where

$$C(X) = \overbrace{C(\mathbf{x}_1)}^{\text{false detections for } \mathbf{x}_1} \cup ... \cup \overbrace{C(\mathbf{x}_n)}^{\text{false detections for } \mathbf{x}_n} \tag{12.167}$$

$$\cup \overbrace{C}^{\text{state-independent false detections}} ; \tag{12.168}$$

where, for $i = 1, ..., n$, $C(\mathbf{x}_i)$ is a Poisson process with mean value and spatial density

$$\lambda_i \stackrel{\text{abbr.}}{=} \lambda(\mathbf{x}_i), \quad c_i(\mathbf{z}) \stackrel{\text{abbr.}}{=} c(\mathbf{z}|\mathbf{x}_i); \tag{12.169}$$

and where C is a Poisson process with mean value λ_0 and spatial distribution $c_0(\mathbf{z})$; and where

$$\Upsilon(\mathbf{x}_1), ...\Upsilon(\mathbf{x}_n), C(\mathbf{x}_1), ..., C(\mathbf{x}_n), C \tag{12.170}$$

are assumed to be statistically independent.

The belief-mass function for the total false alarm process is

$$\begin{aligned}\beta_{C(X)}(T) &= \Pr\left(C(X) \subseteq T|X\right) &(12.171)\\ &= \Pr(C(\mathbf{x}_1),...,C(\mathbf{x}_n),C \subseteq T|X) &(12.172)\\ &= \Pr(C(\mathbf{x}_1) \subseteq T|\mathbf{x}_1)\cdots\Pr(C(\mathbf{x}_n) \subseteq T|\mathbf{x}_n) &(12.173)\\ &\quad\cdot\Pr(C \subseteq T) &(12.174)\\ &= \beta_{C(\mathbf{x}_1)}(T)\cdots\beta_{C(\mathbf{x}_n)}(T)\cdot\beta_C(T). &(12.175)\end{aligned}$$

From (12.52) we know that

$$\beta_{C(\mathbf{x}_i)}(T) = e^{\lambda_i p_i(T)-\lambda_i}, \qquad \beta_C(T) = e^{\lambda_0 p_0(T)-\lambda_0} \tag{12.176}$$

where for $i = 0, 1, ..., n$,

$$p_i(T) \triangleq \int_T c_i(\mathbf{z})d\mathbf{z}. \tag{12.177}$$

Thus

$$\begin{aligned}\beta_C(T|X) &= e^{\lambda_0 p_0(T)-\lambda_0}\cdot e^{\lambda_1 p_1(T)-\lambda_1}\cdots e^{\lambda_n p_n(T)-\lambda_n} &(12.178)\\ &= e^{\lambda_0 p_0(T)-\lambda_0+\lambda_1 p_1(T)-\lambda_1+...+\lambda_n p_n(T)-\lambda_n} &(12.179)\\ &= e^{\lambda(X)\cdot p(T|X)-\lambda(X)} &(12.180)\end{aligned}$$

where we have defined

$$\begin{aligned}\lambda(X) &\triangleq \lambda_0+\lambda_1+...+\lambda_n = \lambda_0 + \sum_{\mathbf{x}\in X}\lambda(\mathbf{x}) &(12.181)\\ c(\mathbf{z}|X) &\triangleq \frac{\lambda_0 c_0(\mathbf{z})+\lambda_1 c_1(\mathbf{z})+...+\lambda_n c_n(\mathbf{z})}{\lambda_0+\lambda_1+...+\lambda_n} &(12.182)\\ &= \frac{\lambda_0 c_0(\mathbf{z})+\lambda(\mathbf{x}_1)p(T|\mathbf{x}_1)+...+\lambda(\mathbf{x}_n)p(T|\mathbf{x}_n)}{\lambda(\mathbf{x}_1)+...+\lambda(\mathbf{x}_n)} &(12.183)\\ p(T|X) &\triangleq \int_T c(\mathbf{z}|X)d\mathbf{z}. &(12.184)\end{aligned}$$

Thus, from (12.51), the probability density function of the false alarm process is

$$f_{C(X)}(Z) = e^{-\lambda(X)}\prod_{\mathbf{z}\in Z}\lambda(X)\cdot c(\mathbf{z}|X). \tag{12.185}$$

In other words, the composite false alarm process $C(X)$ is also Poisson with mean value $\lambda(X)$ and spatial distribution $c(\mathbf{z}|X)$. Consequently, (12.139)-(12.141) become

$$f_{k+1}(Z|X) = e^{\lambda(X)} \cdot f_{C(X)}(Z) \cdot f_{k+1}(\emptyset|X) \tag{12.186}$$

$$\cdot \sum_\theta \prod_{i:\theta(i)>0} \frac{p_D(\mathbf{x}_i) \cdot f_{k+1}(\mathbf{z}_{\theta(i)}|\mathbf{x}_i)}{(1-p_D(\mathbf{x}_i)) \cdot \lambda(X) \cdot c(\mathbf{z}_{\theta(i)}|X)} \tag{12.187}$$

where

$$f_{k+1}(\emptyset|X) = e^{-\lambda(X)} \prod_{\mathbf{x}\in X} (1-p_D(\mathbf{x}))\,. \tag{12.188}$$

12.5.1 p.g.fl. for State-Dependent False Alarms

The multitarget likelihood $f_{k+1}(Z|X)$ for state-dependent false alarms has exactly the same form as the multitarget likelihood for state-independent false alarms, except for the fact that the composite false alarm process has an average of $\lambda(X)$ false alarms, distributed according to $c(\mathbf{z}|X)$. Consequently, if we define

$$c[g|X] \triangleq \int g(\mathbf{z}) \cdot c(\mathbf{z}|X) d\mathbf{z} \tag{12.189}$$

then (12.151) becomes

$$G_{k+1}[g|X] = (1 - p_D + p_D p_g)^X \cdot e^{\lambda(X)\cdot c[g|X] - \lambda(X)}. \tag{12.190}$$

12.6 TRANSMISSION DROP-OUTS

Even though a sensor may have collected observations from targets that are located in its field of view, these observations may not actually be available to the collection site.[5] This may be because of transmission drop-outs caused by atmospheric interference, weather, terrain blockage, transmission latency, and so on. Many of these effects can be modeled using *transmission probabilities of detection.*

In Section 12.2.2 I pointed out that sensors, no less than targets, have states. Suppose that we have a single sensor with state vector $\overset{*}{\mathbf{x}}$. Let

$$\overset{*}{p}_T(\overset{*}{\mathbf{x}}) \tag{12.191}$$

5 This material originally appeared in [139, pp. 254-258].

be the probability that an observation will survive transmission to the processing site, given that the sensor currently has state $\mathring{\mathbf{x}}$. (Note that $\mathring{p}_T(\mathring{\mathbf{x}})$ also has an implicit functional dependence upon the state of the data-collection site.) Consequently, the total probability of detection of the sensor has the form

$$\mathring{p}_D(\mathbf{x}, \mathring{\mathbf{x}}) \triangleq \mathring{p}_T(\mathring{\mathbf{x}}) \cdot p_D(\mathbf{x}, \mathring{\mathbf{x}}) \tag{12.192}$$

where $p_D(\mathbf{x}, \mathring{\mathbf{x}})$ is the usual probability of detection. This means that if $f_{k+1}(Z|X) \overset{\text{abbr.}}{=} f(Z|X, \mathring{\mathbf{x}}_{k+1})$ is the usual multitarget likelihood function for the sensor, then the complete multitarget likelihood is

$$\mathring{f}_{k+1}(Z|X) = \begin{cases} 1 - \mathring{p}_T + \mathring{p}_T \cdot f_{k+1}(\emptyset|X) & \text{if} \quad Z = \emptyset \\ \mathring{p}_T \cdot f_{k+1}(Z|X) & \text{if} \quad \text{otherwise} \end{cases}. \tag{12.193}$$

where $\mathring{p}_T \overset{\text{abbr.}}{=} \mathring{p}_T(\mathring{\mathbf{x}}_{k+1})$ and $\mathring{f}_{k+1}(Z|X) \overset{\text{abbr.}}{=} \mathring{f}(Z|X, \mathring{\mathbf{x}}_{k+1})$.

That is, a non-null measurement set will be collected at the reception site only if the following are true:

- A non-null measurement set was collected by the sensor;
- The measurements successfully survived transmission.

It is left to the reader as Exercise 52 to verify that $\mathring{f}_{k+1}(Z|X)$ is a likelihood function in the sense that $\int \mathring{f}_{k+1}(Z|X)\delta Z = 1$.

12.6.1 p.g.fl. for Transmission Drop-Outs

It is left to the reader as Exercise 53 to show that if $G_{k+1}[g|X]$ is the p.g.fl. of $f_{k+1}(Z|X)$, then the p.g.fl. of $\mathring{f}_{k+1}(Z|X)$ is [139, p. 293]:

$$\mathring{G}_{k+1}[g|X] = 1 - \mathring{p}_T + \mathring{p}_T \cdot G_{k+1}[g|X]. \tag{12.194}$$

12.7 EXTENDED TARGETS

The standard multitarget measurement model can be directly extended to incorporate extended-target measurements of the kind described in Section 9.2.3. As was noted there, any observation generated by an extended target will be a finite set Z

of vectors. The sensing phenomenology is, however, different than that assumed in the standard model. In the standard model, most of the elements of Z will be false alarms that do not provide any information about the target state. In the extended-target case, most or at least a large number of the observation-vectors will provide target-related information.

In what follows I first consider the single-target case (Section 12.7.1), followed by the multitarget case (Section 12.7.2).

An approximate approach proposed by Gilholm et al. is briefly summarized in Section 12.7.3.

12.7.1 Single Extended Target

I begin by specifying a measurement model of standard type for single extended targets. Because this turns out to be formally identical to the standard model in the multitarget case, we will be able to derive the corresponding true likelihood function quite easily.

12.7.1.1 Measurement Model for a Single Extended Target

As in the standard model, we assume that we have a single sensor with state $\overset{*}{\mathbf{x}}$; with probability of detection $p_D(\mathbf{x}, \overset{*}{\mathbf{x}})$; and with likelihood function $f_{k+1}(\mathbf{z}|\mathbf{x}) = f_{\mathbf{W}_{k+1}}(\mathbf{z} - \eta_{k+1}(\mathbf{x}, \overset{*}{\mathbf{x}}))$. The state of an extended target will have the general form

$$\mathbf{x} = (x, y, z, v_x, v_y, v_z, \theta, \varphi, \psi) \qquad (12.195)$$

where x, y, z are the coordinates of its centroid, where v_x, v_y, v_z are the coordinates of the centroid's velocity, where θ, φ, ψ are body-frame coordinates, and where c is the target type.

We model the extended target as a collection of point scatterer sites

$$\breve{\mathbf{x}}^1 + \mathbf{x} \quad ,..., \quad \breve{\mathbf{x}}^L + \mathbf{x} \qquad (12.196)$$

located on the target surface, regardless of orientation relative to the sensor. The $\breve{\mathbf{x}}^1$,...,$\breve{\mathbf{x}}^L$ (implicitly) depend on target type c and have the form

$$\breve{\mathbf{x}}^\ell = (x_\ell, y_\ell, z_\ell, 0, 0, 0, 0, 0, 0) \qquad (12.197)$$

for $\ell = 1, ..., L$.

For any given sensor state $\overset{*}{\mathbf{x}}$, some sites will be visible to the sensor and others will not. Define the *visibility function* $e^{\ell}(\mathbf{x}, \overset{*}{\mathbf{x}})$ for the sensor by

$$e^{\ell}(\mathbf{x}, \overset{*}{\mathbf{x}}) = \begin{cases} 1 & \text{if} \quad \text{the site } \breve{\mathbf{x}}_{\ell} + \mathbf{x} \text{ is visible to sensor} \\ 0 & \text{if} \quad \text{otherwise} \end{cases} . \tag{12.198}$$

Given these preliminaries,

$$p_D^{\ell}(\mathbf{x}, \overset{*}{\mathbf{x}}) \triangleq e^{\ell}(\mathbf{x}, \overset{*}{\mathbf{x}}) \cdot p_D(\breve{\mathbf{x}}^{\ell} + \mathbf{x}, \overset{*}{\mathbf{x}}) \tag{12.199}$$

is the probability of detection of the site $\breve{\mathbf{x}}^{\ell} + \mathbf{x}$, given that it is visible to the sensor. Likewise,

$$\eta^{\ell}(\mathbf{x}, \overset{*}{\mathbf{x}}) \triangleq \eta_{k+1}(\breve{\mathbf{x}}^{\ell} + \mathbf{x}, \overset{*}{\mathbf{x}}) \tag{12.200}$$

is the deterministic measurement model at the site $\breve{\mathbf{x}}^{\ell} + \mathbf{x}$ and

$$f_{k+1}^{\ell}(\mathbf{z}|\mathbf{x}) \triangleq f_{\mathbf{W}_{k+1}}(\mathbf{z} - \eta^{\ell}(\mathbf{x}, \overset{*}{\mathbf{x}})) \tag{12.201}$$

is the corresponding likelihood function.

Let $\emptyset^{\ell}(\mathbf{x}, \overset{*}{\mathbf{x}})$ be the random subset of $\mathfrak{Z}_0$ defined by

$$\emptyset^{\ell}(\mathbf{x}, \overset{*}{\mathbf{x}}) = \emptyset^{p_D^{\ell}(\mathbf{x}, \overset{*}{\mathbf{x}})} \tag{12.202}$$

where $\emptyset^p$ was defined in (11.21) or (12.47). That is,

$$\Pr(\emptyset^{\ell}(\mathbf{x}, \overset{*}{\mathbf{x}}) = T) \triangleq \begin{cases} p_D^{\ell}(\mathbf{x}, \overset{*}{\mathbf{x}}) & \text{if} \quad T = \mathfrak{Z}_0 \\ 1 - p_D^{\ell}(\mathbf{x}, \overset{*}{\mathbf{x}}) & \text{if} \quad T = \emptyset \\ 0 & \text{if} \quad \text{otherwise} \end{cases} . \tag{12.203}$$

Then the random set measurement model at site $\breve{\mathbf{x}}^{\ell} + \mathbf{x}$ is

$$\Upsilon^{\ell}(\mathbf{x}, \overset{*}{\mathbf{x}}) = \{\eta^{\ell}(\mathbf{x}, \overset{*}{\mathbf{x}}) + \mathbf{W}_{\ell}\} \cap \emptyset^{\ell}(\mathbf{x}, \overset{*}{\mathbf{x}}) \tag{12.204}$$

for all $\ell = 1, ..., L$, where $\mathbf{W}_1, ..., \mathbf{W}_L$ are independent, identically distributed random vectors with density function $f_{\mathbf{W}_{k+1}}(\mathbf{z})$. Consequently, the measurement model over all sites is

$$\Sigma_{k+1} = \overbrace{\Upsilon(\mathbf{x})}^{\text{target-generated detections}} \cup \overbrace{C}^{\text{false detections}} \tag{12.205}$$

where

$$\Upsilon(\mathbf{x}) = \overbrace{\Upsilon^1(\mathbf{x})}^{\text{detection set at site 1}} \cup ... \cup \overbrace{\Upsilon^L(\mathbf{x})}^{\text{detection set at site } L} . \tag{12.206}$$

In what follows I will suppress the sensor state $\overset{*}{\mathbf{x}}$ (as well as the target type) for the sake of notational clarity, regarding it as implicit.

12.7.1.2 True Likelihood for a Single Extended Target

From (12.205) and (12.206) we see that the measurement model for a single target is an exact analog of the multitarget standard model of (12.40) and (12.41), with the following.

- $p_D^\ell(\mathbf{x}, \mathring{\mathbf{x}})$ in place of $p_D(\mathbf{x}_i, \mathring{\mathbf{x}})$;
- $\eta^\ell(\mathbf{x}, \mathring{\mathbf{x}})$ in place of $\eta_{k+1}(\mathbf{x}_i, \mathring{\mathbf{x}})$;
- $f_{k+1}^\ell(\mathbf{z}|\mathbf{x})$ in place of $f_{k+1}(\mathbf{z}|\mathbf{x}_i)$.

We can therefore just "transcribe" the belief-mass function at a single site from (12.40)-(12.41). That is, the true likelihood function for a single extended target is, if $Z = \emptyset$,

$$f_{k+1}(\emptyset|\mathbf{x}) = e^{-\lambda} \prod_{\ell=1}^{L} (1 - p_D^\ell(\mathbf{x})) \tag{12.207}$$

and, if $Z \neq \emptyset$,

$$f_{k+1}(Z|\mathbf{x}) = e^\lambda f_C(Z) f_{k+1}(\emptyset|\mathbf{x}) \sum_\theta \prod_{\theta(\ell)>0} \frac{p_D^\ell(\mathbf{x}) \cdot f_{k+1}^\ell(\mathbf{z}_{\theta(\ell)}|\mathbf{x})}{(1 - p_D^\ell(\mathbf{x})) \cdot \lambda c(\mathbf{z}_{\theta(\ell)})}. \tag{12.208}$$

The summation is taken over all associations $\theta : \{1, ..., L\} \to \{0, 1, ..., m\}$ as defined in Section 10.5.4 or Figures 10.2(b) and 10.2(c). Also, $f_C(Z) = e^{-\lambda} \prod_{\mathbf{z} \in Z} \lambda c(\mathbf{z})$.

12.7.2 Multiple Extended Targets

Suppose that n extended targets with states $X = \{\mathbf{x}_1, ..., \mathbf{x}_n\}$ are present.

$$\Sigma_{k+1} = \overbrace{\Upsilon(X)}^{\text{target-generated detections}} \cup \overbrace{C}^{\text{false detections}} \tag{12.209}$$

where

$$\Upsilon(X) = \overbrace{\Upsilon(\mathbf{x}_1)}^{\text{detection set for state 1}} \cup ... \cup \overbrace{\Upsilon(\mathbf{x}_n)}^{\text{detection set for state } n} \tag{12.210}$$

and where each of the $\Upsilon(\mathbf{x}_1), ..., \Upsilon(\mathbf{x}_n)$ has the form of (8.73).

From the fundamental convolution formula, (11.252), the likelihood function of Σ_{k+1} will be

$$f_{k+1}(Z|X) \tag{12.211}$$

$$= \sum_{W_0 \uplus W_1 \uplus \ldots \uplus W_m = Z} f_C(W_0) \cdot f_{\Upsilon(\mathbf{x}_1)}(W_1) \cdots f_{\Upsilon(\mathbf{x}_n)}(W_n) \tag{12.212}$$

$$= \sum_{W_0 \uplus W_1 \uplus \ldots \uplus W_m = Z} f_C(W_0) \tag{12.213}$$

$$\cdot f_{k+1}(W_1|\mathbf{x}_1) \cdots f_{k+1}(W_n|\mathbf{x}_n) \tag{12.214}$$

where the summation is taken over all mutually disjoint subsets $W_0, W_1, \ldots, W_m$ of Z such that $W_0 \cup W_1 \cup \ldots \cup W_m = Z$.

12.7.3 Poisson Approximation

Drawing upon earlier work by Hue, Le Cadre, and Perez [93], Gilholm, Godsill, Maskell, and Salmond have proposed an approximate multitarget likelihood for extended targets [66]. In this section we briefly summarize their approach.

Equation (12.208) is the exact point scatterer multitarget likelihood, assuming the existence of a Poisson false alarm background. If one excludes the false alarm model from (12.208) then from (12.120) one gets:

$$f_{k+1}(Z|\mathbf{x}) = f_{k+1}(\emptyset|\mathbf{x}) \sum_{\theta} \prod_{\theta(\ell)>0} \frac{p_D^\ell(\mathbf{x}) \cdot f_{k+1}^\ell(\mathbf{z}_{\theta(\ell)}|\mathbf{x})}{(1 - p_D^\ell(\mathbf{x}))}. \tag{12.215}$$

In essence, Gilholm et al. propose the replacement of this equation by a Poisson approximation:

$$f_{k+1}(Z|\mathbf{x}) = e^{-\tilde{\lambda}(\mathbf{x})} \cdot \prod_{\mathbf{z} \in Z} \tilde{\lambda}(\mathbf{x}) \cdot \tilde{f}_{\mathbf{x}}(\mathbf{z}). \tag{12.216}$$

That is, the observations generated by the extended target are distributed spatially according to $\tilde{f}_{\mathbf{x}}(\mathbf{z}) \overset{\text{abbr.}}{=} \tilde{f}_{k+1}(\mathbf{z}|\mathbf{x})$; and an average number $\tilde{\lambda}(\mathbf{x}) \overset{\text{abbr.}}{=} \tilde{\lambda}_{k+1}(\mathbf{x})$ of observations are produced per data collection.

Expressed in p.g.fl. form this becomes

$$\tilde{G}_{k+1}[g|\mathbf{x}] = e^{\tilde{\lambda}(\mathbf{x}) \cdot \tilde{f}_{\mathbf{x}}[g] - \tilde{\lambda}(\mathbf{x})}. \tag{12.217}$$

If we restore the Poisson false alarm background model, the corresponding p.g.fl. is

$$G_{k+1}[g|\mathbf{x}] = \tilde{G}_{k+1}[g|\mathbf{x}] \cdot G_C[g] \tag{12.218}$$
$$= e^{\tilde{\lambda}(\mathbf{x}) \cdot \tilde{f}_{\mathbf{x}}[g] - \tilde{\lambda}(\mathbf{x})} \cdot e^{\lambda c[g] - \lambda} \tag{12.219}$$
$$= e^{\lambda(\mathbf{x}) \cdot f_{\mathbf{x}}[g] - \lambda(\mathbf{x})} \tag{12.220}$$

where

$$\lambda(\mathbf{x}) \triangleq \tilde{\lambda}(\mathbf{x}) + \lambda \tag{12.221}$$
$$f_{\mathbf{x}}[g] = \int g(\mathbf{z}) \cdot f_{\mathbf{x}}(\mathbf{z}) d\mathbf{z} \tag{12.222}$$
$$f_{\mathbf{x}}(\mathbf{z}) \triangleq \frac{\tilde{\lambda}(\mathbf{x}) \cdot \tilde{f}(\mathbf{z}|\mathbf{x}) + \lambda \cdot c(\mathbf{z})}{\tilde{\lambda}(\mathbf{x}) + \lambda}. \tag{12.223}$$

Thus the approximate likelihood corresponding to (12.208) is also Poisson:

$$f_{k+1}(Z|\mathbf{x}) = e^{-\lambda(\mathbf{x})} \cdot \prod_{\mathbf{z} \in Z} \lambda(\mathbf{x}) \cdot f_{\mathbf{x}}(\mathbf{z}). \tag{12.224}$$

12.8 UNRESOLVED TARGETS

The standard model is based on the presumption that only single targets generate signal detections.[6] This is, however, a significant limitation. As I remarked in Section 9.2.1, this assumption will not be valid when at least some targets are closely spaced compared to sensor resolution. In this case, they will tend to look like a single target.

The purpose of this section is to propose a conceptually parsimonious generalization of the standard model that will permit the modeling of signal detections originating from multiple targets. This generalization is based on the idea illustrated in Figure 12.2.

The concept of a *point target* is based on the presumption that targets are sufficiently distant that they appear to be mathematical points. What this assumption implies, however, is that unresolved targets are multiple targets that are so much further distant that they appear to be a single point rather than a cluster of closely spaced points. Only when the targets move closer to the sensor do they "resolve"—that is, appear to separate into distinguishably distinct point targets.

6 The material in this section originally appeared in [137, 141, 133].

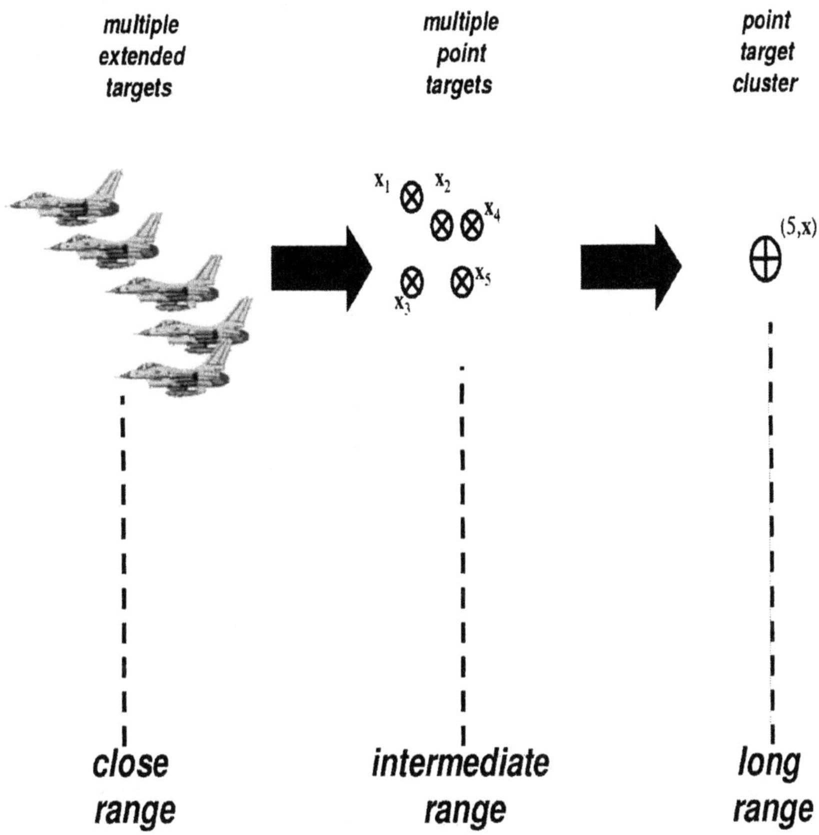

Figure 12.2 Targets sufficiently close to a sensor may have detectable physical extent. When further away, they diminish sufficiently in extent that they can be mathematically idealized as a group of "point targets." When at even greater remove, multiple point targets are diminished even further and so can be further idealized as a "point target cluster" or "point cluster."

I therefore propose the concept of *point target cluster* (*point cluster* for short). On the basis of this concept, I will derive the true likelihood function for problems in which both point targets and point clusters can be present.

The section is organized as follows. I formally define point clusters and their states in Section 12.8.1. I derive formulas for the likelihood functions for single clusters and for multiple clusters in Sections 12.8.2 and 12.8.3, respectively. Finally, in Section 12.8.4, I show that multicluster likelihood functions are *continuous with respect to target number*—that is,

$$\lim_{a \searrow 0} f_{k+1}(Z|\mathring{X} \cup \{(a, \mathbf{x})\}) = f_{k+1}(Z|\mathring{X}). \tag{12.225}$$

See (12.290)-(12.291).

12.8.1 Point Target Clusters

A point cluster is modeled as an *augmented state vector* of the form $\mathring{\mathbf{x}} = (n, \mathbf{x})$ where $\mathbf{x} \in \mathfrak{X}_0$ is a conventional single-target state vector and where n is a positive integer. That is, $\mathring{\mathbf{x}}$ models a cluster of n targets colocated at the single state $\mathbf{x}$. Since the state variable n is not continuous, it is necessary to extend this definition by modeling a point cluster as an augmented vector of the form

$$\mathring{\mathbf{x}} = (a, \mathbf{x}) \in \mathring{\mathfrak{X}}_0 \tag{12.226}$$

where $\mathbf{x} \in \mathfrak{X}_0$ is a conventional single-target state vector and where a is a positive real number.

- That is, $\mathring{\mathbf{x}}$ models a cluster of targets colocated at the single state $\mathbf{x}$, the average number of which is $a > 0$.

Just as a state set $X = \{\mathbf{x}_1, ..., \mathbf{x}_n\}$ is used to represent a multitarget state (a system of many targets), so an augmented state set

$$\mathring{X} = \{\mathring{\mathbf{x}}_1, ..., \mathring{\mathbf{x}}_n\} = \{(a_1, \mathbf{x}_1), ..., (a_n, \mathbf{x}_n)\} \tag{12.227}$$

represents a multicluster state (a system of many point clusters). It is assumed that the point clusters are distinct, in the sense that the $\mathbf{x}_1, ..., \mathbf{x}_n$ are distinct. Thus pairs such as $(a, \mathbf{x}), (a', \mathbf{x})$ cannot occur in $\mathring{X}$.

Given this model the goal of a Bayesian analysis is to determine the following:

- The number n of point clusters;

- The expected numbers $a_1, ..., a_n$ of targets in the point clusters (in which case the expected number of targets in all clusters is $|\mathring{X}| \triangleq a_1 + ... + a_n$);
- The joint states $\mathbf{x}_1, ..., \mathbf{x}_n$ of the point clusters.

In what follows I define the concept of likelihood $f_{k+1}(Z|a, \mathbf{x})$ for a single point cluster (Section 12.8.2) and then the likelihood $f_{k+1}(Z|\mathring{X})$ for multiple point clusters obscured in Poisson false alarms (Section 12.8.3). In Section 12.8.4, I show that likelihoods for point clusters are continuous in the variable $\mathring{X}$.

12.8.2 Single-Cluster Likelihoods

What is the proper definition of the likelihood function $f_{k+1}(Z|\mathring{\mathbf{x}})$ of a point cluster $\mathring{\mathbf{x}} = (a, \mathbf{x})$? Equation (12.263) will show that it is

$$f(Z|a, \mathbf{x}) \triangleq \begin{cases} B_{a,p_D(\mathbf{x})}(0) & \text{if } Z = \emptyset \\ m! \cdot B_{a,p_D(\mathbf{x})}(m) \cdot f(\mathbf{z}_1|\mathbf{x}) \cdots f(\mathbf{z}_m|\mathbf{x}) & \text{if } Z \neq \emptyset \end{cases} \tag{12.228}$$

where $B_{a,q}(m)$, defined in (12.243), is a generalization of the binomial distribution that is continuous (in fact, infinitely differentiable) as a function $a \mapsto B_{a,q}(m)$ for each fixed choice of m and q.

We proceed through three stages: when a is a positive integer; when $0 < a < 1$; and when $a > 0$.

12.8.2.1 Case I: Integer Number of Targets

If $a = n$ is a positive integer, it is clear what we must do. From (12.119) and (12.120) we know that the true likelihood $f_{k+1}(Z|X)$ for a collection $X = \{\mathbf{x}_1, ..., \mathbf{x}_n\}$ of distinct target states is

$$f_{k+1}(\emptyset|X) = \prod_{i=1}^{n} (1 - p_D(\mathbf{x}_i)) \tag{12.229}$$

if $Z = \emptyset$ and, if $Z = \{\mathbf{z}_1, ..., \mathbf{z}_m\}$ with $0 < |Z| = m \leq n$,

$$f_{k+1}(Z|X) = f_{k+1}(\emptyset|X) \sum_{\theta} \prod_{i:\theta(i)>0} \frac{p_D(\mathbf{x}_i) \cdot f_{k+1}(\mathbf{z}_{\theta(i)}|\mathbf{x}_i)}{1 - p_D(\mathbf{x}_i)}. \tag{12.230}$$

where the summation is taken over all associations $\theta : \{1, ..., n\} \to \{0, 1, ..., m\}$ as defined in Section 10.5.4 and Figure 10.2(b). Let $\mathbf{x}_1 \to \mathbf{x}, ..., \mathbf{x}_n \to \mathbf{x}$ as a limit. Then, if $Z = \emptyset$,

$$f_{k+1}(\emptyset|\mathring{\mathbf{x}}) = (1 - p_D(\mathbf{x}))^n = 0! \cdot B_n(0) \tag{12.231}$$

and, if otherwise,

$$f_{k+1}(Z|\mathring{\mathbf{x}}) = f_{k+1}(Z|n, \mathbf{x}) = m! \cdot B_{n,p_D(\mathbf{x})}(m) \cdot \prod_{\mathbf{z} \in Z} f_{k+1}(\mathbf{z}|\mathbf{x}). \tag{12.232}$$

where $B_{n,q}(m)$ is the binomial distribution

$$B_{n,q}(m) \triangleq \begin{cases} C_{n,m}\, q^m\, (1-q)^{n-m} & \text{if} \quad 0 \leq m \leq n \\ 0 & \text{if} \quad \text{otherwise} \end{cases}. \tag{12.233}$$

To see this, note that

$$\begin{aligned} & f_{k+1}(Z|n, \mathbf{x}) && (12.234) \\ = \; & (1 - p_D(\mathbf{x}))^n \sum_\theta \prod_{i:\theta(i)>0} \frac{p_D(\mathbf{x}) \cdot f_{k+1}(\mathbf{z}_{\theta(i)}|\mathbf{x})}{1 - p_D(\mathbf{x})} && (12.235) \\ = \; & p_D(\mathbf{x})^m \cdot (1 - p_D(\mathbf{x}))^{n-m} \cdot \sum_\theta \prod_{i:\theta(i)>0} f_{k+1}(\mathbf{z}_j|\mathbf{x}). && (12.236) \end{aligned}$$

Now, as noted in Section 12.3.5, under current assumptions the set of all $\mathbf{z}_{\theta(i)}$ with $\theta(i) > 0$ is all of Z. Also, the number of non-null θ is equal to the number of selections of m out of n objects. So,

$$\begin{aligned} & f_{k+1}(Z|n, \mathbf{x}) && (12.237) \\ = \; & p_D(\mathbf{x})^m \cdot (1 - p_D(\mathbf{x}))^{n-m} \cdot \sum_{\theta \neq 0} \prod_{j=1}^m f_{k+1}(\mathbf{z}_j|\mathbf{x}) && (12.238) \\ = \; & m! \cdot C_{n,m} \cdot p_D(\mathbf{x})^m \cdot (1 - p_D(\mathbf{x}))^{n-m} \cdot \prod_{j=1}^m f_{k+1}(\mathbf{z}_j|\mathbf{x}) && (12.239) \\ = \; & m! \cdot B_{n,p_D(\mathbf{x})}(m) \cdot \prod_{\mathbf{z} \in Z} f_{k+1}(\mathbf{z}|\mathbf{x}). && (12.240) \end{aligned}$$

12.8.2.2 Case II: Partially Existing Targets

By this I mean $0 < a \leq 1$. In this case it is natural to provisionally define

$$f_{k+1}(Z|a,\mathbf{x}) \cong \begin{cases} 1 - a \cdot p_D(\mathbf{x}) & \text{if} \quad Z = \emptyset \\ a \cdot p_D(\mathbf{x}) \cdot f_{k+1}(\mathbf{z}|\mathbf{x}) & \text{if} \quad Z = \{\mathbf{z}\} \\ 0 & \text{if} \quad \text{otherwise} \end{cases} . \tag{12.241}$$

That is, if a target has less than unity degree of existence a then the probability that an observation $\mathbf{z}$ will be collected from such a target is at least approximately proportional to a.

12.8.2.3 Case III: Target Number Is Continuous

Recall from (11.10) that the probability-generating function (p.g.f.) of a probability distribution $p(m)$ on $m = 0, 1, ...$ is $G(y) = \sum_{m=0}^{\infty} p(m) y^m$. The p.g.f. of the binomial distribution $B_{n,q}(m)$ is

$$G_{n,q}(y) = (1 - q + qy)^n . \tag{12.242}$$

Demonstration of this fact is left to the reader as Exercise 49. Our goal is to construct an approximate p.g.f. $G_{a,q}(y)$ for the case when a is not an integer. Then from it we can construct the corresponding probability distribution $B_{a,q}(m)$ using (11.11):

$$B_{a,q}(m) \triangleq \frac{1}{m!} \frac{d^m G_{a,q}}{dy^m}(0). \tag{12.243}$$

Let n be the largest integer smaller than a. Then, intuitively speaking, the point cluster consists of n ordinary targets and one partially existing target with degree of existence $a - n$. So, again intuitively speaking and judging from (12.241), we should have

$$f_{k+1}(Z|a,\mathbf{x}) = m! \cdot B_{a,p_D(\mathbf{x})}(m) \cdot f(\mathbf{z}_1|\mathbf{x}) \cdots f(\mathbf{z}_m|\mathbf{x}) \tag{12.244}$$

where $B_{a,q}(m)$ is the probability distribution corresponding to the p.g.f.

$$G_{a,q}(y) \cong (1 - q + qy)^n \cdot (1 - (a-n)q + (a-n)qy) . \tag{12.245}$$

However, this naïve definition will not do because the functions $a \mapsto G_{a,q}(y)$ for fixed y, q are not differentiable.

Instead let $\sigma(a)$ be a "sigmoidal" function with the following properties:

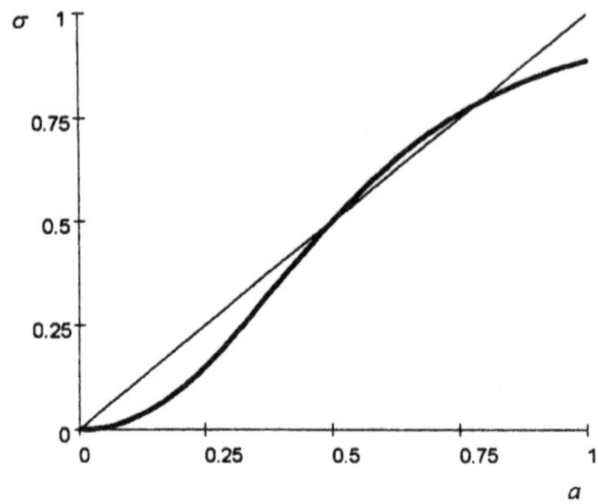

Figure 12.3 An exemplar of a sigmoidal function $x = \sigma(a)$ (thick curve) plotted together with the line $x = a$ (thin line). Here $\sigma(a)$ has been approximated by the function in (12.246).

- $\sigma(a)$ is infinitely differentiable;
- $\sigma(a) = 0$ for $a \leq 0$;
- $\sigma(\frac{1}{2}) = \frac{1}{2}$;
- $\sigma(a) \cong 1$ if $a \geq 1$;
- $(d\sigma/da)(\frac{1}{2}) \cong 1$.

That is, $\sigma(a) = 0$ for $a \leq 0$, $\sigma(a) \cong a$ for $0 < a < 1$, and $\sigma(a) \cong 1$ for $a \geq 1$. Figure 12.3 illustrates one possible exemplar for σ, in this case with $\sigma(a)$ approximated as the function

$$\sigma(a) \cong \frac{(2a)^2}{(2a)^2 + 2^{1-2a}}. \tag{12.246}$$

Let $\sigma_i(a) \triangleq \sigma(a - i)$. Define the p.g.f. $G_{a,q}(y)$ by

$$G_{a,q}(y) \triangleq \prod_{i=0}^{\infty} (1 - \sigma_i(a)q + \sigma_i(a)qy)\,. \tag{12.247}$$

The right-hand side is actually a *finite* product since $\sigma(a-i)=0$ for all $i>a$. Also, it has the intuitive form of (12.245), as we now see.

Example 70 *For* $0<a<1$,

$$\begin{aligned} G_{a,q}(y) &= [1-\sigma(a)q+\sigma(a)qy]\cdot[1-\sigma_1(a)q+\sigma_1(a)qy] \quad (12.248)\\ &\quad \cdot[1-\sigma_2(a)q+\sigma_2(a)qy]\cdots \quad (12.249)\\ &= 1-\sigma(a)q+\sigma(a)qy \quad (12.250)\\ &\cong 1-aq+aqy. \quad (12.251)\end{aligned}$$

For $1<a<2$,

$$\begin{aligned} G_{a,q}(y) &= [1-\sigma(a)q+\sigma(a)qy]\cdot[1-\sigma_1(a)q+\sigma_1(a)qy]\\ &\quad \cdot[1-\sigma_2(a)q+\sigma_2(a)qy]\cdots \quad (12.252)\\ &= [1-\sigma(a)q+\sigma(a)qy]\cdot[1-\sigma_1(a)q+\sigma_1(a)qy] \quad (12.253)\\ &\cong (1-q+qy)\cdot[1-(a-1)q+(a-1)qy]. \quad (12.254)\end{aligned}$$

For $2<a<3$,

$$\begin{aligned} G_{a,q}(y) &= [1-\sigma(a)q+\sigma(a)qy]\cdot[1-\sigma_1(a)q+\sigma_1(a)qy] \quad (12.255)\\ &\quad \cdot[1-\sigma_2(a)q+\sigma_2(a)qy]\cdots \quad (12.256)\\ &\cong (1-q+qy)^2\cdot[1-(a-2)q+(a-2)qy]. \quad (12.257)\end{aligned}$$

In general, if n *is the largest integer smaller than* a, *then for* $n<a<n+1$,

$$G_{a,q}(y)\cong(1-q+qy)^n\cdot[1-(a-n)q+(a-n)qy]. \quad (12.258)$$

Thus the definition of $G_{a,q}(y)$ *is consistent with the intuitive definition of (12.245).*

Also note that

$$\begin{aligned} G_{a,q}(1) &= 1 \quad (12.259)\\ \frac{d^mG_{a,q}}{dy^m}(0) &\geq 0 \quad \text{for all } m\geq 0 \quad (12.260)\\ B_{a,q}(0) &= G_{a,q}(0)=\prod_{i=0}^{\infty}(1-\sigma_i(a)q) \quad (12.261)\end{aligned}$$

and that

$$a \mapsto \frac{d^m G_{a,q}}{dy^m}(0) \tag{12.262}$$

is infinitely differentiable for each m and each y.

Thus we define the likelihood of a single point cluster to be

$$f(Z|a,\mathbf{x}) \triangleq \begin{cases} B_{a,p_D(\mathbf{x})}(0) & \text{if} \quad Z = \emptyset \\ m! B_{a,p_D(\mathbf{x})}(m) \cdot f(\mathbf{z}_1|\mathbf{x}) \cdots f(\mathbf{z}_m|\mathbf{x}) & \text{if} \quad Z \neq \emptyset \end{cases}. \tag{12.263}$$

It is left to the reader as Exercise 50 to verify that $f(Z|a,\mathbf{x})$ is a likelihood in the sense that

$$\int f(Z|a,\mathbf{x})\delta Z = 1. \tag{12.264}$$

Example 71 *We plot the functions* $a \mapsto B_{a,0.8}(0)$ *and* $a \mapsto B_{a,0.8}(1)$. *For the function* $a \mapsto B_{a,0.8}(0)$, *notice that for* $0 < a \leq 1$,

$$B_{a,q}(0) = G_{a,q}(0) = 1 - q\sigma(a) \tag{12.265}$$

and that for $1 < a < 2$,

$$B_{a,q}(0) = G_{a,q}(0) = (1 - \sigma(a)q) \cdot (1 - \sigma_1(a)q) \tag{12.266}$$

and so on. Using the approximate $\sigma(a)$ *of (12.246) we have, ignoring all* $a > 2$,

$$B_{a,0.8}(0) = \begin{cases} 1 - 0.8\frac{(2a)^2}{(2a)^2 + 2^{1-2a}} & \text{if} \quad 0 < a \leq 1 \\ \left[\begin{array}{c} \left(1 - 0.8\frac{(2a)^2}{(2a)^2+2^{1-2a}}\right) \\ \cdot \left(1 - 0.8\frac{(2a-2)^2}{(2a-2)^2+2^{3-2a}}\right) \end{array} \right] & \text{if} \quad 1 < a \leq 2 \end{cases}. \tag{12.267}$$

This is the curve plotted in Figure 12.4. Now turn to the function $a \mapsto B_{a,0.8}(1)$. *For* $0 < a < 1$,

$$B_{a,q}(1) = \frac{dG_{a,q}}{dy}(0) = \left[\frac{d}{dy}\left(1 - \sigma(a)q + \sigma(a)qy\right)\right]_{y=0} \tag{12.268}$$

$$= q\sigma(a). \tag{12.269}$$

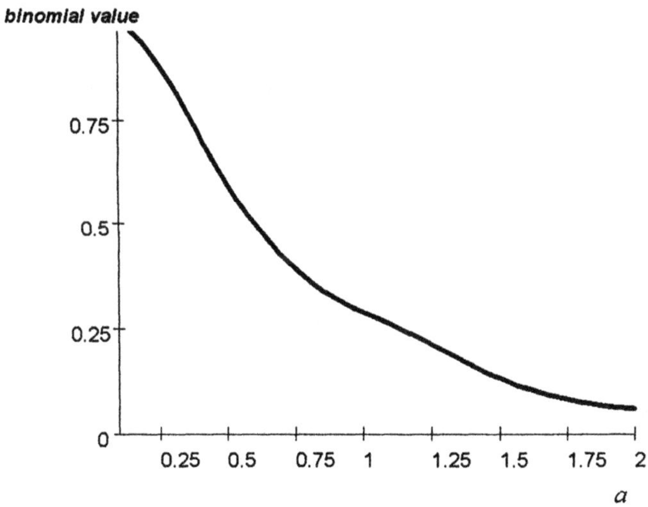

Figure 12.4 A plot of the continuous-variable binomial function $a \mapsto B_{a,q}(0)$ for $q = 0.8$ and $0 < a < 2$.

For $1 < a < 2$,

$$
\begin{aligned}
& B_{a,q}(1) && (12.270)\\
= \; & \frac{dG_{a,q}}{dy}(0) && (12.271)\\
= \; & \left[\frac{d}{dy}[1-\sigma(a)q+\sigma(a)qy]\cdot[1-\sigma_1(a)q+\sigma_1(a)qy]\right]_{y=0} && (12.272)\\
= \; & \left[\begin{array}{c} \sigma(a)q\cdot[1-\sigma_1(a)q+\sigma_1(a)qy] \\ +[1-\sigma(a)q+\sigma(a)qy]\cdot\sigma_1(a)q \end{array}\right]_{y=0} && (12.273)\\
= \; & \sigma(a)q\cdot[1-\sigma_1(a)q]+[1-\sigma(a)q]\cdot\sigma_1(a)q. && (12.274)
\end{aligned}
$$

For $2 < a < 3$,

$$
\begin{aligned}
& B_{a,q}(1) && (12.275)\\
= \; & \frac{dG_{a,q}}{dy}(0) && (12.276)\\
= \; & \left[\begin{array}{c} \frac{d}{dy}[1-\sigma(a)q+\sigma(a)qy]\cdot[1-\sigma_1(a)q+\sigma_1(a)qy] \\ \cdot[1-\sigma_2(a)q+\sigma_2(a)qy] \end{array}\right]_{y=0} && (12.277)\\
= \; & \sigma(a)q\cdot[1-\sigma_1(a)q]\cdot[1-\sigma_2(a)q] && (12.278)\\
& +[1-\sigma(a)q]\cdot\sigma_1(a)q\cdot[1-\sigma_2(a)q] && (12.279)\\
& +[1-\sigma(a)q]\cdot[1-\sigma_1(a)q]\cdot\sigma_2(a)q && (12.280)
\end{aligned}
$$

and so on. Using the approximate $\sigma(a)$ *of (12.246) we get the curve plotted in Figure 12.5.*

12.8.3 Multicluster Likelihoods

Suppose now that we have multiple point clusters with state $\mathring{X} = \{\mathring{\mathbf{x}}_1, ..., \mathring{\mathbf{x}}_n\}$ with $\mathring{\mathbf{x}}_i = (a_i, \mathbf{x}_i)$ for all $i = 1, ..., n$. In this section I generalize the standard multitarget measurement model of Section 12.3.6 to the multicluster case. The measurement model is the obvious generalization of (12.44):

$$
\overbrace{\Sigma_{k+1}}^{\text{measurement set}} = \overbrace{\Upsilon(\mathring{X})}^{\text{cluster detection set}} \cup \overbrace{C}^{\text{false detections}} \tag{12.281}
$$

where C is a Poisson false alarm process, where

$$
\Upsilon(\mathring{X}) = \overbrace{\Upsilon(\mathring{\mathbf{x}}_1)}^{\text{detections from } \mathring{\mathbf{x}}_1} \cup ... \cup \overbrace{\Upsilon(\mathring{\mathbf{x}}_n)}^{\text{detections from } \mathring{\mathbf{x}}_n} \tag{12.282}
$$

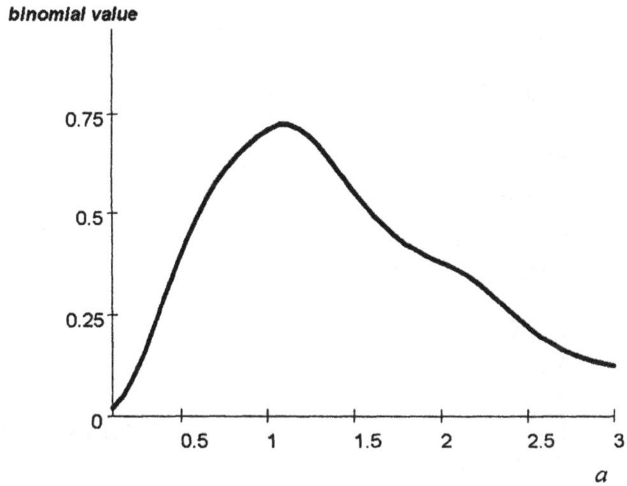

Figure 12.5 A plot of the continuous-variable binomial function $a \mapsto B_{a,q}(1)$ for $q = 0.8$ and $0 < a < 3$.

and where $\Upsilon(\mathring{\mathbf{x}}_1), ..., \Upsilon(\mathring{\mathbf{x}}_n), C$ are statistically independent.

From the fundamental convolution formula, (11.252),

$$f_{k+1}(Z|\mathring{X}) = \sum_{W_0 \uplus W_1 \uplus ... \uplus W_n = Z} f_C(W_0) \cdot f(W_1|\mathring{\mathbf{x}}_1) \cdots f(W_n|\mathring{\mathbf{x}}_n) \tag{12.283}$$

where the summation is taken over all mutually disjoint subsets $W_0, W_1, ..., W_n$ of Z such that $W_0 \cup W_1 \cup ... \cup W_n = Z$. However, by (12.263) the likelihood $f(Z|\mathring{\mathbf{x}}_i)$ for $\Upsilon(\mathring{\mathbf{x}}_i)$ is

$$f(Z|a_i, \mathbf{x}_i) \triangleq \left\{ \begin{array}{ll} B_{a_i, p_D(\mathbf{x}_i)}(0) & \text{if} \quad Z = \emptyset \\ m! B_{a_i, p_D(\mathbf{x}_i)}(m) \cdot f(\mathbf{z}_1|\mathbf{x}_i) \cdots f(\mathbf{z}_m|\mathbf{x}_i) & \text{if} \quad Z \neq \emptyset \end{array} \right. . \tag{12.284}$$

Also, the likelihood $f_C(Z)$ is $f_C(Z) = e^{-\lambda} \prod_{\mathbf{z} \in Z} \lambda c(\mathbf{z})$. Thus

$$\begin{aligned} f_{k+1}(Z|\mathring{X}) &= e^{-\lambda} \sum_{W_0 \uplus W_1 \uplus ... \uplus W_n = Z} |W_1|! \cdots |W_n|! & (12.285) \\ &\cdot \lambda^{|W_0|} \cdot B_{a_1, p_D(\mathbf{x}_1)}(|W_1|) & (12.286) \\ &\cdots B_{a_n, p_D(\mathbf{x}_n)}(|W_n|) & (12.287) \\ &\cdot \left(\prod_{\mathbf{z} \in W_0} \lambda c(\mathbf{z}) \right) & (12.288) \\ &\cdot \left(\prod_{\mathbf{z} \in W_1} f_{k+1}(\mathbf{z}|\mathbf{x}_1) \right) \cdots \left(\prod_{\mathbf{z} \in W_n} f_{k+1}(\mathbf{z}|\mathbf{x}_n) \right) & (12.289) \end{aligned}$$

12.8.4 Continuity of Multicluster Likelihoods

Let the likelihood $f_{k+1}(Z|\mathring{X})$ be as defined in the previous section. In Appendix G.21, I demonstrate that $f_{k+1}(Z|\mathring{X})$ is continuous in the variable $\mathring{X}$ in the following sense:

$$\begin{aligned} \lim_{a \searrow 0} f_{k+1}(Z|a, \mathbf{x}) &= f_{k+1}(Z|\emptyset) & (12.290) \\ \lim_{a \searrow 0} f_{k+1}(Z|\mathring{X} \cup \{(a, \mathbf{x})\}) &= f_{k+1}(Z|\mathring{X}). & (12.291) \end{aligned}$$

12.9 MULTISOURCE MEASUREMENT MODELS

In previous sections we have derived multitarget likelihood functions for *single-sensor* measurement models only. The purpose of this section is to describe what multitarget likelihood functions look like for multisensor scenarios, at least given suitable independence assumptions [70, pp. 225-228].

Readers should reacquaint themselves with the discussion of sensor state spaces in Section 12.2.2.

12.9.1 Conventional Measurements

Suppose that we have s sensors in a scene. The states of the sensors are denoted by $\overset{*}{X} = \{\overset{*1}{\mathbf{x}}, ..., \overset{*s}{\mathbf{x}}\}$. The multisensor measurement space is a disjoint union

$$\mathfrak{Z}_0 = \overset{1}{\mathfrak{Z}}_0 \uplus ... \uplus \overset{s}{\mathfrak{Z}}_0 \tag{12.292}$$

where $\overset{j}{\mathfrak{Z}}_0$ is the measurement space for the jth sensor. The reason for the disjointness is that any measurement in $\overset{j}{\mathfrak{Z}}_0$ has the general form $\overset{j}{\mathbf{z}} = (\overset{j}{\mathbf{y}}, j)$ where $j = 1, ..., s$ is the unique identifying tag of the sensor that originated observation $\overset{j}{\mathbf{y}}$. Thus even if the sensors have identical statistics, their corresponding measurement spaces are distinct. The total observation set collected by all sensors therefore has the form

$$Z = \overset{1}{Z} \uplus ... \uplus \overset{s}{Z} \tag{12.293}$$

where $\overset{j}{Z} = Z \cap \overset{j}{\mathfrak{Z}}_0$ is the observation set collected by the jth sensor.

What is the true likelihood function $f_{k+1}(Z|X, \overset{*}{X})$ for the multisensor case? Assuming that measurements are conditionally independent of target states, in Appendix G.20, I show that, as one would expect,

$$f_{k+1}(Z|X, \overset{*}{X}) = f_{k+1}(\overset{1}{Z}|X, \overset{*1}{\mathbf{x}}) \cdots f_{k+1}(\overset{s}{Z}|X, \overset{*s}{\mathbf{x}}) \tag{12.294}$$

$$= \overset{1}{f}_{k+1}(\overset{1}{Z}|X) \cdots \overset{s}{f}_{k+1}(\overset{s}{Z}|X) \tag{12.295}$$

where $f_{k+1}(\overset{j}{Z}|X, \overset{*j}{\mathbf{x}})$ is the multitarget likelihood for the jth sensor.

Example 72 *Suppose that two sensors interrogate two targets* $X = \{x_1, x_2\}$ *that are moving on the real line. The first sensor is position-measuring and the second* *is velocity-measuring. Neither sensor is affected by missed detections or false alarms. Measurements are assumed conditionally independent of target state. What is the two sensor likelihood function? The measurement spaces are, respectively,* $\overset{1}{\mathfrak{Z}}_0 = \mathbb{R} \times \{1\}$ *and* $\overset{2}{\mathfrak{Z}}_0 = \mathbb{R} \times \{2\}$. *Target states have the form* $\mathbf{x} = (x, v)$ *and the measurement models of the two sensors have the respective forms*

$$\overset{1}{z} = x + \overset{1}{W}, \qquad \overset{2}{z} = v + \overset{2}{W} \tag{12.296}$$

where $\overset{1}{W}$ *and* $\overset{2}{W}$ *are zero-mean random numbers. The corresponding single-target likelihoods are*

$$\overset{1}{f}_{k+1}(\overset{1}{z}|x, v) = f_{\overset{1}{W}}(\overset{1}{z} - x), \qquad \overset{2}{f}_{k+1}(\overset{2}{z}|x, v) = f_{\overset{2}{W}}(\overset{2}{z} - v). \tag{12.297}$$

Typical single-sensor measurement sets have the form $\overset{1}{Z} = \{\overset{1}{z}_1, \overset{1}{z}_2\}$ *and* $\overset{2}{Z} = \{\overset{2}{z}_1, \overset{2}{z}_2\}$. *Therefore the corresponding single-sensor multitarget likelihoods are, by (12.91) or by Example 69,*

$$\overset{1}{f}_{k+1}(\overset{1}{Z}|X) = \overset{1}{f}_{k+1}(\{\overset{1}{z}_1, \overset{1}{z}_2\}|\{(x_1, v_1), (x_2, v_2)\}) \tag{12.298}$$

$$= f_{\overset{1}{W}}(\overset{1}{z}_1 - x_1) \cdot f_{\overset{1}{W}}(\overset{1}{z}_2 - x_2) \tag{12.299}$$

$$+ f_{\overset{1}{W}}(\overset{1}{z}_2 - x_1) \cdot f_{\overset{1}{W}}(\overset{1}{z}_1 - x_2) \tag{12.300}$$

$$\overset{2}{f}_{k+1}(\overset{2}{Z}|X) = \overset{2}{f}_{k+1}(\{\overset{2}{z}_1, \overset{2}{z}_1\}|\{(x_1, v_1), (x_2, v_2)\}) \tag{12.301}$$

$$= f_{\overset{2}{W}}(\overset{2}{z}_1 - v_1) \cdot f_{\overset{2}{W}}(\overset{2}{z}_2 - v_2) \tag{12.302}$$

$$+ f_{\overset{2}{W}}(\overset{2}{z}_2 - v_1) \cdot f_{\overset{2}{W}}(\overset{2}{z}_1 - v_2) \tag{12.303}$$

where $\overset{1}{f}_{k+1}(\overset{1}{Z}|X) = 0$ *and* $\overset{2}{f}_{k+1}(\overset{2}{Z}|X) = 0$ *if* $|X| \neq 2$. *A typical multisensor measurement set has the form*

$$Z = \overset{1}{Z} \cup \overset{2}{Z} = \{\overset{1}{z}_1, \overset{1}{z}_2, \overset{2}{z}_1, \overset{2}{z}_2\}. \tag{12.304}$$

Since measurements are conditionally independent of target states, it follows that the multisensor-multitarget likelihood is

$$\begin{aligned}
& f_{k+1}(\{\overset{1}{z}_1, \overset{1}{z}_2, \overset{2}{z}_1, \overset{2}{z}_2\}|\{(x_1, v_1), (x_2, v_2)\}) && (12.305) \\
= \; & \overset{1}{f}_{k+1}(\{\overset{1}{z}_1, \overset{1}{z}_2\}|\{(x_1, v_1), (x_2, v_2)\}) && (12.306) \\
& \cdot \overset{2}{f}_{k+1}(\{\overset{2}{z}_1, \overset{2}{z}_2\}|\{(x_1, v_1), (x_2, v_2)\}) && (12.307) \\
= \; & \left[f_{\overset{1}{W}}(\overset{1}{z}_1 - x_1) f_{\overset{1}{W}}(\overset{1}{z}_2 - x_2) + f_{\overset{1}{W}}(\overset{1}{z}_2 - x_1) f_{\overset{1}{W}}(\overset{1}{z}_1 - x_2) \right] && (12.308) \\
& \cdot \left[\begin{array}{c} f_{\overset{2}{W}}(\overset{2}{z}_1 - v_1) f_{\overset{2}{W}}(\overset{2}{z}_2 - v_2) \\ + f_{\overset{2}{W}}(\overset{2}{z}_2 - v_1) f_{\overset{2}{W}}(\overset{2}{z}_1 - v_2) \end{array} \right]. && (12.309)
\end{aligned}$$

12.9.2 Generalized Measurements

Until now, we have implicitly assumed that our information source is a conventional sensor that can be modeled using a conventional likelihood function. Clearly, however, identical reasoning can be applied to general information sources modeled using generalized likelihood functions, as described in Chapters 5-7.

Example 73 (Generalized Measurements, Single-Target) *Consider the multitarget likelihood function for the single-target standard model, (12.68):*

$$f_{k+1}(Z|\mathbf{x}) = f_C(Z) \cdot \left(1 - p_D(\mathbf{x}) + p_D(\mathbf{x}) \sum_{\mathbf{z} \in Z} \frac{f(\mathbf{z}|\mathbf{x})}{\lambda c(\mathbf{z})}\right) \quad (12.310)$$

$$f_C(Z) = e^{-\lambda} \prod_{\mathbf{z} \in Z} \lambda c(\mathbf{z}). \quad (12.311)$$

Suppose that the source instead supplies unambiguously generated ambiguous (UGA) fuzzy Dempster-Shafer measurements. The corresponding generalized likelihood function, see (5.73), is:

$$f_{k+1}(o|\mathbf{x}) = \sum_g o(g) \cdot g(\eta(\mathbf{x})). \quad (12.312)$$

Assuming that the usual independence assumptions apply, $f(o|\mathbf{x}) \overset{\text{abbr.}}{=} f_{k+1}(o|\mathbf{x})$ *can be used in place of* $f_{k+1}(\mathbf{z}|\mathbf{x})$ *to get a corresponding multitarget generalized likelihood function* $f_{k+1|k}(Z|X)$*. In this case any measurement set has the form*

$Z = \{o_1, ..., o_m\}$ rather than $Z = \{\mathbf{z}_1, ..., \mathbf{z}_m\}$. Assume that we have a Poisson clutter process λ, $c(o)$ defined on fuzzy DS measurements. Then (12.310) and (12.311) become

$$f_{k+1}(Z|\mathbf{x}) = f_C(Z) \cdot \left(1 - p_D(\mathbf{x}) + p_D(\mathbf{x}) \sum_{o \in Z} \frac{f(o|\mathbf{x})}{\lambda c(o)}\right) \quad (12.313)$$

$$f_C(Z) = e^{-\lambda} \prod_{o \in Z} \lambda c(o). \quad (12.314)$$

12.10 A MODEL FOR BEARING-ONLY MEASUREMENTS

Certain types of bearing-only target tracking problems are substantially different than the standard detection-type problems thus far considered.

Consider the following situation. Several bearing-only sensors scan a surveillance region. A significant amount of time may be required for any given sensor to complete a single scan. Each sensor scans until it obtains a detection and then reports it. The detection can have originated with a target or it can be a false alarm. Each sensor continues in this fashion, reporting back a detection each time one is encountered. If nothing was detected during a scan, the sensor reports this fact.

The measurement reported by a sensor will have the form $Z = \emptyset$ (nothing detected during an entire scan) or $Z = \{a\}$ (a target was detected at scan-angle a). These measurements will be reported in time order, but the label of the sensor delivering a report will be a random variable.

Using FISST modeling techniques, Vihola [234, 232, 233] has derived multitarget likelihood functions for such sensors and employed them in a multitarget particle filter (see Section 15.2.4). I present a slightly generalized version of Vihola's likelihood:[7]

If $X = \emptyset$, then

$$f_{k+1}(Z|\emptyset) = \begin{cases} 1 - p_{FA} & \text{if} \quad Z = \emptyset \\ p_{FA} \cdot \kappa_{k+1}(a) & \text{if} \quad Z = \{a\} \\ 0 & \text{if} \quad |Z| > 1 \end{cases}. \quad (12.315)$$

On the other hand let $X \neq \emptyset$ with $X = \{\mathbf{x}_1, ..., \mathbf{x}_n\}$ and $|X| = n$. Then $f_{k+1}(Z|X) = 0$ whenever $|Z| > 1$. If $Z = \emptyset$

$$f_{k+1}(\emptyset|X) = 1 - q_X. \quad (12.316)$$

7 Vihola has actually proposed two versions of this likelihood. The one described here is from [234].

If $Z = \{a\}$,

$$f_{k+1}(\{a\}|X) = q_X \cdot \left(p_{FA}\kappa(a) + (1 - p_{FA}) \sum_{i=1}^{n} \hat{p}_D^X(\mathbf{x}_i) \cdot L_a(\mathbf{x}_i) \right). \quad (12.317)$$

Here

$$\hat{p}_D^X(\mathbf{x}_i) \triangleq \frac{p_D(\mathbf{x}_i)}{p_D(\mathbf{x}_1) + ... + p_D(\mathbf{x}_n)}. \quad (12.318)$$

for $i = 1, ..., n$ and the following are true:

- $p_D(\mathbf{x})$ is the probability that a target with state $\mathbf{x}$ will generate a detection.
- $L_a(\mathbf{x}) \overset{\text{abbr.}}{=} f_{k+1}(a|\mathbf{x})$ is the likelihood that a detection a will be generated if the target has state $\mathbf{x}$.
- p_{FA} is the (constant) probability that the detection is a false alarm rather than a target detection.
- $\kappa(a) \overset{\text{abbr.}}{=} \kappa_{k+1}(a)$ is the spatial distribution of the false alarm process.
- All measurements are conditionally independent of target state.
- The probability that a non-null measurement will be collected, given that targets with state set X are present, is:

$$q_X \triangleq \begin{cases} p_{FA} & \text{if} \quad X = \emptyset \\ 1 - (1 - p_{FA}) \cdot \prod_{i=1}^{n} (1 - p_D(\mathbf{x}_i)) & \text{if} \quad X \neq \emptyset \end{cases}. \quad (12.319)$$

Consider two special cases. First, if p_D is constant then $\hat{p}_D^X(\mathbf{x}_i) = 1/n$ for all $i = 1, ..., n$. In this case (12.317) reduces to Vihola's formula in [234]:

$$f_{k+1}(\{a\}|X) = q_X \cdot \left(p_{FA}\kappa(a) + \frac{1 - p_{FA}}{n} \sum_{i=1}^{n} L_a(\mathbf{x}_i) \right). \quad (12.320)$$

Second, assume that there are no false alarms and that $\mathbf{x}_2, ..., \mathbf{x}_n$ are not in the sensor field of view: $p_D(\mathbf{x}_i) = 0$ for $i = 2, ..., n$. Then $q_X = p_D(\mathbf{x}_1)$ and $\hat{p}_D^X(\mathbf{x}_1) = 1$ and thus (12.317) reduces to:

$$f_{k+1}(\{a\}|X) = p_D(\mathbf{x}_1) \cdot L_a(\mathbf{x}_1). \quad (12.321)$$

That is, if only a single target is visible to the sensor then $f_{k+1}(\{a\}|X)$ reduces to the usual single-target likelihood with nonunity probability of detection.

In this section I summarize the derivation of (12.315)-(12.317).

12.10.1 Multitarget Measurement Model

Let Σ be the random measurement set. Note that $\Sigma = \emptyset$ when both of the following are true: (1) no target is detected (if present) and (2) no false alarm is collected. The probability of this occurring is

$$(1 - p_{FA}) \cdot \prod_{i=1}^{n} (1 - p_D(\mathbf{x}_i)) = 1 - q_X. \quad (12.322)$$

This tells us that the measurement model must have the form

$$\Sigma = \Sigma' \cap \emptyset^{q_X} \quad (12.323)$$

where $|\Sigma'| = 1$ and where $\emptyset^q$ is defined as in (12.47). (That is, $\emptyset^q = \mathfrak{Z}_0$ with probability q and $\emptyset^q = \emptyset$ with probability $1 - q$.) Our task is to determine the form of the random singleton set Σ'.

Assume first that $X = \emptyset$, so that $q_X = p_{FA}$. Then every non-null measurement will be a false alarm and Σ has the form

$$\Sigma = \{A_0\} \cap \emptyset^{q_{FA}} \quad (12.324)$$

where $A_0 \in [0, 2\pi]$ is a random radian angle with distribution $\kappa(a)$. That is, $\Sigma = \emptyset$ (no detection) with probability $1 - p_{FA}$ and $\Sigma = \{a\}$ with probability p_{FA}, with $\kappa(a)$ being the likelihood that the specific bearing angle a will be collected.

On the other hand, let $X = \{\mathbf{x}_1, ..., \mathbf{x}_n\} \neq \emptyset$. Let A_i be a random angle with distribution $f_{k+1}(a|\mathbf{x}_i)$ and assume that $A_1, ..., A_n$ are statistically independent. If a is generated by $\mathbf{x}_i$ then the measurement set would be $\Sigma = \{A_i\}$ with $A_i = a$. If a is a false alarm then it would be $\Sigma = \{A_0\}$ with $A_0 = a$. What are the probabilities that these respective events will occur?

First note that false alarms dominate target detections in the following sense. If $p_{FA} = 1$ then every a is a false alarm. Thus the probability that a will be a false alarm is p_{FA}. The probability that it will not be a false alarm (i.e, a target detection) is $1 - p_{FA}$. Thus a originates with $\mathbf{x}_i$ only if it is not a false alarm and if it is detected. This probability is $(1 - p_{FA}) \cdot p_D(\mathbf{x}_i)$. The frequencies with which the event A_i will occur, given that it is not a false alarm, are

$$p_D^X(\mathbf{x}_i) = \frac{p_D(\mathbf{x}_i)}{p_D(\mathbf{x}_1) + ... + p_D(\mathbf{x}_n)}. \quad (12.325)$$

So let J be a random integer on $\{0, 1, ..., n\}$ such that

$$\Pr(I = 0) = q_0 \triangleq p_{FA} \tag{12.326}$$
$$\Pr(I = i) = q_i \triangleq (1 - p_{FA}) \cdot p_D^X(\mathbf{x}_i). \tag{12.327}$$

for $i = 1, ..., n$. Then the measurement model for a bearing-only sensor is

$$\Sigma = \{A_I\} \cap \emptyset^{q_X}. \tag{12.328}$$

That is, if a detection is collected then any of the $A_0, A_1, ..., A_m$ are the possible models, with respective probabilities $p_0, p_1, ..., p_n$ that they will occur.

In what follows we assume that A_0, $A_1, ..., A_n$, I, $\emptyset^{q_X}$ are statistically independent.

12.10.2 Belief-Mass Function

Consider the case when $X = \emptyset$ first, so that every non-null measurement is a false alarm. The belief-mass function is

$$\beta_{k+1}(T|X) = \Pr(\Sigma \subseteq T|X) = \Pr(\{A_0\} \cap \emptyset^{p_{FA}} \subseteq T) \tag{12.329}$$
$$= \Pr(\emptyset^{p_{FA}} = \emptyset) + \Pr(\emptyset^{p_{FA}} = \emptyset, A_0 \in T) \tag{12.330}$$
$$= 1 - p_{FA} + p_{FA} \cdot p_0(T) \tag{12.331}$$

where $p_0(T) \overset{\text{abbr.}}{=} \int_T \kappa(a)da$.

On the other hand if $X \neq \emptyset$, then

$$\beta_{k+1}(T|X) = \Pr(\Sigma \subseteq T|X) = \Pr(\{A_I\} \cap \emptyset^{q_X} \subseteq T|X) \tag{12.332}$$
$$= \Pr(\emptyset^{q_X} = \emptyset|X) + \Pr(\emptyset^{q_X} \neq \emptyset|X) \tag{12.333}$$
$$\cdot \Pr(A_I \in T) \tag{12.334}$$
$$= 1 - q_X + q_X \sum_{i=0}^{n} \Pr(A_I \in T, I = i) \tag{12.335}$$
$$= 1 - q_X + q_X \sum_{i=0}^{n} q_i \cdot \Pr(A_I \in T) \tag{12.336}$$
$$= 1 - q_X + q_X \cdot \left(q_0 \cdot p_0(T) + \sum_{i=1}^{n} q_i \cdot p_i(T) \right) \tag{12.337}$$

where $p_i(T) \overset{\text{abbr.}}{=} \int_T f_{k+1}(a|\mathbf{x}_i)da$ for $i = 1, ..., n$. Thus

$$\beta_{k+1}(T|X) = 1 - q_X + q_X \cdot p_{FA} p_0(T) \quad (12.338)$$

$$+ q_X \cdot (1 - p_{FA}) \sum_{i=1}^{n} \hat{p}_D^X(\mathbf{x}_i) \cdot p_i(T). \quad (12.339)$$

12.10.3 Multitarget Likelihood Function

Begin with $X = \emptyset$. The first set derivative of $\beta_{k+1}(T|\emptyset)$ is

$$\frac{\delta\beta_{k+1}}{\delta a}(T|\emptyset) = p_{FA} \cdot \kappa_{k+1}(a). \quad (12.340)$$

Thus

$$f_{k+1}(\emptyset|\emptyset) = \beta_{k+1}(\emptyset|X) = 1 - p_{FA} \quad (12.341)$$

$$f_{k+1}(\{a\}|\emptyset) = \frac{\delta\beta_{k+1}}{\delta a}(\emptyset|X) = p_{FA} \cdot \kappa(a). \quad (12.342)$$

Now let $X \neq \emptyset$. The first set derivative of $\beta_{k+1}(T|X)$ is

$$\frac{\delta\beta_{k+1}}{\delta a}(T|X) = q_X p_{FA} \kappa(a) + q_X (1 - p_{FA}) \sum_{i=1}^{n} \hat{p}_D^X(\mathbf{x}_i) \cdot L_a(\mathbf{x}_i). \quad (12.343)$$

Since this no longer has any functional dependence on T, all higher-order set derivatives vanish. Consequently for $n \geq 1$,

$$f_{k+1}(\emptyset|X) = \beta_{k+1}(\emptyset|X) = 1 - q_X \quad (12.344)$$

$$f_{k+1}(\{a\}|X) = \frac{\delta\beta_{k+1}}{\delta a}(\emptyset|X) \quad (12.345)$$

$$= q_X \cdot p_{FA} \cdot \kappa(a) \quad (12.346)$$

$$+ q_X (1 - p_{FA}) \sum_{i=1}^{n} \hat{p}_D^X(\mathbf{x}_i) \cdot L_a(\mathbf{x}_i). \quad (12.347)$$

12.11 A MODEL FOR DATA-CLUSTER EXTRACTION

"Data clustering" (often also called "data classification" or "data categorization") is the problem of partitioning a finite set $Z = \{\mathbf{z}_1, ..., \mathbf{z}_m\}$ of data into a union Z

$= C_1 \uplus ... \uplus C_\gamma$ of some undetermined number γ of undetermined and mutually disjoint "classes" ("clusters," "categories") $C_1, ..., C_\gamma$ according to some criteria of similarity or closeness.[8] Once this has been accomplished, the items in any C_j are identified as being similar to, related to, or nearby each other.

This process is illustrated in Figure 12.6. It is called "hard clustering" since the classes are sharply defined, disjoint, and any element of Z is unequivocally assigned to one and only one class. "Soft clustering," as illustrated in Figure 12.7, treats the classes $C_1, ..., C_\gamma$ as having "fuzzy," overlapping boundaries.

In this section, I generalize a Bayesian data clustering approach due to Cheeseman et al. [24, 25]. When used as part of a multitarget Bayes filter (Chapter 14) my modeling approach produces a rigorous, systematic, and fully Bayesian means of dynamically partitioning time-evolving data-sets into soft clusters. This includes estimating the number of clusters simultaneously with the shapes of the clusters.

In what follows I describe the finite-mixture model approach to data clustering (Section 12.11.1), the construction of a likelihood function for data clustering (Section 12.11.2), and estimation of soft classes (Section 12.11.3).

12.11.1 Finite-Mixture Models

I begin by assuming that we have drawn an instantiation $Z = \{\mathbf{z}_1, ..., \mathbf{z}_m\}$ from some random subset Σ of a measurement space $\mathfrak{Z}_0$. The statistics of Σ are unknown and must be determined. I assume that the statistics of any single datum $\mathbf{z}$ can be approximately modeled as a mixture distribution of the form

$$f(\mathbf{z}|Q) = \frac{a_1 f(\mathbf{z}|\mathbf{u}_1) + ... + a_\gamma f(\mathbf{z}|\mathbf{u}_\gamma)}{a_1 + ... + a_\gamma} \tag{12.348}$$

where

$$Q = \{(a_1, \mathbf{u}_1), ..., (a_\gamma, \mathbf{u}_\gamma)\} \tag{12.349}$$

and where $a_1, ..., a_\gamma \geq 0$ and where Q is called a "model."

For any $\mathbf{u}_j$, the *class distribution* $f(\mathbf{z}|\mathbf{u}_j)$ defines the statistics of the class whose associated parameter vector is $\mathbf{u}_j$.[9] The mixing parameter

$$A_j \triangleq \frac{a_j}{a_1 + ... + a_\gamma} \tag{12.350}$$

8 The material in this section was first reported in [123].

9 In [24, 25] $f(\mathbf{z}|\mathbf{u})$ is a Gaussian: $f(\mathbf{z}|\mathbf{z}_0, R) = N_R(\mathbf{z} - \mathbf{z}_0)$.

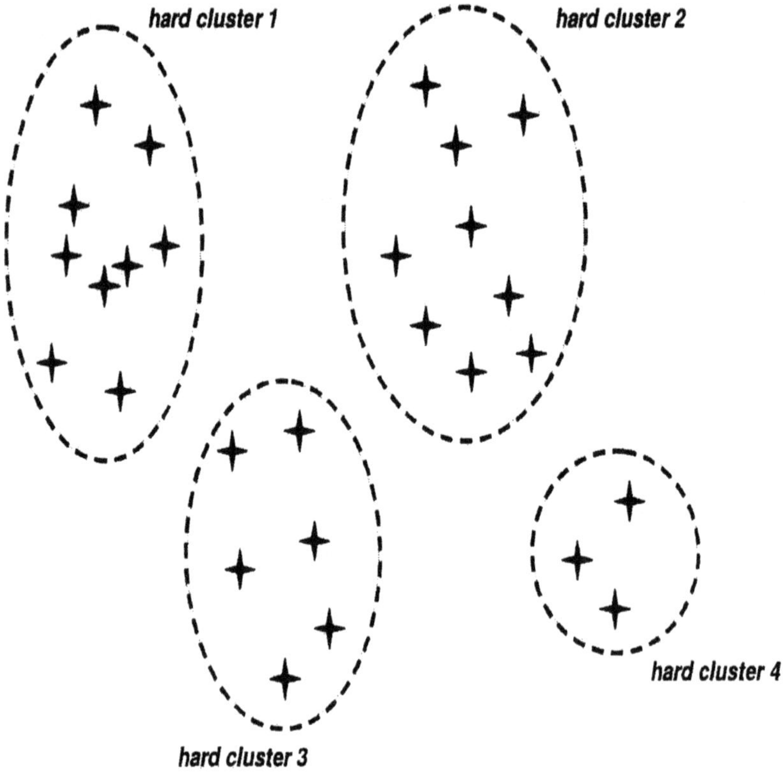

Figure 12.6 Hard clustering: The figure illustrates a data set (the black stars) that has been partitioned into clusters/classes on the basis of spatial distance. The clustering is "hard" because every datum has been uniquely associated with a single cluster/class.

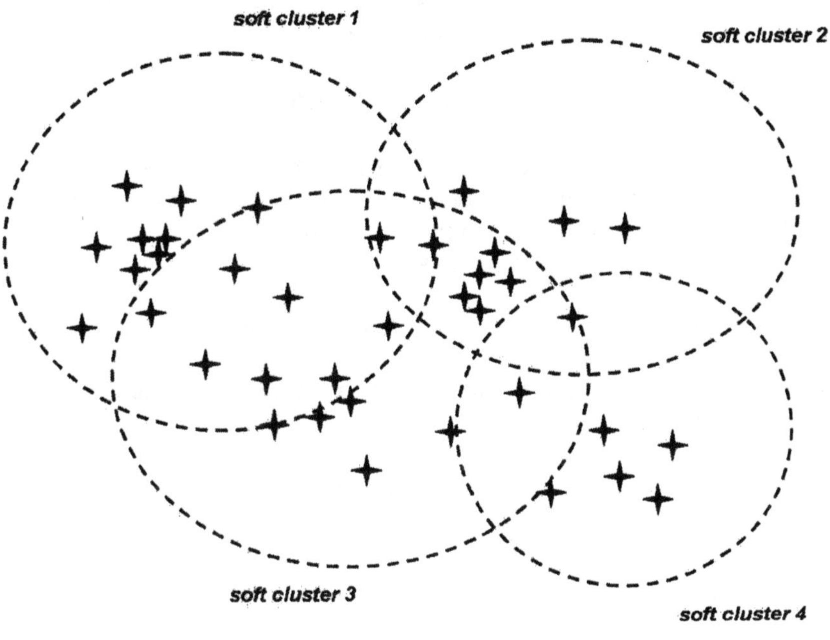

Figure 12.7 Soft clustering: A data set has been "partitioned" into soft clusters. The clustering is "soft" because it is not possible to uniquely assign every datum to a particular cluster.

determines the degree to which the jth class contributes to the observed statistics of $\mathbf{z}$. The integer γ is the number of classes. Hereafter I abbreviate

$$\mathbf{q}_j \triangleq (a_j, \mathbf{u}_j) \tag{12.351}$$

for all $j = 1, ..., \gamma$, so that $Q = \{\mathbf{q}_1, ..., \mathbf{q}_\gamma\}$.

To cluster the data Z, we are to determine which choice of γ, $\mathbf{u}_1, ..., \mathbf{u}_\gamma$, $a_1, ..., a_\gamma$ best explains the generation of Z. In other words, we are to estimate the optimal parameter set (also known as optimal model) $\hat{Q} = \{\hat{\mathbf{q}}_1, ..., \hat{\mathbf{q}}_{\hat{\gamma}}\}$.

12.11.2 A Likelihood for Finite-Mixture Modeling

The data in Z is assumed to be conditionally independent given the model Q. Also, the probability of drawing m pieces of data is given by some probability distribution $p(m)$, which we assume to be parametrized: $p(m) = p_{\mathbf{p}}(m) = p(m|\mathbf{p})$. The probability of drawing a vector $(\mathbf{z}_1, ..., \mathbf{z}_m)$ is

$$f(\mathbf{z}_1, ..., \mathbf{z}_m|Q, \mathbf{p}) = p_{\mathbf{p}}(m) \cdot f(\mathbf{z}_1|Q) \cdots f(\mathbf{z}_m|Q). \tag{12.352}$$

The probability of drawing a finite set Z is, because there are $m!$ vectors corresponding to Z,

$$\begin{aligned} f(Z|Q, \mathbf{p}) &= m! \cdot f(\mathbf{z}_1, ..., \mathbf{z}_m|Q) & (12.353)\\ &= m! \cdot p_{\mathbf{p}}(m) \cdot f(\mathbf{z}_1|Q) \cdots f(\mathbf{z}_m|Q). & (12.354) \end{aligned}$$

In other words,

$$\begin{aligned} & f(Z|Q, \mathbf{p}) & (12.355)\\ = \; & m! \cdot p_{\mathbf{p}}(m) & (12.356)\\ & \cdot \frac{\begin{array}{c}(a_1 f(\mathbf{z}_1|\mathbf{u}_1) + ... + a_\gamma f(\mathbf{z}_1|\mathbf{u}_\gamma)) \\ \cdots (a_1 f(\mathbf{z}_m|\mathbf{u}_1) + ... + a_\gamma f(\mathbf{z}_m|\mathbf{u}_\gamma))\end{array}}{(a_1 + ... + a_\gamma)^m}. & (12.357) \end{aligned}$$

This equation is well defined since it is invariant with respect to reorderings of both the $\mathbf{z}_1, ..., \mathbf{z}_m$ and the $\mathbf{q}_1, ..., \mathbf{q}_\gamma$.[10]

10 Note that (12.355) defines an i.i.d. cluster process as introduced in (11.121).

To be a likelihood function, $f(Z|Q,\mathbf{p})$ must integrate to unity:

$$\int f(Z|Q,\mathbf{p})\delta Z \tag{12.358}$$

$$= \sum_{m=0}^{\infty} \frac{1}{m!} \int f(\{\mathbf{z}_1,...,\mathbf{z}_m\}|Q,\mathbf{p})d\mathbf{z}_1 \cdots d\mathbf{z}_m \tag{12.359}$$

and so

$$= \sum_{m=0}^{\infty} \frac{p_{\mathbf{p}}(m)}{(a_1+...+a_\gamma)^m} \tag{12.360}$$

$$\cdot \int (a_1 f(\mathbf{z}_1|\mathbf{u}_1)+...+a_\gamma f(\mathbf{z}_1|\mathbf{u}_\gamma)) \tag{12.361}$$

$$\cdots (a_1 f(\mathbf{z}_m|\mathbf{u}_1)+...+a_\gamma f(\mathbf{z}_m|\mathbf{u}_\gamma))\, d\mathbf{z}_1 \cdots d\mathbf{z}_m \tag{12.362}$$

and so

$$= \sum_{m=0}^{\infty} p_{\mathbf{p}}(m) = 1. \tag{12.363}$$

In general, solving the clustering problem requires estimation of $\mathbf{p}$ as well as of Q. To avoid this additional complexity, in [123] I chose $p_{\mathbf{p}}(m)$ to be a Poisson distribution of the form

$$p(m|a) = \frac{e^{-a}a^m}{m!}, \qquad a = a_1+...+a_\gamma. \tag{12.364}$$

Since a can be determined from the model parameters in Q, no additional parameters need be computed.

12.11.3 Extraction of Soft Data Classes

Suppose that, through a multitarget state-estimation process such as those described in Section 14.5, we have derived an optimal model

$$\hat{Q} = \{(\hat{a}_1,\hat{\mathbf{u}}_1),...,(\hat{a}_\gamma,\hat{\mathbf{u}}_{\hat{\gamma}})\}, \quad \hat{\mathbf{p}}$$

where, in particular $\hat{\gamma}$ is the estimated number of classes. How do we determine which class any given datum $\mathbf{z}$ belongs to?

The quantity $f(\mathbf{z}|\hat{Q},\hat{\mathbf{p}})$ is the likelihood that $\mathbf{z}$ will be generated given the complete mixture model. On the other hand, $\hat{A}_j f(\mathbf{z}|\hat{\mathbf{u}}_j)$ is the likelihood that $\mathbf{z}$ will be generated by the jth model alone, given the proportion $\hat{A}_j \triangleq \hat{a}_j/(\hat{a}_1 + ... + \hat{a}_{\hat{\gamma}})$ that this model contributes to the total model. Consequently, the degree to which the jth model contributed to the generation of $\mathbf{z}$ is[11]

$$0 \leq \kappa_j(\mathbf{z}) \triangleq \frac{\hat{A}_j \cdot f(\mathbf{z}|\hat{\mathbf{u}}_j)}{f(\mathbf{z}|\hat{Q})} = \frac{\hat{a}_j f(\mathbf{z}|\hat{\mathbf{u}}_j)}{\hat{a}_1 f(\mathbf{z}|\hat{\mathbf{u}}_1) + ... + \hat{a}_{\hat{\gamma}} f(\mathbf{z}|\hat{\mathbf{u}}_{\hat{\gamma}})} \leq 1 \tag{12.365}$$

for $j = 1, ..., \hat{\gamma}$.

If the value of $\kappa_j(\mathbf{z})$ is small then $\mathbf{z}$ almost certainly did not originate with the jth class. If it is close to 1 then it almost certainly did so.

Thus $\kappa_j(\mathbf{z})$ is a *fuzzy membership function* on $\mathfrak{Z}_0$ that specifies the jth class in the form of a *fuzzy subset* of $\mathfrak{Z}_0$. The data classes $\kappa_1(\mathbf{z}), ..., \kappa_{\hat{\gamma}}(\mathbf{z})$ are therefore *soft* data classes.

12.12 CHAPTER EXERCISES

Exercise 49 *Show that the probability-generating function* $G_{n,q}(y)$ *of the binomial distribution* $B_{n,q}(m)$ *is* $G_{n,q}(y) = (1 - q + qy)^n$.

Exercise 50 *Show that (12.264) is true:* $\int f(Z|a, \mathbf{x})\delta Z = 1$.

Exercise 51 *Show that the p.g.fl. of the likelihood for a single extended target is*

$$G_{k+1}[g|\mathbf{x}] = e^{\lambda c[g]-\lambda} \prod_{\ell=1}^{L} \left[1 - p_D^\ell(\mathbf{x}) + p_D^\ell(\mathbf{x}) p_g^\ell(\mathbf{x})\right] \tag{12.366}$$

where $p_g^\ell(\mathbf{x}) \triangleq \int g(\mathbf{z}) \cdot f_{k+1}^\ell(\mathbf{z}|\mathbf{x})d\mathbf{z}$. *Show that the p.g.fl. for multiple extended targets is*

$$G_{k+1}[g|X] = e^{\lambda c[g]-\lambda} \prod_{\ell=1}^{L} \left(1 - p_D^\ell + p_D^\ell p_g^\ell\right)^X . \tag{12.367}$$

Exercise 52 *Verify that* $\mathring{f}_{k+1}(Z|X)$ *as defined in (12.193) is a likelihood function. That is, show that* $\int \mathring{f}_{k+1}(Z|X)\delta Z = 1$.

11 Actually, $\kappa_j(\mathbf{z})$ is the posterior probability $p(j|\mathbf{z})$ that the jth class is the model that best explains the generation of $\mathbf{z}$—see [123].

Exercise 53 *Verify (12.194):* $\overset{*}{G}_{k+1}[g|X] = 1 - \overset{*}{p}_T + \overset{*}{p}_T \cdot G_{k+1}[g|X]$.

Exercise 54 *Show that the p.g.fl. of the finite-mixture model likelihood of (12.353) is*

$$G[g|Q, \mathbf{p}] = G_{\mathbf{p}}(p_Q[g]) \tag{12.368}$$

where

$$p_Q[g] \triangleq \int g(\mathbf{z}) \cdot f(\mathbf{z}|Q) d\mathbf{z} \tag{12.369}$$

and where

$$G_{\mathbf{p}}(x) \triangleq \sum_{m=0}^{\infty} p_{\mathbf{p}}(m) x^m \tag{12.370}$$

is the p.g.f. of the probability distribution $p_{\mathbf{p}}(m)$. *In particular, let* $p_{\mathbf{p}}(m) = e^{-\lambda(Q)} \lambda(Q)^m / m!$ *be a Poisson distribution where* $\lambda(Q) = a_1 + ... + a_\gamma$ *if* $Q = \{(a_1, \mathbf{u}_1), ..., (a_\gamma, \mathbf{u}_\gamma)\}$. *Then show that*

$$G[g|Q] = e^{\lambda(Q) \cdot p[g|Q] - \lambda(Q)} \tag{12.371}$$

where $p[g|Q] \triangleq \int g(\mathbf{z}) \cdot f(\mathbf{z}|Q) d\mathbf{z}$.

Chapter 13

Multitarget Markov Densities

13.1 INTRODUCTION TO THE CHAPTER

Single-target motion modeling is relatively simple because only two types of temporal dynamicism need be considered:

- The presumed physical motion of individual targets;
- Less commonly, the transition of targets from one *mode* to another.[1]

Multitarget motion is considerably more complex. Target number can be constantly varying because targets can appear and disappear. For example, completely new targets can enter a scene spontaneously, as when a missile or jet aircraft suddenly appears over the horizon. Existing targets can give rise to new targets through *spawning*—as when a jet fighter launches a missile; when an ICBM re-entry vehicle (RV) launches decoys or other countermeasures; when an apparently single target at a distance resolves into multiple, closely spaced targets; and so on. Targets can likewise leave a scene, as when disappearing over the horizon or behind some other occlusion, or they can be damaged or destroyed.

A more subtle kind of multitarget dynamicism is best exemplified by the temporal behavior of *group targets* [247]. Many target groupings are ad hoc formations of targets. Such targets just happen to be relatively near each other and they move largely independently of each other. Consequently, knowing that one

1 This occurs, for example, when a diesel-electric submarine transitions between submerged versus snorkeling modes; when a mobile missile launcher transitions between transport and launching modes; or when a variable-wing jet fighter transitions between swept-wing and extended-wing modes.

target is moving in a certain way will tell us little or nothing about how other nearby targets will be moving. By contrast, group targets are target formations (brigades, aircraft sorties, tank columns, and so on) that are tactically organized. Individual targets in a group target will exhibit some degree of independent movement. Overall, however, they move as elements of a single, coordinated mass. Thus, knowing how a single target is moving will provide information about how other targets in a group target are moving.

What may not be apparent is that:

- *All of these instances of multitarget dynamicism should be explicitly modeled.*

This is because multitarget filters that assume uncorrelated motion and constant target number may perform poorly against dynamic multitarget environments. This will be for the same reason that single-target trackers that assume straight-line motion may perform poorly against maneuvering targets. In either case, data is "wasted" in trying to overcome—successfully or otherwise—the effects of motion-model mismatch. This situation is illustrated in Figure 13.1.

The purpose of this chapter is to describe how multitarget motion models and their corresponding true multitarget Markov densities are constructed. In this we have a bit of luck. Both goals can be accomplished in exact analogy to the construction of multitarget measurement models and their corresponding likelihood functions. This is because of the following:

- Target disappearance is mathematically analogous to missed detection.
- Target appearance is mathematically analogous to the generation of state-independent false alarms.
- Target spawning is mathematically analogous to the generation of state-dependent false alarms.
- Coordinated versus uncoordinated motion is mathematically analogous to conditionally dependent versus conditionally independent measurements.

Because of these formal similarities, the formal modeling procedure that we shall adopt is exactly analogous to that introduced in Chapter 12:

- Suppose that, at time step $k+1$, we have succeeded in constructing a model $\Xi_{k+1|k} = \Gamma_k(X')$ of the randomly varying state set, given the targets had state set X' at time step k.

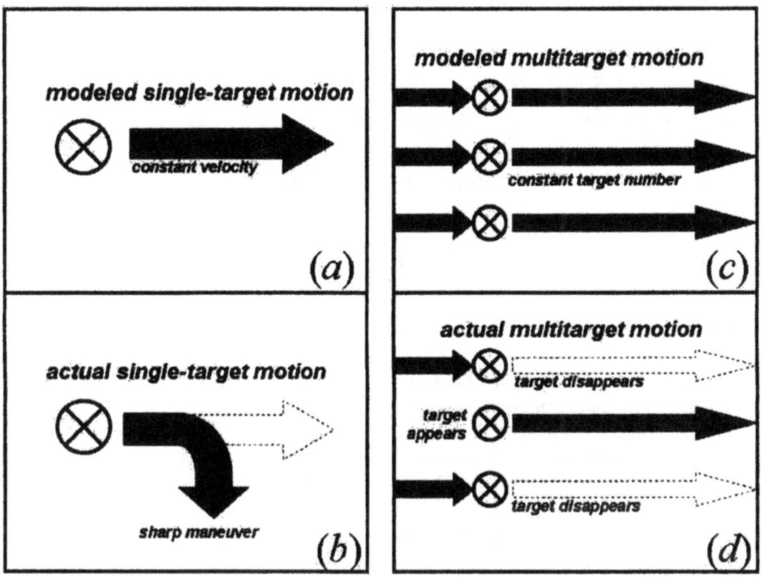

Figure 13.1 The need for multitarget motion modeling is illustrated using an analogy with single-target motion modeling. (a) Assumed target motion is constant-velocity: the target is expected to continue in the same direction at the same speed. However, actual target motion is a sharp maneuver as in (b). A tracking filter will either require many measurements to reacquire the target, or will lose it altogether. (c) An illustration is shown of a multitarget motion model that assumes unknown but constant target number. (d) The actual multitarget motion is shown. A multitarget tracking filter will presume the existence of two targets after two have disappeared and one has appeared. Such a filter will either require many measurements to reacquire correct target number (in this case $n = 1$) or may completely lose one or more targets.

- The statistics of $\Xi_{k+1|k}$ are completely characterized by the belief-mass function $\beta_{k+1|k}(S|X') = \Pr\left(\Xi_{k+1|k} \subseteq S|X'\right)$.
- From (11.249) we know that the probability density function of $\Xi_{k+1|k}$ can be constructed as a set derivative

$$f_{k+1|k}(X|X') \triangleq \frac{\delta\beta_{k+1|k}}{\delta X}(\emptyset|X') = \left[\frac{\delta\beta_{k+1|k}}{\delta X}(S|X')\right]_{S=\emptyset}. \tag{13.1}$$

From Section 11.5 we also know that $f_{k+1|k}(X|X')$ and $\beta_{k+1|k}(S|X')$ contain exactly the same information about $\Xi_{k+1|k}$. This is because $f_{k+1|k}(X|X')$ and $\beta_{k+1|k}(S|X')$ can be derived from each other using the fundamental formulas of multiobject calculus of Section 11.5.1:

$$\beta_{k+1|k}(S|X') = \int_S f_{k+1|k}(X|X')\delta X'. \tag{13.2}$$

In other words, the following are true.

- $f_{k+1|k}(X|X')$ as defined in (13.1) is the *true multitarget Markov density* for presumed multitarget motion.
- That is, it faithfully encapsulates the information in the measurement model $\Xi_{k+1|k} = \Gamma_k(X')$ without introducing extraneous information.
- Equation (13.1) is the *central formula of multitarget motion modeling*.

Multitarget Markov densities $f_{k+1|k}(X|X')$ are unlike conventional single-target Markov densities $f_{k+1|k}(\mathbf{x}|\mathbf{x}')$ in that they *comprehensively* describe the temporal dynamicism of target groupings. They incorporate not just individual target motion models, but any other modeling necessary to describe the dynamics of a scene. This includes models of the following.

- Target appearance, including spawning of new targets by other targets;
- Target disappearance;
- The degree to which targets jointly move in a coordinated manner.

Example 74 (Simple Example) *Suppose that two moving targets in one dimension have statistically independent motions, with no target appearance or disappearance. The motion of both targets individually is governed by the Markov density $f_{k+1|k}(x|x') = N_{\sigma^2}(x - 2x')$. The multitarget Markov density $f_{k+1|k}(X|X')$*

for this problem is nonzero only if $|X| = |X'| = 2$*, in which case for all* $z \neq z'$ *and* $x \neq x'$,

$$
\begin{aligned}
& f(\{x,y\}|\{x',y'\}) && (13.3)\\
= \; & N_{\sigma^2}(x-x') \cdot N_{\sigma^2}(y-y') + N_{\sigma^2}(y-x') \cdot N_{\sigma^2}(x-y') && (13.4)\\
= \; & \frac{1}{2\pi\sigma^2} \left[\begin{array}{c} \exp\left(-\frac{(x-x')^2+(y-y')^2}{2\sigma^2} \right) \\ + \exp\left(-\frac{(y-x')^2+(x-y')^2}{2\sigma^2} \right) \end{array} \right]. && (13.5)
\end{aligned}
$$

13.1.1 Summary of Major Lessons Learned

The following are the major ideas and concepts to be encountered in this chapter:

- Careful multitarget motion models should include models for target appearance and disappearance, not just the motion of individual targets.
- Care must be exercised in the use of target birth and death models. For example, if more targets appear than disappear from the scene, then the motion model biases a multitarget filter to expect a net target gain over time. This is fine if more targets tend to appear than to disappear. Otherwise, enough measurements must be available to allow the filter to overcome the bias induced by the incorrect motion model.
- A formula for the standard multitarget motion model without target spawning; see (13.42):

$$
\begin{aligned}
f_{k+1|k}(X|X') &= e^{\mu_0} f_B(X) \cdot f_{k+1|k}(\emptyset|X') && (13.6)\\
& \quad \cdot \sum_{\theta} \prod_{i:\theta(i)>0} \frac{p_S(\mathbf{x}'_i) \cdot f_{k+1|k}(\mathbf{x}_{\theta(i)}|\mathbf{x}'_i)}{(1-p_S(\mathbf{x}'_i)) \cdot \mu_0 b(\mathbf{x}_{\theta(i)})}. && (13.7)
\end{aligned}
$$

- A formula for the standard multitarget motion model with target spawning; see (13.46):

$$
\begin{aligned}
& f_{k+1|k}(X|X') = e^{\mu(X)} f_{B(X')}(X) \cdot f_{k+1|k}(\emptyset|X') && (13.8)\\
& \cdot \sum_{\theta} \prod_{i:\theta(i)>0} \frac{p_S(\mathbf{x}'_i) \cdot f_{k+1|k}(\mathbf{x}_{\theta(i)}|\mathbf{x}'_i)}{(1-p_S(\mathbf{x}'_i)) \cdot \mu(X') \cdot b(\mathbf{x}_{\theta(i)}|X')}. && (13.9)
\end{aligned}
$$

- A formula for the motion of single point clusters; see (13.73):

$$f_{k+1|k}(a,\mathbf{x}|a',\mathbf{x}') = f_{k+1|k}(a|a',\mathbf{x}') \cdot f_{k+1|k}(\mathbf{x}|\mathbf{x}') \quad (13.10)$$

$$= f_A(a - \varphi_k^0(a',\mathbf{x}')) \cdot f_{\mathbf{V}}(\mathbf{x} - \varphi_k^1(\mathbf{x}')). \quad (13.11)$$

- A formula for the motion of multiple point clusters; see (13.77):

$$f_{k+1|k}(\{\mathring{\mathbf{x}}_1, ..., \mathring{\mathbf{x}}_n\}|\mathring{X}') = e^{-\mu(\mathring{X}')} \cdot \mu(\mathring{X}')^n \quad (13.12)$$

$$\cdot f_{k+1|k}(\mathring{\mathbf{x}}_1|\mathring{X}') \cdots f_{k+1|k}(\mathring{\mathbf{x}}_n|\mathring{X}'). \quad (13.13)$$

- A formula for virtual leader-follower motion without target appearance or disappearance; see (13.95):

$$f_{k+1|k}(X|X') = \sum_\sigma f_{k+1|k}^{\sigma 1}(\mathbf{x}_1|X') \cdots f_{k+1|k}^{\sigma n}(\mathbf{x}_n|X'). \quad (13.14)$$

- A formula for virtual leader-follower motion with target disappearance and appearance; see (13.105):

$$f_{k+1|k}(X|X') = e^{\mu_0} \cdot f_B(X) \cdot f_{k+1|k}(\emptyset|X') \quad (13.15)$$

$$\cdot \sum_\theta \prod_{i:\theta(i)>0} \frac{p_S(\mathbf{x}_i') \cdot f_{k+1|k}^i(\mathbf{x}_{\theta(i)}|X')}{(1 - p_S(\mathbf{x}_i')) \cdot \mu_0 b(\mathbf{x}_{\theta(i)})}. \quad (13.16)$$

13.1.2 Organization of the Chapter

I begin by developing a "standard" multitarget motion model in Section 13.2. This is an exact analog of the standard measurement model introduced in Section 13.2. I indicate the modifications that are required for extended targets (Section 13.3) and for unresolved targets (Section 13.4).

In Section 13.5, I conclude with a simple model of coordinated multitarget motion: the *virtual leader-follower* model. This includes two versions of the model: no target birth or death (Section 13.5.1) and target birth and death (Section 13.5.2). Exercises for the chapter can be found in Section 13.6.

13.2 "STANDARD" MULTITARGET MOTION MODEL

In this section, I devise a model for multitarget motion that I call "standard" because it is identical in mathematical form to the standard single-sensor, multitarget measurement model considered in Chapter 12. Because of this formal similarity, formulas for multitarget motion models and Markov densities can be essentially "transcribed" from the corresponding formulas for multitarget measurement models and likelihood functions.

The standard multitarget motion model is based on the following assumptions (see Figure 13.2):

- The likelihood that a single target will have state vector $\mathbf{x}$ at time step $k+1$ if it had state vector $\mathbf{x}'$ at time step k is described by a Markov transition density $f_{k+1|k}(\mathbf{x}|\mathbf{x}')$ corresponding to a single-target motion model $\mathbf{X}_{k+1|k} = \varphi_k(\mathbf{x}', \mathbf{V}_k)$.[2] Hereafter this is assumed for notational simplicity to be additive: $\mathbf{X}_{k+1|k} = \varphi_k(\mathbf{x}') + \mathbf{V}_k$.
- A single target with state $\mathbf{x}'$ at time step k has a probability

$$p_S(\mathbf{x}) \overset{\text{abbr.}}{=} p_S^{k+1|k}(\mathbf{x}') \tag{13.17}$$

  of surviving into time step $k+1$;
- A single target with state $\mathbf{x}'$ at time step k has a probability (density)

$$b(X|\mathbf{x}') \overset{\text{abbr.}}{=} b_{k+1|k}(X|\mathbf{x}') \tag{13.18}$$

  of spawning a set X of new targets at time step $k+1$.
- The probability (density) that new targets with state set X will spontaneously appear at time step $k+1$ is

$$b(X) \overset{\text{abbr.}}{=} b_{k+1|k}(X). \tag{13.19}$$

- Target birth, target death, and target motion are conditionally independent of the previous multitarget state.

In what follows I will derive the *true multitarget Markov density function* $f_{k+1|k}(X|X')$ for this proposed standard model, where $X = \{\mathbf{x}_1, ..., \mathbf{x}_n\}$ and $X' = \{\mathbf{x}'_1, ..., \mathbf{x}'_{n'}\}$. My proposed standard model has the mathematical form

$$\overset{\text{predicted state set}}{\Xi_{k+1|k}} = \overset{\text{persisting targets}}{\Gamma(X')} \cup \overset{\text{spawned targets}}{B(X')} \tag{13.20}$$

$$\cup \overset{\text{spontaneous targets}}{B} \tag{13.21}$$

2 Since $\mathbf{x}$ and $\mathbf{x}'$ can include target-identity state variables, in principle this allows modeling of different target motions for different target types.

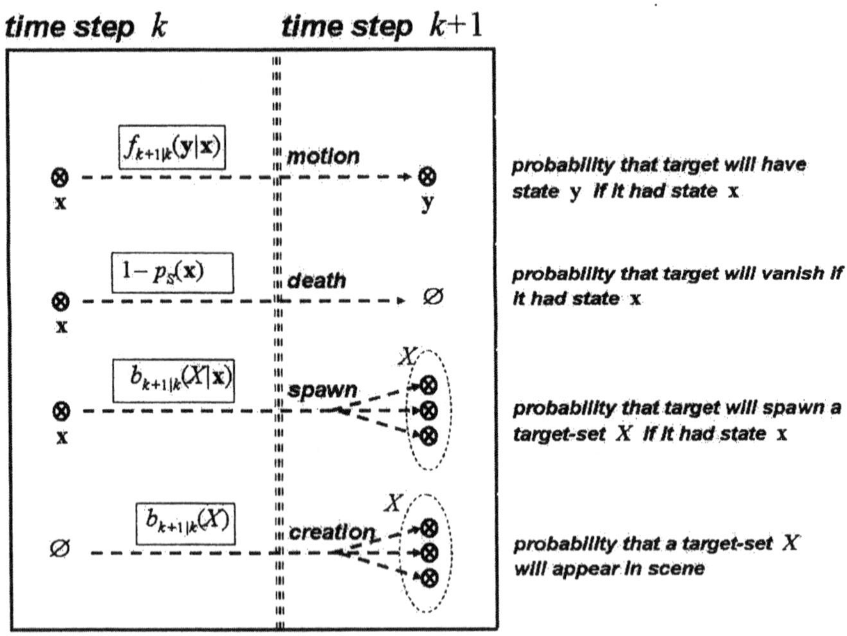

Figure 13.2 A pictoral summary of the major assumptions of the "standard" multitarget motion model.

where $\Gamma(X')$ and $B(X')$ have the form

$$\Gamma(X') = \overset{\text{prediction of } \mathbf{x}'_1}{\Gamma(\mathbf{x}'_1)} \cup ... \cup \overset{\text{prediction of } \mathbf{x}'_{n'}}{\Gamma(\mathbf{x}'_{n'})} \tag{13.22}$$

$$B(X') = \overset{\text{spawned by } \mathbf{x}'_1}{B(\mathbf{x}'_1)} \cup ... \cup \overset{\text{spawned by } \mathbf{x}'_{n'}}{B(\mathbf{x}'_{n'})}. \tag{13.23}$$

We also assume that

$$\Gamma(\mathbf{x}'_1), ..., \Gamma(\mathbf{x}'_{n'}), B(\mathbf{x}'_1), ..., B(\mathbf{x}'_{n'}), B(X'), B \tag{13.24}$$

are statistically independent.

We proceed through five progressively more complex cases:

- Case I: At most one target is present (Section 13.2.1);
- Case II: Arbitrary number of targets, no target birth or death (Section 13.2.2);
- Case III: Arbitrary number of targets, target deaths, but no target birth (Section 13.2.3);
- Case IV: Arbitrary number of targets, target deaths, and spontaneous births (Section 13.2.4);
- Case V: Arbitrary number of targets, target deaths, spontaneous births, and spawning (Section 13.2.5).

13.2.1 Case I: At Most One Target Is Present

We begin with a scenario in which a single target can enter an empty scene, can persist in the scene, or can disappear from the scene.

We assume that a target with state $\mathbf{x}'$ at time step k has a probability $p_S(\mathbf{x}')$ of surviving into time step $k+1$. We assume that a target currently out of the scene has a probability p_R of re-entering the scene, and that its distribution of re-entry is $f_R(\mathbf{x}) \triangleq f_{\mathbf{V}'_k}(\mathbf{x})$. As usual, the single-target Markov density is $f_{k+1|k}(\mathbf{x}|\mathbf{x}') = f_{\mathbf{V}_k}(\mathbf{x} - \varphi_k(\mathbf{x}'))$.

The multitarget motion model in this case is as follows:

$$\overset{\text{predicted state set}}{\Xi_{k+1|k}} = \left\{ \begin{array}{ccc} \overset{\text{prediction of } \mathbf{x}'}{\Gamma(\mathbf{x}')} & \text{if} & X' = \{\mathbf{x}'\} \\ \overset{\text{possibly re-entering target}}{B} & \text{if} & X' = \emptyset \end{array} \right\}; \tag{13.25}$$

where

$$\Gamma(\mathbf{x}') = \{\varphi_k(\mathbf{x}') + \mathbf{V}_k\} \cap \emptyset^{p_S(\mathbf{x}')} \tag{13.26}$$

$$B = \{\mathbf{V}'_k\} \cap \emptyset^{p_R} \tag{13.27}$$

and where the random set $\emptyset^p$ was defined in (12.47) or (11.22). From (13.27) and (11.23),

$$\beta_{k+1|k}(S|\emptyset) = \Pr(B \subseteq S) = 1 - p_R + p_R \cdot p_{\mathbf{V}'_k}(S). \tag{13.28}$$

Thus from (11.29)

$$f_{k+1|k}(X|\emptyset) = \left\{ \begin{array}{ccc} 1 - p_R & \text{if} & X = \emptyset \\ p_R \cdot f_R(\mathbf{x}) & \text{if} & X = \{\mathbf{x}\} \\ 0 & \text{if} & \text{otherwise} \end{array} \right. . \tag{13.29}$$

On the other hand, from (13.26) and (11.23)

$$\beta_{k+1|k}(S|\{\mathbf{x}'\}) = \Pr(\Gamma(\mathbf{x}') \subseteq S) \tag{13.30}$$

$$= 1 - p_S(\mathbf{x}') + p_S(\mathbf{x}') \cdot p_{\mathbf{V}_k}(S|\mathbf{x}') \tag{13.31}$$

and thus by (11.29)

$$f_{k+1|k}(X|\{\mathbf{x}'\}) = \left\{ \begin{array}{ccc} 1 - p_S(\mathbf{x}') & \text{if} & X = \emptyset \\ p_S(\mathbf{x}') \cdot f_{k+1|k}(\mathbf{x}|\mathbf{x}') & \text{if} & X = \{\mathbf{x}\} \\ 0 & \text{if} & \text{otherwise} \end{array} \right. . \tag{13.32}$$

13.2.2 Case II: No Target Death or Birth

In this case $n' = n$ and (13.20) becomes

$$\overset{\text{predicted state set}}{\Xi_{k+1|k}} = \overset{\text{prediction of } \mathbf{x}'_1}{\{\mathbf{X}^1_{k+1|k}\}} \cup \ldots \cup \overset{\text{prediction of } \mathbf{x}'_n}{\{\mathbf{X}^n_{k+1|k}\}} \tag{13.33}$$

$$= \{\varphi_k(\mathbf{x}'_1) + \mathbf{V}^1_k, \ldots, \varphi_k(\mathbf{x}'_n) + \mathbf{V}^n_k\} \tag{13.34}$$

with $\mathbf{V}^1_k, \ldots, \mathbf{V}^n_k$ being independent, identically distributed (i.i.d.) zero-mean random vectors with common density $f_{\mathbf{V}_k}(\mathbf{x} - \varphi_k(\mathbf{x}'_i))$.

This motion model is exactly analogous to the single-sensor, multitarget measurement model considered in Section 12.3.4—that is, no missed detections or false alarms. Consequently, we can immediately transcribe the true Markov density for this model from (12.91) of that section:

$$f_{k+1|k}(X|X') = \sum_{\sigma} f_{k+1|k}(\mathbf{x}_1|\mathbf{x}'_{\sigma 1}) \cdots f_{k+1|k}(\mathbf{x}_n|\mathbf{x}'_{\sigma n}) \tag{13.35}$$

where the summation is taken over all permutations σ on the numbers $1, ..., n$.

13.2.3 Case III: Target Death, No Birth

In this case $n' \geq n$ and (13.20) has the form

$$\overset{\text{predicted state set}}{\Xi_{k+1|k}} = \overset{\text{prediction of } \mathbf{x}'_1}{\Gamma(\mathbf{x}'_1)} \cup ... \cup \overset{\text{prediction of } \mathbf{x}'_{n'}}{\Gamma(\mathbf{x}'_{n'})} \tag{13.36}$$

where for all $i = 1, ..., n'$, $\Gamma(\mathbf{x}'_i)$ are as in (13.26):

$$\Gamma(\mathbf{x}'_i) = \{\varphi_k(\mathbf{x}'_i) + \mathbf{V}^i_k\} \cap \emptyset^{p_S(\mathbf{x}'_i)}. \tag{13.37}$$

This motion model is exactly analogous to the multitarget measurement model considered in Section 12.3.5 (i.e., missed detections but no false alarms). The corresponding true Markov density can be transcribed from (12.119) and (12.120). If $X = \emptyset$, then

$$f_{k+1|k}(\emptyset|X') = (1 - p_S(\mathbf{x}'_1)) \cdots (1 - p_S(\mathbf{x}'_{n'})) \tag{13.38}$$

and, otherwise, if $X \neq \emptyset$ where $X = \{\mathbf{x}_1, ..., \mathbf{x}_n\}$ with $|X| = n \leq n'$,

$$f_{k+1|k}(X|X') = f_{k+1|k}(\emptyset|X') \sum_{\theta} \prod_{i:\theta(i)>0} \frac{p_S(\mathbf{x}'_i) \cdot f_{k+1|k}(\mathbf{x}_{\theta(i)}|\mathbf{x}'_i)}{1 - p_S(\mathbf{x}'_i)} \tag{13.39}$$

where the summation is taken over all associations $\theta : \{1, ..., n'\} \to \{0, 1, ..., n\}$ as defined in Section 10.5.4 or Figure 10.2(b).

13.2.4 Case IV: Target Death and Birth

In this case (13.20) becomes

$$\overset{\text{predicted state set}}{\Xi_{k+1|k}} = \overset{\text{prediction of } \mathbf{x}'_1}{\Gamma(\mathbf{x}'_1)} \cup ... \cup \overset{\text{prediction of } \mathbf{x}'_{n'}}{\Gamma(\mathbf{x}'_{n'})} \tag{13.40}$$

$$\cup \overset{\text{spontaneous targets}}{B} . \tag{13.41}$$

If the appearance of new targets is additionally assumed to be Poisson, then this motion model is exactly analogous to the measurement model considered in Section 12.3.6 (i.e., missed detections and state-independent false alarms). Let μ_0 be the expected number of new targets and let $b(\mathbf{x})$ be their physical distribution. Then the corresponding true Markov density can be read off from (12.139):

$$f_{k+1|k}(X|X') = e^{\mu_0} f_B(X) \cdot f_{k+1|k}(\emptyset|X') \tag{13.42}$$

$$\cdot \sum_\theta \prod_{i:\theta(i)>0} \frac{p_S(\mathbf{x}'_i) \cdot f_{k+1|k}(\mathbf{x}_{\theta(i)}|\mathbf{x}'_i)}{(1-p_S(\mathbf{x}'_i)) \cdot \mu_0 b(\mathbf{x}_{\theta(i)})}. \tag{13.43}$$

The summation is taken over all association hypotheses $\theta : \{1, ..., n'\} \rightarrow \{0, 1, ..., n\}$ as defined in Section 10.5.4 or Figure 10.2(c); and where

$$f_B(X) = e^{-\mu_0} \prod_{\mathbf{x} \in X} \mu_0 b(\mathbf{x}) \tag{13.44}$$

$$f_{k+1|k}(\emptyset|X') = e^{-\mu_0} \prod_{\mathbf{x}' \in X'} (1 - p_S(\mathbf{x}')). \tag{13.45}$$

13.2.5 Case V: Target Death and Birth with Spawning

Equation (13.20) defines the full standard motion model. If both spontaneous target birth and target spawning are assumed to be Poisson, then this motion model is exactly analogous to the measurement model considered in Section 12.5 (i.e., missed detections, and both state-dependent and state-independent false alarms). The corresponding true multitarget Markov density can be read off from (12.186)-(12.188):

$$f_{k+1|k}(X|X') = e^{\mu(X')} \cdot f_{B(X')}(X) \cdot f_{k+1|k}(\emptyset|X') \tag{13.46}$$

$$\cdot \sum_\theta \prod_{i:\theta(i)>0} \frac{p_S(\mathbf{x}'_i) \cdot f_{k+1|k}(\mathbf{x}_{\theta(i)}|\mathbf{x}'_i)}{(1-p_S(\mathbf{x}'_i)) \cdot \mu(X') \cdot b(\mathbf{x}_{\theta(i)}|X')}. \tag{13.47}$$

Here, the summation is taken over all associations $\theta : \{1, ..., n'\} \rightarrow \{0, 1, ..., n\}$ as defined in Section 10.5.4 or Figure 10.2(c). Also,

$$f_{k+1|k}(\emptyset|X') = e^{-\mu(X')} \prod_{\mathbf{x}' \in X'} (1 - p_S(\mathbf{x}')) \tag{13.48}$$

and

$$\mu(X') = \mu_0 + \mu_1(\mathbf{x}') + ... + \mu_{n'}(\mathbf{x}') \quad (13.49)$$

$$b(\mathbf{x}|X') \quad (13.50)$$

$$= \frac{\mu_0 b_0(\mathbf{x}) + \mu_1(\mathbf{x}') \cdot b(\mathbf{x}|\mathbf{x}'_1) + ... + \mu_{n'}(\mathbf{x}') \cdot b(\mathbf{x}|\mathbf{x}'_{n'})}{\mu_0 + \mu_1(\mathbf{x}') + ... + \mu_{n'}(\mathbf{x}')}. \quad (13.51)$$

Here μ_0 is the expected number of spontaneously generated new targets and $b_0(\mathbf{x})$ is their physical distribution. Also, $\mu(\mathbf{x}'_i)$ is the expected number of new targets spawned by a target with previous state $\mathbf{x}'_i$, and $b(\mathbf{x}|\mathbf{x}'_i)$ is their physical distribution.

13.2.6 p.g.fl.s for the Standard Motion Model

In this section, I derive the p.g.fl. form of the standard multitarget motion model. This will prove useful in Chapter 16.

The p.g.fl. of a multiobject probability density function was defined in (11.154). Given a multitarget Markov density $f_{k+1|k}(X|X')$, its p.g.fl. is

$$G_{k+1|k}[h|X'] \triangleq \int h^X \cdot f_{k+1|k}(X|X')\delta X \quad (13.52)$$

where, as defined in (11.153), $h^X = 1$ if $X = \emptyset$ and, otherwise, $h^X \triangleq \prod_{\mathbf{x} \in X} h(\mathbf{x})$. Define

$$b_0[h] \triangleq \int h(\mathbf{x}) \cdot b_0(\mathbf{x})d\mathbf{x} \quad (13.53)$$

$$b[h|X'] \triangleq \int h(\mathbf{x}) \cdot b(\mathbf{x}|X')d\mathbf{x} \quad (13.54)$$

$$p_h(\mathbf{x}') \triangleq \int h(\mathbf{x}) \cdot f_{k+1|k}(\mathbf{x}|\mathbf{x}')d\mathbf{x} \quad (13.55)$$

$$f_R[h] \triangleq \int h(\mathbf{x}) \cdot f_R(\mathbf{x})d\mathbf{x}. \quad (13.56)$$

I will show that the p.g.fl.s for the five cases of the standard multitarget model are the following:

- *Case I: At most one target is present*:

$$G_{k+1|k}[h|\emptyset] = 1 - p_R + p_R \cdot f_R[h] \quad (13.57)$$
$$G_{k+1|k}[h|\mathbf{x}'] = 1 - p_S(\mathbf{x}') + p_S(\mathbf{x}') \cdot p_h(\mathbf{x}'). \quad (13.58)$$

- *Case II: No target death or birth*:

$$G_{k+1|k}[h|X'] = p_h^{X'}. \quad (13.59)$$

- *Case III: Target death, no birth*:

$$G_{k+1|k}[h|X'] = (1 - p_S + p_S p_h)^{X'}. \quad (13.60)$$

- *Case IV: Target death and Poisson spontaneous birth*:

$$G_{k+1|k}[h|X'] = (1 - p_S + p_S p_h)^{X'} \cdot e^{\mu_0 b_0[h] - \mu_0}. \quad (13.61)$$

- *Case V: Target death and Poisson spontaneous birth and spawning*:

$$G_{k+1|k}[h|X'] = (1 - p_S + p_S p_h)^{X'} \cdot e^{\mu(X') \cdot b[h|X'] - \mu(X')}. \quad (13.62)$$

Equations (13.58)-(13.62) follow directly as direct analogs of (12.147)-(12.151), respectively.

13.3 EXTENDED TARGETS

Measurement models for single extended targets were discussed in Section 12.7.1 and for multiple extended targets in Section 12.7.2. Extended targets are mathematically modeled as point targets $\mathbf{x}$ with finite-set likelihoods $f_{k+1}(Z|\mathbf{x})$. Consequently, motion models for extended targets are no different than motion models for point targets.

The motion of a single extended target is governed by a Markov density of the form $f_{k+1|k}(\mathbf{x}|\mathbf{x}')$. The motion of multiple extended targets are governed by multitarget Markov densities identical to those just considered in Sections 13.2.1-13.2.5.

13.4 UNRESOLVED TARGETS

A measurement modeling methodology for unresolved targets, based on the concept of a point target cluster, was introduced in Section 12.8. The purpose of this section is to derive Markov densities $f_{k+1|k}(\mathring{X}|\mathring{X}')$, which give the likelihood that there is a cluster set

$$\mathring{X} = \{(a_1, \mathbf{x}_1), ..., (a_n, \mathbf{x}_n)\} \tag{13.63}$$

at time step $k+1$, given that the cluster set at time step k was

$$\mathring{X}' = \{(a'_1, \mathbf{x}'_1), ..., (a'_{n'}, \mathbf{x}'_{n'})\}. \tag{13.64}$$

In particular, we must have $\int f_{k+1|k}(\mathring{X}|\mathring{X}')\delta\mathring{X} = 1$.[3]

I begin in Section 13.4.1 by describing the dynamic behavior of point clusters from an intuitive point of view. I then construct formal motion models for single point clusters (Section 13.4.2) and multiple point clusters (Section 13.4.3).

13.4.1 Intuitive Dynamic Behavior of Point Clusters

I begin with a discussion of the dynamic behavior that one might intuitively expect point clusters to exhibit. As noted in Figure 12.2, a point cluster models a target group that is so distant from the sensor that it cannot easily be distinguished from a single point target. Consider the following dynamical possibilities:

- *Resolution*: If a point cluster with n targets moves toward the sensor, it eventually resolves into a group of closely spaced point targets:

$$(n', \mathbf{x}') \to \{(1, \mathbf{x}_1), ..., (1, \mathbf{x}_n)\}. \tag{13.65}$$

- *Intraresolution*: During the resolution process, a point cluster may first resolve into multiple point clusters:

$$(n', \mathbf{x}') \to \{(a_1, \mathbf{x}_1), ..., (a_n, \mathbf{x}_n)\} \tag{13.66}$$

3 Since the state space now has the form $\mathfrak{X}_0 = \mathbb{R}^+ \times \mathbb{R}^N$, where $\mathbb{R}^+$ denotes the positive integers, the set integral now has the form

$$\int f(\mathring{X})\delta\mathring{X} \triangleq \sum_{n=0}^{\infty} \frac{1}{n!} \int f(\{\mathring{\mathbf{x}}_1, ..., \mathring{\mathbf{x}}_n\}) d\mathring{\mathbf{x}}_1 \cdots d\mathring{\mathbf{x}}_n$$

where

$$\int f(\mathring{\mathbf{x}})d\mathring{\mathbf{x}} \triangleq \int_{0+}^{\infty} \int f(a, \mathbf{x})d\mathbf{x}da \triangleq \lim_{\varepsilon \searrow 0} \int_{\varepsilon}^{\infty} \int f(a, \mathbf{x})d\mathbf{x}da.$$

with $n' \cong a_1 + ... + a_n$.

- *Deresolution:* If a group of n closely spaced point targets moves away from the sensor, they eventually deresolve into a point cluster:

$$\{(1, \mathbf{x}'_1), ..., (1, \mathbf{x}'_n)\} \to (n, \mathbf{x}). \tag{13.67}$$

- *Intraderesolution:* During deresolution, closely spaced point targets may first deresolve into a smaller number of point clusters:

$$\{(1, \mathbf{x}'_1), ..., (1, \mathbf{x}'_{n'})\} \to \{(a_1, \mathbf{x}_1), ..., (a_n, \mathbf{x}_n)\}, \tag{13.68}$$

with $n' \cong a_1 + ... + a_n$.

- *Preresolution:* If a point cluster with n targets (1) moves away from the sensor, (2) moves laterally with respect to it, or (3) moves toward it but without resolving, then it remains a point cluster:

$$(n', \mathbf{x}') \to (n, \mathbf{x}). \tag{13.69}$$

In what follows I will not consider all of these exhaustive possibilities. I will derive Markov densities for single point clusters and multiple point clusters, in Sections 13.4.2 and 13.4.3, respectively.

13.4.2 Markov Densities for Single Point Clusters

I begin by positing the existence of a Markov density

$$\begin{aligned} f_{k+1|k}(a, \mathbf{x}|a', \mathbf{x}') &= f_{k+1|k}(\mathring{\mathbf{x}}|\mathring{\mathbf{x}}') & (13.70) \\ &= f_{\mathring{\mathbf{V}}}(\mathring{\mathbf{x}} - \mathring{\varphi}_k(\mathring{\mathbf{x}}')) & (13.71) \\ &= f_{(A,\mathbf{V})}(a - \varphi_k^0(a', \mathbf{x}'),\ \mathbf{x} - \varphi_k^1(a', \mathbf{x}')) & (13.72) \end{aligned}$$

that models transitions from point clusters to point clusters. We will assume that geokinematic dynamics are independent of target-number dynamics:

$$\begin{aligned} f_{k+1|k}(a, \mathbf{x}|a', \mathbf{x}') &= f_{k+1|k}(a|a', \mathbf{x}') \cdot f_{k+1|k}(\mathbf{x}|\mathbf{x}') & (13.73) \\ &= f_A(a - \varphi_k^0(a', \mathbf{x}')) \cdot f_{\mathbf{V}}(\mathbf{x} - \varphi_k^1(\mathbf{x}')). & (13.74) \end{aligned}$$

Thus $f_{k+1|k}(\mathbf{x}|\mathbf{x}')$ is the Markov density for a conventional single-target motion model; and $f_{k+1|k}(a|a',\mathbf{x}')$ describes the likelihood that the predicted cluster will have a targets if, previously, it had a targets with state $\mathbf{x}'$.

For example, if the point cluster is still distant then it should remain a single point cluster without significant change in target number: $\mu(a',\mathbf{x}') \cong 1$ and $f_{k+1|k}(a|a',\mathbf{x}') \cong \delta_{a'}(a)$. If on the other hand it is on the verge of resolving into individual point targets, then the expected number of point targets should be approximately a': $\mu(a',\mathbf{x}') \cong a'$ and $f_{k+1|k}(a|a',\mathbf{x}') \cong \delta_1(a)$.

13.4.3 Markov Densities for Multiple Point Clusters

In Section 13.2.5 we assumed that spawned targets are Poisson distributed. We make the same assumption here. The point cluster analog of the standard multitarget motion model of Section 13.2.5 is

$$\overbrace{\mathring{\Xi}_{k+1|k}}^{\text{predicted cluster set}} = \overbrace{\mathring{\Gamma}(\mathring{\mathbf{x}}'_1)}^{\text{prediction of } \mathring{\mathbf{x}}'_1} \cup \ldots \cup \overbrace{\mathring{\Gamma}(\mathring{\mathbf{x}}'_{n'})}^{\text{prediction of } \mathring{\mathbf{x}}'_{n'}} \tag{13.75}$$

$$\cup \overbrace{\mathring{B}}^{\text{spontaneous clusters}} . \tag{13.76}$$

This model differs from the standard model in one respect. Since a mechanism already exists by which point clusters can "lose" targets, one need not provide an explicit probability of disappearance for point clusters. So, we assume that $\mathring{\Gamma}'(\mathring{\mathbf{x}}'_i)$ is a Poisson spawning process with distribution $f_{k+1|k}(\mathring{X}|\mathring{\mathbf{x}}'_i)$; and that $\mathring{B}$ is a Poisson target-birth process with expected value μ_0 and distribution $b_0(\mathring{\mathbf{x}})$.

The situation we are in is exactly analogous to that of Section 13.2.5. In other words, $\mathring{\Xi}_{k+1|k}$ is also a Poisson process

$$f_{k+1|k}(\{\mathring{\mathbf{x}}_1,...,\mathring{\mathbf{x}}_n\}|\mathring{X}') = e^{-\mu(\mathring{X}')} \cdot \mu(\mathring{X}')^n \tag{13.77}$$

$$\cdot f_{k+1|k}(\mathring{\mathbf{x}}_1|\mathring{X}') \cdots f_{k+1|k}(\mathring{\mathbf{x}}_n|\mathring{X}') \tag{13.78}$$

where its expected value $\mu(\mathring{X}')$ and distribution function $f_{k+1|k}(\mathring{\mathbf{x}}|\mathring{X}')$ are defined by

$$\mu(\mathring{X}') \triangleq \mu_0 + \mu(\mathring{\mathbf{x}}'_1) + ... + \mu(\mathring{\mathbf{x}}'_{n'}) \tag{13.79}$$

and

$$f_{k+1|k}(\mathring{\mathbf{x}}|\mathring{X}') \triangleq \frac{\mu_0 b_0(\mathring{\mathbf{x}}) + \mu(\mathring{\mathbf{x}}'_1) f_{k+1|k}(\mathring{\mathbf{x}}|\mathring{\mathbf{x}}'_1) + ... + \mu(\mathring{\mathbf{x}}'_{n'}) f_{k+1|k}(\mathring{\mathbf{x}}|\mathring{\mathbf{x}}'_{n'})}{\mu_0 + \mu(\mathring{\mathbf{x}}'_1) + ... + \mu(\mathring{\mathbf{x}}'_{n'})}. \tag{13.80}$$

13.5 COORDINATED MULTITARGET MOTION

In this section, I briefly consider the problem of modeling coordinated motion. I define and derive true multitarget Markov densities for two simple models:

- Virtual leader-follower without target death or birth (Section 13.5.1);
- Virtual leader-follower with target death and birth (Section 13.5.2).

13.5.1 Simple Virtual Leader-Follower

This is the simplest type of coordinated motion. It has been applied by Salmond and Gordon to group target tracking [200, 201]. The two basic assumptions underlying the virtual leader-follower model are the following:

- Targets do not appear or disappear from the group (i.e., $n = n'$);
- The deterministic state of any target is a translational offset of the average state (centroid) of the group. Figure 13.3 illustrates the concept.

Let $X' = \{\mathbf{x}'_1, ..., \mathbf{x}'_n\}$ be the multitarget state at time step k and let

$$\bar{\mathbf{x}}' \triangleq \frac{1}{|X'|} \sum_{\mathbf{x}' \in X'} \mathbf{x}' = \frac{1}{n} (\mathbf{x}'_1 + ... + \mathbf{x}'_n) \tag{13.81}$$

be the average single-target state—the centroid—at time step k. Let

$$\Delta\mathbf{x}'_i \triangleq \mathbf{x}'_i - \bar{\mathbf{x}}' \tag{13.82}$$

be the translational offset of each target from the centroid, so that $\bar{\mathbf{x}}' + \Delta\mathbf{x}'_i = \mathbf{x}'_i$. Then the predicted state set has the form

$$\begin{aligned} \Xi_{k+1|k} &= \Gamma_1(X') \cup ... \cup \Gamma_n(X') & (13.83) \\ &= \{\varphi_k(\bar{\mathbf{x}}') + \Delta\mathbf{x}'_1 + \mathbf{V}_1, ..., \varphi_k(\bar{\mathbf{x}}') + \Delta\mathbf{x}'_n + \mathbf{V}_n\} & (13.84) \end{aligned}$$

where $\mathbf{V}_1, ..., \mathbf{V}_n$ are independent, identically distributed (i.i.d.), zero-mean random state vectors and where

$$\Gamma_i(X') \triangleq \{\varphi_k(\bar{\mathbf{x}}') + \Delta\mathbf{x}'_i + \mathbf{V}_i\} \tag{13.85}$$

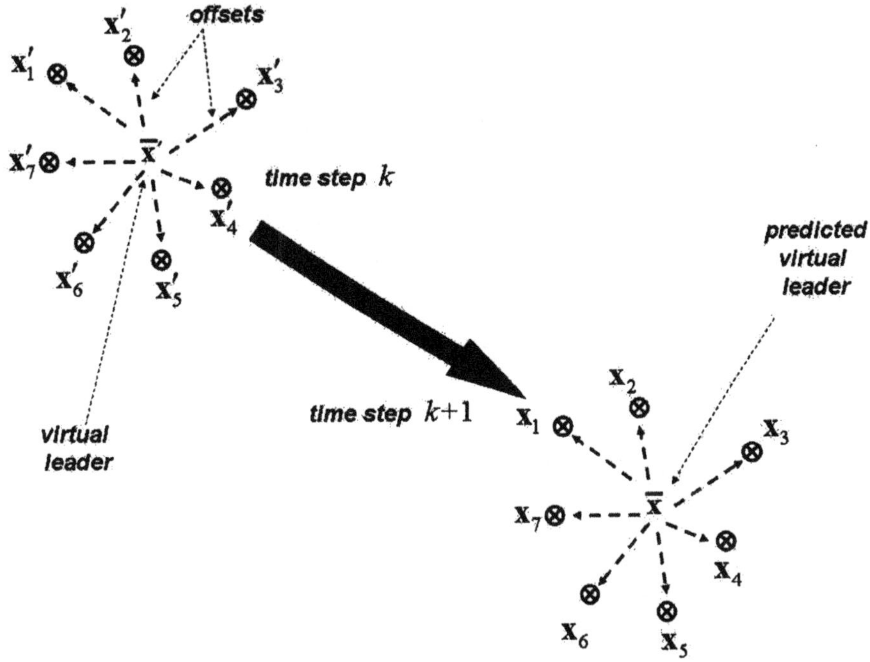

Figure 13.3 The simple virtual leader-follower model of coordinated motion is illustrated. At time step k, the offsets of the state vectors $\mathbf{x}'_1, ..., \mathbf{x}'_7$ from their centroid $\bar{\mathbf{x}}'$ (their "virtual leader") are computed. A motion model is used to predict the centroid at time step $k+1$. The predicted state vectors $\mathbf{x}_1, ..., \mathbf{x}_7$ keep the same offsets as before, except that now they are with respect to the predicted centroid.

for $i = 1, ..., n$.

That is, the targets in the group are predicted to have (except for random variation) the same offsets from the predicted centroid as from the previous centroid. Such targets behave as though they are following a virtual leader whose state is the centroid.

I derive the multitarget Markov density for this model. The belief-mass function for the motion model is

$$\beta_{k+1|k}(S|X') = \Pr(\Xi_{k+1|k} \subseteq S|X') \tag{13.86}$$

$$= \Pr\begin{pmatrix} \varphi_k(\bar{\mathbf{x}}') + \Delta\mathbf{x}'_1 + \mathbf{V}_1 \in S, ..., \\ \varphi_k(\bar{\mathbf{x}}') + \Delta\mathbf{x}'_n + \mathbf{V}_n \in S|X' \end{pmatrix} \tag{13.87}$$

$$= \Pr(\varphi_k(\bar{\mathbf{x}}') + \Delta\mathbf{x}'_1 + \mathbf{V}_1 \in S|X') \tag{13.88}$$

$$\cdots \Pr(\varphi_k(\bar{\mathbf{x}}') + \Delta\mathbf{x}'_n + \mathbf{V}_n \in S|X') \tag{13.89}$$

$$= \Pr(\mathbf{V}_1 \in S - \varphi_k(\bar{\mathbf{x}}') - \Delta\mathbf{x}'_1|X') \tag{13.90}$$

$$\cdots \Pr(\mathbf{V}_n \in S - \varphi_k(\bar{\mathbf{x}}') - \Delta\mathbf{x}'_n|X') \tag{13.91}$$

$$= p^1_{k+1|k}(S|X') \cdots p^n_{k+1|k}(S|X') \tag{13.92}$$

where

$$p^i_{k+1|k}(S|X') \triangleq \int_S f^i_{k+1|k}(S|X')d\mathbf{x} \tag{13.93}$$

$$f^i_{k+1|k}(\mathbf{x}|X') \triangleq f_{\mathbf{V}}(\mathbf{x} - \varphi_k(\bar{\mathbf{x}}') - \Delta\mathbf{x}'_i). \tag{13.94}$$

Using the reasoning of Section 13.2.2, we can transcribe the formula for the multitarget Markov density directly from (12.91). It is $f_{k+1|k}(X|X') = 0$ if $|X| \neq |X'|$ and, otherwise,

$$f_{k+1|k}(X|X') = \sum_\sigma f^{\sigma 1}_{k+1|k}(\mathbf{x}_1|X') \cdots f^{\sigma n}_{k+1|k}(\mathbf{x}_n|X') \tag{13.95}$$

where the summation is taken over all permutations of the numbers $1, ..., n$.

Example 75 (Dead Reckoning) *Assume two-dimensional state space with position-velocity state vectors of the form* $\mathbf{x} = (x, y, v_x, v_y)^T$. *Let* $X' = \{\mathbf{x}'_1, ..., \mathbf{x}'_n\}$ *with* $\mathbf{x}'_i = (x^i, y^i, v^i_x, v^i_y)^T$ *for* $i = 1, ..., n$. *It follows that the average state (group state) is* $\bar{\mathbf{x}}' = (\bar{x}, \bar{y}, \bar{v}_x, \bar{v}_y)^T$ *where* $\bar{x}$ *is the average value of* $x^1, ..., x^n$; *where* $\bar{y}$ *is the average value of* $y^1, ..., y^n$; *and so on. Assume that single-target motion* *follows a dead-reckoning model (also known as constant-velocity model). That is,*

$$\varphi_k(\mathbf{x}') = (x + v_x \Delta t,\ y + v_y \Delta t,\ v_x, v_y)^T \tag{13.96}$$

where Δt *is the time elapsed between time steps* k *and* $k+1$. *It follows that*

$$\varphi_k(\bar{\mathbf{x}}') = (\bar{x} + \bar{v}_x \Delta t,\ \bar{y} + \bar{v}_y \Delta t,\ \bar{v}_x, \bar{v}_y)^T \tag{13.97}$$

and therefore that

$$\begin{aligned}
\varphi_k(\bar{\mathbf{x}}') + \Delta\mathbf{x}'_i &= (\bar{x} + \bar{v}_x \Delta t,\ \bar{y} + \bar{v}_y \Delta t,\ \bar{v}_x, \bar{v}_y)^T & (13.98)\\
&\quad + (x^i, y^i, v^i_x, v^i_y)^T - (\bar{x}, \bar{y}, \bar{v}_x, \bar{v}_y)^T & (13.99)\\
&= (x^i + \bar{v}_x \Delta t,\ y^i + \bar{v}_y \Delta t,\ v^i_x, v^i_y)^T. & (13.100)
\end{aligned}$$

That is, the predicted velocities of individual targets are the same as their previous velocities. Their predicted positions, however, are determined via dead reckoning based on the average velocity (group velocity).

13.5.2 General Virtual Leader-Follower

The simple virtual leader-follower model just described is easily generalized to encompass target disappearance and target appearance. Assume that no targets appear and that the probability that a target with state $\mathbf{x}'$ at time step k will survive into time step $k+1$ is $p_S(\mathbf{x}')$. Then the multitarget motion model has the form

$$\Xi_{k+1|k} = \Gamma_1(X') \cup ... \cup \Gamma_{n'}(X') \tag{13.101}$$

where, for all $i = 1, ..., n'$, $\Gamma_i(X')$ is as in (13.26):

$$\Gamma_i(X') = \{\varphi_k(\bar{\mathbf{x}}') + \Delta\mathbf{x}'_i + \mathbf{V}_i\} \cap \emptyset^{p_S(\mathbf{x}'_i)}. \tag{13.102}$$

This motion model is analogous to that of Section 13.2.3. Thus we can transcribe the true multitarget Markov density directly from (13.38) and (13.39):

$$f_{k+1|k}(\emptyset|X') = (1 - p_S(\mathbf{x}'_1)) \cdots (1 - p_S(\mathbf{x}'_n)) \tag{13.103}$$

and, for $X \neq \emptyset$ where $X = \{\mathbf{x}_1, ..., \mathbf{x}_n\}$ with $|X| = n \leq n'$,

$$f_{k+1|k}(X|X') = f_{k+1|k}(\emptyset|X') \sum_\theta \prod_{i:\theta(i)>0} \frac{p_S(\mathbf{x}'_i) \cdot f^i_{k+1|k}(\mathbf{x}_{\theta(i)}|X')}{(1 - p_S(\mathbf{x}'_i))}. \tag{13.104}$$

Incorporating a birth model can be addressed similarly. From (13.42)-(13.45) we can transcribe:

$$f_{k+1|k}(X|X') = e^{\mu_0} \cdot f_B(X) \cdot f_{k+1|k}(\emptyset|X') \tag{13.105}$$

$$\cdot \sum_\theta \prod_{i:\theta(i)>0} \frac{p_S(\mathbf{x}'_i) \cdot f^i_{k+1|k}(\mathbf{x}_{\theta(i)}|X')}{(1-p_S(\mathbf{x}'_i)) \cdot \mu_0 b(\mathbf{x}_{\theta(i)})} \tag{13.106}$$

where the birth process is Poisson with mean value μ_0 and spatial distribution $b(\mathbf{x})$ and where

$$f_{k+1|k}(\emptyset|X') = e^{-\mu_0} \cdot \prod_{\mathbf{x}' \in X'} (1-p_S(\mathbf{x}')) \tag{13.107}$$

$$f_B(X) = e^{-\mu_0} \cdot \prod_{\mathbf{x} \in X} \mu_0 \cdot b(\mathbf{x}). \tag{13.108}$$

13.6 CHAPTER EXERCISES

Exercise 55 *Show that the p.g.fl. for the simple virtual leader-follower model of (13.95) is*

$$G_{k+1|k}[h|X'] = p_h^{X' - \varphi_k(\bar{\mathbf{x}}') - \bar{\mathbf{x}}'} \tag{13.109}$$

where

$$p_h(\mathbf{x}') \triangleq \int h(\mathbf{x}) \cdot f_{\mathbf{V}}(\mathbf{x}+\mathbf{x}')d\mathbf{x} \tag{13.110}$$

and where, for any $\mathbf{x} \in \mathfrak{X}_0$, $X' - \mathbf{x} \triangleq \{\mathbf{x}' - \mathbf{x} |\ \mathbf{x}' \in X'\}$ *if* $X' \neq \emptyset$ *and* $X' - \mathbf{x} \triangleq \emptyset$ *otherwise.*

Exercise 56 *Assuming that* p_S *is constant, show that the p.g.fl. for the virtual leader-follower with target death of (13.104) is*

$$G_{k+1|k}[h|X'] = (1 - p_S + p_S p_h)^{X' - \varphi_k(\bar{\mathbf{x}}') - \bar{\mathbf{x}}'} \tag{13.111}$$

where p_h *and* $X' - \varphi_k(\bar{\mathbf{x}}') - \bar{\mathbf{x}}'$ *are as defined in Exercise 55.*

Chapter 14

The Multitarget Bayes Filter

14.1 INTRODUCTION TO THE CHAPTER

At this point it is worth summarizing what we have learned thus far. In Chapter 2, I introduced, in the context of a single sensor observing a single moving target, one of the fundamental concepts of the book: the Bayes recursive filter. There we learned that the Bayes filter can process conventional statistical measurements of essentially any type, given that they can be mediated by a likelihood function.

Chapters 3-7 showed that the Bayes filter can, in addition, process increasingly unconventional measurement types provided that they can be mediated by generalized likelihood functions.

Chapter 8 showed that, within a carefully specified sense, certain kinds of unconventional state-estimate data can be processed by the Bayes filter.

Finally, Chapters 9-13 laid the mathematical foundations for constructing true multitarget Markov densities and for true multisource-multitarget measurements of finite-set type.

This chapter represents the culmination of Chapters 9 through 13: the systematic generalization of the single-sensor, single-target Bayes filter to multisource-multitarget problems.

- This filter is the theoretical foundation for multitarget fusion, detection, tracking, and identification.

In brief, let $Z^{(k)} : Z_1, ..., Z_k$ be a time sequence of measurement sets. Let $f_{k+1|k}(X|X')$ be the multitarget Markov density (Chapter 13) and $f_{k+1}(Z|X)$ the multisource likelihood function (Chapter 12). Then the multitarget Bayes filter has the form

$$f_{k+1|k}(X|Z^{(k)}) = \int f_{k+1|k}(X|X') \cdot f_{k|k}(X'|Z^{(k)})\delta X' \quad (14.1)$$

$$f_{k+1|k+1}(X|Z^{(k+1)}) = \frac{f_{k+1}(Z_{k+1}|X) \cdot f_{k+1|k}(X|Z^{(k)})}{f_{k+1}(Z_{k+1}|Z^{(k)})} \quad (14.2)$$

where the Bayes normalization factor is

$$f_{k+1}(Z_{k+1}|Z^{(k)}) = \int f_{k+1}(Z_{k+1}|X) \cdot f_{k+1|k}(X|Z^{(k)})\delta X. \quad (14.3)$$

The multitarget filter propagates through time a sequence

$$f_{0|0}(X|Z^{(0)}) \rightarrow f_{1|0}(X|Z^{(0)}) \rightarrow f_{1|1}(X|Z^{(1)}) \rightarrow \quad (14.4)$$

$$\ldots \rightarrow f_{k|k}(X|Z^{(k)}) \rightarrow f_{k+1|k}(X|Z^{(k)}) \quad (14.5)$$

$$\rightarrow f_{k+1|k+1}(X|Z^{(k+1)}) \rightarrow \ldots \quad (14.6)$$

of *Bayes multitarget posterior probability distributions* (Chapter 11).

The purpose of this chapter is to introduce the multitarget Bayes filter in sufficient detail to permit its application. Subsequent chapters in Part III will be devoted to approximate methods for implementing it.

14.1.1 Summary of Major Lessons Learned

The following are the major concepts to be encountered in this chapter:

- In principle, data from information sources of highly disparate types can be fused using the multisource-multitarget Bayes filter (Sections 14.4.2 and 14.4.3).
- A systematic, unified, and probabilistic approach to multitarget-multisource integration is the consequence.
- The naïve multitarget generalizations of the most common Bayes-optimal state estimators (the expected a posteriori and maximum a posteriori estimators) are mathematically undefined in general (Section 14.5.1).
- Consequently, new Bayes-optimal multitarget state estimators must be devised and shown to be well behaved. These include the following:

- The *marginal multitarget (MaM)* estimator (Section 14.5.2). If $\hat{n} \triangleq \arg\sup_n f_{k|k}(n|Z^{(k)})$, then

$$X_{k|k}^{\text{MaM}} \triangleq \arg \sup_{\mathbf{x}_1,...,\mathbf{x}_{\hat{n}}} f_{k|k}(\{\mathbf{x}_1, ..., \mathbf{x}_{\hat{n}}\}|Z^{(k)}). \tag{14.7}$$

- The *joint multitarget (JoM)* estimator (Section 14.5.3). Given a small number $c \geq 0$ with the same units of measurement as $\mathbf{x}$:

$$X_{k|k}^{\text{JoM}} \triangleq \arg \sup_{n,\mathbf{x}_1,...,\mathbf{x}_n} f_{k|k}(\{\mathbf{x}_1, ..., \mathbf{x}_n\}|Z^{(k)}) \cdot \frac{c^n}{n!}. \tag{14.8}$$

- The JoM estimator is a limiting case of the conventional maximum a posteriori (MAP) estimator (Remark 22).
- The issue of *track labeling* (i.e., maintaining connected sequences of state-estimates) presents special computational difficulty for general multitarget Bayes filtering (Section 14.5.6).
- Measures of the uncertainty in multitarget estimates: track covariances (Section 14.6.2); global mean deviation (Section 14.6.3); and central multitarget entropy (Section 14.6.4);
- The *joint target-detection and tracking* (JoTT) *filter*, a single-sensor, single-target track-before-detect filter that admits essentially arbitrary false alarm processes (Section 14.7).
- The multitarget Bayes filter can be reformulated in terms of probability-generating functionals; see (14.265) and (14.280):[1]

$$G_{k+1|k}[h] = \int G_{k+1|k}[h|X'] \cdot f_{k|k}(X'|Z^{(k)})\delta X' \tag{14.9}$$

$$G_{k+1|k+1}[h] = \frac{\frac{\delta F}{\delta Z_{k+1}}[0, h]}{\frac{\delta F}{\delta Z_{k+1}}[0, 1]}. \tag{14.10}$$

- The concept of ranked target prioritization or ranked tactical significance can be mathematically codified in the form of a *tactical importance function* or TIF $\rho_{k|k}(\mathbf{x})$ (Section 14.9.1).

1 This reformulation permits, in turn, the development of the probability hypothesis density (PHD) and cardinalized probability hypothesis density (CPHD) filters of Chapter 16.

- Target prioritization can be incorporated into the fundamental statistics of a multitarget problem as follows; see (14.291) and (14.292):

$$f_{k|k}^{\rho_{k|k}}(X|Z^{(k)}) = \rho_{k|k}^X \cdot \frac{\delta G_{k|k}}{\delta X}[1 - \rho_{k|k}]. \tag{14.11}$$

14.1.2 Organization of the Chapter

The first few sections of the chapter describe the individual processing steps of the multitarget Bayes filter in greater detail: initialization (Section 14.2); predictor (Section 14.3); corrector (Section 14.4); multitarget state estimation (Section 14.5); and multitarget error estimation (Section 14.6). In particular, in Section 14.5 I describe three multitarget state estimators: the marginal multitarget (MaM) estimator; the joint multitarget (JoM) estimator; and the multitarget maximum likelihood estimator (MMLE).

Section 14.7 applies the Bayes filter to the case when at most a single target is known to be present. (A special case of this "JoTT filter" was the basis of the single-target nonlinear filtering example presented in Section 2.4.1.)

In preparation for Chapter 16, I introduce the p.g.fl. form of the multitarget Bayes filter in Section 14.8. In Section 14.9, I show how concepts of *target prioritization* or *tactical significance* can be incorporated into the fundamental statistics of a multitarget problem. Exercises for the chapter are in Section 14.10.

14.2 MULTITARGET BAYES FILTER: INITIALIZATION

Initialization of the multitarget Bayes filter consists of selection of the initial distribution $f_{0|0}(X) \triangleq f_{0|0}(X|Z^{(0)})$. I describe two methods.

14.2.1 Initialization: Multitarget Poisson Process

If initial knowledge of the number and states of targets is very limited, one could choose a *multitarget Poisson process*, (11.122), with a large mean/variance μ and a very high-variance spatial distribution $f(\mathbf{x})$:

$$f_{0|0}(\{\mathbf{x}_1, ..., \mathbf{x}_n\}) = e^{-\mu}\mu^n \cdot f(\mathbf{x}_1) \cdots f(\mathbf{x}_n). \tag{14.12}$$

14.2.2 Initialization: Target Number Known

Suppose that we know that exactly n statistically independent targets are present, with $f_1(\mathbf{x}), ..., f_\nu(\mathbf{x})$ being their respective a priori spatial distributions. Then $\nu = n$ and $q_1 = ... = q_\nu = 1$ and the multitarget multi-Bernoulli process reduces to $f_{0|0}(X) = 0$ if $|X| \neq n$ and, otherwise,

$$f_{0|0}(\{\mathbf{x}_1, ..., \mathbf{x}_n\}) = \sum_\sigma f_{\sigma 1}(\mathbf{x}_1) \cdots f_{\sigma n}(\mathbf{x}_n) \tag{14.13}$$

where the summation is taken over all permutations σ on the numbers $1, ..., n$.

14.3 MULTITARGET BAYES FILTER: PREDICTOR

The predictor equation for the multitarget Bayes filter is the analog of (2.121), with a set integral (see Section 11.3.3) in place of the vector integral:

$$f_{k+1|k}(X|Z^{(k)}) = \int f_{k+1|k}(X|X') \cdot f_{k|k}(X'|Z^{(k)}) \delta X'. \tag{14.14}$$

With the set integral explicitly written out, this is

$$\begin{aligned} & f_{k+1|k}(X|Z^{(k)}) & (14.15)\\ = \; & f_{k+1|k}(X|\emptyset) \cdot f_{k|k}(\emptyset|Z^{(k)}) & (14.16)\\ + \; & \int f_{k+1|k}(X|\{\mathbf{x}'\}) \cdot f_{k|k}(\{\mathbf{x}'\}|Z^{(k)}) d\mathbf{x}' & (14.17)\\ + \; & \frac{1}{2} \int f_{k+1|k}(X|\{\mathbf{x}'_1, \mathbf{x}'_2\}) \cdot f_{k|k}(\{\mathbf{x}'_1, \mathbf{x}'_2\}|Z^{(k)}) d\mathbf{x}'_1 d\mathbf{x}'_2 + ... & (14.18) \end{aligned}$$

Example 76 (Predictor Step for the 0-or-1 Target Case) *Assume that at most one target is in a scene, and that targets can disappear if present but not appear. The corresponding multitarget Markov density has the form*

$$f_{k+1|k}(X|\emptyset) = \begin{cases} 1 & \text{if} \quad X = \emptyset \\ 0 & \text{if} \quad \text{otherwise} \end{cases}. \tag{14.19}$$

$$f_{k+1|k}(X|\{\mathbf{x}'\}) = \begin{cases} 1 - p_S & \text{if} \quad X = \emptyset \\ p_S \cdot f_{k+1|k}(\mathbf{x}|\mathbf{x}') & \text{if} \quad X = \{\mathbf{x}\} \\ 0 & \text{if} \quad |X| \geq 2 \end{cases}. \tag{14.20}$$

where p_S *is the probability that a target will remain in the scene. The prior density must have the form*

$$f_{k|k}(X) = \begin{cases} 1-p & \text{if} \quad X = \emptyset \\ p \cdot f_{k|k}(\mathbf{x}) & \text{if} \quad X = \{\mathbf{x}\} \\ 0 & \text{if} \quad |X| \geq 2 \end{cases}. \tag{14.21}$$

Thus the predicted posterior distribution has the form

$$\begin{aligned} f_{k+1|k}(\emptyset) &= \int f_{k+1|k}(X|X') \cdot f_{k|k}(X')\delta X' && (14.22) \\ &= f_{k+1|k}(\emptyset|\emptyset) \cdot f_{k|k}(\emptyset) && (14.23) \\ &\quad + \int f_{k+1|k}(\emptyset|\{\mathbf{x}'\}) \cdot f_{k|k}(\{\mathbf{x}'\})d\mathbf{x}' && (14.24) \\ &= 1 - p + p \int (1-p_S) \cdot f_{k|k}(\mathbf{x}')d\mathbf{x}' && (14.25) \\ &= 1 - p_S \cdot p && (14.26) \end{aligned}$$

and

$$\begin{aligned} f_{k+1|k}(\{\mathbf{x}\}) &= \int f_{k+1|k}(\{\mathbf{x}\}|X') \cdot f_{k|k}(X')\delta X' && (14.27) \\ &= f_{k+1|k}(\{\mathbf{x}\}|\emptyset) \cdot f_{k|k}(\emptyset) && (14.28) \\ &\quad + \int f_{k+1|k}(\{\mathbf{x}\}|\{\mathbf{x}'\}) \cdot f_{k|k}(\{\mathbf{x}'\})d\mathbf{x}' && (14.29) \\ &= 0 + p_S \cdot p \int f_{k+1|k}(\mathbf{x}|\mathbf{x}') \cdot f_{k|k}(\mathbf{x}')d\mathbf{x}' && (14.30) \\ &= p_S \cdot p \cdot f_{k+1|k}(\mathbf{x}) && (14.31) \end{aligned}$$

where $f_{k+1|k}(\mathbf{x})$ *is the conventional single-target predicted posterior distribution. Note that* $f_{k+1|k}(X)$ *actually is a multitarget probability density function:*

$$\begin{aligned} \int f_{k+1|k}(X)\delta X &= f_{k+1|k}(\emptyset) + \int f_{k+1|k}(\{\mathbf{x}\})d\mathbf{x} && (14.32) \\ &= 1 - p_S \cdot p + p_S \cdot p \int f_{k+1|k}(\mathbf{x})d\mathbf{x} = 1. && (14.33) \end{aligned}$$

14.3.1 Predictor: No Target Birth or Death

Assume that target motions are independent, that targets neither appear nor disappear, and that target number is known to be n. Then in vector notation the multitarget predictor has the form

$$f_{k+1|k}(\mathbf{x}_1, ..., \mathbf{x}_n|Z^{(k)}) = \int f_{k+1|k}(\mathbf{x}_1|\mathbf{x}'_1) \cdots f_{k+1|k}(\mathbf{x}_n|\mathbf{x}'_n) \quad (14.34)$$

$$\cdot f_{k|k}(\mathbf{x}'_1, ..., \mathbf{x}'_n|Z^{(k)}) d\mathbf{x}'_1 \cdots d\mathbf{x}'_n. \quad (14.35)$$

This multitarget predictor equation is by far the one most often presumed in papers on multitarget filtering. It is obtained by predicting all individual target states in parallel.

To see this, according to (13.35), the corresponding multitarget Markov density is

$$f_{k+1|k}(X|X') = \sum_{\sigma} f_{k+1|k}(\mathbf{x}_1|\mathbf{x}'_{\sigma 1}) \cdots f_{k+1|k}(\mathbf{x}_n|\mathbf{x}'_{\sigma n}) \quad (14.36)$$

where the summation is taken over all permutations σ on the numbers $1, ..., n$ and where $X = \{\mathbf{x}_1, ..., \mathbf{x}_n\}$, $X' = \{\mathbf{x}'_1, ..., \mathbf{x}'_n\}$ with $|X| = |X'| = n$. Substituting this into (14.14) we get

$$f_{k+1|k}(X|Z^{(k)}) \quad (14.37)$$

$$= \frac{1}{n!} \int f_{k+1|k}(\{\mathbf{x}_1, ..., \mathbf{x}_n\}|\{\mathbf{x}'_1, ..., \mathbf{x}'_n\}) \quad (14.38)$$

$$\cdot f_{k|k}(\{\mathbf{x}'_1, ..., \mathbf{x}'_n\}|Z^{(k)}) d\mathbf{x}'_1 \cdots d\mathbf{x}'_n \quad (14.39)$$

$$= \frac{1}{n!} \int \left(\sum_{\sigma} f_{k+1|k}(\mathbf{x}_{\sigma 1}|\mathbf{x}'_1) \cdots f_{k+1|k}(\mathbf{x}_{\sigma n}|\mathbf{x}'_n) \right) \quad (14.40)$$

$$\cdot f_{k|k}(\{\mathbf{x}'_1, ..., \mathbf{x}'_n\}|Z^{(k)}) d\mathbf{x}'_1 \cdots d\mathbf{x}'_n \quad (14.41)$$

$$= \frac{1}{n!} \sum_{\sigma} \int f_{k+1|k}(\mathbf{x}_{\sigma 1}|\mathbf{x}'_1) \cdots f_{k+1|k}(\mathbf{x}_{\sigma n}|\mathbf{x}'_n) \quad (14.42)$$

$$\cdot f_{k|k}(\{\mathbf{x}'_1, ..., \mathbf{x}'_n\}|Z^{(k)}) d\mathbf{x}'_1 \cdots d\mathbf{x}'_n \quad (14.43)$$

$$= \int f_{k+1|k}(\mathbf{x}_1|\mathbf{x}'_1) \cdots f_{k+1|k}(\mathbf{x}_n|\mathbf{x}'_n) \quad (14.44)$$

$$\cdot f_{k|k}(\{\mathbf{x}'_1, ..., \mathbf{x}'_n\}|Z^{(k)}) \delta X'. \quad (14.45)$$

Or, using the fact that vector and set notations for the posteriors are related by (11.116),

$$n! \cdot f_{k|k}(\{\mathbf{x}'_1, ..., \mathbf{x}'_n\}|Z^{(k)}) = f_{k|k}(\mathbf{x}'_1, ..., \mathbf{x}'_n|Z^{(k)}) \quad (14.46)$$

$$n! \cdot f_{k+1|k}(\{\mathbf{x}'_1, ..., \mathbf{x}'_n\}|Z^{(k)}) = f_{k+1|k}(\mathbf{x}'_1, ..., \mathbf{x}'_n|Z^{(k)}). \quad (14.47)$$

Thus we have, as claimed,

$$f_{k+1|k}(\mathbf{x}_1, ..., \mathbf{x}_n|Z^{(k)}) = \int f_{k+1|k}(\mathbf{x}_1|\mathbf{x}'_1) \cdots f_{k+1|k}(\mathbf{x}_n|\mathbf{x}'_n) \quad (14.48)$$

$$\cdot f_{k|k}(\mathbf{x}'_1, ..., \mathbf{x}'_n|Z^{(k)}) d\mathbf{x}'_1 \cdots d\mathbf{x}'_n. \quad (14.49)$$

14.4 MULTITARGET BAYES FILTER: CORRECTOR

The corrector equations for the multitarget Bayes filter are the analogs of the single-target filter equations (2.128) and (2.129):

$$f_{k+1|k+1}(X|Z^{(k+1)}) = \frac{f_{k+1}(Z_{k+1}|X) \cdot f_{k+1|k}(X|Z^{(k)})}{f_{k+1}(Z_{k+1}|Z^{(k)})} \quad (14.50)$$

where the Bayes normalization factor is

$$f_{k+1}(Z_{k+1}|Z^{(k)}) = \int f_{k+1}(Z_{k+1}|X) \cdot f_{k+1|k}(X|Z^{(k)} \delta X. \quad (14.51)$$

The integrals in these equations are set integrals as defined in Section 11.3.3.

In what follows I consider the multitarget filtering of conventional single-sensor measurements (Section 14.4.1), of single-source generalized measurements (Section 14.4.2), and of multisource measurements (Section 14.4.3).

14.4.1 Conventional Measurements

The following simple example illustrates the process of multitarget filtering using conventional measurements.[2]

2 This example was first presented in [70, pp. 186, 187].

Example 77 (Corrector Step for the 0-or-1 Target Case) *A scene containing at most a single target is observed by a single sensor with constant probability of detection p_D and no false alarms. This sensor collects a single observation, that must therefore be either a missed detection $Z = \emptyset$ or a detection $Z = \{\mathbf{z}\}$. Thus the multitarget likelihood function is*

$$f_{k+1}(Z|\emptyset) = \begin{cases} 1 & \text{if} \quad Z = \emptyset \\ 0 & \text{if} \quad \text{otherwise} \end{cases}. \tag{14.52}$$

$$f_{k+1}(Z|\{\mathbf{x}\}) = \begin{cases} 1 - p_D & \text{if} \quad Z = \emptyset \\ p_D \cdot f_{k+1|k}(\mathbf{z}|\mathbf{x}) & \text{if} \quad Z = \{\mathbf{x}\} \\ 0 & \text{if} \quad |Z| \geq 2 \end{cases}. \tag{14.53}$$

Since at most one target is present, the predicted multitarget density must have the form

$$f_{k+1|k}(X) = \begin{cases} 1 - q & \text{if} \quad X = \emptyset \\ q \cdot f_{k+1|k}(\mathbf{x}) & \text{if} \quad X = \{\mathbf{x}\} \\ 0 & \text{if} \quad |X| \geq 2 \end{cases}. \tag{14.54}$$

That is, there is predicted probability q that a target actually exists. The predicted density for a single target with state $\mathbf{x}$ is $q \cdot f_{k+1|k}(\mathbf{x})$. If $Z = \emptyset$ the Bayes factor is

$$f_{k+1}(\emptyset) = \int f_{k+1}(\emptyset|X) \cdot f_{k+1|k}(X)\delta X \tag{14.55}$$

$$= f_{k+1}(\emptyset|\emptyset) \cdot f_{k+1|k}(\emptyset) \tag{14.56}$$

$$+ \int f_{k+1}(\emptyset|\{\mathbf{x}\}) \cdot f_{k+1|k}(\{\mathbf{x}\})d\mathbf{x} \tag{14.57}$$

$$= 1 - q + q \int (1 - p_D) \cdot f_{k+1|k}(\mathbf{x})d\mathbf{x} \tag{14.58}$$

$$= 1 - q \cdot p_D. \tag{14.59}$$

So from Bayes' rule the nonvanishing values of the corresponding data-updated posterior are:

$$f_{k+1|k+1}(\emptyset) = \frac{f_{k+1}(\emptyset|\emptyset) \cdot f_{k+1|k}(\emptyset)}{f_{k+1}(\emptyset)} \tag{14.60}$$

$$= \frac{1 \cdot (1 - q)}{1 - q \cdot p_D} = \frac{1 - q}{1 - q \cdot p_D} \tag{14.61}$$

and

$$f_{k+1|k+1}(\{\mathbf{x}\}) = \frac{f_{k+1}(\emptyset|\mathbf{x}) \cdot f_{k+1|k}(\mathbf{x})}{f_{k+1}(\emptyset)} \tag{14.62}$$

$$= \frac{(1-p_D) \cdot q \cdot f_{k+1|k}(\mathbf{x})}{1 - q \cdot p_D} \tag{14.63}$$

$$= \frac{q - q \cdot p_D}{1 - q \cdot p_D} \cdot f_{k+1|k}(\mathbf{x}). \tag{14.64}$$

On the other hand, if $Z = \{\mathbf{z}\}$, *then the Bayes factor is*

$$f_{k+1}(\{\mathbf{z}\}) = \int f_{k+1}(\{\mathbf{z}\}|X) \cdot f_{k+1|k}(X)\delta X \tag{14.65}$$

$$= f_{k+1}(\{\mathbf{z}\}|\emptyset) \cdot f_{k+1|k}(\emptyset) \tag{14.66}$$

$$+ \int f_{k+1}(\{\mathbf{z}\}|\{\mathbf{x}\}) \cdot f_{k+1|k}(\{\mathbf{x}\})d\mathbf{x} \tag{14.67}$$

$$= 0 + q \cdot p_D \int f_{k+1}(\mathbf{z}|\mathbf{x}) \cdot f_{k+1|k}(\mathbf{x})d\mathbf{x} \tag{14.68}$$

$$= q \cdot p_D \cdot f_{k+1}(\mathbf{z}) \tag{14.69}$$

where $f_{k+1}(\mathbf{z})$ *is the conventional Bayes' factor. So from Bayes' rule*

$$f_{k+1|k+1}(\emptyset) = \frac{f_{k+1}(\{\mathbf{z}\}|\emptyset) \cdot f_{k+1|k}(\emptyset)}{f_{k+1}(\{\mathbf{z}\})} = 0 \tag{14.70}$$

and

$$f_{k+1|k+1}(\{\mathbf{x}\}) = \frac{f_{k+1}(\{\mathbf{z}\}|\{\mathbf{x}\}) \cdot f_{k+1|k}(\{\mathbf{x}\})}{f_{k+1}(\{\mathbf{z}\})} \tag{14.71}$$

$$= \frac{p_D \cdot f_{k+1}(\mathbf{z}|\mathbf{x}) \cdot q \cdot f_{k+1|k}(\mathbf{x})}{q \cdot p_D \cdot f_{k+1}(\mathbf{z})} \tag{14.72}$$

$$= f_{k+1|k+1}(\mathbf{x}) \tag{14.73}$$

where $f_{k+1|k+1}(\mathbf{x})$ *is the conventional data-updated posterior. Equations (14.61) and (14.70) can be interpreted as follows. When we observe nothing* ($Z = \emptyset$) *this may be due to the fact that no target is actually present (with probability* $f_{k+1|k+1}(\emptyset)$*); or it is present but we just did not see it because of a missed detection (with probability* $1 - f_{k+1|k+1}(\emptyset)$*). If we collect an observation* $\mathbf{z}$*, however, a target is known to exist with certainty.*

14.4.2 Generalized Measurements

Until now, we have implicitly assumed that our information source is a conventional sensor that can be modeled using a conventional likelihood function. Clearly, however, identical reasoning can be applied to general information sources modeled using generalized likelihood functions, as described in Chapters 5-7.

To see this, as noted in Section 12.9.2, generalized likelihoods can be as easily used as conventional measurements in multitarget measurement models. For example, there I presented an example of a single target observed by a source supplying fuzzy DS measurements, noting that the generalized multitarget likelihood would have the form

$$f_{k+1}(Z|\mathbf{x}) = f_C(Z) \cdot \left(1 - p_D(\mathbf{x}) + p_D(\mathbf{x}) \sum_{o \in Z} \frac{f(o|\mathbf{x})}{\lambda c(o)}\right) \tag{14.74}$$

where

$$f(o|\mathbf{x}) = \sum_g o(g) \cdot g(\eta(\mathbf{x})). \tag{14.75}$$

When both conventional and generalized multitarget likelihoods are used in the multitarget Bayes filter, diverse evidence types can be fused using the same underlying probabilistic paradigm.

14.4.3 Unified Multitarget-Multisource Integration

I described the construction of likelihood functions for multisource data in Section 12.9. For example, I showed that when measurements are conditionally independent of state, the multitarget likelihood function is

$$f_{k+1}(\overset{1}{Z}, ..., \overset{s}{Z}|X) = \overset{1}{f}_{k+1}(\overset{1}{Z}|X) \cdots \overset{s}{f}_{k+1}(\overset{s}{Z}|X) \tag{14.76}$$

where

$$\overset{j}{f}_{k+1}(\overset{j}{Z}|X) \overset{\text{abbr.}}{=} f_{k+1}(\overset{j}{Z}|X, \overset{*j}{\mathbf{x}}) \tag{14.77}$$

is the multitarget likelihood function for the jth sensor. Simultaneously arriving multisource data can thus be processed in the usual manner using the multitarget corrector equations.

Suppose that the statistical correlations between measurements are known well enough to construct a joint likelihood $f_{k+1}(\overset{1}{Z}, ..., \overset{s}{Z}|X)$. Then sources of very general types can be fused. The result is

- *A systematic, rigorous, and theoretically unified methodology for fusing multisource-multitarget measurements of all types.*

14.5 MULTITARGET BAYES FILTER: STATE ESTIMATION

Multitarget state estimation poses unexpected difficulties. This is primarily because the naïve generalizations of the most commonly used Bayes-optimal state estimators, the expected a posteriori and maximum a posteriori estimators—see (2.133) and (2.134)—are not defined in general. Consequently, new multitarget state estimators must be devised and shown to be well behaved. This is the purpose of this section.

I use simple examples to illustrate the failure of the multitarget EAP and MAP estimators in Section 14.5.1. In Section 14.5.2, I introduce two new Bayes-optimal multitarget state estimators, the marginal multitarget (MaM) and joint multitarget (JoM) estimators, and compare them in simple examples. Computational issues are addressed in Section 14.5.5 and the question of track labeling in Section 14.5.6.

14.5.1 The Failure of the Classical State Estimators

The purpose of this section is to demonstrate, using examples, the following:

- The multitarget MAP is mathematically well defined only when single-target state space has no units of measurement (e.g., when it is discretized).
- A multitarget posterior expectation is meaningless even when it happens to be mathematically well defined.

Example 78 (Naïve Multitarget EAP/MAP: 0 or 1 Target) *Suppose that targets are in the one-dimensional interval [0,2] and that distance is measured in meters.*[3] *Assume that, according to current evidence, there is a 50-50 chance that no target exists. However, if it does, it is a single target that is equally likely to be anywhere in [0,2]. In this case the multitarget posterior distribution is*

$$f(X) = \begin{cases} \frac{1}{2} & \textit{if} \quad X = \emptyset \\ \frac{1}{4}\, m^{-1} & \textit{if} \quad \left\{ \begin{array}{c} X = \{x\} \\ 0 \leq x \leq 2 \end{array} \right. \\ 0 & \textit{if} \quad \textit{otherwise} \end{cases} . \tag{14.78}$$

3 This material originally appeared in [134, p. 41].

First consider the naïve EAP estimate:

$$\hat{X}^{EAP} = \int X \cdot f(X)\delta X \tag{14.79}$$

$$= \emptyset \cdot f(\emptyset) + \int_0^2 x \cdot f(\{x\})dx = \frac{1}{2}\emptyset + \frac{1}{2} \cdot 2m \tag{14.80}$$

$$= \frac{1}{2}(\emptyset + 2m). \tag{14.81}$$

This presents us with a two-fold problem. First, how do we add the unitless quantity $\emptyset$ *to the unit-ed quantity* $2m$*? More to the point, how do we add the empty set* $\emptyset$ *to the number* 2*? We cannot, because addition and subtraction of finite sets cannot be usefully defined.*[4] *In other words, the naïve EAP estimate is not mathematically well defined. Now consider the naïve MAP estimate. Since* $f(\emptyset) = \frac{1}{2} > \frac{1}{4} = f(\{x\})$,

$$\hat{X}^{MAP} = \arg\sup_X f(X) = \emptyset. \tag{14.82}$$

From this we conclude that the scene contains no targets. Now change units of measurement from meters to kilometers. Then the multitarget posterior is

$$f(X) = \begin{cases} \frac{1}{2} & \text{if} \quad X = \emptyset \\ 250\,km^{-1} & \text{if} \quad \left\{ \begin{array}{c} X = \{x\} \\ 0 \leq x \leq 0.002 \end{array} \right. \\ 0 & \text{if} \quad \text{otherwise} \end{cases} . \tag{14.83}$$

In this case, $f(\emptyset) = \frac{1}{2} < 250 = f(\{x\})$ *and so*

$$\arg\sup_X f(X) = \{x\} \tag{14.84}$$

for any $0 \leq x \leq 0.002$*. We now conclude that a target is present, even though we have done nothing more than change units of measurement! The paradox arises because the naïve MAP estimator prescribes a mathematically undefined procedure: comparing a unitless quantity* $f(X)$ *(when* $X = \emptyset$*) to a quantity* $f(X)$ *with units (when* $X = \{x\}$*). The naïve MAP estimate is thus mathematically undefined in general. It is defined only if units-mismatch difficulties do not arise. This occurs, for example, if state space has been discretized into cells.*

4 Addition of finite subsets can be defined in various ways, but none are suitable for our purposes. The Minkowski sum $X + X' \triangleq \{\mathbf{x} + \mathbf{x}' | \mathbf{x} \in X, \mathbf{x}' \in X'\}$ has no subtraction concept. Symmetric difference $X \dot{+} X' \triangleq (X - (X \cap X')) \cup (X' - (X \cap X'))$ has nilpotent subtraction: $-X = X$. There are many other examples of this kind.

The following example shows that the naïve EAP estimator is meaningless even when it happens to be mathematically well defined.

Example 79 (Naïve Multitarget EAP: Two Targets) *Revisit Example 56 of Section 11.3.3. Recall that in that example, two statistically independent targets are located at* $x = a$ *and* $x = b$ *with* $a \neq b$*; both have track variance* σ^2*; and the units are meters. The multitarget posterior is* $f(X) = 0$ *if* $|X| \neq 2$ *and, for all* $x \neq y$*,*

$$
\begin{aligned}
& f(\{x, y\}) && (14.85)\\
= \; & N_{\sigma^2}(x-a) \cdot N_{\sigma^2}(y-b) + N_{\sigma^2}(x-b) \cdot N_{\sigma^2}(y-a) && (14.86)\\
= \; & \frac{1}{2\pi\sigma^2} \exp\left(-\frac{(x-a)^2 + (y-b)^2}{2\sigma^2}\right) && (14.87)\\
& + \frac{1}{2\pi\sigma^2} \exp\left(-\frac{(x-b)^2 + (y-a)^2}{2\sigma^2}\right). && (14.88)
\end{aligned}
$$

The naïve multitarget MAP estimate does exist in this case:

$$
\begin{aligned}
\hat{X}^{EAP} &= \int (x, y) \cdot f(x, y) dx dy && (14.89)\\
&= \frac{1}{2} \cdot \begin{pmatrix} \int x \left(N_{\sigma^2}(x-a) + N_{\sigma^2}(x-b)\right) dx, \\ \int y \left(N_{\sigma^2}(y-b) + N_{\sigma^2}(y-a)\right) dy \end{pmatrix} && (14.90)\\
&= \frac{1}{2} \cdot (a+b, \; b+a). && (14.91)
\end{aligned}
$$

Unfortunately, this naïve EAP estimate always yields the same counterintuitive answer: two targets colocated at the same state $x = \frac{1}{2}(a+b)$*.*

Examples 78 and 79 lead us to the following conclusions:

- *The classical Bayes-optimal state estimators do not exist in general multitarget situations.*
- *We therefore must define new ones and demonstrate that they are statistically well behaved.*

This is the purpose of Sections 14.5.2 and 14.5.3.

Remark 21 (Unitless Multitarget Posterior Distributions?) *One might try to sidestep the failure of the multitarget MAP by using Stieltjes integrals.*[5] *To keep the discussion simple, let* $g(x)$ *be an arbitrary density function on single-target, one-dimensional state space; let* $P(x) = \int_{-\infty}^{x} g(y)dy$ *be its cumulative probability function; and let* $f(x_1, ..., x_n)$ *be a multitarget posterior. One could define a unitless multitarget posterior*

$$\breve{f}(x_1, ..., x_n) \triangleq \frac{f(x_1, ..., x_n)}{g(x_1) \cdots g(x_n)} \tag{14.92}$$

where set integrals are now computed using the Stieltjes integrals

$$\int \breve{f}(x_1, ..., x_n) dP(x_1) \cdots dP(x_n) \tag{14.93}$$

$$\triangleq \int \breve{f}(x_1, ..., x_n) \cdot g(x_1) \cdots g(x_n) dx_1 \cdots dx_n \tag{14.94}$$

$$= \int f(x_1, ..., x_n) dx_1 \cdots dx_n. \tag{14.95}$$

A multitarget MAP estimate computed using $\breve{f}(x_1, ..., x_n)$ *will no longer have difficulties with incommensurability of units. The price, however, is the unacceptable introduction of an arbitrary and extraneous factor—the density g—into the concept of a multitarget posterior. In particular, one could simply choose* $g(x) = c^{-1}$ *to be constant, where* c *has the same units as* x. *In this case*

$$\breve{f}(x_1, ..., x_n) = c^n \cdot f(x_1, ..., x_n). \tag{14.96}$$

14.5.2 Marginal Multitarget (MaM) Estimator

The MaM estimator is defined as follows.[6] Recall that the cardinality distribution was defined in (11.113) as:

$$f_{k|k}(n|Z^{(k)}) \triangleq \frac{1}{n!} \int f_{k|k}(\{\mathbf{x}_1, ..., \mathbf{x}_n\}|Z^{(k)}) d\mathbf{x}_1 \cdots d\mathbf{x}_n. \tag{14.97}$$

The first step of MaM estimation is computation of the MAP estimate of target number:

$$\hat{n} \triangleq \arg\sup_n \ f_{k|k}(n|Z^{(k)}). \tag{14.98}$$

5 This material originally appeared in [134, p. 42].

6 This estimator was originally called the "GMAP-I" estimator in [70, pp. 191, 192].

The second step is computation of a MAP estimate of the states of the individual targets, given that $n = \hat{n}$:

$$\hat{X}^{\text{MaM}} \triangleq \arg \sup_{\mathbf{x}_1,...,\mathbf{x}_{\hat{n}}} f_{k|k}(\{\mathbf{x}_1, ..., \mathbf{x}_{\hat{n}}\}|Z^{(k)}). \tag{14.99}$$

The MaM estimator is Bayes-optimal [70, pp. 192-194]. It is not known whether it is statistically consistent (i.e., always converges to the correct state in the static case).

14.5.3 Joint Multitarget (JoM) Estimator

The JoM estimator is defined as

$$\hat{X}^{\text{JoM}} \triangleq \arg \sup_{n,\mathbf{x}_1,...,\mathbf{x}_n} f_{k|k}(\{\mathbf{x}_1, ..., \mathbf{x}_n\}|Z^{(k)}) \cdot \frac{c^n}{n!} \tag{14.100}$$

where c is a fixed constant, the units of measurement of which are the same as those for a single-target state $\mathbf{x}$.[7] Expressed in more compact notation,

$$\hat{X}^{\text{JoM}} \triangleq \arg\sup_{X} f_{k|k}(X|Z^{(k)}) \cdot \frac{c^{|X|}}{|X|!}. \tag{14.101}$$

The following two step process is an alternative formulation of the JoM estimate. First, for each choice of $n \geq 0$, determine the MAP estimate

$$\hat{X}_n = \arg \sup_{\mathbf{x}_1,...,\mathbf{x}_n} f_{k|k}(\{\mathbf{x}_1, ..., \mathbf{x}_n\}|Z^{(k)}). \tag{14.102}$$

Then

$$\hat{X}^{\text{JoM}} = \hat{X}_{\hat{n}} \text{ where } \hat{n} = \arg\sup_{n} f_{k|k}(\hat{X}_n|Z^{(k)}) \cdot \frac{c^n}{n!}. \tag{14.103}$$

The JoME is Bayes-optimal and will converge to the correct answer if provided with enough data (that is, it is statistically consistent [70, pp. 192-205]).

- *The JoM estimator determines the number $\hat{n}$ and the identities/kinematics $\hat{\mathbf{x}}_1, ..., \hat{\mathbf{x}}_{\hat{n}}$ of targets optimally and simultaneously without resort to optimal report-to-track association.*

7 This estimator was originally called the "GMAP-II" estimator in [70, pp. 191, 192], though the current definition is slightly different in that it contains the factor $1/n!$

Stated differently:

- *The JoM estimator optimally resolves the conflicting objectives of detection, tracking, and identification in a single, unified statistical operation.*

The following question naturally arises:

- *In any given problem, how do we know which value of c to choose?*

The answer to this question will become clear as a result of the following analysis.

Remark 22 (JoM Is a Limiting Case of MAP) *The JoM estimator with very small c is a limiting case of the MAP estimator on a discrete multitarget space. To see this, choose some bounded region D of state space $\mathfrak{X}_0$. Partition D into ν equal-sized rectangular cells $d_1, ..., d_\nu$ of very small (hyper)volume ε, so that $\nu\varepsilon = |D|$ is the total (hyper)volume of D. Let $\hat{\mathbf{x}}_i$ denote the center of cell d_i. Given a multitarget probability distribution $f(\{\mathbf{x}_1, ..., \mathbf{x}_n\})$ on D, define*

$$f_n(\mathbf{x}_1, ..., \mathbf{x}_n) \triangleq (n!)^{-1} \cdot f(\{\mathbf{x}_1, ..., \mathbf{x}_n\}). \tag{14.104}$$

Then the corresponding discrete probability distribution on the discretized space is

$$\begin{aligned}
f(d_{i_1}, ..., d_{i_n}) &\triangleq \int_{d_{i_1} \times ... \times d_{i_n}} f_n(\mathbf{x}_1, ..., \mathbf{x}_n) d\mathbf{x}_1 \cdots d\mathbf{x}_n && (14.105)\\
&\cong f_n(\hat{\mathbf{x}}_{i_1}, ..., \hat{\mathbf{x}}_{i_n}) \cdot \varepsilon^n && (14.106)\\
&= f(\{\hat{\mathbf{x}}_{i_1}, ..., \hat{\mathbf{x}}_{i_n}\}) \cdot \frac{\varepsilon^n}{n!}. && (14.107)
\end{aligned}$$

Here the approximate equality results from elementary calculus and the fact that ε is small. So

$$f(\{\hat{\mathbf{x}}_{i_1}, ..., \hat{\mathbf{x}}_{i_n}\}) \cong f(d_{i_1}, ..., d_{i_n}) \cdot \frac{n!}{\varepsilon^n}. \tag{14.108}$$

Setting $c = \varepsilon$,

$$\begin{aligned}
X^{JoM} &= \arg \sup_{n, \mathbf{x}_1, ..., \mathbf{x}_n} f(\{\mathbf{x}_1, ..., \mathbf{x}_n\}) \cdot \frac{c^n}{n!} && (14.109)\\
&\cong \arg \sup_{n, i_1, .., i_n} f(\{\hat{\mathbf{x}}_{i_1}, ..., \hat{\mathbf{x}}_{i_n}\}) \cdot \frac{\varepsilon^n}{n!} && (14.110)\\
&\cong \arg \sup_{n, i_1, .., i_n} f(d_{i_1}, ..., d_{i_n}). && (14.111)
\end{aligned}$$

This confirms that the JoM estimate is a limiting case of the MAP estimate.

We can now return to the question of choosing the proper value of the JoM estimation constant c. The value of c should be small enough that the approximation of (14.108) is valid for all n and all $i_1, ..., i_n$. However, how small is that?

At minimum,

$$f(\{\hat{\mathbf{x}}_{i_1}, ..., \hat{\mathbf{x}}_{i_n}\}) \cdot \varepsilon^n \leq n! \tag{14.112}$$

since

$$\sum_{i_1,...,i_n} f(d_{i_1}, ..., d_{i_n}) \leq 1. \tag{14.113}$$

Actually, we need $f(\{\hat{\mathbf{x}}_{i_1}, ..., \hat{\mathbf{x}}_{i_n}\}) \cdot \varepsilon^n \leq 1$, see (14.165). Suppose that $\hat{\mathbf{x}}_{i_1}, ..., \hat{\mathbf{x}}_{i_n}$ are located in a part of D where $f(X)$ is very "peaky." Then ε must be very small if (14.108) is to be valid for this choice of $i_1, ..., i_n$.[8] On the other hand, ε need be no smaller than what is necessary for (14.108) to be valid.

Moreover, in any practical problem it is not necessary to estimate the state beyond some minimal degree of accuracy. (Localizing the position of an aircraft carrier to within a centimeter would be pointless, for example.) So as a general rule of thumb:

- *The value of c should be approximately equal to the accuracy to which the state is to be estimated, as long as $f(\{\hat{\mathbf{x}}_{i_1}, ..., \hat{\mathbf{x}}_{i_n}\}) \cdot \varepsilon^n \leq 1$.*

It should also be noted that there is a trade-off between optimality and the rate of convergence of the JoM estimator. The smaller the value of c, the greater the optimality but the slower the convergence [70, pp. 192-205].

14.5.3.1 Multitarget Maximum Likelihood Estimator (MMLE)

The conventional maximum likelihood state estimator (MLE) was introduced in (2.136) as a (in general non-Bayesian) special case of the MAP estimator. In the multitarget case this relationship no longer holds true.

If $f_{k+1}(Z|X)$ is a multitarget likelihood function then the units of measurement for $f(Z_{k+1}|X)$ are determined by the observation set Z (which is fixed) rather than by the multitarget state X. Consequently:

8 For example, if $f(X)$ is so "peaky" that virtually all of its probability mass is located there, then c should take the smallest value such that $f(d_{i_1}, ..., d_{i_n}) \cong 1$. In this case, $\varepsilon \cong n! \cdot f(\{\hat{\mathbf{x}}_{i_1}, ..., \hat{\mathbf{x}}_{i_n}\})^{-1}$.

- *The naïve extension of the classical MLE to the multitarget case is always well defined despite the fact that the classical MAP is not*:

$$\hat{X}^{\text{MMLE}} \triangleq \arg \sup_{n, \mathbf{x}_1, ..., \mathbf{x}_n} f(Z|\{\mathbf{x}_1, ..., \mathbf{x}_n\}). \tag{14.114}$$

In condensed notation,

$$\hat{X}^{\text{MMLE}} \triangleq \arg \sup_{X} f(Z|X). \tag{14.115}$$

As with the conventional single-target MLE, the MMLE will converge to the correct answer given enough data [70, pp. 199-205]. See [70, pp. 256-259] for a numerical example.

14.5.4 JoM and MaM Estimators Compared

I illustrate the computation of MaM and JoM estimates with three simple examples: (1) at most, one target is present; (2) the multitarget distribution is an i.i.d. cluster process; and (3) the multitarget distribution is uniform.

Example 80 (JoM Versus MaM: At Most One Target Present) *Suppose that a multitarget posterior has the form*

$$f(X) = \begin{cases} 1-q & \text{if} \quad X = \emptyset \\ q \cdot N_{\sigma^2}(x) & \text{if} \quad X = \{x\} \\ 0 & \text{if} \quad \text{otherwise} \end{cases}.$$

where $N_{\sigma^2}(x)$ is a normal distribution with very small variance σ^2. That is, at most one target is present in the scene at $x = 0$; and if it exists it is very accurately localized. The MaM estimator will decide in favor of the presence of a target if

$$1 - q = f(0) < f(1) = q \tag{14.116}$$

that is, if $q > \frac{1}{2}$. The JoM estimator will decide in favor of the presence of a target if, given the small-magnitude estimation constant c,

$$1 - q = f(\emptyset) < c \cdot \sup_x f(\{x\}) = \frac{cq}{\sqrt{2\pi}\sigma} \tag{14.117}$$

or, rewriting, if

$$\sigma \cdot \left(\frac{1}{q} - 1\right) < \frac{c}{\sqrt{2\pi}}.$$

In summary: MaM ignores relevant information—what is, σ^2—whereas JoM balances the information contained in q against the information contained in σ. Information supporting a no-target decision (q is small) can be countermanded by sufficiently compelling information supporting a yes-target decision (the target is very well-localized since σ is small, and therefore is much more likely to exist). For example, let $c = 0.1$ and $q = 0.2$. Then the target must be exceedingly well-localized ($\sigma < 0.001$) if a yes-target decision is to be supported. However, if $q = 0.999$, then the target can be quite unlocalized ($\sigma < 40.0$).

Example 81 (JoM Versus MaM: i.i.d. Cluster Process) *Recall that the i.i.d. cluster process was defined in (11.121) by*

$$f(X) = |X|! \cdot p(|X|) \cdot \prod_{\mathbf{x} \in X} f(\mathbf{x}). \tag{14.118}$$

In particular, $f(\emptyset) = p(0)$. To compute the MaM estimate first compute the cardinality distribution as defined in (11.113):

$$\begin{aligned} f(n) &= \frac{1}{n!} \int f(\{\mathbf{x}_1, ..., \mathbf{x}_n\}) d\mathbf{x}_1 \cdots d\mathbf{x}_n & (14.119) \\ &= p(n) \cdot \int f(\mathbf{x}_1) \cdots f(\mathbf{x}_n) d\mathbf{x}_1 \cdots d\mathbf{x}_n & (14.120) \\ &= p(n). & (14.121) \end{aligned}$$

The MAP estimate of target number is

$$\hat{n} = \arg\sup_n \, p(n) \tag{14.122}$$

and the MAP estimate of target states for this choice of target number is

$$\begin{aligned} \hat{X}^{MaM} &= \arg \sup_{\mathbf{x}_1, ..., \mathbf{x}_{\hat{n}}} f(\mathbf{x}_1) \cdots f(\mathbf{x}_{\hat{n}}) & (14.123) \\ &= \underbrace{\{\hat{\mathbf{x}}, ..., \hat{\mathbf{x}}\}}_{\hat{n}} = \{\hat{\mathbf{x}}\}. & (14.124) \end{aligned}$$

That is, the MaM estimator determines that there are $\hat{n}$ targets, all colocated at the same state $\hat{\mathbf{x}} = \arg\sup_{\mathbf{x}} f(\mathbf{x})$. To compute the JoM estimate, note that for

$n > 0$,

$$s(n) \triangleq \sup_{\mathbf{x}_1,...,\mathbf{x}_{\mathring{n}}} f(\{\mathbf{x}_1,...,\mathbf{x}_n\}) \cdot \frac{c^n}{n!} \tag{14.125}$$

$$= \sup_{\mathbf{x}_1,...,\mathbf{x}_{\mathring{n}}} n! \cdot p(n) \cdot f(\mathbf{x}_1) \cdots f(\mathbf{x}_n) \cdot \frac{c^n}{n!} \tag{14.126}$$

$$= p(n) \cdot \left(c \cdot \sup_{\mathbf{x}} f(\mathbf{x}) \right)^n. \tag{14.127}$$

As with the MaM estimate, the JoM estimate will consist of some number of targets, all colocated at the same state $\hat{\mathbf{x}} = \arg\sup_{\mathbf{x}} f(\mathbf{x})$. *The primary problem is to determine the estimated number of targets* n_0. *We must assume that* c *is small enough that* $c \cdot \sup_{\mathbf{x}} f(\mathbf{x}) \leq 1$. *Thus the peak value of* $s(n)$ *will occur for a smaller value of* n *than would be the case with the MaM estimator.*

Example 82 (JoM Versus MaM: Multitarget Uniform Distribution) *Recall that the multitarget uniform distribution* $u_{\mathring{n},D}(X)$ *was defined in (11.126) as:*

$$u_{\mathring{n},D}(X) = \begin{cases} \frac{1}{\mathring{n}+1} & \text{if} \quad X = \emptyset \\ \frac{n!}{|D|^n \cdot (\mathring{n}+1)} & \text{if} \quad \begin{cases} X = \{\mathbf{x}_1,...,\mathbf{x}_n\} \\ 1 \leq |X| = n \leq \mathring{n} \end{cases} \\ 0 & \text{if} \quad \text{otherwise} \end{cases}. \tag{14.128}$$

If the distribution $u_{\mathring{n},D}(X)$ *is to be "uniform" in a meaningful sense, it should not display any preference for target number or individual target states. Thus any Bayes-optimal state estimator should return every possible (allowable) state as an estimate. In the case of the MaM estimator, we first compute the cardinality distribution of* $u_{\mathring{n},D}(X)$:

$$u_{\mathring{n},D}(n) = \frac{1}{n!} \int u_{\mathring{n},D}(X)(\{\mathbf{x}_1,...,\mathbf{x}_n\}) d\mathbf{x}_1 \cdots d\mathbf{x}_n \tag{14.129}$$

$$= \frac{1}{n!} \int \frac{n!}{|D|^n \cdot (\mathring{n}+1)} d\mathbf{x}_1 \cdots d\mathbf{x}_n \tag{14.130}$$

$$= \int_{\underbrace{D \times ... \times D}_{n}} \frac{1}{|D|^n \cdot (\mathring{n}+1)} d\mathbf{x}_1 \cdots d\mathbf{x}_n \tag{14.131}$$

$$= \frac{1}{\mathring{n}+1}. \tag{14.132}$$

With respect to the MaM estimator, therefore, $u_{\mathring{n},D}(X)$ *behaves like a uniform distribution. In the case of the JoM estimator this is not true. Note that for* $X = \{\mathbf{x}_1, ..., \mathbf{x}_n\}$ *with* $|X| = n$,

$$u_{\mathring{n},D}(X) \cdot \frac{c^n}{n!} = \left(\frac{c}{|D|}\right)^n \cdot \frac{1}{\mathring{n}+1}. \tag{14.133}$$

If c *is small then the maximal value occurs for* $X = \emptyset$. *With respect to the JoM estimator, therefore, the multitarget uniform distribution does not behave like a uniform distribution.*

14.5.5 Computational Issues

Both the MaM and JoM estimators require solution of optimization problems of the general form

$$\hat{X} = \arg \sup_{\mathbf{x}_1,...,\mathbf{x}_n} f(\{\mathbf{x}_1, ..., \mathbf{x}_n\}) \tag{14.134}$$

for fixed n. How might such problems be addressed computationally?

14.5.5.1 Direct Gaussian-Mixture Approximation

One approach is to approximate $f(\{\mathbf{x}_1, ..., \mathbf{x}_n\})$ as a Gaussian mixture:

$$f(\{\mathbf{x}_1, ..., \mathbf{x}_n\}) \cong \sum_{\sigma} N_{P_{\sigma 1}}(\mathbf{x}_1 - \hat{\mathbf{x}}_{\sigma 1}) \cdots N_{P_{\sigma n}}(\mathbf{x}_n - \hat{\mathbf{x}}_{\sigma n}) \tag{14.135}$$

where the vectors $\hat{\mathbf{x}}_1, ..., \hat{\mathbf{x}}_n$ and covariance matrices $P_1, ..., P_n$ are to be determined. Given this approximation, the approximate solution to (14.134) is $\hat{X} = \{\hat{\mathbf{x}}_1, ..., \hat{\mathbf{x}}_n\}$. As we noted in (2.239), Dempsters' expectation-maximization (EM) algorithm is well-suited for determining Gaussian-mixture approximations of the form of (14.135) [15, 162].

14.5.5.2 Indirect Gaussian-Mixture Approximation

The essential ideas underlying this approach were proposed by M. Vihola [232, 233, 234] and Sidenbladh and Wirkander [210]. Unfortunately, to describe it I must use some concepts that will not be extensively discussed until Chapter 16.

The *probability hypothesis density* (also known as *first-order multitarget moment density*) of the multitarget posterior $f_{k|k}(X|Z^{(k)})$ is defined as

$$D_{k|k}(\mathbf{x}|Z^{(k)}) \triangleq \int f_{k|k}(\{\mathbf{x}\} \cup X|Z^{(k)})\delta X. \tag{14.136}$$

It is not a probability distribution since $N_{k|k} \triangleq \int D_{k|k}(\mathbf{x}|Z^{(k)})d\mathbf{x}$ is the expected number of targets in the scene. Let $\hat{n}$ be the integer nearest in value to $N_{k|k}$. Then the $\hat{n}$ highest peaks of $D_{k|k}(\mathbf{x}|Z^{(k)})$ correspond to the states of the targets. That is, let $\hat{\mathbf{x}}_1, ..., \hat{\mathbf{x}}_{\hat{n}}$ be such that $D_{k|k}(\hat{\mathbf{x}}_1|Z^{(k)})$,...,$D_{k|k}(\hat{\mathbf{x}}_{\hat{n}}|Z^{(k)})$ are the $\hat{n}$ largest values of $D_{k|k}(\cdot|Z^{(k)})$. Then $\hat{X}_{k|k} = \{\hat{\mathbf{x}}_1, ..., \hat{\mathbf{x}}_{\hat{n}}\}$ is an estimate of the multitarget state.

This estimate can be extracted by using the EM algorithm [15, 162] to approximate $D_{k|k}(\mathbf{x}|Z^{(k)})$ as a mixture of $\hat{n}$ Gaussians:

$$D_{k|k}(\mathbf{x}|Z^{(k)}) \cong \sum_{i=1}^{\hat{n}} \alpha_i \cdot N_{C_i}(\mathbf{x} - \hat{\mathbf{x}}_i) \tag{14.137}$$

Alternatively, in some cases peak detection, clustering, or histogramming techniques can be used to determine the $\hat{n}$ largest suprema of $D_{k|k}(\mathbf{x}|Z^{(k)})$ [232, 233, 234].

14.5.6 State Estimation and Track Labeling

Track labeling does not arise as an issue in single-target filtering. If only a single target is present, its "track" is the time sequence $\mathbf{x}_{0|0}$, $\mathbf{x}_{1|1}$,..., $\mathbf{x}_{k|k}$,... of instantaneous state-estimates computed using some state estimator.

Matters are not so simple in the multitarget case. State estimation will produce a time sequence of state sets $X_{0|0}$, $X_{1|1}$,..., $X_{k|k}$,... Since the elements of sets have no inherent order, for any k which state-estimates in $X_{k+1|k+1}$ correspond to which state-estimates in $X_{k|k}$? Stated differently, how do we "connect the dots" to produce the time-evolving tracks of the targets?

Because they are constructed in a bottom-up fashion, the SHC, MHC, and CHC approaches of Chapter 10 all can accommodate track labeling. However, what about the multitarget Bayes filter?

The easy answer to this question is to note, as we did in Remark 19 of Section 12.2.1, that careful multitarget state representation requires that individual target state vectors include an identity parameter: $\mathbf{x} = (\mathbf{u}, u)$ where $\mathbf{u}$ is the geokinematic state and u is the identity state.

Even in tracking problems that do not involve target identity as such, we must be able to distinguish one track from another. At minimum, target states must include *track labels* $\ell = 1, ..., \nu$ that distinguish one target from another. Complete target states therefore have the form $(\mathbf{x}_1, \ell_1)$,...,$(\mathbf{x}_\nu, \ell_\nu)$. Given this, the process of multitarget state estimation leads to a time sequence $X_{0|0}$, $X_{1|1}$,..., $X_{k|k}$,... in which

$$X_{k|k} = \{(\hat{\mathbf{x}}^1_{k|k}, \hat{\ell}^1_{k|k}), ..., (\hat{\mathbf{x}}^{\hat{n}_{k|k}}_{k|k}, \hat{\ell}^{\hat{n}_{k|k}}_{k|k})\}. \tag{14.138}$$

This means that at time step k, the target with label $\hat{\ell}^i_{k|k}$ is estimated to have state vector $\hat{\mathbf{x}}^i_{k|k}$. Continuity for any given track in a scene is achieved by simply linking time sequences of state-estimates that have the same label. This process is illustrated in Figure 14.1.

This approach to track labeling is theoretically optimal from a Bayesian point of view. Unfortunately, it also significantly increases computational load.

Suppose, for example, that exactly n targets are known to be present. Then computing a MaM or JoM estimate would require solving the labeled optimization problem

$$\arg \sup_{\ell_1 \neq ... \neq \ell_n, \mathbf{x}_1, ..., \mathbf{x}_{\hat{n}}} f(\{(\mathbf{x}_1, \ell_1), ..., (\mathbf{x}_n, \ell_n)\}). \tag{14.139}$$

If a operations are required to solve the unlabeled optimization problem

$$\arg \sup_{\mathbf{x}_1, ..., \mathbf{x}_n} f(\{\mathbf{x}_1, ..., \mathbf{x}_n\}), \tag{14.140}$$

then $n! \cdot a$ operations are required to solve the labeled one. This is because each choice of the list $\ell_1 \neq ... \neq \ell_n$ (i.e., each choice of a permutation of the numbers $1, ..., n$) requires a separate solution of

$$\arg \sup_{\mathbf{x}_1, ..., \mathbf{x}_n} f(\{(\mathbf{x}_1, \ell_1), ..., (\mathbf{x}_n, \ell_n)\}). \tag{14.141}$$

This may be computationally unacceptable for problems involving more than a small number of targets, where the definition of "small" will greatly depend on SNR.

With the exception of the particle labeling approach described later, most solutions for track labeling problem for the full multitarget filter have been less than satisfactory.

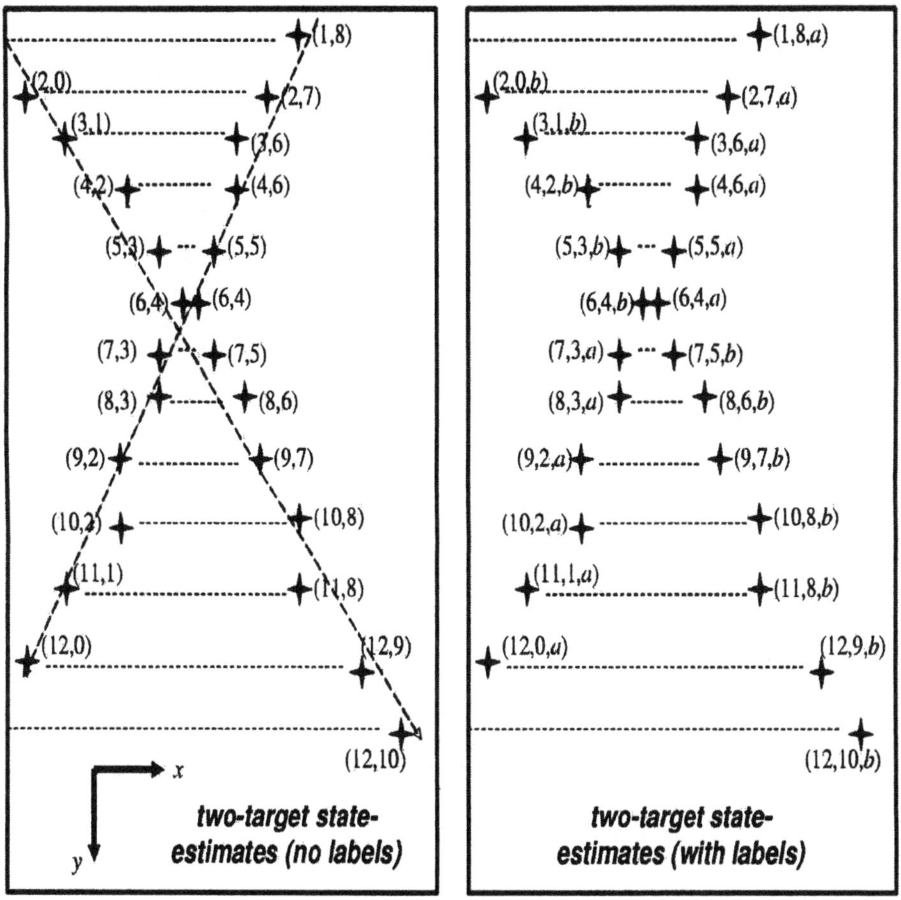

Figure 14.1 The concept of expressing a track label as a discrete state parameter is illustrated. The left-hand figure displays a time sequence $X_{1|1}, ..., X_{12|12}$ of instantaneous multitarget state-estimates in an x-y plane. (The direction of time is downward.) Human "eyeball track labeling" is required to infer the existence of two targets moving from left to right and from right to left and that cross at the location $x = 4, y = 6$. The right-hand figure illustrates the same scene if a track label ℓ with values $\ell = a, b$ is included as an additional state parameter. In this case the instantaneous multitarget state-estimates produce sequences of clearly labeled tracks: track a and track b.

14.5.6.1 "Eyeball Track Labeling"

The great majority of papers on multitarget filtering simply sidestep the issue of track labeling. In such papers the time sequence $X_{0|0}, X_{1|1},..., X_{k|k},...$ is visually displayed and readers must use their human visual integrative powers to connect the dots.

14.5.6.2 Heuristic Labeling Predicated on Favorable Sensing Conditions

Assume that signal-to-noise ratio (SNR) is very large, in the sense that sensor noise power is small and false detections and clutter observations are nearly nonexistent. Then ad hoc pattern recognition approaches can be applied to scenarios containing many tens of targets. In practice, this approach has required a number of unrealistic assumptions besides high SNR: target number is already known; tracks have already been initialized; and target motions are not too nonlinear.

Conventional multitarget tracking techniques such as multihypothesis tracking (MHT) are already capable of successfully addressing most such scenarios at much smaller computational expense. Consequently, approaches of this kind will probably be applicable only in special situations—for example, high-SNR problems with very closely spaced targets and many confusing track crossings.

14.5.6.3 Estimate-to-Track Association

As the name implies, such techniques adapt familiar measurement-to-track association techniques (of the sort described in Section 10.5) to recursively associate state vectors with evolving tracks. Since such techniques have been employed primarily with the probability hypothesis density (PHD) filter of Chapter 16, we will postpone a more detailed discussion until Section 16.3.4.

14.5.6.4 Particle Labeling

Ma, Vo, Singh, and Baddeley [122, 239] have applied the multitarget Bayes filter to the problem of detecting and tracking time-varying numbers of individuals in rooms, based on processing of speech utterances collected by multiple fixed microphones. Their implementation has the ability to maintain track continuity through the device of attaching labels to individual single-target states in the multitarget particles. The authors report that this procedure also greatly simplifies the process of multitarget state estimation.

14.6 MULTITARGET BAYES FILTER: ERROR ESTIMATION

In addition to a multitarget state-estimate $\hat{X}$, we would like to have some assessment of the accuracy or error associated with this estimate. In this section, I describe a few techniques for estimating degree of error. These include: RMS deviation of target number (Section 14.6.1), covariances of the state-estimates of individual targets (Section 14.6.2), global mean deviation (Section 14.6.3), and information-theoretic concepts of multitarget statistical dispersion (Section 14.6.4).

14.6.1 Target Number RMS Deviation

Let

$$p_{k|k}(n|Z^{(k)}) = \frac{1}{n!}\int f_{k|k}(\{\mathbf{x}_1, ..., \mathbf{x}_n\}|Z^{(k)})d\mathbf{x}_1 \cdots d\mathbf{x}_n \tag{14.142}$$

be the cardinality distribution of $f_{k|k}(X|Z^{(k)})$—see (11.113)—and let $\hat{n}$ be the target number as generated by some state estimator. The RMS deviation $\sigma_{k|k}$ of the estimate is given by

$$\sigma_{k|k}^2 = \sum_{n=0}^{\infty}(n - \hat{n})^2 \cdot p_{k|k}(n|Z^{(k)}). \tag{14.143}$$

14.6.2 Track Covariances

Suppose that $\hat{X} = \{\hat{\mathbf{x}}_1, ..., \hat{\mathbf{x}}_{\hat{n}}\}$ is the multitarget estimate produced by some multitarget state estimator.[9] We would like to estimate the error associated with each of the state-estimates $\hat{\mathbf{x}}_1, ..., \hat{\mathbf{x}}_{\hat{n}}$. Computational questions aside, a theoretically reasonable solution is as follows. Given a fixed i, construct the induced posterior distribution

$$f_{k|k}^{(i)}(\mathbf{x}|Z^{(k)}) \triangleq \frac{1}{K_i} f_{k|k}(\{\hat{\mathbf{x}}_1, ..., \hat{\mathbf{x}}_{i-1}, \mathbf{x}, \hat{\mathbf{x}}_{i+1}, ..., \hat{\mathbf{x}}_{\hat{n}}\}|Z^{(k)}) \tag{14.144}$$

where

$$K_i \triangleq \int f_{k|k}(\{\hat{\mathbf{x}}_1, ..., \hat{\mathbf{x}}_{i-1}, \mathbf{x}, \hat{\mathbf{x}}_{i+1}, ..., \hat{\mathbf{x}}_{\hat{n}}\}|Z^{(k)})d\mathbf{x}. \tag{14.145}$$

9 The material in this section originally appeared in [134, pp. 44, 45].

The single-target probability density $f^{(i)}_{k|k}(\mathbf{x}|Z^{(k)})$ is the current spatial distribution of the track $\hat{\mathbf{x}}_i$. For $i = 1, ..., \hat{n}$ the uncertainty in $\hat{\mathbf{x}}_i$ is described by its corresponding covariance (strictly speaking, RMS deviation) matrix:

$$C^{(i)}_{k|k} \triangleq \int (\mathbf{x} - \hat{\mathbf{x}}_i)(\mathbf{x} - \hat{\mathbf{x}}_i)^T \, f^{(i)}_{k|k}(\mathbf{x}|Z^{(k)}) d\mathbf{x}. \tag{14.146}$$

How would one actually compute these track covariances? In Section 14.5 we noted that the EM algorithm can be used to approximate

$$f(\{\mathbf{x}_1, ..., \mathbf{x}_{\hat{n}}\}) \cong \sum_\sigma N_{P_{\sigma 1}}(\mathbf{x}_1 - \hat{\mathbf{x}}_{\sigma 1}) \cdots N_{P_{\sigma n}}(\mathbf{x}_{\hat{n}} - \hat{\mathbf{x}}_{\sigma \hat{n}}). \tag{14.147}$$

Using this approximation, (14.144) becomes

$$f^{(i)}_{k|k}(\mathbf{x}|Z^{(k)}) \tag{14.148}$$

$$\cong \frac{1}{K_i} \sum_\sigma N_{P_{\sigma 1}}(\hat{\mathbf{x}}_1 - \hat{\mathbf{x}}_{\sigma 1}) \cdots N_{P_{\sigma n}}(\hat{\mathbf{x}}_{i-1} - \hat{\mathbf{x}}_{\sigma(i-1)}) \tag{14.149}$$

$$\cdot N_{P_{\sigma i}}(\mathbf{x} - \hat{\mathbf{x}}_{\sigma i}) \tag{14.150}$$

$$\cdot N_{P_{\sigma n}}(\hat{\mathbf{x}}_{i+1} - \hat{\mathbf{x}}_{\sigma(i+1)}) \cdots N_{P_{\sigma n}}(\hat{\mathbf{x}}_{\hat{n}} - \hat{\mathbf{x}}_{\sigma \hat{n}}) \tag{14.151}$$

where

$$K_i = \sum_\sigma N_{P_{\sigma 1}}(\hat{\mathbf{x}}_1 - \hat{\mathbf{x}}_{\sigma 1}) \cdots N_{P_{\sigma n}}(\hat{\mathbf{x}}_{i-1} - \hat{\mathbf{x}}_{\sigma(i-1)}) \tag{14.152}$$

$$\cdot N_{P_{\sigma n}}(\hat{\mathbf{x}}_{i+1} - \hat{\mathbf{x}}_{\sigma(i+1)}) \cdots N_{P_{\sigma n}}(\hat{\mathbf{x}}_{\hat{n}} - \hat{\mathbf{x}}_{\sigma \hat{n}}) \tag{14.153}$$

The term with $\sigma = \iota$ (identity permutation) will often dominate other terms. Therefore $K_i \cong \int N_{P_i}(\mathbf{x} - \hat{\mathbf{x}}_i) d\mathbf{x} = 1$ and

$$f^{(i)}_{k|k}(\mathbf{x}|Z^{(k)}) \cong N_{P_i}(\mathbf{x} - \hat{\mathbf{x}}_i). \tag{14.154}$$

Thus approximately, the $P_1, ..., P_{\hat{n}}$ are the covariance matrices respectively corresponding to the target state-estimates $\hat{\mathbf{x}}_1, ..., \hat{\mathbf{x}}_{\hat{n}}$.

14.6.3 Global Mean Deviation

Let $\hat{X}_{k|k}$ be a multitarget state-estimate. Suppose further that we have a "global miss distance" metric $d(X, X')$ that measures the distance between nonempty finite sets X, X'. Then one could in principle compute the *global mean deviation* from $\hat{X}_{k|k}$ as follows:

$$\Delta_{k|k} = \int d(\hat{X}_{k|k}, X) \cdot \hat{f}_{k|k}(X) \delta X \tag{14.155}$$

where $\hat{f}_{k|k}(X) \overset{\text{abbr.}}{=} f_{k|k}(X|Z^{(k)}, X \neq \emptyset)$ is the probability of the multitarget state X given that it is nonempty. That is, $\hat{f}_{k|k}(\emptyset) \triangleq 0$ and, if $X \neq \emptyset$,

$$\hat{f}_{k|k}(X) \triangleq \frac{f_{k|k}(X|Z^{(k)})}{1 - f_{k|k}(\emptyset|Z^{(k)})}. \tag{14.156}$$

This is indeed possible in principle. The most obvious choice for $d(X, X')$ is the *Hausdorff distance* [87]

$$d^H(X, X') \triangleq \max\left\{\max_{\mathbf{x}\in X} \min_{\mathbf{x}'\in X} d(\mathbf{x}, \mathbf{x}'),\ \max_{\mathbf{x}'\in X'} \min_{\mathbf{x}\in X} d(\mathbf{x}, \mathbf{x}')\right\}. \tag{14.157}$$

Hausdorff distance has a possible disadvantage: It is relatively insensitive to deviation in target number: $|X| \neq |\hat{X}_{k|k}|$.

Hoffman and Mahler [87] have introduced the family of *Wasserstein distances*, that do account for deviations in cardinality. Let $X = \{\mathbf{x}_1, ..., \mathbf{x}_n\}$ and $X' = \{\mathbf{x}'_1, ..., \mathbf{x}'_{n'}\}$ where $|X| = n$ and $|X'| = n'$. Then the Wasserstein distance of power p is defined by

$$d_p^W(X, X') \triangleq \inf_C \sqrt[p]{\sum_{i=1}^{n} \sum_{i'=1}^{n'} C_{i,i'} \cdot d(\mathbf{x}_i, \mathbf{x}'_{i'})^p} \tag{14.158}$$

where the infimum is taken over all $n \times n'$ "transportation matrices" C. A matrix C is a transportation matrix if, for all $i = 1, ..., n$ and $i' = 1, ..., n'$,

$$\sum_{i=1}^{n} C_{i,i'} = \frac{1}{n'}, \qquad \sum_{i'=1}^{n'} C_{i,i'} = \frac{1}{n}. \tag{14.159}$$

The larger the value of p, the more that $d_p^W(X, X')$ penalizes deviations in target number. Efficient algorithms exist for computing the $d_W^p(X, X')$ [87].

Given this, we can define the following measures of mean deviation:

$$\Delta^H_{k|k} \triangleq \int d^H(\hat{X}_{k|k}, X) \cdot \hat{f}_{k|k}(X)\delta X \tag{14.160}$$

$$\Delta^{W,p}_{k|k} \triangleq \left(\int d^W_p(\hat{X}_{k|k}, X)^p \cdot \hat{f}_{k|k}(X)\delta X\right)^{1/p} \tag{14.161}$$

In general, these quantities will be computationally intractable because of the set integral used to define them. However, under certain circumstances (e.g., multitarget SMC approximation as described in Chapter 15) they will become potentially computable in real time. See Section 15.7.1.

Example 83 *[87, p. 330] Let* $X' = \{\mathbf{x}'\}$. *Then* $d^W_p(X, X')$ *reduces to the root-mean-power of the distance between* X *and* $\mathbf{x}'$*:*

$$d^W_p(X, \{\mathbf{x}'\}) \triangleq \sqrt[p]{\frac{1}{n}\sum_{i=1}^{n} d(\mathbf{x}_i, \mathbf{x}')^p} \tag{14.162}$$

On the other hand, assume that $n = n'$. *Then*

$$d^W_p(X, X') \triangleq \min_\sigma \sqrt[p]{\frac{1}{n}\sum_{i=1}^{n} d(\mathbf{x}_i, \mathbf{x}'_{\sigma i})^p} \tag{14.163}$$

where the minimum is taken over all permutations σ *on the numbers* $1, ..., n$.

14.6.4 Information Measures of Multitarget Dispersion

In this section, I briefly describe two information-theoretic measures of dispersion: multitarget Kullback-Leibler discrimination and multitarget central entropy.

14.6.4.1 Multitarget Kullback-Leibler Discrimination

Let $f(X)$ and $f_0(X)$ be multitarget probability density functions.[10] Then the Kullback-Leibler discrimination (or cross-entropy) of f with respect to the

10 This material originally appeared in [70, p. 206]. It generalizes the discussion of single-target measures of dispersion in Section 2.4.9.2.

reference density f_0 is defined as:[11]

$$K(f;f_0) \triangleq \int f(X) \cdot \log\left(\frac{f(X)}{f_0(X)}\right) \delta X. \tag{14.164}$$

14.6.4.2 Multitarget Central Entropy

Let $\hat{X}_{k|k} = \{\hat{\mathbf{x}}_1, ..., \hat{\mathbf{x}}_{\hat{n}}\}$ be an estimate of the multitarget state as derived from $f_{k|k}(X|Z^{(k)})$. Let $E_1, ..., E_{\hat{n}}$ be small neighborhoods of $\hat{\mathbf{x}}_1, ..., \hat{\mathbf{x}}_{\hat{n}}$, respectively, all of which have the same (hyper)volume ε. Then by analogy with (2.150), the *central entropy* of the estimate $\hat{X}_{k|k}$ is

$$\kappa_{k|k} \triangleq -\log\left(\varepsilon^{\hat{n}} \cdot f_{k|k}(\hat{X}_{k|k}|Z^{(k)})\right). \tag{14.165}$$

Note that ε must be small enough that $\varepsilon^{\hat{n}} \cdot f_{k|k}(\hat{X}_{k|k}|Z^{(k)}) \leq 1$. As in the single-target case, if $f_{k|k}(\hat{X}_{k|k}|Z^{(k)})$ is very "peaky" at $X = \hat{X}_{k|k}$ then the central entropy $\kappa_{k|k}$ will be small.

Equation (14.165) follows from the same reasoning that led to (2.150). Let $u_{\hat{\mathbf{x}}_i}(\mathbf{x})$ be the uniform distribution defined by $u_{\hat{\mathbf{x}}_i}(\mathbf{x}) = \varepsilon^{-1}$ if $\mathbf{x} \in E_i$ and $u_{\hat{\mathbf{x}}_i}(\mathbf{x}) = 0$ otherwise. Define the multitarget probability density function $u_{k|k}(X)$ by $u_{k|k}(X) = 0$ if $|X| \neq \hat{n}$ and, otherwise,[12]

$$u_{k|k}(\{\mathbf{x}_1, ..., \mathbf{x}_{\hat{n}}\}) \triangleq \sum_{\sigma} u_{\hat{\mathbf{x}}_{\sigma 1}}(\mathbf{x}_1) \cdots u_{\hat{\mathbf{x}}_{\sigma \hat{n}}}(\mathbf{x}_{\hat{n}}) \tag{14.166}$$

11 Multitarget Kullback-Leibler discrimination, and generalizations of it called Csiszár discrimination measures, have been used as part of a systematic approach to scientific performance evaluation of multitarget information fusion algorithms. See [70, 51, 89].

12 This distribution is an approximation of the multitarget Dirac delta distribution introduced in (11.124). It represents the multitarget state $\hat{X}_{k|k}$ in the same way that the distribution $u_{k|k}(\mathbf{x})$ represented the single-target state $\hat{\mathbf{x}}_{kk}$ in (2.154).

where the summation is taken over all permutations σ on the numbers $1, ..., \hat{n}$. If we set $f = f_{k|k}$ and $f_0 = u_{k|k}$, then (14.164) reduces to

$$K(u_{k|k}; f_{k|k}) \tag{14.167}$$

$$= \int u_{k|k}(X) \cdot \log \left(\frac{u_{k|k}(X)}{f_{k|k}(X|Z^{(k)})} \right) \delta X \tag{14.168}$$

$$= \frac{1}{\hat{n}!} \int u_{k|k}(\{\mathbf{x}_1, ..., \mathbf{x}_{\hat{n}}\}) \tag{14.169}$$

$$\cdot \log \left(\frac{u_{k|k}(\{\mathbf{x}_1, ..., \mathbf{x}_{\hat{n}}\})}{f_{k|k}(\{\mathbf{x}_1, ..., \mathbf{x}_{\hat{n}}\}|Z^{(k)})} \right) d\mathbf{x}_1 \cdots d\mathbf{x}_{\hat{n}} \tag{14.170}$$

and so

$$= \int_{E_1 \times ... \times E_{\hat{n}}} \varepsilon^{-\hat{n}} \tag{14.171}$$

$$\cdot \log \left(\frac{\varepsilon^{-\hat{n}}}{f_{k|k}(\{\mathbf{x}_1, ..., \mathbf{x}_{\hat{n}}\}|Z^{(k)})} \right) d\mathbf{x}_1 \cdots d\mathbf{x}_{\hat{n}} \tag{14.172}$$

$$\cong \varepsilon^{-\hat{n}} \cdot \log \left(\frac{\varepsilon^{-\hat{n}}}{f_{k|k}(\{\hat{\mathbf{x}}_1, ..., \hat{\mathbf{x}}_{\hat{n}}\}|Z^{(k)})} \right) \cdot \varepsilon^{\hat{n}} \tag{14.173}$$

$$= -\log \left(\varepsilon^{\hat{n}} \cdot f_{k|k}(\{\hat{\mathbf{x}}_1, ..., \hat{\mathbf{x}}_{\hat{n}}\}|Z^{(k)}) \right) \tag{14.174}$$

as claimed.

14.7 THE JOTT FILTER

In the simple single-target Bayes filtering example of Section 2.4.1 we encountered the following problem. When clutter backgrounds are very dense, it cannot be assumed that a target is actually present in the volume of space currently being interrogated by the sensor. Indeed, because of the contaminating background it is entirely possible for a target to pass in and out of this volume and not be evidently present in any individual frame or scan. An effective filter must not only track a target but decide if one is present to begin with and then extract it from the background. It must also account for the possibility that the target may leave and re-enter the search volume.

This means that an additional target state denoting target nonpresence, $X = \emptyset$, must be included along with the usual states, $X = \{\mathbf{x}\}$, in which target presence is assumed. In addition, the filter should be able to process data even when the clutter background is complex and time-varying.

In 1994 Musicki, Evans, and Stankovic proposed an algorithm potentially capable of dealing with many problems of this type. They introduced an exact Bayesian solution to the single-target joint detection and tracking problem called the *integrated probabilistic data association* (IPDA) filter [172]. It optimally estimates the probability that a target is present, and estimates its state if it is present.

The IPDA filter is based on the following assumptions:

- At most one target is present in the scene at any time;
- If a target is present it can disappear;
- The observations of a single sensor are corrupted by missed detections (constant p_D) and by false alarms that are spatially uniform and have Poisson-distributed time-arrivals.[13]

The simple example presented in Section 2.4.1 was actually an implementation of an IPDA-type nonlinear filter.

The IPDA filter was originally derived using a bottom-up methodology. Challa, Vo, and Wang subsequently showed [23] that identical filtering formulas result when the FISST top-down methodology (see Chapters 12 and 13) is applied to the same underlying models.

In this section, I describe a generalization of the IPDA filter, which I call the *joint target-detection and tracking* (JoTT) filter. It generalizes the IPDA filter in such a manner that the following occur:

- It explicitly models target appearance at time step $k+1$ if no target was present at time step k.
- Probability of detection can be state-dependent.
- The false alarm process is independent of target state but is otherwise *arbitrary*.

In this section, I describe the JoTT filter step-by-step. It is organized as follows: measurement and motion models (Section 14.7.1); initialization (Section 14.7.2), prediction (Section 14.7.3), correction (Section 14.7.4), state estimation (Section 14.7.5), and error estimation (Section 14.7.6). I conclude in Section 14.7.7

13 Musicki et al. also assumed that the motion and sensor-noise models are linear-Gaussian. However, their derivation does not presume this fact, and so the IPDA filter theory subsumes nonlinear models.

by briefly discussing implementation of the JoTT filter using sequential Monte Carlo (SMC) methods.

In what follows I abbreviate

$$f_{k|k}(X) = f_{k|k}(X|Z^{(k)}) \tag{14.175}$$
$$f_{k+1|k}(X) = f_{k+1|k}(X|Z^{(k)}) \tag{14.176}$$
$$f_{k+1|k+1}(X) = f_{k+1|k+1}(X|Z^{(k+1)}). \tag{14.177}$$

Since we know a priori that at most one target is present, all multitarget posterior distributions for $k = 0, 1, ...$ have the form

$$f_{k|k}(X) = \begin{cases} 1 - p_{k|k} & \text{if} \quad X = \emptyset \\ p_{k|k} \cdot f_{k|k}(\mathbf{x}) & \text{if} \quad X = \{\mathbf{x}\} \\ 0 & \text{if} \quad |X| > 1 \end{cases} . \tag{14.178}$$

where $f_{k|k}(\mathbf{x}) \overset{\text{abbr.}}{=} f_{k|k}(\mathbf{x}|Z^{(k)})$ and $\int f_{k|k}(\mathbf{x})d\mathbf{x} = 1$.

In other words, if the target exists at time step k then the following are true:

- $f_{k|k}(\mathbf{x})$ is its track density;
- $p_{k|k}$ is the probability that it exists.

Remark 23 (Caution Regarding Notation) *The expressions on the right-hand side of (14.178) should be treated with care. It is tempting to abbreviate $f_{k|k}(\{\mathbf{x}\})$ as $f_{k|k}(\mathbf{x})$ and then erroneously presume that $\int f_{k|k}(\mathbf{x})d\mathbf{x} = 1$. In actuality, $\int f_{k|k}(\{\mathbf{x}\})d\mathbf{x} = p_{k|k}$.*

14.7.1 JoTT Filter: Models

In this section, I will briefly review the motion and measurement models assumed for the joint target-detection and tracking filter. These models are special cases of the "standard" multitarget motion and measurement models introduced in Sections 12.3.2, and 12.3.3, and 13.2.1.

14.7.1.1 Motion Model

We assume that if there are no targets in the scene, then there is a probability p_B that a target will enter. If this occurs, we assume that the track distribution of the "birth target" is $b_{k+1|k}(\mathbf{x})$. In this case from (13.29) we know that the multitarget Markov density must have the form

$$f_{k+1|k}(X|\emptyset) = \begin{cases} 1 - p_B & \text{if} \quad X = \emptyset \\ p_B \cdot b_{k+1|k}(\mathbf{x}) & \text{if} \quad X = \{\mathbf{x}\} \\ 0 & \text{if} \quad |X| \geq 2 \end{cases}. \tag{14.179}$$

On the other hand, if a target with state $\mathbf{x}'$ is in the scene at time step k, we assume that there is a probability $p_S(\mathbf{x}')$ that it will remain in the scene. If it does, its motion between time steps k and $k+1$ is governed by the conventional single-target Markov density $f_{k+1|k}(\mathbf{x}|\mathbf{x}')$. In this case, (13.32) informs us that the multitarget Markov density must have the form

$$f_{k+1|k}(X|\{\mathbf{x}'\}) = \begin{cases} 1 - p_S(\mathbf{x}') & \text{if} \quad X = \emptyset \\ p_S(\mathbf{x}') \cdot f_{k+1|k}(\mathbf{x}|\mathbf{x}') & \text{if} \quad X = \{\mathbf{x}\} \\ 0 & \text{if} \quad |X| \geq 2 \end{cases}. \tag{14.180}$$

14.7.1.2 Measurement Model

We assume that the false alarm process $\kappa(Z) \overset{\text{abbr.}}{=} \kappa_{k+1}(Z)$ at time step $k+1$ is independent of target state, but otherwise arbitrary. We also assume that false alarms are independent of the detection and target-observation processes. Let $f_{k+1}(Z|X)$ denote the multitarget likelihood that is assumed to conform to the standard model of Section 12.3.5.

Given this, it follows that

$$f_{k+1}(Z|\emptyset) = \kappa(Z) \tag{14.181}$$

if there is no target in the scene; and if otherwise by

$$f_{k+1}(Z|\{\mathbf{x}\}) = \kappa(Z) \cdot \left(1 - p_D(\mathbf{x}) + p_D(\mathbf{x}) \sum_{\mathbf{z} \in Z} \frac{L_{\mathbf{z}}(\mathbf{x}) \cdot \kappa(Z - \{\mathbf{z}\})}{\kappa(Z)}\right) \tag{14.182}$$

with $f(\emptyset|\{\mathbf{x}\}) = \kappa(Z) \cdot (1 - p_D(\mathbf{x}))$ and $L_{\mathbf{z}}(\mathbf{x}) \triangleq f_{k+1|k}(\mathbf{z}|\mathbf{x})$.

The easiest way to demonstrate (14.181) and (14.182) is to use the fundamental convolution formula, (11.253). From that theorem, the likelihood for the complete observation process is

$$f_{k+1}(Z|X) = \sum_{W \subseteq Z} f(W|X) \cdot \kappa(Z - W) \tag{14.183}$$

$$= f(\emptyset|X) \cdot \kappa(Z) + \sum_{\mathbf{z} \in Z} f(\{\mathbf{z}\}|X) \tag{14.184}$$

$$\cdot \kappa(Z - \{\mathbf{z}\}). \tag{14.185}$$

If $X = \emptyset$, then

$$f_{k+1}(Z|\emptyset) = f(\emptyset|\emptyset) \cdot \kappa(Z) + \sum_{\mathbf{z} \in Z} f(\{\mathbf{z}\}|\emptyset) \cdot \kappa(Z - \{\mathbf{z}\}) \tag{14.186}$$

$$= 1 \cdot \kappa(Z) + 0 = \kappa(Z). \tag{14.187}$$

If $X = \{\mathbf{x}\}$, then

$$f_{k+1}(Z|\{\mathbf{x}\}) \tag{14.188}$$

$$= f(\emptyset|\{\mathbf{x}\}) \cdot \kappa(Z) + \sum_{\mathbf{z} \in Z} f(\{\mathbf{z}\}|\{\mathbf{x}\}) \cdot \kappa(Z - \{\mathbf{z}\}), \tag{14.189}$$

and so

$$= (1 - p_D(\mathbf{x})) \cdot \kappa(Z) \tag{14.190}$$

$$+ p_D(\mathbf{x}) \sum_{\mathbf{z} \in Z} L_{\mathbf{z}}(\mathbf{x}) \cdot \kappa(Z - \{\mathbf{z}\}) \tag{14.191}$$

$$= \kappa(Z) \tag{14.192}$$

$$\cdot \left(1 - p_D(\mathbf{x}) + p_D(\mathbf{x}) \sum_{\mathbf{z} \in Z} L_{\mathbf{z}}(\mathbf{x}) \cdot \frac{\kappa(Z - \{\mathbf{z}\})}{\kappa(Z)} \right). \tag{14.193}$$

14.7.2 JoTT Filter: Initialization

The initial distribution must have the general form

$$f_{0|0}(X) = \begin{cases} 1 - p_{0|0} & \text{if} \quad X = \emptyset \\ p_{0|0} \cdot f_{0|0}(\mathbf{x}) & \text{if} \quad X = \{\mathbf{x}\} \\ 0 & \text{if} \quad \text{otherwise} \end{cases} . \tag{14.194}$$

where $f_{0|0}(\mathbf{x})$ is the a priori distribution of the target and where $p_{0|0}$ is the a priori probability that the target exists.

If $p_{0|0} = 0.5$ and if $f_{0|0}(\mathbf{x}) = |D|^{-1}$ for $\mathbf{x} \in D$ but $f_{0|0}(\mathbf{x}) = 0$ otherwise, then this reduces to a uniform distribution in the sense of (11.126).

14.7.3 JoTT Filter: Predictor

We are given $p_{k|k}$ and $f_{k|k}(\mathbf{x})$—and thus $f_{k|k}(X)$—from the previous time step. We are to determine formulas for $p_{k+1|k}$ and $f_{k+1|k}(\mathbf{x})$—that is, for $f_{k+1|k}(X)$. In Appendix G.23, I demonstrate that these are

$$p_{k+1|k} = p_B \cdot (1 - p_{k|k}) + p_{k|k} \int p_S(\mathbf{x}') \cdot f_{k|k}(\mathbf{x}') d\mathbf{x}' \tag{14.195}$$

and

$$f_{k+1|k}(\mathbf{x}) \tag{14.196}$$

$$= \frac{\begin{array}{c} p_B \cdot b_{k+1|k}(\mathbf{x}) \cdot (1 - p_{k|k}) \\ +p_{k|k} \int p_S(\mathbf{x}') \cdot f_{k+1|k}(\mathbf{x}|\mathbf{x}') \cdot f_{k|k}(\mathbf{x}') d\mathbf{x}' \end{array}}{p_B \cdot (1 - p_{k|k}) + p_{k|k} \int p_S(\mathbf{x}') \cdot f_{k|k}(\mathbf{x}') d\mathbf{x}'}. \tag{14.197}$$

If p_S is constant then these equations become

$$p_{k+1|k} = p_B \cdot (1 - p_{k|k}) + p_S \cdot p_{k|k} \tag{14.198}$$

$$f_{k+1|k}(\mathbf{x}) \tag{14.199}$$

$$= \frac{p_B \cdot b_{k+1|k}(\mathbf{x}) \cdot (1 - p_{k|k}) + p_S \cdot p_{k|k} \cdot \tilde{f}_{k+1|k}(\mathbf{x})}{p_B \cdot (1 - p_{k|k}) + p_S \cdot p_{k|k}} \tag{14.200}$$

where

$$\tilde{f}_{k+1|k}(\mathbf{x}) \triangleq \int f_{k+1|k}(\mathbf{x}|\mathbf{x}') \cdot f_{k|k}(\mathbf{x}') d\mathbf{x}' \tag{14.201}$$

is the conventional prediction of $f_{k|k}(\mathbf{x}')$ to time step $k+1$. Note that if $p_B = 0$ or if $p_{k|k} = 1$ then $f_{k+1|k}(\mathbf{x}) = \tilde{f}_{k+1|k}(\mathbf{x})$.

14.7.4 JoTT Filter: Corrector

We are given $p_{k+1|k}$ and $f_{k+1|k}(\mathbf{x})$—and thus the predicted multitarget posterior $f_{k+1|k}(X)$. We are to determine formulas for $p_{k+1|k+1}$ and $f_{k+1|k+1}(\mathbf{x})$—that is, for $f_{k+1|k+1}(X)$. In Appendix G.24, I show that

$$p_{k+1|k+1} = \frac{1 - f_{k+1|k}[p_D] + \sum_{\mathbf{z}\in\mathbf{Z}} f_{k+1|k}[p_D L_{\mathbf{z}}] \cdot \frac{\kappa(Z-\{\mathbf{z}\})}{\kappa(Z)}}{p_{k+1|k}^{-1} - f_{k+1|k}[p_D] + \sum_{\mathbf{z}\in\mathbf{Z}} f_{k+1|k}[p_D L_{\mathbf{z}}] \cdot \frac{\kappa(Z-\{\mathbf{z}\})}{\kappa(Z)}} \quad (14.202)$$

and

$$f_{k+1|k+1}(\mathbf{x}) \quad (14.203)$$

$$= f_{k+1|k}(\mathbf{x}) \quad (14.204)$$

$$\cdot \frac{1 - p_D(\mathbf{x}) + p_D(\mathbf{x}) \sum_{\mathbf{z}\in\mathbf{Z}} L_{\mathbf{z}}(\mathbf{x}) \cdot \frac{\kappa(Z-\{\mathbf{z}\})}{\kappa(Z)}}{1 - f_{k+1|k}[p_D] + \sum_{\mathbf{z}\in\mathbf{Z}} f_{k+1|k}[p_D L_{\mathbf{z}}] \cdot \frac{\kappa(Z-\{\mathbf{z}\})}{\kappa(Z)}} \quad (14.205)$$

where

$$f_{k+1|k}[h] \triangleq \int h(\mathbf{x}) \cdot f_{k+1|k}(\mathbf{x}) d\mathbf{x}. \quad (14.206)$$

If p_D is constant, these equations become

$$p_{k+1|k+1} = \frac{1 - p_D + p_D \sum_{\mathbf{z}\in Z_{k+1}} f_{k+1|k}[L_{\mathbf{z}}] \cdot \frac{\kappa(Z-\{\mathbf{z}\})}{\kappa(Z)}}{p_{k+1|k}^{-1} - p_D + p_D \sum_{\mathbf{z}\in Z_{k+1}} f_{k+1|k}[L_{\mathbf{z}}] \cdot \frac{\kappa(Z-\{\mathbf{z}\})}{\kappa(Z)}} \quad (14.207)$$

and

$$f_{k+1|k+1}(\mathbf{x}) = f_{k+1|k}(\mathbf{x}) \quad (14.208)$$

$$\cdot \frac{1 - p_D + p_D \sum_{\mathbf{z}\in Z_{k+1}} L_{\mathbf{z}}(\mathbf{x}) \cdot \frac{\kappa(Z-\{\mathbf{z}\})}{\kappa(Z)}}{1 - p_D + p_D \sum_{\mathbf{z}\in Z_{k+1}} f_{k+1|k}[L_{\mathbf{z}}] \cdot \frac{\kappa(Z-\{\mathbf{z}\})}{\kappa(Z)}}. \quad (14.209)$$

14.7.5 JoTT Filter: Estimation

In Example 80, I examined some of the subtleties involving the MaM and JoM estimators when at most one target is present. This section describes the MaM and JoM estimators in the context of JoTT filtering.

14.7.5.1 JoTT Filter: MaM Estimate

The marginal multitarget (MaM) estimator was introduced in (14.98) and (14.99). Since under current assumptions the scene contains at most one target, target existence can be declared if the yes-target probability is greater than 0.5 (or some larger-valued threshold $0.5 < \tau < 1$):

$$p_{k+1|k+1} > \frac{1}{2}. \tag{14.210}$$

From (14.202) this becomes

$$\frac{1 - f_{k+1|k}[p_D] + \sum_{\mathbf{z}\in Z_{k+1}} f_{k+1|k}[p_D L_{\mathbf{z}}] \cdot \frac{\kappa(Z-\{\mathbf{z}\})}{\kappa(Z)}}{p_{k+1|k}^{-1} - f_{k+1|k}[p_D] + \sum_{\mathbf{z}\in Z_{k+1}} f_{k+1|k}[p_D L_{\mathbf{z}}] \cdot \frac{\kappa(Z-\{\mathbf{z}\})}{\kappa(Z)}} > \frac{1}{2}. \tag{14.211}$$

Solving this inequality for $p_{k+1|k}$ yields the following test for target existence:

$$p_{k+1|k} > \frac{1}{2 - p_D + \sum_{\mathbf{z}\in Z_{k+1}} f_{k+1|k}[p_D L_{\mathbf{z}}] \cdot \frac{\kappa(Z-\{\mathbf{z}\})}{\kappa(Z)}}. \tag{14.212}$$

If target existence is declared, the estimate of the target's state is

$$\hat{\mathbf{x}}_{k+1|k+1}^{\text{MaM}} = f_{k+1|k}(\mathbf{x}) \tag{14.213}$$

$$\cdot \arg\sup_{\mathbf{x}} \left(1 - p_D(\mathbf{x}) + p_D(\mathbf{x}) \sum_{\mathbf{z}\in Z_{k+1}} L_{\mathbf{z}}(\mathbf{x}) \cdot \frac{\kappa(Z - \{\mathbf{z}\})}{\kappa(Z)} \right). \tag{14.214}$$

14.7.5.2 JoTT Filter: JoM Estimate

The joint multitarget (JoM) estimator was defined in (14.100) and (14.101). Let $c > 0$ be small. We assume that p_D is constant. Then target existence can be declared on the basis of a JoM estimate if

$$\frac{p_{k+1|k}^{-1} - 1}{c \cdot \sup_{\mathbf{x}} f_{k+1|k}(\mathbf{x})} < 1 - p_D + p_D \sum_{\mathbf{z}\in Z_{k+1}} L_{\mathbf{z}}(\hat{\mathbf{x}}) \cdot \frac{\kappa(Z - \{\mathbf{z}\})}{\kappa(Z)} \tag{14.215}$$

where $\hat{\mathbf{x}} \triangleq \arg\sup_{\mathbf{x}} f_{k+1|k+1}(\{\mathbf{x}\})$. In this case the JoM estimate of target state is the same as the corresponding MaM estimate:

$$\hat{\mathbf{x}}^{\text{JoM}}_{k+1|k+1} = f_{k+1|k}(\mathbf{x}) \tag{14.216}$$

$$\cdot \arg\sup_{\mathbf{x}} \left(1 - p_D(\mathbf{x}) + p_D(\mathbf{x}) \sum_{\mathbf{z} \in Z_{k+1}} L_{\mathbf{z}}(\mathbf{x}) \cdot \frac{\kappa(Z - \{\mathbf{z}\})}{\kappa(Z)}\right). \tag{14.217}$$

To verify these equations, note that target existence can be declared if

$$f_{k+1|k+1}(\emptyset) < c \cdot \sup_{\mathbf{x}} f_{k+1|k+1}(\{\mathbf{x}\}) = c \cdot f_{k+1|k+1}(\{\hat{\mathbf{x}}\}) \tag{14.218}$$

or

$$p^{-1}_{k+1|k+1} < 1 + c \cdot f_{k+1|k+1}(\hat{\mathbf{x}}). \tag{14.219}$$

Substituting (14.202) and (14.203) into this, we get

$$\frac{p^{-1}_{k+1|k} - p_D + p_D \sum_{\mathbf{z} \in Z_{k+1}} f_{k+1|k}[L_{\mathbf{z}}] \cdot \frac{\kappa(Z-\{\mathbf{z}\})}{\kappa(Z)}}{1 - p_D + p_D \sum_{\mathbf{z} \in Z_{k+1}} f_{k+1|k}[L_{\mathbf{z}}] \cdot \frac{\kappa(Z-\{\mathbf{z}\})}{\kappa(Z)}} \tag{14.220}$$

$$< 1 + c \cdot f_{k+1|k}(\hat{\mathbf{x}}) \tag{14.221}$$

$$\cdot \frac{1 - p_D + p_D \sum_{\mathbf{z} \in Z_{k+1}} L_{\mathbf{z}}(\hat{\mathbf{x}}) \cdot \frac{\kappa(Z-\{\mathbf{z}\})}{\kappa(Z)}}{1 - p_D + p_D \sum_{\mathbf{z} \in Z_{k+1}} f_{k+1|k}[L_{\mathbf{z}}] \cdot \frac{\kappa(Z-\{\mathbf{z}\})}{\kappa(Z)}}. \tag{14.222}$$

After a bit of algebra, this reduces to the inequality of (14.215).

If target existence is declared, then the JoM estimate of target state is, as claimed,

$$\hat{\mathbf{x}}^{\text{JoM}}_{k+1|k+1} \tag{14.223}$$

$$= \arg\sup_{\mathbf{x}} \; c \cdot f_{k+1|k+1}(\{\mathbf{x}\}) \tag{14.224}$$

$$= \arg\sup_{\mathbf{x}} \; c \cdot p_{k+1|k+1} \cdot f_{k+1|k+1}(\mathbf{x}) \tag{14.225}$$

$$= \arg\sup_{\mathbf{x}} \left(1 - p_D + p_D \sum_{\mathbf{z} \in Z_{k+1}} f_{k+1}(\mathbf{z}|\mathbf{x}) \cdot \frac{\kappa(Z - \{\mathbf{z}\})}{\kappa(Z)}\right) \tag{14.226}$$

$$\cdot f_{k+1|k}(\mathbf{x}). \tag{14.227}$$

14.7.6 JoTT Filter: Error Estimation

Error estimation for the JoTT filter is somewhat more complex than for the conventional single-target Bayes filter, because error must be estimated not only for track state but also target number. I consider only the most obvious such estimates: variance of target number and track covariance.

14.7.6.1 Target Number Variance

The cardinality distribution for the JoTT data update is

$$p_{k+1|k+1}(n) = p_{k+1|k+1}^n \cdot \left(1 - p_{k+1|k+1}\right)^{1-n} \tag{14.228}$$

for $0 \leq n \leq 1$ and is therefore a binomial distribution. The variance on target number is, therefore,

$$\sigma_{k+1|k+1}^2 = p_{k+1|k+1}\left(1 - p_{k+1|k+1}\right). \tag{14.229}$$

It is left to the reader as Exercise 57 to show that, if p_D is constant, this can be rewritten as

$$\frac{\sigma_{k+1|k+1}^2}{\sigma_{k+1|k}^2} \tag{14.230}$$

$$= \frac{1 - p_D + p_D \sum_{\mathbf{z} \in Z_{k+1}} f_{k+1}(\mathbf{z}) \cdot \frac{\kappa(Z-\{\mathbf{z}\})}{\kappa(Z)}}{\left(1 - p_{k+1|k} p_D + p_{k+1|k} p_D \sum_{\mathbf{z} \in Z_{k+1}} f_{k+1}(\mathbf{z}) \cdot \frac{\kappa(Z-\{\mathbf{z}\})}{\kappa(Z)}\right)^2}. \tag{14.231}$$

14.7.6.2 Track Error Covariance

In the event that a target is declared to exist, the covariance matrix of its track distribution is

$$P_{k+1|k+1} = \int (\mathbf{x} - \hat{\mathbf{x}}_{k+1|k+1})(\mathbf{x} - \hat{\mathbf{x}}_{k+1|k+1})^T \cdot f_{k+1|k+1}(\mathbf{x}) d\mathbf{x}. \tag{14.232}$$

14.7.7 SMC Implementation of JoTT Filter

As a prelude to Chapter 15 and Part III, I discuss one possible implementation technique for the JoTT filter. Sequential Monte Carlo (SMC) approximation was introduced in Section 2.5.3 in connection with the conventional single-target Bayes filter. In this section, I summarize the basic ideas of SMC implementation of the JoTT filter. As in Section 2.5.3, I describe only a generic particle filter that employs the dynamic prior.

SMC implementation of the JoTT filter differs from the conventional case. This is because a particle system representation of $f_{k|k}(X)$ must include more than the particles $\mathbf{x}^1_{k|k},...\mathbf{x}^\nu_{k|k}$ modeling the target state (in the event that the target exists). It must also include an additional particle—the particle $X = \emptyset$, which represents the *absence of targets*.

A particle system approximation for the JoTT filter at time step k is, therefore, a collection

$$\emptyset, \mathbf{x}^1_{k|k},...\mathbf{x}^\nu_{k|k} \tag{14.233}$$

of states and importance weights $w^0_{k|k}, w^1_{k|k},...w^\nu_{k|k}$ with $\sum_{i=0}^\nu w^i_{k|k} = 1$, such that

$$\int \theta(X) \cdot f_{k|k}(X)\delta X \cong w^0_{k|k} \cdot \theta(\emptyset) + \sum_{i=1}^{\nu} w^i_{k|k} \cdot \theta(\mathbf{x}^i_{k|k}) \tag{14.234}$$

for any unitless function $\theta(X)$ of a finite-set variable X with $|X| \leq 1$. The $\emptyset, \mathbf{x}^1_{k|k},...\mathbf{x}^\nu_{k|k}$ are interpreted as samples drawn from $f_{k|k}(X)$:

$$\emptyset, \{\mathbf{x}^1_{k|k}\},..., \{\mathbf{x}^\nu_{k|k}\} \sim f_{k|k}(\cdot). \tag{14.235}$$

In (14.234) set $\theta(X) = 1$ if $|X| = 1$ and $\theta(X) = 0$ otherwise. Then

$$p_{k|k} = \int f_{k|k}(\{\mathbf{x}\})d\mathbf{x} \cong \sum_{i=1}^{\nu} w^i_{k|k} = 1 - w^0_{k|k}. \tag{14.236}$$

This means that caution is necessary when applying equal particle weighting. Naïve equal weighting would set $w^0_{k|k}, w^1_{k|k},...w^\nu_{k|k}$ all equal to $(\nu+1)^{-1}$. The probability of target existence would then always be $p_{k|k} \cong 1 - (\nu+1)^{-1} = \nu/(\nu+1)$. When the number of particles is large, this would mean that $p_{k|k} \cong 1$ for all $k \geq 1$, even if no target were actually in the scene.

The apparent paradox arises from the fact that $\emptyset, \mathbf{x}^1_{k|k},...\mathbf{x}^\nu_{k|k}$ are drawn from a *mixed discrete-continuous* distribution $f_{k|k}(X|Z^{(k)})$ with $X = \emptyset$ or $X = \{\mathbf{x}\}$. Any two samples drawn from a purely continuous distribution are very unlikely to be the same. This is not the case when drawing from the distribution

$p(j) = p^j_{k|k} \cdot (1 - p_{k|k})^{1-j}$ on the finite set $\{0, 1\}$, for $j = 0, 1$. If one draws $\nu + 1$ samples from $f_{k|k}(X|Z^{(k)})$ and ν is large, then on average there will be $(1 - p_{k|k}) \cdot (\nu + 1)$ duplicates of the null particle $\emptyset$ and $p_{k|k} \cdot (\nu + 1)$ distinct non-null particles.

The duplicate null particles can be consolidated into a single null particle $\emptyset$, the weight of which is the sum of the weights of all of the duplicates:

$$w^0_{k|k} = (1 - p_{k|k}) \cdot (\nu + 1) \cdot \frac{1}{\nu + 1} = 1 - p_{k|k}. \tag{14.237}$$

The non-null particles will have equal weights

$$w^i_{k|k} = \frac{p_{kk}}{\nu_{k|k}} \tag{14.238}$$

for $i = 1, ..., \nu_{k|k}$, where $\nu_{k|k}$ is the number of non-null particles. This weighting is what we shall assume hereafter.

14.7.7.1 SMC-JoTT Filter Initialization

If an initial distribution $p_{0|0}$, $f_{0|0}(\mathbf{x})$ is available, then the filter is initialized by drawing ν samples from $f_{0|0}(\mathbf{x})$:

$$\mathbf{x}^1_{0|0}, ...\mathbf{x}^\nu_{0|0} \sim f_{0|0}(\cdot). \tag{14.239}$$

Then, weights are set as

$$w^0_{0|0} = 1 - p_{0|0} \tag{14.240}$$

$$w^i_{0|0} = \frac{p_{0|0}}{\nu}, \qquad i = 1, ..., \nu. \tag{14.241}$$

14.7.7.2 SMC-JoTT Filter Predictor

Assume that the Bayes posterior at the previous time step k has been approximated by a particle system $\emptyset, \mathbf{x}^1_{k|k}, ..., \mathbf{x}^\nu_{k|k}$ with weights $w^0_{k|k} = 1 - p_{k|k}$ and $w^i_{k|k} = p_{k|k}/\nu$ for $i = 1, ..., \nu$. Here I have abbreviated $\nu \overset{\text{abbr.}}{=} \nu_{k|k}$.

Following the procedure described in Section 2.5.3, we draw samples

$$X^0_{k+1|k} \sim f_{k+1|k}(\cdot|\emptyset) \tag{14.242}$$

and

$$X^i_{k+1|k} \sim f_{k+1|k}(\cdot|\{\mathbf{x}^i_{k|k}\}) \tag{14.243}$$

for $i = 1, ..., \nu$ from the respective multitarget dynamic priors

$$f_{k+1|k}(X|\emptyset), \quad f_{k+1|k}(X|\{\mathbf{x}^1_{k|k}\}) \ ,..., \ f_{k+1|k}(X|\{\mathbf{x}^\nu_{k|k}\}). \tag{14.244}$$

We will have either $X^0_{k+1|k} = \emptyset$ or $X^0_{k+1|k} = \{\mathbf{x}^0_{k+1|k}\}$ for some $\mathbf{x}^0_{k+1|k}$. Likewise, $X^i_{k+1|k} = \emptyset$ or $X^i_{k+1|k} = \{\mathbf{x}^i_{k+1|k}\}$. The total weight of the null particle is therefore

$$1 - p_{k+1|k} = w^0_{k+1|k} = \sum_{i:X^i_{k+1|k}=\emptyset} w^i_{k|k}. \tag{14.245}$$

Let $\nu_{k+1|k}$ be the number of non-null predicted particles. Then their weight will be

$$w^i_{k+1|k} = \frac{1 - w^0_{k+1|k}}{\nu_{k+1|k}} \tag{14.246}$$

for all $i = 1, ..., \nu_{k+1|k}$.

14.7.7.3 SMC-JoTT Filter Corrector

Assume that the predicted posterior is represented by $\emptyset, \mathbf{x}^1_{k+1|k}, ..., \mathbf{x}^\nu_{k+1|k}$ with weights $w^0_{k+1|k} = 1 - p_{k+1|k}$ and $w^i_{k+1|k} = p_{k+1|k}/\nu$ for $i = 1, ..., \nu$. Once again, I have abbreviated $\nu \overset{\text{abbr.}}{=} \nu_{k+1|k}$. In this case the Bayes normalization factor becomes:

$$
\begin{aligned}
& f_{k+1}(Z|Z^k) && (14.247)\\
= & \int f_{k+1}(Z|X) \cdot f_{k+1|k}(X)\delta X && (14.248)\\
\cong & \ (1 - p_{k+1|k}) \cdot f_{k+1}(Z|\emptyset) && (14.249)\\
+ & \ \frac{p_{k+1|k}}{\nu} \sum_{i=1}^{\nu} f_{k+1}(Z|\{\mathbf{x}^i_{k+1|k}\}) && (14.250)\\
= & \ (1 - p_{k+1|k}) \cdot \kappa(Z) + \frac{\kappa(Z) \cdot p_{k+1|k}}{\nu} && (14.251)\\
& \cdot \sum_{i=1}^{\nu} \left(\begin{array}{c} 1 - p_D(\mathbf{x}^i_{k+1|k}) \\ + p_D(\mathbf{x}^i_{k+1|k}) \sum_{\mathbf{z}\in Z} \frac{L_{\mathbf{z}}(\mathbf{x}^i_{k+1|k}) \cdot \kappa(Z-\{\mathbf{z}\})}{\kappa(Z)} \end{array} \right). && (14.252)
\end{aligned}
$$

Abbreviate

$$\Phi_i \overset{\text{abbr.}}{=} 1 - p_D(\mathbf{x}^i_{k+1|k}) + p_D(\mathbf{x}^i_{k+1|k}) \sum_{\mathbf{z} \in Z_{k+1}} \frac{L_{\mathbf{z}}(\mathbf{x}^i_{k+1|k}) \cdot \kappa(Z_{k+1} - \{\mathbf{z}\})}{\kappa(Z_{k+1})}. \tag{14.253}$$

The Bayes corrector equation, (2.82), in particle system form becomes

$$\int \theta(X) \cdot f_{k+1|k+1}(X) \delta X \tag{14.254}$$

$$\cong \frac{(1 - p_{k+1|k}) \cdot \kappa(Z) \cdot \theta(\emptyset) + \frac{\kappa(Z) \cdot p_{k+1|k}}{\nu} \sum_{i=1}^{\nu} \theta(\mathbf{x}^i_{k+1|k}) \Phi_i}{(1 - p_{k+1|k}) \cdot \kappa(Z) + \frac{\kappa(Z) \cdot p_{k+1|k}}{\nu} \sum_{e=1}^{\nu} \Phi_e} \tag{14.255}$$

$$= \frac{(1 - p_{k+1|k}) \cdot \theta(\emptyset) + \frac{p_{k+1|k}}{\nu} \sum_{i=1}^{\nu} \theta(\mathbf{x}^i_{k+1|k}) \cdot \Phi_i}{1 - p_{k+1|k} + \frac{p_{k+1|k}}{\nu} \sum_{e=1}^{\nu} \Phi_e} \tag{14.256}$$

and so

$$\int \theta(X) \cdot f_{k+1|k+1}(X) \delta X \tag{14.257}$$

$$= \frac{1 - p_{k+1|k}}{1 - p_{k+1|k} + \frac{p_{k+1|k}}{\nu} \sum_{e=1}^{\nu} \Phi_e} \cdot \theta(\emptyset) \tag{14.258}$$

$$+ \sum_{i=1}^{\nu} \frac{\frac{p_{k+1|k}}{\nu} \cdot \Phi_i}{1 - p_{k+1|k} + \frac{p_{k+1|k}}{\nu} \sum_{e=1}^{\nu} \Phi_e} \cdot \theta(\mathbf{x}^i_{k+1|k}). \tag{14.259}$$

From this we conclude that the new particle system is $\emptyset, \mathbf{x}^1_{k+1|k+1}, ..., \mathbf{x}^{\nu}_{k+1|k+1}$ with $\mathbf{x}^i_{k+1|k+1} = \mathbf{x}^i_{k+1|k}$ for $i = 1, ..., \nu$. The weights are

$$\int \theta(X) \cdot f_{k+1|k+1}(X) \delta X \cong w^0_{k+1|k+1} \cdot \theta(\emptyset) + \sum_{i=1}^{\nu} w^i_{k+1|k+1} \cdot \theta(\mathbf{x}^i_{k+1|k+1}) \tag{14.260}$$

where

$$w^0_{k+1|k+1} \triangleq \frac{1 - p_{k+1|k}}{1 - p_{k+1|k} + \frac{p_{k+1|k}}{\nu} \sum_{e=1}^{\nu} \Phi_e} \tag{14.261}$$

and

$$w^i_{k+1|k+1} \triangleq \frac{\frac{p_{k+1|k}}{\nu} \cdot \Phi_i}{1 - p_{k+1|k} + \frac{p_{k+1|k}}{\nu} \sum_{e=1}^{\nu} \Phi_e} \tag{14.262}$$

for $i = 1, ..., \nu$. The weights are then equalized by applying some resampling technique. Given this, the new weights are $\tilde{w}^0_{k+1|k+1} = w^0_{k+1|k+1} \triangleq p_{k+1|k+1}$ and $\tilde{w}^i_{k+1|k+1} = p_{k+1|k+1}/\nu_{k+1|k+1}$ for $i = 1, ..., \nu_{k+1|k+1}$, where $\nu_{k+1|k+1}$ is the new number of non-null particles.

14.7.7.4 SMC-JoTT Filter State and Error Estimation

Implementation of state and error estimates for the JoTT filter proceeds essentially as in (2.239) and (2.241).

14.8 THE P.G.FL. MULTITARGET BAYES FILTER

In this section, I introduce advanced theoretical material that is central to the derivation of the PHD and CPHD filters in Chapter 16 and the para-Gaussian filter in Chapter 17. The less theoretically inclined reader may choose to pass on directly to Chapter 15.

The probability-generating functional (p.g.fl.) was defined in (11.154):

$$G_f[h] = \int h^Y \cdot f(Y)\delta Y \tag{14.263}$$

where $h^Y = 1$ if $Y = \emptyset$ and $h^Y = \prod_{\mathbf{y}\in\mathbf{Y}} h(\mathbf{y})$ if otherwise. In Sections 11.1.1, 11.1.2, and 11.3.1, I noted that p.g.fl.s are important because they permit many difficult mathematical derivations to be simplified. I offered (11.87) and (11.88) as examples of such simplifications.

This section lays the groundwork for a third simplification. Recasting the multitarget predictor and corrector equations in p.g.fl. form makes it possible to derive concrete multitarget predictor and corrector equations for real-world problems.

In what follows I describe the p.g.fl. form of the multitarget predictor equation (Section 14.8.1); and multitarget corrector equation (Section 14.8.2).

14.8.1 The p.g.fl. Multitarget Predictor

The multitarget predictor equation, (14.14), is

$$f_{k+1|k}(X) = \int f_{k+1|k}(X|X') \cdot f_{k|k}(X')\delta X'. \tag{14.264}$$

The p.g.fl. form of this equation is

$$G_{k+1|k}[h] = \int G_{k+1|k}[h|X'] \cdot f_{k|k}(X'|Z^{(k)})\delta X' \tag{14.265}$$

where $G_{k+1|k}[h]$ is the p.g.fl. of $f_{k+1|k}(X|Z^{(k)})$ and where

$$G_{k+1|k}[h|X'] \triangleq \int h^X \cdot f_{k+1|k}(X|X')\delta X \tag{14.266}$$

is the p.g.fl. of $f_{k+1|k}(X|X')$.

Equation (14.265) is easily established as follows:

$$\begin{aligned} G_{k+1|k}[h] &\triangleq \int h^X \cdot f_{k+1|k}(X)\delta X && (14.267) \\ &= \int h^X \cdot \left(\int f_{k+1|k}(X|X') \cdot f_{k|k}(X')\delta X' \right) \delta X && (14.268) \\ &= \int \left(\int h^X \cdot f_{k+1|k}(X|X')\delta X \right) \cdot f_{k|k}(X')\delta X' && (14.269) \\ &= \int G_{k+1|k}[h|X'] \cdot f_{k|k}(X')\delta X'. && (14.270) \end{aligned}$$

Example 84 (p.g.fl. Form of Standard Multitarget Motion) *In (13.61) we derived the p.g.fl. for the multitarget Markov density $f_{k+1|k}(X|X')$ for the standard multitarget motion model without spawning:*

$$G_{k+1|k}[h|X'] = (1 - p_S + p_S p_h)^{X'} \cdot e^{\mu \cdot b[h] - \mu} \tag{14.271}$$

where

$$b[h] \triangleq \int h(\mathbf{x}) \cdot b(\mathbf{x})d\mathbf{x}, \qquad p_h(\mathbf{x}') \triangleq \int h(\mathbf{x}) \cdot f_{k+1|k}(\mathbf{x}|\mathbf{x}')d\mathbf{x}. \tag{14.272}$$

This fact and (14.265) allow us to derive the following concrete formula for the p.g.fl. $G_{k+1|k}[h]$ of the predicted multitarget posterior distribution [136, p. 1172]:

$$G_{k+1|k}[h] = e^{\mu \cdot b[h] - \mu} \cdot G_{k|k}[1 - p_S + p_S p_h] \tag{14.273}$$

where $G_{k|k}[h] = \int h^X \cdot f_{k|k}(X)\delta X$ *is the p.g.fl. of* $f_{k|k}(X')$. *To see this, note that*

$$\begin{aligned} & G_{k+1|k}[h] && (14.274) \\ = & \int G_{k+1|k}[h|X'] \cdot f_{k|k}(X')\delta X' && (14.275) \\ = & e^{\mu \cdot b[h]-\mu} \cdot \int (1 - p_S + p_S p_h)^{X'} \cdot f_{k|k}(X')\delta X' && (14.276) \\ = & e^{\mu \cdot b[h]-\mu} \cdot G_{k|k}[1 - p_S + p_S p_h]. && (14.277) \end{aligned}$$

From this one can, in principle, compute the predicted multitarget posterior $f_{k+1|k}(X|Z^{(k)})$ *as the functional derivative of* $G_{k+1|k}[h]$:

$$f_{k+1|k}(X|Z^{(k)}) = \frac{\delta G_{k+1|k}}{\delta X}[0]. \tag{14.278}$$

14.8.2 The p.g.fl. Multitarget Corrector

The multitarget corrector equation, (14.50), is

$$f_{k+1|k+1}(X) = \frac{f_{k+1}(Z_{k+1}|X) \cdot f_{k+1|k}(X)}{\int f_{k+1}(Z_{k+1}|X') \cdot f_{k+1|k}(X')\delta X'}. \tag{14.279}$$

The p.g.fl. form of this equation is

$$G_{k+1|k+1}[h] = \frac{\frac{\delta F}{\delta Z_{k+1}}[0, h]}{\frac{\delta F}{\delta Z_{k+1}}[0, 1]}. \tag{14.280}$$

Here

$$F[g, h] \triangleq \int h^X \cdot G_{k+1}[g|X] \cdot f_{k+1|k}(X)\delta X \tag{14.281}$$

and

$$G_{k+1}[g|X] \triangleq \int g^Z \cdot f_{k+1}(Z|X)\delta Z \tag{14.282}$$

is the p.g.fl. of the multitarget likelihood $f_{k+1}(Z|X)$. The functional derivatives of $F[g, h]$ are taken with respect to the variable g. This result is established in Appendix G.25.

Example 85 *In (12.151) we derived the p.g.fl. for the multitarget likelihood function for the standard measurement model:*

$$G_{k+1}[g|X] = (1 - p_D + p_D p_g)^X \cdot e^{\lambda c[g] - \lambda} \tag{14.283}$$

where

$$c[g] \triangleq \int g(\mathbf{z}) \cdot c(\mathbf{z}) d\mathbf{z}, \qquad p_g(\mathbf{x}) \triangleq \int g(\mathbf{z}) \cdot f_{k+1}(\mathbf{z}|\mathbf{x}) d\mathbf{z}. \tag{14.284}$$

From this and (14.281) it is easy to derive the following concrete formula for $F[g,h]$ *[136, p. 1173]:*

$$F[g,h] = e^{\lambda c[g] - \lambda} \cdot G_{k+1|k}[h(q_D + p_D p_g)]. \tag{14.285}$$

where $q_D(\mathbf{x}) \triangleq 1 - p_D(\mathbf{x})$. *To see this, note that*

$$F[g,h] \tag{14.286}$$

$$= \int h^X \cdot G_{k+1}[g|X] \cdot f_{k+1|k}(X) \delta X \tag{14.287}$$

$$= e^{\lambda c[g] - \lambda} \int h^X \cdot (1 - p_D + p_D p_g)^X \cdot f_{k+1|k}(X) \delta X \tag{14.288}$$

$$= e^{\lambda c[g] - \lambda} \int [h(1 - p_D + p_D p_g)]^X \cdot f_{k+1|k}(X) \delta X \tag{14.289}$$

$$= e^{\lambda c[g] - \lambda} \cdot G_{k+1|k}[h(q_D + p_D p_g)]. \tag{14.290}$$

14.9 TARGET PRIORITIZATION

As noted in Section 11.3.1, *targets of interest* (ToIs) are targets that have greater tactical importance than others. This may be because they have high immediacy (e.g., are threateningly near friendly installations or forces), high intrinsic value (e.g., tanks and missile launchers), and so on. In many applications, one must neglect non-ToIs to gain better information about ToIs.

One approach would be to wait until accumulated information strongly suggests that particular targets are probable ToIs and then bias computation toward these targets. However, ad hoc techniques of this sort have inherent limitations:

- Information about target type accumulates incrementally, not suddenly. Preferential biasing of sensors toward targets should likewise be accomplished incrementally, only to the degree supported by accumulated evidence.
- Information about target type may be erroneous, and may be reversed by additional data. It may not be possible to recover from an erroneous hard-and-fast decision to ignore a target, since the target has been lost because of that decision.
- Target preference may not be an either-or choice, since ToIs themselves may be ranked in order of tactical importance. For example, missile launchers and tanks both have high tactical value, but the former even more so than the latter.

Rather than resorting to ad hoc techniques with inherent limitations, one should

- *Integrally incorporate target preference into the fundamental statistical representation of multisensor-multitarget systems.*

The purpose of this section is to describe how this can be done.[14]

In Section 14.9.1, I shall show how concepts of tactical significance at any given time can be captured mathematically in the form of a *tactical importance function* (TIF) $\rho(\mathbf{x}) \overset{\text{abbr.}}{=} \rho_{k|k}(\mathbf{x})$. In Section 14.9.2, I will show that if $G_{k|k}[h]$ is the p.g.fl. of the multitarget posterior distribution $f_{k|k}(X|Z^{(k)})$, then the p.g.fl.

$$G^{\rho}_{k|k}[h] \triangleq G_{k|k}[1 - \rho + h\rho] \tag{14.291}$$

has the effect of rank-biasing $G_{k|k}[h]$ in favor of ToIs. Finally, in Section 14.9.3, I will show that its corresponding multitarget posterior is

$$f^{\rho}_{k|k}(X|Z^{(k)}) = \rho^X \cdot \frac{\delta G_{k|k}}{\delta X}[1 - \rho] \tag{14.292}$$

$$= \rho^X \cdot \int (1 - \rho)^W \cdot f_{k|k}(X \cup W|Z^{(k)})\delta W \tag{14.293}$$

where as usual $\rho^X = 1$ if $X = \emptyset$ and $\rho^X = \prod_{\mathbf{x} \in X} \rho(\mathbf{x})$ otherwise.

14 This material originally appeared in [150] and [139, pp. 284, 285]. It is a fundamental aspect of the sensor management approach described in [139], see [49, 55, 54, 53].

14.9.1 Tactical Importance Functions (TIFs)

The tactical significance of a target depends on multiple factors. One factor is target type c. Any target that is capable of causing massive damage or disruption, such as an enemy mobile missile launcher, is of inherent tactical interest. A second factor is position $\mathbf{p}$. Any target of given type may have greater or lesser tactical significance depending on where it is located relative to other objects in the scene. Likewise for speed and heading $\mathbf{v}$. Any target of undetermined type that is heading toward a friendly command and control center at high speed has inherent tactical interest. A fourth factor is threat level τ. A target with an active fire control radar has greater potential tactical significance than one that does not. There may be other factors as well.

The current tactical significance of a target is a function of both its current state vector $\mathbf{x} = (\mathbf{e}, \mathbf{v}, c, \tau, ...)$ and the states of the other objects, fixed or mobile, in the scene. Since the scene is dynamically changing, the definition of tactical significance must also be dynamic.

Tactical significance at the current time step k can be modeled as a relative-ranking function

$$0 \leq \rho_{k|k}(\mathbf{x}) \leq 1. \tag{14.294}$$

It states that the relative tactical significance of a target of given type c, with current position $\mathbf{p}$, velocity $\mathbf{v}$, and threat-state τ, is $\rho_{k|k}(\mathbf{x})$. If $\rho_{k|k}(\mathbf{x}) = 0$, then the target has no tactical significance whatsoever. If $\rho_{k|k}(\mathbf{x}) = 1$, it is of the highest possible tactical significance.

I call $\rho_{k|k}(\mathbf{x})$ a *tactical importance function* (TIF). Figure 14.2 illustrates the concept of a TIF.

14.9.2 The p.g.fl. for a TIF

Consider the simplest possible situation first: any given target is unequivocally either a ToI or a non-ToI. In this case, there is a specific subset $S \subseteq \mathfrak{X}_0$ of all possible ToIs. The multitarget posterior density $f_{k|k}(X|Z^{(k)})$ is the probability distribution of a random finite set $\Xi_{k|k}$ of target-states, without target preference being taken into account. Consequently, $\Xi_{k|k} \cap S$ is the random finite set of all targets at time step k that are of current tactical significance. It can be shown (see [70, p. 164, Prop. 23] or (11.332) that the p.g.fl. of $\Xi_{k|k} \cap S$ is

$$G^S_{k|k}[h] \triangleq G_{k|k}[1 - \mathbf{1}_S + \mathbf{1}_S h]. \tag{14.295}$$

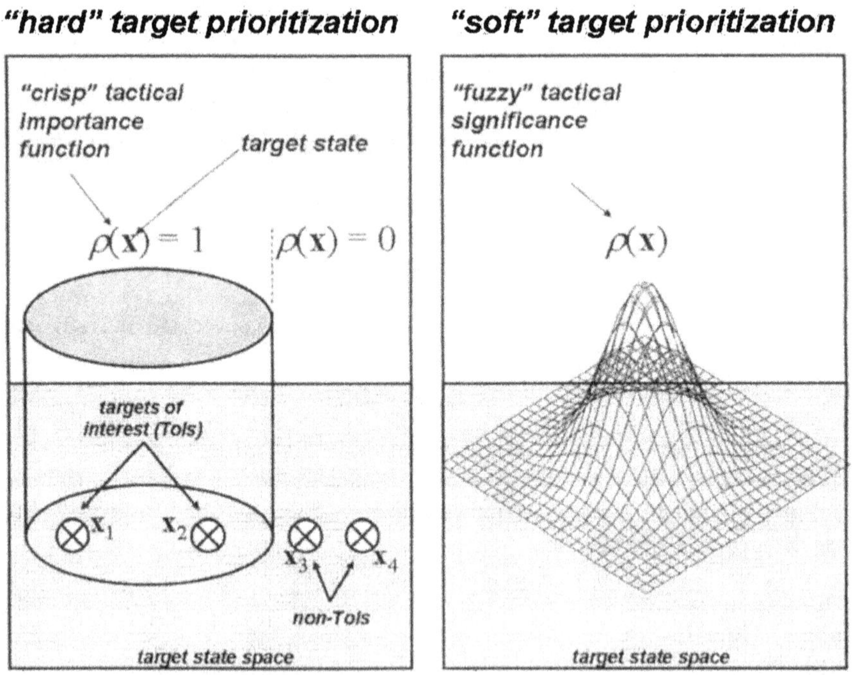

Figure 14.2 The concept of a tactical importance function (TIF) $\rho(\mathbf{x})$. The value of $\rho(\mathbf{x})$ is a relative ranking of the tactical prioritization of a target that has state $\mathbf{x}$. In the left-hand figure, targets are either of total interest, $\rho(\mathbf{x}) = 1$, or of no interest at all, $\rho(\mathbf{x}) = 0$. In the figure at right, intermediate rankings or degrees of importance can be expressed as a fuzzy membership function.

Let $\rho(\mathbf{x}) = \rho_{k|k}(\mathbf{x})$ be the tactical importance function at time step k. By analogy, the p.g.fl. corresponding to all targets of tactical interest is

$$G^{\rho}_{k|k}[h] \triangleq G_{k|k}[1 - \rho + \rho h]. \tag{14.296}$$

14.9.3 The Multitarget Posterior for a TIF

Note that $G^{\rho}_{k|k}[h] = G_{k|k}[T[h]]$ where $T[h] \triangleq 1 - \rho + h\rho$ is a functional transformation. This is an affine functional transformation as defined in Example 64 of Chapter 11. From (11.215) of that example we know that

$$\frac{\delta G^{\rho}_{k|k}}{\delta \mathbf{x}}[h] = \frac{\delta}{\delta \mathbf{x}} G_{k|k}[\rho h + 1 - \rho] \tag{14.297}$$

$$= \rho(\mathbf{x}) \cdot \frac{\delta G_{k|k}}{\delta \mathbf{x}}[\rho h + 1 - \rho]. \tag{14.298}$$

If $X = \{\mathbf{x}_1, ..., \mathbf{x}_n\}$ with $|X| = n$, it follows by induction that

$$\frac{\delta G^{\rho}_{k|k}}{\delta X}[h] = \frac{\delta^n G^{\rho}_{k|k}}{\delta \mathbf{x}_1 \cdots \delta \mathbf{x}_n}[h] \tag{14.299}$$

$$= \rho(\mathbf{x}_1) \cdots \rho(\mathbf{x}_n) \cdot \frac{\delta^n G_{k|k}}{\delta \mathbf{x}_1 \cdots \delta \mathbf{x}_n}[1 - \rho + h\rho] \tag{14.300}$$

$$= \rho^X \cdot \frac{\delta G_{k|k}}{\delta X}[1 - \rho + h\rho]. \tag{14.301}$$

Therefore

$$f^{\rho}_{k|k}(X|Z^{(k)}) \triangleq \frac{\delta G^{\rho}_{k|k}}{\delta X}[0] = \rho^X \cdot \frac{\delta G_{k|k}}{\delta X}[1 - \rho]. \tag{14.302}$$

Finally, from the Radon-Nikodým theorem for functional derivatives—see 11.251)—we know that

$$\frac{\delta G_{k|k}}{\delta X}[1 - \rho] = \int (1 - \rho)^W \cdot f_{k|k}(X \cup W|Z^{(k)}) \delta W. \tag{14.303}$$

Example 86 (Target Preference for JoTT Filter) *We know that the p.g.fl. of the data-updated JoTT filter posterior (Section 14.7.4) must be*

$$
\begin{aligned}
& G_{k+1|k+1}[h] && (14.304)\\
= & \int h^X \cdot f_{k+1|k+1}(X)\delta X && (14.305)\\
= & f_{k+1|k+1}(\emptyset) + \int h(\mathbf{x}) \cdot f_{k+1|k+1}(\{\mathbf{x}\})d\mathbf{x} && (14.306)\\
= & p_{k+1|k+1} + p_{k+1|k+1} \int h(\mathbf{x}) \cdot f_{k+1|k+1}(\mathbf{x})d\mathbf{x} && (14.307)\\
= & 1 - p_{k+1|k+1} + p_{k+1|k+1} \cdot f_{k+1|k+1}[h]. && (14.308)
\end{aligned}
$$

Suppose that we wish to bias the JoTT filter away from uninteresting targets. We encapsulate our definition of "uninteresting" into a tactical importance function $\rho(\mathbf{x}) \overset{\text{abbr.}}{=} \rho_{k+1|k+1}(\mathbf{x})$. *Then from (14.296) we get*

$$
\begin{aligned}
G^\rho_{k+1|k+1}[h] &= G_{k+1|k+1}[1 - \rho + \rho h] && (14.309)\\
&= 1 - p_{k+1|k+1} + p_{k+1|k+1} && (14.310)\\
&\quad \cdot f_{k+1|k+1}[1 - \rho + \rho h] && (14.311)\\
&= 1 - p_{k+1|k+1} + p_{k+1|k+1} && (14.312)\\
&\quad + p_{k+1|k+1} \cdot f_{k+1|k+1}[-\rho + \rho h] && (14.313)\\
&= 1 + p_{k+1|k+1} \cdot f_{k+1|k+1}[\rho(h-1)]. && (14.314)
\end{aligned}
$$

Thus

$$\frac{\delta G^\rho_{k+1|k+1}}{\delta \mathbf{x}}[h] = p_{k+1|k+1} \cdot \rho(\mathbf{x}) \cdot f_{k+1|k+1}(\mathbf{x}). \tag{14.315}$$

Since

$$
\begin{aligned}
f^\rho_{k+1|k+1}(\emptyset) &= G^\rho_{k+1|k+1}[0] && (14.316)\\
&= 1 - p_{k+1|k+1} \cdot f_{k+1|k+1}[\rho] && (14.317)\\
f^\rho_{k+1|k+1}(\{\mathbf{x}\}) &= \frac{\delta G^\rho_{k+1|k+1}}{\delta \mathbf{x}}[0] && (14.318)\\
&= p_{k+1|k+1} \cdot \rho(\mathbf{x}) \cdot f_{k+1|k+1}(\mathbf{x}) && (14.319)
\end{aligned}
$$

it follows that the multitarget distribution for targets of interest is, in JoTT notation, exactly what one would intuitively expect:

$$p^{\rho}_{k+1|k+1} = p_{k+1|k+1} \cdot f_{k+1|k+1}[\rho] \quad (14.320)$$

$$f^{\rho}_{k+1|k+1}(\mathbf{x}) = \frac{\rho(\mathbf{x}) \cdot f_{k+1|k+1}(\mathbf{x})}{f_{k+1|k+1}[\rho]}. \quad (14.321)$$

14.10 CHAPTER EXERCISES

Exercise 57 *Verify (14.230)—that is, the formula for the variance on target number for the JoTT filter.*

Exercise 58 *Suppose that* $G[h] = e^{\mu s[h]-\mu}$ *is multitarget Poisson and let* $G^{\rho}[h]$ *be its restriction to targets of interest (ToIs) as defined in (14.296). Show that* $G^{\rho}[h] = e^{\mu^{\rho} s^{\rho}[h]-\mu^{\rho}}$ *is also multitarget Poisson with* $\mu^{\rho} = \mu \cdot s[\rho]$ *and* $s^{\rho}(\mathbf{x}) = \rho(\mathbf{x}) \cdot s(\mathbf{x})/s[\rho]$.

Exercise 59 *Suppose that* $G[h] = G(s[h])$ *is the p.g.fl. of a multitarget i.i.d. cluster process. Let* $G^{\rho}[h]$ *be its restriction to targets of interest (ToIs) as defined in (14.296). Show that* $G^{\rho}[h] = G^{\rho}(s^{\rho}[h])$ *is also a multitarget i.i.d. cluster process with* $\mu^{\rho} = \mu \cdot s[\rho]$ *and* $s^{\rho}(\mathbf{x}) = s(\mathbf{x})/s[\rho]$.

Part III

Approximate Multitarget Filtering

Chapter 15

Multitarget Particle Approximation

15.1 INTRODUCTION TO THE CHAPTER

The multitarget Bayes filter of Chapter 14 is the theoretically optimal mathematical foundation for multitarget detection, tracking, localization, and target identification. This having been stated, an obvious question looms:

- How might one actually implement the full multitarget Bayes filter in a computationally tractable manner?

This is no small problem. In Section 15.2.1, I will illustrate the challenging combinatorial complexity of the multitarget Bayes filter via operation counts for fixed-grid and particle system approximations. From these examples I will draw the following conclusion:

- For most situations in which the multitarget Bayes filter is actually appropriate (i.e., when conventional approaches fail) it will likely be implementable in real time for only a small number of targets.

Consequently, principled approximation techniques must be devised with the aim of making the multitarget filter practical. This chapter is the first of three that address this issue:

- Multitarget sequential Monte Carlo approximation (this chapter);
- Multitarget-moment approximation (Chapter 16);
- Multitarget multi-Bernoulli approximation (Chapter 17).

The last approach is speculative, in the sense that it has never been implemented even in simple simulations. The second approach, while still in the research and development stage of maturity, has been implemented by numerous researchers and at this time appears promising. The second and third approaches are based on a common strategy: *approximations at the p.g.fl. level* that render the p.g.fl. form of the multitarget Bayes filter (Section 14.8) mathematically tractable.

In this chapter, I describe the most familiar approach: multitarget SMC approximation. Though it also is still in a research and development stage of maturity, it is based on increasingly familiar and well-understood techniques.[1]

In (2.211) we noted that a single-target particle filter propagates a particle system through time:

$$\{\mathbf{x}^i_{0|0}\}_{1\leq i\leq \nu} \rightarrow \{\mathbf{x}^i_{1|0}\}_{1\leq i\leq \nu} \rightarrow \{\mathbf{x}^i_{1|1}\}_{1\leq i\leq \nu} \rightarrow \cdots \tag{15.1}$$

$$\rightarrow \{\mathbf{x}^i_{k|k}\}_{1\leq i\leq \nu} \rightarrow \{\mathbf{x}^i_{k+1|k}\}_{1\leq i\leq \nu} \tag{15.2}$$

$$\rightarrow \{\mathbf{x}^i_{k+1|k+1}\}_{1\leq i\leq \nu} \rightarrow \cdots \tag{15.3}$$

where $\{\mathbf{x}^i_{k|k}\}_{1\leq i\leq \nu} \overset{\text{abbr.}}{=} \{\mathbf{x}^1_{k|k}, ..., \mathbf{x}^\nu_{k|k}\}$ and where the $\mathbf{x}^i_{k|k}$ are single-target states in $\mathfrak{X}_0$. Similarly, a multitarget particle filter propagates a *multitarget particle system* through time:

$$\{X^i_{0|0}\}_{0\leq i\leq \nu} \rightarrow \{X^i_{1|0}\}_{0\leq i\leq \nu} \rightarrow \{X^i_{1|1}\}_{0\leq i\leq \nu} \rightarrow \cdots \tag{15.4}$$

$$\rightarrow \{X^i_{k|k}\}_{0\leq i\leq \nu} \rightarrow \{X^i_{k+1|k}\}_{0\leq i\leq \nu} \tag{15.5}$$

$$\rightarrow \{X^i_{k+1|k+1}\}_{0\leq i\leq \nu} \rightarrow \cdots \tag{15.6}$$

where $\{X^i_{k|k}\}_{0\leq i\leq \nu} \overset{\text{abbr.}}{=} \{X^0_{k|k}, X^1_{k|k}, ..., X^\nu_{k|k}\}$ and where the $X^i_{k|k}$ are multitarget state sets in the multitarget state space $\mathfrak{X}$.

15.1.1 Summary of Major Lessons Learned

The following are the major concepts to be encountered in this chapter:

- Even with highly unrealistic assumptions, fixed-grid implementation of the multitarget Bayes filter is inherently intractable.
- This remains true even when using the simplest possible model of targets moving between discrete locations in one dimension (Section 15.2.1).

1 See, for example, [10, 19, 92, 93, 106, 110, 111].

- Sequential Monte Carlo (particle system) approximations can be generalized to the multitarget Bayes filter.
- Even with multitarget SMC implementation, the multitarget Bayes filter is not necessarily tractable in those applications for which it is actually appropriate (Section 15.2.3).
- From a practitioner's point of view, the multitarget Bayes filter is appropriate primarily for those applications that cannot be adequately addressed using conventional techniques (Section 15.2.3).
- Multitarget state and error estimation is not easily accomplished for multitarget SMC approximation (Section 15.7.1).
- State and error estimation based on the probability hypothesis density (PHD) techniques of Chapter 16 is one possible approach (Section 15.7.1).
- The issue of computationally tractable track labeling appears to be especially difficult in SMC implementations (Section 15.7.3).

15.1.2 Organization of the Chapter

The chapter is organized as follows. In Section 15.2, I investigate the computational complexity of the multitarget Bayes filter, using fixed-grid and particle system implementations as examples. In this section, I also summarize a few FISST-based implementations of the multitarget Bayes filter. The concept of multitarget particle systems approximation is described in Section 15.3.

In subsequent sections, I describe the steps of SMC implementation of the multitarget filter: initialization (Section 15.4); prediction (Section 15.5); correction (Section 15.6); and multitarget state and error estimation (Section 15.7).

15.2 THE MULTITARGET FILTER: COMPUTATION

The purpose of this section is to spotlight the severe computational challenges associated with real-time implementation of the multitarget recursive Bayes filter. Section 15.2.1 generalizes my discussion of single-target fixed grid approximation (Section 2.4.11) into the multitarget realm. Section 15.2.2 similarly generalizes my discussion in Section 2.4.11 of single-target sequential Monte Carlo approximation. Section 15.2.3 addresses the issue, *When is the multitarget Bayes filter appropriate?*

Finally, in Section 15.2.4, I summarize some existing implementations of the FISST multitarget Bayes filter.

15.2.1 Fixed-Grid Approximation

Recall from Section 2.4.11 that in fixed-grid approximation one first chooses bounded regions of the measurement space $\mathbb{R}^M$ and the state space $\mathbb{R}^N$. One then discretizes these regions into collections of $\mu = \mu_0^M$ measurement cells and $\nu = \nu_0^N$ state cells, where μ_0 and ν_0 are the respective numbers of single-dimensional cells. Target and sensor constraints are modeled by heuristically specifying Markov transitions $f_{k+1|k}(x|x')$ from each state cell x' to all others x; and sensor noise by heuristically specifying the likelihood $f_{k+1}(z|x)$ that the sensor will collect in observation-cell z if the target is present at a given state cell x.

Suppose that if x, x', z are fixed, then a operations are required to compute $f_{k+1|k}(x|x')$ and b operations are required to compute $f_{k+1}(z|x)$. Also assume that a single sensor interrogates a scene containing n targets.

Begin with the multitarget predictor step, (14.14), and assume the "standard" multitarget motion model introduced in Section 13.2.2. That is, simplify by supposing that target number n is known and target motions are independent. As was shown in (14.34), the multitarget Markov density $f_{k+1|k}(X|X')$ in vector notation has the form

$$f_{k+1|k}(x_1, ..., x_n|x'_1, ..., x'_n) = f_{k+1|k}(x_1|x'_1) \cdots f_{k+1|k}(x_n|x'_n). \tag{15.7}$$

Thus in vector notation (14.14) becomes

$$f_{k+1|k}(x_1, ..., x_n|Z^{(k)}) \tag{15.8}$$

$$= \sum_{x'_1,...,x'_n} f_{k+1|k}(x_1|x'_1) \cdots f_{k+1|k}(x_n|x'_n) \tag{15.9}$$

$$\cdot f_{k|k}(x'_1, ..., x'_n|Z^{(k)}). \tag{15.10}$$

For each choice of $x_1, ..., x_n$ and $x'_1, ..., x'_n$, the number of operations required to compute $f_{k+1|k}(x_1|x'_1) \cdots f_{k+1|k}(x_n|x'_n)$ is na. Consequently, the number of operations required to compute the sum in (15.8) is at least $\nu^n na$. Thus the number of operations required to compute the entire predicted multitarget posterior distribution $f_{k+1|k}(\cdot|Z^{(k)})$ is at least $\nu^n \cdot \nu^n na = \nu^{2n} na$.

Next, consider the multitarget corrector equation, (14.50). We simplify by assuming that the sensor has the "standard" multitarget measurement model introduced in Section 12.3.4—that is, no missed detections or false alarms. In this case the multitarget likelihood $f_{k+1}(Z|X)$ is, in vector notation,

$$f_{k+1}(z_1, ..., z_n|x_1, ..., x_n) = \frac{1}{n!} \sum_{\sigma} f_{k+1}(z_1|x_{\sigma 1}) \cdots f_{k+1}(z_n|x_{\sigma n}) \tag{15.11}$$

where the summation is taken over all permutations σ on $1, ..., n$. Thus the multitarget corrector, (14.50), is

$$f_{k+1|k+1}(x_1, ..., x_n|Z^{(k+1)}) \tag{15.12}$$

$$\propto f_{k+1}(z_1, ..., z_n|x_1, ..., x_n) \cdot f_{k+1|k}(x_1, ..., x_n|Z^{(k)}). \tag{15.13}$$

For each choice of $x_1, ..., x_n$, (15.11) requires at least $n!nb$ operations to compute and thus so does (15.12).

At least $\nu^n n!nb$ operations are required to compute the entire data-updated multitarget posterior $f_{k+1|k+1}(\cdot|Z^{(k+1)})$. Consequently, at least

$$\nu^{2n} na + \nu^n n!nb = n\nu^n (\nu^n a + n!b) \tag{15.14}$$

operations are required for each recursive cycle of (15.8) and (15.12).

We can attempt to increase tractability by stripping off even more application realism than is already the case. We get operation counts of the following:

- $n\nu^n(1+n!b)$ if in addition, states transition only to immediately neighboring cells;
- $n\nu^n(1 + b)$ if in addition targets are well-separated;
- $2n\nu^n$ if in addition observations are binary;
- $2n\nu_0^n$ if in addition state space is one-dimensional—that is, if $\nu = \nu_0$.

In other words, even assuming a drastic, cumulative loss of application realism, fixed-grid implementation is:

- *Inherently intractable even when using the simplest possible model of targets moving between discrete locations in one dimension.*

15.2.2 SMC Approximation

As with the single-target Bayes filter, matters can be improved if we employ SMC techniques such as the ones to be described later in the chapter. Roughly speaking, in particle systems approximation the number ν^n of multitarget cells is replaced by the number Π of multitarget particles. In this case at least $n(a+n!b)\Pi$ operations are required—or $n(a+1)\Pi$ if observations are binary and targets are well-separated. Significant increases in computational efficiency can be achieved since Π will usually be substantially smaller than ν^n.

However, this still does not mean that the computational savings will be enough to permit real-time application. First, there is the troublesome factor $n!$. Second, in general the number Π of multitarget particles must be large under realistic sensing conditions, such as lower SNR, in which conventional methods such as MHT fail. Third, simple implementations based on the dynamic prior will be intractable for larger numbers of targets. More sophisticated proposal densities must be devised.

15.2.3 When Is the Multitarget Filter Appropriate?

In Section 10.7.1 I sketched typical situations in which the performance of conventional multitarget correlation approaches, such as MHC, will deteriorate. It is in *these* applications (i.e., applications in which even the best MHC filters may perform poorly) that the multitarget Bayes filter becomes potentially important.

This point is not universally recognized. Some researchers have, for example, claimed that multitarget SMC filters can tractably track many tens of targets. However, to achieve such performance they have found it necessary to make a number of assumptions:

- SNR is very large.
- Targets are preinitialized.
- Target number is known.
- Target-motion nonlinearities are moderate.

Under such conditions the multitarget Bayes filter is probably unnecessary to begin with, regardless of what implementation is used. This is because multihypothesis trackers already perform well against much larger numbers of targets, with much smaller computational loading, without any such assumptions.

Indeed, it can be stated with confidence that:

- *Hundreds of targets are being successfully tracked at any given time using multitarget SMC filters and real data.*

This is because the alpha-beta filter of Section 2.2.6 (the tracker most commonly used in commercial air traffic control) is actually a primitive single-target particle filter. Effective signal-to-noise ratio in the air traffic control application is extremely large. Also, targets tend to follow straight-line paths infrequently punctuated by shallow ascents, descents, and turns. Thus *an alpha-beta filter can track a single target with just a single particle*—namely, the filter's evolving state-estimate $\mathbf{x}_{k|k}$. If we concatenate a large number of single-target alpha-beta filters, we get a "virtual multitarget SMC filter" that tracks large numbers of targets using a single (multitarget) particle.

As in the single-target case, the following questions should be fundamental to any application of the multitarget Bayes filter in real-world scenarios:

- Is the multitarget Bayes filter *inappropriate* for the application at hand from a practitioner's point of view?
- *That is, can the application be adequately addressed using conventional techniques?*

15.2.4 Implementations of the Multitarget Filter

Only a few implementations of the full FISST multitarget Bayes filter have been reported in the literature. I describe five: a bearings-only multitarget tracker devised by Vihola; a passive-acoustic tracking system due to Ma, Vo, Singh, and Baddeley; a ground-target tracker reported by Sidenbladh and Wirkander; a MITRE Corporation road-network tracking and sensor management system; and a communications network multiuser detector due to Biglieri and Lops.[2]

15.2.4.1 Bearings-Only Tracking

In Section 12.10, I described a multitarget likelihood function for bearings-only sensors devised by Vihola. Vihola has incorporated this measurement model into SMC multitarget Bayes filters and used it to track multiple targets using multiple bearings-only sensors whose observations are corrupted by missed detections and false alarms. He has reported two such implementations: one using a standard

2 Many papers have reported SMC implementations of the multitarget Bayes filter without resort to FISST methods. In most of these, target number has either been assumed known a priori, or ad hoc methods have been employed to incorporate unknown target number. For this reason I have limited my discussion to FISST-based approaches, since these attempt to deal with unknown target number in the systematic fashion advocated in this book.

sequential importance resampling (SIR) technique [232, 233] and another using a more efficient Rao-Blackwellized approach [234].

Vihola tested the SIR implementation on a simple two-dimensional scenario containing three bearings-only sensors and one or two targets. Measurements were drawn randomly from the three sensors. Probability of detection was state-dependent and the false alarm probability p_{FA} was varied. Because of the difficulty of implementing MaM or JoM estimators (Sections 14.5.2, 14.5.3) in an SMC filter context, a "PHD visualization" scheme (Section 14.5.5) was used to extract estimates of target number and state.

In single-target tests, the multitarget SIR filter performed well given false alarm rates of $p_{FA} = 0.3$ and $p_{FA} = 0.7$. Its performance began to degrade when the false alarm rate was increased to $p_{FA} = 0.8$. Performance degraded even more in the two target tests with $p_{FA} = 0.1$, 0.3, and 0.5.

The author implemented three different versions of the Rao-Blackwellized (R-B) multitarget particle filter [234].[3] This filter also employed a symmetrization scheme, in which the filter target birth and target death models were balanced to achieve no net gain or loss of targets per recursive cycle.

Monte Carlo simulations consisting of 100 scenarios were used to determine average performance. Baseline comparisons with a linear-Gaussian sensor and up to seven appearing and disappearing targets in moderate clutter showed that the basic SIR filter required two orders of magnitude more particles to achieve the same performance as the best of the three R-B filters. In bearings-only simulations, the three R-B filters exhibited promising behavior.

15.2.4.2 Passive-Acoustic Tracking

Ma, Vo, Singh, and Baddeley [122, 239] have applied the multitarget Bayes filter to the problem of detecting and tracking time-varying numbers of individuals in rooms, based on processing of speech utterances collected by multiple fixed microphones. This constitutes a time direction-of-arrival (TDOA) problem. The authors used a Langevin motion model and a generalized cross-correlation (GCC) measurement processing technique.

The authors implemented the multitarget Bayes filter using the motion and measurement models. This implementation incorporated track labeling, through the device of attaching labels to individual single-target states in the multitarget particles.

3 See also [202].

The authors successfully tested their algorithm in simulations. The first simulation involved four microphone pairs and two moving, appearing and disappearing individuals. The M-SMC filter successfully detected and tracked these individuals throughout the scenario. In a second simulation, the robustness of the method against motion model mismatch was tested. In this case the simulation involved eight microphone pairs and three nonmoving, but appearing and disappearing, individuals. The model mismatch resulted in occasional localization errors, but overall tracking performance was reasonable.

15.2.4.3 Ground-Target Tracking in Terrain

Tracking targets in terrain is, in general, a highly nonlinear problem because of terrain constraints such as lakes and rivers, cliffs, roads, intersections, and so on. An extended Kalman filter often performs poorly in such applications: tanks seemingly vault cliffs and submarines apparently tunnel inland. In [210] Sidenbladh and Wirkander describe the application of an SMC implementation of the full multitarget Bayes filter to this problem.

The authors tested the approach in two-dimensional simulations. Vehicles were assumed to be of the same type and have three travel options: roads, fields, and forest, with a given a priori probability of traveling in each. A terrain map was used to construct an (in general) nonlinear motion model. The targets were observed by simulated human ground observers who reported estimates of position, speed, and direction. Such observations tend to have lower probabilities of detection but also low false alarm rates. Thus the simulations did not include false alarms.

In these tests, three vehicles traveled over roads, with one vehicle temporarily going off-road into a field. The multitarget filter performed well when probability of detection was $p_D = 0.5$, using only 200 particles. Performance drastically deteriorated when $p_D = 0.1$.

15.2.4.4 Tracking and Sensor Management in Road Networks

Kuklinski, Otero, Stein, Theophanis, and Witkoskie [219, 249, 250] of MITRE Corporation, McLean, Virginia, are applying FISST techniques to the problem of tracking moving vehicles in road networks. They have asserted that FISST-based filters are preferable in such problems because, by taking all possible measurement-to-track associations into account, such filters can more robustly track target direction-switches at road intersections.

The multitarget filter was implemented using a Gaussian-mixture approximation, while a "restless bandit" approach was used for the sensor management implementation.

The authors successfully tested their algorithms on real vehicles in a real road network adjacent to their facility, consisting of six roads. Data was collected from six sensors that included radars, optical cameras, and acoustic sensors. A single vehicle was outfitted with GPS for the purposes of ground truthing, and then driven through the network amidst "confuser" targets (i.e., normal local traffic).

15.2.4.5 Multiuser Detection (MUD) in Communications Networks

Biglieri and Lops [14] have argued that FISST techniques "provide a fairly natural approach" for dynamic, mobile, multiple-access wireless digital communications networks. The problem addressed is that of detecting and identifying active users in a dynamic network in which users are continually entering and leaving the system.

The received signal is a superposition of the signals $s(\mathbf{x}^1), ..., s(\mathbf{x}^n)$ generated by the current users with states $\mathbf{x}^1, ..., \mathbf{x}^n$, as well as the signal $s(\mathbf{x}^0)$ generated by a reference user with state $\mathbf{x}^0$:

$$\mathbf{z} = s(\mathbf{x}^0) + s(\mathbf{x}^1) + ... + s(\mathbf{x}^n) + \mathbf{W} \tag{15.15}$$

where $\mathbf{W}$ is a noise vector with probability density $f_{\mathbf{W}}(\mathbf{z})$. This leads to the following multitarget likelihood function $f(Z|X)$:

$$f(Z|X) = \begin{cases} f_{\mathbf{W}}(\mathbf{z} - s(\mathbf{x}^0) - s(X)) & \text{if} \quad Z = \{\mathbf{z}\} \\ 0 & \text{if} \quad \text{otherwise} \end{cases}. \tag{15.16}$$

where $s(X) \triangleq \sum_{\mathbf{x} \in X} s(\mathbf{x})$. The authors apply FISST techniques to construct a multitarget motion model and, from it, a multitarget Markov density.

The authors implemented a multitarget filter with a MAP multiuser state estimator, and tested it in simulations with a small maximum number of users. They compared their algorithm with a standard MUD algorithm (ML multiuser detection), that assumes that all users are simultaneously active. They found that the new algorithm provided a detector much more robust to variations in user activity when compared to the classic MUD technique.

15.3 MULTITARGET PARTICLE SYSTEMS

In this section, I summarize the basic ideas of SMC approximation of the multitarget Bayes filter. As in earlier descriptions of SMC implementations in this book, for the sake of clarity we limit ourselves to describing a generic SMC multitarget filter based on the dynamic prior. The computational difficulties of such methods greatly increase in the multitarget case as target number increases. As a consequence, the importance of a good proposal density also increases (see Remark 26). As Vo, Singh, and Doucet note [237, p. 490]:

> The main practical problem with the multitarget particle filter is the need to perform importance sampling in very high dimensional spaces if many targets are present. Moreover, it can be difficult to find an efficient importance density...

For greater detail see [238, 45, 237, 232, 234, 210]. Also, the reader may find it useful to review the material on SMC approximation of the JoTT filter (Section 14.7).

I summarized the basic elements of single-target sequential Monte Carlo (SMC) approximation in Section 2.5.3. There I defined a particle system approximation of a single-target posterior distribution $f_{k|k}(\mathbf{x}|Z^k)$ to be a collection $\mathbf{x}^1_{k|k}, ...\mathbf{x}^\nu_{k|k}$ of state vectors together with a collection of positive importance weights $w^1_{k|k}, ...w^\nu_{k|k}$ with $\sum_{i=1}^\nu w^i_{k|k} = 1$, such that

$$\int \theta(\mathbf{x}) \cdot f_{k|k}(\mathbf{x}|Z^k) d\mathbf{x} \cong \sum_{i=1}^{\nu} w^i_{k|k} \cdot \theta(\mathbf{x}^i_{k|k}) \tag{15.17}$$

for any unitless function $\theta(\mathbf{x})$ of a state vector variable $\mathbf{x}$.

Multitarget SMC approximation follows fundamentally the same recipe. Now, however, particles are *state sets* rather than *state vectors*. A multitarget particle system approximation of the multitarget posterior distribution $f_{k|k}(X|Z^{(k)})$ is a collection

$$X^0_{k|k}, X^1_{k|k}, ..., X^\nu_{k|k} \tag{15.18}$$

of state sets and positive importance weights $w^0_{k|k}, ...w^\nu_{k|k}$ with $\sum_{i=0}^\nu w^i_{k|k} = 1$, such that

$$\int \theta(X) \cdot f_{k|k}(X|Z^{(k)}) \delta X \cong \sum_{i=0}^{\nu} w^i_{k|k} \cdot \theta(X^i_{k|k}) \tag{15.19}$$

for any unitless function $\theta(X)$ of a state set variable X.[4] As in the single-target case, the following convergence property must be satisfied:[5]

$$\lim_{\nu \to \infty} \sum_{i=0}^{\nu} w^i_{k|k} \cdot \theta(X^i_{k|k}) = \int \theta(X) \cdot f_{k|k}(X|Z^{(k)}) \delta X. \tag{15.20}$$

The multitarget particles are interpreted as random samples drawn from the multitarget posterior distribution $f_{k|k}(X|Z^{(k)})$:

$$X^0_{k|k}, X^1_{k|k}, ..., X^\nu_{k|k} \sim f_{k|k}(\cdot|Z^{(k)}). \tag{15.21}$$

Again as in the single-target case, it is common practice to abbreviate (15.19) using the following abuse of notation:[6]

$$f_{k|k}(X|Z^{(k)}) \cong \sum_{i=0}^{\nu} w^i_{k|k} \cdot \delta_{X^i_{k|k}}(X). \tag{15.22}$$

where the multitarget Dirac density was defined in (11.124).

Multitarget particle systems are far more complex than single-target ones. It is often convenient to order $X^0_{k|k}, X^1_{k|k}, ..., X^\nu_{k|k}$ by increasing target number. That is, $w^0_{k|k}$ is the weight of the no-target particle

$$X^0_{k|k} = \emptyset. \tag{15.23}$$

The next ν_1 particles $X^1_{k|k}, ..., X^{\nu_1}_{k|k}$ represent single-target samples:

$$X^1_{k|k} = \{\mathbf{x}^1_1\} \quad ,....., \quad X^{\nu_1}_{k|k} = \{\mathbf{x}^{\nu_1}_1\}. \tag{15.24}$$

The next ν_2 particles $X^{\nu_1+1}_{k|k}, ..., X^{\nu_2}_{k|k}$ represent two target samples:

$$X^{\nu_1+1}_{k|k} = \{\mathbf{x}^{\nu_1+1}_1, \mathbf{x}^{\nu_1+1}_2\} \quad ,....., \quad X^{\nu_2}_{k|k} = \{\mathbf{x}^{\nu_2}_1, \mathbf{x}^{\nu_2}_2\}. \tag{15.25}$$

4 Note that if $\theta(X)$ were not unitless, then the set integral in (15.19) might be mathematically undefined because of units-mismatch problems.

5 See, for example, [238, p. 1230, Prop. 2].

6 The multitarget Dirac delta function $\delta_Y(X)$ concentrated at Y was defined in (11.124). Some researchers, going by analogy with the notation $\delta(\mathbf{x}-\mathbf{y}) = \delta_\mathbf{y}(\mathbf{x})$ for the single-target Dirac delta, have employed the notation $\delta(X-Y)$ rather than $\delta_Y(X)$. This practice invites careless mistakes. Since vectors can be added and subtracted, '$\mathbf{x}-\mathbf{y}$' is mathematically well defined. Since finite sets cannot be meaningfully added or subtracted, however, '$X - Y$' is mathematically undefined.

In general, the n-target states are represented by ν_n multitarget particles

$$X_{k|k}^{\nu_{n-1}+1} = \{\mathbf{x}_1^{\nu_{n-1}+1}, ..., \mathbf{x}_n^{\nu_{n-1}+1}\} \quad ,...., \quad X_{k|k}^{\nu_n} = \{\mathbf{x}_1^{\nu_n}, ..., \mathbf{x}_n^{\nu_n}\} \tag{15.26}$$

all the way up to some largest number $\breve{n}$ of targets:

$$X_{k|k}^{\nu_{\breve{n}-1}+1} = \{\mathbf{x}_1^{\nu_{\breve{n}-1}+1}, ..., \mathbf{x}_n^{\nu_{\breve{n}-1}+1}\} \quad ,...., \quad X_{k|k}^{\nu_{\breve{n}}} = \{\mathbf{x}_1^{\nu_{\breve{n}}}, ..., \mathbf{x}_n^{\nu_{\breve{n}}}\}. \tag{15.27}$$

Thus

$$\nu = 1 + \nu_1 + ... + \nu_{\breve{n}} \tag{15.28}$$

where $\nu_0 = 1$ is the number of zero-target particles.

Define $\theta(X)$ by $\theta(X) = 1$ if $|X| = n$ and $\theta(X) = 0$ otherwise. Then (15.19) reduces to

$$f_{k|k}(n|Z^{(k)}) \cong \sum_{i:|X_{k|k}^i|=n} w_{k|k}^i = \sum_{i=\nu_{n-1}+1}^{\nu_n} w_{k|k}^i \tag{15.29}$$

where $f_{k|k}(n|Z^{(k)})$ is the cardinality distribution of $f_{k|k}(X|Z^{(k)})$ as defined in (11.113). That is, $f_{k|k}(n|Z^{(k)})$ is the probability that n targets are present in the scene, for $n = 0, 1, ...$

Remark 24 *Since the target states in multitarget state sets are inherently order-independent (permutation-symmetric), so are the target states in multitarget particles.*

Remark 25 *In what follows we will follow the common simplifying practice of assuming that multitarget particles are equally weighted:* $w_{k|k}^i = 1/\nu$ *for* $i = 1, ..., n$. *However, and as was noted in (14.237) and (14.238) in the context of the JoTT filter, some care may be necessary regarding the weight* $w_{k|k}^0$ *of the null particle* $X_{k|k}^0 = \emptyset$. *As before, this is because* $\emptyset$ *is a discrete state and thus sampling from a multitarget density* $f_{k|k}(X|Z^{(k)})$ *will generally result in multiple copies of it. When these copies are amalgamated into a single particle,* $w_{k|k}^0$ *will not be the same as the other weights. Indeed,* $w_{k|k}^i = (1 - w_{k|k}^0) \cdot \nu^{-1}$ *for* $i = 1, ..., \nu$. *It is this sense of "equal weighting" that will be presumed in what follows.*

15.4 M-SMC FILTER INITIALIZATION

If an initial multitarget distribution $f_{0|0}(X)$ is available, then ν multitarget particles are drawn from it:

$$\emptyset, X^1_{0|0}..., X^\nu_{0|0} \sim f_{0|0}(\cdot). \tag{15.30}$$

I consider some ways of accomplishing this.

15.4.1 Target Number is Known

Suppose that the initial number n of targets is known a priori. Then the multitarget distribution of (14.13) applies: $f_{0|0}(X) = 0$ if $|X| \neq n$ and, otherwise,

$$f_{0|0}(\{\mathbf{x}_1, ..., \mathbf{x}_n\}) = \sum_\sigma f_{\sigma 1}(\mathbf{x}_1) \cdots f_{\sigma n}(\mathbf{x}_n) \tag{15.31}$$

where $f_1(\mathbf{x}) \cdots f_n(\mathbf{x})$ are the initial distributions of the targets. In this case the initialization procedure is as follows. (1) Decide what number ν of particles is required to maintain track on each target. (2) Draw ν single-target particles from the respective distributions:

$$\mathbf{x}^{1,1}_{0|0}, ..., \mathbf{x}^{1,\nu}_{0|0} \sim f_1(\cdot) \quad ,....., \quad \mathbf{x}^{n,1}_{0|0}, ..., \mathbf{x}^{n,\nu}_{0|0} \sim f_n(\cdot). \tag{15.32}$$

The initial particle system is chosen by grouping these samples into ν finite subsets, each of which has n elements:

$$X^1_{0|0} = \{\mathbf{x}^{1,1}_{0|0}, ..., \mathbf{x}^{n,1}_{0|0}\} \tag{15.33}$$

$$\vdots \tag{15.34}$$

$$X^\nu_{0|0} = \{\mathbf{x}^{1,\nu}_{0|0}, ..., \mathbf{x}^{n,\nu}_{0|0}\}. \tag{15.35}$$

They are equally weighted in the sense that $w^1_{0|0} = ... = w^\nu_{0|0} = (1 - w^0_{0|0}) \cdot \nu^{-1}$. Though there are $\nu + 1$ multitarget particles, these correspond to $n \cdot \nu$ single-target particles.

15.4.2 Null Multitarget Prior

If the multitarget filter incorporates a birth model, then the initial distribution could be chosen to be [234]

$$f_{0|0}(X) = \begin{cases} 1 & \text{if} \quad X = \emptyset \\ 0 & \text{if} \quad X \neq \emptyset \end{cases}. \tag{15.36}$$

That is, our initial guess is that no targets are present and the initial particle system is just $X^0_{0|0} = \emptyset$ with $w^0_{0|0} = 1$. The multitarget filter must subsequently detect the presence of any targets that might be present.

15.4.3 Poisson Multitarget Prior

Suppose that the initial distribution is multitarget Poisson as in (14.12):

$$f_{0|0}(\{\mathbf{x}_1, ..., \mathbf{x}_n\}) = e^{-\mu} \cdot \mu^n \cdot f_0(\mathbf{x}_1) \cdots f_0(\mathbf{x}_n). \tag{15.37}$$

Choose some maximum number $\breve{n}$ of targets. Note that the weight of the null particle is $w^0_{0|0} = e^{-\mu}$. Draw $\breve{n}$ nonzero samples from the Poisson distribution

$$\nu_1, ..., \nu_{\breve{n}} \sim \frac{e^{-\mu}\mu^\nu}{\nu!} \tag{15.38}$$

with $\nu = \nu_1 + ... + \nu_{\breve{n}}$. The number ν_i tells us how many multitarget particles with i targets should be drawn from $f_{0|0}(\{\mathbf{x}_1, ..., \mathbf{x}_{\nu_i}\})$. That is, $X^1_{0|0}$ consists of a single sample drawn from $f_0(\mathbf{x})$:

$$\mathbf{x}^1_{0|0} \sim f_0(\cdot). \tag{15.39}$$

Repeat this procedure until ν_1 single-target particles $X^1_{0|0}, ..., X^{\nu_1}_{0|0}$ have been created. Likewise, $X^{\nu_1+1}_{0|0}$ consists of two samples drawn from $f_0(\mathbf{x})$:

$$\mathbf{x}^1_{0|0}, \mathbf{x}^2_{0|0} \sim f_0(\cdot). \tag{15.40}$$

Repeat this procedure until ν_2 two target particles $X^{\nu_1+1}_{0|0}, ..., X^{\nu_1+\nu_2}_{0|0}$ have been created. And so on. At each stage we construct $X^{\nu_i+1}_{0|0}$ by drawing i samples from $f_0(\mathbf{x})$:

$$\mathbf{x}^1_{0|0}, ..., \mathbf{x}^i_{0|0} \sim f_0(\cdot). \tag{15.41}$$

Repeat this procedure until $\nu_{\breve{n}}$ $\breve{n}$-target particles

$$X^{\nu_1+...+\nu_{\breve{n}-1}+1}_{0|0}, ..., X^{\nu_1+...+\nu_{\breve{n}-1}+\nu_{\breve{n}}}_{0|0} \tag{15.42}$$

have been created.

15.5 M-SMC FILTER PREDICTOR

As in the single-target case we draw a single multitarget random sample from the multitarget dynamic prior: $X^i_{k+1|k} \sim f_{k+1|k}(\cdot|X^i_{k|k})$. Thus for any unitless $\theta(X)$,

$$\int \theta(X) \cdot f_{k+1|k}(X|Z^{(k)})\delta X \tag{15.43}$$

$$\cong \quad w^0_{k+1|k} \cdot \theta(\emptyset) + \frac{1 - w^0_{k+1|k}}{\nu} \sum_{i=1}^{\nu} \theta(X^i_{k+1|k}). \tag{15.44}$$

The predicted particle system is $X^0_{k+1|k}, ..., X^\nu_{k+1|k}$.

There is more to the expression $X^i_{k+1|k} \sim f_{k+1|k}(\cdot|X^i_{k|k})$ than at first meets the eye. SMC implementations of the multitarget Bayes filter typically presume the simplest possible multitarget motion model, described in Section 13.2.2. That is, target motions are statistically independent and targets do not appear or disappear from the scene. In this case, from (14.34) we know that the multitarget prediction integral of (14.14) reduces to

$$f_{k+1|k}(\{\mathbf{x}_1, ..., \mathbf{x}_n\}|Z^{(k)}) \tag{15.45}$$

$$= \int f_{k+1|k}(\mathbf{x}_1|\mathbf{x}'_1) \cdots f_{k+1|k}(\mathbf{x}_n|\mathbf{x}'_n) \tag{15.46}$$

$$\cdot f_{k|k}(\{\mathbf{x}'_1, ..., \mathbf{x}'_n\}|Z^{(k)})d\mathbf{x}'_1 \cdots d\mathbf{x}'_n. \tag{15.47}$$

In this case, for each i and with $X^i_{k|k} = \{\mathbf{x}^{i,1}_{k|k}, ..., \mathbf{x}^{i,n(i)}_{k|k}\}$ it is enough to sample $n(i)$ particles from the single-target Markov densities

$$\mathbf{x}^{i,1}_{k+1|k} \sim f_{k+1|k}(\cdot|\mathbf{x}^{i,1}_{k|k}) \quad ,..., \quad \mathbf{x}^{i,n(i)}_{k+1|k} \sim f_{k+1|k}(\cdot|\mathbf{x}^{i,n(i)}_{k|k}) \tag{15.48}$$

and define the predicted multitarget particles to be

$$X^i_{k+1|k} \triangleq \{\mathbf{x}^{i,1}_{k+1|k}, ..., \mathbf{x}^{i,n(i)}_{k+1|k}\} \tag{15.49}$$

for $i = 1, ..., \nu$.

When the multitarget Markov density incorporates target appearance and disappearance models, matters are complicated. Sampling from $f_{k+1|k}(X|X^i_{k|k})$ must account not only for time evolution of target states but also for the birth and death models.

Assume, for example, the standard motion model introduced in Section 13.2. Then the predicted multitarget particle will have the form

$$X^i_{k+1|k} \triangleq X^{i,\text{persist}}_{k+1|k} \cup X^{i,\text{spawn}}_{k+1|k} \cup X^{i,\text{appear}}_{k+1|k}. \tag{15.50}$$

We address the terms in the union in the following order: $X^{i,\text{persist}}_{k+1|k}$, $X^{i,\text{appear}}_{k+1|k}$, and $X^{i,\text{spawn}}_{k+1|k}$.

Remark 26 (Caution Regarding the Dynamic Prior) *The choice of the dynamic prior as proposal density is even more problematic in the multitarget case than in the single-target case. As Vo, Singh, and Doucet note, "...the choice of a naïve importance density like [the dynamic prior] will typically lead to an algorithm whose efficiency decreases exponentially with the number of targets for a fixed number of particles" [237, p. 490]. As a consequence, simple bootstrap methods work best when only a small number of targets are present.*

15.5.1 Persisting and Disappearing Targets

According to the standard multitarget motion model of Section 13.2, a target with state $\mathbf{x}'$ at time step k has a probability $p_S(\mathbf{x}')$ of persisting into time step $k+1$. This means that the multitarget probability distribution of $X^{i,\text{persist}}_{k+1|k}$ is a multitarget multi-Bernoulli distribution as introduced in (11.131) and (11.132). Let $X^i_{k|k} = \{\mathbf{x}'_1, ..., \mathbf{x}'_{n'}\}$ be a particle from time step k, where the index i has been suppressed in the $\mathbf{x}'_1, ..., \mathbf{x}'_{n'}$ for the sake of notational clarity.

Let $p^i_S \triangleq p_S(\mathbf{x}'_i)$. The probability that all of the $\mathbf{x}'_1, ..., \mathbf{x}'_{n'}$ will persist is $p^1_S \cdots p^{n'}_S$. The probability that none will persist is $(1 - p^1_S) \cdots (1 - p^{n'}_S)$. The probability that the subset $\{\mathbf{x}'_{j_1}, ..., \mathbf{x}'_{j_n}\}$ with $1 \leq j_1 < ... < j_n \leq n'$ will persist is[7]

$$p_{n'}(n, j_1, ..., j_n) = \frac{p_S^{j_1} \cdots p_S^{j_n}}{(1 - p_S^{j_1}) \cdots (1 - p_S^{j_n})} \cdot (1 - p_S^1) \cdots (1 - p_S^{n'}). \tag{15.51}$$

Consequently, draw a sample from this distribution:

$$(\hat{n}, \hat{j}_1, ..., \hat{j}_{\hat{n}}) \sim p_{n'}(\cdot, \cdot, ..., \cdot). \tag{15.52}$$

Then for each $\hat{j}_1, ..., \hat{j}_{\hat{n}}$ draw particles

$$\begin{aligned} \mathbf{x}_{k+1|k}^{\hat{j}_1} &\sim f_{k+1|k}(\cdot|\mathbf{x}_{\hat{j}_1}) \\ &\vdots \\ \mathbf{x}_{k+1|k}^{\hat{j}_{\hat{n}}} &\sim f_{k+1|k}(\cdot|\mathbf{x}_{\hat{j}_{\hat{n}}}). \end{aligned}$$

Then define

$$X_{k+1|k}^{i,\text{persist}} = \{\mathbf{x}_{k+1|k}^{\hat{j}_1}, ..., \mathbf{x}_{k+1|k}^{\hat{j}_{\hat{n}}}\}.$$

This process is repeated for each $i = 1, ..., \nu$.

15.5.2 Appearing Targets

According to the standard multitarget motion model, a multitarget probability density $B_{k+1|k}(X)$ models those targets that appear but that are not spawned by existing targets. The easiest way to draw $X_{k+1|k}^{i,\text{appear}}$ from $B_{k+1|k}(X)$ is as follows. Let $p_{k+1|k}^{\text{appear}}(n)$ be the cardinality distribution of $B_{k+1|k}(X)$ as defined in (11.113). Draw a sample

$$n_{k+1|k} \overset{\text{abbr.}}{=} n_{k+1|k}^{\text{appear}} \sim p_{k+1|k}^{\text{appear}}(\cdot) \tag{15.53}$$

from this distribution. Then draw a sample from the distribution

$$\hat{B}_{k+1|k}(\mathbf{x}_1, ..., \mathbf{x}_{n_{k+1|k}}) \triangleq \frac{1}{n_{k+1|k}!} \cdot B_{k+1|k}(\{\mathbf{x}_1, ..., \mathbf{x}_{n_{k+1|k}}\}). \tag{15.54}$$

That is,

$$(\mathbf{x}_{k+1|k}^1, ..., \mathbf{x}_{k+1|k}^{n_{k+1|k}}) \sim \hat{B}_{k+1|k}(\cdot, ..., \cdot). \tag{15.55}$$

7 From (11.141)-(11.147) we know that $\sum_{n=0}^{n'} \sum_{1 \leq j_1 < ... < j_n \leq n'} p_{n'}(n, j_1, ..., j_n) = 1$.

Then

$$X_{k+1|k}^{i,\text{appear}} = \{\mathbf{x}_{k+1|k}^{1}, ..., \mathbf{x}_{k+1|k}^{n_{k+1|k}}\}. \tag{15.56}$$

This process is repeated for each $i = 1, ..., \nu$.

Example 87 *Suppose that*

$$B_{k+1|k}(X) = e^{-\lambda_{k+1|k}} \prod_{\mathbf{x} \in X} \lambda_{k+1|k} b_{k+1|k}(\mathbf{x}) \tag{15.57}$$

is Poisson. Draw $n_{k+1|k} \sim p_{k+1|k}^{appear}(\cdot)$ *from*

$$p_{k+1|k}^{appear}(n) = \frac{e^{-\lambda_{k+1|k}} \lambda_{k+1|k}^{n}}{n!} \tag{15.58}$$

and then draw $n_{k+1|k}$ *samples from* $b_{k+1|k}(\mathbf{x})$,

$$\mathbf{x}_{k+1|k}^{1}, ..., \mathbf{x}_{k+1|k}^{n_{k+1|k}} \sim b_{k+1|k}(\cdot). \tag{15.59}$$

15.5.2.1 Spawning Targets

The probability density $B_{k+1|k}(X|\mathbf{x}')$ models those targets that are spawned by an existing target that had state $\mathbf{x}'$ at time step k. The subparticle $X_{k+1|k}^{i,\text{spawn}}$ is drawn from $B_{k+1|k}(X|\mathbf{x}')$ in the same way that $X_{k+1|k}^{i,\text{appear}}$ was drawn from $B_{k+1|k}(X)$.

That is, let $p_{k+1|k}^{\text{spawn}}(n|\mathbf{x}')$ be the cardinality distribution of $B_{k+1|k}(X|\mathbf{x}')$. Let $X_{k|k}^{i} = \{\mathbf{x}'_1, ..., \mathbf{x}'_{n'}\}$, where the index i has been suppressed in the $\mathbf{x}'_1, ..., \mathbf{x}'_{n'}$ for the sake of notational simplicity. Draw samples

$$n_{k+1|k}^{1} \overset{\text{abbr.}}{=} n_{k+1|k}(\mathbf{x}'_1) \sim p_{k+1|k}^{\text{spawn}}(\cdot|\mathbf{x}'_1) \tag{15.60}$$

$$\vdots \tag{15.61}$$

$$n_{k+1|k}^{n'} \overset{\text{abbr.}}{=} n_{k+1|k}(\mathbf{x}'_{n'}) \sim p_{k+1|k}^{\text{spawn}}(\cdot|\mathbf{x}'_{n'}). \tag{15.62}$$

Then for $j = 1, ..., n'$ draw samples from the distributions

$$\hat{B}_{k+1|k}^{j}(\mathbf{x}_1, ..., \mathbf{x}_{n_{k+1|k}^{j}}) \triangleq \frac{1}{n_{k+1|k}^{j}!} \cdot B_{k+1|k}(\{\mathbf{x}_1, ..., \mathbf{x}_{n_{k+1|k}^{j}}\}|\mathbf{x}'_j). \tag{15.63}$$

That is:

$$(\mathbf{x}_{k+1|k}^{1,1}, ..., \mathbf{x}_{k+1|k}^{1,n_{k+1|k}^{1}}) \sim \hat{B}_{k+1|k}^{1}(\cdot, ..., \cdot) \tag{15.64}$$

$$\vdots \tag{15.65}$$

$$(\mathbf{x}_{k+1|k}^{n',1}, ..., \mathbf{x}_{k+1|k}^{n',n_{k+1|k}^{n'}}) \sim \hat{B}_{k+1|k}^{n'}(\cdot, ..., \cdot). \tag{15.66}$$

Then

$$X_{k+1|k}^{i,\text{spawn}}(\mathbf{x}') = \{\mathbf{x}_{k+1|k}^{1}, ..., \mathbf{x}_{k+1|k}^{n_{k+1|k}}, ..., \mathbf{x}_{k+1|k}^{n',1}, ..., \mathbf{x}_{k+1|k}^{n',n_{k+1|k}}\}. \tag{15.67}$$

This process is repeated for each $i = 1, ..., \nu$.

15.6 M-SMC FILTER CORRECTOR

Recall that the multitarget corrector was defined in (14.50) and (14.51). Substituting

$$f_{k+1|k}(X|Z^k) \cong w_{k+1|k}^0 \cdot \delta_\emptyset(X) + \frac{1 - w_{k+1|k}^0}{\nu} \sum_{i=1}^{\nu} \delta_{X_{k+1|k}^i}(X) \tag{15.68}$$

into the Bayes normalization factor, (14.51), yields:

$$f_{k+1}(Z_{k+1}|Z^k) = \int f_{k+1}(Z_{k+1}|X) \cdot f_{k+1|k}(X|Z^{(k)})\delta X \tag{15.69}$$

$$\cong w_{k+1|k}^0 \cdot f_{k+1}(Z_{k+1}|\emptyset) \tag{15.70}$$

$$+\frac{1 - w_{k+1|k}^0}{\nu} \sum_{i=0}^{\nu} f_{k+1}(Z_{k+1}|X_{k+1|k}^i). \tag{15.71}$$

The multitarget Bayes corrector equation, (14.50), then becomes

$$\int \theta(X) \cdot f_{k+1|k+1}(X|Z^{(k+1)})\delta X \tag{15.72}$$

$$= \frac{\int \theta(X) \cdot f_{k+1}(Z_{k+1}|X) \cdot f_{k+1|k}(X|Z^{(k)})\delta X}{f_{k+1}(Z_{k+1}|Z^{(k)})} \tag{15.73}$$

$$\cong \sum_{i=0}^{\nu} w_{k+1|k+1}^i \cdot \theta(X_{k+1|k+1}^i) \tag{15.74}$$

where $X^i_{k+1|k+1} \triangleq X^i_{k+1|k}$ and where

$$w^i_{k+1|k+1} \triangleq \frac{f_{k+1}(Z_{k+1}|X^i_{k+1|k})}{\sum_{e=0}^{\nu} f_{k+1}(Z_{k+1}|X^e_{k+1|k})}. \tag{15.75}$$

Techniques such as resampling and roughening can be applied much as in the single-target case.

15.7 M-SMC FILTER STATE AND ERROR ESTIMATION

The number of targets can be estimated easily using (15.29):

$$f_{k|k}(n|Z^{(k)}) \cong \sum_{i:|X^i_{k|k}|=n} = \sum_{i=\nu_{n-1}+1}^{\nu_n} w^i_{k|k} \tag{15.76}$$

where the first summation is taken over all indices i such that $|X^i_{k|k}| = n$.

As for estimating the target states, multitarget state estimation based on particle systems is difficult. In particular, the MaM and JoM estimators of Sections 14.5.2 and 14.5.3 are not easily implemented in this context. Most reported implementations sidestep the issue by displaying particle clouds and letting the reader sort things out visually.

Here I describe two approaches: "PHD visualization" (Section 15.7.1) and global mean deviation (Section 15.7.2). The issue of track labeling is briefly discussed in Section 15.7.3.

15.7.1 PHD-Based State and Error Estimation

Equation (14.137) introduced the multitarget first-moment approach to multitarget state and error estimation, due to Vihola and to Sidenbladh and Wirkander. This approach is based on concepts to be more fully explored in Chapter 16.

Recall our discussion of (14.137). The approach is based on the probability hypothesis density (PHD) $D_{k|k}(\mathbf{x}|Z^{(k)})$ of $f_{k|k}(X|Z^{(k)})$:

$$D_{k|k}(\mathbf{x}|Z^{(k)}) \triangleq \int f_{k|k}(\{\mathbf{x}\} \cup X|Z^{(k)})\delta X. \tag{15.77}$$

The PHD is not a probability distribution since $N_{k|k} \triangleq \int D_{k|k}(\mathbf{x}|Z^{(k)})d\mathbf{x}$ is the expected number of targets in the scene.

Let $\hat{n}$ be the integer nearest in value to $N_{k|k}$. Then the $\hat{n}$ highest peaks of $D_{k|k}(\mathbf{x}|Z^{(k)})$ correspond to the states of the targets. That is, let $\hat{\mathbf{x}}_1, ..., \hat{\mathbf{x}}_{\hat{n}}$ be such that $D_{k|k}(\hat{\mathbf{x}}_1|Z^{(k)})$,...,$D_{k|k}(\hat{\mathbf{x}}_{\hat{n}}|Z^{(k)})$ are the $\hat{n}$ largest values of $D_{k|k}(\cdot|Z^{(k)})$. Then $\hat{X}_{k|k} = \{\hat{\mathbf{x}}_1, ..., \hat{\mathbf{x}}_{\hat{n}}\}$ is an estimate of the multitarget state.

Vo, Singh, and Doucet [238, p. 1230] have shown that the particle approximation of $D_{k|k}(\mathbf{x}|Z^{(k)})$ easily arises from the multitarget particle approximation of $f_{k|k}(X|Z^{(k)})$:

$$D_{k|k}(\mathbf{x}|Z^{(k)}) \cong \sum_{i=0}^{\nu} w^i_{k|k} \sum_{\mathbf{w}\in X^i_{k|k}} \delta_{\mathbf{w}}(\mathbf{x}). \tag{15.78}$$

This fact is verified in Appendix G.26.

State and error estimates can be extracted from (15.78) by histogramming or by using the EM algorithm [15, 162] to approximate $D_{k|k}(\mathbf{x}|Z^{(k)})$ as a mixture of $\hat{n}$ Gaussians:

$$D_{k|k}(\mathbf{x}|Z^{(k)}) \cong \sum_{i=1}^{\hat{n}} \alpha_i \cdot N_{C_i}(\mathbf{x} - \hat{\mathbf{x}}_i). \tag{15.79}$$

15.7.2 Global Mean Deviation

Another approach is to compute *global mean deviation*, as introduced in Section 14.6.3—see (14.160) and (14.161). There I defined

$$\Delta^H_{k|k} \triangleq \int d^H(\hat{X}_{k|k}, X) \cdot \hat{f}_{k|k}(X)\delta X \tag{15.80}$$

$$\Delta^{W,p}_{k|k} \triangleq \sqrt[p]{\int d^W_p(\hat{X}_{k|k}, X)^p \cdot \hat{f}_{k|k}(X)\delta X} \tag{15.81}$$

where $\Delta^H_{k|k}$ and $\Delta^{W,p}_{k|k}$ are the global mean deviations defined in terms of Hausdorff distance and Wasserstein distance, respectively; and where $\hat{f}_{k|k}(\emptyset) \triangleq 0$ and, if $X \neq \emptyset$,

$$\hat{f}_{k|k}(X) \triangleq \frac{f_{k|k}(X|Z^{(k)})}{1 - f_{k|k}(\emptyset|Z^{(k)})}. \tag{15.82}$$

Consequently, suppose that $f_{k|k}(X|Z^{(k)})$ is approximated by the particle system $X^0_{k|k}, X^1_{k|k}, ..., X^\nu_{k|k}$ with weights $w^0_{k|k}, w^1_{k|k}, ..., w^\nu_{k|k}$. Also assume that

$X^0_{k|k} = \emptyset$ is the only null particle. Then

$$\Delta^H_{k|k} \cong \frac{1}{1 - w^0_{k|k}} \cdot \sum_{i=1}^{\nu} w^i_{k|k} \cdot d^H(\hat{X}_{k|k}, X^i_{k|k}) \tag{15.83}$$

$$\Delta^{W,p}_{k|k} \cong \frac{1}{\sqrt[p]{1 - w^0_{k|k}}} \cdot \sqrt[p]{\sum_{i=1}^{\nu} w^i_{k|k} \cdot d^W_p(\hat{X}_{k|k}, X^i_{k|k})^p} \tag{15.84}$$

15.7.3 Track Labeling for the Multitarget SMC Filter

In Section 14.5.6, I observed that, in most attempts to incorporate track labeling, computationally tractable labeling was possible only because of highly unrealistic assumptions. The only promising approach at the current time, due to Ma, Vo, Singh, and Baddeley [122, 239], is based on attaching labels to individual single-target states in the multitarget particles.

Chapter 16

Multitarget-Moment Approximation

16.1 INTRODUCTION TO THE CHAPTER

In many applications, individual targets are of less importance than the target groupings to which they belong. Such applications include the following:

- *Group target processing*: The goal is to detect, track, and identify tactically significant target formations such as brigades or battalions, fighter aircraft sorties, naval convoys, and so on [247]. In such cases one may require only gross parameters such as group centroid, group average velocity, group size, group expanse, and group shape;
- *Tracking in high target density*: There are far too many targets to detect and track individually. One may have no other choice than to track the entire target formation(s).
- *Detecting and tracking a few targets of interest (ToIs) obscured in a dense multitarget background*: It is often preferable to track a target formation ("cluster tracking") until subsequent target-identity information permits extraction of the ToI(s) from the background.
- *Tracking of closely spaced targets*: In this case targets are too closely spaced (compared to sensor resolution) to permit effective tracking of individual targets. Tracking of the group—and whatever group parameters one can determine—may be the only feasible option.

The conventional approach to such problems would be to attempt to track individual targets and assemble the group from them. Such an approach is most infeasible in the densest parts of formations, where targets of interest are most likely to be but where confusion is greatest. Thus in such applications standard techniques, such as those sketched in Chapter 10, may begin to fail.

In this chapter we turn to two advanced approximations of the multitarget recursive Bayes filter that can potentially address such applications:

- The *probability hypothesis density* (PHD) *filter;*
- The *cardinalized PHD* (CPHD) *filter.*

The strategy employed in these filters is opposite to that used in conventional approaches. They first track only the overall group behavior, and then attempt to detect and track individual targets only as the quantity and quality of data permits. Derivation of them requires the probability-generating functional (p.g.fl.) and multitarget calculus techniques introduced in Chapter 11.

Remark 27 (Innovative Claims for the PHD/CPHD Filters) *As we shall see in Sections 16.3.4.1 and 16.5.1, preliminary research has suggested that the PHD and CPHD filters may be more effective than MHC-type filters in some conventional multitarget detection and tracking problems. Whether such claims hold up is for future research to determine. Here we emphasize that the PHD and CPHD approaches were originally devised to address nontraditional tracking problems such as those just described.*

In the remainder of this section, I review the concepts of first-moment and second-moment filtering in the single-target case (Section 16.1.1). Then, in Sections 16.1.2 and 16.1.3, respectively, I briefly summarize how I intend to generalize first- and second-moment approximation to the multitarget case. Lessons learned may be found in Section 16.1.4 and the organization of the chapter in Section 16.1.5.

16.1.1 Single-Target Moment-Statistic Filters

I begin by revisiting familiar single-target filtering approaches previously described in Section 2.2: the Kalman filter and its special case, the constant-gain Kalman filter.

16.1.1.1 Second-Order Moment Approximation: Kalman Filter

The Kalman filter presumes that signal-to-noise ratio (SNR) is high enough that all time-evolving posteriors $f_{k|k}(\mathbf{x}|Z^k)$ are not too complicated. That is, the

$f_{k|k}(\mathbf{x}|Z^k)$ are essentially unimodal and not too skewed for all $k \geq 0$. In this case one can *(lossily) compress* the posterior into *sufficient statistics* (first and second statistical moments) and propagate these statistics in place of the posterior itself.

In the case of the Kalman filter the relevant statistics are the first-moment vector and second-moment matrix

$$\begin{aligned}\mathbf{x}_{k|k} &= \int \mathbf{x}\, f_{k|k}(\mathbf{x}|Z^k)d\mathbf{x} \\ Q_{k|k} &= \int \mathbf{x}\mathbf{x}^T\, f_{k|k}(\mathbf{x}|Z^k)d\mathbf{x}\end{aligned}$$

where "^{T}" denotes matrix transpose. If SNR is high enough that higher-order moments can be neglected, $\mathbf{x}_{k|k}$ and $Q_{k|k}$ are approximate sufficient statistics in the sense that

$$f_{k|k}(\mathbf{x}|Z^k) \cong f_{k|k}(\mathbf{x}|\mathbf{x}_{k|k}, Q_{k|k}) = N_{P_{k|k}}(\mathbf{x} - \mathbf{x}_{k|k}) \tag{16.1}$$

where $N_{P_{k|k}}(\mathbf{x}-\mathbf{x}_{k|k})$ is a multidimensional Gaussian distribution with covariance matrix $P_{k|k} = Q_{k|k} - \mathbf{x}_{k|k}\mathbf{x}_{k|k}^T$. Given this assumption, we can propagate $\mathbf{x}_{k|k}$ and $P_{k|k}$ instead of the full distribution $f_{k|k}(\mathbf{x}|Z^k)$ using a Kalman filter. That is, we have a diagram of the form

$$\begin{array}{ccccccc}
\cdots \rightarrow & f_{k|k}(\mathbf{x}) & \overset{\text{predictor}}{\rightarrow} & f_{k+1|k}(\mathbf{x}) & \overset{\text{corrector}}{\rightarrow} & f_{k+1|k+1}(\mathbf{x}) & \rightarrow \cdots \\
& \downarrow & & \downarrow & & \downarrow & \\
\cdots \rightarrow & \begin{matrix}\mathbf{x}_{k|k}\\ P_{k|k}\end{matrix} & \overset{\text{KF predictor}}{\rightarrow} & \begin{matrix}\mathbf{x}_{k+1|k}\\ P_{k+1|k}\end{matrix} & \overset{\text{KF corrector}}{\rightarrow} & \begin{matrix}\mathbf{x}_{k+1|k+1}\\ P_{k+1|k+1}\end{matrix} & \rightarrow \cdots
\end{array}$$

Here the top row portrays the prediction and correction steps of the single-target Bayes filter. The downward-pointing arrows indicate the compression of the posteriors into their corresponding first- and second-order moments. The bottom row portrays the prediction and correction steps for the Kalman filter, as described in Section 2.2.

16.1.1.2 First-Moment Approximation: Constant-Gain Kalman Filter

If SNR is even higher, the second-order moment can be neglected as well. In this case the first moment is an approximate sufficient statistic:

$$f_{k|k}(\mathbf{x}|Z^k) \cong f_{k|k}(\mathbf{x}|\hat{\mathbf{x}}_{k|k}) = N_P(\mathbf{x} - \mathbf{x}_{k|k}) \tag{16.2}$$

where now the covariance P is *constant.* We can then propagate $\mathbf{x}_{k|k}$ alone using a constant-gain Kalman filter (CGKF) such as the alpha-beta filter discussed in Section 2.2.6. In this case, the previous diagram reduces to

$$\begin{array}{ccccccccc} \cdots\rightarrow & f_{k|k}(\mathbf{x}) & \stackrel{\text{predictor}}{\rightarrow} & f_{k+1|k}(\mathbf{x}) & \stackrel{\text{corrector}}{\rightarrow} & f_{k+1|k+1}(\mathbf{x}) & \rightarrow\cdots \\ & \downarrow & & \downarrow & & \downarrow & \\ \cdots\rightarrow & \mathbf{x}_{k|k} & \stackrel{\substack{\text{CGKF} \\ \text{predictor}}}{\rightarrow} & \mathbf{x}_{k+1|k} & \rightarrow\stackrel{\substack{\text{CGKF} \\ \text{corrector}}}{\rightarrow} & \mathbf{x}_{k+1|k+1} & \rightarrow\cdots \end{array}$$

The bottom row now portrays the prediction and correction steps for the constant-gain Kalman filter, as described in Section 2.2.6.

16.1.2 First-Order Multitarget-Moment Filtering

Our goal is to extend this reasoning to the multitarget case. The approach will be described in detail in subsequent sections of the chapter. Here we only sketch the basic ideas.

We assume that SNR is high enough that a first-order statistical moment $D_{k|k}$ of the *multitarget system* is an approximate sufficient statistic:

$$f_{k|k}(X|Z^{(k)}) \cong f_{k|k}(X|D_{k|k}). \tag{16.3}$$

We must then "fill in the question marks" in the following diagram,

$$\begin{array}{ccccccccc} \cdots\rightarrow & f_{k|k}(X) & \stackrel{\text{predictor}}{\rightarrow} & f_{k+1|k}(X) & \stackrel{\text{corrector}}{\rightarrow} & f_{k+1|k+1}(X) & \rightarrow\cdots \\ & \downarrow & & \downarrow & & \downarrow & \\ \cdots\rightarrow & D_{k|k}(\mathbf{x}) & \stackrel{\substack{\text{1st-order} \\ \text{predictor?}}}{\rightarrow} & D_{k+1|k}(\mathbf{x}) & \stackrel{\substack{\text{1st-order} \\ \text{corrector?}}}{\rightarrow} & D_{k+1|k+1}(\mathbf{x}) & \rightarrow\cdots \end{array}$$

Here the top row portrays the multitarget Bayes filter and the downward-pointing arrows indicate the compression of the multitarget posteriors into their corresponding *first-order multitarget moments.* The bottom row portrays the (as yet undetermined) time prediction and data-update steps for the (as yet undefined) first-order multitarget-moment filter.

16.1.2.1 The Probability Hypothesis Density (PHD)

What is the first-order multitarget moment $D_{k|k}$ of $f_{k|k}$? It is not a vector, as in the single-target case. Rather, it is a *density function*

$$D_{k|k} \stackrel{\text{abbr.}}{=} D_{k|k}(\mathbf{x}) \stackrel{\text{abbr.}}{=} D_{k|k}(\mathbf{x}|Z^{(k)}) \tag{16.4}$$

defined on single-target states $\mathbf{x} \in \mathfrak{X}_0$. In point process theory $D_{k|k}(\mathbf{x})$ is called the *first-moment density* or *intensity density* [36, p. 130]. The author has, for historical and correct authorial-attribution reasons [136, p. 1154] adopted the name *probability hypothesis density* (PHD).

The PHD is not a probability density. In fact, it is uniquely characterized by the following property. Given any region S of single-target state space $\mathfrak{X}_0$, the integral $\int_S D_{k|k}(\mathbf{x})d\mathbf{x}$ is the expected number of targets in S. In particular, if $S = \mathfrak{X}_0$ is the entire state space then

$$N_{k|k} \triangleq \int D_{k|k}(\mathbf{x})d\mathbf{x} \tag{16.5}$$

is the total expected number of targets in the scene.

Why is the PHD uniquely determined by this property? Let $g(\mathbf{x})$ be any other density that gives the expected number of targets in S when integrated over S. Then since $\int_S g(\mathbf{x})d\mathbf{x} = \int_S D_{k|k}(\mathbf{x})d\mathbf{x}$ for all measurable S, it follows that $g = D_{k|k}$ almost everywhere.

In Section 16.3 we will derive formulas for the predictor and corrector steps of a "PHD filter":

$$\cdots \rightarrow \quad D_{k|k}(\mathbf{x}) \quad \overset{\text{PHD predictor}}{\rightarrow} \quad D_{k+1|k}(\mathbf{x}) \quad \overset{\text{PHD corrector}}{\rightarrow} \quad D_{k+1|k+1}(\mathbf{x}) \quad \rightarrow \cdots$$

This filter is more kindred in spirit to composite-hypothesis correlation (JPDA) approaches (Section 10.6) than to multihypothesis correlation approaches (Section 10.5). This is because it simultaneously associates all measurements with all tracks, rather than attempting to enumerate and rank a list of possible measurement-to-track associations.

Example 88 (Simple Example of a PHD) *In Example 56 of Section 11.3.3 we presented the following simple illustration of a multitarget posterior distribution for two one-dimensional targets located at* $x = a, b$*:*

$$f(\{x, y\}) = N_{\sigma^2}(x-a) \cdot N_{\sigma^2}(y-b) + N_{\sigma^2}(x-b) \cdot N_{\sigma^2}(y-a). \tag{16.6}$$

The corresponding PHD can be shown to be:

$$\begin{aligned} D(x) &= N_{\sigma^2}(x-a) + N_{\sigma^2}(x-b) && (16.7) \\ &= \frac{1}{\sqrt{2\pi}\sigma}\left[\exp\left(-\frac{(x-a)^2}{2\sigma^2}\right) + \exp\left(-\frac{(x-b)^2}{2\sigma^2}\right)\right]. && (16.8) \end{aligned}$$

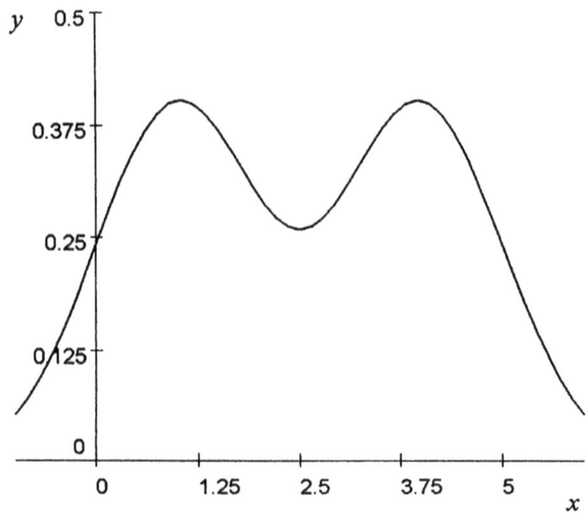

Figure 16.1 The graph of the probability hypothesis density (PHD) of the multitarget probability distribution of (16.6) and Figure 11.1.

The graph of this function is plotted in Figure 16.1 for $\sigma = 1$, $a = 1$, *and* $b = 4$. *The peaks of* $D(x)$ *occur near the target locations* $x = 1$ *and* $x = 4$. *The integral of* $D(x)$ *is the actual number of targets:*

$$N = \int D(x)dx = \int N_{\sigma^2}(x-a)dx + \int N_{\sigma^2}(x-b)dx \quad (16.9)$$

$$= 1 + 1 = 2. \quad (16.10)$$

16.1.2.2 Strengths and Weaknesses of the PHD Filter

As Erdinc, Willett, and Bar-Shalom have noted, the PHD filter "is attracting increasing but cautious attention" [56]. It has been implemented and tested by many researchers, primarily in simulations but to a lesser extent with real data. This research will be summarized in Section 16.5.1. What potential advantages have spurred this interest in the PHD filter?

- It has potentially desirable computational characteristics. At any moment its computational complexity is order $O(mn)$ where n is the number of targets in the scene and m is the number of observations in the current measurement set Z.[1]
- It admits explicit statistical models for missed detections, sensor field of view (FoV), and false alarms (which, as is the case with conventional techniques, are assumed to be Poisson).
- It admits explicit statistical models of the major aspects of multitarget dynamicism: target disappearance, target appearance, and the spawning of new targets by prior targets.
- It can be implemented using both sequential Monte Carlo and Gaussian-mixture approximation techniques.
- It does not require measurement-to-track association.
- It has potentially desirable clutter-rejection properties. At every step it computes an estimate $N_{k|k}$ of target number *directly from the data.*

The last point should be contrasted with conventional multitarget filtering techniques, which either assume target number is known or infer it indirectly. Generally speaking, having a good estimate of target number is half of the battle in multitarget tracking. If one has $1,000$ measurements but we know that roughly 900 of them are false alarms, then the problem of detecting the actual targets has been greatly simplified. As Lin, Bar-Shalom, and Kirubarajan have noted [121, p. 464]:

> ...when multiple targets are closely spaced and subject to birth, death, spawning, merging, and so on, this conventional approach (MHT / assignment) may not give satisfactory results. This is mainly because of the difficulty in deciding what the number of targets is.

The PHD filter also has potential disadvantages:

1 Note, however, that the PHD filter involves multidimensional integrals. Thus it must be implemented using various computational approximations—see Section 16.5.

- Though it provides instantaneous estimates of target number, these estimates tend to be very unstable (i.e., high variance) in the presence of false alarms and, especially, missed detections.
- One can work around this by averaging $N_{k|k}$ for $k = i, ..., i+T$ in a running time window of fixed length T. This tends to produce accurate estimates of target number. However, if the multitarget scene is highly dynamic (e.g., rapid entry and exit of targets) then this "trick" will not be effective.
- Replacement of a full multitarget distribution $f_{k|k}(X)$ by its PHD loses a tremendous amount of information. This loss can be overcome only if SNR is high enough—and individual measurement sets are therefore information-rich enough—to compensate for the shortfall.[2]

16.1.3 Second-Order Multitarget-Moment Filtering

The potential limitations of the PHD filter should be evident to anyone who is aware of the superior performance of Kalman (second-order) compared to alpha-beta (first-order) filters in the single-target case. The potential desirability of second-order extensions of the PHD filter should also be evident.

16.1.3.1 Full Second-Order Multitarget-Moment Filtering

The possibility of second-order multitarget-moment filters was investigated at an early stage [137, pp. 155, 156]. Such a filter would propagate not only a PHD $D_{k|k}(\mathbf{x})$ but also a second-order multitarget moment—for example, a multitarget covariance density $C_{k|k}(\mathbf{x}, \mathbf{x}') \overset{\text{abbr.}}{=} C_{k|k}(\mathbf{x}, \mathbf{x}'|Z^{(k)})$. The resulting filter would look like this:

$$\begin{array}{ccccccc} \cdots \rightarrow & f_{k|k}(X) & \overset{\text{predictor}}{\rightarrow} & f_{k+1|k}(X) & \overset{\text{corrector}}{\rightarrow} & f_{k+1|k+1}(X) & \rightarrow \cdots \\ & \downarrow & & \downarrow & & \downarrow & \\ \cdots \rightarrow & \begin{array}{c} D_{k|k} \\ C_{k|k} \end{array} & \overset{\substack{\text{2nd-order} \\ \text{predictor?}}}{\rightarrow} & \begin{array}{c} D_{k+1|k} \\ C_{k+1|k} \end{array} & \overset{\substack{\text{2nd-order} \\ \text{corrector?}}}{\rightarrow} & \begin{array}{c} D_{k+1|k+1} \\ C_{k+1|k+1} \end{array} & \rightarrow \cdots \end{array}$$

It is possible in principle to construct the predictor and corrector equations for such a filter. However, these are unlikely to be computationally tractable for problems involving more than a small number of targets.

2 Despite this fact, research has demonstrated that the PHD filter performs perhaps surprisingly well.

16.1.3.2 Partial Second-Order Multitarget-Moment Filtering

The CPHD filter steers a middle ground between the information loss of first-order multitarget-moment approximation and the intractability of a full second-order approximation. In addition to propagating the PHD $D_{k|k}(\mathbf{x})$ of $f_{k|k}(X)$, it also propagates the cardinality distribution $p_{k|k}(n)$ of $f_{k|k}(X)$—where, recall, the cardinality distribution was defined in (11.113). That is, it propagates the *entire probability distribution on target number*.

Propagation of $p_{k|k}(n)$ is drastically less computationally challenging than the propagation of a covariance density $C_{k|k}(\mathbf{x}, \mathbf{x}')$. As a result, the CPHD filter sidesteps most of the implementation difficulties of a full second-order multitarget-moment approximation.

As with the PHD filter, we must "fill in the question marks" in the following diagram:

$$
\begin{array}{ccccccc}
\cdots \to & f_{k|k}(X) & \overset{\text{predictor}}{\longrightarrow} & f_{k+1|k}(X) & \overset{\text{corrector}}{\longrightarrow} & f_{k+1|k+1}(X) & \to \cdots \\
 & \downarrow & & \downarrow & & \downarrow & \\
\cdots \to & \begin{array}{c} p_{k|k}(n) \\ D_{k|k}(\mathbf{x}) \end{array} & \overset{\substack{\text{CPHD} \\ \text{predictor?}}}{\longrightarrow} & \begin{array}{c} p_{k+1|k}(n) \\ D_{k+1|k}(\mathbf{x}) \end{array} & \overset{\substack{\text{CPHD} \\ \text{corrector?}}}{\longrightarrow} & \begin{array}{c} p_{k+1|k+1}(n) \\ D_{k+1|k+1}(\mathbf{x}) \end{array} & \to \cdots
\end{array}
$$

In Section 16.7 we will derive formulas for the predictor and corrector steps of the "CPHD filter" indicated in the bottom row.

16.1.3.3 Strengths and Weaknesses of the CPHD Filter

Currently the CPHD filter has been implemented and tested (successfully) in a small number of reduced-complexity simulations (see Section 16.9.1). What are the potential strengths of the CPHD filter that have inspired such efforts?

- It generates low-variance (accurate and stable) instantaneous estimates of target number, directly from the data.
- This is partially due to the fact that it permits MAP estimates $\hat{n}_{k|k} = \arg\sup_n p_{k|k}(n)$ of target number rather than (as with the PHD filter) less accurate EAP estimates $N_{k|k} = \sum_{n=0}^{\infty} n \cdot p_{k|k}(n)$.
- As a consequence, tracking and localization of targets is improved.
- It admits more general false alarm processes than does the PHD filter: i.i.d. cluster processes as defined in (11.121).

- It reduces to the PHD filter if the i.i.d. false alarm process is Poisson and if explicit modeling of target spawning is neglected.
- It can be implemented using both sequential Monte Carlo and Gaussian-mixture approximation techniques.

What are the potential disadvantages of the CPHD filter?

- It is more computationally demanding than the PHD filter: order $O(m^3n)$ complexity rather than $O(mn)$.
- It does not admit explicit models for the spawning of new targets by prior targets.
- The corrector step appears algebraically complex.

16.1.4 Summary of Major Lessons Learned

The following are the other major ideas and concepts to be encountered in this chapter:

- The probability hypothesis density (PHD) is a first-order moment of a multitarget system, in somewhat the same way that an expected value is a first-order moment of a single-target system (Section 16.2.1).
- PHDs can be computed from the multitarget statistical descriptors—multitarget probability densities, belief-mass functions, and p.g.fl.s—using the multitarget calculus (Section 16.2.3):

$$D_\Xi(\mathbf{x}) = \frac{\delta\beta_\Xi}{\delta\mathbf{x}}(\mathfrak{X}_0) = \frac{\delta G_\Xi}{\delta\mathbf{x}}[1] = \int f_\Xi(\{\mathbf{x}\} \cup W)\delta W. \tag{16.11}$$

- A PHD can be interpreted as (1) a target density, (2) an information-theoretic best-fit approximation, and (3) a generalization of a fuzzy membership function (Section 16.2.1).
- The PHD filter permits explicit modeling of missed detections, Poisson false alarm processes, target appearance, target disappearance, and spawning of targets by other targets (Section 16.2).
- The PHD filter can be computationally implemented using sequential Monte Carlo (particle system) approximations (Section 16.5.2).

- The PHD filter can be computationally implemented using Gaussian-mixture approximations (Section 16.5.3).
- When it is applicable, Gaussian-sum approximation appears to be superior because it is much computationally faster, and easily incorporates track labeling and state and error estimation (Section 16.5.3).
- The CPHD filter permits explicit modeling of missed detections, i.i.d. cluster false alarm processes, target appearance, and target disappearance (Section 16.7).
- The CPHD filter can be implemented using SMC approximations (Section 16.9.2).
- The CPHD filter can be implemented using Gaussian-sum approximations (Section 16.9.3).

16.1.5 Organization of the Chapter

The chapter is organized as follows. The PHD is formally introduced in Section 16.2, along with examples and its relationship to the multitarget calculus. The PHD filter is described in detail in Section 16.3. A physical interpretation of the PHD filter, due to Erdinc, Willett, and Bar-Shalom, is discussed in Section 16.4.

Section 16.5 describes two different computational implementation approaches for the PHD filter: sequential Monte Carlo (SMC) approximation, and Gaussian-mixture (GM) approximation. Section 16.7 introduces the CPHD filter. SMC and GM methods for implementing the CPHD filter are discussed in Section 16.9. Section 16.10 briefly summarizes the p.g.fl.-based approach used to derive the PHD and CPHD filters. The chapter concludes, in Section 16.11, with the following challenging research question:

- Can the CPHD filter approach be further modified to produce a PHD-type filter that is *second-order* in target number and has computational complexity $O(mn)$?

Chapter exercises may be found in Section 16.12.

16.2 THE PROBABILITY HYPOTHESIS DENSITY (PHD)

The purpose of this section is to introduce the concept of a probability hypothesis density (PHD), and to examine some of its major properties.

The section is organized as follows. I introduce the concept of a first-order multitarget moment in Section 16.2.1. This includes three interpretations of the PHD: as a target density, as an information-theoretic best-fit approximation, and as a generalized fuzzy membership function. In Section 16.2.3, I show how to use multitarget calculus to construct PHDs from the fundamental multitarget statistical descriptors (multitarget distributions, belief-mass functions, and p.g.fl.s). I provide several examples of PHDs in Section 16.2.4. The section concludes, in Section 16.2.5, with a brief introduction to higher-order multitarget moments.

16.2.1 First-Order Multitarget Moments

The first question that confronts us is,

- *What is the multitarget counterpart of an expected value*?

Let $f_\Xi(X)$ be the multitarget probability distribution of a random finite set Ξ. Then a naïve definition of its expected value would be

$$\Xi^{\text{naïve}} \triangleq \int X \cdot f_\Xi(X)\delta X.$$

However, this integral is mathematically undefined since addition $X + X'$ of finite subsets X, X' cannot be usefully defined—see Section 14.5.1.

Consequently, one must instead resort to a different strategy. Select a transformation $X \mapsto T_X$ that converts finite subsets X into vectors T_X in some vector space. This transformation should preserve basic set-theoretic structure by transforming unions into sums: $T_{X \cup X'} = T_X + T_{X'}$ whenever $X \cap X' = \emptyset$. Given this, one can define an *indirect expected value* as

$$\Xi^{\text{indirect}} \triangleq \mathrm{E}[T_\Xi] = \int T_X \cdot f_\Xi(X)\delta X. \tag{16.12}$$

What should the transformation $X \mapsto T_X$ be? Common practice in the point process literature [36] is to select $T_X = \delta_X$ where $\delta_X(\mathbf{x}) \triangleq 0$ if $X = \emptyset$ and, otherwise,

$$\delta_X(\mathbf{x}) \triangleq \sum_{\mathbf{w} \in X} \delta_{\mathbf{w}}(\mathbf{x}) \tag{16.13}$$

where $\delta_{\mathbf{w}}(\mathbf{x})$ is the Dirac delta density concentrated at $\mathbf{w}$. Given this,

$$D_{\Xi}(\mathbf{x}) \triangleq \mathrm{E}[\delta_{\Xi}(\mathbf{x})] = \int \delta_X(\mathbf{x}) \cdot f_{\Xi}(X)\delta X \qquad (16.14)$$

is a multitarget analog of the concept of expected value. It is a density function on single-target state space $\mathfrak{X}_0$. It is called the *probability hypothesis density* (PHD) (also known as *intensity density, first-moment density*) of Ξ or of $f_{\Xi}(X)$.[3]

16.2.1.1 PHD as a Target Density

The PHD is a density function but not a probability density. To see this, note that the integral of $D_{\Xi}(\mathbf{x})$ in any region S is the expected number of objects of Ξ in that region (see [136] or Appendix G.27):

$$\int_S D_{\Xi}(\mathbf{x})d\mathbf{x} = \mathrm{E}[|S \cap \Xi|]. \qquad (16.15)$$

In (16.15) let $S = E_{\mathbf{x}'}$ be an arbitrarily small region around a state vector $\mathbf{x}'$ with (hyper)volume $\varepsilon = |E_{\mathbf{x}'}|$. Then from elementary calculus, the expected number of targets in $E_{\mathbf{x}'}$ is

$$D_{\Xi}(\mathbf{x}') \cdot \varepsilon \cong \int_{E_{\mathbf{x}'}} D_{\Xi}(\mathbf{x})d\mathbf{x}. \qquad (16.16)$$

So, approximately, $D_{\Xi}(\mathbf{x}')$ is the expected number of targets in $E_{\mathbf{x}'}$, divided by the (hyper)volume of $E_{\mathbf{x}'}$. In other words:

- For any state vector $\mathbf{x}$, the value $D_{\Xi}(\mathbf{x})$ can be interpreted as the *density of targets* (target density) at $\mathbf{x}$.

This interpretation of the PHD is illustrated in Figure 16.2, which portrays a snapshot of an evolving PHD in a two-dimensional simulation.[4]

16.2.1.2 PHD as an Information-Theoretic Best Approximation

It can be shown that the PHD $D_{k|k}(\mathbf{x})$ is the function that is an information-theoretic best fit to the multitarget density $f_{k|k}(X) = f_{k|k}(X|Z^{(k)})$.

3 The fact that the "historical" PHD is a point process first-order multitarget moment was first demonstrated by the author in [70, pp. 168, 169].

4 The implementation of the PHD filter that was used to generate Figure 16.2 is based on a wavelet approximation. It is not a computationally efficient approach, but effectively illustrates the dynamical behavior of PHDs.

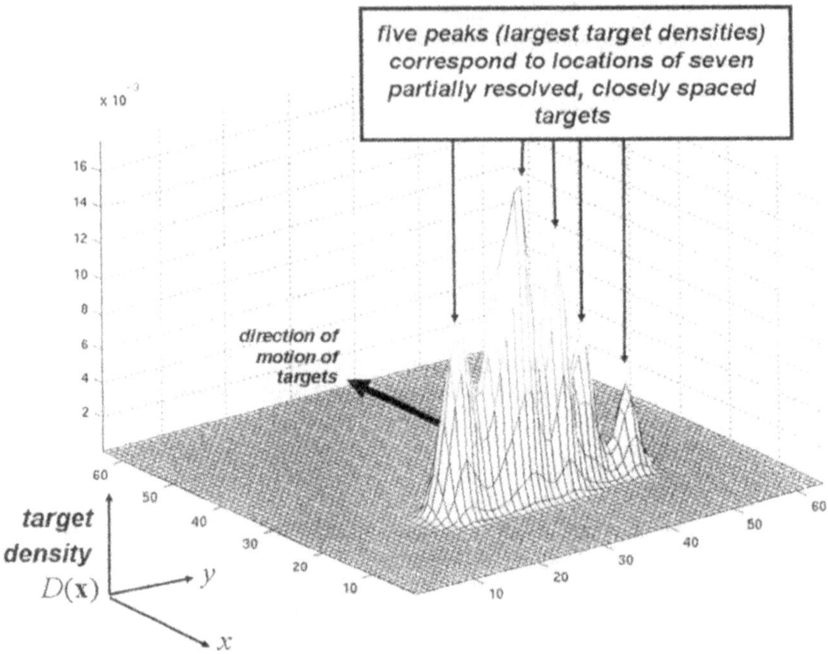

Figure 16.2 A typical probability hypothesis density (PHD) is pictured for targets moving in the x-y plane. Seven targets are moving abreast in the plane in parallel, from the lower right to the upper left. The mass of the PHD is seven, corresponding to the seven targets. Generally speaking, the seven highest peaks of the PHD should correspond to the locations of the targets. Only five targets are apparent. This is because filter resolution is too coarse, given the magnitude of sensor noise power. Seven, rather than five, peaks would be apparent if filter resolution were increased. This would also require increased computational load.

Let $f_I(X)$ denote the multitarget Poisson density with intensity function $I(\mathbf{x}) = \mu \cdot s(\mathbf{x})$ as defined in (11.122) and (11.123):

$$f_I(X) \triangleq e^{-\mu} \prod_{\mathbf{x} \in X} I(\mathbf{x}). \tag{16.17}$$

In (14.164) we introduced the concept of multitarget Kullback-Leibler discrimination. Let

$$K(I) \triangleq \int f_{k|k}(X) \cdot \log \left(\frac{f_{k|k}(X)}{f_I(X)} \right) \delta X \tag{16.18}$$

be the Kullback-Leibler cross-entropy between $f_{k|k}(X)$ and $f_I(X)$. I ask: What choice of I minimizes $K(I)$? The answer is $I(\mathbf{x}) = D_{k|k}(\mathbf{x})$ [136, p. 1166, Theorem 4]. That is:

- $I(\mathbf{x}) = D_{k|k}(\mathbf{x})$ is the intensity of the Poisson process $f_I(X)$ that most closely approximates $f_{k|k}(X)$ in an information-theoretic sense.

16.2.2 PHD as a Continuous Fuzzy Membership Function

Let the state space $\mathfrak{X}_0$ be *finite*. Then the PHD of a random subset Ξ of $\mathfrak{X}_0$ is a *fuzzy membership function*. That is, let $\Xi \overset{\text{abbr.}}{=} \Xi_{k|k}$ be the randomly varying track set and note that

$$f_\Xi(X) \overset{\text{abbr.}}{=} f_{k|k}(X|Z^{(k)}) = \Pr(\Xi = X). \tag{16.19}$$

Then

$$D_{k|k}(x) = \Pr(x \in \Xi) = \mu_\Xi(x) \tag{16.20}$$

where $\mu_\Xi(x)$ is the one point covering function of Ξ as defined in (4.20).

To see this, note that from (16.26):

$$D_\Xi(x) = \int \delta_X(x) \cdot f_\Xi(X) \delta X = \sum_X \delta_X(x) \cdot f_\Xi(X). \tag{16.21}$$

However, $\delta_X(x) \cdot f_\Xi(X)$ is nonzero only if $x \in X$, so

$$\begin{aligned} D_\Xi(x) &= \sum_{X \ni x} \Pr(\Xi = X) = \sum_X \Pr(x \in X, \Xi = X) & (16.22)\\ &= \sum_X \Pr(x \in \Xi, \Xi = X) & (16.23)\\ &= \Pr(x \in \Xi). & (16.24) \end{aligned}$$

This relationship shows that in the continuous case, the PHD represents the zero-probability event $\Pr(\mathbf{x} \in \Xi)$ in the same way that the probability density $f_{\mathbf{X}}(\mathbf{x})$ of a continuous random vector $\mathbf{X}$ represents the zero-probability event "$\Pr(\mathbf{X} = \mathbf{x})$".

Finally, note that

$$\sum_{x} D_{\Xi}(x) = \sum_{x} \mu_{\Xi}(x). \tag{16.25}$$

In the fuzzy logic literature the sum on the right is called the "sigma-count" of the fuzzy membership function μ_{Ξ}. It is interpreted as the "number of elements" in the fuzzy set corresponding to μ_{Ξ}.

Accordingly, in the continuous-space case:

- The PHD can be interpreted as a density-function analog of the concept of a fuzzy membership function.

Remark 28 (Visual Portrayal of Discrete PHD) *Recall that, in Figure 4.1, I illustrated the concept of the one point covering function of a random set. Given the results of this section, Figure 4.1 is also an illustration of a PHD on a discrete space.*

16.2.3 PHDs and Multitarget Calculus

PHDs can be constructed from fundamental multitarget descriptors in three ways: (1) from multitarget densities via set integration, (2) from belief-mass functions via set differentiation, and (3) from p.g.fl.s via functional differentiation.

16.2.3.1 Integration Formula for the PHD

Equation (16.14) gives rise to a specific formula for the PHD in terms of set integrals [136, p. 1166]:

$$D_{\Xi}(\mathbf{x}) = \int f_{\Xi}(\{\mathbf{x}\} \cup W)\delta W = \int_{X \ni \mathbf{x}} f_{\Xi}(X)\delta X. \tag{16.26}$$

To see this, note that

$$
\begin{aligned}
& D_\Xi(\mathbf{x}) && (16.27) \\
= & \int \delta_X(\mathbf{x}) \cdot f_\Xi(X)\delta X && (16.28) \\
= & \sum_{n=1}^{\infty} \frac{1}{n!} \int [\delta_{\mathbf{x}_1}(\mathbf{x}) + ... + \delta_{\mathbf{x}_n}(\mathbf{x})] && (16.29) \\
& f_\Xi(\{\mathbf{x}_1, ..., \mathbf{x}_n\}) d\mathbf{x}_1 \cdots d\mathbf{x}_n && (16.30) \\
= & \sum_{n=1}^{\infty} \frac{n}{n!} \int f_\Xi(\{\mathbf{x}, \mathbf{w}_1, ..., \mathbf{w}_{n-1}\}) d\mathbf{w}_1 \cdots d\mathbf{w}_{n-1} && (16.31) \\
= & \sum_{n=1}^{\infty} \frac{1}{(n-1)!} \int f_\Xi(\{\mathbf{x}, \mathbf{w}_1, ..., \mathbf{w}_{n-1}\}) d\mathbf{w}_1 \cdots d\mathbf{w}_{n-1} && (16.32) \\
= & \sum_{i=0}^{\infty} \frac{1}{i!} \int f_\Xi(\{\mathbf{x}, \mathbf{w}_1, ..., \mathbf{w}_i\}) d\mathbf{w}_1 \cdots d\mathbf{w}_i && (16.33) \\
= & \int f_\Xi(\{\mathbf{x}\} \cup W) \delta W. && (16.34)
\end{aligned}
$$

16.2.3.2 Differentiation Formulas for the PHD

The PHD $D_\Xi(\mathbf{x})$ can be computed from the p.g.fl. $G_\Xi[h]$ or the belief-mass function $\beta_\Xi(S)$ of $f_\Xi(X)$ as follows:

$$
\begin{aligned}
D_\Xi(\mathbf{x}) &= \frac{\delta G_\Xi}{\delta \mathbf{x}}[1] = \frac{\delta \beta_\Xi}{\delta \mathbf{x}}(\mathfrak{X}_0) && (16.35) \\
&= \frac{\delta \log G_\Xi}{\delta \mathbf{x}}[1] = \frac{\delta \log \beta_\Xi}{\delta \mathbf{x}}(\mathfrak{X}_0). && (16.36)
\end{aligned}
$$

The second of the two equations follows from the first, since from (11.223)

$$\frac{\delta \beta_\Xi}{\delta \mathbf{x}}(\mathfrak{X}_0) = \frac{\delta G_\Xi}{\delta \mathbf{x}}[\mathbf{1}_{\mathfrak{X}_0}] = \frac{\delta G_\Xi}{\delta \mathbf{x}}[1]. \tag{16.37}$$

As for the first one, note that for $X = \{\mathbf{x}_1, ..., \mathbf{x}_n\}$ with $|X| = n$,

$$\frac{\delta}{\delta \mathbf{x}} h^X = \frac{h^X}{h(\mathbf{x})} \cdot \delta_X(\mathbf{x}) \tag{16.38}$$

since

$$\frac{\delta}{\delta \mathbf{x}} h^X = \frac{\delta}{\delta \mathbf{x}} [h(\mathbf{x}_1) \cdots h(\mathbf{x}_n)] \tag{16.39}$$

$$= \sum_{i=1}^{n} h(\mathbf{x}_1) \cdots \frac{\delta}{\delta \mathbf{x}} h(\mathbf{x}_i) \cdots h(\mathbf{x}_n) \tag{16.40}$$

$$= \sum_{i=1}^{n} h(\mathbf{x}_1) \cdots \delta_{\mathbf{x}}(\mathbf{x}_i) \cdots h(\mathbf{x}_n) \tag{16.41}$$

$$= h^X \cdot \sum_{\mathbf{w} \in X} \frac{\delta_{\mathbf{x}}(\mathbf{w})}{h(\mathbf{w})} \tag{16.42}$$

$$= \frac{h^X}{h(\mathbf{x})} \cdot \sum_{\mathbf{w} \in X} \delta_{\mathbf{x}}(\mathbf{w}) \tag{16.43}$$

$$= \frac{h^X}{h(\mathbf{x})} \cdot \delta_X(\mathbf{x}). \tag{16.44}$$

Thus from the definition of a p.g.fl., (11.154),

$$\frac{\delta G_\Xi}{\delta \mathbf{x}}[h] = \int \left(\frac{\delta}{\delta \mathbf{x}} h^X \right) \cdot f_\Xi(X) \delta X \tag{16.45}$$

$$= \int \left(\frac{h^X}{h(\mathbf{x})} \cdot \delta_X(\mathbf{x}) \right) \cdot f_\Xi(X) \delta X. \tag{16.46}$$

If we set $h = 1$, then the definition of the PHD, (16.14), yields the desired result:

$$\frac{\delta G_\Xi}{\delta \mathbf{x}}[1] = \int \delta_X \cdot f_\Xi(X) \delta X = D_\Xi(\mathbf{x}). \tag{16.47}$$

The second of the two equations follows from the first. As for the first, note that from the first chain rule for functional derivatives, (11.280),

$$\frac{\delta \log G_\Xi}{\delta \mathbf{x}}[h] = \frac{1}{G_\Xi[h]} \cdot \frac{\delta G_\Xi}{\delta \mathbf{x}}[h]. \tag{16.48}$$

Setting $h = 1$ we get

$$D_\Xi(\mathbf{x}) = \frac{\delta G_\Xi}{\delta \mathbf{x}}[1] = \frac{1}{G_\Xi[1]} \cdot \frac{\delta G_\Xi}{\delta \mathbf{x}}[1] = \frac{\delta \log G_\Xi}{\delta \mathbf{x}}[1]. \tag{16.49}$$

Remark 29 (Log-p.g.fl. Construction of PHD) *Equation (16.49) can greatly simplify the derivation of PHDs since* $\log G_\Xi$ *may be algebraically simpler than* G_Ξ.

16.2.4 Examples of PHDs

In this section, I provide concrete examples of the PHDs of specific multitarget processes: multitarget Poisson, i.i.d. cluster, multitarget multi-Bernoulli, multitarget Dirac, and multitarget uniform.

Example 89 (PHD of a Poisson Process) *The Poisson process*

$$f(X) = e^{-\mu} \prod_{\mathbf{x} \in X} \mu s(\mathbf{x}) \tag{16.50}$$

was introduced in (11.122). Its PHD is

$$D(\mathbf{x}) = \mu \cdot s(\mathbf{x}). \tag{16.51}$$

To see this, note that

$$\begin{aligned} D(\mathbf{x}) &= \int f(\{\mathbf{x}\} \cup X) \delta X && (16.52) \\ &= \sum_{n=0}^{\infty} \frac{1}{n!} \int f(\{\mathbf{x}, \mathbf{x}_1, ..., \mathbf{x}_n\}) d\mathbf{x}_1 \cdots d\mathbf{x}_n && (16.53) \\ &= \sum_{n=0}^{\infty} \frac{1}{n!} \int e^{-\mu} \mu^{n+1} \cdot s(\mathbf{x}) \cdot s(\mathbf{x}_1) \cdots s(\mathbf{x}_n) d\mathbf{x}_1 \cdots d\mathbf{x}_n && (16.54) \\ &= e^{-\mu} \mu \cdot s(\mathbf{x}) \sum_{n=0}^{\infty} \frac{\mu^n}{n!} = e^{-\mu} \mu f(\mathbf{x}) \cdot e^{\mu} && (16.55) \\ &= \mu \cdot s(\mathbf{x}). && (16.56) \end{aligned}$$

As an alternative derivation, recall from (11.172) that the p.g.fl. of this multitarget Poisson process is $G[h] = e^{\mu s[h] - \mu}$. *Its first functional derivative is*

$$\frac{\delta G}{\delta \mathbf{x}}[h] = e^{\mu s[h] - \mu} \cdot \frac{\delta}{\delta \mathbf{x}} \mu s[h] = e^{\mu s[h] - \mu} \cdot \mu \cdot s(\mathbf{x}). \tag{16.57}$$

Thus from (16.35) its PHD is

$$D(\mathbf{x}) = \frac{\delta G}{\delta \mathbf{x}}[1] = \mu \cdot s(\mathbf{x}). \tag{16.58}$$

Example 90 (PHD of an i.i.d. Cluster Process) *The i.i.d. cluster process* $f(X) = |X|! \cdot p(|X|) \cdot \prod_{\mathbf{x} \in X} s(\mathbf{x})$ *was introduced in (11.121). Its PHD is*

$$D(\mathbf{x}) = \bar{n} \cdot s(\mathbf{x}) \tag{16.59}$$

where $\bar{n}$ *is the expected value of the probability distribution* $p(n)$*. To see this, note that*

$$\begin{aligned}
D(\mathbf{x}) &= \int f(\{\mathbf{x}\} \cup X)\delta X & (16.60)\\
&= \sum_{n=0}^{\infty} \frac{1}{n!} \int f(\{\mathbf{x}, \mathbf{x}_1, ..., \mathbf{x}_n\})d\mathbf{x}_1 \cdots d\mathbf{x}_n & (16.61)\\
&= \sum_{n=0}^{\infty} \frac{1}{n!} \int (n+1)! \cdot p(n+1) \cdot s(\mathbf{x}) & (16.62)\\
&\quad \cdot s(\mathbf{x}_1) \cdots s(\mathbf{x}_n) d\mathbf{x}_1 \cdots d\mathbf{x}_n & (16.63)\\
&= s(\mathbf{x}) \cdot \sum_{n=0}^{\infty} (n+1) \cdot p(n+1) = s(\mathbf{x}) \cdot \sum_{n=1}^{\infty} n \cdot p(n) & (16.64)\\
&= \bar{n} \cdot s(\mathbf{x}). & (16.65)
\end{aligned}$$

Example 91 (PHD of a Multi-Bernoulli Process) *The multitarget multi-Bernoulli process was introduced in (11.131)-(11.135). Its PHD is*

$$D(\mathbf{x}) = q_1 f_1(\mathbf{x}) + ... + q_\nu f_\nu(\mathbf{x}). \tag{16.66}$$

To see this, note that from (11.175) we know that the p.g.fl. of the multitarget multi-Bernoulli process is

$$G[h] = (1 - q_1 + q_1 f_1[h]) \cdots (1 - q_\nu + q_\nu f_\nu[h]). \tag{16.67}$$

Then from (16.35) we know that its PHD is given by

$$D(\mathbf{x}) = \frac{\delta G}{\delta \mathbf{x}}[1]. \tag{16.68}$$

However,

$$\frac{\delta G}{\delta \mathbf{x}}[h] = \sum_{i=1}^{\nu} (1 - q_1 + q_1 f_1[h]) \cdots q_i f_i(\mathbf{x}) \cdots (1 - q_\nu + q_\nu f_\nu[h]) \tag{16.69}$$

and so, setting $h = 1$, *we get*

$$D(\mathbf{x}) = \frac{\delta G}{\delta \mathbf{x}}[1] = \sum_{i=1}^{\nu} q_i f_i(\mathbf{x}). \tag{16.70}$$

Example 92 (PHD of a Multitarget Dirac Delta Function) *The multitarget Dirac delta function* $\delta_{X'}(X)$ *was introduced in (11.124). If* $X' = \{\mathbf{x}'_1, ..., \mathbf{x}'_n\}$ *with* $|X'| = n$, *then its PHD is*

$$D_{X'}(\mathbf{x}) = \delta_{\mathbf{x}'_1}(\mathbf{x}) + ... + \delta_{\mathbf{x}'_n}(\mathbf{x}) \triangleq \delta_{X'}(\mathbf{x}). \tag{16.71}$$

To see this, note that

$$D_{X'}(\mathbf{x}_1) \tag{16.72}$$

$$= \int \delta_{X'}(\{\mathbf{x}_1\} \cup X)\delta X \tag{16.73}$$

$$= \frac{1}{(n-1)!} \int \delta_{X'}(\{\mathbf{x}_1, \mathbf{x}_2, ..., \mathbf{x}_n\}) d\mathbf{x}_2 \cdots d\mathbf{x}_n \tag{16.74}$$

$$= \frac{1}{(n-1)!} \int \left(\sum_{\sigma} \delta_{\mathbf{x}'_{\sigma 1}}(\mathbf{x}_1) \cdot \delta_{\mathbf{x}'_{\sigma 2}}(\mathbf{x}_2) \cdots \delta_{\mathbf{x}'_{\sigma n}}(\mathbf{x}_n) \right) \tag{16.75}$$

$$\cdot d\mathbf{x}_2 \cdots d\mathbf{x}_n \tag{16.76}$$

$$= \frac{1}{(n-1)!} \sum_{\sigma} \delta_{\mathbf{x}'_{\sigma 1}}(\mathbf{x}_1) = \sum_{i=1}^{n} \delta_{\mathbf{x}'_i}(\mathbf{x}_1) = \delta_{X'}(\mathbf{x}_1). \tag{16.77}$$

Example 93 (PHD of a Multitarget Uniform Distribution) *The multitarget uniform distribution* $u_{\mathring{n},D}(X)$ *was introduced in (11.126). Its PHD is*

$$D_{\mathring{n},D}(\mathbf{x}) = \frac{\mathring{n}}{2} \cdot \frac{\mathbf{1}_D(\mathbf{x})}{|D|}. \tag{16.78}$$

For,

$$D_{\mathring{n},D}(\mathbf{x}) = \int u_{\mathring{n},D}(\{\mathbf{x}\} \cup X)\delta X \tag{16.79}$$

and so

$$= \sum_{n=0}^{\infty} \frac{1}{n!} \int u_{\mathring{n},D}(\{\mathbf{x}, \mathbf{x}_1, ..., \mathbf{x}_n\}) d\mathbf{x}_1 \cdots d\mathbf{x}_n \tag{16.80}$$

$$= \sum_{n=0}^{\mathring{n}-1} \frac{1}{n!} \int_{D^n} \frac{(n+1)! \cdot \mathbf{1}_D(\mathbf{x})}{|D|^{n+1} \cdot (\mathring{n}+1)} d\mathbf{x}_1 \cdots d\mathbf{x}_n \tag{16.81}$$

and so

$$= \frac{\mathbf{1}_D(\mathbf{x})}{|D|} \cdot \sum_{n=0}^{\mathring{n}-1} \frac{n+1}{\mathring{n}+1} \tag{16.82}$$

$$= \frac{\mathbf{1}_D(\mathbf{x})}{|D|} \cdot \frac{1}{\mathring{n}+1} \cdot \sum_{j=1}^{\mathring{n}} j \tag{16.83}$$

and so

$$= \frac{\mathbf{1}_D(\mathbf{x})}{|D|} \cdot \frac{1}{\mathring{n}+1} \cdot \frac{\mathring{n} \cdot (\mathring{n}+1)}{2} \tag{16.84}$$

$$= \frac{\mathbf{1}_D(\mathbf{x})}{|D|} \cdot \frac{\mathring{n}}{2}. \tag{16.85}$$

16.2.5 Higher-Order Multitarget Moments

In this section I briefly point out that it is possible to define multitarget moments of all orders. The *multitarget factorial moment density* of a multitarget probability distribution $f_\Xi(X)$ is defined in analogy with (16.26) [136, p. 1162]:

$$D_\Xi(X) \triangleq \int_{W \supseteq X} f_\Xi(W) \delta W = \int f_\Xi(X \cup W) \delta W. \tag{16.86}$$

Notice that the set integral is well defined in the sense that $f_\Xi(X \cup W)\delta W$ always has the same units of measurement as X.

The moment density $D_{k|k}(X|Z^{(k)})$ can also be defined in terms of set and functional derivatives [136, p.1162,(60)]:

$$D_\Xi(X) = \frac{\delta G_\Xi}{\delta X}[1] = \frac{\delta \beta_\Xi}{\delta X}(\mathfrak{X}_0). \tag{16.87}$$

The multitarget probability distribution $f_\Xi(X)$ can be recovered from its multitarget-moment density $D_\Xi(X)$ via the following inversion formula [136, p. 1164, Theorem 3]:

$$f_\Xi(X) = \int (-1)^{|W|} \cdot D_\Xi(X \cup W)\delta W. \tag{16.88}$$

16.3 THE PHD FILTER

I now turn to the specification of the PHD filter described in Section 16.1.2. I describe the basic steps of the PHD filter, each in turn: initialization (Section 16.3.1); prediction (Section 16.3.2); correction (Section 16.3.3); and state and error estimation (Section 16.3.4). Then I address the issues of track labeling with the PHD (Section 16.3.4.1) and target identification with the PHD (Section 16.3.5).

16.3.1 PHD Filter Initialization

Initialization of the PHD filter consists of choosing an a priori PHD:

$$D_{0|0}(\mathbf{x}) = D_{0|0}(\mathbf{x}|Z^{(0)}) = n_0 \cdot s_0(\mathbf{x}) \tag{16.89}$$

where $s_0(\mathbf{x})$ is a probability density whose peaks correspond to a priori target positions, and where n_0 is an initial estimate of the expected number of targets.

For example, one could choose $D_{0|0}$ to be a sum of Gaussians

$$D_{0|0}(\mathbf{x}) = N_{P^1_{0|0}}(\mathbf{x} - \mathbf{x}^1_{0|0}) + ... + N_{P^{n_0}_{0|0}}(\mathbf{x} - \mathbf{x}^{n_0}_{0|0}) \tag{16.90}$$

in which case $\int D_{0|0}(\mathbf{x})d\mathbf{x} = n_0$.

If very little is known about the initial target positions, one can choose a "uniform PHD." That is, $D_{0|0}(\mathbf{x}) = n_0 \cdot s_0(\mathbf{x})$ where $s_0(\mathbf{x})$ is uniform in some region D and where n_0 is a guess about how many targets might be present.

16.3.2 PHD Filter Predictor

I describe the multitarget motion assumptions underlying the PHD filter predictor, and then give the PHD predictor equations.

16.3.2.1 PHD Filter Predictor Assumptions

The PHD filter presumes the "standard" multitarget motion model of Section 13.2. More precisely, target motions are statistically independent; targets can disappear from the scene; new targets can be spawned by existing targets; and new targets can appear in the scene independently of existing targets. These possibilities are described as follows.

- *Motion of individual targets:* $f_{k+1|k}(\mathbf{x}|\mathbf{x}')$ is the single-target Markov transition density.
- *Disappearance of existing targets:* $p_S(\mathbf{x}') \overset{\text{abbr.}}{=} p_{S,k+1|k}(\mathbf{x}')$ is the probability that a target with state $\mathbf{x}'$ at time step k will survive in time step $k+1$.
- *Spawning of new targets by existing targets:* $b_{k+1|k}(X|\mathbf{x}')$ is the likelihood that a group of new targets with state set X will be spawned at time step $k+1$ by a single target that had state $\mathbf{x}'$ at time step k; and its PHD is

$$b_{k+1|k}(\mathbf{x}|\mathbf{x}') \triangleq \int b_{k+1|k}(\{\mathbf{x}\} \cup W|\mathbf{x}')\delta W. \tag{16.91}$$

- *Appearance of completely new targets:* $b_{k+1|k}(X)$ is the likelihood that new targets with state set X will enter the scene at time step $k+1$, and its PHD is

$$b_{k+1|k}(\mathbf{x}) \triangleq \int b_{k+1|k}(\{\mathbf{x}\} \cup W)\delta W. \tag{16.92}$$

16.3.2.2 PHD Filter Predictor Equations

From time step k we have in hand the PHD

$$D_{k|k}(\mathbf{x}) \overset{\text{abbr.}}{=} D_{kk}(\mathbf{x}|Z^{(k)}). \tag{16.93}$$

We are to derive a formula for the predicted PHD

$$D_{k+1|k}(\mathbf{x}) \overset{\text{abbr.}}{=} D_{k+1|k}(\mathbf{x}|Z^{(k+1)}). \tag{16.94}$$

This can be shown to be [136, (75)]:

$$D_{k+1|k}(\mathbf{x}) = \overbrace{b_{k+1|k}(\mathbf{x})}^{\text{birth targets}} + \int F_{k+1|k}(\mathbf{x}|\mathbf{x}') \cdot D_{k|k}(\mathbf{x}')d\mathbf{x}' \tag{16.95}$$

where the PHD "pseudo-Markov transition density" is

$$F_{k+1|k}(\mathbf{x}|\mathbf{x}') \triangleq \overbrace{p_S(\mathbf{x}') \cdot f_{k+1|k}(\mathbf{x}|\mathbf{x}')}^{\text{persisting targets}} + \overbrace{b_{k+1|k}(\mathbf{x}|\mathbf{x}')}^{\text{spawned targets}}. \tag{16.96}$$

The predicted expected number of targets is therefore

$$N_{k+1|k} = \int D_{k+1|k}(\mathbf{x}|Z^{(k)})d\mathbf{x} \tag{16.97}$$

$$= N^{\text{birth}}_{k+1|k} + N^{\text{persist}}_{k+1|k} + N^{\text{spawn}}_{k+1|k} \tag{16.98}$$

where

$$N^{\text{birth}}_{k+1|k} \triangleq \int b_{k+1|k}(\mathbf{x})d\mathbf{x} \tag{16.99}$$

$$N^{\text{persist}}_{k+1|k} \triangleq \int p_S(\mathbf{x}) \cdot D_{k|k}(\mathbf{x})d\mathbf{x} \tag{16.100}$$

$$N^{\text{spawn}}_{k+1|k} \triangleq \int b_{k+1|k}(\mathbf{x}|\mathbf{x}') \cdot D_{k|k}(\mathbf{x}')d\mathbf{x}d\mathbf{x}' \tag{16.101}$$

are the predicted expected numbers of, respectively, newly appearing targets, persisting targets, and spawned targets.

Consider the following special cases. If there are no missed detections ($p_D = 1$) or false alarms ($\lambda = 0$) then (16.95) and (16.96) reduce to the general form of the single-target Bayes filter predictor, (2.81):

$$D_{k+1|k}(\mathbf{x}) = \int f_{k+1|k}(\mathbf{x}|\mathbf{x}') \cdot D_{k|k}(\mathbf{x}')d\mathbf{x}'. \tag{16.102}$$

With birth but without spawning, (16.95) and (16.96) reduce to

$$D_{k+1|k}(\mathbf{x}) = b_{k+1|k}(\mathbf{x}) + \int f_{k+1|k}(\mathbf{x}|\mathbf{x}') \cdot D_{k|k}(\mathbf{x}')d\mathbf{x}'. \tag{16.103}$$

16.3.3 PHD Filter Corrector

I describe the multitarget measurement assumptions underlying the PHD filter corrector, and then give the PHD corrector equations for both single and multiple sensors.

16.3.3.1 PHD Filter Single-Sensor Corrector Assumptions

The PHD filter presumes the "standard" multitarget measurement model of Section 12.3. That is: no target generates more than one measurement and each measurement is generated by no more than a single target, all measurements are conditionally independent of target state, missed detections, and a multiobject Poisson false alarm process. This is more precisely described as follows.

- *Single-target measurement generation*: $L_{\mathbf{z}}(\mathbf{x}) \overset{\text{abbr.}}{=} f_{k+1}(\mathbf{z}|\mathbf{x}, \mathbf{\mathring{x}})$ is the sensor likelihood function.
- *Probability of detection (sensor field of view):* $p_D(\mathbf{x}) \overset{\text{abbr.}}{=} p_{D,k+1}(\mathbf{x}, \mathbf{\mathring{x}})$ is the probability that an observation will be collected at time step $k+1$ from a target with state $\mathbf{x}$, if the sensor has state $\mathbf{\mathring{x}}$ at that time step.
- *Poisson false alarms:* at time step $k+1$ the sensor collects an average number $\lambda \overset{\text{abbr.}}{=} \lambda_{k+1}(\mathbf{\mathring{x}})$ of Poisson-distributed false alarms, the spatial distribution of which is governed by the probability density $c(\mathbf{z}) \overset{\text{abbr.}}{=} c_{k+1}(\mathbf{z}|\mathbf{\mathring{x}})$.

An additional simplifying assumption is required if we are to derive closed-form formulas for the corrector step:

- *Poisson multitarget prior:* the predicted multitarget distribution is approximately Poisson with average number μ of targets and target spatial distribution $f_{k+1|k}(\mathbf{x}|Z^{(k)})$:

$$f_{k+1|k}(X|Z^{(k)}) \cong e^{-\mu} \cdot \prod_{\mathbf{x} \in X} \mu \cdot f_{k+1|k}(\mathbf{x}|Z^{(k)}). \tag{16.104}$$

16.3.3.2 PHD Filter Single-Sensor Corrector Equations

From the predictor step we have in hand the predicted PHD

$$D_{k+1|k}(\mathbf{x}) \overset{\text{abbr.}}{=} D_{k+1|k}(\mathbf{x}|Z^{(k)}). \tag{16.105}$$

At time step $k+1$ we collect a new observation set $Z_{k+1} = \{\mathbf{z}_1, ..., \mathbf{z}_m\}$ with m elements. We require a formula for the data-updated PHD

$$D_{k+1|k+1}(\mathbf{x}) \stackrel{\text{abbr.}}{=} D_{k+1|k+1}(\mathbf{x}|Z^{(k+1)}). \tag{16.106}$$

Abbreviate

$$D_{k+1|k}[h] \stackrel{\text{abbr.}}{=} \int h(\mathbf{x}) \cdot D_{k+1|k}(\mathbf{x})d\mathbf{x}. \tag{16.107}$$

Then the PHD corrector step is [136, (87-88)]:

$$D_{k+1|k+1}(\mathbf{x}) \cong L_{Z_{k+1}}(\mathbf{x}) \cdot D_{k+1|k}(\mathbf{x}) \tag{16.108}$$

where for any measurement set Z the "PHD pseudolikelihood" $L_Z(\mathbf{x}) \stackrel{\text{abbr.}}{=} L_Z(\mathbf{x}|Z^{(k)})$ is[5]

$$L_Z(\mathbf{x}) \triangleq 1 - p_D(\mathbf{x}) + p_D(\mathbf{x}) \sum_{\mathbf{z} \in Z} \frac{L_{\mathbf{z}}(\mathbf{x})}{\lambda c(\mathbf{z}) + D_{k+1|k}[p_D L_{\mathbf{z}}]}. \tag{16.109}$$

The data-updated expected number of targets is

$$\begin{aligned} N_{k+1|k+1} &= \int D_{k+1|k+1}(\mathbf{x})d\mathbf{x} && (16.110) \\ &= N_{k+1|k} - D_{k+1|k}[p_D] && (16.111) \\ &\quad + \sum_{\mathbf{z} \in Z_{k+1}} \frac{D_{k+1|k}[p_D L_{\mathbf{z}}]}{\lambda c(\mathbf{z}) + D_{k+1|k}[p_D L_{\mathbf{z}}]}. && (16.112) \end{aligned}$$

In particular, if p_D is constant then this reduces to

$$\begin{aligned} N_{k+1|k+1} &= (1 - p_D) \cdot N_{k+1|k} && (16.113) \\ &\quad + p_D \sum_{\mathbf{z} \in Z_{k+1}} \frac{D_{k+1|k}[L_{\mathbf{z}}]}{\lambda c(\mathbf{z}) + p_D D_{k+1|k}[L_{\mathbf{z}}]}. && (16.114) \end{aligned}$$

5 *Caution*: In some early PHD filter papers, the term $1 - p_D$ was erroneously written as

$$\frac{1 - p_D}{1 - (1 - p_D) \cdot N_{k+1|k}}.$$

16.3.3.3 Computational Complexity of the PHD Filter

Equation (16.109) is linear in the number m of measurements and does not explicitly depend on the number n of targets at all. This has caused some authors to argue that the PHD filter has computational order of complexity $O(m)$. This is true only for applications in which tracking individual targets is not of primary importance. Otherwise, in practice a certain minimum number of operations will be required to achieve sufficient resolution of any given target. Thus in general the complexity is of order $O(mn)$.

16.3.3.4 Interpretation of the PHD Corrector

Equations (16.108) and (16.109) may appear mysterious, but they are intuitively reasonable.

First we examine *behavior with respect to false alarms*. Suppose that $\mathbf{z}$ is one of the collected observations in Z_{k+1}. It could have originated with a target or as a false alarm. Suppose that probability of detection p_D is constant, that the average number λ of false alarms is large, and that the false alarm spatial distribution $c(\mathbf{z})$ is uniform in some region D of $\mathfrak{Z}_0$:

$$c(\mathbf{z}) = \frac{\mathbf{1}_D(\mathbf{z})}{|D|}. \tag{16.115}$$

If the measurement $\mathbf{z}_0 \in Z_{k+1}$ is not in D, then it cannot be a false alarm—it must be an actual target detection. Thus

$$\frac{D_{k+1|k}[p_D L_{\mathbf{z}_0}]}{\lambda c(\mathbf{z}_0) + D_{k+1|k}[p_D L_{\mathbf{z}_0}]} = \frac{D_{k+1|k}[p_D L_{\mathbf{z}_0}]}{D_{k+1|k}[p_D L_{\mathbf{z}_0}]} = 1 \tag{16.116}$$

and so $\mathbf{z}_0$ contributes one target to the total target number.

On the other hand, suppose that $\mathbf{z}_0$ is in D. Then this term becomes

$$\frac{D_{k+1|k}[p_D L_{\mathbf{z}_0}]}{\lambda c(\mathbf{z}_0) + D_{k+1|k}[p_D L_{\mathbf{z}}]} = \frac{D_{k+1|k}[p_D \cdot L_{\mathbf{z}_0}]}{\frac{\lambda}{|D|} + D_{k+1|k}[p_D L_{\mathbf{z}_0}]} \tag{16.117}$$

$$\cong \frac{1}{\lambda/|D|} \cdot D_{k+1|k}[p_D L_{\mathbf{z}_0}] \cong 0 \tag{16.118}$$

where the approximations are valid if λ is so large that it dominates the value of the denominator. Thus a clear false alarm contributes, as it should, almost nothing to the total posterior target count.

Next we examine *behavior with respect to probability of detection.* Assume that

$$p_D(\mathbf{x}) = \mathbf{1}_E(\mathbf{x}) \tag{16.119}$$

for some region E of $\mathfrak{Z}_0$. If $\mathbf{x}_0$ is not in E then it could not have been observed: $p_D(\mathbf{x}_0) = 0$. No additional information is available about targets with state $\mathbf{x}_0$ and so $D_{k+1|k+1}(\mathbf{x}_0)$ reduces, as it should, to the predicted PHD value $D_{k+1|k}(\mathbf{x}_0)$:

$$D_{k+1|k+1}(\mathbf{x}) \cong (1 - 0 + 0) \cdot D_{k+1|k}(\mathbf{x}) = D_{k+1|k}(\mathbf{x}). \tag{16.120}$$

On the other hand, suppose that $\mathbf{x}_0$ is in E and thus was unequivocally detected. In this case

$$D_{k+1|k+1}(\mathbf{x}_0) = \left(\sum_{\mathbf{z} \in Z_{k+1}} \frac{L_\mathbf{z}(\mathbf{x}_0)}{\lambda c(\mathbf{z}) + D_{k+1|k}[L_\mathbf{z}]} \right) \cdot D_{k+1|k}(\mathbf{x}_0). \tag{16.121}$$

Then the likelihood that $\mathbf{x}_0$ is the state of an actual target depends on the magnitude of $L_\mathbf{z}(\mathbf{x}_0)$ for some $\mathbf{z}$ in Z_{k+1}.

Finally, we examine *behavior with respect to prior information.* Suppose that sensor resolution is good, so that $L_\mathbf{z}(\mathbf{x})$ is very "peaky" for that $\mathbf{x}$ corresponding to $\mathbf{z}$. Let $\mathbf{z}_0 \in Z_{k+1}$ and suppose $\mathbf{z}_0$ is consistent with prior information. Then $D_{k+1|k}[p_D L_{\mathbf{z}_0}]$ will tend to dominate $\lambda c(\mathbf{z}_0) + D_{k+1|k}[p_D L_{\mathbf{z}_0}]$, and so the term

$$\frac{D_{k+1|k}[p_D L_{\mathbf{z}_0}]}{\lambda c(\mathbf{z}_0) + D_{k+1|k}[p_D L_{\mathbf{z}_0}]} \cong 1 \tag{16.122}$$

tends to contribute one target to the total target count.

Conversely, assume that $\mathbf{z}_0$ is very inconsistent with prior information. Then $D_{k+1|k}[p_D L_{\mathbf{z}_0}]$ will be small and its corresponding term in the summation will tend to be ignored.

Remark 30 (PHD "Self-Gating" Property) *This fact highlights a "self-gating" feature of the PHD corrector equation. The corrector tends to discount observations that are not associated with a prior track.*

The behavior of the PHD corrector is thus reasonable. However, (16.121) shows that, when probability of detection is constant and unity, any memory it has of previous information resides in $D_{k+1|k}[p_D L_\mathbf{z}]$. Whether or not $\mathbf{z}_0$ is (implicitly) designated as a false alarm is in part determined by how large $\lambda c(\mathbf{z}_0)$ is when compared to $D_{k+1|k}[L_{\mathbf{z}_0}]$. In this case the effect of past measurements is muted. Stated differently:

- The PHD filter has a "poor memory" and is thus more responsive to new measurements than to predicted tracks.

The sometimes peculiar behavior attributable to the term $1 - p_D(\mathbf{x})$ in (16.109) will be revisited in Section 16.6.

16.3.3.5 PHD Filter Corrector: Multiple Sensors

Suppose that at time step $k+1$ two measurement sets $\overset{1}{Z}_{k+1}$, $\overset{2}{Z}_{k+1}$ are collected from two different sensors. The rigorous formula for the PHD corrector step appears to be too complicated to be of practical use [136, p. 1169].

The most common heuristic approach is to apply the PHD corrector step twice in succession, once for $\overset{1}{Z}_{k+1}$ and once for $\overset{2}{Z}_{k+1}$. That is, apply the PHD corrector equation for the first sensor,

$$D^1_{k+1|k+1}(\mathbf{x}) \cong \overset{1}{L}_{\overset{1}{Z}_{k+1}}(\mathbf{x}) \cdot D_{k+1|k}(\mathbf{x}) \tag{16.123}$$

where

$$\overset{1}{L}_{\overset{1}{Z}}(\mathbf{x}) \triangleq 1 - \overset{1}{p}_D(\mathbf{x}) + \sum_{\overset{1}{\mathbf{z}} \in \overset{1}{Z}} \frac{\overset{1}{p}_D(\mathbf{x}) \cdot \overset{1}{L}_{\overset{1}{\mathbf{z}}}(\mathbf{x})}{\overset{1}{\lambda}\overset{1}{c}(\overset{1}{\mathbf{z}}) + D_{k+1|k}[\overset{1}{p}_D L_{\overset{1}{\mathbf{z}}}]} \tag{16.124}$$

is the pseudolikelihood for the first sensor. Then apply the corrector again for the second sensor,

$$D_{k+1|k+1}(\mathbf{x}) \cong \overset{2}{L}_{\overset{2}{Z}_{k+1}}(\mathbf{x}) \cdot D^1_{k+1|k}(\mathbf{x}) \tag{16.125}$$

where

$$\overset{2}{L}_{\overset{2}{Z}}(\mathbf{x}) \triangleq 1 - \overset{2}{p}_D(\mathbf{x}) + \sum_{\overset{2}{\mathbf{z}} \in \overset{2}{Z}} \frac{\overset{2}{p}_D(\mathbf{x}) \cdot \overset{2}{L}_{\overset{2}{\mathbf{z}}}(\mathbf{x})}{\overset{2}{\lambda}\overset{2}{c}(\overset{2}{\mathbf{z}}) + D_{k+1|k}[\overset{2}{p}_D L_{\overset{2}{\mathbf{z}}}]} \tag{16.126}$$

is the pseudolikelihood for the second sensor.[6]

6 Note that this approach will produce different results than the following heuristic approach, which is independent of the order of sensor processing:

$$D_{k+1|k+1}(\mathbf{x}) \cong \overset{1}{L}_{\overset{1}{Z}_{k+1}}(\mathbf{x}) \cdot \overset{2}{L}_{\overset{2}{Z}_{k+1}}(\mathbf{x}) \cdot D_{k+1|k}(\mathbf{x}).$$

The heuristic approach is not entirely satisfactory since changing the order in which the sensor outputs are processed will produce different numerical results. In practice, however, simulations seem to show that sensor order does not actually result in observable differences in the behavior of the PHD filter. This may be because the PHD approximation itself loses so much information that any information loss due to heuristic multisensor fusion is essentially irrelevant.

A more rigorous approximate "pseudosensor" approach for the multiple-sensor PHD corrector has been described in [139, pp. 271-276].

16.3.4 PHD Filter State and Error Estimation

In principle, extracting multitarget state-estimates from a PHD is conceptually simple. The PHD provides an instantaneous estimate $N_{k|k}$ of the number of targets. After rounding $N_{k|k}$ to the nearest integer n, one looks for the n largest local suprema $D_1, ..., D_n$ of the PHD and declares the corresponding $\mathbf{x}_1, ..., \mathbf{x}_n$ such that $D_i = D_{k|k}(\mathbf{x}_i|Z^{k)})$ to be the state-estimates of the targets.

One difficulty has already been pointed out: the instantaneous target number estimates $N_{k|k}$ are unstable. Consequently, a running average

$$\bar{N}_{k|k} = \frac{1}{T}\sum_{j=0}^{T-1} N_{k+j|k+j} \tag{16.127}$$

must be computed over a moving time window of some length T. The quantity $\bar{N}_{k|k}$ turns out to be a reasonably accurate and stable estimator of target number as long as target births and deaths do not occur too quickly in succession.

Given this disclaimer, state estimation is easy or difficult depending on the technique that has been employed to implement the PHD filter. It is very easy if one employs the closed-form Gaussian-sum approximation approach of Vo and Ma [235], to be explained in Section 16.5.3.

It is considerably more difficult if sequential Monte Carlo (also known as particle system) approximation is adopted. In this case, one common approach is to use the expectation-maximization (EM) algorithm [15, 162] to find the best approximation of the PHD using a weighted sum of n Gaussian distributions:

$$D_{k|k}(\mathbf{x}) \cong \sum_{i=1}^{e} w_i \cdot N_{P_i}(\mathbf{x} - \mathbf{x}_i). \tag{16.128}$$

The centroids of these Gaussians are then chosen as the target state estimates, and their covariances as measures of track uncertainty. Clustering techniques have also been applied.

16.3.4.1 Track Labeling and the PHD Filter

In Section 14.5.6, I noted that in multitarget filtering, the issue of track labeling is important. Multitarget state estimation will produce a time sequence of state sets $X_{0|0}$, $X_{1|1}$,..., $X_{k|k}$,... Since the elements of sets have no inherent order, which state-estimates in $X_{k+1|k+1}$ correspond to which state-estimates in $X_{k|k}$? That is, how do we "connect the dots"?

I also noted that the easy and optimal answer—including track label as an additional state variable—is computationally infeasible in general. It increases the computational complexity of the multitarget filter by a factor of $n!$ where n is the current number of targets.

Similar problems arise with the PHD filter. Once again, the easy answer is computationally troublesome. If n targets are in the scene, then including track label as an additional state variable has the effect of increasing the computational load of the PHD filter by a factor of n.

Consequently, some authors have proposed alterations of the PHD filter to include track labeling capability. In this section, I summarize three approaches. The first two employ hybridizations of the SMC-PHD and MHC filter methods. The third is based on attaching track labels to particles in an SMC-PHD filter. A fourth approach, due to Vo and Ma, integrates track labeling with the PHD filter itself. It will be described in detail in Section 16.5.3.

16.3.4.2 MHC Filter as a Declutterer for a PHD Filter

The track labeling approach of Lin, Bar-Shalom, and Kirubarajan [121, 120] is based on discretization of an SMC-PHD filter. The authors note that a major difficulty in maintaining labeled tracks with an SMC-PHD filter can be attributed to two factors:

- Peak extraction in an SMC implementation is difficult in and of itself.
- In the presence of false alarms and missed detections, the evolving PHD has many randomly varying "oscillation peaks."

Lin et al. address this problem by discretizing the (single-target) state space into cells and then evaluating the PHD in the cells. This procedure tends to both smooth out the oscillatory peaks and to make peak extraction easier. They call the resulting SMC-PHD filter the PHD-RC ("PHD in a resolution cell") filter.

The authors incorporate track labeling capability into this filter by using "peak-to-track association." That is, an MHC filter is used to maintain and propagate labeled tracks. At each recursive step, the PHD-RC technique is used to extract number of targets and state-estimates (peaks) from the evolving PHD. These peaks are used as the "data" in an MHC data-to-track association approach. The PHD peaks are thereby associated with labeled tracks.[7] The authors call the resulting hybrid PHD-MHC filter the DA-PHD-RC filter.

The authors compared three filters—the PHD-RC filter, the hybrid PHD-MHC filter, and the MHC filter alone—in one-dimensional simulations. In these simulations, a position-measuring sensor observed three targets moving in a moderate false alarm environment (an average of eight per frame). Two of the targets were present throughout the scenario, while the third appeared and disappeared during it.

The simulations verified that the PHD-RC filter provides better estimates of both target number and target state. Furthermore, the hybrid PHD-MHC filter provided much better performance than the PHD-RC filter with less computation time.

Finally, the authors compared the hybrid filter with the MHC filter. The hybrid filter significantly outperformed the MHC filter in the sense of creating far fewer false tracks. The authors attribute this difference to the fact that the hybrid filter does a better job of estimating the actual number of targets.

16.3.4.3 PHD Filter as a Declutterer for a MHC Filter

Panta, Vo, Singh, and Doucet [182, 184] employ a different approach for track labeling: using an SMC-PHD filter as a decluttering preprocessor for a track-oriented MHC filter. They propose two versions of their approach.

In the first scheme, measurements are used as the inputs of an SMC-PHD filter. The state-estimate outputs of this filter are treated as decluttered pseudomeasurements and fed into the MHC filter, which performs peak-to-track association. Because of the decluttering, the MHC filter not only propagates track labels but also requires far smaller hypothesis tables. The outputs of the MHC filter can also be used to construct better proposal densities for the SMC-PHD declutterer at the next stage.

7 This information can be used to create better proposal distributions for the SMC-PHD filter. It also aids peak extraction at the next recursive step, since particles can be preferentially placed in regions of space more likely to contain targets.

In a second scheme, the authors use the SMC-PHD filter to perform "global gating" on measurements. The SMC-PHD filter track estimates are used to define validation gates around the estimates, and measurements that do not fall within these gates are discarded. The resulting clutter-censored measurement set is then fed into the MHC filter. One limitation of this approach is that gating tends to reduce the filter's ability to detect new targets.

The authors compare three filters—the peak-to-track association filter, the global-gating filter, and an MHC filter alone—in one-dimensional simulations. A position-measuring sensor observed five appearing, spawning, and disappearing targets in a low false alarm environment (average two per frame). The MHC and peak-to-track association filters showed comparable ability to detect and track actual targets, but the MHC filter tended to produce more false tracks. Similar performance was obtained when the MHC and global-gating filters were compared.

16.3.4.4 Track Labeling of Particles

Clark and Bell [30] and Panta, Vo, and Singh [183] have proposed similar approaches to track labeling for SMC-PHD filters.

In both approaches track labels are treated as additional state parameters, with the result that each particle in a particle PHD has an additional label or "tag" indicating its association with a track having the same label. The basic idea is that labeled particles with small weights will be eliminated, along with their labels, whereas labeled particles with large weights will be duplicated, along with their labels. Thus labels will tend to be associated with the particle clusters that represent individual targets. Since particles must be clustered in some fashion in order to extract state-estimates, correspondences between particle labels and cluster labels can be reinforced through suitable association procedures.

In [30], after the particle system is partitioned into clusters in each recursive cycle, the particles in each cluster are assigned the same label. In resampling, the children of a particle inherit their parents' labels. After resampling, the particle system is repartitioned into new clusters. If the majority of the particles in a cluster have the same label then all particles in the cluster are assigned that label.

The approach of [183] is similar, except that special pruning and merging procedures are used to address closely spaced clusters.

16.3.4.5 GM-PHD Filter

The Gaussian-mixture (GM-PHD) filter of Vo and Ma [235] includes an explicit procedure for maintaining track labels [33, 34]. The price is that the sensor likelihood function and target Markov motion density must both be linear-Gaussian and the probability of detection constant.

Since the GM-PHD filter is discussed in detail in Section 16.5.3, I will not go into the matter further here.

16.3.5 Target ID and the PHD Filter

Suppose that target identity or class c is included as a state variable, and that there are e possible classes $c_1, ..., c_e$. The inclusion of this discrete state variable will have the same effect computationally as including track label as an additional state variable. That is, it will have the effect of increasing the computational load of the PHD filter by a factor of e. New techniques are required that allow target identification to be incorporated with the PHD filter, but with less computational cost. It is possible that the track labeling techniques for the PHD filter can be extended to include a more general kind of label: target type.

16.4 PHYSICAL INTERPRETATION OF PHD FILTER

In [56], Erdinc, Willett, and Bar-Shalom proposed an interpretation of the PHD filter based on physical reasoning. Because their findings provide an intuitive perspective of the PHD filter, I describe them here.

The model of Erdinc et al. begins with a state space $\mathfrak{X}_0$ that has been discretized into a space $\check{\mathfrak{X}}_0$ of near-infinitesimal cells c_i indexed by integers $i = ... - 1, 0, 1, ...$ (see Figure 16.3). Let $\Xi_{k+1|k}$ be the random finite set of predicted target-states at time steps $k+1$, and $\Xi_{k|k}$ the random finite set of target states at time step k. Let Ξ and Ξ' denote the respective restrictions of $\Xi_{k+1|k}$ and $\Xi_{k|k}$ to $\check{\mathfrak{X}}_0$.

According to Figure 4.1 or (16.20), the prior and predicted PHDs on $\check{\mathfrak{X}}_0$ are

$$D_{k|k}(c) \triangleq \int_c D_{k|k}(\mathbf{x})d\mathbf{x} = \Pr(c \in \Xi') \tag{16.129}$$

$$D_{k+1|k}(c) \triangleq \int_c D_{k+1|k}(\mathbf{x})d\mathbf{x} = \Pr(c \in \Xi). \tag{16.130}$$

If the number of cells approaches infinity, if the size of cells approaches zero, and if $\mathbf{x}$ is held inside of a given cell $c_{\mathbf{x}}$, then

$$\frac{D_{k|k}(c_{\mathbf{x}})}{|c_{\mathbf{x}}|} \rightarrow D_{k|k}(\mathbf{x}) \quad (16.131)$$

$$\frac{D_{k+1|k}(c_{\mathbf{x}})}{|c_{\mathbf{x}}|} \rightarrow D_{k+1|k}(\mathbf{x}). \quad (16.132)$$

Note that the probability that cell c is occupied by some target at time step $k+1$ is

$$\Pr(\{c\} \cap \Xi \neq \emptyset) = D_{k+1|k}(c). \quad (16.133)$$

The probability that cell c is occupied by some target at time step k is, similarly,

$$\Pr(\{c\} \cap \Xi' \neq \emptyset) = D_{k|k}(c). \quad (16.134)$$

16.4.1 Physical Interpretation of PHD Predictor

Let $B_{k+1|k}$ be the random finite set of spontaneous birth-targets at time step $k+1$, and let $B_{k+1|k}(\mathbf{x}')$ be the RFS of targets spawned by a prior target with state $\mathbf{x}'$. Let B and $B_{c'}$ denote their respective restrictions to $\breve{\mathfrak{X}}_0$. Then just as in (16.129)-(16.132), the PHDs of $B_{k+1|k}$ and $B_{k+1|k}(\mathbf{x}')$ in $\breve{\mathfrak{X}}_0$ are

$$p_B(c) \triangleq \int_c b(\mathbf{x})d\mathbf{x} = \Pr(c \in B) \quad (16.135)$$

$$p_G(c|c') \triangleq \frac{1}{|c'|}\int_{c\times c'} b(\mathbf{x}|\mathbf{x}')d\mathbf{x}d\mathbf{x}' = \Pr(c \in B_{c'}) \quad (16.136)$$

$$p_S(c) \triangleq \frac{1}{|c|}\int_c p_S(\mathbf{x})d\mathbf{x} \quad (16.137)$$

$$p_{k+1|k}(c|c') \triangleq \frac{1}{|c'|}\int_{c\times c'} f_{k+1|k}(\mathbf{x}|\mathbf{x}')d\mathbf{x}d\mathbf{x}' \quad (16.138)$$

where $|c|$ is the (hyper)volume of c. Thus as cell size goes to zero and the number of cells goes to infinity,

$$\frac{p_B(c_{\mathbf{x}})}{|c_{\mathbf{x}}|} \rightarrow b(\mathbf{x}), \quad \frac{p_G(c_{\mathbf{x}}|c_{\mathbf{x}'})}{|c_{\mathbf{x}}|} \rightarrow b(\mathbf{x}|\mathbf{x}') \quad (16.139)$$

$$p_S(c) \rightarrow p_S(\mathbf{x}), \quad \frac{p_{k+1|k}(c_{\mathbf{x}}|c_{\mathbf{x}'})}{|c_{\mathbf{x}}|} \rightarrow f_{k+1|k}(\mathbf{x}|\mathbf{x}'). \quad (16.140)$$

Also, $b(c) = \Pr(\{c\} \cap B \neq \emptyset)$ is the probability that cell c contains some target spontaneously appearing in the scene; and $b(c|c') = \Pr(\{c\} \cap B_{c'} \neq \emptyset)$ is the probability that c contains some target spawned by a prior target in cell c'. The situation is portrayed in Figure 16.3.

We are now in a position to interpret the PHD predictor equation in terms of Figure 16.3. If the cell c_i contains a target, this occurred because of one of the following three possibilities:

- A target spontaneously appeared in c_i, with probability $p_B(c)$;
- A target moved from some cell c_e into c_i with total probability

$$\overbrace{D_{k|k}(c_e)}^{\text{prob. targ. in } c_e} \cdot \overbrace{p_{k+1|k}(c_i|c_e)}^{\text{prob. targ. moved}} \cdot \overbrace{p_S(c_e)}^{\text{prob. targ. survived}} \tag{16.141}$$

- A target in some cell c_e spawned a new target in c_i with total probability

$$\overbrace{D_{k|k}(c_e)}^{\text{prob. targ. in } c_e} \cdot \overbrace{p_G(c_i|c_e)}^{\text{prob. } c_e \text{ spawned } c_i} . \tag{16.142}$$

These possibilities are mutually exclusive because the cell size is near-infinitesimal. Thus the total probability that the cell c_i contains a target is

$$D_{k+1|k}(c_i) \tag{16.143}$$

$$= \overbrace{p_B(c_i)}^{\text{birth}} + \overbrace{\sum_e p_S(c_e) \cdot p_{k+1|k}(c_i|c_e) \cdot D_{k|k}(c_e)}^{\text{survival}} \tag{16.144}$$

$$+ \overbrace{\sum_e p_G(c_i|c_e) \cdot D_{k|k}(c_e)}^{\text{spawning}} \tag{16.145}$$

$$= p_B(c_i) \tag{16.146}$$

$$+ \sum_e \left[p_S(c_e) \cdot p_{k+1|k}(c_i|c_e) + p_G(c_i|c_e)\right] \cdot D_{k|k}(c_e). \tag{16.147}$$

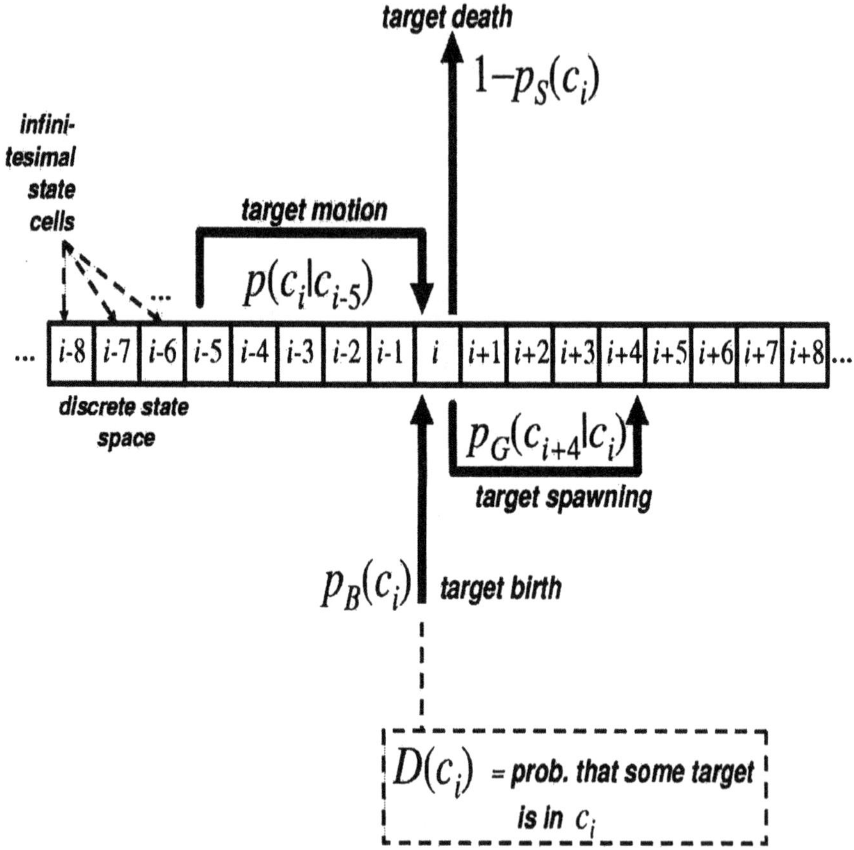

Figure 16.3 The physical interpretation of the PHD filter predictor step due to Erdinc, Willett, and Bar-Shalom [56]. State space is discretized into an arbitrarily large number of arbitrarily small cells c. A PHD $D(c)$ is the probability that some target is contained in c. The PHD predictor is derived and then shown to converge to the continuous-space PHD filter equations as the number of cells increases without bound and their size approaches zero.

Thus

$$\frac{D_{k+1|k}(c_i)}{|c_i|} = \frac{p_B(c_i)}{|c_i|} \tag{16.148}$$

$$+\sum_e \left[p_S(c_e) \cdot \frac{p_{k+1|k}(c_i|c_e)}{|c_i|} + \frac{p_G(c_i|c_e)}{|c_i|}\right] \tag{16.149}$$

$$\cdot \frac{D_{k|k}(c_e)}{|c_e|} \cdot |c_e|. \tag{16.150}$$

Taking the limit as cell size goes to zero and the number of cells approaches infinity, this equation converges to the PHD predictor equation of (16.95):

$$D_{k+1|k}(\mathbf{x}) = b(\mathbf{x}) + \int \left[p_S(\mathbf{x}') \cdot f_{k+1|k}(\mathbf{x}|\mathbf{x}') + b(\mathbf{x}|\mathbf{x}')\right] \cdot D_{k|k}(\mathbf{x}')d\mathbf{x}'. \tag{16.151}$$

16.4.2 Physical Interpretation of PHD Corrector

The physical interpretation of the PHD corrector step is more complicated than that for the PHD predictor. Let

$$p_D(c) \triangleq \frac{1}{|c|}\int_c p_D(\mathbf{x})d\mathbf{x} \tag{16.152}$$

$$L_{\mathbf{z}}(c) \triangleq \frac{1}{|c|}\int_c L_{\mathbf{z}}(\mathbf{x})d\mathbf{x} \tag{16.153}$$

be, respectively, the restrictions of the probability of detection $p_D(\mathbf{x})$ and the likelihood function $L_{\mathbf{z}}(\mathbf{x}) = f_{k+1}(\mathbf{z}|\mathbf{x})$ to $\breve{\mathfrak{X}}_0$, where $|c|$ denotes the (hyper)volume of c. In the limit as the number of cells approaches infinity and their size approaches zero,

$$p_S(c_{\mathbf{x}}) \to p_S(\mathbf{x}'), \quad L_{\mathbf{z}}(c_{\mathbf{x}}) \to L_{\mathbf{z}}(\mathbf{x}). \tag{16.154}$$

The situation is shown in Figure 16.4.

The physical derivation of the corrector step proceeds through the following stages:

- Poisson approximation of the measurement process $f(Z|Z^{(k)})$;
- Derivation of a data-update equation for the discrete PHD;

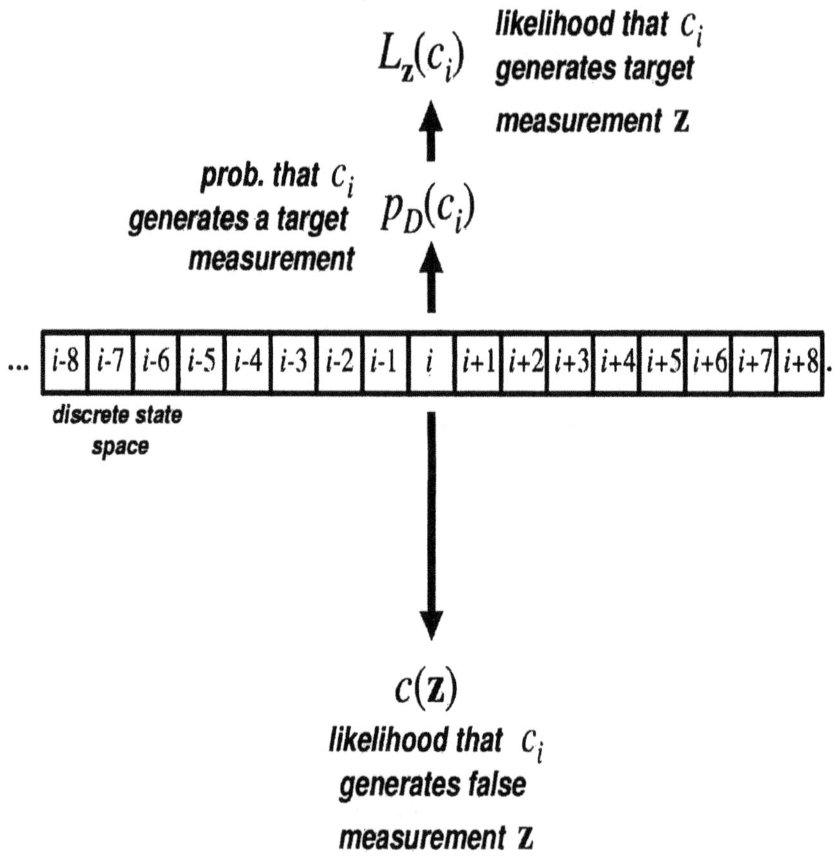

Figure 16.4 The physical interpretation of the PHD filter corrector step due to Erdinc, Willett, and Bar-Shalom [56]. The PHD corrector is derived in the discrete case. It is then shown to converge to the continuous-space PHD filter equations as the number of cells increases without bound and their size approaches zero.

- Approximation of the discrete data-update equation assuming that cell size is arbitrarily small;
- Demonstration that the discrete data-update equation converges to the usual PHD predictor equation.

16.4.2.1 Poisson Approximation

Since the average number of targets in cell c is $D_{k+1|k}(c)$, it follows that $p_D(c) \cdot D_{k+1|k}(c)$ is the average number of target detections arising from c. Consequently, the total average number of measurements is

$$\mu = \overbrace{\sum_e p_D(c_e) \cdot D_{k+1|k}(c_e)}^{\text{target detections}} + \overbrace{\lambda}^{\text{false alarms}} . \tag{16.155}$$

We assume that the measurement process is approximately Poisson with parameter μ. That is, for any measurement set with $|Z| = m$,

$$f_{k+1}(Z|Z^{(k)}) \cong \frac{e^{-\mu}\mu^m}{m!} \cdot \prod_{\mathbf{z}\in Z} f_{k+1}(\mathbf{z}|Z^{(k)}) \tag{16.156}$$

where $f_{k+1}(\mathbf{z}|Z^{(k)})$ is the Poisson spatial density.

What should $f_{k+1}(\mathbf{z}|Z^{(k)})$ be? We have:

$$f(\mathbf{z}|Z^{(k)}) = c(\mathbf{z}) \cdot \Pr(\mathbf{z} \text{ is false alarm}|Z^{(k)}) \tag{16.157}$$

$$+\sum_i \Pr(\mathbf{z} \text{ generated by } c_i|Z^{(k)}) \tag{16.158}$$

where

$$\Pr(\mathbf{z} \text{ is false alarm}|Z^{(k)}) = \frac{\lambda}{\mu} \tag{16.159}$$

$$\Pr(\mathbf{z} \text{ generated by } c_i|Z^{(k)}) = L_{\mathbf{z}}(c_i) \cdot \frac{p_D(c_i) \cdot D_{k+1|k}(c_i)}{\mu}. \tag{16.160}$$

Thus

$$f(\mathbf{z}|Z^{(k)}) = \frac{\lambda c(\mathbf{z})}{\mu} + \sum_i \frac{p_D(c_i) \cdot L_{\mathbf{z}}(c_i) \cdot D_{k+1|k}(c_i)}{\mu} \tag{16.161}$$

$$= \frac{1}{\mu} \cdot \left(\lambda c(\mathbf{z}) + \sum_i p_D(c_i) \cdot L_{\mathbf{z}}(c_i) \cdot D_{k+1|k}(c_i) \right). \tag{16.162}$$

Note that $\int f(\mathbf{z}|Z^{(k)}) d\mathbf{z} = 1$, so that $f(\mathbf{z}|Z^{(k)})$ actually is the spatial distribution of the Poisson process $f(Z|Z^{(k)})$.

16.4.2.2 Discrete-PHD Corrector Equation

An application of Bayes' rule yields the following data-update equation for the discrete-space PHD:

$$D_{k+1|k+1}(c) = \frac{f(Z|c \in \Xi, Z^{(k)})}{f(Z_{k+1}|Z^{(k)})} \cdot D_{k+1|k}(c). \tag{16.163}$$

Noting that

$$\Pr(c \notin \Xi|Z^{(k)}) = 1 - D_{k+1|k}(c) \tag{16.164}$$

and using Bayes' rule and the additivity of probability, we can write the denominator of (16.163) as

$$f(Z_{k+1}|Z^{(k)}) = f(Z_{k+1}|c \in \Xi, Z^{(k)}) \cdot D_{k+1|k}(c) \tag{16.165}$$

$$+f(Z_{k+1}|c \notin \Xi, Z^{(k)}) \cdot (1 - D_{k+1|k}(c)). \tag{16.166}$$

Thus (16.163) becomes

$$D_{k+1|k+1}(c) = \frac{D_{Z_{k+1}}(c) \cdot D_{k+1|k}(c)}{\left(\begin{array}{c} f(Z|c \in \Xi, Z^{(k)}) \cdot D_{k+1|k}(c) \\ +f(Z_{k+1}|c \notin \Xi, Z^{(k)}) \cdot (1 - D_{k+1|k}(c)) \end{array} \right)}. \tag{16.167}$$

The fraction on the right-hand side of this equation is a function of $D_{k+1|k}(c)$. Since c is arbitrarily small, $D_{k+1|k}(c)$ is arbitrarily small by (16.130). We expand the fraction to first order in a Taylor series about $D_{k+1|k}(c) = 0$:

$$D_{k+1|k+1}(c) \cong \frac{f(Z_{k+1}|c \in \Xi, Z^{(k)})}{f(Z_{k+1}|c \notin \Xi, Z^{(k)})} \cdot D_{k+1|k}(c). \tag{16.168}$$

We must determine formulas for the numerator and denominator of the fraction.

Denominator of corrector: If $c \notin \Xi$ then c contains no targets. Let $Z = \{\mathbf{z}_1, ..., \mathbf{z}_m\}$ with $|Z| = m$. Then

$$f(Z_{k+1}|c \notin \Xi, Z^{(k)}) = \frac{e^{-\mu}\mu^m}{m!} \cdot \prod_{j=1}^{m} f(\mathbf{z}_j|Z^{(k)}) = f(Z_{k+1}|Z^{(k)}). \quad (16.169)$$

Numerator of corrector. Let ϖ be a binary random variable such that $\varpi = 1$ if a target in c generates a detection, and $\varpi = 0$ otherwise. Then

$$\begin{aligned} f(Z_{k+1}|c \in \Xi, Z^{(k)}) &= f(Z_{k+1}|c \in \Xi, \varpi = 1, Z^{(k)}) \cdot \Pr(\varpi = 1|c \in \Xi, Z^{(k)}) \\ &+ f(Z_{k+1}|c \in \Xi, \varpi = 0, Z^{(k)}) \cdot \Pr(\varpi = 0|c \in \Xi, Z^{(k)}) \end{aligned} \quad (16.170)$$

or, since $p_D(c) = \Pr(\varpi = 1|c \in \Xi, Z^{(k)})$,

$$\begin{aligned} f(Z_{k+1}|c \in \Xi, Z^{(k)}) &= p_D(c) \cdot f(Z_{k+1}|c \in \Xi, \varpi = 1, Z^{(k)}) \quad (16.171) \\ &+ (1 - p_D(c)) \cdot f(Z_{k+1}|c \in \Xi, \varpi = 0, Z^{(k)}). \end{aligned}$$

Consider the factor $f(Z_{k+1}|c \in \Xi, \varpi = 1, Z^{(k)})$ in (16.171). This is the probability (density) that a target is in c and that it generated one of the measurements in Z_{k+1}, say $\mathbf{z}_j$. The other measurements in Z_{k+1} originated from other sources. The likelihood that $\mathbf{z}_j$ originated from a target in cell c and that the other measurements in Z_{k+1} originated from other sources is

$$\begin{aligned} & L_{\mathbf{z}_j}(c) \quad (16.172) \\ & \cdot \left(\frac{e^{-\mu}\mu^{m-1}}{(m-1)!} \cdot f(\mathbf{z}_1|Z^{(k)}) \cdots \widehat{f(\mathbf{z}_j|Z^{(k)})} \cdots f(\mathbf{z}_m|Z^{(k)}) \right) \quad (16.173) \\ = \;& f(Z_{k+1}|Z^{(k)}) \cdot \frac{m}{\mu} \cdot \frac{L_{\mathbf{z}_j}(c)}{f(\mathbf{z}_j|Z^{(k)})} \quad (16.174) \end{aligned}$$

where '$\widehat{f(\mathbf{z}_j|Z^{(k)})}$' indicates that the factor $f(\mathbf{z}_j|Z^{(k)})$ is omitted from the product. Averaging over all possibilities $\mathbf{z}_j \in Z_{k+1}$ we get

$$f(Z_{k+1}|c \in \Xi, \varpi = 1, Z^{(k)}) = f(Z_{k+1}|Z^{(k)}) \cdot \frac{1}{\mu} \sum_{j=1}^{m} \frac{L_{\mathbf{z}_j}(c)}{f(\mathbf{z}_j|Z^{(k)})}. \quad (16.175)$$

Next consider the factor $f(Z_{k+1}|c \in \Xi, \varpi = 0, Z^{(k)})$ in (16.171). This is the probability (density) that a target is in c but does not generate a measurement. Consequently all measurements in c originate with other sources and so

$$f(Z_{k+1}|c \in \Xi, \varpi = 0, Z^{(k)}) = \frac{e^{-\mu}\mu^m}{m!} \prod_{j=1}^{m} f(\mathbf{z}_j|Z^{(k)}). \tag{16.176}$$

Thus

$$f(Z_{k+1}|c \in \Xi, \varpi = 0, Z^{(k)}) = f(Z_{k+1}|Z^{(k)}). \tag{16.177}$$

Substituting (16.175) and (16.177) into (16.171) we get

$$f(Z_{k+1}|c \in \Xi, Z^{(k)}) = p_D(c) \cdot f(Z_{k+1}|Z^{(k)}) \cdot \frac{1}{\mu} \sum_{j=1}^{m} \frac{L_{\mathbf{z}_j}(c)}{f(\mathbf{z}_j|Z^{(k)})} \tag{16.178}$$

$$+ (1 - p_D(c)) \cdot f(Z_{k+1}|Z^{(k)}) \tag{16.179}$$

and so

$$f(Z_{k+1}|c \in \Xi, Z^{(k)}) = f(Z_{k+1}|Z^{(k)}) \tag{16.180}$$

$$\cdot \left(\frac{1}{\mu} \sum_{j=1}^{m} \frac{p_D(c) \cdot L_{\mathbf{z}_j}(c)}{f(\mathbf{z}_j|Z^{(k)})} + 1 - p_D(c) \right). \tag{16.181}$$

16.4.2.3 Continuous-PHD Corrector Equation

Substituting this result and (16.169) into (16.168)

$$D_{k+1|k+1}(c) \cong \cdot f(Z_{k+1}|Z^{(k)}) \cdot D_{k+1|k}(c) \tag{16.182}$$

$$\cdot \frac{\frac{1}{\mu} \sum_{j=1}^{m} \frac{p_D(c) \cdot L_{\mathbf{z}_j}(c)}{f(\mathbf{z}_j|Z^{(k)})} + 1 - p_D(c)}{f(Z_{k+1}|Z^{(k)})} \tag{16.183}$$

and so

$$= \left(\frac{1}{\mu} \sum_{j=1}^{m} \frac{p_D(c) \cdot L_{\mathbf{z}_j}(c)}{f(\mathbf{z}_j|Z^{(k)})} + 1 - p_D(c) \right) \tag{16.184}$$

$$\cdot D_{k+1|k}(c) \tag{16.185}$$

and so

$$= (1 - p_D(c)) \cdot D_{k+1|k}(c) \tag{16.186}$$

$$+ \left(\sum_{j=1}^{m} \frac{p_D(c) \cdot L_{\mathbf{z}_j}(c)}{\lambda c(\mathbf{z}_j) + \sum_i p_D(c_j) \cdot L_{\mathbf{z}_j}(c_i) \cdot D_{k+1|k}(c_i)} \right) \tag{16.187}$$

$$\cdot D_{k+1|k}(c). \tag{16.188}$$

Dividing both sides by the (hyper)volume $|c|$ of c, we get

$$\frac{D_{k+1|k+1}(c)}{|c|} \tag{16.189}$$

$$= (1 - p_D(c)) \cdot \frac{D_{k+1|k}(c)}{|c|} \tag{16.190}$$

$$+ \left(\sum_{j=1}^{m} \frac{p_D(c) \cdot L_{\mathbf{z}_j}(c)}{\lambda c(\mathbf{z}_j) + \sum_i p_D(c_j) \cdot L_{\mathbf{z}_j}(c_i) \cdot \frac{D_{k+1|k}(c_i)}{|c_i|} \cdot |c_i|} \right) \tag{16.191}$$

$$\cdot \frac{D_{k+1|k}(c)}{|c|}. \tag{16.192}$$

As the number of cells increases without bound and their size goes to zero, this equation converges to the PHD corrector equation

$$D_{k+1|k+1}(\mathbf{x}) = (1 - p_D(\mathbf{x})) \cdot D_{k+1|k}(\mathbf{x}) \tag{16.193}$$

$$+ \left(\sum_{j=1}^{m} \frac{p_D(\mathbf{x}) \cdot L_{\mathbf{z}_j}(\mathbf{x})}{\lambda c(\mathbf{z}_j) + \int p_D(\mathbf{x}) \cdot L_{\mathbf{z}_j}(\mathbf{x}) \cdot D_{k+1|k}(\mathbf{x}) d\mathbf{x}} \right) \tag{16.194}$$

$$\cdot D_{k+1|k}(\mathbf{x}). \tag{16.195}$$

16.5 IMPLEMENTING THE PHD FILTER

The PHD filter drastically reduces the computational load associated with the full multitarget Bayes filter. Despite this fact, it involves multidimensional integrals that must be approximated using some computational implementation technique. The purpose of this section is to describe the major approximation methods that have been used.

Section 16.5.1 begins with a survey of several implementations and applications to which the PHD filter has been applied. Most of these are based on simulations, though a few do use real data. I then describe particle system approximation and Gaussian-mixture approximation in Sections 16.5.2 and 16.5.3.

16.5.1 Survey of PHD Filter Implementations

In this section, I briefly summarize some of the applications to which PHD filter implementations have been applied. These include: tracking in terrain; multiple moving-target tracking; distributed tracking; direction of arrival (DOA) tracking; active-acoustic tracking; bistatic radio frequency (RF) tracking; tracking in images; sensor management; and group target tracking.

In what follows, "SMC-PHD" will refer to particle systems implementations of the PHD filter (Section 16.5.2), whereas "GM-PHD" will refer to the Gaussian-mixture implementation of Vo and Ma (Section 16.5.3).

16.5.1.1 Tracking Multiple Moving Targets in Terrain

As noted in Section 15.2.4, tracking targets in terrain is, in general, a highly nonlinear problem because of terrain constraints. In [208], Sidenbladh describes an application of an SMC-PHD filter to the simulations previously described in Section 15.2.4. The author compared the SMC-PHD filter with the multitarget SMC filter described in that section.

As before, vehicles are of the same type and have three travel options: roads, fields, and forest, with a given a priori probability of traveling in each. The terrain map is used to construct a nonlinear motion model. The targets are observed by human observers who report estimates of position, speed, and direction. Such observations tend to have lower probabilities of detection but also low false alarm rates. In the simulations, three vehicles travel over roads, with one vehicle temporarily going off-road into a field.

Since the PHD filter is a first-order approximation of the full multitarget Bayes filter, the latter would be expected to have better performance than the former. This proved to be true in the case of target number, which was more accurately estimated by the full filter. Position accuracy was comparable, though the PHD filter was more prone to temporary mistakes followed by rapid recovery.

16.5.1.2 Multiple Moving-Target Tracking

Punithakumar, Kirubarajan, and Sinha [190] generalized the SMC-PHD filter to include multiple motion model techniques using jump-Markov switching. This is accomplished by sampling from the model transition probabilities as well as a conventional proposal density. Their jump-Markov model presumes two motion models: constant-velocity and coordinated turn.

The authors tested their multiple-model SMC-PHD filter against the two SMC-PHD filters that result from disabling one or the other of the two models. The two-dimensional simulation consisted of two targets moving along straight lines, occasionally punctuated by sharp turns. The targets were observed by a single range-bearing sensor with uniformly distributed false alarms. The multiple-model filter was significantly better at following the turns than either of the single-model filters.

The multiple motion model problem has also been addressed by Pasha, Vo, Tuan, and Ma [185, 186]. They generalize their GM-PHD filter technique to include linear jump-Markov motion models. Their multiple-model GM-PHD filter presumes three motion models: constant-velocity, clockwise constant turn, and counterclockwise constant turn.

They successfully tested the algorithm in two-dimensional simulations involving rapidly intertwining targets in a dense false alarm environment. In one simulation, five targets underwent rapid maneuvers with occasional tracking crossings and track osculations. A second simulation involved four targets, two of which have rapidly intertwining trajectories with frequent track crossings. The multiple-model GM–PHD filter successfully tracked the targets in both simulations.

16.5.1.3 Distributed Tracking

Punithakumar, Kirubarajan, and Sinha [189] devised and implemented a *distributed SMC-PHD filter*. This filter addresses the problem of communicating and fusing track information from a distributed network of sensor-carrying platforms, that are detecting and tracking a time-varying number of targets. They successfully demonstrated the performance of their approach in a simulation consisting of four computational nodes, sixteen bearings-only sensors, and multiple ground targets.

16.5.1.4 Direction of Arrival (DOA) Tracking of Dynamic Sources

Balakumar, Sinha, Kirubarajan, and Reilly [9] applied an SMC-PHD filter to the problem of tracking an unknown and time-varying number of narrowband, far-field signal sources, using a uniform linear array of passive sensors, in a highly nonstationary sensing environment. The authors note that conventional DOA techniques such as MUSIC fail in nonstationary environments, and that difficulties are only compounded when the number of sources can vary.

The actual signal consists of outputs of the different sensors in the array, each of which is a superposition of the signals generated by the individual sources. Since this is a very different measurement phenomenology than that presumed by the PHD filter, the authors had to convert the actual measurement model to a detection-type model. They used discrete Fourier transform (DFT) techniques to determine coarse estimates of the DOAs, and these estimates were then employed as measurement set inputs to the SMC-PHD filter. The filter was used to estimate the number of sources, as well as their DOAs and intensities.

The SMC-PHD filter was compared to another SMC filter [116] that was specifically designed to detect and track a varying number of narrowband signals in a nonstationary environment. It uses a random-jump Markov chain Monte Carlo (RJMCMC) technique to estimate target number and for resampling. It had previously been shown to outperform a conventional root-MUSIC approach.

The two filters were compared in two simulations: one with fixed and one with varying target number. In the first simulation, two sources move near to each other without actually crossing. The authors found that the PHD filter "instantaneously" estimated the number of targets correctly without initialization, whereas the RJMCMC filter required 25 iterations to converge to the actual target number even with initialization. Furthermore, the SMC-PHD filter successfully separated the tracks as they moved closer together, whereas the RJMCMC filter estimated target number to be unity.

The second simulation involved two appearing and disappearing targets. The first target entered the scene and disappeared after three quarters of the scenario had elapsed. The second target appeared in the scene after one quarter of the scenario had elapsed and remained thereafter. The two targets approach each other at approximately midscenario. The RJMCMC filter required 25 iterations to detect the first target, easily detected the second target when it appeared, but once again estimated target number as one when the targets neared each other. The SMC-PHD filter "instantly" estimated the correct number of targets even when they were closely spaced, and also correctly tracked the DOAs.

16.5.1.5 Active-Acoustic Tracking

Clark, Vo, and Bell [28, 31] applied the GM-PHD filter to detection and tracking of underwater objects using forward-looking active sonar mounted on an autonomous underwater vehicle (AUV). Such objects can include marine animals and objects on the sea floor. Because an AUV is moving, even motionless objects on the bottom must be detected and tracked for purposes of collision avoidance and registration. The authors use image segmentation techniques to extract high-reflectivity areas in sonar images. The centroids of the segmented regions are extracted as features and fed into the GM-PHD algorithm.

The authors tested their algorithm on simulated sonar data and on real data collected by a sonar on an AUV. They also compared the performance of the GM-PHD and particle PHD filters. The former significantly outperformed the latter. The authors attribute this to the better computational characteristics of the GM-PHD filter, as well as the greater ease of state estimation and track labeling associated with it.

Clark and Bell have extended their approach to include three-dimensional active sonar [32]. They successfully tested their SMC-PHD filter on real data from an Echoscope forward-looking sonar with 64×64 beams. The data contained a single bottom-located target. They also compared a number of state estimation schemes, including the EM algorithm and k-means. They found that the k-means algorithm was not only significantly faster than the EM algorithm, but also outperformed it.

16.5.1.6 Bistatic Radio Frequency (RF) Tracking

Conventional radar tracking is based on radar systems in which the transmitter(s) and receiver(s) are located in a single apparatus. Bistatic radar tracking exploits the fact that radar transmissions reflect from airborne targets in all directions—not just back in the direction of the originating radar. Multiple receiving sensors, located at considerable distances from the actual transmitters, can parasitically collect RF reflections from airborne targets and use them to detect targets and infer their trajectories.

Clutter objects called "ghost targets" result from spurious triangulations and must be eliminated as part of the tracking process. The more transmitter-receiver pairs that can be exploited and the more advantageous the transmitter-receiver geometries, the more effective the tracking and localization.

A closely related bistatic RF technique, passive coherent localization (PCL), employs continuous-wave sources (TV and FM radio stations) rather than pulsed-wave sources (radar).

Tobias and Lanterman [225, 224, 226] have applied SMC-PHD filters to the PCL problem in reduced-complexity simulations. The EM algorithm was used to extract multitarget state-estimates. Some success was achieved when both range and Doppler information were used.

16.5.1.7 Tracking in Images

Ikoma, Uchino, and Maeda [96] applied an SMC-PHD filter to the problem of tracking the trajectories of feature points in time-varying optical images. A corner detector is applied to each frame in the video sequence to extract feature points, which are then processed using the SMC-PHD filter. The authors successfully tested their algorithm on actual video images of a scene containing a walking person and a moving, radio-controlled toy car.

Wang, Wu, Kassim, and Huang [246] applied SMC-PHD methods to the problem of tracking groups of humans in digital video. They use an adaptive background-subtraction method to extract objects of interest. The video background is adaptively estimated, and then subtracted from the image pixels to produce foreground images. Morphological operations (dilation and erosion) are performed on the foregrounds to remove noise. The resulting foreground consists of a number of "blobs" associated with the targets. The centroids of these blobs are fed to the SMC-PHD filter as its data.

The authors tested their algorithms on real video images consisting of pedestrians entering, moving within, and exiting from a fixed camera field of view. They observe that their method can successfully track human groups in video.

16.5.1.8 Sensor Management

The author has proposed a systematic approach to multisensor-multitarget sensor management [139, 149, 159] based on the PHD filter. The approach includes the ability to preferentially bias sensor collections toward targets of tactical interest, based on their relative rank order of priority [139, 150].

El-Fallah et al. have successfully demonstrated SMC-PHD implementations of the approach in reduced-complexity simulations [53, 54, 55, 49].

16.5.1.9 Joint Tracking and Classification with HRRR

Zajic et al. [255] report an algorithm in which an SMC-PHD filter is integrated with a robust classifier algorithm that identifies airborne targets from high range-resolution radar (HRRR) signatures [194]. The classifier algorithm utilizes wavelet features ϕ extracted from the HRRR signatures to produce a feature likelihood function $L_\phi(u)$ where u is target type. The classifier was trained on real HRRR data-sets drawn from seven possible airborne target types. Assuming that target identities and kinematics are independent, the total likelihood function has the form $L_{\mathbf{z},\phi}(c, \mathbf{x}) = L_\phi(c) \cdot L_\mathbf{z}(\mathbf{x})$. This was then incorporated into an SMC-PHD filter algorithm.

The joint identification and tracking algorithm was tested on real HRRR signatures and simulated two-dimensional kinematics. The dynamics measurements included moderately dense false alarms (120 false alarms per frame). The classifier was not turned on until after the PHD filter had achieved good track estimates. The algorithm successfully detected, tracked, and classified the two targets.

16.5.1.10 Group Target Detection and Tracking

Ahlberg, Hörling, Kjellström, Jöred, Mårtenson, Neider, Schubert, Svenson, Svensson, Undén, and Walter of the Swedish Defence Research Agency (FOI) have employed PHD filters for group target tracking in an ambitious situation assessment simulator system called "IFD03" [2, 209]. A force aggregation algorithm based on Dempster-Shafer methods is used to detect and classify force structure. Separate PHD filters are used to track group targets at each force structure level. The PHD filters can take account of *a priori* information derived from terrain simulators.

It should also be mentioned here that the author has proposed [130, 131] a generalization of the PHD filter, the "group PHD filter," for detecting and tracking group objects such as squads, platoons, and brigades. To the author's knowledge, there are no published implementations of this filter.

16.5.2 SMC-PHD Approximation

Beginning with Sidenbladh [208] and Zajic and Mahler [254], most researchers have implemented the PHD filter using sequential Monte Carlo (SMC) methods. This section summarizes the basic ideas behind such "particle PHD" filtering techniques, using a generic methodology based on the dynamic prior. For more detailed discussions, see [238, 121, 208, 54].

SMC techniques (Section 2.5.3) were originally devised to approximate probability densities and the single-target Bayes filter. Consequently, they must be modified for use with the PHD. This is because the PHD is not a probability density, and because the PHD filter equations are more complex than the single-target Bayes filter equations.

Despite these differences, SMC approximation carries over to the PHD filter relatively easily (with the exception of regularization). In this section we sketch the basic elements of particle PHD filtering. The basic concept is pictured in Figure 16.5, which is the PHD filter analog of Figure 2.12.

A particle system approximation of the PHD $D_{k|k}(\mathbf{x}|Z^k)$ is a collection $\mathbf{x}^1_{k|k}, ...\mathbf{x}^\nu_{k|k}$ of state vectors and importance weights $w^1_{k|k}, ...w^\nu_{k|k}$, such that

$$\int \theta(\mathbf{x}) \cdot D_{k|k}(\mathbf{x}|Z^k) d\mathbf{x} \cong \sum_{i=1}^{\nu} w^i_{k|k} \cdot \theta(\mathbf{x}^i_{k|k}) \qquad (16.196)$$

for any unitless function $\theta(\mathbf{x})$ of $\mathbf{x}$. As in Section 2.5.3, for the sake of clarity we will assume that the particles are equally weighted: $w^i_{k|k} = 1/\nu$ for all $i = 1, ..., \nu$.

The usual convergence property, the obvious analog of (2.208), must be satisfied:[8]

$$\lim_{\nu\to\infty} \sum_{i=1}^{\nu} w^i_{k|k} \cdot \theta(\mathbf{x}^i_{k|k}) = \int \theta(\mathbf{x}) \cdot D_{k|k}(\mathbf{x}|Z^k) d\mathbf{x}. \qquad (16.197)$$

As an abbreviation,we also employ the obvious analog of (2.210).

The major difference between the PHD particle system approximation, (16.196) and the conventional single-target one, (2.207), is that the importance weights sum not to unity but to the expected number of targets:

$$\sum_{i=1}^{\nu} w^i_{k|k} \cong N_{k|k} \,. \qquad (16.198)$$

Another significant difference is that when targets have been adequately localized, a single-target particle system $\mathbf{x}^1_{k|k}, ...\mathbf{x}^\nu_{k|k}$ consists of a single particle cluster. The corresponding PHD-particle system, on the other hand, will consist of several particle clusters—one corresponding to each detected target.

In what follows I summarize each of the particle PHD filter processing steps in turn: initialization, prediction, correction, and state and error estimation.

8 See, for example, [238, p. 1234, Prop. 3] and [29].

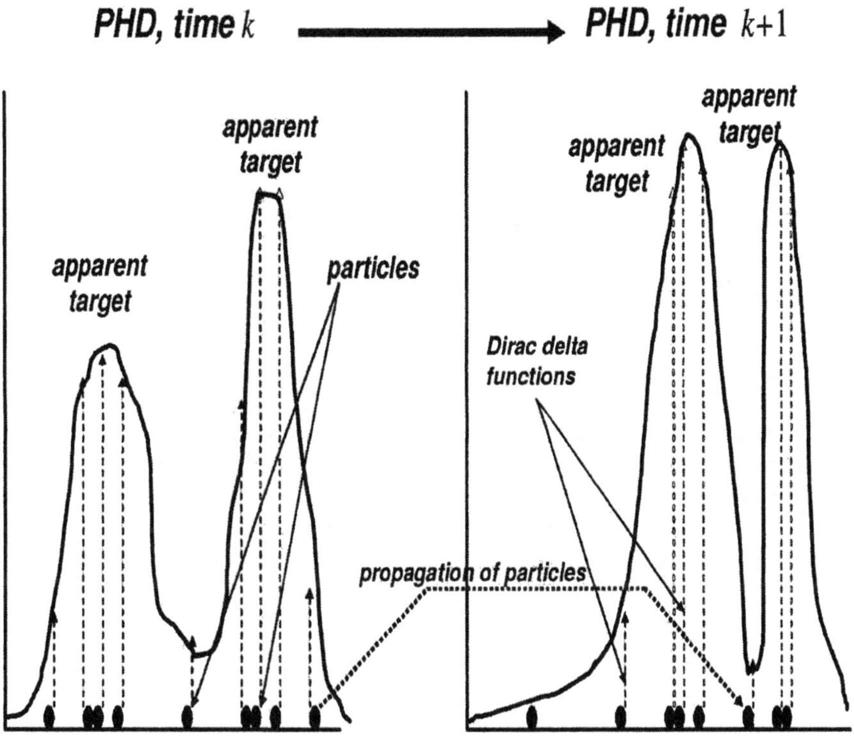

Figure 16.5 The concept of particle PHD (SMC-PHD) filtering is illustrated. Particles represent random samples drawn from a posterior PHD. Particles are supposed to be more densely located where targets are most likely to be present. As with single-target SMC filtering, the basic idea is to propagate particles from time step to time step so that this assumption remains valid.

16.5.2.1 SMC-PHD Filter: Initialization

The SMC-PHD filter can be initialized by applying the reasoning of Section 2.5.3 to Section 16.3.1. That is, one can choose an a priori PHD

$$D_{0|0}(\mathbf{x}) = D_{0|0}(\mathbf{x}|Z^{(0)}) = n_0 \cdot s_0(\mathbf{x}) \tag{16.199}$$

and then draw ν samples from $s_0(\mathbf{x})$:

$$\mathbf{x}^1_{0|0}, \ldots \mathbf{x}^\nu_{0|0} \sim s_0(\cdot). \tag{16.200}$$

Here ν should be chosen large enough to permit adequate particle representation for every target. Intuitively speaking, if ρ particles are required to adequately maintain track on any individual target, then one chooses $\nu \cong \rho \cdot n_0$ [238, p. 1233].

16.5.2.2 SMC-PHD Filter: Predictor

If neither target appearance nor target disappearance were modeled, then the particle PHD predictor would be essentially identical to the single-target particle predictor, (2.222) and (2.223). With the inclusion of target birth and death models, however, the derivation of the SMC-PHD is complicated by the incorporation of these two additional processes.

Assume that the prior PHD has been approximated by an equally weighted particle system $\mathbf{x}^1_{k|k}, \ldots, \mathbf{x}^\nu_{k|k}$. The total expected number of predicted targets will be

$$N_{k+1|k} = N^{\text{birth}}_{k+1|k} + N^{\text{persist}}_{k+1|k} + N^{\text{spawn}}_{k+1|k} \tag{16.201}$$

where

$$N^{\text{birth}}_{k+1|k} = \int b_{k+1|k}(\mathbf{x})d\mathbf{x} \tag{16.202}$$

$$N^{\text{persist}}_{k+1|k} = \int p_S(\mathbf{x}') \cdot D_{k+1|k}(\mathbf{x}')d\mathbf{x}' \tag{16.203}$$

$$N^{\text{spawn}}_{k+1|k} = \int N^{\text{spawn}}_{k+1|k}(\mathbf{x}') \cdot D_{k+1|k}(\mathbf{x}')d\mathbf{x}' \tag{16.204}$$

and where

$$N^{\text{spawn}}_{k+1|k}(\mathbf{x}') = \int b_{k+1|k}(\mathbf{x}|\mathbf{x}')d\mathbf{x} \tag{16.205}$$

is the expected number of targets spawned by a prior target with state $\mathbf{x}'$.

Let $\rho = \nu / N_{k|k}$ be the number of particles required to effectively track a single target. Then as noted in Section 16.3.1, approximately

$$\nu_{k+1|k} = \frac{N_{k+1|k}}{N_{k|k}} = \nu^{\text{birth}}_{k+1|k} + \nu^{\text{persist}}_{k+1|k} + \nu^{\text{spawn}}_{k+1|k} \tag{16.206}$$

particles are required to track $N_{k+1|k}$ targets, where

$$\nu^{\text{birth}}_{k+1|k} \triangleq \frac{N^{\text{birth}}_{k+1|k}}{N_{k|k}}, \quad \nu^{\text{persist}}_{k+1|k} \triangleq \frac{N^{\text{persist}}_{k+1|k}}{N_{k|k}}, \quad \nu^{\text{spawn}}_{k+1|k} \triangleq \frac{N^{\text{spawn}}_{k+1|k}}{N_{k|k}}. \tag{16.207}$$

Persisting particles: To model persisting targets, note that the particle approximation of $D_{k|k}(\mathbf{x}|Z^{(k)})$ yields

$$N^{\text{persist}}_{k+1|k} \cong \frac{1}{\nu} \sum_{i=1}^{\nu} p_S(\mathbf{x}^i_{k|k}). \tag{16.208}$$

Define the discrete probability distribution $p_S(i)$ on $i \in \{1, ..., \nu\}$ by

$$p_{\text{persist}}(i) \triangleq \frac{p_S(\mathbf{x}^i_{k|k})}{\sum_{e=1}^{\nu} p_S(\mathbf{x}^e_{k|k})}. \tag{16.209}$$

Let $\breve{\nu}_p$ be the nearest integer to $\nu^{\text{persist}}_{k+1|k}$ and draw $\breve{\nu}_p$ samples from $p_{\text{persist}}(i)$:

$$i_1, ..., i_{\breve{\nu}_p} \sim p_{\text{persist}}(\cdot). \tag{16.210}$$

Then the particles $\mathbf{x}^{i_1}_{k|k}, ..., \mathbf{x}^{i_{\breve{\nu}_p}}_{k|k}$ are chosen to represent targets that persist into time step $k+1$. For each of these persisting particles, draw a single sample from the dynamic prior:

$$\mathbf{x}^1_{k+1|k} \sim f_{k+1|k}(\cdot|\mathbf{x}^{i_1}_{k|k}) \quad ,...., \quad \mathbf{x}^{\breve{\nu}_p}_{k+1|k} \sim f_{k+1|k}(\cdot|\mathbf{x}^{i_{\breve{\nu}_p}}_{k|k}). \tag{16.211}$$

Then the predicted particles $\mathbf{x}^1_{k+1|k}, ..., \mathbf{x}^{\breve{\nu}_p}_{k+1|k}$ represent the persisting targets.

Appearing particles: We need approximately $\nu^{\text{birth}}_{k+1|k}$ particles to represent spontaneously appearing targets. Let $\breve{\nu}_b$ be the nearest integer to $\nu^{\text{birth}}_{k+1|k}$. Define the probability density $\hat{b}_{k+1|k}(\mathbf{x})$ by

$$\hat{b}_{k+1|k}(\mathbf{x}) \triangleq \frac{b_{k+1|k}(\mathbf{x})}{N^{\text{birth}}_{k+1|k}}. \tag{16.212}$$

Draw $\breve{\nu}_b$ samples from $\hat{b}_{k+1|k}(\mathbf{x})$:

$$\mathbf{x}^{\breve{\nu}_p+1}_{k+1|k}, ..., \mathbf{x}^{\breve{\nu}_p+\breve{\nu}_b}_{k+1|k} \sim \hat{b}_{k+1|k}(\cdot). \tag{16.213}$$

Then the predicted particles $\mathbf{x}^{\breve{\nu}_p+1}_{k+1|k}, ..., \mathbf{x}^{\breve{\nu}_p+\breve{\nu}_b}_{k+1|k}$ represent the spontaneously appearing targets.

Spawning particles: We need approximately $\nu^{\text{spawn}}_{k+1|k}$ particles to represent all spawned targets. From the particle approximation of $D_{k|k}(\mathbf{x}|Z^{(k)})$ we know that

$$N^{\text{spawn}}_{k+1|k} \cong \frac{1}{\nu}\sum_{i=1}^{\nu} N^{\text{spawn}}_{k+1|k}(\mathbf{x}^i_{k|k}). \tag{16.214}$$

Define the probability distribution $p_{\text{spawn}}(i)$ on $i \in \{1, ..., \nu\}$ by

$$p_{\text{spawn}}(i) \triangleq \frac{N^{\text{spawn}}_{k+1|k}(\mathbf{x}^i_{k|k})}{\sum_{e=1}^{\nu} N^{\text{spawn}}_{k+1|k}(\mathbf{x}^e_{k|k})}. \tag{16.215}$$

Let $\breve{\nu}_s$ be the nearest integer to $\nu^{\text{spawn}}_{k+1|k}$ and draw $\breve{\nu}_s$ samples from $p_{\text{spawn}}(i)$:

$$j_1, ..., j_{\breve{\nu}_s} \sim p_{\text{spawn}}(\cdot). \tag{16.216}$$

Then $\mathbf{x}^{i_1}_{k|k}, ..., \mathbf{x}^{i_{\breve{\nu}_p}}_{k|k}$ are chosen to represent targets that spawned new targets.

We have to determine predicted particles that represent targets spawned by each of these targets. Let $\breve{\nu}^i_s$ be the nearest integer to $\nu^{\text{spawn}}_{k+1|k}$. Define the probability density functions $\hat{b}^i_{k+1|k}(\mathbf{x})$ by

$$\hat{b}^i_{k+1|k}(\mathbf{x}) \triangleq \frac{b_{k+1|k}(\mathbf{x}|\mathbf{x}^i_{k|k})}{N^{\text{spawn}}_{k+1|k}} \tag{16.217}$$

for $i = 1, ..., \nu$. Then for each $i = 1, ..., \breve{\nu}_s^i$, draw $\breve{\nu}_s^i$ samples from $\hat{b}_{k+1|k}^i(\mathbf{x})$:

$$\mathbf{x}_{k+1|k}^{\breve{\nu}_p+\breve{\nu}_b+1}, ..., \mathbf{x}_{k+1|k}^{\breve{\nu}_p+\breve{\nu}_b+\breve{\nu}_s^1} \sim \hat{b}_{k+1|k}^1(\cdot) \tag{16.218}$$

$$\mathbf{x}_{k+1|k}^{\breve{\nu}_p+\breve{\nu}_b+\breve{\nu}_s^1+1}, ..., \mathbf{x}_{k+1|k}^{\breve{\nu}_p+\breve{\nu}_b+\breve{\nu}_s^1+\breve{\nu}_s^2} \sim \hat{b}_{k+1|k}^1(\cdot) \tag{16.219}$$

$$\vdots \tag{16.220}$$

$$\mathbf{x}_{k+1|k}^{\breve{\nu}_p+\breve{\nu}_b+\breve{\nu}_s^1+...+\breve{\nu}_s^{\nu-1}+1}, ..., \mathbf{x}_{k+1|k}^{\breve{\nu}_p+\breve{\nu}_b+\breve{\nu}_s^1+...+\breve{\nu}_s^{\nu-1}+\breve{\nu}_s^{\nu}} \sim \hat{b}_{k+1|k}^1(\cdot). \tag{16.221}$$

Then the $\mathbf{x}_{k+1|k}^{\breve{\nu}_p+\breve{\nu}_b+1}, ..., \mathbf{x}_{k+1|k}^{\breve{\nu}_p+\breve{\nu}_b+\breve{\nu}_s^1}$ are the predicted particles representing targets spawned by $\mathbf{x}_{k|k}^1$; and $\mathbf{x}_{k+1|k}^{\breve{\nu}_p+\breve{\nu}_b+\breve{\nu}_s^1+1}, ..., \mathbf{x}_{k+1|k}^{\breve{\nu}_p+\breve{\nu}_b+\breve{\nu}_s^1+\breve{\nu}_s^2}$ are the predicted particles representing targets spawned by $\mathbf{x}_{k|k}^2$; and so on.

16.5.2.3 SMC-PHD Filter: Corrector

Assume that the predicted PHD has been approximated by a predicted particle system $\mathbf{x}_{k+1|k}^1, ..., \mathbf{x}_{k+1|k}^{\nu}$. The integrals $D_{k+1|k}[p_D L_{\mathbf{z}}]$ in the PHD corrector equation become

$$D_{k+1|k}[p_D L_{\mathbf{z}}] \cong \frac{1}{\nu} \sum_{i=1}^{\nu} p_D(\mathbf{x}_{k+1|k}^i) \cdot L_{\mathbf{z}}(\mathbf{x}_{k+1|k}^i). \tag{16.222}$$

Consequently, the Bayes corrector equation, (2.82), becomes

$$\int \theta(\mathbf{x}) \cdot D_{k+1|k+1}(\mathbf{x}|Z^{k+1}) d\mathbf{x} \tag{16.223}$$

$$= \int \theta(\mathbf{x}) \cdot L_{Z_{k+1}}(\mathbf{x}) \cdot D_{k+1|k}(\mathbf{x}|Z^{k+1}) d\mathbf{x} \tag{16.224}$$

$$\cong \frac{1}{\nu} \sum_{i=1}^{\nu} \theta(\mathbf{x}_{k+1|k}^i) \cdot L_{Z_{k+1}}(\mathbf{x}_{k+1|k}^i) \tag{16.225}$$

or

$$= \frac{1}{\nu}\sum_{i=1}^{\nu}\theta(\mathbf{x}^i_{k+1|k}) \tag{16.226}$$

$$\cdot\left(\begin{array}{c} 1-p_D(\mathbf{x}^i_{k+1|k}) \\ +\sum_{\mathbf{z}\in Z_{k+1}}\frac{p_D(\mathbf{x}^i_{k+1|k})\cdot L_{Z_{k+1}}(\mathbf{x}^i_{k+1|k})}{\lambda c(\mathbf{z})+\frac{1}{\nu}\sum_{e=1}^{\nu}p_D(\mathbf{x}^e_{k+1|k})\cdot L_{\mathbf{z}}(\mathbf{x}^e_{k+1|k})}\end{array}\right) \tag{16.227}$$

$$= \sum_{i=1}^{\nu} w^i_{k+1|k+1}\cdot\theta(\mathbf{x}^i_{k+1|k+1}) \tag{16.228}$$

where $\mathbf{x}^i_{k+1|k+1} \triangleq \mathbf{x}^i_{k+1|k}$ with corresponding weights

$$w^i_{k+1|k+1} \triangleq \frac{1}{\nu}\left(\begin{array}{c} 1-p_D(\mathbf{x}^i_{k+1|k}) \\ +\sum_{\mathbf{z}\in Z_{k+1}}\frac{p_D(\mathbf{x}^i_{k+1|k})\cdot L_{Z_{k+1}}(\mathbf{x}^i_{k+1|k})}{\lambda c(\mathbf{z})+\frac{1}{\nu}\sum_{e=1}^{\nu}p_D(\mathbf{x}^e_{k+1|k})\cdot L_{\mathbf{z}}(\mathbf{x}^e_{k+1|k})}\end{array}\right). \tag{16.229}$$

The data-updated expected number of targets is, therefore,

$$N_{k+1|k+1} = \sum_{i=1}^{\nu} w^i_{k+1|k+1}. \tag{16.230}$$

Since the particles $\mathbf{x}^1_{k+1|k+1},...,\mathbf{x}^\nu_{k+1|k+1}$ have unequal weights, we must use resampling techniques to replace them with new, equally-weighted particles. Any conventional resampling technique can be applied if we replace the $w^i_{k+1|k+1}$ by the normalized weights

$$\hat{w}^i_{k+1|k+1} = \frac{w^i_{k+1|k+1}}{N_{k+1|k+1}} \tag{16.231}$$

and then apply the technique to the $\hat{w}^1_{k+1|k+1},...,\hat{w}^\nu_{k+1|k+1}$. We then end up with a new particle system $\tilde{\mathbf{x}}^1_{k+1|k+1},...,\tilde{\mathbf{x}}^\nu_{k+1|k+1}$ with equal weights $\tilde{w}^i_{k+1|k+1} = N_{k+1|k+1}/\nu$ for all $i = 1,...,\nu$. See [54, Section 8] or [238, p. 1233, Step 3] for examples of resampling techniques.

Remark 31 (Regularization of Particle PHDs) *In connection with (2.236) we mentioned "regularization." This is a technique for improving the statistical* *diversity of particle systems by approximating them with a smooth distribution and then sampling from that distribution. The usual regularization techniques cannot be applied to particle PHD approximation because they presume that the distribution to be smoothed is a unimodal probability distribution. Zajic et al. [254, 255] proposed a regularization scheme based on clustering, in which clusters are extracted and regularization performed on the clusters. El-Fallah et al. [54, Section 8] proposed a similar scheme for the SMC-PHD filter, but with clustering achieved using the EM algorithm.*

16.5.2.4 SMC-PHD Filter: State and Error Estimation

Estimation of the instantaneous expected target number is easy since this quantity is recursively maintained throughout the operation of the particle PHD filter. However, as was noted in Section 16.3.4, this number tends to be unstable and must be averaged over a running time window.

State and error estimation for the particle PHD filter is not a simple matter. As previously noted in Section 16.3.4, techniques such as clustering, the EM algorithm, or histogramming are commonly employed.

Clark and Bell [32, pp. 266, 267] have conducted a comparative analysis of different state estimation techniques for SMC-PHD filters. Let ν be the number of particles, n the number of targets, and i the number of iterations required to converge to a solution. They considered three approaches with known orders of computational complexity:

- Hierarchical clustering, order $O(\nu^2 \log \nu)$;
- k-means, order $O(i\nu n)$;
- The EM algorithm, order $O(i\nu n^2)$.

The k-means approach is computationally attractive, but it is statistically biased and inconsistent, and nonrobust. The EM algorithm is unbiased and more robust, but also more computationally intensive. In tests using real three-dimensional active sonar data, the authors determined that k-means actually outperformed the EM algorithm.

16.5.3 GM-PHD Approximation

The Gaussian-Mixture (GM) PHD filter was devised in 2005 by Vo and Ma [235, 236] and has since been extended to multiple motion models [185] and applied to active-sonar tracking [31]. It has also been extended so that it can maintain track labels [33, 34]. It generalizes the Gaussian-mixture methods of Section 2.5.2 to PHD filtering. It propagates through time components of the form

$$(w^i_{k|k}, \mathbf{x}^i_{k|k}, P^i_{k|k}, \ell^i_{k|k}), \tag{16.232}$$

where $(w^i_{k|k}, \mathbf{x}^i_{k|k}, P^i_{k|k})$ is a conventional Gaussian component as in Section 2.5.2, and where $\ell^i_{k|k}$ is a track label:

$$
\begin{aligned}
& (w^i_{0|0}, \mathbf{x}^i_{0|0}, P^i_{0|0}, \ell^i_{0|0})_{1 \le i \le n_{0|0}} && (16.233)\\
\rightarrow\; & (w^i_{1|0}, \mathbf{x}^i_{1|0}, P^i_{1|0}, \ell^i_{1|0})_{1 \le i \le n_{1|0}} && (16.234)\\
\rightarrow\; & (w^i_{1|1}, \mathbf{x}^i_{1|1}, P^i_{1|1}, \ell^i_{1|1})_{1 \le i \le n_{1|1}} \rightarrow \cdots && (16.235)\\
\rightarrow\; & (w^i_{k|k}, \mathbf{x}^i_{k|k}, P^i_{k|k}, \ell^i_{k|k})_{1 \le i \le n_{k|k}} && (16.236)\\
\rightarrow\; & (w^i_{k+1|k}, \mathbf{x}^i_{k+1|k}, P^i_{k+1|k}, \ell^i_{k+1|k})_{1 \le i \le n_{k+1|k}} && (16.237)\\
\rightarrow\; & (w^i_{k+1|k+1}, \mathbf{x}^i_{k+1|k+1}, P^i_{k+1|k+1}, \ell^i_{k+1|k+1},)_{1 \le i \le n_{k+1|k+1}} && (16.238)\\
\rightarrow\; & \cdots && (16.239)
\end{aligned}
$$

At any given time step k, the evolving PHD has the form

$$D_{k|k}(\mathbf{x}) = \sum_{i=1}^{n_{k|k}} w^i_{k|k} \cdot N_{P^i_{k|k}}(\mathbf{x} - \mathbf{x}^i_{k|k}). \tag{16.240}$$

Clark and Vo [34] have proved a strong L_1 uniform convergence property for the GM-PHD filter: "each step in time of the PHD filter will maintain a suitable approximation error that converges to zero as the number of Gaussians in the mixture tends to infinity." This includes the pruning and merging stages.

The underlying assumptions of the GM-PHD filter are more restrictive than those for the SMC-PHD filter. This fact having been granted, however, the GM-PHD approximation method has potentially major advantages over SMC-PHD approximation:

- It is exact, in the sense that it provides a true closed-form algebraic solution to the PHD filter equations.
- The predictor and corrector equations are much less computationally demanding.

- State and error estimation are easily accomplished;
- Track labels can be maintained without resort to complicated state-to-track correlation techniques.
- It is more easily implemented.

In what follows I summarize the processing steps of the GM-PHD filter: assumptions; initialization, prediction, correction, pruning and merging, and state and error estimation.

16.5.3.1 GM-PHD Filter Assumptions

In addition to the usual PHD filter assumptions, the GM-PHD filter makes the following additional ones:

- The single-target Markov transition density is linear-Gaussian:

$$f_{k+1|k}(\mathbf{x}|\mathbf{x}') = N_{Q_k}(\mathbf{x} - F_k\mathbf{x}'). \tag{16.241}$$

- The probability p_S that a target will persist is constant.
- The single-target likelihood function is linear-Gaussian:

$$f_{k+1}(\mathbf{z}|\mathbf{x}) = N_{R_{k+1}}(\mathbf{z} - H_{k+1}\mathbf{x}). \tag{16.242}$$

- The probability of detection p_D is constant.[9]
- The PHDs of the target-birth and target-spawning processes are both Gaussian mixtures

$$b_{k+1|k}(\mathbf{x}) = \sum_{i=1}^{a_k} \beta_k^i \cdot N_{B_k^i}(\mathbf{x} - \mathbf{x}_b^i) \tag{16.243}$$

$$b_{k+1|k}(\mathbf{x}|\mathbf{x}') = \sum_{i=1}^{b_k} \gamma_k^i \cdot N_{G_k^j}(\mathbf{x} - E_k^i\mathbf{x}'). \tag{16.244}$$

9 This restriction is required since both $p_D(\mathbf{x})$ and $1-p_D(\mathbf{x})$ occur in the PHD corrector equation. For example, one could define $p_D(\mathbf{x}) = \hat{N}_D(L\mathbf{x}-\overset{*}{L}\overset{*}{\mathbf{x}})$ where $\hat{N}_D(\mathbf{x})$ is a normalized Gaussian as in (5.207). This would result in a non-Gaussian component $1 - p_D(\mathbf{x})$ in the corrector.

- The PHDs $D_{k|k}(\mathbf{x}|Z^{(k)})$ and $D_{k+1|k}(\mathbf{x}|Z^{(k)})$ are Gaussian mixtures:

$$D_{k|k}(\mathbf{x}|Z^{(k)}) = \sum_{i=1}^{n_{k|k}} w_{k|k}^i \cdot N_{P_{k|k}^i}(\mathbf{x} - \mathbf{x}_{k|k}^i) \quad (16.245)$$

$$D_{k+1|k}(\mathbf{x}|Z^{(k)}) = \sum_{i=1}^{n_{k+1|k}} w_{k+1|k}^i \quad (16.246)$$

$$\cdot N_{P_{k+1|k}^i}(\mathbf{x} - \mathbf{x}_{k+1|k}^i). \quad (16.247)$$

Given this representation, expected numbers of targets are computed as:

$$N_{k|k} = \sum_{i=1}^{n_{k|k}} w_{k|k}^i, \qquad N_{k+1|k} = \sum_{i=1}^{n_{k+1|k}} w_{k+1|k}^i. \quad (16.248)$$

16.5.3.2 GM-PHD Filter Initialization

The initial PHD is chosen to be a Gaussian mixture:

$$D_{0|0}(\mathbf{x}|Z^{(0)}) = \sum_{i=1}^{n_{0|0}} w_{0|0}^i \cdot N_{P_{0|0}^i}(\mathbf{x} - \mathbf{x}_{0|0}^i). \quad (16.249)$$

The initial expected number of targets is, therefore,

$$N_{0|0} = \sum_{i=1}^{n_{0|0}} w_{0|0}^i. \quad (16.250)$$

Furthermore, each Gaussian component $(w_{0|0}^i, \mathbf{x}_{0|0}^i, P_{0|0}^i)$ in (16.249) is assigned an *identifying label* $\ell_{0|0}^i$ to aid in labeling of tracks as the processing proceeds. This results in an initial *label table*

$$\mathcal{L}_{0|0} = \{\ell_{0|0}^1, ..., \ell_{0|0}^{n_{0|0}}\}. \quad (16.251)$$

16.5.3.3 GM-PHD Filter Predictor

We assume that we have a prior GM-PHD

$$D_{k|k}(\mathbf{x}|Z^{(k)}) = \sum_{i=1}^{n_{k|k}} w_{k|k}^i \cdot N_{P_{k|k}^i}(\mathbf{x} - \mathbf{x}_{k|k}^i) \quad (16.252)$$

and a prior label table $\mathcal{L}_{k|k}$. In Appendix G.28 we verify that the predicted PHD has the form

$$D_{k+1|k}(\mathbf{x}) = \sum_{i=1}^{a_k} v^i_{k+1|k} \cdot N_{B^i_{k+1|k}}(\mathbf{x} - \mathbf{b}^i_{k+1|k}) \tag{16.253}$$

$$+\sum_{i=1}^{n_{k|k}} w^i_{k+1|k} \cdot N_{P^i_{k+1|k}}(\mathbf{x} - \mathbf{x}^i_{k+1|k}) \tag{16.254}$$

$$+\sum_{j=1}^{b_k}\sum_{i=1}^{n_{k|k}} w^{i,j}_{k+1|k} \cdot N_{P^{i,j}_{k+1|k}}(\mathbf{x} - \mathbf{x}^{i,j}_{k+1|k}) \tag{16.255}$$

where

$$v^i_{k+1|k} = \beta^i_k, \quad \mathbf{b}^i_{k+1|k} = \mathbf{x}^i_b, \quad B^i_{k+1|k} = B_i \tag{16.256}$$

$$w^i_{k+1|k} = p_S \cdot w^i_{k|k}, \quad \mathbf{x}^i_{k+1|k} = F\mathbf{x}^i, \tag{16.257}$$

$$P^i_{k+1|k} = Q + FP_iF^T \tag{16.258}$$

$$w^{i,j}_{k+1|k} = \gamma^j_k \cdot w^i_{k|k}, \quad \mathbf{x}^{i,j}_{k+1|k} = E_j\mathbf{x}^i, \tag{16.259}$$

$$P^{i,j}_{k+1|k} = G_j + E_iP_iE_i^T \tag{16.260}$$

and where we have abbreviated

$$F \overset{\text{abbr.}}{=} F_k, \quad Q \overset{\text{abbr.}}{=} Q_k, \tag{16.261}$$

$$P_i \overset{\text{abbr.}}{=} P^i_{k|k}, \quad \mathbf{x}^i \overset{\text{abbr.}}{=} \mathbf{x}^i_{k|k} \tag{16.262}$$

$$G_i \overset{\text{abbr.}}{=} G^i_k, \quad B_i \overset{\text{abbr.}}{=} B^i_k, \quad E_i \overset{\text{abbr.}}{=} E^i_k. \tag{16.263}$$

The predicted GM-PHD has

$$n_{k+1|k} = a_k + n_{k|k} + b_k \cdot n_{k|k} \tag{16.264}$$

$$= a_k + (1 + b_k) \cdot n_{k|k} \tag{16.265}$$

Gaussian components. The number of components will therefore increase exponentially. Standard techniques, such as pruning the components with smallest weights or merging similar components [31], can be used to limit this computational increase to a tractable level—see (16.286)-(16.289).

We must also construct the predicted label table $\mathcal{L}_{k+1|k}$. The prediction

$$(w^i_{k+1|k}, \mathbf{x}^i_{k+1|k}, P^i_{k+1|k}) \tag{16.266}$$

of the prior ith component is assigned the same label as its predecessor $(w^i_{k|k}, \mathbf{x}^i_{k|k}, P^i_{k|k})$. Each birth component $(v^i_{k+1|k}, \mathbf{b}^i_{k+1|k}, B^i_{k+1|k})$ is assigned a new label, and likewise for each spawned component

$$(w^{i,j}_{k+1|k}, \mathbf{x}^{i,j}_{k+1|k}, P^{i,j}_{k+1|k}). \tag{16.267}$$

The complete predicted label table thus has the form

$$\mathcal{L}_{k+1|k} = \mathcal{L}^{\text{birth}}_{k+1|k} \cup \mathcal{L}^{\text{prior}}_{k+1|k} \cup \mathcal{L}^{\text{spawn}}_{k+1|k}. \tag{16.268}$$

16.5.3.4 GM-PHD Filter Corrector

Assume that from the previous time step we have a predicted GM-PHD

$$D_{k+1|k}(\mathbf{x}) = \sum_{i=1}^{n_{k+1|k}} w^i_{k+1|k} \cdot N_{P^i_{k+1|k}}(\mathbf{x} - \mathbf{x}^i_{k+1|k}) \tag{16.269}$$

and a predicted label table $\mathcal{L}_{k+1|k}$. We collect a new measurement set $Z_{k+1} = \{\mathbf{z}_1, ..., \mathbf{z}_{m_{k+1}}\}$ with $|Z_{k+1}| = m_{k+1}$. In Appendix G.29 we verify that the data-updated PHD has the form

$$D_{k+1|k+1}(\mathbf{x}) \tag{16.270}$$

$$= \sum_{i=1}^{n_{k+1|k}} w^i_{k+1|k+1} \cdot N_{P^i_{k+1|k+1}}(\mathbf{x} - \mathbf{x}^i_{k+1|k+1}) \tag{16.271}$$

$$+ \sum_{i=1}^{n_{k+1|k}} \sum_{j=1}^{m_{k+1}} w^{i,j}_{k+1|k+1} \cdot N_{P^i_{k+1|k+1}}(\mathbf{x} - \mathbf{x}^{i,j}_{k+1|k+1}) \tag{16.272}$$

where

$$\mathbf{x}^i_{k+1|k+1} = \mathbf{x}^i, \quad P^i_{k+1|k+1} = P_i \tag{16.273}$$

$$w^i_{k+1|k+1} = (1 - p_D) \cdot w^i_{k+1|k+1} \tag{16.274}$$

and

$$\mathbf{x}^{i,j}_{k+1|k+1} = \mathbf{x}_{k+1|k} + K_i\left(\mathbf{z}_j - H\mathbf{x}^i\right) \quad (16.275)$$

$$P^{i,j}_{k+1|k+1} = \left(I - K_i H\right) P_i \quad (16.276)$$

$$K_i = P_i H^T \left(H P_i H^T + R\right)^{-1} \quad (16.277)$$

$$w^{i,j}_{k+1|k+1} \quad (16.278)$$

$$= \frac{w^i_{k+1|k} \cdot p_D \cdot N_{R+HP_iH^T}(\mathbf{z}_j - H\mathbf{x}^i) \cdot N_{C_i}(\mathbf{x} - \mathbf{c}_i)}{\lambda c(\mathbf{z}_j) + p_D \sum_{e=1}^{n_{k+1|k}} w^e_{k+1|k} \cdot N_{R+HP_eH^T}(\mathbf{z}_j - H\mathbf{x}^e)} \quad (16.279)$$

and where we have abbreviated

$$H \overset{\text{abbr.}}{=} H_{k+1}, \quad R \overset{\text{abbr.}}{=} R_{k+1}, \quad P_i \overset{\text{abbr.}}{=} P^i_{k+1|k} \quad (16.280)$$

$$\mathbf{x}^i \overset{\text{abbr.}}{=} \mathbf{x}^i_{k+1|k}, \quad K_i \overset{\text{abbr.}}{=} K^i_{k+1}. \quad (16.281)$$

The updated GM-PHD has

$$n_{k+1|k+1} = n_{k+1|k+1} + n_{k+1|k+1} \cdot m_{k+1} \quad (16.282)$$

$$= n_{k+1|k+1} \cdot (1 + m_{k+1}) \quad (16.283)$$

components. As with the predictor step, the number of components will increase exponentially. Techniques such as pruning and merging must be used to limit combinatorial explosion.

The predicted label table $\mathcal{L}_{k+1|k}$ must be updated to $\mathcal{L}_{k+1|k+1}$. Each single-index component

$$(w^i_{k+1|k+1}, \mathbf{x}^i_{k+1|k+1}, P^i_{k+1|k+1}) \quad (16.284)$$

retains the same label as its predecessor $(w^i_{k+1|k}, \mathbf{x}^i_{k+1|k}, P^i_{k+1|k})$. Likewise, each two-index component

$$(w^{i,j}_{k+1|k+1}, \mathbf{x}^{i,j}_{k+1|k+1}, P^i_{k+1|k+1}) \quad (16.285)$$

is assigned the same label as its predecessor $(w^i_{k+1|k}, \mathbf{x}^i_{k+1|k}, P^i_{k+1|k})$. This last step will result in different components that have the same label. Such duplication is rectified after pruning and merging.

16.5.3.5 GM-PHD Filter Pruning and Merging

Suppose that the current set of components is $(w^i_{k|k}, \mathbf{x}^i_{k|k}, P^i_{k|k})$ for $i = 1, ..., n_{k|k}$. Those components for which $w^i_{k|k}$ is sufficiently small are eliminated. The weights of the remaining components are renormalized to sum to unity.

Clark, Panta, and Vo [33, 34] employ the following merging approach. They use the merging distance

$$d_{i,j} \triangleq (\mathbf{x}^i_{k|k} - \mathbf{x}^j_{k|k})^T (P^i_{k|k})^{-1} (\mathbf{x}^i_{k|k} - \mathbf{x}^j_{k|k}). \tag{16.286}$$

If this is sufficiently small then the two components $(w^i_{k|k}, \mathbf{x}^i_{k|k}, P^i_{k|k})$ and $(w^j_{k|k}, \mathbf{x}^j_{k|k}, P^j_{k|k})$ are merged into a single component $(w_{i,j}, \mathbf{c}_{i,j}, C_{i,j})$ with

$$w_{i,j} \triangleq w^i_{k|k} + w^j_{k|k} \tag{16.287}$$

$$\mathbf{c}_{i,j} = \frac{w_i \mathbf{x}^i_{k|k} + w_j \mathbf{x}^j_{k|k}}{w_i + w_j} \tag{16.288}$$

$$C_{i,j} = \frac{w_i P^i_{k|k} + w_j P^j_{k|k}}{w_i + w_j} + (\mathbf{x}^i_{k|k} - \mathbf{x}^j_{k|k})(\mathbf{x}^i_{k|k} - \mathbf{x}^j_{k|k})^T. \tag{16.289}$$

The merged component is assigned the label of the premerged component that had the largest weight.

Also, because of the corrector step it will often be the case that two or more components will have the same label even after pruning and merging. In this case the component with the largest weight keeps the label and new labels are assigned to the others. This is repeated until all components have unique labels.

16.5.3.6 GM-PHD Filter State and Error Estimation

Suppose that the current set of components is $(w^i_{k|k}, \mathbf{x}^i_{k|k}, P^i_{k|k})$ for $i = 1, ..., n_{k|k}$. Those components for which $w^i_{k|k}$ is sufficiently large are declared to be live tracks. Their corresponding covariance matrices are chosen to be their error estimates. The declared tracks are identified by their labels. By comparing these labels to the labels in the previous state and error estimate, we therefore get time-connected tracks.

16.6 LIMITATIONS OF THE PHD FILTER

Because it is a first-order multitarget moment, the PHD $D_{k|k}(\mathbf{x}|Z^{(k)})$ represents a major loss of information compared to the multitarget distribution $f_{k|k}(X|Z^{(k)})$. The Poisson assumption, necessary to produce closed-form formulas for the PHD corrector, produces a still greater loss of information. The purpose of this section is to gain a better understanding of this information loss via examination of a special case originally proposed by Erdinc, Willett, and Bar-Shalom [57].

Suppose that at most a single target is present. Then it is possible to develop an *exact* formula for the PHD corrector equation. Note that, under the same conditions, the JoTT filter (Section 14.7) is not only exact but theoretically optimal. In this case, and using the notation of that section, prior and predicted PHDs can be written as

$$D_{k+1|k}(\mathbf{x}) = p_{k+1|k} \cdot f_{k+1|k}(\mathbf{x}) \tag{16.290}$$

$$D_{k+1|k+1}(\mathbf{x}) = p_{k+1|k+1} \cdot f_{k+1|k+1}(\mathbf{x}) \tag{16.291}$$

where the respected expected target numbers are

$$N_{k+1|k} = p_{k+1|k}, \qquad N_{k+1|k+1} = p_{k+1|k+1}. \tag{16.292}$$

From the JoTT corrector equations (14.202) and (14.203), we can derive an *exact* equation for the PHD corrector. In Appendix G.30, I show that it is

$$D_{k+1|k+1}(\mathbf{x}) = \ell_Z(\mathbf{x}) \cdot D_{k+1|k}(\mathbf{x}) \tag{16.293}$$

where the exact PHD pseudolikelihood $\ell_Z(\mathbf{x}) \overset{\text{abbr.}}{=} \ell_Z(\mathbf{x}|Z^{(k)})$ is

$$\ell_Z(\mathbf{x}) = \frac{1 - p_D(\mathbf{x}) + p_D(\mathbf{x}) \sum_{\mathbf{z}\in Z} \frac{L_{\mathbf{z}}(\mathbf{x})}{\lambda c(\mathbf{z})}}{1 - D_{k+1|k}[p_D] + \sum_{\mathbf{z}\in Z} \frac{D_{k+1|k}[p_D L_{\mathbf{z}}]}{\lambda c(\mathbf{z})}}. \tag{16.294}$$

The corresponding PHD pseudolikelihood is, of course,

$$L_Z(\mathbf{x}) = 1 - p_D(\mathbf{x}) + \sum_{\mathbf{z}\in Z} \frac{p_D(\mathbf{x}) \cdot L_{\mathbf{z}}(\mathbf{x})}{\lambda c(\mathbf{z}) + D_{k+1|k}[p_D L_{\mathbf{z}}]}. \tag{16.295}$$

Compare (16.294) and (16.295). Suppose that probability of detection is constant and that no measurements are collected: $Z_{k+1} = \emptyset$. Then the two equations reduce to:

$$\ell_Z(\mathbf{x}) = \frac{1-p_D}{1-p_{k+1|k}\cdot p_D} \tag{16.296}$$

$$L_Z(\mathbf{x}) = 1-p_D. \tag{16.297}$$

Accordingly, the expected number of targets in the two cases becomes

$$N_{k+1|k+1} = \frac{(1-p_D)\cdot N_{k+1|k}}{1-N_{k+1|k}\cdot p_D} \tag{16.298}$$

$$N_{k+1|k+1} = (1-p_D)\cdot N_{k+1|k}. \tag{16.299}$$

That is, the first-order approximation assumed in the PHD filter corrector equation linearizes the exact target number to $N_{k+1|k+1} = (1-p_D)\cdot N_{k+1|k}$.

Thus, for example, if $p_{k+1|k} = 1$ and $p_D = 0.9$ then the two filters arrive at very different instantaneous estimates for the data-updated target number:

$$\text{exact PHD filter:} \qquad N_{k+1|k+1} = 1.0 \tag{16.300}$$

$$\text{approx. PHD filter:} \qquad N_{k+1|k+1} = 0.1. \tag{16.301}$$

In [57], Erdinc, Willett, and Bar-Shalom presented (16.298) and (16.299) to illustrate information loss in the PHD filter. As a consequence they posed the following question:

- Can the PHD filter be generalized to include higher-order information on target number such as variance?

This query motivated the development of the CPHD filter, to be introduced next in Section 16.7. See also Section 16.11.

16.7 THE CARDINALIZED PHD (CPHD) FILTER

The purpose of this section is to construct the CPHD filter, which was summarized in Section 16.1.3. The CPHD filter propagates not only the PHD $D_{k|k}(\mathbf{x})$ but also the cardinality distribution $p_{k|k}(n)$ and its probability-generating function

$G_{k|k}(x)$:

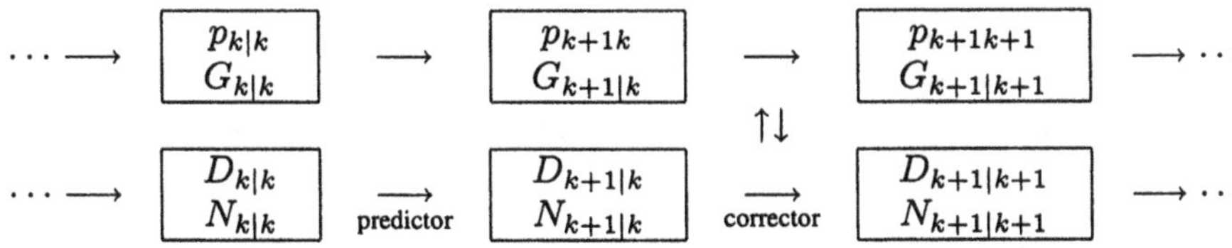

The pair of vertical arrows at the corrector step indicates that the upper and lower filters are mutually coupled. On the one hand, the formula for $D_{k+1|k+1}(\mathbf{x})$ requires $G_{k+1|k}(x)$ and its derivatives of all orders $G^{(i)}_{k+1|k}(x)$, $i \geq 0$. On the other hand, the formulas for $G_{k+1|k+1}(x)$ and $p_{k+1k+1}(n)$ require $D_{k+1|k}(\mathbf{x})$ and its integral $N_{k+1|k} = \int D_{k+1|k}(\mathbf{x})d\mathbf{x}$.

In what follows I describe the basic steps of the CPHD filter: initialization (Section 16.7.1), prediction (Section 16.7.2), correction (Section 16.7.3), and state and error estimation (Section 16.7.4). I discuss computational issues in Section 16.7.5 and the relationship between the CPHD and JoTT filters in Section 16.7.6.

16.7.1 CPHD Filter Initialization

Initialization of the CPHD filter consists of choosing an initial PHD and an initial cardinality distribution:

$$D_{0|0}(\mathbf{x}) = D_{0|0}(\mathbf{x}|Z^{(0)}) = n_0 \cdot s_0(\mathbf{x}) \tag{16.302}$$

$$p_{0|0}(n) = p_{0|0}(n|Z^{(0)}) = p_0(n) \tag{16.303}$$

where $s_0(\mathbf{x})$ is a probability density with peaks corresponding to a priori target positions; and where $p_0(n)$ is a probability distribution on n, the expected value of which is n_0:

$$n_0 = \sum_{n=0}^{\infty} n \cdot p_0(n). \tag{16.304}$$

As with the PHD filter, one can choose $D_{0|0}$ to be a sum of Gaussians

$$D_{0|0}(\mathbf{x}) = N_{P^1_{0|0}}(\mathbf{x} - \mathbf{x}^1_{0|0}) + ... + N_{P^{n_0}_{0|0}}(\mathbf{x} - \mathbf{x}^{n_0}_{0|0}). \tag{16.305}$$

The initial distribution could be chosen to be binomial:

$$p_0(n) = B_{\nu_0,q_0}(n) = C_{\nu_0,n} \cdot q_0^n \cdot (1 - q_0)^{\nu_0 - n} \tag{16.306}$$

where $n_0 = \nu_0 q_0$.

If very little is known about the initial target positions, one can choose a uniform PHD and a uniform target number distribution. That is, $D_{0|0}(\mathbf{x}) = n_0 \cdot s_0(\mathbf{x})$ where $s_0(\mathbf{x})$ is uniform in some region D and where n_0 is a guess about how many targets might be present. Also, $p_0(n)$ is uniform on some interval $[0, \nu]$ with $n_0 = \nu_0/2$.

16.7.2 CPHD Filter Predictor

I describe the multitarget motion assumptions underlying the CPHD filter predictor, and then list the CPHD predictor equations.

16.7.2.1 CPHD Predictor Assumptions

The CPHD filter presumes the same multitarget motion model as the PHD filter, with the exception that target spawning can no longer be explicitly modeled. That is, it presumes target motions are statistically independent; targets can disappear from the scene; and new targets can appear in the scene independently of existing targets. These possibilities are described as follows.

- *Motion of individual targets:* $f_{k+1|k}(\mathbf{x}|\mathbf{x}')$ is the single-target Markov transition density;
- *Disappearance of existing targets:* $p_S(\mathbf{x}) \overset{\text{abbr.}}{=} p_{S,k+1|k}(\mathbf{x})$ is the probability that any target with state $\mathbf{x}$ at time step k will survive to time step $k+1$;
- *Appearance of completely new targets:* $b(X) \overset{\text{abbr.}}{=} b_{k+1|k}(X)$ is the likelihood that new targets with state set X will enter the scene at time step $k+1$, and its PHD, cardinality distribution, and p.g.f. are

$$b(\mathbf{x}) \overset{\text{abbr.}}{=} \int b_{k+1|k}(\{\mathbf{x}\} \cup W)\delta W \tag{16.307}$$

$$p_B(n) \overset{\text{abbr.}}{=} \int_{|X|=n} b_{k+1|k}(X)\delta X \tag{16.308}$$

$$B(x) \overset{\text{abbr.}}{=} \sum_{n=0}^{\infty} p_B(n) \cdot x^n. \tag{16.309}$$

Closed-form formulas for the predicted p.g.f. require the following simplifying assumption:

- *I.i.d. cluster process multitarget prior:* The p.g.fl. of $f_{k|k}(X|Z^{(k)})$ is an i.i.d. cluster process—that is, has the form $G[h] = G(s[h])$ where $G(x) \overset{\text{abbr.}}{=} G_{k|k}(x)$ is the p.g.f. of the cardinality distribution of $f_{k|k}(X|Z^{(k)})$ and where $s[h]$ is defined later.

16.7.2.2 CPHD Predictor Equations

From time step k we have in hand the following items:

- The PHD $D_{k|k}(\mathbf{x}) \overset{\text{abbr.}}{=} D_{k|k}(\mathbf{x}|Z^{(k)})$;
- The expected number of targets $N_{k|k}$;
- The cardinality distribution $p(n) \overset{\text{abbr.}}{=} p_{k|k}(n|Z^{(k)})$;
- The p.g.f. $G(x) \overset{\text{abbr.}}{=} G_{k|k}(x|Z^{(k)})$.

We are to specify formulas for the following things:

- The predicted PHD $D_{k+1|k}(\mathbf{x}) \overset{\text{abbr.}}{=} D_{k+1|k}(\mathbf{x}|Z^{(k)})$;
- The predicted expected number of targets $N_{k+1|k}$;
- The predicted cardinality distribution $p_{k+1|k}(n) \overset{\text{abbr.}}{=} p_{k+1|k}(n|Z^{(k)})$;
- The predicted p.g.f. $G_{k+1|k}(x) \overset{\text{abbr.}}{=} G_{k+1|k}(x|Z^{(k)})$.

Abbreviate

$$s(\mathbf{x}) \overset{\text{abbr.}}{=} \frac{D_{k|k}(\mathbf{x})}{N_{k|k}}, \qquad s[h] \overset{\text{abbr.}}{=} \int h(\mathbf{x}) \cdot s(\mathbf{x}) d\mathbf{x}. \tag{16.310}$$

Then the CPHD filter corrector equations are as follows [142]:

$$G_{k+1|k}(x) \cong B(x) \cdot G(1 - s[p_S] + s[p_S] \cdot x) \tag{16.311}$$

$$D_{k+1|k}(\mathbf{x}) = b(\mathbf{x}) + \int p_S(\mathbf{x}') \cdot f_{k+1|k}(\mathbf{x}|\mathbf{x}') \cdot D_{k|k}(\mathbf{x}') d\mathbf{x}' \tag{16.312}$$

$$p_{k+1|k}(n) \cong \sum_{i=0}^{n} \frac{1}{i!} \cdot p_B(n-i) \cdot G^{(i)}(1 - s[p_S]) \cdot s[p_S]^i \tag{16.313}$$

$$N_{k+1|k} = N^{\text{birth}}_{k+1|k} + N^{\text{persist}}_{k+1|k} \tag{16.314}$$

where

$$N^{\text{birth}}_{k+1|k} = \int b(\mathbf{x})d\mathbf{x} \tag{16.315}$$

$$N^{\text{persist}}_{k+1|k} = \int p_S(\mathbf{x}) \cdot D_{k|k}(\mathbf{x})d\mathbf{x} \tag{16.316}$$

are, respectively, the expected number of new and persisting targets.

16.7.3 CPHD Filter Single-Sensor Corrector

I describe the multitarget measurement assumptions underlying the CPHD filter corrector, and then give the CPHD corrector equations for both single and multiple sensors.

16.7.3.1 CPHD Single-Sensor Corrector Assumptions

The CPHD filter presumes a multitarget measurement model more general than that for the PHD filter. That is: no target generates more than one measurement and each measurement is generated by no more than a single target, all measurements are conditionally independent of target state, there are missed detections, and the false alarm process is an i.i.d. cluster process as defined in Section 11.121. More precisely, these factors are stated as follows.

- *Single-target measurement generation*: $L_{\mathbf{z}}(\mathbf{x}) \overset{\text{abbr.}}{=} f_{k+1}(\mathbf{z}|\mathbf{x}, \mathring{\mathbf{x}})$ is the sensor likelihood function.
- *Probability of detection (sensor field of view):* $p_D(\mathbf{x}) \overset{\text{abbr.}}{=} p_{D,k+1}(\mathbf{x}, \mathring{\mathbf{x}})$ is the probability that an observation will be collected at time step $k+1$ from a target with state $\mathbf{x}$, if the sensor has state $\mathring{\mathbf{x}}$ at that time step.
- *I.i.d. cluster process false alarms:* at time step $k+1$ the sensor collects false alarms whose spatial distribution is given by the probability density $c(\mathbf{z}) \overset{\text{abbr.}}{=} c_{k+1}(\mathbf{z}|\mathring{\mathbf{x}})$ and whose cardinality distribution is $\kappa(m) \overset{\text{abbr.}}{=} \kappa_{k+1}(m|\mathring{\mathbf{x}})$. Also, the p.g.f. of $\kappa(m)$ is $C(z) \overset{\text{abbr.}}{=} C_{k+1}(z)$.

We cannot derive closed-form formulas for the corrector step without making an additional simplifying assumption:

- *I.i.d. cluster process multitarget prior:* the predicted multitarget distribution $f_{k+1|k}(X|Z^{(k)})$ is approximately an i.i.d. cluster process with p.g.fl. $G[h] =$

$G(s[h])$. Here $G(x) \overset{\text{abbr.}}{=} G_{k+1|k}(x|Z^{(k)})$ is the p.g.f. of the predicted cardinality distribution $p(n) \overset{\text{abbr.}}{=} p_{k+1|k}(n|Z^{(k)})$ and $s[h]$ is defined later.

The ith derivatives of $G(x)$ and of $C(z)$ are $G^{(i)}(x)$ and $C^{(i)}(z)$, respectively.

16.7.3.2 CPHD Single-Sensor Corrector Equations

From time step k we have in hand the following things:

- The predicted PHD $D_{k+1|k}(\mathbf{x}) \overset{\text{abbr.}}{=} D_{k+1|k}(\mathbf{x}|Z^{(k)})$;
- The predicted expected number of targets $N_{k+1|k}$;
- The predicted cardinality distribution $p(n) \overset{\text{abbr.}}{=} p_{k+1|k}(n|Z^{(k)})$;
- The predicted p.g.f. $G(x) \overset{\text{abbr.}}{=} G_{k+1|k}(x|Z^{(k)})$.

We collect a new measurement set $Z_{k+1} = \{\mathbf{z}_1, ..., \mathbf{z}_m\}$ with $|Z_{k+1}| = m$. We are to specify formulas for the following items:

- The data-updated PHD

$$D_{k+1|k+1}(\mathbf{x}) \overset{\text{abbr.}}{=} D_{k+1|k+1}(\mathbf{x}|Z^{(k+1)}); \tag{16.317}$$

- The data-updated expected number of targets $N_{k+1|k+1}$;
- The data-updated cardinality distribution

$$p_{k+1|k+1}(n) \overset{\text{abbr.}}{=} p_{k+1|k+1}(n|Z^{(k+1)}); \tag{16.318}$$

- The data-updated p.g.f.

$$G_{k+1|k+1}(x) \overset{\text{abbr.}}{=} G_{k+1|k+1}(x|Z^{(k+1)}). \tag{16.319}$$

Let $G^{(i)}(x)$ be the ith derivative of the predicted p.g.f. $G(x)$. Abbreviate

$$q(\mathbf{x}) \overset{\text{abbr.}}{=} 1 - p_D(\mathbf{x}) \tag{16.320}$$

$$s(\mathbf{x}) \overset{\text{abbr.}}{=} \frac{D_{k+1|k}(\mathbf{x})}{N_{k+1|k}} \tag{16.321}$$

$$s[h] \overset{\text{abbr.}}{=} \int h(\mathbf{x}) \cdot s(\mathbf{x}) d\mathbf{x} \tag{16.322}$$

and for any $Z = \{\mathbf{z}_1, ..., \mathbf{z}_m\}$ further abbreviate[10]

$$\sigma_i(Z) \triangleq \sigma_{m,i}\left(\frac{D_{k+1|k}[p_D L_{\mathbf{z}_1}]}{c(\mathbf{z}_1)}, ..., \frac{D_{k+1|k}[p_D L_{\mathbf{z}_m}]}{c(\mathbf{z}_m)}\right) \tag{16.323}$$

$$\hat{G}^{(i)}(x) \triangleq \frac{G^{(i)}(x)}{G^{(1)}(1)^i} = \frac{G^{(i)}(x)}{N^i_{k+1|k}} \tag{16.324}$$

where $\sigma_{m,i}(x_1, ..., x_m)$ is the elementary symmetric function as defined in (11.141).

Then the first two corrector equations for the CPHD filter are [142]:

$$G_{k+1|k+1}(x) \tag{16.325}$$

$$\cong \frac{\sum_{j=0}^m x^j \cdot C^{(m-j)}(0) \cdot \hat{G}^{(j)}(xs[q_D]) \cdot \sigma_j(Z_{k+1})}{\sum_{i=0}^m C^{(m-i)}(0) \cdot \hat{G}^{(i)}(s[q_D]) \cdot \sigma_i(Z_{k+1})} \tag{16.326}$$

and

$$D_{k+1|k+1}(\mathbf{x}) \cong L_{Z_{k+1}}(\mathbf{x}) \cdot D_{k+1|k}(\mathbf{x}) \tag{16.327}$$

where

$$L_Z(\mathbf{x}) \tag{16.328}$$

$$\triangleq \frac{\sum_{j=0}^m C^{(m-j)}(0) \cdot \hat{G}^{(j+1)}(s[q_D]) \cdot \sigma_j(Z)}{\sum_{i=0}^m C^{(m-i)}(0) \cdot \hat{G}^{(i)}(s[q_D]) \cdot \sigma_i(Z)} \cdot (1 - p_D(\mathbf{x})) \tag{16.329}$$

$$+p_D(\mathbf{x}) \cdot \sum_{\mathbf{z} \in Z} \frac{L_{\mathbf{z}}(\mathbf{x})}{c(\mathbf{z})} \tag{16.330}$$

$$\cdot \frac{\sum_{j=0}^{m-1} C^{(m-j-1)}(0) \cdot \hat{G}^{(j+1)}(s[q_D]) \cdot \sigma_j(Z - \{\mathbf{z}\})}{\sum_{i=0}^m C^{(m-i)}(0) \cdot \hat{G}^{(i)}(s[q_D]) \cdot \sigma_i(Z)}. \tag{16.331}$$

10 The notation $\hat{G}^{(i)}(x)$ requires cautious handling. Note that

$$\frac{d^e}{dx^e}\hat{G}^{(i)}(x) = \hat{G}^{(i+e)}(x) \cdot G^{(1)}(1)^e$$

and not

$$\frac{d^e}{dx^e}\hat{G}^{(i)}(x) = \hat{G}^{(i+e)}(x).$$

For this reason we write

$$\hat{G}^{(i)(e)}(x) \triangleq \frac{d^e}{dx^e}\hat{G}^{(i)}(x).$$

The second two corrector equations are [142]:

$$N_{k+1|k+1} \cong \frac{\sum_{j=0}^{m} C^{(m-j)}(0) \cdot \alpha_j \cdot \sigma_j(Z_{k+1})}{\sum_{i=0}^{m} C^{(m-i)}(0) \cdot \hat{G}^{(i)}(s[q_D]) \cdot \sigma_i(Z_{k+1})} \tag{16.332}$$

and

$$p_{k+1|k+1}(n) = \frac{\sum_{j=0}^{m} C^{(m-j)}(0) \cdot \sigma_j(Z_{k+1}) \cdot \frac{1}{(n-j)!} \cdot \hat{G}^{(j)(n-j)}(0) \cdot s[q_D]^{n-j}}{\sum_{i=0}^{m} C^{(m-i)}(0) \cdot \hat{G}^{(i)}(s[q_D]) \cdot \sigma_i(Z_{k+1})} \tag{16.333}$$

where

$$\alpha_j \triangleq j \cdot \hat{G}^{(j)}(s[q_D]) + \hat{G}^{(j+1)}(s[q_D]) \cdot s[q_D] \cdot G^{(1)}(1) \tag{16.334}$$

$$= j \cdot \hat{G}^{(j)}(s[q_D]) + \hat{G}^{(j+1)}(s[q_D]) \cdot D_{k+1|k}[q_D]. \tag{16.335}$$

It is also possible to derive closed-form formulas for the $G^{(i)}_{k+1|k+1}(x)$. These are

$$G^{(n)}_{k+1|k+1}(x) = \frac{\left(\begin{array}{c} \sum_{e=0}^{m} \sum_{j=0}^{n} C_{n,j} \cdot C^{(m-e)}(0) \cdot \sigma_e(Z_{k+1}) \\ \cdot \frac{e!}{(e-j)!} \cdot x^{e-j} \cdot \hat{G}^{(e)(n-j)}(xs[q_D]) \cdot s[q_D]^{n-j} \end{array} \right)}{\sum_{i=0}^{m} C^{(m-i)}(0) \cdot \hat{G}^{(i)}(s[q_D]) \cdot \sigma_i(Z_{k+1})}. \tag{16.336}$$

Note that $i! = \Gamma(i+1) = \pm\infty$ for all integers $i < 0$ where $\Gamma(x)$ is the gamma function. Thus any terms in these equations with $i > n$ vanish. The $\sigma_{M,i}$ can be computed using the method of (16.340) and (16.341).

16.7.3.3 CPHD Multisensor Corrector Equations

In (16.123) and (16.125) I described a heuristic technique for fusing simultaneous observation sets from different sensors: applying the PHD corrector step in succession, one time for each sensor. The same technique can be used with the CPHD filter corrector.

16.7.4 CPHD Filter State and Error Estimation

The PHD's EAP estimate $N_{k|k}$ of the posterior expected number of targets may be both unstable and inaccurate under lower-SNR conditions. One reason for this is that EAP estimation often produces unstable and inaccurate state-estimates under lower-SNR conditions. The minor modes of the posterior distribution are induced by false alarms and thus tend to be highly random, and they can erratically perturb the expected value away from the target-induced primary mode. The maximum a posteriori (MAP) estimator, by way of contrast, ignores minor modes and locks onto the more stable and accurate major mode.

Consequently, for the CPHD filter the MAP estimate

$$\hat{n}_{k|k} \triangleq \arg\sup_{n} \; p_{k|k}(n|Z^{(k)}) \tag{16.337}$$

may produce more accurate and stable estimates of target number.

Target state-estimates are computed in the same manner as for the PHD filter (Section 16.3.4).

16.7.5 Computational Complexity of the CPHD Filter

The cardinality distributions $p_{k|k}(n|Z^{(k)})$ and $p_{k|k}(n|Z^{(k)})$ can be nonvanishing for all $n \geq 0$. Consequently, the CPHD filter will be inherently computationally intractable in the event that these distributions have heavy tails.

Suppose on the other hand that these distributions vanish, at least approximately, for all n larger than some largest value ν. From (16.311) we see that their p.g.f.s are polynomials of degrees not exceeding ν. It can be shown [142] that in the corrector step,

$$\deg G_{k+1|k+1}(x) \leq \deg G_{k+1|k}(x). \tag{16.338}$$

For the predictor step, however, it can be shown [142] that

$$\deg G_{k+1|k}(x) \leq \nu_B + \deg G_{k|k}(x) \tag{16.339}$$

where v_B is the maximum number of new targets. Thus the computability of the CPHD filter is partially limited by what target-birth model is adopted.

Beyond these contributing effects, the primary source of computational complexity will be the sums of the elementary symmetric functions $\sigma_1(Z)$,....,$\sigma_m(Z)$ in the corrector step. Luckily, these sums are not as combinatorially daunting as they might at first seem. The reason is that $\sigma_{m,i}(x_1, ..., x_m)$ for $i = 1, ..., m$ can be computed with only order m^2 complexity using the following double recursion on $j = 1, ..., m$ and $i = 1, ..., j$ [70, p. 40]:

$$\sigma_{j,1}(x_1, ..., x_m) = x_1 + ... + x_j \tag{16.340}$$
$$\sigma_{j,i}(x_1, ..., x_m) = \sigma_{j-1,i}(x_1, ..., x_{m-1}) \tag{16.341}$$
$$+x_j \cdot \sigma_{j-1,i-1}(x_1, ..., x_{m-1}). \tag{16.342}$$

Consequently, taken as a whole the CPHD filter will have computational complexity $O(m^3 n)$.[11]

Vo, Vo, and Cantoni [242, Section IV-B] have proposed an alternative method, based on Vieta's theorem, for computing the $\sigma_i(Z)$. Suppose that

$$a_M x^M + ... + a_1 x + a_0 = a_M(x - r_1) \cdots (x - r_M) \tag{16.343}$$

is a polynomial with M known roots $r_1, ..., r_M$. Then for $i = 0, 1, ..., M$

$$\sigma_{M,i}(r_1, ..., r_M) = (-1)^i \cdot \frac{a_{M-i}}{a_M}. \tag{16.344}$$

16.7.6 CPHD and JoTT Filters Compared

In this section, I point out the following fact: the CPHD filter reduces to the JoTT filter of Section 14.7 when the following are true.

- The number n of targets is no larger than one.
- The probability of detection p_D is constant.
- The false alarm process $\kappa(Z)$ is an i.i.d. cluster process.

This means that *under these assumptions, the CPHD filter is a special case of the multitarget Bayes filter, rather than just an approximation of it.*

The CPHD predictor and corrector equations are predicated on the following assumption: the multitarget posteriors $f_{k|k}(X|Z^{(k)})$ and $f_{k|k}(X|Z^{(k)})$ are, approximately, the distributions of i.i.d. cluster processes. That is, their respective p.g.fl.s have the form

$$G_{k|k}[h] \cong G_{k|k}(s_{k|k}[h]) \tag{16.345}$$
$$G_{k+1|k}[h] \cong G_{k+1|k}(s_{k+1|k}[h]) \tag{16.346}$$

11 The tractability of this double recursion was one impetus behind the development of Kamen's symmetric measurement equation (SME) filter [103, 170].

where $G_{k|k}(x)$ and $G_{k+1|k}(x)$ are p.g.f.s and $s_{k|k}(\mathbf{x})$ and $s_{k+1|k}(\mathbf{x})$ are probability density functions.

When the maximum number of targets is one, this assumption is not an approximation—it is exact. To see why, note that in this case all multitarget posteriors must have the form

$$f_{k|k}(X|Z^{(k)}) = \begin{cases} 1 - p_{k|k} & \text{if} \quad X = \emptyset \\ p_{k|k} \cdot s_{k|k}(\mathbf{x}) & \text{if} \quad X = \{\mathbf{x}\} \\ 0 & \text{if} \quad |X| \geq 2 \end{cases} . \tag{16.347}$$

As a result, all of the p.g.fl.s of the multitarget posteriors must have the form

$$G_{k|k}[h] = \int h^X \cdot f_{k|k}(X|Z^{(k)})\delta X \tag{16.348}$$

$$= f_{k+1|k}(X|Z^{(k)}) + \int h(\mathbf{x}) \cdot f_{k+1|k}(\{\mathbf{x}\}|Z^{(k)})d\mathbf{x} \tag{16.349}$$

$$= 1 - p_{k|k} + p_{k|k} \cdot s_{k|k}[h] \tag{16.350}$$

where $s_{k|k}[h] \triangleq \int h(\mathbf{x}) \cdot s_{k|k}(\mathbf{x})d\mathbf{x}$. The p.g.f. of the cardinality distribution of $f_{k+1|k}(X|Z^{(k)})$ is, therefore,

$$G_{k|k}(x) = G_{k|k}[x] = 1 - p_{k|k} + p_{k|k} \cdot x. \tag{16.351}$$

Thus in this special case $G_{k|k}[h]$ actually is an i.i.d. cluster process with $N_{k+1|k} = G^{(1)}_{k|k}(1) = p_{k|k}$. So, the CPHD filter must be exact, rather than an approximation.

16.8 PHYSICAL INTERPRETATION OF CPHD FILTER

In Section 16.4, I described a physical interpretation of the PHD filter due to Erdinc, Willett, and Bar-Shalom [56]. In the same paper, these authors derived a similar physical interpretation of the CPHD filter. I will not discuss this further here, instead referring the interested reader to the original paper.

16.9 IMPLEMENTING THE CPHD FILTER

As with the PHD filter, the CPHD filter can be approximated using both sequential Monte Carlo and Gaussian-mixture methods. The purpose of this section is to summarize how this is accomplished. I begin in Section 16.9.1 with an overview of implementations of CPHD filter currently in existence. In Section 16.9.2, I summarize particle approximation of the CPHD filter (SMC-CPHD filter). In Section 16.9.3, I summarize Gaussian-mixture approximation of the CPHD filter (GM-CPHD filter).

16.9.1 Survey of CPHD Filter Implementations

The CPHD filter was first introduced in 2006. As a result, only a few implementations have appeared in the literature.

In [243], Vo, Vo, and Cantoni announced the Gaussian-mixture implementation of the CPHD filter. They also described two versions of the GM-CPHD filter: one using an extended Kalman filter (EKF) and another using the unscented Kalman filter (UKF), which is capable of operating under nonlinear conditions.

They tested and compared these two versions in two-dimensional simulations. In one simulation, five targets appeared and disappeared while observed by a linear-Gaussian sensor in a dense Poisson false alarm environment. The EKF version of the GM-CPHD filter correctly detected all target births and deaths and successfully tracked the targets during the times they were present in the scene. It also successfully negotiated a track crossing that occurred at midscenario.

A second scenario also involved five targets with target appearances and disappearances. This time, however, the sensor was nonlinear (range-bearing) and nonlinear target dynamics occurred. Both the EKF and UKF versions of the GM-CPHD filter were tested on the simulated data. The two versions both successfully negotiated the simulation with similar performance.

In [240] Vo, Vo, and Cantoni described detailed performance comparisons of the GM-PHD and GM-CPHD filters. In the first scenario, up to 10 targets could appear randomly, with track crossings. The GM-PHD and GM-CPHD filters were both successful at identifying target births and deaths, at tracking the targets, and at negotiating track crossings. As expected, for any individual sample path (measurement sequence), the GM-CPHD filter's estimates of instantaneous target number were far more accurate and stable (small variance) than those of the GM-PHD filter.

However, the authors also noted unexpected differences between the sample-path and the Monte Carlo behaviors of the two filters. The GM-PHD filter's Monte Carlo instantaneous estimates of target number were essentially as accurate as those for the GM-CPHD filter. (For any particular sample path, of course, the PHD's instantaneous target number estimates would be much worse.) Moreover, the GM-PHD filter's Monte Carlo response to target appearance and disappearance was "almost instantaneous"whereas the GM-CPHD filter's Monte Carlo response was sluggish.

Vo, Vo, and Cantoni speculated that this behavior was due to the fact that the PHD filter has a weak memory and thus is easily influenced by new measurements. I suspect, however, that this only partially explains the results. Because the CPHD filter has a better memory—that is, its effectiveness as a filter is better—its behavior is more heavily influenced by its underlying multitarget motion model for birth and death. If actual target birth and death in a scenario deviates from this model, then the CPHD filter will tend to respond sluggishly in an averaged, Monte Carlo sense. The PHD filter, which is less influenced by the motion model because of its more limited memory, will have better transient response in a Monte Carlo (but not a sample-path) sense.

It is possible that the GM-CPHD filter would exhibit better Monte Carlo transient response if it were capable of adaptively choosing its internal birth-death models using jump-Markov techniques.

16.9.2 Particle Approximation (SMC-CPHD)

The basic concepts of particle system approximation of the CPHD filter are essentially the same as for the SMC-PHD filter of Section 16.5.2. In what follows I summarize the basic steps: initialization, prediction, correction, and state and error estimation.

16.9.2.1 SMC-CPHD Filter Initialization

Initialization of the SMC-CPHD filter is essentially the same as that for the SMC-PHD filter of Section 16.5.2. The only difference is that the target-number distribution must also be initialized as in Section 16.7.1.

16.9.2.2 SMC-CPHD Filter Predictor

The SMC-CPHD filter predictor is essentially the same as that for the SMC-PHD filter predictor of Section 16.5.2. The only differences are that target spawning is not modeled and that a predicted p.g.f. must be computed as in Section 16.7.2.

16.9.2.3 SMC-CPHD Filter Corrector

Assume that the predicted PHD has been approximated by a predicted particle system $\mathbf{x}^1_{k+1|k}, ..., \mathbf{x}^\nu_{k+1|k}$. The integral $D_{k+1|k}[p_D L_\mathbf{z}]$ in the PHD corrector equation becomes

$$D_{k+1|k}[p_D L_\mathbf{z}] \cong \frac{1}{\nu}\sum_{i=1}^{\nu} p_D(\mathbf{x}^i_{k+1|k}) \cdot L_\mathbf{z}(\mathbf{x}^i_{k+1|k}). \tag{16.352}$$

Consequently, the Bayes corrector equation, (2.82), becomes

$$\int \theta(\mathbf{x}) \cdot D_{k+1|k+1}(\mathbf{x}|Z^{k+1})d\mathbf{x} \tag{16.353}$$

$$= \int \theta(\mathbf{x}) \cdot L_{Z_{k+1}}(\mathbf{x}) \cdot D_{k+1|k}(\mathbf{x}|Z^{k+1})d\mathbf{x} \tag{16.354}$$

$$\cong \frac{1}{\nu}\sum_{i=1}^{\nu} \theta(\mathbf{x}^i_{k+1|k}) \cdot L_{Z_{k+1}}(\mathbf{x}^i_{k+1|k}) \tag{16.355}$$

or

$$= \frac{1}{\nu}\sum_{i=1}^{\nu} \theta(\mathbf{x}^i_{k+1|k}) \tag{16.356}$$

$$\cdot \frac{\sum_{j=0}^{m} C^{(m-j)}(0) \cdot \hat{G}^{(j+1)}(s[q_D]) \cdot \sigma_j(Z)}{\sum_{i=0}^{m} C^{(m-i)}(0) \cdot \hat{G}^{(i)}(s[q_D]) \cdot \sigma_i(Z)} \tag{16.357}$$

$$\cdot \left(1 - p_D(\mathbf{x}^i_{k+1|k})\right) \tag{16.358}$$

$$+ \frac{1}{\nu}\sum_{i=1}^{\nu} \theta(\mathbf{x}^i_{k+1|k}) \tag{16.359}$$

$$\cdot p_D(\mathbf{x}^i_{k+1|k}) \cdot \sum_{\mathbf{z}\in Z} \frac{L_\mathbf{z}(\mathbf{x}^i_{k+1|k})}{c(\mathbf{z})} \tag{16.360}$$

$$\cdot \frac{\sum_{j=0}^{m-1} C^{(m-j-1)}(0) \cdot \hat{G}^{(j+1)}(s[q_D]) \cdot \sigma_j(Z - \{\mathbf{z}\})}{\sum_{i=0}^{m} C^{(m-i)}(0) \cdot \hat{G}^{(i)}(s[q_D]) \cdot \sigma_i(Z)}. \tag{16.361}$$

Abbreviate

$$\Lambda_0 \overset{\text{abbr.}}{=} \frac{\sum_{j=0}^{m} C^{(m-j)}(0) \cdot \hat{G}^{(j+1)}(s[q_D]) \cdot \sigma_j(Z)}{\sum_{i=0}^{m} C^{(m-i)}(0) \cdot \hat{G}^{(i)}(s[q_D]) \cdot \sigma_i(Z)} \tag{16.362}$$

$$\Lambda_e \overset{\text{abbr.}}{=} \frac{\sum_{j=0}^{m-1} C^{(m-j-1)}(0) \cdot \hat{G}^{(j+1)}(s[q_D]) \cdot \sigma_j(Z - \{\mathbf{z}_e\})}{c(\mathbf{z}_e) \cdot \sum_{i=0}^{m} C^{(m-i)}(0) \cdot \hat{G}^{(i)}(s[q_D]) \cdot \sigma_i(Z)} \tag{16.363}$$

Then we conclude that $\mathbf{x}^i_{k+1|k+1} \triangleq \mathbf{x}^i_{k+1|k}$ where

$$w^i_{k+1|k+1} \triangleq \frac{\Lambda_0}{\nu} \cdot \left(1 - p_D(\mathbf{x}^i_{k+1|k})\right) \tag{16.364}$$

$$+\frac{p_D(\mathbf{x}^i_{k+1|k})}{\nu} \cdot \sum_{e=1}^{m_{k+1}} \Lambda_e \cdot L_{\mathbf{z}_e}(\mathbf{x}^i_{k+1|k}). \tag{16.365}$$

16.9.2.4 SMC-CPHD Filter State and Error Estimation

State and error estimation for the SMC-CPHD filter is essentially the same as that for the SMC-PHD filter predictor of Section 16.5.2. The only difference is that target number and error in target number must be estimated as in Section 16.7.4.

16.9.3 Gaussian-Mixture Approximation (GM-CPHD)

The GM-CPHD filter was devised by Vo, Vo, and Cantoni [242, 240, 241]. It presumes the same underlying GM-PHD assumptions listed in (16.241) and (16.246).

In what follows I summarize the basic steps: initialization, prediction, correction, and state and error estimation.

16.9.3.1 GM-CPHD Filter Initialization

Initialization of the GM-CPHD filter, including initialization of component labels, is the same as that for the GM-PHD filter, except for the fact that an initial cardinality distribution must also be chosen:

$$D_{0|0}(\mathbf{x}|Z^{(0)}) = \sum_{i=1}^{n_{0|0}} w^i_{0|0} \cdot N_{P^i_{0|0}}(\mathbf{x} - \mathbf{x}^i_{0|0}) \tag{16.366}$$

$$p_{0|0}(n) = p_{0|0}(n|Z^{(0)}) = p_0(n). \tag{16.367}$$

16.9.3.2 GM-CPHD Filter Predictor

The GM-CPHD predictor is essentially the same as the GM-PHD predictor, except that target spawning is not modeled and a predicted p.g.f. must be computed. For completeness we list out the complete predictor equations.

Assume that we have a prior GM-PHD

$$D_{k|k}(\mathbf{x}|Z^{(k)}) = \sum_{i=1}^{n_{k|k}} w^i_{k|k} \cdot N_{P^i_{k|k}}(\mathbf{x} - \mathbf{x}^i_{k|k}) \tag{16.368}$$

and a prior label table $\mathcal{L}_{k|k}$. Under current assumptions the predicted p.g.f. is

$$G_{k+1|k}(x) \cong B(x) \cdot G(1 - p_S + p_S \cdot x). \tag{16.369}$$

In Appendix G.31 we verify that the predicted PHD has the form

$$D_{k+1|k}(\mathbf{x}) = \sum_{i=1}^{a_k} b^i_{k+1|k} \cdot N_{B^i_{k+1|k}}(\mathbf{x} - \mathbf{b}^i_{k+1|k}) \tag{16.370}$$

$$+p_S \sum_{i=1}^{n_{k|k}} w^i_{k|k} \cdot N_{Q+FP_iF^T}(\mathbf{x} - F\mathbf{x}^i) \tag{16.371}$$

where

$$b^i_{k+1|k} = \beta^i_k \tag{16.372}$$

$$\mathbf{b}^i_{k+1|k} = \mathbf{x}^i_b, \quad B^i_{k+1|k} = B_i \tag{16.373}$$

$$w^i_{k+1|k} = p_S \cdot w^i_{k|k} \tag{16.374}$$

$$\mathbf{x}^i_{k+1|k} = F\mathbf{x}^i, \quad P^i_{k+1|k} = Q + FP_iF^T \tag{16.375}$$

and where as usual we have abbreviated

$$F \overset{\text{abbr.}}{=} F_k, \quad Q \overset{\text{abbr.}}{=} Q_k, \tag{16.376}$$

$$P_i \overset{\text{abbr.}}{=} P^i_{k|k}, \quad \mathbf{x}^i \overset{\text{abbr.}}{=} \mathbf{x}^i_{k|k} \tag{16.377}$$

$$G_i \overset{\text{abbr.}}{=} G^i_k, \quad B_i \overset{\text{abbr.}}{=} B^i_k, \quad E_i \overset{\text{abbr.}}{=} E^i_k. \tag{16.378}$$

16.9.3.3 GM-CPHD Filter Corrector

Assume that from the previous time step we have a predicted GM-PHD

$$D_{k+1|k}(\mathbf{x}) = \sum_{i=1}^{n_{k+1|k}} w^i_{k+1|k} \cdot N_{P^i_{k+1|k}}(\mathbf{x} - \mathbf{x}^i_{k+1|k}) \tag{16.379}$$

and a predicted label table $\mathcal{L}_{k+1|k}$. We collect a new measurement set $Z_{k+1} = \{\mathbf{z}_1, ..., \mathbf{z}_{m_{k+1}}\}$ with $|Z_{k+1}| = m_{k+1}$. In Appendix G.32 we verify that the data-updated PHD has the form

$$D_{k+1|k+1}(\mathbf{x}) \tag{16.380}$$

$$= \sum_{i=1}^{n_{k+1|k}} w^i_{k+1|k+1} \cdot N_{P^i_{k+1|k+1}}(\mathbf{x} - \mathbf{x}^i_{k+1|k+1}) \tag{16.381}$$

$$+ \sum_{j=1}^{m_{k+1}} \sum_{i=1}^{n_{k+1|k}} w^{i,j}_{k+1|k+1} \cdot N_{P^i_{k+1|k+1}}(\mathbf{x} - \mathbf{x}^{i,j}_{k+1|k+1}) \tag{16.382}$$

where

$$w^i_{k+1|k+1} = \Lambda_0 \cdot w^i_{k+1|k} \tag{16.383}$$

$$\mathbf{x}^i_{k+1|k+1} = \mathbf{x}^i \tag{16.384}$$

$$P^i_{k+1|k+1} = P^i_{k+1|k} \tag{16.385}$$

$$w^{i,j}_{k+1|k+1} = w^i_{k+1|k} \cdot \Lambda_e \cdot N_{R+HP_iH^T}(\mathbf{z}_j - H\mathbf{x}^i) \tag{16.386}$$

$$\mathbf{x}^{i,j}_{k+1|k+1} = \mathbf{x}^i_{k+1|k} + K_i \left(\mathbf{z}_j - H\mathbf{x}^i\right) \tag{16.387}$$

$$P^i_{k+1|k+1} = (I - K_iH)\, P_i \tag{16.388}$$

$$K_i = P_iH^T \left(HP_iH^T + R\right)^{-1} \tag{16.389}$$

and where

$$\Lambda_0 \triangleq \frac{\sum_{j=0}^m C^{(m-j)}(0) \cdot \hat{G}^{(j+1)}(q_D) \cdot \sigma_j(Z)}{\sum_{i=0}^m C^{(m-i)}(0) \cdot \hat{G}^{(i)}(q_D) \cdot \sigma_i(Z)} \cdot (1 - p_D) \tag{16.390}$$

$$\Lambda_e \triangleq \frac{p_D \sum_{j=0}^{m-1} C^{(m-j-1)}(0) \cdot \hat{G}^{(j+1)}(q_D) \cdot \sigma_j(Z - \{\mathbf{z}_e\})}{c(\mathbf{z}_e) \cdot \sum_{i=0}^m C^{(m-i)}(0) \cdot \hat{G}^{(i)}(q_D) \cdot \sigma_i(Z)} \tag{16.391}$$

and where we have abbreviated

$$H \overset{\text{abbr.}}{=} H_{k+1}, \quad R \overset{\text{abbr.}}{=} R_{k+1}, \quad P_i \overset{\text{abbr.}}{=} P^i_{k+1|k} \tag{16.392}$$

$$\mathbf{x}^i \overset{\text{abbr.}}{=} \mathbf{x}^i_{k+1|k}, \quad K_i \overset{\text{abbr.}}{=} K^i_{k+1}. \tag{16.393}$$

Also, the updated p.g.f. is

$$G_{k+1|k+1}(x) \cong \frac{\sum_{j=0}^{m} x^j \cdot C^{(m-j)}(0) \cdot \hat{G}^{(j)}(q_D x) \cdot \sigma_j(Z_{k+1})}{\sum_{i=0}^{m} C^{(m-i)}(0) \cdot \hat{G}^{(i)}(q_D) \cdot \sigma_i(Z_{k+1})} \tag{16.394}$$

where

$$\sigma_i(Z) \triangleq p_D^i \cdot \sigma_{m,i}\left(\frac{D_{k+1|k}[L_{\mathbf{z}_1}]}{c(\mathbf{z}_1)}, \ldots, \frac{D_{k+1|k}[L_{\mathbf{z}_m}]}{c(\mathbf{z}_m)}\right) \tag{16.395}$$

and where

$$D_{k+1|k}[L_{\mathbf{z}_j}] = \sum_{i=1}^{n_{k+1|k}} w_{k+1|k}^i \cdot N_{R+HP_iH^T}(\mathbf{z}_j - H\mathbf{x}^i). \tag{16.396}$$

16.9.3.4 GM-CPHD Filter State and Error Estimation

State estimation and error estimation for the GM-CPHD filter is the same as that for the GM-PHD filter with one difference. Target number is determined using a MAP estimate of the cardinality distribution, as in Section 16.7.4.

16.10 DERIVING THE PHD AND CPHD FILTERS

In this section, I summarize the methodology used to derive the predictor and corrector equations for the PHD and CPHD filters. My purpose is to provide a brief overview of the two most important ideas underlying multitarget moment approximation:

- Approximate at the p.g.fl. level to promote mathematical tractability;
- Then use the multitarget calculus to construct the actual approximation formulas.

For complete proofs, the reader is directed to [136, 142].

16.10.1 Derivation of PHD and CPHD Predictors

In Section 14.8.1, I showed that the multitarget predictor equation, (14.14), can be written in p.g.fl. form as follows:

$$G_{k+1|k}[h] = \int G_{k+1|k}[h|X'] \cdot f_{k|k}(X'|Z^{(k)})\delta X' \tag{16.397}$$

where $G_{k+1|k}[h]$ is the p.g.fl. of $f_{k+1|k}(X|Z^{(k)})$ and where

$$G_{k+1|k}[h|X'] \triangleq \int h^X \cdot f_{k+1|k}(X|X')\delta X \tag{16.398}$$

is the p.g.fl. of $f_{k+1|k}(X|X')$. Then in (14.273) I showed that, assuming the PHD filter motion model, we can derive the following formula for $G_{k+1|k}[h]$ [136, p. 1172]:

$$G_{k+1|k}[h] = e_h \cdot G[(1 - p_S + p_S p_h) \cdot b_h] \tag{16.399}$$

where $G[h] \overset{\text{abbr.}}{=} G_{k|k}[h]$ is the p.g.fl. of $f_{k|k}(X'|Z^{(k)})$ and where

$$p_h(\mathbf{x}') \triangleq \int h(\mathbf{x}) \cdot f_{k+1|k}(\mathbf{x}|\mathbf{x}')d\mathbf{x} \tag{16.400}$$

$$b_h(\mathbf{x}') \triangleq \int h^X \cdot b_{k+1|k}(X|\mathbf{x}')\delta X \tag{16.401}$$

$$e_h \triangleq \int h^X \cdot b_{k+1|k}(X)\delta X. \tag{16.402}$$

From (16.35) we know that the predicted PHD $D_{k+1|k}(\mathbf{x}|Z^{(k)})$ can be determined as a functional derivative:

$$D_{k+1|k}(\mathbf{x}|Z^{(k)}) = \frac{\delta G_{k+1|k}}{\delta \mathbf{x}}[1]. \tag{16.403}$$

Applying the basic formulas for the functional derivative (Section 11.6) to this formula, one arrives at the PHD filter predictor equations.

The CPHD predictor can be derived in the same fashion, with two primary differences:

- $b_h(\mathbf{x}') = 1$ identically (i.e., no target spawning);
- $G[h] = G(s[h])$ is assumed to be an i.i.d. cluster process.

16.10.2 Derivation of PHD and CPHD Correctors

In Section 14.8.2, I showed that the multitarget corrector, (14.50), can be written in p.g.fl. form as

$$G_{k+1|k+1}[h] = \frac{\frac{\delta F}{\delta Z_{k+1}}[0,h]}{\frac{\delta F}{\delta Z_{k+1}}[0,1]} \tag{16.404}$$

where

$$F[g,h] \triangleq \int h^X \cdot G_{k+1}[g|X] \cdot f_{k+1|k}(X|Z^{(k)})\delta X \tag{16.405}$$

$$G_{k+1}[g|X] \triangleq \int g^Z \cdot f_{k+1}(Z|X)\delta Z. \tag{16.406}$$

Given the PHD filter multitarget measurement model, in (14.285), I showed that [136, p. 1173]:

$$F[g,h] = e^{\lambda c[g]-\lambda} \cdot G_{k+1|k}[h(q_D + p_D p_g)] \tag{16.407}$$

where $q_D(\mathbf{x}) \triangleq 1 - p_D(\mathbf{x})$ and where

$$c[g] \triangleq \int g(\mathbf{z}) \cdot c(\mathbf{z})d\mathbf{z} \tag{16.408}$$

$$p_g(\mathbf{x}) \triangleq \int g(\mathbf{z}) \cdot f_{k+1}(\mathbf{z}|\mathbf{x})d\mathbf{z}. \tag{16.409}$$

To arrive at a closed-form formula for the PHD corrector equation, we must assume that $G_{k+1|k}[h]$ is Poisson. That is,

$$G_{k+1|k}[h] = e^{\mu s[h]-\mu} \tag{16.410}$$

where $s[h] \triangleq \int h(\mathbf{x}) \cdot s(\mathbf{x})d\mathbf{x}$ and where $N_{k+1|k} = \mu$ and $D_{k+1|k}(\mathbf{x}) = \mu s(\mathbf{x})$. In this case (16.407) simplifies to

$$F[g,h] = \exp\left(\lambda c[g] - \lambda + \mu s[h(q_D + p_D p_g)] - \mu\right). \tag{16.411}$$

Using the product rule and chain rule for functional derivatives (see Section 11.6), we then derive closed-form formulas for the numerator and denominator of (16.404). This gives us a closed-form formula for $G_{k+1|k+1}[h]$. From (16.35) we know that the data-updated PHD $D_{k+1|k+1}(\mathbf{x}|Z^{(k)})$ can be computed as a functional derivative,

$$D_{k+1|k+1}(\mathbf{x}|Z^{(k+1)}) = \frac{\delta G_{k+1|k+1}}{\delta \mathbf{x}}[1] = \frac{\frac{\delta F}{\delta Z_{k+1}\delta \mathbf{x}}[0,1]}{\frac{\delta F}{\delta Z_{k+1}}[0,1]}. \tag{16.412}$$

The closed-form formula for the PHD corrector is then derived from this equation using the basic rules of functional differentiation in Section 11.6.

The corrector equations for the CPHD filter are derived using identical reasoning (but considerably more algebra). The major differences:

- The false alarm p.g.fl. $C[g] = e^{\lambda c[g]-\lambda}$ is replaced by an i.i.d. cluster process: $C[g] = C(c[g])$;
- The predicted p.g.fl. $G_{k+1|k}[h] = e^{\mu s[h]-\mu}$ is replaced by an i.i.d. cluster process: $G_{k+1|k}[h] = G(s[h])$.

16.11 PARTIAL SECOND-ORDER FILTERS?

I anticipate that the CPHD filter will be appropriate for many applications. Nevertheless, the fact that it has computational complexity of order $O(m^3 n)$ means that it will become intractable when the number of measurements becomes too large.

A major reason for the cubic dependence on target number is the fact that the CPHD filter must compute the entire probability distribution on target number at each recursive step. One possible way of addressing this challenge would be to devise a modification of the CPHD filter that *propagates only the first and second moments (mean* $n_{k|k}$ *and variance* $\sigma^2_{k|k}$*) of target number.* In this manner, it might be possible to devise a multitarget-moment filter that is second-order in target number and has computational complexity $O(mn)$.

Such a filter would have the form

$$\begin{array}{ccccccc}
\cdots \to & f_{k|k}(X) & \overset{\text{predictor}}{\longrightarrow} & f_{k+1|k}(X) & \overset{\text{corrector}}{\longrightarrow} & f_{k+1|k+1}(X) & \to \cdots \\
 & \downarrow & & \downarrow & & \downarrow & \\
 & n_{k|k} & & n_{k+1|k} & & n_{k+1|k+1} & \\
\cdots \to & \sigma^2_{k|k} & \overset{\text{predictor?}}{\longrightarrow} & \sigma^2_{k+1|k} & \overset{\text{corrector?}}{\longrightarrow} & \sigma^2_{k+1|k+1} & \to \cdots \\
 & D_{k|k}(\mathbf{x}) & & D_{k+1|k}(\mathbf{x}) & & D_{k+1|k+1}(\mathbf{x}) &
\end{array}$$

The author has reported an only partially successful attack on this problem in [143]. A second-order CPHD filter was devised by approximating the cardinality distribution $p_{k|k}(n)$ as a binomial distribution. The resulting filter does indeed have computational complexity $O(mn)$. However, the binomial approximation requires the approximate inequality

$$\lambda + p_D n < n \tag{16.413}$$

where λ is the expected number of Poisson false alarms; where p_D is probability of detection; and where n is the actual number of targets. In other words, the average number of false alarms must be smaller than a generally small fraction of the actual target number:

$$\lambda < (1 - p_D) \cdot n. \tag{16.414}$$

This, clearly, constitutes a major limitation.

16.12 CHAPTER EXERCISE

Exercise 60 *Let $G[h]$ be the p.g.fl. of a multitarget system. In (14.291) it was shown that $G^\rho[h] = G[1 - \rho + \rho h]$ is the p.g.fl. of the corresponding system of targets of interest (ToIs). Let $D(\mathbf{x})$ and $G(x)$ be, respectively, the PHD and p.g.f. of $G[h]$. Then show that the PHD of $G^\rho[h]$ is $D^\rho(\mathbf{x}) = \rho(\mathbf{x}) \cdot D(\mathbf{x})$; that the p.g.f. of $G^\rho[h]$ is $G^\rho(x) = G(1 - s[\rho] + s[\rho] \cdot x)$; and that the cardinality distribution of $G^\rho[h]$ is*

$$p^\rho(n) = \frac{1}{n!} \cdot s[\rho]^n \cdot G^{(n)}(1 - s[\rho]) \tag{16.415}$$

where $G^{(n)}(x)$ denotes the nth derivative of $G(x)$.

Chapter 17

Multi-Bernoulli Approximation

17.1 INTRODUCTION TO THE CHAPTER

In this concluding chapter, I introduce a final multitarget approximation technique. It is speculative in the sense that, to date, it has never been implemented even in simulations. One version of it was first proposed by the author in 1999 [138], where it was called "para-Gaussian approximation." A different version was independently formulated by Moreland and Challa in 2003 [168]. In what follows I will do the following:

- Remind the reader of my fundamental multitarget approximation strategy (Section 17.1.1).
- Motivate the use of the multitarget multi-Bernoulli approximation (Section 17.1.2).
- Summarize the theoretical foundation of the chapter, the *multitarget multi-Bernoulli (MeMBer) filter* (Section 17.1.3).
- Describe a potentially tractable Gaussian-mixture implementation, the (refined) *para-Gaussian filter* (Section 17.1.4).

17.1.1 p.g.fl.-Based Multitarget Approximation

Three of the multitarget filters described so far in this book have been based on approximations of the evolving multitarget posteriors $f_{k|k}(X|Z^{(k)})$ by simpler multitarget densities. Assume that $X = \{\mathbf{x}_1, ..., \mathbf{x}_n\}$ with $|X| = n$. Then the following are true:

- *JoTT filter* (Section 14.7): the $f_{k|k}(X|Z^{(k)})$ are multitarget multi-Bernoulli processes with the number ν of hypothesized tracks being equal to one; see (11.150):

$$f_{k|k}(X|Z^{(k)}) = \begin{cases} 1-p_{k|k} & \text{if} \quad X=\emptyset \\ p_{k|k}\cdot f_{k|k}(\mathbf{x}) & \text{if} \quad X=\{\mathbf{x}\} \\ 0 & \text{if} \quad |X|\geq 2 \end{cases} \tag{17.1}$$

$$G_{k|k}[h] = 1-p_{k|k}+p_{k|k}\cdot f_{k|k}[h]. \tag{17.2}$$

- *PHD filter* (Section 16.3): the $f_{k|k}(X|Z^{(k)})$ are approximate multitarget Poisson processes; see (11.122):

$$f_{k|k}(X|Z^{(k)}) \cong e^{-\mu_{k|k}}\cdot\prod_{i=1}^{n}\mu_{k|k}f_{k|k}(\mathbf{x}_i) \tag{17.3}$$

$$G_{k|k}[h] \cong e^{\mu_{k|k}\cdot f_{k|k}[h]-\mu_{k|k}}. \tag{17.4}$$

- *CPHD filter* (Section 16.7): the $f_{k|k}(X|Z^{(k)})$ are approximate i.i.d. cluster processes; see (11.121):

$$f_{k|k}(X|Z^{(k)}) \cong n!\cdot p(n)\cdot\prod_{i=1}^{n}f_{k|k}(\mathbf{x}_i) \tag{17.5}$$

$$G_{k|k}[h] \cong G_{k|k}(f_{k|k}[h]). \tag{17.6}$$

This chapter investigates filters based on a less restrictive assumption:

- The $f_{k|k}(X|Z^{(k)})$ are approximate multitarget multi-Bernoulli processes with arbitrary ν; see (11.131)-(11.135):

$$f_{k|k}(X|Z^{(k)}) \cong n!\cdot Q \tag{17.7}$$

$$\cdot \sum_{1\leq i_1<\ldots<i_n\leq\nu_{k|k}}\frac{q_{i_1}f_{i_1}(\mathbf{x}_1)}{1-q_{i_1}} \tag{17.8}$$

$$\cdots\frac{q_{i_n}f_{i_n}(\mathbf{x}_n)}{1-q_{i_n}} \tag{17.9}$$

$$G_{k|k}[h] \cong \prod_{i=1}^{\nu_{k|k}}\left(1-q_i+q_i\cdot f_i[h]\right). \tag{17.10}$$

where $Q \triangleq \prod_{i=1}^{n}(1 - q_i)$.

As with the PHD and CPHD filters, derivation of the equations for the multitarget Bayes filter based on this last approximation requires extensive application of the probability-generating functional (p.g.fl.) and multitarget calculus techniques introduced in Chapter 11.

17.1.2 Why Multitarget Multi-Bernoulli Processes?

The prediction set integral and the Bayes-factor set integral in (14.1)-(14.3) are major computational obstacles for multitarget Bayes filtering:

$$f_{k+1|k}(X|Z^{(k)}) = \int f_{k+1|k}(X|X') \cdot f_{k|k}(X'|Z^{(k)})\delta X' \quad (17.11)$$

$$f_{k+1|k+1}(X|Z^{(k+1)}) = \frac{f_{k+1}(Z_{k+1}|X) \cdot f_{k+1|k}(X|Z^{(k)})}{f_{k+1}(Z_{k+1}|Z^{(k)})} \quad (17.12)$$

$$f_{k+1}(Z_{k+1}|Z^{(k)}) = \int f_{k+1}(Z_{k+1}|X) \quad (17.13)$$

$$\cdot f_{k+1|k}(X|Z^{(k)})\delta X. \quad (17.14)$$

In *Mathematics of Data Fusion* [70, pp. 239-244] I pointed out a potential analogy between Kalman filtering and multitarget filtering. One reason for the tractability of the Kalman filter is the fact that the prediction and Bayes-factor integrals in the single-target Bayes filter can be solved in closed form when all relevant distributions—likelihoods, Markov densities, posterior densities—are linear-Gaussian.

Similarly, *the set integrals in the multitarget Bayes filter can be solved in closed form* if all relevant distributions have a specific form: They are, at least approximately, multitarget multi-Bernoulli processes.

17.1.3 The Multitarget Multi-Bernoulli Filter

The multitarget multi-Bernoulli (MeMBer) filter propagates a *track table*:

$$\mathcal{T}_{0|0} \to \mathcal{T}_{1|0} \to \mathcal{T}_{1|1} \to \cdots \to \mathcal{T}_{k|k} \to \mathcal{T}_{k+1|k} \to \mathcal{T}_{k+1|k+1} \to \cdots \quad (17.15)$$

For a given value of k, $\mathcal{T}_{k|k}$ has the form

$$\mathcal{T}_{k|k} = \{(\ell^1_{k|k}, q^1_{k|k}, f^1_{k|k}(\mathbf{x})), \ldots, (\ell^{\nu_{k|k}}_{k|k}, q^{\nu_{k|k}}_{k|k}, f^{\nu_{k|k}}_{k|k}(\mathbf{x}))\}. \quad (17.16)$$

Each 3-tuple $(\ell^i_{k|k}, q^i_{k|k}, f^i_{k|k}(\mathbf{x}))$ for $i = 1, ..., \nu$ is called a "track" where the following are true:

- $\ell_i \overset{\text{abbr.}}{=} \ell^i_{k|k}$ is the identifying label of the ith hypothesized track.
- $q_i \overset{\text{abbr.}}{=} q^i_{k|k} < 1$ is the probability that the ith hypothesized track is an actual target (probability of existence).
- $f_i(\mathbf{x}) \overset{\text{abbr.}}{=} f^i_{k|k}(\mathbf{x})$ is the track density of the ith hypothesized track.
- $\nu \overset{\text{abbr.}}{=} \nu_{k|k}$ is the current number of hypothesized tracks.

As we shall see, given a track table $\mathcal{T}_{k|k}$ the expected number of targets is

$$N_{k|k} = q^1_{k|k} + ... + q^{\nu_{k|k}}_{k|k} < \underbrace{1 + ... + 1}_{\nu_{k|k}} = \nu_{k|k}. \tag{17.17}$$

This means that $\nu_{k|k}$ must be larger than the actual number of targets.

17.1.4 The Para-Gaussian Filter

The original "para-Gaussian" form of the multitarget multi-Bernoulli filter, as announced in [138], had significant limitations. It provided no means for modeling target appearance, and was algebraically complex for more than a few targets [144, p. 2].

In Section 17.3 I describe a more general and tractable approach, which I also call the "para-Gaussian filter." We will assume that the $f_1(\mathbf{x}), ..., f_\nu(\mathbf{x})$ in (17.7) are Gaussian mixtures:

$$f_i(\mathbf{x}) = \sum_{e=1}^{n_i} N_{P^{i,e}_{k|k}}(\mathbf{x} - \mathbf{x}^{i,e}_{k|k}). \tag{17.18}$$

With this assumption, (17.7) becomes a representation of the multitarget posterior $f_{k|k}(X|Z^{(k)})$ as a multitarget Gaussian mixture density.[1] The result is a more general and tractable version of the para-Gaussian filter originally described in [138] or in [134, pp. 51, 52].

The following are anticipated advantages of the refined para-Gaussian filter:

1 In this sense the para-Gaussian filter is a multitarget Gaussian-mixture approximation. It is quite different, however, from the one originally proposed by Alspach in [5].

- Easily implemented birth model;
- Formal Poisson false alarm model;
- Target number is estimated directly rather than inferred, either through MAP or EAP estimation;
- No measurement-to-track association;
- More accurate than a PHD or CPHD filter (though also more computationally demanding).

The para-Gaussian filter can also be viewed as a generalization of the Vo-Ma GM-PHD filter of Section 16.9.3.

17.1.5 Summary of Major Lessons Learned

The following are the major concepts to be encountered in this chapter:

- Multitarget multi-Bernoulli approximation (Section 17.2);
- Equations for the para-Gaussian filter predictor; see (17.92)-(17.97):

$$\ell_i = \ell'_i \tag{17.19}$$

$$q_i = p_S \cdot q'_i \tag{17.20}$$

$$f_i(\mathbf{x}) = \sum_{e=1}^{n_i} w_{i,e} \cdot N_{P_{i,e}}(\mathbf{x} - \mathbf{x}_{i,e}) \tag{17.21}$$

where

$$\mathbf{x}_{i,e} = F'\mathbf{x}_{i,e} \tag{17.22}$$

$$P_{i,e} = Q + FP'_{i,e}F^T \tag{17.23}$$

$$w_{i,e} = w'_{i,e}; \tag{17.24}$$

- Equations for the para-Gaussian filter corrector; see (17.105)-(17.112):

$$q_i = \frac{q'_i \cdot (1 - p_D)}{1 - p_D \cdot q'_i} \leq q'_i, \quad f_i(\mathbf{x}) = f'_i(\mathbf{x}) \tag{17.25}$$

$$Q_j \tag{17.26}$$

$$= \frac{\sum_{i,e=1}^{\nu'} p_D \rho_i w'_{i,e} N_{R+HP'_{i,e}H^T}(\mathbf{z}_j - H\mathbf{x}'_{i,e})}{\lambda c(\mathbf{z}_j) + \sum_{i,e=1}^{\nu'} p_D \rho_i w'_{i,e} N_{R+HP'_i H^T}(\mathbf{z}_j - H\mathbf{x}'_{i,e})} \tag{17.27}$$

$$F_j(\mathbf{x}) = \sum_{i,e=1}^{\nu'} w_{i,e,j} \cdot N_{C_{i,e}}(\mathbf{x} - \mathbf{c}_{i,e,j}) \tag{17.28}$$

$$w_{i,e,j} \tag{17.29}$$

$$= \frac{\rho_i \cdot w'_{i,e} \cdot N_{R+HP'_{i,e}H^T}(\mathbf{z}_j - H\mathbf{x}'_{i,e})}{\sum_{i',e'=1}^{\nu'} p_D \rho_{i'} w'_{i',e'} N_{R+HP'_{i',e'}H^T}(\mathbf{z}_j - H\mathbf{x}'_{i',e'})}; \tag{17.30}$$

- Equations for estimates of target number and states; see (17.72)-(17.75):

$$N_{k|k} = \sum_{i=1}^{\nu} q_i, \qquad \sigma^2_{k|k} = \sum_{i=1}^{\nu} q_i(1-q_i) \tag{17.31}$$

$$\hat{n}_{k|k} = \arg\max_n B_{q_1,\ldots,q_{\nu_{k|k}}}(n), \tag{17.32}$$

$$\hat{\sigma}^2_{k|k} = \sigma^2_{k|k} + N^2_{k|k} - n^2_{k|k}. \tag{17.33}$$

17.1.6 Organization of the Chapter

The chapter is organized as follows. The multitarget multi-Bernoulli (MeMBer) filter is described in Section 17.2. Its Gaussian-mixture special case, the para-Gaussian filter, is then described in Section 17.3. The basic ideas behind the derivation of the MeMBer filter are summarized in Section 17.4. Chapter exercises are in Section 17.5.

17.2 MULTITARGET MULTI-BERNOULLI FILTER

As already noted, the MeMBer filter is based on the assumption that every multitarget posterior is the probability law of a multitarget multi-Bernoulli process. That is, the p.g.fl. of $f_{k|k}(X|Z^{(k)})$ is assumed to have the form

$$G_{k|k}[h] \cong \prod_{i=1}^{\nu_{k|k}} \left(1 - q^i_{k|k} + q^i_{k|k} \cdot f^i_{k|k}[h]\right). \tag{17.34}$$

This is interpreted to mean that the filter is carrying $\nu_{k|k}$ hypotheses about possible real-world targets, where $q^i_{k|k}$ is the probability that the ith hypothesized track is valid and where $f^i_{k|k}(\mathbf{x})$ is its spatial distribution if it does exist.

In what follows I describe the individual stages of the MeMBer filter: initialization (Section 17.2.1), prediction (Section 17.2.2), correction (Section 17.2.3), pruning and merging (Section 17.2.4), and state and error estimation (Section 17.2.5). Finally, in Section 17.2.6, I discuss the differences between the MeMBer and Moreland-Challa approaches to multi-Bernoulli approximation.

17.2.1 MeMBer Filter Initialization

Initialization of the MeMBer filter consists of specifying an initial track table:

$$\mathcal{T}_{0|0} = \{(\ell^1_{0|0}, q^1_{0|0}, f^1_{0|0}(\mathbf{x})) , ..., (\ell^{\nu_{0|0}}_{0|0}, q^{\nu_{0|0}}_{0|0}, f^{\nu_{0|0}}_{0|0}(\mathbf{x}))\}. \tag{17.35}$$

That is, the maximum number $\nu_{0|0}$ of hypothesized targets is specified. In general this should be significantly larger than the actual expected number of targets. The spatial distributions $f^1_{0|0}(\mathbf{x})$ of these hypothesized targets are chosen. Finally, one designates the degree $q^i_{0|0}$ to which the ith hypothesized target is an actual target. The (usually integer) labels $\ell^i_{0|0}$ are selected to enable propagation of continuously identifiable tracks in later steps.

If very little is known about target numbers and states, one can choose $q^i_{0|0} = 1/2$ for all $i = 1, ..., \nu_{k|k}$ and make the distributions $f^1_{0|0}(\mathbf{x})$ essentially uniform.

17.2.2 MeMBer Filter Predictor

We assume that:

- Between time steps k and $k+1$, $\alpha \overset{\text{abbr.}}{=} \alpha_{k+1|k}$ new tracks

$$(l_1, b_1, B_1(\mathbf{x})), ..., (l_\alpha, b_\alpha, B_\alpha(\mathbf{x})) \tag{17.36}$$

appear with new labels $l_1, ..., l_\alpha$.

From time step k we are given a track table

$$\mathcal{T}_{k|k} = \{(\ell'_1, q'_1, f'_1(\mathbf{x})), ..., (\ell'_{\nu'}, q'_{\nu'}, f'_{\nu'}(\mathbf{x}))\}. \tag{17.37}$$

We are to define the predicted track table $\mathcal{T}_{k+1|k}$.

In Section 17.4.1, I show that it is

$$\mathcal{T}_{k+1|k} = \mathcal{T}^{\text{persist}}_{k+1|k} \cup \mathcal{T}^{\text{birth}}_{k+1|k} \tag{17.38}$$

where

$$\mathcal{T}^{\text{persist}}_{k+1|k} = \{(\ell_1, q_1, f_1(\mathbf{x})), ..., (\ell_{\nu'}, q_{\nu'}, f_{\nu'}(\mathbf{x}))\} \tag{17.39}$$

$$\mathcal{T}^{\text{birth}}_{k+1|k} = \{(l_1, b_1, B_1(\mathbf{x})), ..., (l_\alpha, b_\alpha, B_\alpha(\mathbf{x}))\} \tag{17.40}$$

and where

$$\ell_i = \ell'_i \tag{17.41}$$

$$q_i = q'_i \cdot f'_i[p_S] \leq q_i \tag{17.42}$$

$$f_i(\mathbf{x}) = \frac{f'_i[p_S M_{\mathbf{x}}]}{f'_i[p_S]} \tag{17.43}$$

where $M_{\mathbf{x}}(\mathbf{x}') = f_{k+1|k}(\mathbf{x}|\mathbf{x}')$ and $f_i[h] = \int h(\mathbf{x}) \cdot f_i(\mathbf{x}) d\mathbf{x}$.

Because of the birth tracks, the number of predicted tracks will increase:

$$\nu_{k+1|k} = \nu_{k|k} + \alpha_{k+1|k}. \tag{17.44}$$

Thus pruning and merging operations must be employed to limit growth in the number of tracks.

17.2.3 MeMBer Filter Corrector

We assume that:

- The expected number λ of false alarms is *nonzero* and "not too large" in some sense that must be determined in implementations.

  We are given a predicted track table

$$\mathcal{T}_{k+1|k} = \{(\ell'_1, q'_1, f'_1(\mathbf{x})), ..., (\ell'_{\nu'}, q'_{\nu'}, f'_{\nu'}(\mathbf{x}))\}. \tag{17.45}$$

We collect a new measurement set $Z \overset{\text{abbr.}}{=} Z_{k+1} = \{\mathbf{z}_1, ..., \mathbf{z}_m\}$ where $m \overset{\text{abbr.}}{=} m_{k+1}$. We are to define the measurement-updated track table $\mathcal{T}_{k+1|k+1}$.

In Section 17.4.2, I show that it is

$$\mathcal{T}_{k+1|k+1} = \mathcal{T}^{\text{legacy}}_{k+1|k+1} \cup \mathcal{T}^{\text{meas}}_{k+1|k+1} \tag{17.46}$$

where the following are true:

- $\mathcal{T}^{\text{legacy}}_{k+1|k+1}$ contains "legacy tracks" (updates of predicted tracks, assuming that no measurements were collected from them);
- $\mathcal{T}^{\text{meas}}_{k+1|k+1}$ contains new tracks (joint updates of all predicted tracks using each of the measurements separately).

Specifically,

$$\mathcal{T}^{\text{legacy}}_{k+1|k+1} = \{(\ell_1, q_1, f_1(\mathbf{x})), ..., (\ell_{\nu'}, q_{\nu'}, f_{\nu'}(\mathbf{x}))\} \tag{17.47}$$

$$\mathcal{T}^{\text{meas}}_{k+1|k+1} = \{(\mathcal{L}_1, Q_1, F_1(\mathbf{x})), ..., (\mathcal{L}_m, Q_m, F_m(\mathbf{x}))\} \tag{17.48}$$

where

$$\ell_i = \ell'_i \tag{17.49}$$

$$q_i = \frac{q'_i \cdot (1 - f'_i[p_D])}{1 - q'_i f_i[p_D]} \leq q'_i \tag{17.50}$$

$$f_i(\mathbf{x}) = \frac{1 - p_D(\mathbf{x})}{1 - f'_i[p_D]} \cdot f'_i(\mathbf{x}) \tag{17.51}$$

$$\mathcal{L}_i = \text{new label} \tag{17.52}$$

$$Q_j = \frac{\tilde{D}[p_D L_{\mathbf{z}_j}]}{\lambda c(\mathbf{z}_j) + \tilde{D}[p_D L_{\mathbf{z}_j}]} < 1 \tag{17.53}$$

$$F_j(\mathbf{x}) = \frac{p_D(\mathbf{x}) \cdot L_{\mathbf{z}_j}(\mathbf{x}) \cdot \tilde{D}(\mathbf{x})}{\tilde{D}[p_D L_{\mathbf{z}_j}]} \tag{17.54}$$

and where the "predicted pseudo-PHD" $\tilde{D}(\mathbf{x})$ is defined by[2]

$$\tilde{D}(\mathbf{x}) \triangleq \sum_{i=1}^{\nu} \frac{q'_i}{1 - q'_i f'_i[p_D]} \cdot f'_i(\mathbf{x}). \tag{17.55}$$

The number of measurement-updated tracks increases in general:

$$\nu_{k+1|k+1} = \nu_{k+1|k} + m_{k+1}. \tag{17.56}$$

Pruning and merging must be used to limit growth in the number of tracks.

2 From (16.66) we know that the actual predicted PHD is $D(\mathbf{x}) = \sum_{i=1}^{\nu} q_i f_i(\mathbf{x})$.

17.2.3.1 Interpretation of the MeMBer Filter Corrector

In what follows I examine more closely some of the properties of the MeMBer filter corrector equations.

- If $p_D = 0$ then (17.50) and (17.51) reduce to $f_i(\mathbf{x}) = q'_i$ and $f_i(\mathbf{x}) = f'_i(\mathbf{x})$ and (17.53) and (17.54) reduce to $Q_j = 0$ and $F_j(\mathbf{x}) = 0$. That is, when no target-generated measurements can be collected then the updated MeMBer corrector reverts to prior information—that is, to the predicted track table.
- If $p_D = 1$ then (17.50) and (17.51) reduce to $q_i f_i(\mathbf{x}) = 0$. That is, when measurements are always collected from all targets then all tracks are measurement-generated—there are no legacy tracks.
- The legacy part of the MeMBer filter corrector resembles the JoTT filter corrector.

To understand this final point, recall from (14.202) and (14.203), that the JoTT corrector equations are

$$p_{k+1|k+1} = \frac{1 - f_{k+1|k}[p_D] + \sum_{\mathbf{z}\in Z} \frac{f_{k+1|k}[p_D L_{\mathbf{z}}]}{\lambda c(\mathbf{z})}}{p_{k+1|k}^{-1} - f_{k+1|k}[p_D] + \sum_{\mathbf{z}\in Z} \frac{f_{k+1|k}[p_D L_{\mathbf{z}}]}{\lambda c(\mathbf{z})}} \tag{17.57}$$

$$f_{k+1|k+1}(\mathbf{x}) = \frac{1 - p_D(\mathbf{x}) + \sum_{\mathbf{z}\in Z} \frac{p_D(\mathbf{x}) L_{\mathbf{z}}(\mathbf{x})}{\lambda c(\mathbf{z})}}{1 - f_{k+1|k}[p_D] + \sum_{\mathbf{z}\in Z} \frac{f_{k+1|k}[p_D L_{\mathbf{z}}]}{\lambda c(\mathbf{z})}} \tag{17.58}$$

$$\cdot f_{k+1|k}(\mathbf{x}) \tag{17.59}$$

where we have assumed that the false alarm process is Poisson: $\kappa(Z) = e^{-\lambda} \prod_{\mathbf{z}\in Z} \lambda c(\mathbf{z})$. In our current notation, $N_{k+1|k} = q'_i$, $N_{k+1|k+1} = q_i$, $f_{k+1|k} = f'_i$, and $f_{k+1|k+1} = f_i$, and so

$$q_i = \frac{q'_i - q'_i f'_i[p_D] + \sum_{\mathbf{z}\in Z} \frac{q'_i f'_i[p_D L_{\mathbf{z}}]}{\lambda c(\mathbf{z})}}{1 - q'_i f'_i[p_D] + \sum_{\mathbf{z}\in Z} \frac{q'_i f'_i[p_D L_{\mathbf{z}}]}{\lambda c(\mathbf{z})}} \tag{17.60}$$

$$f_i(\mathbf{x}) = \frac{q'_i \cdot (1 - p_D(\mathbf{x})) + \sum_{\mathbf{z}\in Z} \frac{q'_i p_D(\mathbf{x}) L_{\mathbf{z}}(\mathbf{x})}{\lambda c(\mathbf{z})}}{q'_i - q'_i f'_i[p_D] + \sum_{\mathbf{z}\in Z} \frac{q'_i f_{k+1|k}[p_D L_{\mathbf{z}}]}{\lambda c(\mathbf{z})}} \cdot f'_i(\mathbf{x}). \tag{17.61}$$

If no measurements were collected then

$$q_i = \frac{q_i' \cdot (1 - f_i'[p_D])}{1 - q_i' f_i'[p_D]} \tag{17.62}$$

$$f_i(\mathbf{x}) = \frac{1 - p_D(\mathbf{x})}{1 - f_i'[p_D]} \cdot f_i'(\mathbf{x}). \tag{17.63}$$

These are the equations (17.50) and (17.51) of the MeMBer corrector.

- The measurement-generated part of the MeMBer filter corrector resembles the PHD filter corrector.

To see this, notice that for a single measurement $Z_{k+1} = \{\mathbf{z}_j\}$ the PHD corrector, (16.108) and (16.109), becomes

$$D_{k+1|k+1}(\mathbf{x}) = \left(1 - p_D(\mathbf{x}) + \frac{p_D(\mathbf{x}) \cdot L_{\mathbf{z}_j}(\mathbf{x})}{\lambda c(\mathbf{z}_j) + D_{k+1|k}[p_D L_{\mathbf{z}_j}]}\right) \cdot D_{k+1|k}(\mathbf{x}) \tag{17.64}$$

and so the expected number of targets is

$$N_{k+1|k+1} = \int D_{k+1|k+1}(\mathbf{x}) d\mathbf{x} \tag{17.65}$$

$$= D_{k+1|k}[1 - p_D] + \frac{D_{k+1|k}[p_D L_{\mathbf{z}_j}]}{\lambda c(\mathbf{z}_j) + D_{k+1|k}[p_D L_{\mathbf{z}_j}]}. \tag{17.66}$$

This has the same form as (17.53) except that the predicted PHD $D_{k+1|k}$ has been replaced by the pseudopredicted PHD $\tilde{D}$ and that the term $D_{k+1|k}[1 - p_D]$ is not present.

17.2.4 MeMBer Filter Pruning and Merging

Suppose that the p.g.fl. at time step k is

$$G_{k|k}[h] = (1 - q_1 + q_1 f_1[h]) \cdots (1 - q_\nu + q_\nu f_\nu[h]).$$

If the probability of existence q_i of a track is smaller than some prespecified threshold, then the track can be eliminated.

On the other hand, suppose that two tracks $(\ell_i, q_i, f_i(\mathbf{x}))$ and $(\ell_j, q_j, f_j(\mathbf{x}))$ are such that $q_i + q_j < 1$. They are eligible for merging if the probability density of association

$$p_{i,j} = \int f_i(\mathbf{x}) \cdot f_j(\mathbf{x}) d\mathbf{x} \leq 1 \tag{17.67}$$

exceeds some threshold. In this case the merged track is $q_{i,j}, f_{i,j}(\mathbf{x})$ where

$$q_{i,j} = q_i + q_j \tag{17.68}$$

$$f_{i,j}(\mathbf{x}) = \frac{f_i(\mathbf{x}) \cdot f_j(\mathbf{x})}{p_{i,j}}. \tag{17.69}$$

As for labels, assign $\ell_{i,j} = \ell_i$ or $\ell_{i,j} = \ell_j$ depending on which is largest, q_i or q_j.

17.2.5 MeMBer Filter State and Error Estimation

Suppose that at time step k the p.g.fl. of the MeMBer filter is

$$G_{k|k}[h] = (1 - q_1 + q_1 \cdot f_1[h]) \cdots (1 - q_\nu + q_\nu \cdot f_\nu[h]). \tag{17.70}$$

17.2.5.1 Target Number State and Error Estimation

From (11.136) we know that the cardinality distribution $p_{k|k}(n)$ is the multi-Bernoulli distribution:

$$p_{k|k}(n) = B_{q_1,\ldots,q_\nu}(n). \tag{17.71}$$

Its expected number and variance are

$$N_{k|k} = \sum_{i=1}^{\nu} q_i \tag{17.72}$$

$$\sigma^2_{k|k} = \sum_{i=1}^{\nu} q_i(1 - q_i). \tag{17.73}$$

On the other hand, the MAP estimate of target number and its mean-square deviation are

$$\hat{n}_{k|k} = \arg\max_n \ B_{q_1,\ldots,q_\nu}(n) \tag{17.74}$$

$$\hat{\sigma}^2_{k|k} = \sigma^2_{k|k} + N^2_{k|k} - n^2_{k|k}. \tag{17.75}$$

These equations are established in Appendix G.33.

17.2.5.2 Multitarget State and Error Estimation

Because state estimation will be difficult in general, one must use approximate techniques. The simplest (and also least accurate) approach is to take the nearest integer $\breve{n}_{k|k}$ of the MAP estimate $\hat{n}_{k|k}$ and then select those $\breve{n}_{k|k}$ tracks that have largest probabilities q_i of existence.

Alternatively, one can construct the PHD

$$D_{k|k}(\mathbf{x}) = \sum_{i=1}^{\nu} q_i \cdot f_i(\mathbf{x}) \tag{17.76}$$

of $G_{k|k}[h]$ and use "PHD visualization" techniques, such as those described in Section 14.5.5).

17.2.6 Relationship with the Moreland-Challa Filter

In this section, I clarify the relationships between the MeMBer filter and the Moreland-Challa (M-C) filter of [168].

Moreland and Challa proposed their filter as a generalization of the IPDA filter [172], noting that it strongly resembled another generalized IPDA filter known as the JIPDA filter [173]. Like the MeMBer filter, it presumes that multitarget posterior distributions have belief-mass functions of the form [168, p. 809, (2)]

$$\beta(S) = (1 - q_1 + q_1 p_1(S)) \cdots (1 - q_\nu + q_\nu p_\nu(S)). \tag{17.77}$$

Here the pairs $q_i, f_i(\mathbf{x})$ are hypothesized ("tentative") tracks with probability of existence q_i and spatial distribution $f_i(\mathbf{x})$, and $p_i(S) = \int_S f_i(\mathbf{x}) d\mathbf{x}$ is the probability-mass function of $f_i(\mathbf{x})$.

The MeMBer and M-C filters both presume Poisson false alarms [168, p. 809, (3)]. They differ, however, in regard to labeling, track initiation and birth, and corrector approximation. The M-C filter does not maintain track labels. In addition, consider the following.

- *Track initiation and birth*: The MeMBer filter assumes that track birth is a multi-Bernoulli process, whereas the M-C filter presumes that it is a truncated Poisson process [168, p. 810]. Also, in the MeMBer filter both initial and birth tracks *must* have nonunity probabilities of existence.[3] The M-C filter presumes the opposite.

3 This can be guaranteed by ensuring that the maximal number ν of targets is larger than the current actual number of targets.

- *Corrector approximation*: In both filters the predicted posterior distribution is assumed to be the probability law of a multitarget multi-Bernoulli process. The Bayes update does not result, however, in another multitarget multi-Bernoulli process. Consequently, approximations must be imposed to arrive at formulas for the Bayes-updated tracks

$$q_1, f_1(\mathbf{x}), ..., q_\nu, f_\nu(\mathbf{x}). \tag{17.78}$$

The M-C filter employs an approximation at the multitarget density level, using measurement-to-track association techniques. Validation gates are established around the persisting tracks and the birth tracks and measurements not contained in any gate are ignored. Association weights between the tracks and the remaining measurements are computed. The $q_i, f_i(\mathbf{x})$ are then computed in terms of these weights [168, p. 813, Eqs. (21, 22)].

Measurement-gating may cause the M-C filter to have difficulty in detecting new tracks. If a new target appears near one of the birth tracks then it will be detected. Otherwise, the measurements it generates will be eliminated by the gating process.

The MeMBer filter corrector does not require gating and thus will not have this difficulty. It employs two successive approximations at the p.g.fl. level, based on the assumption that false alarm density is not too great, see (17.170) to (17.172) and (17.176). The success or failure of the MeMBer filter depends on the degree to which these approximations are valid in practice.

17.3 PARA-GAUSSIAN FILTER

In this section we derive a computationally tractable version of the MeMBer filter. Given that the p.g.fl. at time step k is

$$G_{k|k}[h] = (1 - q_1 + q_1 f_1[h]) \cdots (1 - q_\nu + q_\nu f_\nu[h])$$

we assume that each $f_i(\mathbf{x})$ is a Gaussian mixture:

$$f_i(\mathbf{x}) = \sum_{e=1}^{n^i_{k|k}} w_{i,e} \cdot N_{P^i_e}(\mathbf{x} - \mathbf{x}_{i,e}) \tag{17.79}$$

where

$$\sum_{e=1}^{n^i_{k|k}} w_{i,e} = 1 \tag{17.80}$$

for each $i = 1, ..., \nu$.

In what follows I describe the individual stages of the para-Gaussian filter: initialization (Section 17.3.1), prediction (Section 17.3.2), correction (Section 17.3.3), pruning and merging (Section 17.3.4), and state and error estimation (Section 17.3.5).

17.3.1 Para-Gaussian Filter Initialization

Initialization of the para-Gaussian filter consists of specifying an initial track table:

$$\mathcal{T}_{0|0} = \{(\ell^1_{0|0}, q^1_{0|0}, f^1_{0|0}(\mathbf{x})) \;, ..., \; (\ell^{\nu_{0|0}}_{0|0}, q^{\nu_{0|0}}_{0|0}, f^{\nu_{0|0}}_{0|0}(\mathbf{x}))\}. \tag{17.81}$$

First specify a maximum number $\nu_{0|0}$ of hypothesized targets, significantly larger than the actual expected number of targets. Then choose the states $\mathbf{x}^i_{0|0}$ and uncertainties $P^i_{0|0}$ of these hypothesized targets, so that

$$f^i_{0|0}(\mathbf{x}) = N_{P^i_{0|0}}(\mathbf{x} - \mathbf{x}^i_{0|0}). \tag{17.82}$$

Finally, designate the degree $q^i_{0|0}$ to which the ith hypothesized target is an actual target. The (usually integer) labels $\ell^i_{0|0}$ are chosen to enable propagation of track labels in later steps.

17.3.2 Para-Gaussian Filter Predictor

17.3.2.1 Para-Gaussian Filter Predictor Assumptions

The para-Gaussian filter predictor step requires the following assumptions:

- The Markov density is linear-Gaussian:

$$f_{k+1|k}(\mathbf{x}|\mathbf{x}') = N_{Q_k}(\mathbf{x} - F_k\mathbf{x}') \tag{17.83}$$

  where we abbreviate $Q \overset{\text{abbr.}}{=} Q_k$, and $F \overset{\text{abbr.}}{=} F_k$.

- The probability p_S that targets will survive into time step $k+1$ is constant.

- The target birth model consists of introducing $\alpha \stackrel{\text{abbr.}}{=} \alpha_{k+1|k}$ new linear-Gaussian tracks at time step $k+1$:

$$(l^1_{k+1|k}, b^1_{k+1|k}, B^1_{k+1|k}(\mathbf{x})) \;,....,\; (l^\alpha_{k+1|k}, b^\alpha_{k+1|k}, B^\alpha_{k+1|k}(\mathbf{x})) \quad (17.84)$$

where

$$B^i_{k+1|k}(\mathbf{x}) = N_{B^i_{k+1|k}}(\mathbf{x} - \mathbf{b}^i_{k+1|k}) \quad (17.85)$$

and where we abbreviate $b_i \stackrel{\text{abbr.}}{=} b^i_{k+1|k}$, $\mathbf{b}_i \stackrel{\text{abbr.}}{=} \mathbf{b}^i_{k+1|k}$, and $B_i \stackrel{\text{abbr.}}{=} B^i_{k+1|k}$.

17.3.2.2 Para-Gaussian Filter Predictor Equations

From time step k we are given a track table

$$\mathcal{T}_{k|k} = \{(\ell'_1, q'_1, f'_1(\mathbf{x})), ..., (\ell'_{\nu'}, q'_{\nu'}, f'_{\nu'}(\mathbf{x}))\} \quad (17.86)$$

where

$$f'_i(\mathbf{x}) = \sum_{e=1}^{n'_i} w'_{i,e} \cdot N_{P'_{i,e}}(\mathbf{x} - \mathbf{x}'_{i,e}). \quad (17.87)$$

and where $n'_i \stackrel{\text{abbr.}}{=} n^i_{k|k}$. We are to define the predicted track table $\mathcal{T}_{k+1|k}k$. It is

$$\mathcal{T}_{k+1|k} = \mathcal{T}^{\text{persist}}_{k+1|k} \cup \mathcal{T}^{\text{birth}}_{k+1|k} \quad (17.88)$$

where

$$\mathcal{T}^{\text{persist}}_{k+1|k} = \{(\ell_1, q_1, f_1(\mathbf{x})), ..., (\ell_{\nu'}, q_{\nu'}, f_{\nu'}(\mathbf{x}))\} \quad (17.89)$$

$$\mathcal{T}^{\text{birth}}_{k+1|k} = \{(l_1, b_1, B_1(\mathbf{x})), ..., (l_\alpha, b_\alpha, B_\alpha(\mathbf{x}))\}. \quad (17.90)$$

The purpose of birth tracks is to cover regions where target appearance might be expected, or simply to cover regions that are not currently covered. Each birth track is assigned a unique label that differentiates it from other birth tracks or the persisting tracks.

The number of tracks in the predictor step is the sum of the number of persisting tracks and the number of birth tracks:

$$\nu_{k+1|k} = \nu_{k|k} + \alpha_{k+1|k}. \quad (17.91)$$

The birth tracks have already been specified in (17.90). Thus we need only specify the time-update formulas for the persisting tracks in (17.89). The predictor equations for the persisting tracks are, for $i = 1, ..., \nu'$,

$$\ell_i = \ell'_i \tag{17.92}$$

$$q_i = p_S \cdot q'_i \tag{17.93}$$

$$f_i(\mathbf{x}) = \sum_{e=1}^{n'_i} w_{i,e} \cdot N_{P_{i,e}}(\mathbf{x} - \mathbf{x}_{i,e}) \tag{17.94}$$

$$\mathbf{x}_{i,e} = F'\mathbf{x}'_{i,e} \tag{17.95}$$

$$P_{i,e} = Q + FP'_{i,e}F^T \tag{17.96}$$

$$w_{i,e} = w'_{i,e}. \tag{17.97}$$

These equations are derived in Appendix G.34.

17.3.3 Para-Gaussian Filter Corrector

17.3.3.1 Para-Gaussian Filter Corrector Assumptions

The para-Gaussian filter predictor step requires the following assumptions:

- The likelihood function is linear-Gaussian:

$$f_{k+1}(\mathbf{z}|\mathbf{x}) = N_{R_{k+1}}(\mathbf{z} - H_{k+1}\mathbf{x}) \tag{17.98}$$

  where we abbreviate $R \overset{\text{abbr.}}{=} R_{k+1}$, and $H \overset{\text{abbr.}}{=} H_{k+1}$;
- The probability of detection p_D is constant;
- The false alarm process is Poisson with nonzero expected value $\lambda \overset{\text{abbr.}}{=} \lambda_{k+1} > 0$ and spatial distribution $c(\mathbf{z}) \overset{\text{abbr.}}{=} c_{k+1}(\mathbf{z})$.

17.3.3.2 Para-Gaussian Filter Corrector Equations

From time step k we are given a predicted track table

$$\mathcal{T}_{k+1|k} = \{(\ell'_1, q'_1, f'_1(\mathbf{x})), ..., (\ell'_{\nu'}, q'_{\nu'}, f'_{\nu'}(\mathbf{x}))\} \tag{17.99}$$

where

$$f'_i(\mathbf{x}) = \sum_{e=1}^{n'_i} w'_{i,e} \cdot N_{P'_{i,e}}(\mathbf{x} - \mathbf{x}'_{i,e}). \tag{17.100}$$

We are to define the measurement-updated track table $\mathcal{T}_{k+1|k+1}$. It is

$$\mathcal{T}_{k+1|k+1} = \mathcal{T}^{\text{legacy}}_{k+1|k+1} \cup \mathcal{T}^{\text{meas}}_{k+1|k+1} \tag{17.101}$$

where

$$\mathcal{T}^{\text{legacy}}_{k+1|k+1} = \{(\ell_1, q_1, f_1(\mathbf{x})), ..., (\ell_{\nu'}, q_{\nu'}, f_{\nu'}(\mathbf{x}))\} \tag{17.102}$$

$$\mathcal{T}^{\text{meas}}_{k+1|k+1} = \{(\mathcal{L}_1, Q_1, F_1(\mathbf{x})), ..., (\mathcal{L}_m, Q_m, F_m(\mathbf{x}))\}. \tag{17.103}$$

Thus the number of tracks in the corrector step is

$$\nu_{k+1|k+1} = \nu_{k+1|k} + m_{k+1}. \tag{17.104}$$

For the legacy tracks,

$$q_i = \frac{q'_i \cdot (1 - p_D)}{1 - p_D \cdot q'_i} \leq q'_i \tag{17.105}$$

$$f_i(\mathbf{x}) = f'_i(\mathbf{x}) \tag{17.106}$$

for $i = 1, ..., \nu'$. In the case of the measurement-generated tracks, for $j = 1, ..., m$,

$$Q_j = \frac{\sum_{i,e=1}^{\nu'} p_D \rho_i w'_{i,e} N_{R+HP'_{i,e}H^T}(\mathbf{z}_j - H\mathbf{x}'_{i,e})}{\lambda c(\mathbf{z}_j) + \sum_{i,e=1}^{\nu'} p_D \rho_i w'_{i,e} N_{R+HP'_i H^T}(\mathbf{z}_j - H\mathbf{x}'_{i,e})} \tag{17.107}$$

$$F_j(\mathbf{x}) = \sum_{i,e=1}^{\nu'} w_{i,e,j} \cdot N_{C_{i,e}}(\mathbf{x} - \mathbf{c}_{i,e,j}) \tag{17.108}$$

$$w_{i,e,j} = \frac{\rho_i \cdot w'_{i,e} \cdot N_{R+HP'_{i,e}H^T}(\mathbf{z}_j - H\mathbf{x}'_{i,e})}{\sum_{i',e'=1}^{\nu'} p_D \rho_{i'} w'_{i',e'} N_{R+HP'_{i',e'}H^T}(\mathbf{z}_j - H\mathbf{x}'_{i',e'})} \tag{17.109}$$

where

$$\begin{aligned} \mathbf{c}_{i,e,j} &= \mathbf{x}'_{i,e} + K_{i,e}(\mathbf{z}_j - H\mathbf{x}'_{i,e}) && (17.110) \\ C_{i,e} &= (1 - K_{i,e}H)P'_{i,e} && (17.111) \\ K_{i,e} &= P'_{i,e}H^T\left(HP'_{i,e}H^T + R\right)^{-1} && (17.112) \\ \rho_i &= \frac{q'_i}{1 - p_D q'_i}. && (17.113) \end{aligned}$$

These equations are proved in Appendix G.35.

17.3.4 Para-Gaussian Filter Pruning and Merging

We follow the same basic procedure outlined in Section 17.2.4. Suppose that the p.g.fl. at time step k is

$$G_{k|k}[h] = (1 - q_1 + q_1 f_1[h]) \cdots (1 - q_\nu + q_\nu f_\nu[h])$$

where the $f_i(\mathbf{x})$ are Gaussian mixtures

$$f_i(\mathbf{x}) = \sum_{e=1}^{n_i} w_{i,e} \cdot N_{P_{i,e}}(\mathbf{x} - \mathbf{x}_{i,e}). \tag{17.114}$$

Those tracks for which the probability of existence q_i is smaller than some threshold can be eliminated.

On the other hand, suppose that two tracks $q_i, f_i(\mathbf{x})$ and $q_j, f_j(\mathbf{x})$ are such that $q_i + q_j < 1$. From, (D.1), the fundamental identity for Gaussian distributions, (17.68) becomes

$$p_{i,j} = \int f_i(\mathbf{x}) \cdot f_j(\mathbf{x}) d\mathbf{x} \tag{17.115}$$

$$= \int \left(\sum_{e=1}^{n_i} w_{i,e} \cdot N_{P_{i,e}}(\mathbf{x} - \mathbf{x}_{i,e}) \right) \tag{17.116}$$

$$\cdot \left(\sum_{e'=1}^{n_j} w_{j,e'} \cdot N_{P_{j,e'}}(\mathbf{x} - \mathbf{x}_{j,e'}) \right) d\mathbf{x} \tag{17.117}$$

$$= \sum_{e=1}^{n_i} \sum_{e'=1}^{n_j} w_{i,e} \cdot w_{j,e'} \tag{17.118}$$

$$\cdot \int N_{P_{i,e}}(\mathbf{x} - \mathbf{x}_{i,e}) \cdot N_{P_{j,e'}}(\mathbf{x} - \mathbf{x}_{j,e'}) d\mathbf{x} \tag{17.119}$$

$$= \sum_{e=1}^{n_i} \sum_{e'=1}^{n_j} w_{i,e} \cdot w_{j,e'} \cdot N_{P_{i,e}+P_{j,e'}}(\mathbf{x}_{j,e'} - \mathbf{x}_{i,e}). \tag{17.120}$$

The two tracks can be merged if this value exceeds some preselected threshold. In this case the merged track is given by

$$q_{i,j} = q_i + q_j$$

$$f_{i,j}(\mathbf{x}) = \sum_{e=1}^{n_i} \sum_{e'=1}^{n_j} w_{i,e,j,e'} \cdot N_{C_{i,e,j,e'}}(\mathbf{x} - \mathbf{c}_{i,e,j,e'})$$

where

$$w_{i,e,j,e'} = \frac{w_{i,e} \cdot w_{j,e'} \cdot N_{P_{i,e}+P_{j,e'}}(\mathbf{x}_{j,e'} - \mathbf{x}_{i,e})}{\sum_{e=1}^{n_i} \sum_{e'=1}^{n_j} w_{i,e} \cdot w_{j,e'} \cdot N_{P_{i,e}+P_{j,e'}}(\mathbf{x}_{j,e'} - \mathbf{x}_{i,e})} \tag{17.121}$$

and where

$$\mathbf{c}_{i,e,j,e'} = \mathbf{x}_{i,e} + K_{i,e,j,e'}(\mathbf{x}_{j,e'} - \mathbf{x}_{i,e}) \tag{17.122}$$

$$C_{i,e,j,e'} = (I - K_{i,e,j,e'}) P_{i,e} \tag{17.123}$$

$$K_{i,e,j,e'} = P_{i,e} H^T (H P_{i,e} H^T + P_{j,e'})^{-1}. \tag{17.124}$$

17.3.5 Para-Gaussian Filter State and Error Estimation

The techniques are the same as described in Section 17.2.5. For example, suppose that at time step k the track table is

$$\mathcal{T}_{k|k} = \{(\ell_1, q_1, f_1[h]), ..., (\ell_\nu, q_\nu, f_\nu[h])\}. \tag{17.125}$$

Then by (16.66) the data-updated PHD is

$$D_{k|k}(\mathbf{x}) = \sum_{i=1}^{\nu} q_i f_i(\mathbf{x}). \tag{17.126}$$

17.4 MEMBER FILTER DERIVATION

The purpose of this section is to summarize the major ideas underlying the multi-target multi-Bernoulli approximations employed in this chapter. I summarize the main points of the MeMBer filter predictor in Section 17.4.1, and the main points of the MeMBer filter corrector in Section 17.4.2.

17.4.1 Derivation of the MeMBer Filter Predictor

Suppose that the prior p.g.fl.—that is, the p.g.fl. of $f_{k|k}(X|Z^{(k)})$—is the multitarget multi-Bernoulli process

$$G_{k|k}[h] = \prod_{i=1}^{\nu'} (1 - q'_i + q'_i \cdot f'_i[h]). \tag{17.127}$$

We want to show that the predicted p.g.fl.—that is, the p.g.fl. of $f_{k+1|k}(X|Z^{(k)})$—is also a multitarget multi-Bernoulli process:

$$G_{k+1|k}[h] = G_B[h] \cdot \overbrace{\prod_{i=1}^{\nu'} (1 - q_i + q_i \cdot f_i[h])}^{\text{predicted targets}}. \tag{17.128}$$

Here,

$$G_B[h] = \overbrace{\prod_{i=1}^{\alpha} (1 - b_i + b_i \cdot b_i[h])}^{\text{birth targets}} \tag{17.129}$$

is the p.g.fl. of the target-birth process (which is itself a multitarget multi-Bernoulli process by assumption). Also,

$$q_i = q'_i \cdot f'_i[p_S] \quad (17.130)$$

$$f_i(\mathbf{x}) = \frac{f'_i[p_S M_{\mathbf{x}}]}{f'_i[p_S]} \quad (17.131)$$

where $M_{\mathbf{x}}(\mathbf{x}') \triangleq f_{k+1|k}(\mathbf{x}|\mathbf{x}')$ and $f_i[h] = \int h(\mathbf{x}) \cdot f_i(\mathbf{x})d\mathbf{x}$.

Equation (17.128) results from the following reasoning. It is left to the reader as Exercise 61 to show that the predicted p.g.fl. $G_{k+1|k}[h]$ is related to the prior p.g.fl. $G_{k|k}[h]$ by

$$G_{k+1|k}[h] = G_B[h] \cdot G_{k|k}[1 - p_S + p_S p_h]. \quad (17.132)$$

Combining (17.127) and (17.132), we get

$$G_{k+1|k}[h] \quad (17.133)$$

$$= G_B[h] \cdot G_{k|k}[1 - p_S + p_S p_h] \quad (17.134)$$

$$= G_B[h] \cdot \prod_{i=1}^{\nu'} (1 - q'_i + q'_i \cdot f'_i[1 - p_S + p_S p_h]) \quad (17.135)$$

$$= G_B[h] \cdot \prod_{i=1}^{\nu'} \left(1 - q'_i \cdot f'_i[p_S] + q'_i \cdot f'_i[p_S] \cdot \frac{f'_i[p_S p_h]}{f'_i[p_S]}\right). \quad (17.136)$$

Equation (17.130) follows immediately. As for (17.131), note that

$$\frac{f'_i[p_S p_h]}{f'_i[p_S]} \quad (17.137)$$

$$= \frac{1}{f'_i[p_S]} \cdot \int p_S(\mathbf{x}') \cdot p_h(\mathbf{x}') \cdot f'_i(\mathbf{x}')d\mathbf{x}' \quad (17.138)$$

$$= \frac{1}{f'_i[p_S]} \cdot \int p_S(\mathbf{x}') \cdot \left(\int h(\mathbf{x}) \cdot f_{k+1|k}(\mathbf{x}|\mathbf{x}')d\mathbf{x}\right) \quad (17.139)$$

$$\cdot f'_i(\mathbf{x}')d\mathbf{x}' \quad (17.140)$$

$$
\begin{aligned}
&= \frac{1}{f_i'[p_S]} \cdot \int h(\mathbf{x}) \cdot \left(\int p_S(\mathbf{x}') \cdot M_{\mathbf{x}}(\mathbf{x}') \cdot f_i'(\mathbf{x}') d\mathbf{x}' \right) d\mathbf{x} &(17.141)\\
&= \frac{1}{f_i'[p_S]} \cdot \int h(\mathbf{x}) \cdot f_i'[p_S M_{\mathbf{x}}] d\mathbf{x} &(17.142)\\
&= \int h(\mathbf{x}) \cdot f_i(\mathbf{x}) d\mathbf{x} = f_i[h]. &(17.143)
\end{aligned}
$$

17.4.2 Derivation of the MeMBer Filter Corrector

Suppose that the predicted p.g.fl.—that is, the p.g.fl. of $f_{k+1|k}(X|Z^{(k)})$— is the multitarget multi-Bernoulli process

$$G_{k+1|k}[h] = \prod_{i=1}^{\nu'} (1 - q_i' + q_i' \cdot f_i'[h]). \tag{17.144}$$

We want to show that the data-updated p.g.fl.—that is, the p.g.fl. of $f_{k+1|k+1}(X|Z^{(k)})$)— is *approximately* a multitarget multi-Bernoulli process:

$$
\begin{aligned}
G_{k+1|k}[h] \;\cong\; & \overbrace{\prod_{i=1}^{\nu'} (1 - q_i + q_i \cdot f_i[h])}^{\text{legacy tracks}} &(17.145)\\
& \cdot \overbrace{\prod_{j=1}^{m} (1 - Q_i + Q_i \cdot F_i[h])}^{\text{data-induced tracks}}. &(17.146)
\end{aligned}
$$

Here,

$$
\begin{aligned}
q_i &= \frac{q_i' f_i'[q_D]}{1 - q_i' f_i[p_D]} &(17.147)\\
f_i(\mathbf{x}) &= \frac{1 - p_D(\mathbf{x})}{1 - f_i'[p_D]} \cdot f_i'(\mathbf{x}) &(17.148)
\end{aligned}
$$

and

$$
\begin{aligned}
Q_j &= \frac{\tilde{D}[p_D L_{\mathbf{z}_j}]}{\lambda c(\mathbf{z}_j) + \tilde{D}[p_D L_{\mathbf{z}_j}]} &(17.149)\\
F_j(\mathbf{x}) &= \frac{p_D(\mathbf{x}) \cdot L_{\mathbf{z}_j}(\mathbf{x}) \cdot \tilde{D}(\mathbf{x})}{\tilde{D}[p_D L_{\mathbf{z}_j}]} &(17.150)
\end{aligned}
$$

where

$$\tilde{D}(\mathbf{x}) \triangleq \sum_{i=1}^{\nu'} \frac{q'_i}{1 - q'_i f_i[p_D]} \cdot f'_i(\mathbf{x}) \tag{17.151}$$

To see why this is true, recall that in Section 14.8.2, I showed that the multitarget corrector, (14.50), can be written in p.g.fl. form as

$$G_{k+1|k+1}[h] = \frac{\frac{\delta F}{\delta Z_{k+1}}[0, h]}{\frac{\delta F}{\delta Z_{k+1}}[0, 1]}. \tag{17.152}$$

In (14.285), I showed that

$$F[g, h] = e^{\lambda c[g] - \lambda} \cdot G_{k+1|k}[h(q_D + p_D p_g)] \tag{17.153}$$

where $q_D(\mathbf{x}) \triangleq 1 - p_D(\mathbf{x})$ and where

$$c[g] \triangleq \int g(\mathbf{z}) \cdot c(\mathbf{z}) d\mathbf{z} \tag{17.154}$$

$$p_g(\mathbf{x}) \triangleq \int g(\mathbf{z}) \cdot f_{k+1}(\mathbf{z}|\mathbf{x}) d\mathbf{z}. \tag{17.155}$$

Combining (17.144) and (17.153) we get:

$$F[g, h] = e^{\lambda c[g] - \lambda} \cdot \prod_{i=1}^{\nu'} \left(1 - q'_i + q'_i f'_i[h q_D] + q'_i f'_i[h p_D p_g]\right). \tag{17.156}$$

17.4.2.1 MeMBer Filter Corrector: First Approximation

It follows that

$$F[0, h] = e^{-\lambda} \cdot \prod_{i=1}^{\nu'} (1 - q'_i + q'_i f'_i[h q_D]) \tag{17.157}$$

and

$$\frac{\delta F}{\delta \mathbf{z}_1}[0, h] = e^{-\lambda} \cdot \left(\prod_{i=1}^{\nu'} (1 - q'_i + q'_i f'_i[h q_D]) \right) \tag{17.158}$$

$$\cdot \left(\lambda c(\mathbf{z}_1) + \sum_{1 \leq i_1 \leq \nu'} \frac{q'_{i_1} f'_{i_1}[h p_D L_{\mathbf{z}_1}]}{1 - q'_{i_1} + q'_{i_1} f'_{i_1}[h q_D]} \right) \tag{17.159}$$

and

$$\frac{\delta^2 F}{\delta \mathbf{z}_2 \delta \mathbf{z}_1}[0,h] \tag{17.160}$$

$$= e^{-\lambda} \cdot \left(\prod_{i=1}^{\nu'} (1 - q'_i + q'_i f'_i[hq_D]) \right) \tag{17.161}$$

$$\cdot \left(\begin{array}{c} \lambda^2 c(\mathbf{z}_1) \cdot c(\mathbf{z}_2) \\ +\lambda c(\mathbf{z}_1) \sum_{1 \leq i_1 \leq \nu'} \frac{q'_{i_1} f'_{i_1}[hp_D L_{\mathbf{z}_2}]}{1 - q'_{i_1} + q'_{i_1} f'_{i_1}[hq_D]} \\ +\lambda c(\mathbf{z}_2) \sum_{1 \leq i_1 \leq \nu'} \frac{q'_{i_1} f'_{i_1}[hp_D L_{\mathbf{z}_1}]}{1 - q'_{i_1} + q'_{i_1} f'_{i_1}[hq_D]} \\ + \sum_{1 \leq i_1 \neq i_2 \leq \nu'} \frac{q'_{i_1} f'_{i_1}[hp_D L_{\mathbf{z}_1}] \cdot q'_{i_2} f'_{i_2}[hp_D L_{\mathbf{z}_2}]}{(1 - q'_{i_1} + q'_{i_1} f'_{i_1}[hq_D]) \cdot (1 - q'_{i_2} + q'_{i_2} f'_{i_2}[hq_D])} \end{array} \right). \tag{17.162}$$

The difference between

$$\left(\sum_{i=1}^{\nu'} \frac{q'_i f'_i[hp_D L_{\mathbf{z}_1}]}{1 - q'_i + q'_i f'_i[hq_D]} \right) \left(\sum_{i=1}^{\nu'} \frac{q'_i f'_i[hp_D L_{\mathbf{z}_2}]}{1 - q'_i + q'_i f'_i[hq_D]} \right) \tag{17.163}$$

and

$$\sum_{1 \leq i_1 \neq i_2 \leq \nu'} \frac{q'_{i_1} f'_{i_1}[hp_D L_{\mathbf{z}_1}] \cdot q'_{i_2} f'_{i_2}[hp_D L_{\mathbf{z}_2}]}{(1 - q'_{i_1} + q'_{i_1} f'_{i_1}[hq_D]) \cdot (1 - q'_{i_2} + q'_{i_2} f'_{i_2}[hq_D])} \tag{17.164}$$

is

$$\sum_{i=1}^{\nu'} \frac{(q'_i)^2 \cdot f'_i[hp_D L_{\mathbf{z}_1}] \cdot f'_i[hp_D L_{\mathbf{z}_2}]}{(1 - q'_i + q'_i f'_i[hq_D])^2}. \tag{17.165}$$

Suppose that the false alarm density is not too large, in the sense that it is unlikely that two typical observations $\mathbf{z}_1, \mathbf{z}_2$ will be near a typical track. Thus the factor $f_i[hp_D L_{\mathbf{z}_1}] \cdot f_i[hp_D L_{\mathbf{z}_2}]$ will generally be small. So, the expressions in (17.163) and (17.164) will be approximately equal. One of them will generally be small as long as clutter is not too dense. Thus (17.162) becomes

$$\frac{\delta^2 F}{\delta \mathbf{z}_2 \delta \mathbf{z}_1}[0, h] \tag{17.166}$$

$$\cong e^{-\lambda} \cdot \left(\prod_{i=1}^{\nu'} (1 - q'_i + q'_i f'_i[hq_D]) \right) \tag{17.167}$$

$$\cdot \left(\frac{\lambda c(\mathbf{z}_1) + \sum_{i=1}^{\nu'} \frac{q'_i f'_i[hp_D L_{\mathbf{z}_1}]}{1-q'_i+q'_i f'_i[hq_D]}}{\lambda c(\mathbf{z}_1) + \sum_{i=1}^{\nu'} \frac{q'_i f'_i[p_D L_{\mathbf{z}_1}]}{1-q_i+q_i f_i[q_D]}} \right) \tag{17.168}$$

$$\cdot \left(\frac{\lambda c(\mathbf{z}_2) + \sum_{i=1}^{\nu'} \frac{q'_i f'_i[hp_D L_{\mathbf{z}_2}]}{1-q'_i+q'_i f'_i[hq_D]}}{\lambda c(\mathbf{z}_2) + \sum_{i=1}^{\nu'} \frac{q'_i f'_i[p_D L_{\mathbf{z}_2}]}{1-q'_i+q'_i f'_i[q_D]}} \right). \tag{17.169}$$

In general, as we take higher functional derivatives the diagonal terms will tend to be small. Thus for a measurement set $Z_{k+1} = \{\mathbf{z}_1, ...\mathbf{z}_m\}$ we can write

$$\frac{\delta^m F}{\delta \mathbf{z}_m \cdots \delta \mathbf{z}_1}[0, h] \tag{17.170}$$

$$\cong e^{-\lambda} \cdot \left(\prod_{i=1}^{\nu'} (1 - q'_i + q'_i f'_i[hq_D]) \right) \tag{17.171}$$

$$\cdot \prod_{j=1}^{m} \frac{\lambda c(\mathbf{z}_j) + \sum_{i=1}^{\nu'} \frac{q'_i f'_i[hp_D L_{\mathbf{z}_j}]}{1-q'_i+q'_i f'_i[hq_D]}}{\lambda c(\mathbf{z}_j) + \sum_{i=1}^{\nu'} \frac{q'_i f'_i[p_D L_{\mathbf{z}_j}]}{1-q'_i+q'_i f'_i[q_D]}}. \tag{17.172}$$

Consequently, the data-updated p.g.fl. will be

$$G_{k+1|k+1}[h] \cong \left(\prod_{i=1}^{\nu} \frac{1 - q'_i + q'_i f'_i[hq_D]}{1 - q'_i + q'_i f'_i[q_D]} \right) \tag{17.173}$$

$$\cdot \left(\prod_{j=1}^{m} \frac{\lambda c(\mathbf{z}_j) + \sum_{i=1}^{\nu} \frac{q'_i f'_i[hp_D L_{\mathbf{z}_j}]}{1-q'_i+q'_i f'_i[hq_D]}}{\lambda c(\mathbf{z}_j) + \sum_{i=1}^{\nu} \frac{q'_i f'_i[p_D L_{\mathbf{z}_j}]}{1-q'_i+q'_i f'_i[q_D]}} \right). \tag{17.174}$$

17.4.2.2 Second Approximation

This does not have the form of a multi-Bernoulli process since the factors

$$\lambda c(\mathbf{z}_j) + \sum_{i=1}^{\nu'} \frac{q_i' f_i'[hp_D L_{\mathbf{z}_j}]}{1 - q_i' + q_i' f_i'[hq_D]} \tag{17.175}$$

are nonlinear in h. An additional approximation is required:

$$\frac{q_i' f_i'[hp_D L_{\mathbf{z}_j}]}{1 - q_i' + q_i' f_i'[hq_D]} \cong \frac{q_i'}{1 - q_i' + q_i' f_i'[q_D]} \cdot f_i'[hp_D L_{\mathbf{z}_j}] \tag{17.176}$$

and thus

$$\begin{aligned} & \lambda c(\mathbf{z}_j) + \sum_{i=1}^{\nu'} \frac{q_i' f_i'[hp_D L_{\mathbf{z}_j}]}{1 - q_i' + q_i' f_i'[hq_D]} && (17.177) \\ \cong\; & \lambda c(\mathbf{z}_j) + \sum_{i=1}^{\nu'} \frac{q_i' f_i'[hp_D L_{\mathbf{z}_j}]}{1 - q_i' + q_i' f_i'[q_D]} && (17.178) \\ =\; & \lambda c(\mathbf{z}_j) + \tilde{D}[hp_D L_{\mathbf{z}_j}] && (17.179) \end{aligned}$$

where

$$\tilde{D}(\mathbf{x}) \triangleq \sum_{i=1}^{\nu'} \frac{q_i'}{1 - q_i' f_i[p_D]} \cdot f_i'(\mathbf{x}). \tag{17.180}$$

17.4.2.3 Final Steps

The first set of factors in (17.173) can be rewritten as:

$$\begin{aligned} \frac{1 - q_i' + q_i' f_i'[hq_D]}{1 - q_i' + q_i' f_i'[q_D]} &= \frac{1 - q_i' + q_i' f_i'[q_D] \cdot \frac{f_i'[hq_D]}{f_i'[q_D]}}{1 - q_i' + q_i' f_i'[q_D]} && (17.181) \\ &= \frac{1 - q_i'}{1 - q_i' + q_i' f_i'[q_D]} && (17.182) \\ &\quad + \frac{q_i' f_i'[q_D]}{1 - q_i' + q_i' f_i'[q_D]} \cdot \frac{f_i'[hq_D]}{f_i'[q_D]} && (17.183) \\ &= 1 - q_i + q_i \cdot f_i[h] && (17.184) \end{aligned}$$

where q_i and $f_i(\mathbf{x})$ were defined in (17.147) and (17.148).

The second set of factors (17.173) can be rewritten as:

$$\frac{\lambda c(\mathbf{z}_j) + \tilde{D}[hp_D L_{\mathbf{z}_j}]}{\lambda c(\mathbf{z}_j) + \tilde{D}[p_D L_{\mathbf{z}_j}]} = \frac{\lambda c(\mathbf{z}_j)}{\lambda c(\mathbf{z}_j) + \sum_{i=1}^{\nu} \rho_i \cdot f_i'[p_D L_{\mathbf{z}_j}]} \tag{17.185}$$

$$+\frac{\tilde{D}[p_D L_{\mathbf{z}_j}]}{\lambda c(\mathbf{z}_j) + \tilde{D}[p_D L_{\mathbf{z}_j}]} \cdot \frac{\tilde{D}[hp_D L_{\mathbf{z}_j}]}{\tilde{D}[p_D L_{\mathbf{z}_j}]} \tag{17.186}$$

$$= 1 - Q_j + Q_j \cdot F_j[h] \tag{17.187}$$

where Q_j and $F_j(\mathbf{x})$ were defined in (17.149) and (17.150).

17.5 CHAPTER EXERCISE

Exercise 61 *Verify (17.132):*

$$G_{k+1|k}[h] = G_B[h] \cdot G_{k|k}[1 - p_S + p_S p_h]. \tag{17.188}$$

Hint: Employ the same reasoning used to derive (14.273) for the PHD filter predictor.

Appendix A

Glossary of Notation

A.1 TRANSPARENT NOTATIONAL SYSTEM

I employ what is sometimes called a "transparent" system of notation. That is, to the greatest extent possible the reader should be able to infer the semantics of mathematical symbols at a glance.

The most familiar example of transparent notation is the nearly universal practice of reserving the symbols i, j, k for integer variables. (In our system, i, j, l, e are integer variables but k always means an integer time-index.) The system employed in this book extends that of the influential Bar-Shalom school.

The letter x and its variants always relate to *target states* and the letter z and its variants always relate to *measurements*. Thus the following are true:

- Fractur $\mathfrak{X}$ (general state space), ξ (state from a general state space), Fractur $\mathfrak{X}_0$ (baseline state space, e.g., of state vectors), $\mathbf{x}, \mathbf{x}'$ (state vector), x (scalar component of a state vector), $\mathbf{X}$ (random state vector), X, X' (finite set of state vectors, including the empty set), Ξ random finite state set. When more than one symbol is required for a finite-state set, we usually use the symbols Y and V. Also, n, n' denote the number of elements in finite-state sets and N usually denotes the dimension of a Euclidean state space $\mathfrak{X}_0 = \mathbb{R}^N$. Nonfinite subsets of $\mathfrak{X}_0$ will be denoted S, S';
- Fractur $\mathfrak{Z}$ (general measurement space), ζ (measurement from a general measurement space), Fractur $\mathfrak{Z}_0$ (baseline measurement space, e.g., of measurement vectors), $\mathbf{z}$, $\mathbf{z}'$ (state vector), z (scalar component of a measurement vector), $\mathbf{Z}$ (random measurement vector), Z^k (time sequence $\mathbf{z}_1, ..., \mathbf{z}_k$ of measurement vectors), Z, Z' (finite set of measurement vectors,

including the empty set), $Z^{(k)}$ (time sequence $Z_1, ..., Z_k$ of measurement sets), Σ random finite measurement set. When more than one symbol is required for a measurement set, we usually use the symbol W. Also, m, m' denote the number of elements in finite-state sets and M usually denotes the dimension of a Euclidean measurement space, $\mathfrak{Z}_0 = \mathbb{R}^M$. Nonfinite subsets of $\mathfrak{Z}_0$ will be denoted T, T'. Nonfinite random subsets of $\mathfrak{Z}_0$ will be denoted Θ;

- I also follow the usual conventions for deterministic versus random quantities: lowercase is deterministic whereas uppercase indicates the corresponding random quantity. Thus a, a' are real numbers whereas A, A' are random real numbers; and $\mathbf{y}$, $\mathbf{y}'$ are vectors in $\mathfrak{Y}_0$ whereas $\mathbf{Y}$, $\mathbf{Y}'$ are random vectors. With few exceptions, capital Greek letters denote random subsets (finite or infinite) of some underlying space.

The symbol ν (lowercase Greek letter nu) usually denotes some fixed integer number.

A.2 GENERAL MATHEMATICS

This section summarizes notation for basic mathematical concepts.

- $A \triangleq B$: A is defined to be equal to B;
- $A \overset{\text{abbr.}}{=} B$: A is an abbreviation of B;
- $A \cong B$: A is approximately equal to B;
- $A \overset{?}{=} B$: Are A and B equal?
- $f(u) \equiv a$: Function $f(u)$ is identically equal to the constant a;
- $n!$: n factorial;
- $[0, 1]$: Unit interval: real numbers between zero and one;
- $C_{n,k} = \frac{n!}{k!(n-k)!}$: Combinatorial coefficient;
- $\mathbb{R}^n$: n-dimensional Euclidean space;
- $\mathbf{y} = (y_1, ..., y_n, a_1, ..., a_{n'})$: Vector with continuous components $y_1, ..., y_n$ and discrete components $a_1, ..., a_{n'}$;

- $[f(\mathbf{y})]_{\mathbf{y}=\mathbf{a}} = f(\mathbf{a})$: Substitute value $\mathbf{y} = \mathbf{a}$ for $\mathbf{y}$ in the function $f(\mathbf{y})$;
- $\langle \mathbf{y}_1, \mathbf{y}_2 \rangle$: Scalar (dot) product of vectors $\mathbf{y}_1, \mathbf{y}_2$;
- $||\mathbf{y}_1 - \mathbf{y}_2||$: Euclidean distance between vectors $\mathbf{y}_1, \mathbf{y}_2$;
- C^T: Transpose of the matrix C;
- $\int f(\mathbf{y})d\mathbf{y} = \sum_{a_1,\ldots,a_{n'}} \int f(y_1, ..., y_n, a_1, ..., a_{n'})dy_1 \cdots dy_n$: Multivariate integral of function $f(\mathbf{y})$;
- $\frac{\partial f}{\partial g}(\mathbf{y})$: Gradient derivative of a vector function $f(\mathbf{y})$, Appendix C;
- $\frac{\delta p}{\delta \mathbf{y}}$: Radon-Nikodým derivative of probability-mass function $p(S)$ at point $\mathbf{y}$;
- $\lim_{a \searrow 0} f(a)$: One sided limit of $f(a)$ as a approaches 0 from above;
- $\sup_{\mathbf{y}} f(\mathbf{y})$: Supremal (maximal) value of function $f(\mathbf{y})$;
- $\arg\sup_{\mathbf{y}} f(\mathbf{y})$: Value of $\mathbf{y}$ that supremizes $f(\mathbf{y})$;
- $\min\{a_1, ..., a_n\}$: Smallest number in the list $a_1, ..., a_n$;
- $\max\{a_1, ..., a_n\}$: Largest number in the list $a_1, ..., a_n$;
- $O(n)$: Order of magnitude is n;
- $\log x$: Natural (Naperian) logarithm of x.

A.3 SET THEORY

This section summarizes basic notation for set theory.

- $\emptyset$: Empty set;
- $\{a, b, c, d\}$: Finite set with elements a, b, c, d (usually assumed to be distinct);
- $\{a | \ P(a)\}$: Set of all elements a of some universe that satisfy some logical proposition $P(a)$;
- X, Y, Z, W: Finite sets;
- S, T, U: Not necessarily finite sets;

- $\mathbf{1}_S(\mathbf{y})$: Indicator function of subset S, (5.38);
- $|X|$: Number of elements in the finite set X;
- $|S|$: Hypervolume (Lebesgue measure) of the Lebesgue-measurable set S;
- $\mathfrak{Z}$: General measurement space;
- $\mathfrak{X}$: General state space;
- $S \cap T$: Intersection of sets S, T;
- $S \cup T$: Union of sets S, T;
- $S \uplus T$: Disjoint union of sets S, T;
- $S \times T$: Cartesian product of sets S, T;
- $S - T$: Difference of sets S, T;
- $S \subseteq T$: Set S is included in (and possibly equal to) set T';
- $S \subset T$: Set S is included in (but not equal to) set T;
- $S \supseteq T$: Set S is included in (and possibly equal to) set T;
- S^c: Complement of set S;
- $a \in S$: a is an element of the set S;
- $a \notin S$: a is not an element of the set S;
- $\delta_{\mathbf{x},\mathbf{y}}$: Kronecker delta;
- $\delta_{\mathbf{x}}(\mathbf{y})$: Dirac delta density concentrated at $\mathbf{y} = \mathbf{x}$, Appendix B;
- $\Delta_{\mathbf{x}}(S)$: Dirac measure concentrated at $\mathbf{x}$, Appendix B;
- $E_{\mathbf{z}}$: Small region surrounding the vector $\mathbf{z}$;
- $B_{\varepsilon,\mathbf{z}}$: (Hyper)ball of radius ε centered at the vector $\mathbf{z}$.

A.4 FUZZY LOGIC AND DEMPSTER-SHAFER THEORY

This section summarizes the basic notation of expert systems theory.

- $|f|$: magnitude of fuzzy set f, (5.213);

- $f \wedge f'$: Fuzzy *AND* of fuzzy membership functions f, f', Section 4.3;
- $f \vee f'$: Fuzzy *OR* of fuzzy membership functions f, f', Section 4.3;
- $f^c = 1 - f$: Fuzzy negation of fuzzy membership function f, Section 4.3;
- $f \wedge_{A,A'} f'$: Copula fuzzy *AND* of fuzzy membership functions f, f', Section 4.3.4;
- $f \vee_{A,A'} f'$: Copula fuzzy *OR* of fuzzy membership functions f, f', Section 4.3.4;
- $m : f_1, ..., f_b; m_1, ..., m_b$: Fuzzy Dempster-Shafer basic mass assignment (b.m.a.), Section 4.6;
- $m * m'$: Dempster's combination of two b.m.a.s m, m', Section 4.5.1;
- $m \cap m'$: Unnormalized Dempster's combination of two b.m.a.s m, m', Section 4.5.1;
- $m *_q m'$: Modified Dempster's combination of m, m', Sections 4.5.1 and 4.6;
- $\pi_m(u)$: Pignistic probability of a b.m.a. m, Section 4.5.3.2;
- $v_m(u)$: Voorbraak probability of a b.m.a. m, Section 4.5.3.1.

A.5 PROBABILITY AND STATISTICS

This section summarizes the basic notations of probability theory.

- p_D: Probability of detection, Section 9.2.1;
- p_{FA}: Probability of false alarm;
- λ: Average number of false alarms (Poisson process);
- $\mathbf{y}_1, ..., \mathbf{y}_m \sim f(\mathbf{y})$: $\mathbf{y}_1, ..., \mathbf{y}_m$ are sampled from the probability distribution $f(\mathbf{y})$;
- $N_C(\mathbf{x} - \hat{\mathbf{x}})$: Gaussian distribution with mean $\hat{\mathbf{x}}$ and covariance matrix C, (2.34);

- $\hat{N}_C(\mathbf{x} - \hat{\mathbf{x}})$: Normalized Gaussian distribution with mean $\hat{\mathbf{x}}$ and covariance matrix C, (5.207);
- $N_{\sigma^2}(x - \hat{x})$: One-dimensional Gaussian distribution with mean $\hat{x}$ and variance σ^2;
- $\mathbf{x} = (x_1, ..., x_n, c_1, ..., c_{n'})$: State vector with continuous components $x_1, ..., x_n$ and discrete components $c_1, ..., c_{n'}$;
- $\mathbf{z} = (z_1, ..., z_n, u_1, ..., u_{n'})$: Measurement vector with continuous components $z_1, ..., z_n$ and discrete components $u_1, ..., u_{n'}$;
- $\overset{s}{\mathbf{z}}$: Measurement vector $\overset{s}{\mathbf{z}}$: originating with a sensor with sensor identifying-tag s, Section 12.2;
- $\overset{s}{\mathfrak{Z}}$: Measurement space for sensor with sensor identifying tag s, Section 12.2;
- $\overset{s}{S}$: Subset of measurement space for sensor with sensor identifying-tag s, Section 12.2;
- $Z^k : \mathbf{z}_1, ..., \mathbf{z}_k$: Time sequence of observations at time step k
- $\mathbf{X}, \mathbf{Y}, \mathbf{Z}, \mathbf{V}, \mathbf{W}$: Random vectors;
- $\mathbf{Z} = \eta(\mathbf{x}, \mathbf{W})$: General nonlinear sensor measurement model, (2.91);
- $\mathbf{X}_{k+1} = \varphi_k(\mathbf{X}_k, \mathbf{V}_k)$: General (discrete-time) nonlinear target motion model, (2.89);
- $\Pr(\mathbf{Z} = \mathbf{z})$: Probability that (discrete-valued) random vector $\mathbf{Z}$ has value $\mathbf{z}$;
- $p_{\mathbf{Z}}(S) = \Pr(\mathbf{Z} \in S)$: Probability-mass function of random vector $\mathbf{Z}$;
- $p_{\mathbf{Z}|\mathbf{X}}(S|\mathbf{x}) = \Pr(\mathbf{Z} \in S)$: Probability-mass function of random vector $\mathbf{Z}$ conditioned on $\mathbf{X} = \mathbf{x}$;
- $\mathrm{E}[\mathbf{X}]$: Expected value of random vector $\mathbf{X}$;
- $f(\mathbf{z}|\mathbf{x}) = f_{\mathbf{Z}|\mathbf{X}}(\mathbf{z}|\mathbf{x})$: Sensor likelihood function, (2.114);
- $f_{\mathbf{Z}}(\mathbf{z})$: Density function corresponding to probability-mass function $p_{\mathbf{Z}}(S)$;

- $\overset{s}{f}(\overset{s}{\mathbf{z}}|\mathbf{x})$: Sensor likelihood function for sensor with identifying sensor tag s 2.4.10;
- $f(\overset{1}{\mathbf{z}}, ..., \overset{s}{\mathbf{z}}|\mathbf{x})$: Joint multisensor likelihood function;
- $f_{k+1|k}(\mathbf{y}|\mathbf{x})$: Markov transition density from time step k to time step $k+1$, Section 2.4.6;
- $f_{k|k}(\mathbf{x}|Z^k)$: Posterior density at time step k conditioned on observation-stream Z^k, (2.84);
- $f_{k+1|k}(\mathbf{x}|Z^k)$: Time-extrapolated posterior density at time step $k+1$, (2.84);
- $f_0(\mathbf{x}) = f_{0|0}(\mathbf{x})$: Prior distribution (posterior at time step $k = 0$);
- $u(\mathbf{x})$: Uniform distribution on target states;
- $\hat{\mathbf{x}}(\mathbf{z}_1, ..., \mathbf{z}_m)$: State estimator based on measurements $\mathbf{z}_1, ..., \mathbf{z}_m$, Section 2.4.8;
- $\hat{\mathbf{x}}^{\text{MAP}}_{k|k}$: Maximum a posteriori (MAP) state estimator, Section 2.4.8;
- $\hat{\mathbf{x}}^{\text{EAP}}_{k|k}$: Expected a posteriori (MAP) state estimator (posterior expectation), Section 2.4.8;
- $K(f;g)$: Kullback-Leibler discrimination (cross-entropy) of two probability density functions f, g, Section 2.4.9.

A.6 RANDOM SETS

This section summarizes basic notations of random set theory.

- $\emptyset^q$: Random subset such that $\emptyset^q = \emptyset$ with probability $1-q$ and $\emptyset^q =$ entire (measurement or state) space with probability q, (12.47);
- A: Uniformly distributed random number on unit interval $[0,1]$, Section 4.3.2;
- $\Sigma_A(f) = \{u|\ A \leq f(u)\}$: Synchronous random-set model of fuzzy membership function $f(u)$ on universe U, Section 4.3.2;
- $\Sigma_\alpha(f) = \{u|\ \alpha(u) \leq f(u)\}$: Asynchronous random-set model of fuzzy membership function $f(u)$ on universe U, Section 4.3.5;

- $\Sigma_A(W)$: Synchronous random-set model of subset W of $U \times [0,1]$, Section 4.4.1;
- $\Sigma_\Phi(X \Rightarrow S)$: Synchronous random-set model of a rule $X \Rightarrow S$, Section 4.7.3;
- $\mu_\Sigma(a) = \Pr(a \in \Sigma)$: One point covering function (fuzzy membership function) of random set Σ, Section 4.3.2;
- $\beta_\Sigma(S) = \Pr(\Sigma \subseteq S)$: Belief-mass function of random set Σ, Section 11.3.2;
- $\beta_{\Sigma|\Xi}(S|X) = \Pr(\Sigma \subseteq S|X)$: Belief-mass function of random set Σ conditioned on $\Xi = X$.

A.7 MULTITARGET CALCULUS

This section summarizes the basic notation of multitarget calculus.

- $\int f(Y)\delta Y$: Set integral of a multiobject density function $f(Y)$, Section 11.3.3;
- $f(n)$: Cardinality distribution of a multiobject density $f(X)$, (11.113);
- $\frac{\delta\beta}{\delta\mathbf{y}}(S)$: Set derivative of a function $\beta(S)$ of a (nonfinite) set variable S, with respect to a vector $\mathbf{y}$, Section 11.4.2;
- $\frac{\delta\beta}{\delta Y}(S)$: Set derivative of a function $\beta(S)$ of a (nonfinite) set variable S, with respect to a finite set Y of vectors, Section 11.4.2;
- $\frac{\delta\beta}{\delta\emptyset}(S) = \beta(S)$: Set derivative of a function $\beta(S)$ of a (nonfinite) set variable S, with respect to empty set $\emptyset$, Section 11.4.2;
- $\frac{\delta F}{\delta\mathbf{y}}[h]$: Functional derivative of a functional $F[h]$ with respect to a vector $\mathbf{y}$, Section 11.4.1;
- $\frac{\delta F}{\delta Y}[h]$: Functional derivative of a functional $F[h]$ with respect to a finite set Y of vectors, Section 11.4.1;
- $\frac{\delta F}{\delta\emptyset}[h] = F[h]$: Set derivative of a function $F[h]$ with respect to empty set $\emptyset$, Section 11.4.1.

A.8 FINITE-SET STATISTICS

This section summarizes the basic notations of FISST.

- $\mathfrak{Z}_0$: Underlying measurement space, Section 12.2.3;
- $\mathfrak{X}_0$: Underlying state space, Section 12.2.1;
- $\mathfrak{Z}$: Space of all finite subsets of $\mathfrak{Z}_0$, Section 12.2.4;
- $\mathfrak{X}$: Space of all finite subsets of $\mathfrak{X}_0$, Section 12.2.1;
- $f(Y) = f(\{\mathbf{y}_1, ..., \mathbf{y}_n\}) = n!\ f(\mathbf{y}_1, ..., \mathbf{y}_n)$: Three different notations for a function f of a finite-set variable $Y = \{\mathbf{y}_1, ..., \mathbf{y}_n\}$;
- $X = \{\mathbf{x}_1, ..., \mathbf{x}_n\}$: Multitarget state (finite set of ordinary state vectors);
- Θ: Random subset of multisensor measurement space (models an "ambiguous" measurement);
- $\overset{s}{\Theta}$: Generalized measurement collected by source with identifying-tag s;
- $Z = \{\mathbf{z}_1, ..., \mathbf{z}_m, \Theta_1, ..., \Theta_m\}$: multitarget-multisensor observation set (finite set of precise and ambiguous observations);
- $\overset{s}{Z}$: Multitarget-multisensor observation set originating with sensor with identifying-tag s;
- $\overset{s}{\Sigma}$: Random multitarget-multisensor observation set originating with sensor with identifying-tag s;
- Σ: Random set of multisensor-multitarget observations;
- $\Xi_{k|k}$: Random set of multitarget state vectors at time step k;
- Ξ: Random set of multitarget state vectors at time step $k = 0$;
- $Z^{(k)} : Z_1, ..., Z_m$: Time sequence of measurement sets, (14.4);
- $Z(X) = T(X) \cup C(X)$: General multisource/target measurement model with target observations $T(X)$ and clutter observations $C(X)$, Section 12.3;
- $X_{k+1} = \Gamma_{k+1}(X) \cup B$: General multitarget motion model, Section 13.2;
- $f_{k+1}(Z|X)$: General multisource-multitarget likelihood function, (14.50);

- $\overset{s}{f}(\overset{s}{Z}|X)$: General multisource-multitarget likelihood function for sensor with identifying-tag s, Section 12.9;
- $f(\overset{1}{Z},...,\overset{s}{Z}|X)$: Joint multisource-multitarget likelihood function, Section 12.9;
- $f_{k+1|k}(Y|X)$: General multitarget Markov transition density, (14.14);
- $f_{k|k}(X|Z^{(k)})$: Multitarget posterior distribution at time step k conditioned on observation set stream $Z^{(k)}$, (14.4);
- $f_{k+1|k}(X|Z^{(k)})$: Time-extrapolated multitarget posterior distribution at time step $k+1$, (14.4);
- $f_0(X) = f_{0|0}(X)$: Prior multitarget distribution (multitarget posterior distribution at time step $k=0$), (14.4);
- $u_{\mathring{n},D}(X)$: Multitarget uniform distribution on multitarget states, (11.126);
- $\hat{X}^{\text{JoM}}$: Joint multitarget estimator, Section 14.5.3;
- $\hat{X}^{\text{MaM}}$: Marginal multitarget (MaM) estimator, Section 14.5.2;
- $K(f;g)$: Multitarget Kullback-Leibler discrimination of two multitarget densities f, g, Section 14.6.4.

A.9 GENERALIZED MEASUREMENTS

This section summarizes the basic notations of generalized measurements and likelihoods.

- $f(\Theta|\mathbf{x})$: Generalized likelihood function for generalized measurement Θ, (3.14);
- $f(\Theta_1,...,\Theta_m|\mathbf{x})$: Joint generalized likelihood function for generalized measurements $\Theta_1...,\Theta_m$, (3.15) and (6.7);
- $\overset{s}{f}(\overset{s}{\Theta}|\mathbf{x})$: Generalized likelihood function for generalized measurement $\overset{s}{\Theta}$ (source with identifying-tag s), Section 3.6;
- $f(\overset{1}{\mathbf{z}},...,\overset{s}{\mathbf{z}},\overset{1}{\Theta},...,\overset{t}{\Theta}|\mathbf{x})$: Joint generalized likelihood function over all sources, Section 3.6.

Appendix B

Dirac Delta Functions

In this appendix, I tabulate the major properties of Dirac delta functions. Let $\delta_{\mathbf{y}_0}(\mathbf{y})$ denote a delta function concentrated at $\mathbf{y}_0 \in \mathfrak{Y}_0$. Then:

$$\delta_{\mathbf{y}_0}(\mathbf{y}) = \begin{cases} 0 & \text{if} \quad \mathbf{y} \neq \mathbf{y}_0 \\ \infty & \text{if} \quad \text{otherwise} \end{cases} . \tag{B.1}$$

The delta function is characterized by the fact that, for any (ordinary) function $h(\mathbf{y})$,

$$\int h(\mathbf{y}) \cdot \delta_{\mathbf{y}_0}(\mathbf{y}) d\mathbf{y} = h(\mathbf{y}_0). \tag{B.2}$$

Setting $h(\mathbf{y}) = 1$ identically and $h(\mathbf{y}) = \mathbf{1}_S(\mathbf{y})$ yields, respectively, the following two special cases:

$$\int_S \delta_{\mathbf{y}_0}(\mathbf{y}) d\mathbf{y} = \mathbf{1}_S(\mathbf{y}_0) \tag{B.3}$$

$$\int \delta_{\mathbf{y}_0}(\mathbf{y}) d\mathbf{y} = 1. \tag{B.4}$$

It also leads to the following multiplicative property. For any $h(\mathbf{y})$,

$$h(\mathbf{y}) \cdot \delta_{\mathbf{y}_0}(\mathbf{y}) = h(\mathbf{y}_0) \cdot \delta_{\mathbf{y}}(\mathbf{y}). \tag{B.5}$$

We also write

$$\Delta_{\mathbf{y}_0}(S) = \int_S \delta_{\mathbf{y}_0}(\mathbf{y}) d\mathbf{y} \tag{B.6}$$

and call $\Delta_{\mathbf{y}_0}(S)$ the *Dirac measure concentrated at* $\mathbf{y}_0$.

The following basic formula is a consequence of the change of variables formula for multidimensional integrals. Let $T : \mathfrak{Y}_0 \rightarrow \mathfrak{Y}_0$ be a nonsingular transformation and let $J_T(\mathbf{y})$ be its Jacobian determinant. Then

$$\delta_{\mathbf{y}_0}(T(\mathbf{y})) = \frac{1}{J_T(\mathbf{y}_0)} \cdot \delta_{T^{-1}(\mathbf{y}_0)}(\mathbf{y}). \tag{B.7}$$

In particular, let $T(\mathbf{y}) = H\mathbf{y} + \mathbf{y}_0$ be an affine transformation, where H is a nonsingular matrix. Then

$$\delta_{\mathbf{y}}(H\mathbf{w} + \mathbf{y}_0) = \delta_{\mathbf{y}-\mathbf{y}_0}(H\mathbf{w}) = \frac{1}{\det H} \cdot \delta_{H^{-1}(\mathbf{y}-\mathbf{y}_0)}(\mathbf{w}). \tag{B.8}$$

Appendix C

Gradient Derivatives

Let $\tau(\mathbf{x})$ be a vector-valued transformation of $\mathbf{x}$. The *gradient derivative* (also known as Frechét derivative) of τ in the direction of $\mathbf{w}$ is

$$\frac{\partial \tau}{\partial \mathbf{w}}(\mathbf{x}) = \lim_{\varepsilon \to 0} \frac{\tau(\mathbf{x} + \varepsilon \mathbf{w}) - \tau(\mathbf{x})}{\varepsilon} \tag{C.1}$$

where, in addition, for each fixed $\mathbf{x}$ the transformation

$$\mathbf{w} \mapsto \frac{\partial \tau}{\partial \mathbf{w}}(\mathbf{x}) \tag{C.2}$$

must be linear and continuous. Thus

$$\frac{\partial \tau}{\partial \mathbf{w}}(\mathbf{x}) = L_{\tau,\mathbf{x}} \mathbf{w} \tag{C.3}$$

for some matrix $L_{\tau,\mathbf{x}}$. The gradient derivative obeys the usual rules of elementary calculus: sum rule, product rule, and so on.

Iterated gradient derivatives are defined in the usual way:

$$\frac{\partial^{i+1} \tau}{\partial \mathbf{w}_{i+1} \cdots \partial \mathbf{w}_1} = \frac{\partial}{\partial \mathbf{w}_{i+1}} \frac{\partial^i \tau}{\partial \mathbf{w}_i \cdots \partial \mathbf{w}_1} \tag{C.4}$$

$$\frac{\partial^i \tau}{\partial \mathbf{w}^i} = \frac{\partial^i \tau}{\partial \mathbf{w} \cdots \partial \mathbf{w}}. \tag{C.5}$$

C.1 RELATIONSHIP WITH PARTIAL DERIVATIVES

Given an orthonormal basis $\hat{\mathbf{e}}_1, ..., \hat{\mathbf{e}}_N$ and $\mathbf{w} = w_1\hat{\mathbf{e}}_1 + ... + w_N\hat{\mathbf{e}}_N$,

$$\frac{\partial \tau}{\partial \mathbf{w}}(\mathbf{x}) = w_1 \frac{\partial \tau}{\partial x_1}(\mathbf{x}) + ... + w_N \frac{\partial \tau}{\partial x_N}(\mathbf{x}) \tag{C.6}$$

where

$$\frac{\partial \tau}{\partial x_i} \triangleq \frac{\partial \tau}{\partial \hat{\mathbf{e}}_i} \tag{C.7}$$

are the usual partial derivatives with respect to the basis $\hat{\mathbf{e}}_1, ..., \hat{\mathbf{e}}_N$.

C.2 MULTIDIMENSIONAL TAYLOR SERIES

Transformations can be expanded in a Taylor series

$$\tau(\mathbf{x}) = \tau(\mathbf{x}_0) + \frac{\partial \tau}{\partial(\mathbf{x} - \mathbf{x}_0)}(\mathbf{x}_0) + \frac{1}{2} \cdot \frac{\partial^2 \tau}{\partial(\mathbf{x} - \mathbf{x}_0)^2}(\mathbf{x}_0) \tag{C.8}$$

$$+ ... + \frac{1}{i!} \cdot \frac{\partial^i \tau}{\partial(\mathbf{x} - \mathbf{x}_0)^i}(\mathbf{x}_0) + ... \tag{C.9}$$

where

$$\frac{\partial^i \tau}{\partial \mathbf{x}^i}(\mathbf{x}) \triangleq \frac{\partial^i \tau}{\underbrace{\partial \mathbf{x} \cdots \partial \mathbf{x}}_{i \text{ times}}}(\mathbf{x}). \tag{C.10}$$

C.3 MULTIDIMENSIONAL EXTREMA

Suppose that $\tau(\mathbf{x})$ is real-valued. Then

$$\frac{\partial \tau}{\partial \mathbf{w}}(\mathbf{x}) = \mathbf{c}_\tau^T \mathbf{w}_2 \tag{C.11}$$

for some unique vector $\mathbf{c}_\tau$ and, similarly,

$$\frac{\partial^2 \tau}{\partial \mathbf{w}_1 \partial \mathbf{w}_2}(\mathbf{x}) = \mathbf{w}_1^T C_{\tau,\mathbf{x}} \mathbf{w}_2 \tag{C.12}$$

for some unique symmetric matrix $C_{\tau,\mathbf{x}}$.

If $\mathbf{x}_0$ is a local maximum or local minimum of τ then

$$\frac{\partial \tau}{\partial \mathbf{x}}(\mathbf{x}_0) = 0 \tag{C.13}$$

for all $\mathbf{x}$. For $\mathbf{x}$ very near $\mathbf{x}_0$ the quadratic term in (C.8) dominates:

$$\tau(\mathbf{x}) \cong \tau(\mathbf{x}_0) + \frac{1}{2}\frac{\partial^2 \tau}{\partial(\mathbf{x}-\mathbf{x}_0)^2}(\mathbf{x}_0). \tag{C.14}$$

Thus $\mathbf{x}_0$ locally minimizes $\tau(\mathbf{z})$ if and only if $\tau(\mathbf{x}) > \tau(\mathbf{x}_0)$ for all $\mathbf{x}$ very near $\mathbf{x}_0$ and thus if and only if

$$\frac{\partial^2 \tau}{\partial \mathbf{x}^2}(\mathbf{x}_0) > 0 \tag{C.15}$$

for all small $\mathbf{x}$ (and thus for all $\mathbf{x}$). Similarly, $\mathbf{x}_0$ locally maximizes $\tau(\mathbf{x})$ if and only if $\tau(\mathbf{x}) < \tau(\mathbf{x}_0)$ for all $\mathbf{x}$ very near $\mathbf{x}_0$ and thus if and only if

$$\frac{\partial^2 \tau}{\partial \mathbf{x}^2}(\mathbf{x}_0) < 0 \tag{C.16}$$

for all $\mathbf{x}$. Consequently, the matrix $C_{\tau,\mathbf{x}_0}$ is positive-definite if and only if $\mathbf{x}_0$ is a unique local maximum of τ.

Appendix D

Fundamental Gaussian Identity

Let P be an $n \times n$ positive-definite matrix and R an $m \times m$ positive-definite matrix with $m \leq n$. Let H be an $m \times n$ matrix. Then I prove the following identity:

$$N_R(\mathbf{r} - H\mathbf{x}) \cdot N_P(\mathbf{x} - \mathbf{p}) = N_C(\mathbf{r} - H\mathbf{p}) \cdot N_E(\mathbf{x} - \mathbf{e}) \tag{D.1}$$

for all $\mathbf{r}$, $\mathbf{x}$, and $\mathbf{p}$, where $\mathbf{e}$ and E are defined by

$$E^{-1} \triangleq P^{-1} + H^T R^{-1} H \tag{D.2}$$

$$E^{-1}\mathbf{e} \triangleq P^{-1}\mathbf{p} + H^T R^{-1}\mathbf{r} \tag{D.3}$$

$$C = R + HPH^T. \tag{D.4}$$

Restated, we have

$$\frac{1}{\sqrt{\det 2\pi R \cdot \det 2\pi P}} \tag{D.5}$$

$$\cdot \exp\left(-\frac{1}{2}(\mathbf{r} - H\mathbf{x})^T R^{-1}(\mathbf{r} - H\mathbf{x}) + \frac{1}{2}(\mathbf{x} - \mathbf{p})^T P^{-1}(\mathbf{x} - \mathbf{p})\right) \tag{D.6}$$

$$= \frac{1}{\sqrt{\det 2\pi(R + HPH^T) \cdot \det 2\pi E}} \tag{D.7}$$

$$\cdot \exp\left(-\frac{1}{2}(\mathbf{r} - H\mathbf{p})^T C^{-1}(\mathbf{r} - H\mathbf{p}) + \frac{1}{2}(\mathbf{x} - \mathbf{e})^T E^{-1}(\mathbf{x} - \mathbf{e})\right) \tag{D.8}$$

and so it is enough to show that

$$
\begin{aligned}
& (\mathbf{r}-H\mathbf{x})^T R^{-1}(\mathbf{r}-H\mathbf{x})+(\mathbf{x}-\mathbf{p})^T P^{-1}(\mathbf{x}-\mathbf{p}) && \text{(D.9)}\\
= \; & (\mathbf{r}-H\mathbf{p})^T(R+HPH^T)^{-1}(\mathbf{r}-H\mathbf{p}) && \text{(D.10)}\\
& +(\mathbf{x}-\mathbf{e})^T E^{-1}(\mathbf{x}-\mathbf{e}) && \text{(D.11)}
\end{aligned}
$$

and that

$$\det R \cdot \det P = \det(R+HPH^T)\cdot \det(P^{-1}+H^T R^{-1} H)^{-1} \tag{D.12}$$

or, equivalently,

$$\det(I_n + PH^T R^{-1} H) = \det(I_m + HPH^T R^{-1}) \tag{D.13}$$

where I_n is the $n \times n$ identity matrix.

Equation (D.9) is proved using matrix completion of the square:

$$
\begin{aligned}
& (\mathbf{r}-H\mathbf{x})^T R^{-1}(\mathbf{r}-H\mathbf{x})+(\mathbf{x}-\mathbf{p})^T P^{-1}(\mathbf{x}-\mathbf{p}) && \text{(D.14)}\\
= \; & \mathbf{x}^T H^T R^{-1} H\mathbf{x}+\mathbf{x}^T P^{-1}\mathbf{x} && \text{(D.15)}\\
& -2\mathbf{r}^T R^{-1} H\mathbf{x}-2\mathbf{p}^T P^{-1}\mathbf{x} && \text{(D.16)}\\
& +\mathbf{r}^T R^{-1}\mathbf{r}+\mathbf{p}^T P^{-1}\mathbf{p} && \text{(D.17)}\\
= \; & \mathbf{x}^T E^{-1}\mathbf{x}-2\mathbf{x}^T\left(H^T R^{-1}\mathbf{r}+P^{-1}\mathbf{p}\right) && \text{(D.18)}\\
& +\mathbf{r}^T R^{-1}\mathbf{r}+\mathbf{p}^T P^{-1}\mathbf{p} && \text{(D.19)}\\
= \; & \mathbf{x}^T E^{-1}\mathbf{x}-2\mathbf{x}^T E^{-1}\mathbf{e}+\mathbf{r}^T R^{-1}\mathbf{r}+\mathbf{p}^T P^{-1}\mathbf{p} && \text{(D.20)}\\
= \; & (\mathbf{x}-\mathbf{e})^T E^{-1}(\mathbf{x}-\mathbf{e})+\mathbf{p}^T P^{-1}\mathbf{p}+\mathbf{r}^T R^{-1}\mathbf{r}-\mathbf{e}^T E^{-1}\mathbf{e}. && \text{(D.21)}
\end{aligned}
$$

We are done if we show that

$$\mathbf{p}^T P^{-1}\mathbf{p}+\mathbf{r}^T R^{-1}\mathbf{r}-\mathbf{e}^T E^{-1}\mathbf{e} = (\mathbf{r}-H\mathbf{p})^T(R+HPH^T)^{-1}(\mathbf{r}-H\mathbf{p}). \tag{D.22}$$

To show this let

$$g(\mathbf{p},\mathbf{r}) \triangleq \mathbf{p}^T P^{-1}\mathbf{p}+\mathbf{r}^T R^{-1}\mathbf{r}-\mathbf{e}^T E^{-1}\mathbf{e} \tag{D.23}$$

First, note that

$$\begin{aligned}
g(\mathbf{p}, H\mathbf{p}) &= \mathbf{p}^T P^{-1}\mathbf{p} + \mathbf{r}^T R^{-1}\mathbf{r} - \mathbf{e}^T E^{-1}\mathbf{e} && \text{(D.24)}\\
& \mathbf{p}^T P^{-1}\mathbf{p} + \mathbf{p}^T H^T R^{-1} H\mathbf{p} && \text{(D.25)}\\
& -\mathbf{p}^T (P^{-1} + H^T R^{-1} H)^T E (P^{-1} + H^T R^{-1} H)\mathbf{p} && \text{(D.26)}\\
& \mathbf{p}^T P^{-1}\mathbf{p} + \mathbf{p}^T H^T R^{-1} H\mathbf{p} && \text{(D.27)}\\
& -\mathbf{p}^T (P^{-1} + H^T R^{-1} H)\mathbf{p} && \text{(D.28)}\\
&= 0. && \text{(D.29)}
\end{aligned}$$

This tells us that $g(\mathbf{p}, \mathbf{r})$ must have the form

$$g(\mathbf{p}, \mathbf{r}) = (\mathbf{r} - H\mathbf{p})^T C^{-1} (\mathbf{r} - H\mathbf{p}) + (\mathbf{r} - H\mathbf{p})^T \mathbf{c} \qquad \text{(D.30)}$$

for some matrix C and some vector $\mathbf{c}$.

Second, take the gradient derivatives with respect to $\mathbf{r}_1$

$$\frac{\partial g}{\partial \mathbf{r}_1}(\mathbf{p}, \mathbf{r}) = 2\mathbf{r}_1^T R^{-1}\mathbf{r} + 2\mathbf{r}_1^T R^{-1} H E (P^{-1}\mathbf{p} + H^T R^{-1}\mathbf{r}) \qquad \text{(D.31)}$$

and set them all to zero. This results in

$$\begin{aligned}
\mathbf{r} &= \left(R + HPH^T\right) R^{-1} H E P^{-1}\mathbf{p} && \text{(D.32)}\\
&= HP\left(P^{-1} + H^T R^{-1} H\right) E P^{-1}\mathbf{p} && \text{(D.33)}\\
&= HPP^{-1}\mathbf{p} = H\mathbf{p}. && \text{(D.34)}
\end{aligned}$$

However, since also

$$\frac{\partial g}{\partial \mathbf{r}_1}(\mathbf{p}, \mathbf{r}) = \mathbf{r}_1^T C^{-1} (\mathbf{r} - H\mathbf{p}) + \mathbf{r}_1^T \mathbf{c} \qquad \text{(D.35)}$$

for all $\mathbf{r}_1$ and since $\frac{\partial g}{\partial \mathbf{r}_1}(\mathbf{p}, H\mathbf{p}) = 0$ we must have $\mathbf{c} = \mathbf{0}$.

Finally, take the second gradient derivatives with respect to $\mathbf{r}$:

$$\begin{aligned}
\frac{\partial^2 g}{\partial \mathbf{r}_2 \partial \mathbf{r}_1}(\mathbf{p}, \mathbf{r}) &= 2\mathbf{r}_1^T R^{-1}\mathbf{r}_2 + 2\mathbf{r}_1^T R^{-1} H E H^T R^{-1}\mathbf{r}_2 && \text{(D.36)}\\
&= 2\mathbf{r}_1^T \left(R^{-1} + R^{-1} H E H^T R^{-1}\right)\mathbf{r}_2 && \text{(D.37)}\\
&= 2\mathbf{r}_1^T \left(P + H^T R H\right)^{-1} \mathbf{r}_2 && \text{(D.38)}
\end{aligned}$$

where we have used the familiar matrix identity [61]

$$R^{-1} + R^{-1}H\left(P^{-1} + H^T R^{-1} H\right)^{-1} H^T R^{-1} = \left(P + H^T R H\right)^{-1}. \quad \text{(D.39)}$$

Since also

$$\frac{\partial^2 g}{\partial \mathbf{r}_2 \partial \mathbf{r}_1}(\mathbf{p}, \mathbf{r}) = 2\mathbf{r}_1^T C^{-1} \mathbf{r}_2 \quad \text{(D.40)}$$

for all $\mathbf{r}_1, \mathbf{r}_2$ it follows that

$$C^{-1} = P + H^T R H. \quad \text{(D.41)}$$

Turning now to (D.13), let $G = PH^T R^{-1}$ be an $n \times m$ matrix. We will be finished if we show that

$$\det(I_n + GH) = \det(I_m + HG). \quad \text{(D.42)}$$

Define the $n \times n$ matrices $\tilde{G}$ and $\tilde{H}$ by

$$\tilde{H} = \begin{pmatrix} H & 0_{n \times m} \end{pmatrix}, \quad \tilde{G} = \begin{pmatrix} G \\ 0_{m \times n} \end{pmatrix}. \quad \text{(D.43)}$$

Then for any real number λ,

$$-\lambda I_n + \tilde{H}\tilde{G} = \begin{pmatrix} -\lambda I_m & 0_{(n-m) \times m} \\ 0_{m \times (n-m)} & -\lambda I_{n-m} \end{pmatrix} \quad \text{(D.44)}$$

$$+ \begin{pmatrix} H & 0_{n \times m} \end{pmatrix} \begin{pmatrix} G \\ 0_{m \times n} \end{pmatrix} \quad \text{(D.45)}$$

$$= \begin{pmatrix} -\lambda I_m + HG & 0 \\ 0 & -\lambda I_{n-m} \end{pmatrix} \quad \text{(D.46)}$$

and, likewise,

$$-\lambda I_n + \tilde{G}\tilde{H} = -\lambda I_n + \begin{pmatrix} G \\ 0_{m \times n} \end{pmatrix} \begin{pmatrix} H & 0_{n \times m} \end{pmatrix} \quad \text{(D.47)}$$

$$= -\lambda I_n + GH. \quad \text{(D.48)}$$

If

$$\chi_M(\lambda) = \det(-\lambda I_n + M) \quad \text{(D.49)}$$

denotes the characteristic polynomial of an $n \times n$ matrix M, it follows that

$$\chi_{\tilde{G}\tilde{H}}(\lambda) = \chi_{GH}(\lambda), \qquad \chi_{\tilde{H}\tilde{G}}(\lambda) = (-\lambda)^{n-m} \cdot \chi_{HG}(\lambda). \tag{D.50}$$

However, for any $n \times n$ matrices $\tilde{G}$ and $\tilde{H}$ the characteristic polynomials of $\tilde{G}\tilde{H}$ and $\tilde{H}\tilde{G}$ are the same [79, p. 141]:

$$\chi_{\tilde{G}\tilde{H}}(\lambda) = \chi_{\tilde{H}\tilde{G}}(\lambda) \tag{D.51}$$

from which it follows that

$$\chi_{GH}(\lambda) = (-\lambda)^{n-m} \cdot \chi_{HG}(\lambda). \tag{D.52}$$

Setting $\lambda = -1$ we get, as claimed,

$$\det(I_n + GH) = \det(I_m + HG). \tag{D.53}$$

Appendix E

Finite Point Processes

In this appendix, I briefly summarize some basic concepts of point process theory [36, 37, 214, 104, 220] and describe how they are related to FISST. For a more detailed discussion see [137]. Specifically, I show that any general point process can be represented as a certain type of random finite set. I argue that, consequently and from the point of view of multitarget tracking theory, general point processes are:

- Unnecessary;
- Result in notational and mathematical obfuscation.

E.1 MATHEMATICAL REPRESENTATIONS OF MULTIPLICITY

Point process theory is the mathematical theory of stochastic multiobject systems. Intuitively speaking, a point process is a *random finite multiset*. A finite multiset (sometimes also called a "bag") is a finite unordered list

$$L = \overbrace{\mathbf{y}_1, \ldots, \mathbf{y}_1}^{\nu_1}, \ldots, \overbrace{\mathbf{y}_n, \ldots, \mathbf{y}_n}^{\nu_n} \tag{E.1}$$

of distinct elements $\mathbf{y}_1, ..., \mathbf{y}_n$ of some space $\mathfrak{Y}$ in which repetition of elements is allowed: there are ν_1 copies of $\mathbf{y}_1$, ν_2 copies of $\mathbf{y}_2$, and so on.

Multisets can be thought of as generalizations of fuzzy sets in which elements have multiple rather than partial membership [3]. That is, they can be represented as membership functions $\mu(\mathbf{y})$ on $\mathfrak{Y}$ whose values are nonnegative integers:

$$\mu(\mathbf{y}) = \begin{cases} \nu_i & \text{if} \quad \mathbf{y} = \mathbf{y}_i \\ 0 & \text{if} \quad \text{otherwise} \end{cases}. \tag{E.2}$$

Intersection and union of multisets are defined in the same manner as Zadeh conjunction and disjunction:

$$(\mu \wedge \mu')(\mathbf{y}) \triangleq \min\{\mu(\mathbf{y}), \mu'(\mathbf{y})\} \tag{E.3}$$

$$(\mu \vee \mu')(\mathbf{y}) \triangleq \max\{\mu(\mathbf{y}), \mu'(\mathbf{y})\} \tag{E.4}$$

respectively [3]. Complementation of multisets cannot be meaningfully defined at all.

The multiset is not the only possible way of mathematically representing the concept of multiplicity of elements. Other equivalent representations are as follows (where in each case the $\mathbf{y}_1, ..., \mathbf{y}_n$ are distinct):

1. Dirac sum:
$$\delta(\mathbf{y}) = \nu_1 \delta_{\mathbf{y}_1}(\mathbf{y}) + ... + \nu_n \delta_{\mathbf{y}_n}(\mathbf{y}). \tag{E.5}$$
2. Counting measure:
$$\Delta(T) = \int_T \delta(\mathbf{y}) d\mathbf{y} = \nu_1 \Delta_{\mathbf{y}_1}(T) + ... + \nu_n \Delta_{\mathbf{y}_n}(T). \tag{E.6}$$
3. Simple counting measure on pairs:
$$\Delta(S) = \Delta_{(\nu_1, \mathbf{y}_1)}(S) + ... + \Delta_{(\nu_n, \mathbf{y}_n)}(S). \tag{E.7}$$
4. Simple Dirac sum on pairs:
$$\delta(\nu, \mathbf{y}) = \delta_{(\nu_1, \mathbf{y}_1)}(\nu, \mathbf{y}) + ... + \delta_{(\nu_n, \mathbf{y}_n)}(\nu, \mathbf{y}). \tag{E.8}$$
5. Finite set of pairs:
$$\{(\nu_1, \mathbf{y}_1), ..., (\nu_n, \mathbf{y}_n)\}. \tag{E.9}$$

Here $\Delta_{\mathbf{y}}(S)$ denotes the Dirac measure:

$$\Delta_{\mathbf{y}}(S) = \begin{cases} 1 & \text{if} \quad \mathbf{y} \in S \\ 0 & \text{if} \quad \mathbf{y} \notin S \end{cases}. \tag{E.10}$$

If $\nu_i = 1$ for all $i = 1, ..., n$ then (E.5)-(E.9) are all equivalent representations of a finite subset of $\mathfrak{Y}$.

E.2 RANDOM POINT PROCESSES

If any of the representations in (E.5)-(E.9) are randomized, the result is a *random point process*. In theoretical statistics, the random counting measure formulation of (E.6) is the norm. It is this formulation that I refer to as "general point processes."

The family of *Janossy measures* $J_{\delta,n}(T_1, ..., T_n)$ of a point process δ has the properties [36, p. 122],

$$J_{\delta,n}(T_{\sigma 1}, ..., T_{\sigma n}) = J_{\delta,n}(T_{\sigma 1}, ..., T_{\sigma n}) \quad \text{(E.11)}$$

$$\sum_{n=0}^{\infty} J_{\delta,n}(\mathfrak{Y}, ..., \mathfrak{Y}) = 1 \quad \text{(E.12)}$$

for all permutations σ on the numbers $1, ..., n,$ for every $n \geq 2$. The family of *Janossy densities* $J_{\delta,n}(\mathbf{y}_1, ..., \mathbf{y}_n)$ of δ are the density functions of the Janossy measures, provided that they exist:

$$J_{\delta,n}(T_1, ..., T_n) = \int_{T_1 \times ... \times T_n} j_{\delta,n}(\mathbf{y}_1, ..., \mathbf{y}_n) d\mathbf{y}_1 \cdots d\mathbf{y}_n. \quad \text{(E.13)}$$

A point process is *simple* if and only if its Janossy densities

- Exist (i.e., are finite-valued functions);
- Vanish on the diagonals: $j_{\delta,n}(\mathbf{y}_1, ..., \mathbf{y}_n) = 0$ whenever $\mathbf{y}_i = \mathbf{y}_j$ with $i \neq j$ [36, p. 134, Prop. 5.4.IV].

If δ is not simple then $j_{\delta,n}(\mathbf{y}_1, ..., \mathbf{y}_n)$ *is not a finite-valued function.* Rather, it is some complicated combinatorial sum of Dirac-type densities, as Example 94 shows.

The reason for this is that a diagonal constitutes a set of measure zero. Consequently, if the Janossy densities are conventional finite-valued functions, then the diagonal values of $j_{\delta,n}(\mathbf{y}_1, ..., \mathbf{y}_n)$ have no effect on the integral in (E.13). The diagonal values of $j_{\delta,n}(\mathbf{y}_1, ..., \mathbf{y}_n)$ can affect the value of this integral only if they are infinite—that is, of Dirac type.

Example 94 *Let* $\mathbf{Y}$ *be a random variable on* $\mathfrak{Y}$. *Define the point process* $\delta_{\mathbf{Y}}$ *by* $\delta_{\mathbf{Y}}(\mathbf{y}) = 2 \cdot \delta_{\mathbf{Y}}$. *That is, it is equivalent to the random finite multiset* $\mathbf{Y}, \mathbf{Y}$ *or, equivalently, the random pair* $(2, \mathbf{Y})$. *The Janossy densities of* $\delta_{\mathbf{Y}}$ *must vanish for* $n \neq 2$. *Likewise,* $j_{\delta,2}(\mathbf{y}_1, \mathbf{y}_2) = 0$ *if* $\mathbf{y}_1 \neq \mathbf{y}_2$ *since the only instantiations of* $\delta_{\mathbf{Y}}$ *have multiplicity* 2. *Thus* $j_{\delta,2}(\mathbf{y}, \mathbf{y})$ *must be infinite-valued for any* $\mathbf{y}$ *if*

$\int_{T_1 \times T_2} j_{\delta,2}(\mathbf{y}_1, \mathbf{y}_2) d\mathbf{y}_1 d\mathbf{y}_2$ *is to be nonzero. Setting* $\mathbf{y}_0 = \mathbf{0}$ *in (11.59) we see that*

$$j_{\delta,2}(\mathbf{y}_1, \mathbf{y}_2) = \delta_{\mathbf{y}_1}(\mathbf{y}_2) \cdot f_{\mathbf{Y}}(\mathbf{y}_1) + \delta_{\mathbf{y}_2}(\mathbf{y}_1) \cdot f_{\mathbf{Y}}(\mathbf{y}_2) \qquad \text{(E.14)}$$

$$= 2 \cdot \delta_{\mathbf{y}_1}(\mathbf{y}_2) \cdot f_{\mathbf{Y}}(\mathbf{y}_1). \qquad \text{(E.15)}$$

To see this, note that $j_{\delta,2}(\mathbf{y}_1, \mathbf{y}_2)$ *is zero outside of the diagonal events* $(\mathbf{y}, \mathbf{y})$*; is infinite-valued on the diagonal events; and* $\int j_{\delta,2}(\mathbf{y}_1, \mathbf{y}) d\mathbf{y}_1 = 2 f_{\mathbf{Y}}(\mathbf{y})$.

Equation (E.9) shows that multiplicity of elements can be formulated in terms of ordinary set theory—and thus that:

- *Any point process can be represented as a certain type of random finite set.*[1]

In particular, let Ψ be a finite random subset of $\mathfrak{Y}$. Then

$$N_\Psi(Y) \triangleq |Y \cap \Psi|, \qquad \delta_\Psi(\mathbf{y}) \triangleq \sum_{\mathbf{w} \in \Psi} \delta_{\mathbf{w}}(\mathbf{y}) \qquad \text{(E.16)}$$

are equivalent mathematical representations of Ψ.

E.3 POINT PROCESSES VERSUS RANDOM FINITE SETS

From a practical point of view, (E.5)-(E.8) are all *essentially just changes of notation* of (E.9) that do the following:

- Increase notational obfuscation and theoretical complexity;
- Add no new substance;
- Lose the simple, intuitive tools of ordinary set theory.

For example, represent multitarget states as Dirac sums as in (E.5). Then a multitarget probability distribution must be a functional $F[\delta]$. That is, it is a scalar-valued function whose argument δ is an ordinary function. If $F[\delta]$ is a probability distribution then we must define a functional integral $\int F[\delta]\mu(d\delta)$ with respect to some measure μ defined on the space of functions $\delta(\mathbf{y})$.

1 More rigorously speaking, (E.8) and (E.9) show that any point process can be represented as a *marked simple point process* in which the baseline or mother process Ψ is a finite random set and the marks on the $\mathbf{y}$'s in Ψ are positive integers [36, pp. 205; 207, Exercise 7.1.6], [104, pp. 5, 6, 22]. For details about marked point processes, see [36, pp. 204-206].

The theory of functional integrals is, in general, a very difficult subject. However, in our particular case functions have the form

$$\delta = \delta_{\mathbf{y}_1} + \ldots + \delta_{\mathbf{y}_n} \tag{E.17}$$

and probability distributions the form

$$F[\delta] = F[\delta_{\mathbf{y}_1} + \ldots + \delta_{\mathbf{y}_n}]. \tag{E.18}$$

So, functional integrals will also take a simpler form. Since δ is a density and since the differential probability mass $F[\delta]d\delta$ must be unitless, $F[\delta]$ must have the same units of measurement as $\mathbf{y}$ since $d\delta$ must have the same units as δ and the units of δ are the inverse of those of $\mathbf{y}$. Given a choice of units of measurement, let ι denote unity in this system of units. Then if the functional integral is to produce intuitively desired results we must have

$$F[\delta_{\mathbf{y}_1} + \ldots + \delta_{\mathbf{y}_n}] = \iota^{n+1} \cdot n! \cdot f(\mathbf{y}_1, \ldots, \mathbf{y}_n) \tag{E.19}$$

and so

$$\int F[\delta]d\delta = \sum_{n=0}^{\infty} \frac{1}{\iota^{n+1} \cdot n!} \int F[\delta_{\mathbf{y}_1} + \ldots + \delta_{\mathbf{y}_n}]d\mathbf{y}_1 \cdots d\mathbf{y}_n \tag{E.20}$$

where $f(\mathbf{y}_1, \ldots, \mathbf{y}_n)$ is a conventional density function in the variables $\mathbf{y}_1, \ldots, \mathbf{y}_n$. If $\mathfrak{Y}$ is continuous then the diagonals $\mathbf{y}_i = \mathbf{y}_j$ for $i \neq j$ are zero-probability events and so we might as well assign $f(\mathbf{y}_1, \ldots, \mathbf{y}_n) = 0$ on the diagonals, in which case

$$n! \cdot f(\mathbf{y}_1, \ldots, \mathbf{y}_n) = f_{\text{FISST}}(\{\mathbf{y}_1, \ldots, \mathbf{y}_n\}) \tag{E.21}$$

and so

$$\int F[\delta]d\delta = \sum_{n=0}^{\infty} \frac{1}{n!} \int f_{\text{FISST}}(\{\mathbf{y}_1, \ldots, \mathbf{y}_n\})d\mathbf{y}_1 \cdots d\mathbf{y}_n \tag{E.22}$$

$$= \int f_{\text{FISST}}(Y)\delta Y. \tag{E.23}$$

In other words:

- $\int F[\delta]d\delta$ is just the set integral of (11.96) written in notationally and mathematically obfuscated form.

The only feature that would recommend multisets over finite sets is the fact that the former account for multiplicity of elements whereas the latter do not. One might argue, for example, that multiplicity of states occurs every time that the trajectories of two distinct targets cross simultaneously. At that instant two targets have the same state vector. However, and as I pointed out in Remark 19 in Section 12.2.1, a careful analysis reveals that this kind of multiplicity is an artifact of incomplete multitarget-state modeling. Thus we conclude the following:

- For the purposes of multitarget tracking theory, general point processes are unnecessary.

Appendix F

FISST and Probability Theory

In Section 9.3.1 and in footnote 1 of Section 12.2.1, I noted that a pure mathematician would approach the subject of this book differently. The purpose of this appendix is to summarize the relationships between FISST and the conventional mathematical approach. In particular, I will clarify the relationships between:[1]

- Belief-mass functions and multiobject probability-mass functions;
- The set integral and the measure theoretic integral;
- The set and functional derivatives and the measure theoretic Radon-Nikodým derivative.

F.1 MULTIOBJECT PROBABILITY THEORY

One of the basic concepts of FISST is that of a random finite set on some space, say single-target state space $\mathfrak{X}_0$. One cannot define a random variable of any kind without, typically, first defining a topology on the space of objects to be randomized and then defining random elements of that space in terms of the Borel-measurable subsets O.

A finite set is a special kind of closed set. In random set theory, the space $\mathfrak{X}^!$ of closed subsets of $\mathfrak{X}_0$ is called a "hyperspace" [175] and is topologized using the Mathéron "hit-or-miss" topology [160]. This topology is defined as follows. The

1 Some of the material in this section was conceived independently by Vo, Singh, and Doucet [238, pp. 1226-1229].

elementary open subsets of $\mathfrak{X}^!$ have the form

$$O_G = \{C \in \mathfrak{X}^! | \, C \cap G \neq \emptyset\} \text{ "hit"} \tag{F.1}$$
$$O^K = \{C \in \mathfrak{X}^! | \, C \cap K = \emptyset\} \text{ "miss"} \tag{F.2}$$

where K and G are arbitrary compact and open subsets of $\mathfrak{X}_0$, respectively. The open sets of $\mathfrak{X}^!$ are generated by taking all possible combinations of infinite intersections and finite unions of the elementary open subsets. Under the resulting topology, $\mathfrak{X}^!$ is compact, Hausdorff, and separable; and the null set $\emptyset$ behaves like a point at infinity.

Let (Ω, S, p_Ω) be the underlying probability space. Then a random (closed) subset of $\mathfrak{X}_0$ is a measurable map $\Gamma : \Omega \to \mathfrak{X}^!$. Its probability-mass function (probability measure) is

$$p_\Gamma(O) = \Pr(\Gamma \in O) \triangleq p_\Omega(\{\omega \in \Omega | \, \Gamma(\omega) \in O\}) \tag{F.3}$$
$$= p_\Omega(\Gamma^{-1}O). \tag{F.4}$$

A random finite set (RFS) results when the Mathéron topology is restricted to the subhyperspace $\mathfrak{X} \subseteq \mathfrak{X}^!$ of finite subsets of $\mathfrak{X}_0$. An RFS is a measurable map $\Xi : \Omega \to \mathfrak{X}$ with probability measure

$$p_\Xi(O) = \Pr(\Xi \in O) \tag{F.5}$$

defined on the measurable subsets O of $\mathfrak{X}$.

Given a measure μ on $\mathfrak{X}$, one can define a corresponding measure theoretic integral $\int f(X)\mu(dX)$ that has the property

$$\mu(O) = \int \mathbf{1}_O(X)\mu(dX) \tag{F.6}$$

where $\mathbf{1}_O(X)$ is the set indicator function for O:

$$\mathbf{1}_O(X) = \left\{ \begin{array}{lll} 1 & \text{if} & X \in O \\ 0 & \text{if} & \text{otherwise} \end{array} \right. . \tag{F.7}$$

If μ' is another measure on $\mathfrak{X}$ that is absolutely continuous with respect to μ, then the Radon-Nikodým theorem states that there is an almost everywhere unique function $f(X)$, such that

$$\mu'(O) = \int \mathbf{1}_O(X) \cdot f(X)\mu(dX). \tag{F.8}$$

In this case $f(X)$ is called the Radon-Nikodým derivative of μ' with respect to μ and is denoted as

$$f(X) = \frac{d\mu'}{d\mu}(X). \tag{F.9}$$

F.2 BELIEF-MASS FUNCTIONS VERSUS PROBABILITY MEASURES

The probability measure $p_\Xi(O)$ characterizes the statistics of Ξ, but forces us to work in $\mathfrak{X}$, the abstract topological hyperspace whose "points" are finite subsets of $\mathfrak{X}_0$. In FISST we instead use the belief-mass function

$$\beta_\Xi(S) = \Pr(\Xi \subseteq S)$$

defined on the closed subsets S of $\mathfrak{X}_0$. It is actually a special case of $p_\Xi(O)$. To see this, note that if S is closed then S^c is open and so

$$O^c_{S^c} = \{X \in \mathfrak{X} | \ X \cap S^c \neq \emptyset\}^c \tag{F.10}$$
$$= \{X \in \mathfrak{X} | \ X \cap S^c = \emptyset\} \tag{F.11}$$
$$= \{X \in \mathfrak{X} | \ X \subseteq S\} \tag{F.12}$$

is an elementary closed subset of $\mathfrak{X}$. Thus

$$p_\Xi(O^c_{S^c}) = \Pr(\Xi \subseteq S) = \beta_\Xi(S). \tag{F.13}$$

Since $\beta_\Xi(S)$ is a restriction of $p_\Xi(O)$ to a much smaller class of measurable subsets of $\mathfrak{X}$, normally it would not be possible to replace $p_\Xi(O)$ with $\beta_\Xi(S)$. However, one consequence of the *Choquet-Mathéron theorem* [160, p. 30], is that $p_\Xi(O)$ is equivalent to the "capacity" or "Choquet functional"

$$\pi_\Xi(C) = \Pr(\Xi \cap C \neq \emptyset) = \Pr(\Xi \not\subseteq C^c) \tag{F.14}$$
$$= 1 - \Pr(\Xi \subseteq C^c) = 1 - \beta_\Xi(C^c). \tag{F.15}$$

Thus $p_\Xi(O)$ is also equivalent to $\beta_\Xi(S)$ and so we can substitute the latter for the former.

If we are to work entirely in $\mathfrak{X}_0$ rather than $\mathfrak{X}$ we need substitutes not only for $p_\Xi(O)$ but also for the measure theoretic integral and the Radon-Nikodým derivative. This is the purpose of the set integral and set derivative.

F.3 SET INTEGRALS VERSUS MEASURE THEORETIC INTEGRALS

Begin by recalling the definition of a general set integral [70, pp. 141-143].[2] Let O be a Borel-measurable subset of the Mathéron topology on the space $\mathfrak{X}$ of finite subsets X of single-target state space $\mathfrak{X}_0$. Let χ_n be the transformation from vectors $(\mathbf{x}_1, ..., \mathbf{x}_n)$ of length n into finite subsets of finite subsets X of cardinality $\leq n$ defined by

$$\chi_n(\mathbf{x}_1, ..., \mathbf{x}_n) = \{\mathbf{x}_1, ..., \mathbf{x}_n\}. \tag{F.16}$$

Let $\chi_n^{-1}(\sigma_n \cap O)$ denote the subset of vectors $(\mathbf{x}_1, ..., \mathbf{x}_n)$ such that

$$\chi_n(\mathbf{x}_1, ..., \mathbf{x}_n) \in \sigma_n \cap O \tag{F.17}$$

where σ_n denotes the set of finite state sets of cardinality n. Then the set integral evaluated on O is

$$\int_O f(X)\delta X \triangleq \sum_{n=0}^{\infty} \frac{1}{n!} \int_{\chi_n^{-1}(\sigma_n \cap O)} f(\{\mathbf{x}_1, ..., \mathbf{x}_n\}) d\mathbf{x}_1 \cdots d\mathbf{x}_n. \tag{F.18}$$

Now note that the set indicator function $\mathbf{1}_O(X)$ of O has no units of measurement. If the set integral were an integral with respect to some measure μ then by (F.6)

$$\mu(O) = \int \mathbf{1}_O(X)\delta X \tag{F.19}$$

$$= \sum_{n=0}^{\infty} \frac{1}{n!} \int_{\chi_n^{-1}(\sigma_n \cap O)} \mathbf{1}_O(\{\mathbf{x}_1, ..., \mathbf{x}_n\}) d\mathbf{x}_1 \cdots d\mathbf{x}_n \tag{F.20}$$

$$= \sum_{n=0}^{\infty} \frac{1}{n!} \cdot \lambda^n(\chi_n^{-1}(\sigma_n \cap O)) \tag{F.21}$$

where λ^n denotes the extension of Lebesgue measure from $\mathfrak{X}_0$ to $\mathfrak{X}_0^n$. However, in general the infinite sum is undefined since each of its terms has different units of measurement. That is:

- *The set integral is not a measure theoretic integral.*

2 This material first appeared in [144, pp. 12, 13].

The problem is that Lebesgue measure cannot be generalized to multitarget state space without introducing an additional factor. Inspection of (F.19) reveals that the only simple generalization of Lebesgue measure from subsets S of $\mathfrak{X}_0$ to subsets O of $\mathfrak{X}$ must have the form

$$\lambda_c(O) = \int_O \frac{1}{c^{|X|}} \delta X = \sum_{i=0}^{\infty} \frac{1}{c^i \cdot i!} \cdot \lambda^n(\chi_n^{-1}(\sigma_n \cap O)) \tag{F.22}$$

where c is some positive constant with the same units of measurement as $\mathbf{x}$. The generalized Lebesgue integral

$$\int_O F(X)\lambda_c(dX) \tag{F.23}$$

$$= \sum_{n=0}^{\infty} \frac{1}{c^n \cdot n!} \int_{\chi_n^{-1}(\sigma_n \cap O)} F(\{\mathbf{x}_1, ..., \mathbf{x}_n\}) d\mathbf{x}_1 \cdots d\mathbf{x}_n \tag{F.24}$$

$$= \int_O c^{-|X|} \cdot F(X)\delta X \tag{F.25}$$

corresponding to the measure λ_c depends on the same constant c; and will be definable only if the function $F(X)$ has no units of measurement for all X.

In point process theory the common practice is to assume that $c = 1\ \iota$ where ι is the unit of measurement in $\mathfrak{X}_0$ [238, pp. 1226-1229], [65, p. 94].

F.4 SET DERIVATIVES VERSUS RADON-NIKODÝM DERIVATIVES

Similarly, let

$$p_\Xi(O) = \Pr(\Xi \in O) \tag{F.26}$$

be the probability measure of the random state set Ξ. Then its Radon-Nikodým derivative

$$F_\Xi(X) = \frac{dp_\Xi}{d\lambda_c}(X) \tag{F.27}$$

is defined by

$$p_\Xi(O) = \int_O F_\Xi(X)\lambda_c(dX). \tag{F.28}$$

Since

$$\int_O \frac{dp_\Xi}{d\lambda_c}(X)\lambda_c(dX) = p_\Xi(O) \tag{F.29}$$

$$= \int_O \frac{\delta\beta_\Xi}{\delta X}(\emptyset)\delta X \tag{F.30}$$

$$= \int_O c^{|X|} \cdot \frac{\delta\beta_\Xi}{\delta X}(\emptyset)\lambda_c(dX) \tag{F.31}$$

for all O, it follows that

$$\frac{dp_\Xi}{d\lambda_c}(X) = c^{|X|} \cdot \frac{\delta\beta_\Xi}{\delta X}(\emptyset) \tag{F.32}$$

almost everywhere. These two densities will be numerically equal if $c = 1$ ι. Thus:

- *The set derivative is not a Radon-Nikodým derivative.*

Stated differently: There are two ways of defining multitarget density functions that are almost but not quite the same:

- As Frechét functional derivatives $f_\Xi(X)$ of $G_\Xi[h]$;
- As Radon-Nikodým derivatives $F_\Xi[h]$ of $p_\Xi(O)$.

Set integrals are antiderivatives of Frechét functional derivatives, not Radon-Nikodým derivatives.

Appendix G

Mathematical Proofs

G.1 LIKELIHOODS FOR FIRST-ORDER FUZZY RULES

We are to prove (5.80). We assume that $\mathfrak{Z}_0$ is finite with M elements. From (5.80) and (4.159) we have

$$f(\xi \Rightarrow \sigma|\mathbf{x}) \triangleq \Pr(\eta(\mathbf{x}) \in \Sigma_{\Phi,A,A'}(\xi \Rightarrow \sigma)) \tag{G.1}$$
$$= \Pr(\eta(\mathbf{x}) \in \Sigma_A(\sigma) \cap \Sigma_{A'}(\xi)) + \Pr(\eta(\mathbf{x}) \in \Sigma_{A'}(\xi)^c \cap \Phi) \tag{G.2}$$

where the second equation results from the fact that $\Sigma_A(\sigma) \cap \Sigma_{A'}(\xi)$ and $\Sigma_{A'}(\xi)^c \cap \Phi$ are disjoint. On the one hand,

$$\begin{aligned}\Pr(\eta(\mathbf{x}) &\in \Sigma_{A'}(\xi)^c \cap \Phi) && \text{(G.3)}\\ &= \sum_{O \subseteq \mathfrak{Z}_0} \Pr(\eta(\mathbf{x}) \in O \cap \Phi) \cdot \Pr(\Sigma_{A'}(\xi)^c = O) && \text{(G.4)}\\ &= \sum_{O} \Pr(\eta(\mathbf{x}) \in O) \cdot \Pr(\eta(\mathbf{x}) \in \Phi) \cdot \Pr(\Sigma_{A'}(\xi)^c = O) && \text{(G.5)}\end{aligned}$$

and so

$$\begin{aligned}&= \Pr(\eta(\mathbf{x}) \in \Phi) \cdot \sum_{O} \Pr(\eta(\mathbf{x}) \in O) \cdot \Pr(\Sigma_{A'}(\xi)^c = O) && \text{(G.6)}\\ &= \frac{1}{2} \cdot \Pr(\eta(\mathbf{x}) \in \Sigma_{A'}(\xi)^c) && \text{(G.7)}\\ &= \frac{1}{2}(1 - \Pr(\eta(\mathbf{x}) \in \Sigma_{A'}(\xi)) = \frac{1}{2}(1 - \xi(\eta(\mathbf{x}))) = \frac{1}{2}\xi^c(\eta(\mathbf{x})) && \text{(G.8)}\end{aligned}$$

where we have assumed that Φ, A' are independent. The fact that $\Pr(\mathbf{z} \in \Phi) = 1/2$ follows from

$$\Pr(\mathbf{z} \in \Phi) = \sum_{O} \Pr(\mathbf{z} \in O) \cdot \Pr(\Phi = O) \tag{G.9}$$

$$= 2^{-M} \cdot \sum_{O \ni \mathbf{z}} 1 = 2^{-M} \cdot 2^{M-1} = 1/2 \tag{G.10}$$

since there are 2^{M-1} subsets that contain $\mathbf{z}$. On the other hand,

$$\Pr(\eta(\mathbf{x}) \in \Sigma_A(\sigma) \cap \Sigma_{A'}(\xi)) \tag{G.11}$$

$$= \Pr(A \leq \sigma(\eta(\mathbf{x})),\ A' \leq \xi(\eta(\mathbf{x}))) \tag{G.12}$$

$$= (\sigma \wedge_{A,A'} \xi)(\eta(\mathbf{x})) \tag{G.13}$$

where $a \wedge_{A,A'} a' \stackrel{\text{def.}}{=} \Pr(A \leq a,\ A' \leq a')$ for $0 \leq a, a' \leq 1$. Equation (5.80) then follows.

G.2 LIKELIHOODS FOR COMPOSITE RULES

We are to prove (5.102):

$$f(\rho \vee \rho' | \mathbf{x}) = \tau(\eta(\mathbf{x})) + \frac{1}{2}\theta^c(\eta(\mathbf{x})) \tag{G.14}$$

where

$$\tau \triangleq \sigma \,\bar{\wedge}\, \xi + \sigma' \,\bar{\wedge}\, \xi' - \sigma \,\bar{\wedge}\, \xi \,\bar{\wedge}\, \sigma' \,\bar{\wedge}\, \xi' \tag{G.15}$$

$$\theta \triangleq \sigma \,\bar{\wedge}\, \xi + \sigma' \,\bar{\wedge}\, \xi' + \xi \,\bar{\wedge}\, \xi' - \sigma \,\bar{\wedge}\, \xi \,\bar{\wedge}\, \xi' - \xi \,\bar{\wedge}\, \sigma' \,\bar{\wedge}\, \xi'. \tag{G.16}$$

By definition,

$$f(\rho \vee \rho' | \mathbf{x}) \triangleq \Pr(\eta(\mathbf{x}) \in \Sigma_{\Phi,A_1,A_1',A_2,A_s'}(\rho_1 \vee \rho_2)) \tag{G.17}$$

where the random set representation of $\rho \vee \rho'$ was defined in (4.163)

$$\Sigma_{\Phi,A_1,A_1',A_2,A_s'}(\rho_1 \vee \rho_2) \triangleq T \cup (Y^c \cap \Phi) \tag{G.18}$$

using simple substitution into the conditional event algebra disjunction formulas, (4.148)-(4.150):

$$T = (\Sigma_{A_1}(\sigma) \cap \Sigma_{A_2}(\xi)) \cup (\Sigma_{A_3}(\sigma') \cap \Sigma_{A_4}(\xi')) \tag{G.19}$$

$$Y = (\Sigma_{A_1}(\sigma) \cap \Sigma_{A_2}(\xi)) \cup (\Sigma_{A_3}(\sigma') \cap \Sigma_{A_4}(\xi')) \tag{G.20}$$

$$\cup(\Sigma_{A_2}(\xi) \cap \Sigma_{A_4}(\xi')). \tag{G.21}$$

Since T and $Y^c \cap \Phi$ are disjoint since $T \subseteq Y$ by assumption,

$$f(\rho \vee \rho'|\mathbf{x}) = \Pr(\eta(\mathbf{x}) \in T) + \Pr(\eta(\mathbf{x}) \in Y^c \cap \Phi) \tag{G.22}$$

$$= \Pr(\eta(\mathbf{x}) \in T) + \frac{1}{2}(\Pr(\eta(\mathbf{x}) \in Y)) \tag{G.23}$$

where the second equation follows from (G.9) and (G.10) in the proof in Section G.1.

On the one hand, from (4.42) we get

$$\Pr(\mathbf{z} \in T) = \Pr(\mathbf{z} \in \Sigma_{A_1}(\sigma) \cap \Sigma_{A_2}(\xi)) \tag{G.24}$$

$$+\Pr(\mathbf{z} \in \Sigma_{A_3}(\sigma') \cap \Sigma_{A_4}(\xi')) \tag{G.25}$$

$$-\Pr(\mathbf{z} \in \Sigma_{A_1}(\sigma) \cap \Sigma_{A_2}(\xi) \cap \Sigma_{A_3}(\sigma') \cap \Sigma_{A_4}(\xi')) \tag{G.26}$$

$$= (\sigma \barwedge \xi)(\mathbf{z}) + (\sigma' \barwedge \xi')(\mathbf{z}) - (\sigma \barwedge \xi \barwedge \sigma' \barwedge \xi')(\mathbf{z}) \tag{G.27}$$

which establishes (G.15).

On the other hand we get

$$\Pr(\mathbf{z} \in Y) = \Pr(\mathbf{z} \in T) + \Pr(\mathbf{z} \in \Sigma_{A_2}(\xi) \cap \Sigma_{A_4}(\xi')) \tag{G.28}$$

$$-\Pr(\mathbf{z} \in T \cap \Sigma_{A_2}(\xi) \cap \Sigma_{A_4}(\xi')) \tag{G.29}$$

$$= (\sigma \barwedge \xi)(\mathbf{z}) + (\sigma' \barwedge \xi')(\mathbf{z}) - (\sigma \barwedge \xi \barwedge \sigma' \barwedge \xi')(\mathbf{z}) \tag{G.30}$$

$$+(\xi \barwedge \xi')(\mathbf{z}) \tag{G.31}$$

$$-\Pr\left(\begin{array}{c} \mathbf{z} \in (\Sigma_{A_1}(\sigma) \cap \Sigma_{A_2}(\xi) \cap \Sigma_{A_4}(\xi')) \\ \cup(\Sigma_{A_2}(\xi) \cap \Sigma_{A_3}(\sigma') \cap \Sigma_{A_4}(\xi')) \end{array}\right) \tag{G.32}$$

and so

$$\Pr(\eta(\mathbf{x}) \in Y) = (\sigma \barwedge \xi)(\mathbf{z}) + (\sigma' \barwedge \xi')(\mathbf{z}) - (\sigma \barwedge \xi \barwedge \sigma' \barwedge \xi')(\mathbf{z}) \tag{G.33}$$

$$+(\xi \barwedge \xi')(\mathbf{z}) \tag{G.34}$$

$$-(\sigma \barwedge \xi \barwedge \xi')(\mathbf{z}) - (\xi \barwedge \sigma' \barwedge \xi')(\mathbf{z}) + (\sigma \barwedge \xi \barwedge \sigma' \barwedge \xi')(\mathbf{z}) \tag{G.35}$$

$$= (\sigma \barwedge \xi)(\mathbf{z}) + (\sigma' \barwedge \xi')(\mathbf{z}) + (\xi \barwedge \xi')(\mathbf{z}) \tag{G.36}$$

$$-(\sigma \barwedge \xi \barwedge \xi')(\mathbf{z}) - (\xi \barwedge \sigma' \barwedge \xi')(\mathbf{z}) \tag{G.37}$$

which establishes (G.16).

G.3 LIKELIHOODS FOR SECOND-ORDER FUZZY RULES

We are to prove (5.111). From (4.170) we know that

$$\Sigma_{\Phi_1,\Phi_2,A_1,A_2,A_3,A_4}((X \Rightarrow S) \Rightarrow (Y \Rightarrow T)) \tag{G.38}$$
$$= (S \cap X \cap T \cap Y) \cup (S \cap X \cap Y^c \cap \Phi_1) \tag{G.39}$$
$$\cup ((X - S) \cap \Phi_1^c) \tag{G.40}$$
$$\cup (X^c \cap T \cap Y \cap (\Phi_1 \cup \Phi_2)) \tag{G.41}$$
$$\cup (X^c \cap Y^c \cap (\Phi_1 \cup \Phi_2)) \tag{G.42}$$
$$\cup (X^c \cap (Y - T) \cap \Phi_1^c \cap \Phi_2) \tag{G.43}$$

where to simplify notation we abbreviate $X \overset{\text{abbr.}}{=} \Sigma_{A_1}(\xi)$, $S \overset{\text{abbr.}}{=} \Sigma_{A_2}(\sigma)$, $Y \overset{\text{abbr.}}{=} \Sigma_{A_3}(\theta)$, $T \overset{\text{abbr.}}{=} \Sigma_{A_4}(\tau)$. Because the six terms in the union on the right-hand side of this equation are mutually disjoint,

$$\Pr(\mathbf{z} \in \Sigma_{\Phi_1,\Phi_2,A_1,A_2,A_3,A_4}((X \Rightarrow S) \Rightarrow (Y \Rightarrow T))) \tag{G.44}$$
$$= \Pr(\mathbf{z} \in S \cap X \cap T \cap Y) + \Pr(\mathbf{z} \in S \cap X \cap Y^c \cap \Phi_1) \tag{G.45}$$
$$+ \Pr(\mathbf{z} \in X \cap S^c \cap \Phi_1^c) \tag{G.46}$$
$$+ \Pr(\mathbf{z} \in X^c \cap T \cap Y \cap (\Phi_1 \cup \Phi_2)) \tag{G.47}$$
$$+ \Pr(\mathbf{z} \in X^c \cap Y^c \cap (\Phi_1 \cup \Phi_2)) \tag{G.48}$$
$$+ \Pr(\mathbf{z} \in X^c \cap T^c \cap Y \cap \Phi_1^c \cap \Phi_2). \tag{G.49}$$

We compute each of these terms in turn, assuming that Φ_1 and Φ_2 are independent, and that Φ_1 and Φ_2 are independent of A_1, A_2, A_3, A_4:

$$\Pr(\mathbf{z} \in S \cap X \cap T \cap Y) = (\sigma \wedge \xi \wedge \tau \wedge \theta)(\mathbf{z}) \tag{G.50}$$

$$\Pr(\mathbf{z} \in S \cap X \cap Y^c \cap \Phi_1) \tag{G.51}$$

$$= \frac{1}{2}(\sigma \wedge \xi)(\mathbf{z}) - \frac{1}{2}(\sigma \wedge \xi \wedge \theta)(\mathbf{z}) \tag{G.52}$$

$$\Pr(\mathbf{z} \in S^c \cap X \cap \Phi_1^c) = \frac{1}{2}\xi(\mathbf{z}) - \frac{1}{2}(\sigma \wedge \xi)(\mathbf{z}) \tag{G.53}$$

$$\Pr(\mathbf{z} \in X^c \cap T \cap Y \cap (\Phi_1 \cup \Phi_2)) \tag{G.54}$$

$$= \frac{3}{4}(\tau \wedge \theta)(\mathbf{z}) - \frac{3}{4}(\xi \wedge \tau \wedge \theta)(\mathbf{z}) \tag{G.55}$$

$$\Pr(\mathbf{z} \in X^c \cap Y^c \cap (\Phi_1 \cup \Phi_2)) \tag{G.56}$$

$$= \frac{3}{4} - \frac{3}{4}\theta(\mathbf{z}) - \frac{3}{4}\xi(\mathbf{z}) + \frac{3}{4}(\xi \wedge \theta)(\mathbf{z}) \tag{G.57}$$

$$\Pr(\mathbf{z} \in X^c \cap T^c \cap Y \cap \Phi_1^c \cap \Phi_2) = \frac{1}{4}\theta(\mathbf{z}) - \frac{1}{4}(\tau \wedge \theta)(\mathbf{z}) \tag{G.58}$$

$$-\frac{1}{4}(\xi \wedge \theta)(\mathbf{z}) + \frac{1}{4}(\xi \wedge \tau \wedge \theta)(\mathbf{z}). \tag{G.59}$$

After combining terms and simplifying we end up with (5.111).

G.4 UNIFICATION OF DS COMBINATIONS

This refers to Section 5.4.1. Abbreviate $\alpha = \alpha_{\mathrm{DS}}(o_1, o_2)$. From (4.87) and (4.88),

$$f(o_1 * o_2|\mathbf{x}) = \sum_O (o_1 * o_2)(O) \cdot \mathbf{1}_O(\eta(\mathbf{x})) \tag{G.60}$$

$$= \alpha^{-1} \sum_O \left(\sum_{O_1 \cap O_2 = O} o_1(O_1) \cdot o_2(O_2) \right) \tag{G.61}$$

$$\cdot \mathbf{1}_O(\eta(\mathbf{x})) \tag{G.62}$$

$$= \alpha^{-1} \sum_{O_1, O_2} o_1(O_1) \cdot o_2(O_2) \cdot \mathbf{1}_{O_1 \cap O_2}(\eta(\mathbf{x})) \tag{G.63}$$

$$
\begin{aligned}
&= \alpha^{-1} \sum_{O_1,O_2} o_1(O_1) \cdot o_2(O_2) \cdot \mathbf{1}_{O_1}(\eta(\mathbf{x})) \cdot \mathbf{1}_{O_2}(\eta(\mathbf{x})) &\text{(G.64)}\\
&= \alpha^{-1} \left(\sum_{O_1} o_1(O_1) \cdot \mathbf{1}_{O_1}(\eta(\mathbf{x})) \right) &\text{(G.65)}\\
&\quad \cdot \left(\sum_{O_2} o_2(O_2) \cdot \mathbf{1}_{O_2}(\eta(\mathbf{x})) \right) &\text{(G.66)}\\
&= \alpha^{-1} \cdot f(o_1|\mathbf{x}) \cdot f(o_2|\mathbf{x}). &\text{(G.67)}
\end{aligned}
$$

Almost identical reasoning shows that $f(o_1 \cap o_2|\mathbf{x}) = f(o_1|\mathbf{x}) \cdot f(o_2|\mathbf{x})$.

Finally, for $O \neq \emptyset$

$$
\begin{aligned}
m_{o \cap o'}(O) &= \frac{(o \cap o')(O)}{\sum_{Q \neq \emptyset} (o \cap o')(Q)} &\text{(G.68)}\\
&= \frac{(m_o \cap m_{o'})(O)}{\sum_{Q \neq \emptyset} (m_o \cap m_{o'})(Q)} &\text{(G.69)}\\
&= \frac{(m_o * m_{o'})(O)}{\sum_{Q \neq \emptyset} (m_o * m_{o'})(Q)} &\text{(G.70)}\\
&= (m_o * m_{o'})(O). &\text{(G.71)}
\end{aligned}
$$

G.5 UNIFICATION OF RULE-FIRING

This refers to Section 5.4.4. The partial firing of a fuzzy rule $\rho = (g \Rightarrow g')$ by a fuzzy event g'' is expressed as the conjunction operation $g'' \wedge \rho$ defined in (4.145). The likelihood of $g'' \wedge \rho$ is

$$f(g'' \wedge \rho|\mathbf{x}) \triangleq \Pr(\eta(\mathbf{x}) \in \Sigma_{\Phi,A'',A,A'}(g'' \wedge \rho)) \tag{G.72}$$

where by (5.143)

$$\Sigma_{\Phi,A'',A,A'}(g'' \wedge \rho) = \Sigma_{A''}(g'') \cap \Sigma_{\Phi,A,A'}(\rho) \tag{G.73}$$

and where by (4.162)

$$\Sigma_{\Phi,A,A'}(\rho) = (\Sigma_{A'}(g') \cap \Sigma_A(g)) \cup (\Sigma_A(g)^c \cap \Phi). \tag{G.74}$$

On the other hand, the joint likelihood of g'' and ρ is

$$f(g'',\rho|\mathbf{x}) \triangleq \Pr(\eta(\mathbf{x}) \in \Sigma_{A''}(g''),\ \eta(\mathbf{x}) \in \Sigma_{\Phi,A,A'}(\rho)) \quad \text{(G.75)}$$
$$= \Pr(\eta(\mathbf{x}) \in \Sigma_{A''}(g'') \cap \Sigma_{\Phi,A,A'}(\rho)) \quad \text{(G.76)}$$

from which immediately follows

$$f(g'' \wedge \rho|\mathbf{x}) = f(g'',\rho|\mathbf{x}). \quad \text{(G.77)}$$

Consider now full firing of the rule, in which case $g'' = g$ and $A'' = A$. Then

$$\Sigma_{\Phi,A,A,A'}(g \wedge \rho) = \Sigma_A(g) \cap \Sigma_{\Phi,A,A'}(\rho) = \Sigma_A(g) \cap \Sigma_{A'}(g'). \quad \text{(G.78)}$$

Since

$$f(g'',\rho|\mathbf{x}) = f(g'' \wedge \rho|\mathbf{x}) \quad \text{(G.79)}$$

then $f(g,\rho|\mathbf{x}) = f(g \wedge \rho|\mathbf{x})$ and so

$$f(g,\rho|\mathbf{x}) = \Pr(\eta(\mathbf{x}) \in \Sigma_{\Phi,A,A,A'}(g \wedge \rho)) \quad \text{(G.80)}$$
$$= \Pr(\eta(\mathbf{x}) \in \Sigma_A(g) \cap \Sigma_{A'}(g')) \quad \text{(G.81)}$$
$$= (g \wedge_{A,A'} g')(\eta(\mathbf{x})) \quad \text{(G.82)}$$
$$= \Pr(\eta(\mathbf{x}) \in \Sigma_{A''}(g \wedge_{A,A'} g')) \quad \text{(G.83)}$$
$$= f(g \wedge_{A,A'} g'|\mathbf{x}) \quad \text{(G.84)}$$

and so

$$f(g, g \Rightarrow g'|\mathbf{x}) = f(g \wedge_{A,A'} g'|\mathbf{x}). \quad \text{(G.85)}$$

G.6 GENERALIZED LIKELIHOODS: $\mathfrak{Z}_0$ IS FINITE

This refers to Section 5.4.5. Let $M' = 2^M - 1$ and $m = 2^M$. There are m subsets of $\mathfrak{Z}_0$ that contain $\eta(\mathbf{x})$ since they have the form $\{\eta(\mathbf{x})\} \cup (\mathfrak{Z}_0 - \{\eta(\mathbf{x})\})$. Order the M' nonempty subsets $O_1, ..., O_{M'}$ of $\mathfrak{Z}_0$ so that the first m are the subsets of $\mathfrak{Z}_0 - \{\eta(\mathbf{x})\}$. Then

$$\int f(o|\mathbf{x})do = \int_S \left(\sum_{i=1}^{M'} a_i \mathbf{1}_{O_i}(\eta(\mathbf{x})) \right) da_1 \cdots da_{M'} \tag{G.86}$$

$$= \int_S \left(\sum_{i=1}^{m} a_i \mathbf{1}_{O_i}(\eta(\mathbf{x})) \right) da_1 \cdots da_{M'} \tag{G.87}$$

$$+ \int_S \left(\sum_{i=m+1}^{M'} a_i \mathbf{1}_{O_i}(\eta(\mathbf{x})) \right) da_1 \cdots da_{M'} \tag{G.88}$$

$$= \int_S \left(\sum_{i=1}^{m} a_i \right) da_1 \cdots da_{M'}. \tag{G.89}$$

Thus $\int f(o|\mathbf{x})do$ is independent of $\mathbf{x}$. Moreover, it is finite and nonzero since

$$\int_S \left(\sum_{i=1}^{m} a_i \right) da_1 \cdots da_{M'} \leq \int_0^1 \cdots \int_0^1 \left(\sum_{i=1}^{m} a_i \right) da_1 \cdots da_m \tag{G.90}$$

$$\cdot da_{m+1} \cdots da_{M'} \tag{G.91}$$

$$= m \cdot \int_0^1 a da = m/2 < \infty. \tag{G.92}$$

G.7 NOTA FOR FUZZY DS MEASUREMENTS

We are to prove (5.193):

$$f(o|v_0) = \sum_g o(g) \cdot g^c(\eta(v_1)) \cdots g^c(\eta(v_n)). \tag{G.93}$$

From Section 4.6.1 we know that for $i = 1, ..., n$ we have $f(o|v_0) \triangleq \Pr(\Sigma_A(W_o)|v_0)$ where W_o is defined by $W_o \triangleq W_1 \cup ... \cup W_e$ and where

$$W_i \triangleq \{(\mathbf{z}, a)|\ o_{i-1}^+ < a \leq o_{i-1}^+ + o_i g_i(\mathbf{z})\} \tag{G.94}$$

and where $o_0^+ \triangleq 0$ and $o_i^+ \triangleq o_1 + ... + o_i$ for all $i = 1, .., e$. Thus $o_i^+ - o_{i-1}^+ = o_i$ for all $i = 1, ..., e$ and $W_i \cap W_j \neq \emptyset$ for $i \neq j$. From (5.184) we know that

$$f(o|v_0) = \int_0^1 \mathbf{1}_{W_o^c}(\eta(v_1), a) \cdots \mathbf{1}_{W_o^c}(\eta(v_n), a) da. \tag{G.95}$$

Now, note that $W_i \subseteq \mathfrak{Z}_0 \times I_i$ where $I_i = (o_{i-1}^+, o_i^+]$. Thus $W_o^c = W_1^c \cap ... \cap W_e^c$ where

$$W_i^c = (\mathfrak{Z}_0 \times I_1) \cup ... \cup \tilde{W}_i \cup ... \cup (\mathfrak{Z}_0 \times I_n) \tag{G.96}$$

and where

$$\tilde{W}_i = \{(\mathbf{z}, a) | \; o_i^+ \geq a > o_{i-1}^+ + o_i g_i(\mathbf{z}) \text{ or } o_{i-1}^+ < a \leq o_{i-1}^+\} \tag{G.97}$$

$$= \{(\mathbf{z}, a) | \; o_i^+ \geq a > o_{i-1}^+ + o_i - o_i g_i^c(\mathbf{z})\} \tag{G.98}$$

$$= \{(\mathbf{z}, a) | \; o_i^+ - o_i g_i^c(\mathbf{z}) < a \leq o_i^+\}. \tag{G.99}$$

Consequently, $W_o^c = \tilde{W}_1 \cup ... \cup \tilde{W}_e$ with $\tilde{W}_i \cap \tilde{W}_j = \emptyset$ for $i \neq j$. So

$$f(o|v_0) = \int_0^1 \left(\sum_{i_1=1}^{e} \mathbf{1}_{\tilde{W}_i}(\eta(v_1), a) \right) \tag{G.100}$$

$$\cdots \left(\sum_{i_n=1}^{e} \mathbf{1}_{\tilde{W}_n}(\eta(v_n), a) \right) da \tag{G.101}$$

$$= \sum_{i=1}^{e} \int_0^1 \mathbf{1}_{\tilde{W}_i}(\eta(v_1), a) \cdots \mathbf{1}_{\tilde{W}_i}(\eta(v_n), a) da \tag{G.102}$$

where the second equation results from the fact that $\mathbf{1}_{\tilde{W}_i}(\eta(v_1), a) \cdot \mathbf{1}_{\tilde{W}_j}(\eta(v_1), a) = 0$ for $i \neq j$. Consequently,

$$f(o|v_0) = \sum_{i=1}^{e} \int_{o_{i-1}^+}^{o_i^+} \mathbf{1}_{\tilde{W}_i}(\eta(v_1), a) \cdots \mathbf{1}_{\tilde{W}_i}(\eta(v_n), a) da \tag{G.103}$$

$$= \sum_{i=1}^{e} \int_{o_{i-1}^+}^{o_i^+} \mathbf{1}_{(o_i^+ - o_i g_i^c(\eta(v_1)), o_i^+]}(a) \tag{G.104}$$

$$\cdots \mathbf{1}_{(o_i^+ - o_i g_i^c(\eta(v_n)), o_i^+]}(a) da \tag{G.105}$$

and so

$$= \sum_{i=1}^{e} \int_0^{o_i} \mathbf{1}_{(0,o_i g_i^c(\eta(v_1))]}(a') \cdots \mathbf{1}_{(0,o_i g_i^c(\eta(v_n))]}(a') da' \tag{G.106}$$

$$= \sum_{i=1}^{e} \int_0^{o_i} \mathbf{1}_{(0,m_i]}(a') da' \tag{G.107}$$

where

$$m_i = \min\{o_i g_i^c(\eta(v_1)), ..., o_i g_i^c(\eta(v_n))\} \tag{G.108}$$

$$= o_i \cdot \min\{o_i g_i^c(\eta(v_1)), ..., g_i^c(\eta(v_n)) \tag{G.109}$$

Thus, as claimed,

$$f(o|v_0) = \sum_{i=1}^{e} o_i \cdot \min\{_i g_i^c(\eta(v_1)), ..., g_i^c(\eta(v_n)). \tag{G.110}$$

G.8 KEF PREDICTOR

This refers to Section 5.6.2. Assume that the previous posterior distribution has the form of (5.214):

$$f_{k|k}(\mathbf{x}|Z^k) = f(\mathbf{x}|\mu_{k|k}) = \frac{1}{|\mu_{k|k}|} \sum_f \mu_{k|k}(f) \cdot f(\mathbf{x}). \tag{G.111}$$

Define $\mu_{k+1|k}$ on $\mathfrak{X}_0$ as follows. Enumerate the focal sets of $\mu_{k|k}(f)$ as $f_1, ..., f_e$. Define the focal sets of $\mu_{k+1|k}$ to be

$$f_i'(\mathbf{x}) \triangleq \frac{1}{\sigma_i} \int f_{k+1|k}(\mathbf{x}|\mathbf{x}') \cdot f_i(\mathbf{x}') d\mathbf{x}'. \tag{G.112}$$

Note that

$$\sigma_i \triangleq \sup_{\mathbf{x}} \int f_{k+1|k}(\mathbf{x}|\mathbf{x}') \cdot f_i(\mathbf{x}') d\mathbf{x}' > 0 \tag{G.113}$$

since by assumption $f_{k+1|k}(\mathbf{x}|\mathbf{x}')$ is linear-Gaussian. Define the mass $\mu_i' = \mu_{k+1|k}(f_i')$ of f_i' to be

$$\mu_i' \triangleq \frac{\mu_i \cdot \sigma_i}{\sum_{i'=1}^{e} \mu_{i'} \cdot \sigma_{i'}}. \tag{G.114}$$

Then $\mu_{k+1|k}$ is bounded since $\mu_{k|k}$ is bounded. To see this, note that

$$\int f'_i(\mathbf{x})d\mathbf{x} = \frac{1}{\sigma_i}\int\int f_{k+1|k}(\mathbf{x}|\mathbf{x}')\cdot f_i(\mathbf{x}')d\mathbf{x}'d\mathbf{x} \quad (G.115)$$

$$= \frac{1}{\sigma_i}\int f_i(\mathbf{x}')d\mathbf{x}' < \infty. \quad (G.116)$$

Our claim is that $f_{k+1|k}(\mathbf{x}|Z^k) = f(\mathbf{x}|\mu_{k+1|k})$. To see this, note that

$$f_{k+1|k}(\mathbf{x}|Z^k) = \int f_{k+1|k}(\mathbf{x}|\mathbf{x}')\cdot f_{k|k}(\mathbf{x}'|Z^k)d\mathbf{x}' \quad (G.117)$$

$$\propto \sum_{i=1}^{e}\mu_i\cdot\int f_{k+1|k}(\mathbf{x}|\mathbf{x}')\cdot f_i(\mathbf{x}')d\mathbf{x}' \quad (G.118)$$

$$\propto \sum_{i=1}^{e}\left(\frac{\mu_i\cdot\sigma_i}{\sum_{j=1}^{e}\mu_j\cdot\sigma_{i'}}\right) \quad (G.119)$$

$$\cdot\left(\frac{1}{\sigma_i}\int f_{k+1|k}(\mathbf{x}|\mathbf{x}')\cdot f_i(\mathbf{x}')d\mathbf{x}'\right) \quad (G.120)$$

$$= \sum_i \mu_i\cdot f'_i(\mathbf{x}). \quad (G.121)$$

Thus

$$K\cdot f_{k+1|k}(\mathbf{x}|Z^k) = \sum_i \mu_i\cdot f'_i(\mathbf{x}) \quad (G.122)$$

for some constant K. Integrating both sides of this we get $K = \sum_i \mu_i\cdot|f'_i|$ and so

$$f_{k+1|k}(\mathbf{x}|Z^{k+1}) = f(\mathbf{x}|\mu_{k+1|k}) \quad (G.123)$$

by (5.214).

Now, (G.113) becomes, because of (2.36),

$$\sigma_i \triangleq \sup_{\mathbf{x}} \int f_{k+1|k}(\mathbf{x}|\mathbf{x}') \cdot f_i(\mathbf{x}')d\mathbf{x}' \tag{G.124}$$

$$= \sup_{\mathbf{x}} \int N_{Q_k}(\mathbf{x} - F_k\mathbf{x}') \cdot \hat{N}_{C_i}(\mathbf{x}' - \mathbf{x}_i)d\mathbf{x}' \tag{G.125}$$

$$= N_{C_i}(\mathbf{0})^{-1} \cdot \sup_{\mathbf{x}} \int N_{Q_k}(\mathbf{x} - F_k\mathbf{x}') \cdot N_{C_i}(\mathbf{x}' - \mathbf{x}_i)d\mathbf{x}' \tag{G.126}$$

$$= N_{C_i}(\mathbf{0})^{-1} \cdot \sup_{\mathbf{x}} N_{Q_k + F_k C_i F_k^T}(\mathbf{x} - F_k\mathbf{x}_i) \tag{G.127}$$

$$= N_{C_i}(\mathbf{0})^{-1} \cdot N_{Q_k + F_k C_i F_k^T}(\mathbf{0}). \tag{G.128}$$

Thus (G.112) becomes

$$f'_i(\mathbf{x}) \triangleq \frac{1}{\sigma_i} \int f_{k+1|k}(\mathbf{x}|\mathbf{x}') \cdot f_i(\mathbf{x}')d\mathbf{x}' \tag{G.129}$$

$$= N_{Q_k + F_k C_i F_k^T}(\mathbf{0})^{-1} \cdot N_{Q_k + F_k C_i F_k^T}(\mathbf{x} - F_k\mathbf{x}_i) \tag{G.130}$$

$$= \hat{N}_{Q_k + F_k C_i F_k^T}(\mathbf{x} - F_k\mathbf{x}_i). \tag{G.131}$$

Likewise, (G.114) becomes

$$\mu'_i = \frac{\mu_i \cdot \sigma_i}{\sum_{i'=1}^{e} \mu_{i'} \cdot \sigma_{i'}} = \frac{\omega_i \cdot \mu_i}{\sum_{i'=1}^{e} \omega_{i'} \cdot \mu_{i'}} \tag{G.132}$$

where

$$\omega_i \triangleq \frac{N_{Q_k + F_k C_i F_k^T}(\mathbf{0})}{N_{C_i}(\mathbf{0})} \tag{G.133}$$

$$= \sqrt{\frac{\det 2\pi C_i}{\det 2\pi(Q_k + F_k C_i F_k^T)}} \tag{G.134}$$

$$= \sqrt{\frac{\det C_i}{\det(Q_k + F_k C_i F_k^T)}}. \tag{G.135}$$

G.9 KEF CORRECTOR (FUZZY DS MEASUREMENTS)

This refers to Section 5.6.3. Assume that the predicted posterior has the form of (5.214): $f_{k+1|k}(\mathbf{x}|Z^k) = f(\mathbf{x}|\mu_{k+1|k})$. From (5.215) recall that

$$(H^{-1}o)(f) \triangleq \sum_{gH=f} o(g). \tag{G.136}$$

Define the fuzzy b.m.a. $\mu_{k+1|k+1}(f)$ on $\mathfrak{X}_0$ by

$$\mu_{k+1|k+1} = H_{k+1}^{-1} o_{k+1} * \mu_{k+1|k}. \tag{G.137}$$

Then $\int f(\mathbf{x})d\mathbf{x} < \infty$ for any focal set f of $\mu_{k+1|k+1}$, even though the same is not necessarily true for $H_{k+1}^{-1} o_{k+1}$. To see this, let μ and s be fuzzy b.m.a.s on $\mathfrak{X}_0$ with $\int f'(\mathbf{x})d\mathbf{x} < \infty$ for any focal set f' of μ. From (4.129),

$$(s * \mu)(f'') = \frac{1}{\alpha_{s,\mu}} \cdot \sum_{f \cdot f' = f''} s(f) \cdot \mu(f') \tag{G.138}$$

where $\alpha_{s,\mu} = \sum_{f \cdot f' \neq 0} s(f) \cdot \mu(f')$. The focal sets of $s * \mu$ are finite in number since they have the form $f''_{i,j} = f_i \cdot f'_j$ where f_i and f'_j are the focal sets of s and μ, respectively. Also,

$$\int f''_{i,j}(\mathbf{x})d\mathbf{x} = \int f_i(\mathbf{x}) \cdot f'_j(\mathbf{x})d\mathbf{x} \leq \int f'_j(\mathbf{x})d\mathbf{x} < \infty. \tag{G.139}$$

We claim that

$$f_{k+1|k+1}(\mathbf{x}|Z^{k+1}) = f(\mathbf{x}|\mu_{k+1|k+1}). \tag{G.140}$$

See (5.214). Note that

$$
\begin{aligned}
f_{k+1|k+1}(\mathbf{x}|Z^{k+1}) &\propto f_{k+1}(o_{k+1}|\mathbf{x}) \cdot f_{k+1|k}(\mathbf{x}|Z^k) && \text{(G.141)} \\
&\propto \left(\sum_g o_{k+1}(g) \cdot g(\eta(\mathbf{x})) \right) && \text{(G.142)} \\
&\quad \cdot \left(\sum_f \mu_{k+1|k}(f) \cdot f(\mathbf{x}) \right) && \text{(G.143)} \\
&= \sum_f \sum_{f'} \left(\sum_{\eta^{-1}g=f'} o_{k+1}(g) \right) && \text{(G.144)} \\
&\quad \cdot \mu_{k+1|k}(f) \cdot f'(\mathbf{x}) \cdot f(\mathbf{x}) && \text{(G.145)} \\
&= \sum_f \sum_{f'} (\eta^{-1} o_{k+1})(f') \cdot \mu_{k+1|k}(f) && \text{(G.146)} \\
&\quad \cdot f'(\mathbf{x}) \cdot f(\mathbf{x}) && \text{(G.147)}
\end{aligned}
$$

and so

$$
\begin{aligned}
& f_{k+1|k+1}(\mathbf{x}|Z^{k+1}) && \text{(G.148)} \\
\propto\; & \sum_{f''} \left(\sum_{f \cdot f' = f''} (\eta^{-1} o_{k+1})(f') \cdot \mu_{k+1|k}(f) \right) \cdot f''(\mathbf{x}) && \text{(G.149)} \\
\propto\; & \sum_{f''} (\eta^{-1} o_{k+1} * \mu_{k+1|k})(f'') \cdot f''(\mathbf{x}) && \text{(G.150)} \\
=\; & \sum_{f''} \mu_{k+1|k+1}(f'') \cdot f''(\mathbf{x}) && \text{(G.151)} \\
\propto\; & f(\mathbf{x}|\mu_{k+1|k+1}). && \text{(G.152)}
\end{aligned}
$$

Thus $f_{k+1|k+1}(\mathbf{x}|Z^{k+1}) = f(\mathbf{x}|\mu_{k+1|k+1})$ by (5.214).

Next, let the focal sets of $\mu_{k+1|k}$ be $f_i(\mathbf{x}) = \hat{N}_{C_i}(\mathbf{x} - \mathbf{x}_i)$ for $i = 1, ..., e$ and the focal sets g_j of o_{k+1} as $g_j(\mathbf{z}) = \hat{N}_{D_j}(\mathbf{z} - \mathbf{z}_j)$ for $j = 1, ..., d$. Abbreviate $\mu_i = \mu_{k+1|k}(f_i)$ and $o_j = o_{k+1}(g_j)$. We can treat $g_d(\mathbf{z}) \equiv 1$ as a limiting case of $g_d(\mathbf{z}) = \hat{N}_{D_d}(\mathbf{z} - \mathbf{z}_d)$, so without loss of generality we can assume that all g_j have the first form. Under our current assumptions (G.146) becomes

$$f_{k+1|k+1}(\mathbf{x}|Z^{k+1}) \propto \sum_{i,j} o_j \cdot \mu_i \cdot g_j(H_{k+1}\mathbf{x}) \cdot f_i(\mathbf{x}) \tag{G.153}$$
$$= \sum_{i,j} o_j \mu_i \cdot \hat{N}_{D_j}(\mathbf{z}_j - H_{k+1}\mathbf{x}) \tag{G.154}$$
$$\cdot \hat{N}_{C_i}(\mathbf{x} - \mathbf{x}_i) \tag{G.155}$$
$$= \sum_{i,j} o_j \mu_i \cdot N_{D_j}(\mathbf{0})^{-1} N_{C_i}(\mathbf{0})^{-1} \tag{G.156}$$
$$\cdot N_{D_j}(\mathbf{z}_j - H_{k+1}\mathbf{x}) \cdot N_{C_i}(\mathbf{x} - \mathbf{x}_i) \tag{G.157}$$

and so

$$f_{k+1|k+1}(\mathbf{x}|Z^{k+1}) \propto \sum_{i,j} o_j \mu_i \cdot N_{D_j}(\mathbf{0})^{-1} N_{C_i}(\mathbf{0})^{-1} \tag{G.158}$$
$$\cdot N_{C_i + H_{k+1}^T D_j H_{k+1}}(\mathbf{z}_j - H_{k+1}\mathbf{x}_i) \tag{G.159}$$
$$\cdot N_{E_{i,j}}(\mathbf{x} - \mathbf{e}_{i,j}) \tag{G.160}$$
$$= \sum_{i,j} o_j \mu_i \cdot N_{D_j}(\mathbf{0})^{-1} N_{C_i}(\mathbf{0})^{-1} \tag{G.161}$$
$$\cdot N_{C_i + H_{k+1}^T D_j H_{k+1}}(\mathbf{0}) \tag{G.162}$$
$$\cdot \hat{N}_{C_i + H_{k+1}^T D_j H_{k+1}}(\mathbf{z}_j - H_{k+1}\mathbf{x}_i) \tag{G.163}$$
$$\cdot N_{E_{i,j}}(\mathbf{0}) \cdot \hat{N}_{E_{i,j}}(\mathbf{x} - \mathbf{e}_{i,j}). \tag{G.164}$$

Or to summarize,

$$f_{k+1|k+1}(\mathbf{x}|Z^{k+1}) \propto \sum_{i,j} \mu_{i,j} \cdot f_{i,j} \tag{G.165}$$

where

$$f_{i,j}(\mathbf{x}) \triangleq \hat{N}_{E_{i,j}}(\mathbf{x} - \mathbf{e}_{i,j}) \tag{G.166}$$
$$E_{i,j}^{-1} \triangleq C_i^{-1} + H_{k+1}^T D_j^{-1} H_{k+1} \tag{G.167}$$
$$E_{i,j}^{-1}\mathbf{e}_{i,j} \triangleq C_i^{-1}\mathbf{x}_i + H_{k+1}^T D_j^{-1} \mathbf{z}_j \tag{G.168}$$

where

$$\mu_{i,j} \triangleq \frac{o_j \mu_i \omega_{i,j} \cdot \hat{N}_{C_i + H_{k+1}^T D_j H_{k+1}}(\mathbf{z}_j - H_{k+1}\mathbf{x}_i)}{\sum_{i'=1}^{e} \sum_{j'=1}^{d} o_{j'} \mu_{i'} \omega_{i',j'} \cdot \hat{N}_{C_{i'} + H_{k+1}^T D_{j'} H_{k+1}}(\mathbf{z}_{j'} - H_{k+1}\mathbf{x}_{i'})} \tag{G.169}$$

and where

$$\omega_{i,j} \triangleq N_{C_i+H_{k+1}^T D_j H_{k+1}}(\mathbf{0}) \cdot N_{E_{i,j}}(\mathbf{0}) \cdot N_{D_j}(\mathbf{0})^{-1} \tag{G.170}$$

$$\cdot N_{C_i}(\mathbf{0})^{-1} \tag{G.171}$$

$$= \sqrt{\frac{\det C_i \cdot \det D_j}{\det E_{i,j} \cdot \det(C_i + H_{k+1}^T D_j H_{k+1})}}. \tag{G.172}$$

Thus define the fuzzy b.m.a. $\mu_{k+1|k+1}$ on $\mathfrak{X}_0$ by $\mu_{k+1|k+1}(f) = 0$ unless $f = f_{i,j}$, in which case $\mu_{k+1|k+1}(f) = \mu_{i,j}$. Applying (5.214) to (G.161), we find that $f_{k+1|k+1}(\mathbf{x}|Z^{k+1}) = f(\mathbf{x}|\mu_{k+1|k+1})$.

G.10 LIKELIHOODS FOR AGA FUZZY MEASUREMENTS

We are to prove (6.11), that is $f(g|\mathbf{x}) = \sup_{\mathbf{z}} \min\{g(\mathbf{z}),\ \eta_{\mathbf{x}}(\mathbf{z})\}$. By definition,

$$f(g|\mathbf{x}) = \Pr(\Sigma_A(g) \cap \Sigma_A(\eta_{\mathbf{x}}) \neq \emptyset) \tag{G.173}$$

$$= \Pr(\Sigma_A(g \wedge \eta_{\mathbf{x}}) \neq \emptyset) \tag{G.174}$$

$$= 1 - \Pr(\Sigma_A(g \wedge \eta_{\mathbf{x}}) = \emptyset) \tag{G.175}$$

where '$\wedge$' is Zadeh conjunction and where we have used (4.24). In general, if g' is a fuzzy membership on $\mathfrak{Z}_0$ when does $\Sigma_A(g') = \emptyset$? For any $0 \leq a \leq 1$, $\Sigma_a(g') = \{\mathbf{z}|\ a \leq g'(\mathbf{z})\}$. So $\Sigma_a(g') = \emptyset$ occurs if and only if $g'(\mathbf{z})$ is always strictly less than one and if a is larger than any possible value of g'—that is, if $a > \sup_{\mathbf{z}} g'(\mathbf{z})$. Consequently

$$\Pr(\Sigma_A(g') = \emptyset) = 1 - \sup_{\mathbf{z}} g'(\mathbf{z}). \tag{G.176}$$

From this we get, as claimed,

$$f(g|\mathbf{x}) = 1 - \Pr(A > \sup_{\mathbf{z}}(g \wedge \eta_{\mathbf{x}})(\mathbf{z})) \tag{G.177}$$

$$= \Pr(A \leq \sup_{\mathbf{z}}(g \wedge \eta_{\mathbf{x}})(\mathbf{z})) \tag{G.178}$$

$$= \sup_{\mathbf{z}}(g \wedge \eta_{\mathbf{x}})(\mathbf{z}) \tag{G.179}$$

$$= \sup_{\mathbf{z}} \min\{g(\mathbf{z}),\ \eta_{\mathbf{x}}(\mathbf{z})\}. \tag{G.180}$$

G.11 LIKELIHOODS FOR AGA GENERALIZED FUZZY MEASUREMENTS

We are to prove (6.17):

$$f(W|\mathbf{x}) = \sup_{\mathbf{z}} \int_0^1 \mathbf{1}_W(\mathbf{x}, a) \cdot \mathbf{1}_{W_\mathbf{x}}(\mathbf{z}, a) da. \tag{G.181}$$

To see this, note that from (4.76)

$$f(W|\mathbf{x}) = \Pr(\Sigma_A(W) \cap \Sigma_A(W_\mathbf{x}) \neq \emptyset) \tag{G.182}$$

$$= \Pr(\Sigma_A(W \cap W_\mathbf{x}) \neq \emptyset). \tag{G.183}$$

Now consider Figure 4.4. For any generalized fuzzy set V,

$$\Pr(\Sigma_A(V) \neq \emptyset) = \Pr(\{\mathbf{z}| \ (\mathbf{z}, A) \in V\}) \neq \emptyset) \tag{G.184}$$

$$= \Pr(V \cap (\mathfrak{Z}_0 \times \{A\}) \neq \emptyset). \tag{G.185}$$

Now, the probability that the "vertical" cross-section through V at $\mathbf{z}$ is nonempty is $\Pr(V \cap (\{\mathbf{z}\} \times \{A\}) \neq \emptyset)$. However, from the figure, it is clear that $\Pr(V \cap (\mathfrak{Z}_0 \times \{A\}) \neq \emptyset)$ is the supremal value of $\Pr(V \cap (\{\mathbf{z}\} \times \{A\}) \neq \emptyset)$ over all $\mathbf{z}$:

$$\Pr(\Sigma_A(V) \neq \emptyset) = \sup_{\mathbf{z}} \Pr(V \cap (\{\mathbf{z}\} \times \{A\}) \neq \emptyset) \tag{G.186}$$

$$= \sup_{\mathbf{z}} \int_0^1 \mathbf{1}_V(\mathbf{z}, a) da \tag{G.187}$$

Consequently, as claimed,

$$f(W|\mathbf{x}) = \Pr(\Sigma_A(W \cap W_\mathbf{x}) \neq \emptyset) \tag{G.188}$$

$$= \sup_{\mathbf{z}} \int_0^1 \mathbf{1}_{W \cap W_\mathbf{x}}(\mathbf{z}, a) da \tag{G.189}$$

$$= \sup_{\mathbf{z}} \int_0^1 \mathbf{1}_W(\mathbf{z}, a) \cdot \mathbf{1}_{W_\mathbf{x}}(\mathbf{z}, a) da.$$

G.12 LIKELIHOODS FOR AGA FUZZY DS MEASUREMENTS

We are to prove (6.24). Since $f(o|\mathbf{x}) = \Pr(\Sigma_A(W_o) \cap \Sigma_{A'}(W_{\sigma_\mathbf{x}}) \neq \emptyset)$ it will be sufficient to prove that

$$\Pr(\Sigma_A(W_o) \cap \Sigma_{A'}(W_{o'}) \neq \emptyset) = \alpha_{\text{FDS}}(o, o'). \tag{G.190}$$

Let $g_1, ..., g_e$ and $o_1, ..., o_e$ be the focal sets and weights for o; and $g'_1, ..., g'_{e'}$ and $o'_1, ..., o'_{e'}$ be the focal sets and weights for o'. Recall from Section 4.6.1 that the generalized fuzzy sets W_o and $W_{o'}$ are defined as $W_o = W_1 \cup ... \cup W_e$ and $W_{o'} = W'_1 \cup ... \cup W'_e$ where

$$W_i \triangleq \{(u,a) | \; o^+_{i-1} < a \leq o^+_{i-1} + o_i g_i(u)\} \tag{G.191}$$
$$W'_{i'} \triangleq \{(u,a) | \; o'^+_{i'-1} < a \leq o'^+_{i'-1} + o'_{i'} g'_{i'}(u)\} \tag{G.192}$$

and where

$$o^+_i = o_1 + ... + o_i, \tag{G.193}$$
$$o'^+_{i'} = o'_1 + ... + o'_{i'} \tag{G.194}$$
$$o^+_0 = 0, \; o'^+_0 = 0. \tag{G.195}$$

Recall that, by construction, $W_i \cap W_j \neq \emptyset$ for $i \neq j$ and $W'_{i'} \cap W'_{j'} \neq \emptyset$ for $i' \neq j'$. Define

$$I_i \triangleq (o^+_{i-1}, o^+_i] \tag{G.196}$$
$$I'_i \triangleq (o'^+_{i-1}, o'^+_i]. \tag{G.197}$$

Then $W_i \subseteq \mathfrak{Z}_0 \times I_i$. Since the I_i are mutually disjoint and the I'_i are mutually disjoint,

$$\Pr(\Sigma_A(W_o) \cap \Sigma_{A'}(W_{o'}) \neq \emptyset) \tag{G.198}$$

$$= \sum_{i=1}^{e} \sum_{i'=1}^{e'} \Pr(A \in I_i, A' \in I'_{i'}, \Sigma_A(W_i) \cap \Sigma_{A'}(W'_{i'}) \neq \emptyset). \tag{G.199}$$

By construction, however, $\Sigma_a(W_i) = \Sigma_{\tilde{a}}(g_i)$ for all $a \in I_i$, where $\tilde{a} \triangleq (a - o^+_{i-1})/o_i$. Likewise, $\Sigma_{a'}(W'_{i'}) = \Sigma_{\tilde{a}'}(g'_{i'})$ for all $a' \in I'_{i'}$, where $\tilde{a}' \triangleq (a' - o'^+_{i'-1})/o_{i'}$. Consequently, $\Sigma_A(W_i) \cap \Sigma_{A'}(W'_{i'}) \neq \emptyset$ if and only if

$\Sigma_{\tilde{a}}(g_i) \cap \Sigma_{\tilde{a}'}(g'_{i'}) \neq \emptyset$. However, if A, A' are independent then the only way that

$$\Pr(\Sigma_{\tilde{A}}(g_i) \cap \Sigma_{\tilde{A}'}(g'_{i'}) = \emptyset) > 0 \tag{G.200}$$

is for g_i and $g'_{i'}$ to be completely disjoint: $(g_i \cdot g'_{i'})(\mathbf{z}) \triangleq g_i(\mathbf{z}) \cdot g'_{i'}(\mathbf{z}) = 0$ for all $\mathbf{z}$.

So

$$\Pr(\Sigma_{\tilde{A}}(g_i) \cap \Sigma_{\tilde{A}'}(g'_{i'}) \neq \emptyset) > 0 \tag{G.201}$$

if and only if $g_i \cdot g'_{i'} \neq 0$. Thus as claimed, from (4.130) we get

$$\Pr(\Sigma_A(W_o) \cap \Sigma_{A'}(W_{o'}) \neq \emptyset) \tag{G.202}$$

$$= \sum_{g_i \cdot g'_{i'} \neq 0} \Pr(A \in I_i) \cdot \Pr(A' \in I'_{i'}) \tag{G.203}$$

$$= \sum_{g_i \cdot g'_{i'} \neq 0} (o_i^+ - o_{i-1}^+) \cdot (o'^+_{i'} - o'^+_{i'-1}) \tag{G.204}$$

$$= \sum_{g_i \cdot g'_{i'} \neq 0} o_i \cdot o'_{i'} = \alpha_{\text{FDS}}(o, o'). \tag{G.205}$$

G.13 INTERVAL ARGSUP FORMULA

We are to prove (7.9): $\hat{S}_{\mathbf{z}} = \{\mathbf{x} | \overline{J_{\mathbf{z}}(\mathbf{x})} \geq \sup_{\mathbf{y}} \underline{J_{\mathbf{z}}(\mathbf{y})}\}$. Let

$$T_{\mathbf{z}} = \{\mathbf{x} | \overline{J_{\mathbf{z}}(\mathbf{x})} \geq \sup_{\mathbf{y}} \underline{J_{\mathbf{z}}(\mathbf{y})}. \tag{G.206}$$

Then $\hat{\mathbf{x}} = \arg\sup_{\mathbf{x}} L_{\hat{\mathbf{x}}}(\mathbf{x})$ for some function $L_{\hat{\mathbf{x}}}$ such that $L_{\hat{\mathbf{x}}}(\mathbf{x}) \in J_{\mathbf{z}}(\mathbf{x})$ for all $\mathbf{x}$. Since $L_{\hat{\mathbf{x}}}(\mathbf{x}) \in J_{\mathbf{z}}(\mathbf{x})$ for all $\mathbf{x}$ it follows that

$$\overline{J_{\mathbf{z}}(\hat{\mathbf{x}})} \geq L_{\hat{\mathbf{x}}}(\hat{\mathbf{x}}) \geq L_{\hat{\mathbf{x}}}(\mathbf{y}) \geq \underline{J_{\mathbf{z}}(\mathbf{y})} \tag{G.207}$$

for all $\mathbf{y}$ and thus

$$\overline{J_{\mathbf{z}}(\hat{\mathbf{x}})} \geq \sup_{\mathbf{y}} \underline{J_{\mathbf{z}}(\mathbf{y})}. \tag{G.208}$$

So $\hat{\mathbf{x}} \in T_{\mathbf{z}}$ and thus $\hat{S}_{\mathbf{z}} \subseteq T_{\mathbf{z}}$.

Conversely let $\mathbf{x}_0 \in T_{\mathbf{z}}$. Then

$$\overline{J_{\mathbf{z}}(\mathbf{x}_0)} \geq \underline{J_{\mathbf{z}}(\mathbf{y})} \tag{G.209}$$

for all $\mathbf{y}$. Define L_0 by

$$L_0(\mathbf{x}) \triangleq \begin{cases} \overline{J_{\mathbf{z}}(\mathbf{x}_0)} & \text{if} \quad \mathbf{x} = \mathbf{x}_0 \\ \underline{J_{\mathbf{z}}(\mathbf{x})} & \text{if} \quad \mathbf{x} \neq \mathbf{x}_0 \end{cases}. \tag{G.210}$$

Then by construction $L_0(\mathbf{x}) \in J_{\mathbf{z}}(\mathbf{x})$ and $L_0(\mathbf{x}_0) \geq L_0(\mathbf{x})$ for all $\mathbf{x}$. From this it follows that $\mathbf{x}_0 \in \hat{S}_{\mathbf{z}}$ and so $\hat{S}_{\mathbf{z}} \supseteq T_{\mathbf{z}}$ and so $\hat{S}_{\mathbf{z}} = T_{\mathbf{z}}$.

G.14 CONSONANCE OF THE RANDOM STATE SET $\hat{\Gamma}_{\mathbf{z}}$

We are to show that the random subset

$$\hat{\Gamma}_{\mathbf{z}} = \{\mathbf{x} | \ \overline{\Sigma_A(\theta_{\mathbf{z},\mathbf{x}})} \geq \sup_{\mathbf{y}} \underline{\Sigma_A(\theta_{\mathbf{z},\mathbf{y}})}\} \tag{G.211}$$

of (7.20) is consonant. That is, its instantiations are linearly ordered under set inclusion. The instantiations of $\hat{\Gamma}_{\mathbf{z}}$ are the

$$S_a \triangleq \{\mathbf{x} | \ \overline{\Sigma_a(\theta_{\mathbf{z},\mathbf{x}})} \geq \sup_{\mathbf{y}} \underline{\Sigma_a(\theta_{\mathbf{z},\mathbf{y}})}\} \tag{G.212}$$

for all $0 \leq a \leq 1$. Suppose that $a \leq a'$. Then $S_{a'} \subseteq S_a$. To see this, note that if $a \leq a'$ then $\Sigma_a(\theta_{\mathbf{z},\mathbf{x}}) \supseteq \Sigma_{a'}(\theta_{\mathbf{z},\mathbf{x}})$ and so

$$\overline{\Sigma_a(\theta_{\mathbf{z},\mathbf{x}})} \quad \geq \quad \overline{\Sigma_{a'}(\theta_{\mathbf{z},\mathbf{x}})} \tag{G.213}$$

$$\underline{\Sigma_a(\theta_{\mathbf{z},\mathbf{x}})} \quad \leq \quad \underline{\Sigma_{a'}(\theta_{\mathbf{z},\mathbf{x}})}. \tag{G.214}$$

Thus for all $\mathbf{x}$,

$$\overline{\Sigma_a(\theta_{\mathbf{z},\mathbf{x}})} \geq \overline{\Sigma_{a'}(\theta_{\mathbf{z},\mathbf{x}})} \geq \underline{\Sigma_{a'}(\theta_{\mathbf{z},\mathbf{x}})} \geq \underline{\Sigma_a(\theta_{\mathbf{z},\mathbf{x}})}. \tag{G.215}$$

Consequently, let $\mathbf{x} \in S_{a'}$. Then by definition $\overline{\Sigma_{a'}(\theta_{\mathbf{z},\mathbf{x}})} \geq \underline{\Sigma_{a'}(\theta_{\mathbf{z},\mathbf{y}})}$ for all $\mathbf{y}$ and so, also,

$$\overline{\Sigma_a(\theta_{\mathbf{z},\mathbf{x}})} \geq \overline{\Sigma_{a'}(\theta_{\mathbf{z},\mathbf{x}})} \geq \underline{\Sigma_{a'}(\theta_{\mathbf{z},\mathbf{y}})} \geq \underline{\Sigma_a(\theta_{\mathbf{z},\mathbf{y}})} \tag{G.216}$$

and thus $\overline{\Sigma_a(\theta_{\mathbf{z},\mathbf{x}})} \geq \underline{\Sigma_a(\theta_{\mathbf{z},\mathbf{y}})}$ for all $\mathbf{y}$ and so $\mathbf{x} \in S_a$.

G.15 SUFFICIENT STATISTICS AND MODIFIED COMBINATION

We are to prove (8.45), that is $f(s *_0 s'|\mathbf{x}) = f(s|\mathbf{x}) \cdot f(s'|\mathbf{x})$. To see this, note that from (8.29)

$$f(s|\mathbf{x}) \triangleq \frac{\sum_S \frac{s(S)}{p_0(S)} \cdot \mathbf{1}_S(\mathbf{x})}{\sum_T \frac{s(T)}{p_0(T)}}. \tag{G.217}$$

So, from (4.96) and (8.29),

$$f(s *_0 s'|\mathbf{x}) \triangleq \frac{\sum_{S''} \frac{(s*_0 s')(S'')}{p_0(S'')} \cdot \mathbf{1}_{S''}(\mathbf{x})}{\sum_{T''} \frac{(s*_0 s')(T'')}{p_0(T'')}} \tag{G.218}$$

$$= \frac{\sum_{S''} \sum_{S\cap S'=S''} \frac{(s*_0 s')(S'')}{p_0(S'')} \cdot \mathbf{1}_{S''}(\mathbf{x})}{\sum_{T''} \sum_{T\cap T'=T''} \frac{(s*_0 s')(T'')}{p_0(T'')}} \tag{G.219}$$

$$= \frac{\alpha_0(s,s')^{-1} \cdot \sum_{S,S'} \frac{s(S)\cdot s'(S')\cdot \alpha_0(S,S')}{p_0(S\cap S')} \cdot \mathbf{1}_{S\cap S'}(\mathbf{x})}{\alpha_0(s,s')^{-1} \cdot \sum_{T,T'} \frac{s(T)\cdot s'(T')}{p_0(T\cap T')}} \tag{G.220}$$

$$= \frac{\sum_{S,S'} \frac{s(S)\cdot s'(S')}{p_0(S)\cdot p_0(S')} \cdot \mathbf{1}_S(\mathbf{x}) \cdot \mathbf{1}_{S'}(\mathbf{x})}{\sum_{T,T'} \frac{s(T)\cdot s'(T')}{p_0(T)\cdot p_0(T')}} \tag{G.221}$$

$$= \left(\frac{\sum_S \frac{s(S)}{p_0(S)} \cdot \mathbf{1}_S(\mathbf{x})}{\sum_T \frac{s(T)}{p_0(T)}}\right) \left(\frac{\sum_{S'} \frac{s'(S')}{p_0(S')} \cdot \mathbf{1}_{S'}(\mathbf{x})}{\sum_{T'} \frac{s'(T')}{p_0(T')}}\right) \tag{G.222}$$

$$= f(s|\mathbf{x}) \cdot f(s'|\mathbf{x}) \tag{G.223}$$

as claimed.

G.16 TRANSFORMATION INVARIANCE

We are to prove (8.53): $\eta^{\sim 1}(o * o') = \eta^{\sim 1}o *_0 \eta^{\sim 1}o'$. To see this, note that

$$\eta^{\sim 1}(o * o')(S'') \tag{G.224}$$

$$= \frac{\sum_{\eta^{-1}O''=S''}(o * o')(O'') \cdot p_0(S'')}{\sum_{O''}(o * o')(O'') \cdot p_0(\eta^{-1}O'')} \tag{G.225}$$

$$= \frac{\begin{array}{c}\alpha_{\text{DS}}(o,o')^{-1} \cdot \sum_{\eta^{-1}O''=S''} \\ \sum_{O\cap O'=O''} o(O) \cdot o'(O') \cdot p_0(\eta^{-1}O \cap \eta^{-1}O')\end{array}}{\alpha_{\text{DS}}(o,o')^{-1} \cdot \sum_{O,O'} o(O) \cdot o'(O') \cdot p_0(\eta^{-1}O \cap \eta^{-1}O')} \tag{G.226}$$

$$= \frac{\sum_{\eta^{-1}O\cap\eta^{-1}O'=S''} o(O) \cdot o'(O') \cdot p_0(\eta^{-1}O \cap \eta^{-1}O')}{\sum_{O,O'} o(O) \cdot o'(O') \cdot p_0(\eta^{-1}O \cap \eta^{-1}O')} \tag{G.227}$$

and so

$$= \frac{\begin{array}{c}\sum_{S\cap S'=S''} \sum_{\eta^{-1}O=S} \\ \sum_{\eta^{-1}O'=S'} o(O) \cdot o'(O') \cdot p_0(\eta^{-1}O \cap \eta^{-1}O')\end{array}}{\sum_{O,O'} o(O) \cdot o'(O') \cdot p_0(\eta^{-1}O \cap \eta^{-1}O')} \tag{G.228}$$

$$= \frac{\left(\begin{array}{c}\sum_{S\cap S'=S''} \left(\sum_{\eta^{-1}O=S} o(O)\right) \\ \cdot \left(\sum_{\eta^{-1}O'=S'} o'(O')\right) \cdot p_0(\eta^{-1}O \cap \eta^{-1}O')\end{array}\right)}{\sum_{O,O'} o(O) \cdot o'(O') \cdot p_0(\eta^{-1}O \cap \eta^{-1}O')} \tag{G.229}$$

$$= \frac{\sum_{S\cap S'=S''}(\eta^{\sim 1}o)(S) \cdot (\eta^{\sim 1}o')(S') \cdot \alpha_0(S,S')}{\sum_{S,S'}(\eta^{\sim 1}o)(S) \cdot (\eta^{\sim 1}o')(S') \cdot \alpha_0(T,T')} \tag{G.230}$$

$$= \frac{\sum_{S\cap S'=S''}(\eta^{\sim 1}o)(S) \cdot (\eta^{\sim 1}o')(S') \cdot \alpha_0(S,S')}{\alpha_0(\eta^{\sim 1}o, \eta^{\sim 1}o')} \tag{G.231}$$

$$= \eta^{\sim 1}o *_0 \eta^{\sim 1}o' \tag{G.232}$$

as claimed.

G.17 MHT HYPOTHESIS PROBABILITIES

We are to establish (10.115) and (10.116):

$$p_0 = \frac{1}{1+\sum_{\theta'\neq 0} K^{n_{\theta'}} \cdot \ell_{\theta'} \cdot \left(\prod_{\mathbf{z}\in W_{\theta'}} c(\mathbf{z})^{-1}\right)} \tag{G.233}$$

$$p_\theta = \frac{K^{n_\theta} \cdot \ell_\theta \cdot \left(\prod_{\mathbf{z}\in W_\theta} c(\mathbf{z})^{-1}\right)}{1+\sum_{\theta'\neq 0} K^{n_{\theta'}} \cdot \ell_{\theta'} \cdot \left(\prod_{\mathbf{z}\in W_{\theta'}} c(\mathbf{z})^{-1}\right)}. \tag{G.234}$$

where

$$K \triangleq \frac{p_D}{\lambda(1-p_D)} \tag{G.235}$$

$$d_\theta^2 \triangleq \sum_{\theta(i)>0} (\mathbf{z}_{\theta(i)} - H\mathbf{x}_i)^T (R + HP_i' H^T)^{-1} (\mathbf{z}_{\theta(i)} - H\mathbf{x}_i) \tag{G.236}$$

$$\ell_\theta \triangleq \frac{1}{\prod_{\theta(i)>0} \sqrt{\det 2\pi S_i^{n_\theta/2}}} \cdot \exp\left(-\frac{1}{2} d_\theta^2\right). \tag{G.237}$$

First note that under current assumptions the likelihood function of (10.53) must have the form

$$f(Z|\mathbf{x}_1, ..., \mathbf{x}_n, \theta) = \overbrace{\prod_{i:\theta(i)>0} p_D f(\mathbf{z}_{\theta(i)}|\mathbf{x}_i)}^{\text{target measurements}} \tag{G.238}$$

$$\cdot \overbrace{\prod_{i:\theta(i)=0} (1-p_D)}^{\text{missed detections}} \tag{G.239}$$

$$\overbrace{\cdot e^{-\lambda} \prod_{\mathbf{z}\in Z-W_\theta} \lambda c(\mathbf{z})}^{\text{false alarms}} \tag{G.240}$$

and so

$$f(Z|\mathbf{x}_1,...,\mathbf{x}_n,\theta) = e^{-\lambda}\lambda^{m-n_\theta}p_D^{n_\theta}(1-p_D)^{n-n_\theta} \tag{G.241}$$
$$\cdot\left(\prod_{i:\theta(i)>0} f(\mathbf{z}_{\theta(i)}|\mathbf{x}_i)\right) \tag{G.242}$$
$$\cdot\left(\prod_{\mathbf{z}\in Z-W_\theta} c(\mathbf{z})\right). \tag{G.243}$$

If $\theta = 0$ then $n_\theta = n_0 = 0$ and $W_\theta = W_0 = \emptyset$ and so

$$f(Z|\mathbf{x}_1,...,\mathbf{x}_n,0) = e^{-\lambda}\lambda^m(1-p_D)^n \cdot \prod_{\mathbf{z}\in Z} c(\mathbf{z}). \tag{G.244}$$

Abbreviate $S_i \overset{\text{abbr.}}{=} R + HP_i'H^T$ and $\tilde{\mathbf{z}}_i = \mathbf{z}_{\theta(i)} - H\mathbf{x}_i'$ for $i = 1,..,n$. Then for $\theta \neq 0$ (10.57) becomes

$$p_\theta \propto \int f(\mathbf{z}_1,...,\mathbf{z}_n|\mathbf{x}_1,...,\mathbf{x}_n,\theta) \tag{G.245}$$
$$\cdot f_1(\mathbf{x})\cdots f_n(\mathbf{x})d\mathbf{x}_1\cdots d\mathbf{x}_n \tag{G.246}$$
$$= e^{-\lambda}\lambda^{m-n_\theta}p_D^{n_\theta}(1-p_D)^{n-n_\theta} \tag{G.247}$$
$$\cdot\left(\prod_{i:\theta(i)>0} N_{S_i}(\tilde{\mathbf{z}}_{\theta(i)})\right)\cdot\left(\prod_{\mathbf{z}\in Z-W_\theta} c(\mathbf{z})\right) \tag{G.248}$$

and so

$$p_\theta \propto e^{-\lambda}\lambda^{m-n_\theta}p_D^{n_\theta}(1-p_D)^{n-n_\theta} \tag{G.249}$$
$$\cdot\frac{1}{\prod_{\theta(i)>0}^{n_\theta}\sqrt{\det 2\pi S_i^{n_\theta/2}}} \tag{G.250}$$
$$\cdot\exp\left(-\frac{1}{2}\sum_{i:\theta(i)>0}\tilde{\mathbf{z}}_i^T S_i^{-1}\tilde{\mathbf{z}}_i\right)\cdot\left(\prod_{\mathbf{z}\in Z-W_\theta} c(\mathbf{z})\right) \tag{G.251}$$

and so

$$p_\theta \propto e^{-\lambda}\lambda^{m-n_\theta}p_D^{n_\theta}(1-p_D)^{n-n_\theta} \cdot \frac{1}{\prod_{\theta(i)>0}^{n_\theta}\sqrt{\det 2\pi S_i^{n_\theta/2}}} \tag{G.252}$$

$$\cdot \exp\left(-\frac{1}{2}d_\theta^2\right) \cdot \left(\prod_{\mathbf{z}\in Z-W_\theta} c(\mathbf{z})\right) \tag{G.253}$$

$$= \frac{e^{-\lambda}\lambda^{m-n_\theta}p_D^{n_\theta}(1-p_D)^{n-n_\theta}}{\prod_{\theta(i)>0}^{n_\theta}\sqrt{\det 2\pi S_i^{n_\theta/2}}} \tag{G.254}$$

$$\cdot \exp\left(-\frac{1}{2}d_\theta^2\right) \cdot \left(\prod_{\mathbf{z}\in Z-W_\theta} c(\mathbf{z})\right) \tag{G.255}$$

$$= e^{-\lambda}\lambda^{m-n_\theta}p_D^{n_\theta}(1-p_D)^{n-n_\theta} \cdot \left(\prod_{\mathbf{z}\in Z-W_\theta} c(\mathbf{z})\right) \cdot \ell_\theta. \tag{G.256}$$

On the other hand, for $\theta = 0$ we get:

$$p_0 \propto \int f(\mathbf{z}_1,...,\mathbf{z}_n|\mathbf{x}_1,...,\mathbf{x}_n,0) \cdot f_1(\mathbf{x}) \cdots f_n(\mathbf{x}) \tag{G.257}$$

$$\cdot d\mathbf{x}_1 \cdots d\mathbf{x}_n \tag{G.258}$$

$$= e^{-\lambda}\lambda^m (1-p_D)^n \cdot \prod_{\mathbf{z}\in Z} c(\mathbf{z}) \tag{G.259}$$

Consequently,

$$p_0 \tag{G.260}$$

$$= \frac{\lambda^m (1-p_D)^n \cdot \prod_{\mathbf{z}\in Z} c(\mathbf{z})}{\left(\begin{array}{c} \lambda^m (1-p_D)^n \cdot \prod_{\mathbf{z}\in Z} c(\mathbf{z}) \\ +\sum_{\theta'\neq 0}\lambda^{m-n_{\theta'}}p_D^{n_{\theta'}}(1-p_D)^{n-n_{\theta'}} \cdot \ell_{\theta'} \\ \cdot \left(\prod_{\mathbf{z}\in Z-W_{\theta'}} c(\mathbf{z})\right) \end{array}\right)} \tag{G.261}$$

$$= \frac{1}{1+\sum_{\theta'\neq 0}\left(\frac{p_D}{\lambda(1-p_D)}\right)^{n_{\theta'}} \cdot \ell_{\theta'} \cdot \left(\prod_{\mathbf{z}\in W_{\theta'}} c(\mathbf{z})^{-1}\right)} \tag{G.262}$$

$$= \frac{1}{1+\sum_{\theta'\neq 0}K^{n_{\theta'}} \cdot \ell_{\theta'} \cdot \left(\prod_{\mathbf{z}\in Z-W_{\theta'}} c(\mathbf{z})^{-1}\right)}. \tag{G.263}$$

Similarly, if $\theta \neq 0$ then

$$p_\theta \tag{G.264}$$

$$= \frac{\lambda^{m-n_\theta} p_D^{n_\theta} (1-p_D)^{n-n_\theta} \cdot \ell_\theta \cdot \left(\prod_{\mathbf{z}\in Z-W_\theta} c(\mathbf{z})\right)}{\left(\begin{array}{c} \lambda^m (1-p_D)^n \cdot \prod_{\mathbf{z}\in Z} c(\mathbf{z}) \\ +\sum_{\theta'\neq 0} \lambda^{m-n_{\theta'}} p_D^{n_{\theta'}} (1-p_D)^{n-n_{\theta'}} \\ \cdot \ell_{\theta'} \cdot \left(\prod_{\mathbf{z}\in Z-W_{\theta'}} c(\mathbf{z})\right) \end{array}\right)} \tag{G.265}$$

$$= \frac{K^{n_\theta} \cdot \ell_\theta \cdot \left(\prod_{\mathbf{z}\in W_\theta} c(\mathbf{z})^{-1}\right)}{1+\sum_{\theta'\neq 0} K^{n_{\theta'}} \cdot \ell_\theta \cdot \left(\prod_{\mathbf{z}\in W_{\theta'}} c(\mathbf{z})^{-1}\right)} \tag{G.266}$$

$$= \frac{K^{n_\theta} \cdot \ell_\theta \cdot \left(\prod_{\mathbf{z}\in W_{\theta'}} c(\mathbf{z})^{-1}\right)}{1+\sum_{\theta'\neq 0} K^{n_{\theta'}} \cdot \ell_{\theta'} \cdot \left(\prod_{\mathbf{z}\in W_{\theta'}} c(\mathbf{z})^{-1}\right)}. \tag{G.267}$$

G.18 LIKELIHOOD FOR STANDARD MEASUREMENT MODEL

We are to prove (12.139):

$$f_{k+1}(Z|X) = e^\lambda f_C(Z) \cdot f_{k+1}(\emptyset|X) \tag{G.268}$$

$$\cdot \sum_\theta \prod_{i:\theta(i)>0} \frac{p_D(\mathbf{x}_i) \cdot f_{k+1}(\mathbf{z}_{\theta(i)}|\mathbf{x}_i)}{(1-p_D(\mathbf{x}_i)) \cdot \lambda c(\mathbf{z}_{\theta(i)})} \tag{G.269}$$

where

$$f_C(Z) = e^{-\lambda} \prod_{\mathbf{z}\in Z} \lambda c(\mathbf{z}), \quad f_{k+1}(\emptyset|X) = e^{-\lambda} \prod_{\mathbf{x}\in X} (1-p_D(\mathbf{x})). \tag{G.270}$$

Begin by noting that the belief-mass function of the model is

$$\beta_{k+1}(T|X) = \Pr(\Upsilon(X) \cup C \subseteq T|X) \tag{G.271}$$

$$= \Pr(\Upsilon(X) \subseteq T, C \subseteq T|X) \tag{G.272}$$

$$= \Pr(\Upsilon(X) \subseteq T|X) \cdot \Pr(C \subseteq T) \tag{G.273}$$

$$= \beta_{\Upsilon(X)}(T) \cdot \beta_C(T). \tag{G.274}$$

From the general product rule (11.270) we get

$$\frac{\delta\beta_{k+1}}{\delta Z}(T|X) = \sum_{W\subseteq Z} \frac{\delta\beta_{\Upsilon(X)}}{\delta W}(T) \cdot \frac{\delta\beta_C}{\delta(Z-W)}(T). \tag{G.275}$$

Setting $T = \emptyset$, this becomes

$$\begin{aligned} f_{k+1}(Z|X) &= \frac{\delta\beta_{k+1}}{\delta Z}(\emptyset|X) \\ &= \sum_{W\subseteq Z} f_{\Upsilon(X)}(W) \cdot f_C(Z-W) \end{aligned} \tag{G.276}$$

where $f_{\Upsilon(X)}(Z)$ and $f_C(Z)$ are the probability density functions of $\Upsilon(X)$ and C, respectively.

From (12.51) we know that

$$f_C(Z-W) = e^{-\lambda} \prod_{\mathbf{z}\in Z-W} \lambda c(\mathbf{z}). \tag{G.277}$$

From (12.120) we know that if $W = \{\mathbf{w}_1, ..., \mathbf{w}_e\}$ with $|W| = e$ then $f_{\Upsilon(X)}(W) = 0$ if $e > n$ and, otherwise,

$$f_{\Upsilon(X)}(W) = f_{\Upsilon(X)}(\emptyset) \sum_{1\leq i_1 \neq ... \neq i_e \leq n} \prod_{j=1}^{e} \frac{p_D(\mathbf{x}_{i_j}) \cdot f_{k+1}(\mathbf{w}_j|\mathbf{x}_{i_j})}{1-p_D(\mathbf{x}_{i_j})} \tag{G.278}$$

where

$$f_{\Upsilon(X)}(\emptyset) = \prod_{\mathbf{x}\in X} (1-p_D(\mathbf{x})). \tag{G.279}$$

We rewrite (G.278) in a more useful form. Given any e-tuple $(i_1, ..., i_e)$ with $1 \leq i_1 \neq ... \neq i_e \leq n$. Define the function $\tau : W \to X$ by $\tau(\mathbf{w}_j) = \mathbf{x}_{i_j}$ for all $j = 1, ...e$. Then τ is a one-to-one function, and any such function $\tau : W \to X$ defines an e-tuple $(i_1, ..., i_e)$ with $1 \leq i_1 \neq ... \neq i_e \leq n$. Thus (G.278) can be rewritten as

$$f_{\Upsilon(X)}(W) = f_{\Upsilon(X)}(\emptyset) \sum_{\tau:W\to X} \prod_{\mathbf{z}\in W} \frac{p_D(\tau(\mathbf{z})) \cdot f_{k+1}(\mathbf{z}|\tau(\mathbf{z}))}{1-p_D(\tau(\mathbf{z}))} \tag{G.280}$$

where the summation is taken over all one-to-one functions $\tau : W \to X$.

Substituting (G.277) and (G.280) into (G.276) we get

$$
\begin{aligned}
& f_{k+1}(Z|X) && \text{(G.281)} \\
= & \sum_{W\subseteq Z:|W|\leq n} f_{\Upsilon(X)}(W)\cdot f_C(Z-W) && \text{(G.282)} \\
= & e^{-\lambda} f_{\Upsilon(X)}(\emptyset) && \text{(G.283)} \\
& \cdot \sum_{W\subseteq Z:|W|\leq n} \left(\sum_{\tau:W\to X} \prod_{\mathbf{z}\in W} \frac{p_D(\tau(\mathbf{z}))\cdot f_{k+1}(\mathbf{z}|\tau(\mathbf{z}))}{1-p_D(\tau(\mathbf{z}))} \right) && \text{(G.284)} \\
& \cdot \left(\prod_{\mathbf{z}\in Z-W} \lambda c(\mathbf{z}) \right) && \text{(G.285)} \\
= & f_C(Z) f_{\Upsilon(X)}(\emptyset) \sum_{W\subseteq Z:|W|\leq n} && \text{(G.286)} \\
& \sum_{\tau:W\to X} \prod_{\mathbf{z}\in W} \frac{p_D(\tau(\mathbf{z}))\cdot f_{k+1}(\mathbf{z}|\tau(\mathbf{z}))}{(1-p_D(\tau(\mathbf{z})))\cdot \lambda c(\tau(\mathbf{z}))}. && \text{(G.287)}
\end{aligned}
$$

From the general product rule (11.270),

$$
\begin{aligned}
f_{k+1}(\emptyset|X) &= \frac{\delta\beta_{\Upsilon(X)}}{\delta\emptyset}(\emptyset)\cdot\frac{\delta\beta_C}{\delta\emptyset}(\emptyset) = \beta_{\Upsilon(X)}(\emptyset)\cdot\beta_C(\emptyset) && \text{(G.288)} \\
&= f_{\Upsilon(X)}(\emptyset)\cdot e^{-\lambda} && \text{(G.289)}
\end{aligned}
$$

and thus

$$
\begin{aligned}
f_{k+1}(Z|X) &= e^{\lambda} f_C(Z) f_{k+1}(\emptyset|X) \sum_{W\subseteq Z:|W|\leq n} && \text{(G.290)} \\
& \sum_{\tau:W\to X} \prod_{\mathbf{z}\in W} \frac{p_D(\tau(\mathbf{z}))\cdot f_{k+1}(\mathbf{z}|\tau(\mathbf{z}))}{(1-p_D(\tau(\mathbf{z})))\cdot \lambda c(\tau(\mathbf{z}))}. && \text{(G.291)}
\end{aligned}
$$

Now, let $Z\cup\{\phi\}$ be the set obtained by appending an additional element ϕ to Z. Then for each choice of W with $|W|\leq n$ and each choice of a one-to-one function $\tau: W\to X$, we get a function $\gamma_{W,\tau}: X\to Z\cup\{\phi\}$ defined by $\gamma_{W,\tau}(\mathbf{x})=\mathbf{z}$ if there is a $\mathbf{z}$ with $\tau(\mathbf{z})=\mathbf{x}$, and $\gamma_{W,\tau}(\mathbf{x})=\phi$ otherwise. The set of all $\gamma_{W,\tau}(\mathbf{x})$ such $\gamma_{W,\tau}(\mathbf{x})\neq\phi$ is W, and $\gamma_{W,\tau}(\tau(\mathbf{z}))=\mathbf{z}$ if $\mathbf{z}\in W$.

Each $\gamma_{W,t}$ comes about in this fashion. Thus we can write

$$f_{k+1}(Z|X) = e^{\lambda} f_C(Z) f_{k+1}(\emptyset|X) \tag{G.292}$$

$$\cdot \sum_{\gamma:X\to Z\cup\{\phi\}} \prod_{\mathbf{x}:\gamma(\mathbf{x})\neq\phi} \frac{p_D(\mathbf{x})\cdot f_{k+1}(\gamma(\mathbf{z})|\mathbf{x})}{(1-p_D(\mathbf{x}))\cdot\lambda c(\gamma(\mathbf{z}))} \tag{G.293}$$

where the summation is taken over all functions $\gamma : X \to Z \cup \{\phi\}$ such that if $\gamma(\mathbf{x}) = \gamma(\mathbf{x}') \neq \phi$ then $\mathbf{x} = \mathbf{x}'$.

Finally, choose some specific order $\mathbf{z}_1, ..., \mathbf{z}_m$ for the elements of Z. Then (30) can be further rewritten as

$$f_{k+1}(Z|X) = e^{\lambda} f_C(Z) f_{k+1}(\emptyset|X) \sum_{\theta} \prod_{i:\theta(i)>0} \frac{p_D(\mathbf{x}_i)\cdot f_{k+1}(\mathbf{z}_{\theta(i)}|\mathbf{x}_i)}{(1-p_D(\mathbf{x}_i))\cdot\lambda c(\mathbf{z}_{\theta(i)})} \tag{G.294}$$

where the summation is taken over all association hypotheses as defined in Section 10.5.4. This establishes the result.

G.19 P.G.FL. FOR STANDARD MEASUREMENT MODEL

We are to prove (12.147)-(12.151). We prove the five cases each in turn:

Case I: No targets are present. We are to show that $G_{k+1}[g|\emptyset] = e^{\lambda c[g]-\lambda}$. However,

$$G_{k+1}[g|\emptyset] = \int g^Z \cdot f_{k+1}(Z|\emptyset)\delta Z \tag{G.295}$$

$$= e^{-\lambda} \sum_{m=0}^{\infty} \frac{1}{m!} \int g(\mathbf{z}_1)\cdots g(\mathbf{z}_m) \tag{G.296}$$

$$\cdot\lambda^m \cdot c(\mathbf{z}_1)\cdots c(\mathbf{z}_m) d\mathbf{z}_1 \cdots d\mathbf{z}_m \tag{G.297}$$

$$= e^{-\lambda} \sum_{m=0}^{\infty} \frac{\lambda^m c[g]^m}{m!} = e^{-\lambda}\cdot e^{\lambda c[g]} = e^{\lambda c[g]-\lambda}. \tag{G.298}$$

Case II: A single target is present. We are to show that $G_{k+1}[g|\mathbf{x}] = (1 - p_D(\mathbf{x}) + p_D(\mathbf{x}) \cdot p_g(\mathbf{x})) \cdot e^{\lambda c[g]-\lambda}$. Since false alarms and target-generated measurements are conditionally independent of state, from (11.166) we know that

$$G_{k+1}[g|X] = G_{\text{targ}}[g|X] \cdot G_{\text{FA}}[g] \tag{G.299}$$

where, from the previous case, we know that $G_{\text{FA}}[g] = e^{\lambda c[g]-\lambda}$. So it is enough to derive $G_{\text{targ}}[g|X]$. In this case we can assume that there are no false alarms. There are only two possibilities: $Z = \emptyset$ or $|Z| = 1$. Thus

$$\begin{aligned} G_{\text{targ}}[g|\mathbf{x}] &= \int g^Z \cdot f_{k+1}(Z|\mathbf{x})\delta Z && \text{(G.300)} \\ &= f_{k+1}(\emptyset|\mathbf{x}) + \int g(\mathbf{z}) \cdot f_{k+1}(\{\mathbf{z}\}|\mathbf{x})d\mathbf{z} && \text{(G.301)} \\ &= 1 - p_D(\mathbf{x}) + \int g(\mathbf{z}) \cdot p_D(\mathbf{x}) \cdot f_{k+1}(\mathbf{z}|\mathbf{x})d\mathbf{z} && \text{(G.302)} \\ &= 1 - p_D(\mathbf{x}) + p_D(\mathbf{x}) \cdot \int g(\mathbf{z}) \cdot f_{k+1}(\mathbf{z}|\mathbf{x})d\mathbf{z} && \text{(G.303)} \\ &= 1 - p_D(\mathbf{x}) + p_D(\mathbf{x}) \cdot p_g(\mathbf{x}). && \text{(G.304)} \end{aligned}$$

Thus we are done.

Case III: No missed detections or false alarms. We are to show that $G_{k+1}[g|X] = p_g^X$. In this case the random measurement set

$$\Sigma_{k+1} = \Upsilon(\mathbf{x}_1) \cup ... \cup \Upsilon(\mathbf{x}_n) \tag{G.305}$$

is the union of n statistically independent random singletons $\Upsilon(\mathbf{x}_1), ..., \Upsilon(\mathbf{x}_n)$. From the previous proof we know that the p.g.fl. of $\Upsilon(\mathbf{x})$ is $G_{\Upsilon(\mathbf{x})}[g] = p_g(\mathbf{x})$. From (11.166) we know that the p.g.fl. of a union of independent processes is the product of their respective p.g.fl.s. Thus

$$G_{k+1}[g|X] = G_{\Upsilon(\mathbf{x}_1)}[g] \cdots G_{\Upsilon(\mathbf{x}_n)}[g] = p_g(\mathbf{x}_1) \cdots p_g(\mathbf{x}_n) = p_g^X. \tag{G.306}$$

Case IV: Missed detections, no false alarms. We are to show that $G_{k+1}[g|X] = (1 - p_D + p_D p_g)^X$. As in the previous section,

$$\Sigma_{k+1} = \Upsilon(\mathbf{x}_1) \cup ... \cup \Upsilon(\mathbf{x}_n) \tag{G.307}$$

is the union of n statistically independent random singletons $\Upsilon(\mathbf{x}_1), ..., \Upsilon(\mathbf{x}_n)$. In this case, $G_{\Upsilon(\mathbf{x})}[g] = 1 - p_D(\mathbf{x}) + p_D(\mathbf{x}) \cdot p_g(\mathbf{x})$. Thus

$$\begin{aligned} G_{k+1}[g|X] &= G_{\Upsilon(\mathbf{x}_1)}[g] \cdots G_{\Upsilon(\mathbf{x}_n)}[g] && \text{(G.308)} \\ &= (1 - p_D(\mathbf{x}_1) + p_D(\mathbf{x}_1) \cdot p_g(\mathbf{x}_1)) && \text{(G.309)} \\ &\quad \cdots (1 - p_D(\mathbf{x}_n) + p_D(\mathbf{x}_n) \cdot p_g(\mathbf{x}_n)) && \text{(G.310)} \end{aligned}$$

$$= (1 - p_D + p_D p_g)(\mathbf{x}_1) \cdots (1 - p_D + p_D p_g)(\mathbf{x}_n) \tag{G.311}$$

$$= (1 - p_D + p_D p_g)^X . \tag{G.312}$$

Case V: Missed detections and false alarms. We are to show that $G_{k+1}[g|X] = (1 - p_D + p_D p_g)^X \cdot e^{\lambda c[g] - \lambda}$. However, since false alarms and target-generated measurements are conditionally independent of state, from (11.166) we know that

$$G_{k+1}[g|X] = G_{\text{targets}}[g|X] \cdot G_{\text{false alarms}}[g]. \tag{G.313}$$

Combining the results for Case I and Case IV, we get the claimed result.

G.20 MULTISENSOR MULTITARGET LIKELIHOODS

We are to verify (12.294). That is, we are to show that if sensor observations are conditionally independent of state, then the multitarget likelihood function $f_{k+1}(Z|X)$ for all sensors is just the product of the multitarget likelihood functions $\overset{1}{f}_{k+1}(\overset{1}{Z}|X), ..., \overset{s}{f}_{k+1}(\overset{s}{Z}|X)$ of the individual sensors:

$$f_{k+1}(Z|X) = \overset{1}{f}_{k+1}(\overset{1}{Z}|X) \cdots \overset{s}{f}_{k+1}(\overset{s}{Z}|X). \tag{G.314}$$

The measurement space for the multisensor problem is a disjoint union

$$\mathfrak{Z}_0 = \overset{1}{\mathfrak{Z}}_0 \uplus ... \uplus \overset{s}{\mathfrak{Z}}_0 \tag{G.315}$$

where $\overset{j}{\mathfrak{Z}}_0$ is the measurement space for the jth sensor. Thus if $T \subseteq \mathfrak{Z}_0$ is any subset of $\mathfrak{Z}_0$, it must have the form

$$T = \overset{1}{T} \uplus ... \uplus \overset{s}{T} \tag{G.316}$$

where $\overset{s}{T} = T \cap \overset{j}{\mathfrak{Z}}_0$ for $j = 1, ..., s$. In particular, the random observation set supplied by all sensors will be a disjoint union

$$\Sigma = \overset{1}{\Sigma} \uplus ... \uplus \overset{s}{\Sigma} \tag{G.317}$$

where $\overset{j}{\Sigma} = \Sigma \cap \overset{j}{\mathfrak{Z}}_0$ is the random observation set supplied by the jth sensor.

The belief measure of Σ is

$$\begin{aligned}
\beta_\Sigma(T|X) &= \Pr(\Sigma \subseteq T|X) && \text{(G.318)}\\
&= \Pr(\overset{1}{\Sigma} \uplus ... \uplus \overset{s}{\Sigma} \subseteq \overset{1}{T} \uplus ... \uplus \overset{s}{T}|X) && \text{(G.319)}\\
&= \Pr(\overset{1}{\Sigma} \subseteq \overset{1}{T}, ..., \overset{s}{\Sigma} \subseteq \overset{s}{T}|X) && \text{(G.320)}\\
&= \Pr(\overset{1}{\Sigma} \subseteq \overset{1}{T}|X) \cdots \Pr(\overset{s}{\Sigma} \subseteq \overset{s}{T}|X) && \text{(G.321)}\\
&= \overset{1}{\beta}(\overset{1}{T}|X) \cdots \overset{s}{\beta}(\overset{s}{T}|X) && \text{(G.322)}
\end{aligned}$$

where

$$\begin{aligned}
\overset{j}{\beta}(\overset{j}{T}|X) &\triangleq \Pr(\overset{j}{\Sigma} \subseteq \overset{j}{T}|X) = \Pr(\overset{j}{\Sigma} \subseteq T|X) && \text{(G.323)}\\
&\triangleq \overset{j}{\beta}(T|X) && \text{(G.324)}
\end{aligned}$$

for $j = 1, ..., s$. Now let $\overset{j}{\mathbf{z}} \in \overset{j}{\mathfrak{Z}}_0$ and let $E_{\overset{j}{\mathbf{z}}} \subseteq \overset{j}{\mathfrak{Z}}_0$ be an arbitrarily small neighborhood of $\overset{j}{\mathbf{z}}$. Then

$$\frac{\delta \overset{j}{\beta}}{\delta \overset{i}{\mathbf{z}}}(T|X) = \lim_{|\overset{j}{E}_{\overset{i}{\mathbf{z}}}| \searrow 0} \frac{\overset{j}{\beta}(T \cup \overset{i}{E}_{\overset{i}{\mathbf{z}}}|X) - \overset{j}{\beta}(T|X)}{|\overset{j}{E}_{\overset{i}{\mathbf{z}}}|}. \tag{G.325}$$

However, this vanishes identically if $i \neq j$. To see this, note that

$$\begin{aligned}
\overset{j}{\beta}(T \cup E_{\overset{i}{\mathbf{z}}}|X) &= \Pr(\overset{j}{\Sigma} \subseteq (T \cup E_{\overset{i}{\mathbf{z}}}) \cap \overset{j}{\mathfrak{Z}}_0|X) && \text{(G.326)}\\
&= \Pr(\overset{j}{\Sigma} \subseteq \overset{j}{T} \cup (E_{\overset{i}{\mathbf{z}}} \cap \overset{j}{\mathfrak{Z}}_0)|X) && \text{(G.327)}\\
&= \Pr(\overset{j}{\Sigma} \subseteq \overset{j}{T} \cup \emptyset|X) = \overset{j}{\beta}(T|X). && \text{(G.328)}
\end{aligned}$$

If $i = j$, on the other hand,

$$\frac{\delta \overset{j}{\beta}}{\delta \overset{j}{\mathbf{z}}}(T|X) = \frac{\delta \overset{j}{\beta}}{\delta \overset{j}{\mathbf{z}}}(\overset{j}{T}|X). \tag{G.329}$$

So,

$$\frac{\delta \beta}{\delta \overset{j}{\mathbf{z}}}(T|X) = \overset{1}{\beta}(\overset{1}{T}|X) \cdots \frac{\delta \overset{j}{\beta}}{\delta \overset{j}{\mathbf{z}}}(\overset{j}{T}|X) \cdots \overset{s}{\beta}(\overset{s}{T}|X) \tag{G.330}$$

and thus

$$\frac{\delta^m \beta}{\delta \overset{j}{Z}}(T|X) = \overset{1}{\beta}(\overset{1}{T}|X)\cdots\frac{\delta \overset{j}{\beta}}{\delta \overset{j}{Z}}(\overset{j}{T}|X)\cdots\overset{s}{\beta}(\overset{s}{T}|X). \tag{G.331}$$

So in general,

$$\frac{\delta^m \beta}{\delta Z}(T|X) = \frac{\delta \overset{1}{\beta}}{\delta \overset{1}{Z}}(\overset{1}{T}|X)\cdots\frac{\delta \overset{s}{\beta}}{\delta \overset{s}{Z}}(\overset{s}{T}|X). \tag{G.332}$$

Thus, as claimed,

$$f_{k+1}(Z|X) = \frac{\delta^m \beta}{\delta Z}(\emptyset|X) \tag{G.333}$$

$$= \frac{\delta \overset{1}{\beta}}{\delta \overset{1}{Z}}(\emptyset|X)\cdots\frac{\delta \overset{s}{\beta}}{\delta \overset{s}{Z}}(\emptyset|X) \tag{G.334}$$

$$= \overset{1}{f}_{k+1}(\overset{1}{Z}|X)\cdots\overset{s}{f}_{k+1}(\overset{s}{Z}|X). \tag{G.335}$$

G.21 CONTINUITY OF LIKELIHOODS FOR UNRESOLVED TARGETS

We are to verify (12.290) and (12.291):

$$\lim_{a\searrow 0} f_{k+1}(Z|a,\mathbf{x}) = f_{k+1}(Z|\emptyset) \tag{G.336}$$

$$\lim_{a\searrow 0} f_{k+1}(Z|\mathring{X}\cup\{(a,\mathbf{x})\}) = f_{k+1}(Z|\mathring{X}). \tag{G.337}$$

We first note that

$$\lim_{a\searrow 0} f(Z|a,\mathbf{x}) = \left\{\begin{array}{ll} 1 & \text{if} \quad Z=\emptyset \\ 0 & \text{if} \quad Z\neq\emptyset \end{array}\right. . \tag{G.338}$$

For on the one hand,

$$\lim_{a\searrow 0} f(\emptyset|a,\mathbf{x}) = \lim_{a\searrow 0} B_{a,q}(0) \tag{G.339}$$

$$= \lim_{a\searrow 0} G_{a,q}(0) \tag{G.340}$$

$$= \lim_{a\searrow 0} \prod_{i=0}^{\infty}(1-\sigma_i(a)q) \tag{G.341}$$

$$= \prod_{i=0}^{\infty}(1-\sigma(-i)q) = \prod_{i=0}^{\infty} 1 = 1. \tag{G.342}$$

On the other hand, if $Z \neq \emptyset$

$$\lim_{a \searrow 0} f(Z|a, \mathbf{x}) = \lim_{a \searrow 0} m! B_{a,q}(m) \prod_{\mathbf{z} \in Z} f_{k+1}(\mathbf{z}|\mathbf{x}) \tag{G.343}$$

$$= m! \left(\lim_{a \searrow 0} B_{a,q}(m) \right) \prod_{\mathbf{z} \in Z} f_{k+1}(\mathbf{z}|\mathbf{x}) \tag{G.344}$$

$$\lim_{a \searrow 0} B_{a,q}(m) = \lim_{a \searrow 0} \left[\frac{1}{m!} \frac{d^m}{dy^m} G_{a,q}(y) \right]_{y=0} \tag{G.345}$$

$$= \lim_{a \searrow 0} \left[\frac{1}{m!} \frac{d^m}{dy^m} (1 - \sigma(a)q + \sigma(a)qy) \right]_{y=0} \tag{G.346}$$

$$= \lim_{a \searrow 0} \sigma(a)q = 0 \quad \text{if } m = 1 \tag{G.347}$$

$$= 0 \text{ identically if } m > 1. \tag{G.348}$$

Now turn to (G.336). We have

$$\lim_{a \searrow 0} f_{k+1}(Z|a, \mathbf{x}) = \sum_{W \subseteq Z} \lim_{a \searrow 0} f(W|a, \mathbf{x}) \cdot f_C(Z - W) \tag{G.349}$$

$$= \lim_{a \searrow 0} f(\emptyset|a, \mathbf{x}) \cdot f_C(Z) \tag{G.350}$$

$$+ \sum_{W \subseteq Z, W \neq \emptyset} \lim_{a \searrow 0} f(W|a, \mathbf{x}) \cdot f_C(Z - W) \tag{G.351}$$

$$= 1 \cdot f_C(Z) + \sum_{W \subseteq Z, W \neq \emptyset} 0 \cdot f_C(Z - W) \tag{G.352}$$

$$= f_C(Z) = f_{k+1}(Z|\emptyset). \tag{G.353}$$

Finally turn to (G.337). Assume that $(a, \mathbf{x}) \notin \mathring{X}$. Then, as claimed,

$$\lim_{a \searrow 0} f_{k+1}(Z|\mathring{X} \cup \{(a, \mathbf{x})\}) \tag{G.354}$$

$$= \lim_{a \searrow 0} \sum_{W_1 \uplus W_2 \uplus W_3 = Z} f(W_1|\mathring{X}) \cdot f(W_2|a, \mathbf{x}) \cdot f_C(W_3) \tag{G.355}$$

$$= \sum_{W_1 \uplus W_2 \uplus W_3 = Z} f(W_1|\mathring{X}) \cdot \lim_{a \searrow 0} f(W_2|a, \mathbf{x}) \cdot f_C(W_3) \tag{G.356}$$

$$= \sum_{W_1 \uplus W_3 = Z} f(W_1|\mathring{X}) \cdot \lim_{a \searrow 0} f(\emptyset|a, \mathbf{x}) \cdot f_C(W_3) \tag{G.357}$$

$$+ \sum_{W_1 \uplus W_2 \uplus W_3 = Z, W_2 = \emptyset} f(W_1|\mathring{X}) \cdot \lim_{a \searrow 0} f(W_2|a, \mathbf{x}) \cdot f_C(W_3) \tag{G.358}$$

$$= \sum_{W_1 \uplus W_3 = Z} f(W_1|\mathring{X}) \cdot 1 \cdot f_C(W_3) \tag{G.359}$$

$$+ \sum_{W_1 \uplus W_2 \uplus W_3 = Z, W_2 = \emptyset} f(W_1|\mathring{X}) \cdot 0 \cdot f_C(W_3) \tag{G.360}$$

$$= \sum_{W_1 \uplus W_3 = Z} f(W_1|\mathring{X}) \cdot f_C(W_3) \tag{G.361}$$

$$= f_{k+1}(Z|\mathring{X}). \tag{G.362}$$

G.22 ASSOCIATION FOR FUZZY DEMPSTER-SHAFER

We are to prove (10.151) and (10.152):

$$\ell(o|\mu) = \sum_{i=1}^{e} \sum_{j=1}^{d} \omega_{i,j} \cdot o_j \cdot \mu_i \cdot \hat{N}_{C_j + HD_iH^T}(H\mathbf{x}'_i - \mathbf{z}_j) \tag{G.363}$$

where

$$\omega_{i,j} \triangleq \sqrt{\frac{\det 2\pi C_j}{\det 2\pi \left(C_j + HD_iH^T\right)}} \cdot \frac{\sqrt{\det 2\pi D_i}}{\sum_{i'=1}^{e} \mu_{i'} \cdot \sqrt{\det 2\pi D_{i'}}}. \tag{G.364}$$

Let o be a fuzzy DS measurement in the sense of Section 5.6. That is, it assigns a number $o(g) \geq 0$ to every fuzzy membership function $g(\mathbf{z})$ on measurement space $\mathfrak{Z}_0$, with $o(g) > 0$ for only a finite number of g and such that $\sum_g o(g) = 1$. The likelihood for a fuzzy DS measurement o was given by (5.73):

$$f(o|\mathbf{x}) = \sum_g o(g) \cdot g(\eta(\mathbf{x}')). \tag{G.365}$$

By assumption (see Section 5.6.3) $\eta(\mathbf{x}') = H\mathbf{x}'$ and $g_j(\mathbf{z}) = \hat{N}_{C_j}(\mathbf{z} - \mathbf{z}_j)$ for the focal fuzzy subsets $g_1, ..., g_m$ of $o(g)$. Recall that the normalized Gaussian functions $\hat{N}_C(\mathbf{z})$ were defined in (5.207). Thus

$$f(o|\mathbf{x}) = \sum_{j=1}^{d} o_j \cdot \hat{N}_{C_j}(H\mathbf{x}' - \mathbf{z}_j) \tag{G.366}$$

where $o_j \triangleq o(g_j)$.

On the other hand, let μ be a fuzzy DS state in the sense of Section 5.6. It assigns a number $\mu(f) \geq 0$ to every fuzzy membership function $f(\mathbf{x}')$ on the state space $\mathfrak{X}_0$, with $\mu(f) > 0$ for only a finite number of f and such that $\sum_f \mu(f) = 1$. By assumption (see Section 5.6.3) $f_i(\mathbf{x}') = \hat{N}_{D_i}(\mathbf{x}' - \mathbf{x}'_i)$ for the focal fuzzy subsets $f_1, ..., f_n$ of $\mu(f)$, and their weights are $\mu_i = \mu(f_i)$. Thus from (5.213) and (5.214) we know that the track distribution for a fuzzy DS state has the form (see Section 5.6.3)

$$f(\mathbf{x}'|i) = f(\mathbf{x}'|\mu) \triangleq \frac{1}{|\mu|} \sum_f \mu(f) \cdot f(\mathbf{x}') \tag{G.367}$$

$$= \frac{1}{|\mu|} \sum_{i=1}^{e} \mu_i \cdot \hat{N}_{D_i}(\mathbf{x}' - \mathbf{x}'_i) \tag{G.368}$$

where

$$|\mu| \triangleq \sum_f \mu(f) \cdot \int f(\mathbf{x}) d\mathbf{x} = \sum_{i=1}^{e} \mu_i \cdot \sqrt{\det 2\pi D_i}. \tag{G.369}$$

Thus (10.149) becomes

$$\ell(o|i) \triangleq \int f(o|\mathbf{x}) \cdot f(\mathbf{x}|\mu) d\mathbf{x} \tag{G.370}$$

$$= \frac{1}{|\mu|} \int \left(\sum_g o(g) \cdot g(H\mathbf{x}) \right) \cdot \left(\sum_f \mu(f) \cdot f(\mathbf{x}) \right) d\mathbf{x} \tag{G.371}$$

or

$$= \frac{1}{|\mu|} \sum_{i=1}^{e} \sum_{j=1}^{d} o_j \cdot \mu_i \cdot \int \hat{N}_{C_j}(H\mathbf{x} - \mathbf{z}_j) \cdot \hat{N}_{D_i}(\mathbf{x} - \mathbf{x}'_i) d\mathbf{x} \tag{G.372}$$

$$= \frac{1}{|\mu|} \sum_{i=1}^{e} \sum_{j=1}^{d} o_j \cdot \mu_i \cdot \sqrt{\det 2\pi C_j} \cdot \sqrt{\det 2\pi D_i} \tag{G.373}$$

$$\cdot \int N_{C_j}(H\mathbf{x}' - \mathbf{z}_j) \cdot N_{D_i}(\mathbf{x} - \mathbf{x}'_i) d\mathbf{x}. \tag{G.374}$$

From the fundamental theorem of Gaussian identities, (D.1), this becomes, as claimed,

$$\ell(o|i) = \frac{1}{|\mu|}\sum_{i=1}^{e}\sum_{j=1}^{d} o_j \cdot \mu_i \cdot \sqrt{\det 2\pi C_j} \cdot \sqrt{\det 2\pi D_i} \tag{G.375}$$

$$\cdot N_{C_j+HD_iH^T}(H\mathbf{x}'_i - \mathbf{z}_j) \tag{G.376}$$

$$= \frac{1}{|\mu|}\sum_{i=1}^{e}\sum_{j=1}^{d} o_j \cdot \mu_i \cdot \sqrt{\frac{\det 2\pi C_j \cdot \det 2\pi D_i}{\det 2\pi\,(C_j + HD_iH^T)}} \tag{G.377}$$

$$\cdot \hat{N}_{C_j+HD_iH^T}(H\mathbf{x}'_i - \mathbf{z}_j) \tag{G.378}$$

$$= \sum_{i=1}^{e}\sum_{j=1}^{d} \omega_{i,j} \cdot o_j \cdot \mu_i \cdot \hat{N}_{C_j+HD_iH^T}(H\mathbf{x}'_i - \mathbf{z}_j) \tag{G.379}$$

where

$$\omega_{i,j} \triangleq \sqrt{\frac{\det 2\pi C_j}{\det 2\pi\,(C_j + HD_iH^T)}} \cdot \frac{\sqrt{\det 2\pi D_i}}{\sum_{i=1}^{e} \mu_i \cdot \sqrt{\det 2\pi D_i}}. \tag{G.380}$$

G.23 JOTT FILTER PREDICTOR

We are to prove (14.195) and (14.196). Equation (14.195) follows by substituting (14.179) and (14.180) into the multitarget predictor integral, (14.14). If $X = \emptyset$ then

$$f_{k+1|k}(\emptyset) = \int f_{k+1|k}(\emptyset|X') \cdot f_{k|k}(X)\delta X' \tag{G.381}$$

$$= f_{k+1|k}(\emptyset|\emptyset) \cdot f_{k|k}(\emptyset) \tag{G.382}$$

$$+ \int f_{k+1|k}(\emptyset|\{\mathbf{x}'\}) \cdot f_{k|k}(\{\mathbf{x}'\})d\mathbf{x}' \tag{G.383}$$

$$= (1 - p_B) \cdot f_{k|k}(\emptyset) \tag{G.384}$$

$$+ \int (1 - p_S(\mathbf{x}')) \cdot f_{k|k}(\{\mathbf{x}'\})d\mathbf{x}'. \tag{G.385}$$

So,

$$1 - p_{k+1|k} = (1 - p_B) \cdot (1 - p_{k|k}) + p_{k|k} \int (1 - p_S(\mathbf{x}')) \cdot f_{k|k}(\mathbf{x}')d\mathbf{x}' \tag{G.386}$$

and thus

$$
\begin{aligned}
& p_{k+1|k} && \text{(G.387)}\\
= \; & 1-(1-p_B)\cdot(1-p_{k|k}) && \text{(G.388)}\\
& -p_{k|k}\int(1-p_S(\mathbf{x}'))\cdot f_{k|k}(\mathbf{x}')d\mathbf{x}' && \text{(G.389)}\\
= \; & p_B+p_{k|k}-p_Bp_{k|k}-p_{k|k}\int(1-p_S(\mathbf{x}'))\cdot f_{k|k}(\mathbf{x}')d\mathbf{x}' && \text{(G.390)}\\
= \; & p_B\cdot(1-p_{k|k})+p_{k|k}\int p_S(\mathbf{x}')\cdot f_{k|k}(\mathbf{x}')d\mathbf{x}'. && \text{(G.391)}
\end{aligned}
$$

Equation (14.196) follows in similar fashion. If $X=\{\mathbf{x}\}$, then

$$
\begin{aligned}
f_{k+1|k}(\{\mathbf{x}\}) &= \int f_{k+1|k}(\{\mathbf{x}\}|X')\cdot f_{k|k}(X)\delta X' && \text{(G.392)}\\
&= f_{k+1|k}(\{\mathbf{x}\}|\emptyset)\cdot f_{k|k}(\emptyset) && \text{(G.393)}\\
&\quad +\int f_{k+1|k}(\{\mathbf{x}\}|\{\mathbf{x}'\})\cdot f_{k|k}(\{\mathbf{x}'\})d\mathbf{x}' && \text{(G.394)}\\
&= p_B\cdot b_{k+1|k}(\mathbf{x})\cdot f_{k|k}(\emptyset) && \text{(G.395)}\\
&\quad +\int p_S(\mathbf{x}')\cdot f_{k+1|k}(\mathbf{x}|\mathbf{x}')\cdot f_{k|k}(\{\mathbf{x}'\})d\mathbf{x}'. && \text{(G.396)}
\end{aligned}
$$

Thus

$$
\begin{aligned}
& p_{k+1|k}\cdot f_{k+1|k}(\mathbf{x}) && \text{(G.397)}\\
= \; & p_B\cdot b_{k+1|k}(\mathbf{x})\cdot(1-p_{k|k}) && \text{(G.398)}\\
& +p_{k|k}\int p_S(\mathbf{x}')\cdot f_{k+1|k}(\mathbf{x}|\mathbf{x}')\cdot f_{k|k}(\mathbf{x}')d\mathbf{x}'. && \text{(G.399)}
\end{aligned}
$$

Substituting (G.387) into this and solving for $f_{k+1|k}(\mathbf{x})$ yields

$$
f_{k+1|k}(\mathbf{x})=\frac{\begin{array}{c}p_B\cdot b_{k+1|k}(\mathbf{x})\cdot(1-p_{k|k})\\ +p_{k|k}\int p_S(\mathbf{x}')\cdot f_{k+1|k}(\mathbf{x}|\mathbf{x}')\cdot f_{k|k}(\mathbf{x}')d\mathbf{x}'\end{array}}{p_B\cdot(1-p_{k|k})+p_{k|k}\int p_S(\mathbf{x}')\cdot f_{k|k}(\mathbf{x}')d\mathbf{x}'}. \tag{G.400}
$$

G.24 JOTT FILTER CORRECTOR

We are to prove (14.202) and (14.203). By (14.50),

$$f_{k+1|k+1}(X) = \frac{f_{k+1}(Z_{k+1}|X) \cdot f_{k+1|k}(X)}{f_{k+1}(Z_{k+1})} \tag{G.401}$$

where

$$
\begin{aligned}
f_{k+1}(Z) &\overset{\text{abbr.}}{=} f_{k+1}(Z|Z^{(k)}) && \text{(G.402)}\\
&= \int f_{k+1}(Z_{k+1}|X) \cdot f_{k+1|k}(X) && \text{(G.403)}\\
&= f_{k+1}(Z|\emptyset) \cdot f_{k+1|k}(\emptyset) && \text{(G.404)}\\
&\quad + \int f_{k+1}(Z|\{\mathbf{x}\}) \cdot f_{k+1|k}(\{\mathbf{x}\}) d\mathbf{x}. && \text{(G.405)}
\end{aligned}
$$

From (14.181) and (14.182)

$$
\begin{aligned}
f_{k+1}(Z) &= \kappa(Z) \cdot (1 - p_{k+1|k}) + \kappa(Z) \cdot p_{k+1|k} && \text{(G.406)}\\
&\quad \cdot \int \left(q_D(\mathbf{x}) + p_D(\mathbf{x}) \sum_{\mathbf{z} \in \mathbf{Z}} L_{\mathbf{z}}(\mathbf{x}) \cdot \frac{\kappa(Z - \{\mathbf{z}\})}{\kappa(Z)} \right) && \text{(G.407)}\\
&\quad \cdot f_{k+1|k}(\mathbf{x}) d\mathbf{x} && \text{(G.408)}\\
&= \kappa(Z) \cdot (1 - p_{k+1|k}) + \kappa(Z) \cdot p_{k+1|k} && \text{(G.409)}\\
&\quad \cdot \left[f_{k+1|k}[q_D] + \sum_{\mathbf{z} \in \mathbf{Z}} f_{k+1|k}[p_D L_{\mathbf{z}}] \cdot \frac{\kappa(Z - \{\mathbf{z}\})}{\kappa(Z)} \right] && \text{(G.410)}
\end{aligned}
$$

and so

$$
\begin{aligned}
&= \kappa(Z) \cdot \left(\begin{array}{c} 1 - p_{k+1|k} + p_{k+1|k} \\ \cdot \left(\begin{array}{c} f_{k+1|k}[q_D] + \sum_{\mathbf{z} \in \mathbf{Z}} f_{k+1|k}[p_D L_{\mathbf{z}}] \\ \cdot \frac{\kappa(Z - \{\mathbf{z}\})}{\kappa(Z)} \end{array} \right) \end{array} \right) && \text{(G.411)}\\
&= \kappa(Z) \cdot \left(\begin{array}{c} 1 - p_{k+1|k} + p_{k+1|k} \cdot f_{k+1|k}[q_D] \\ + p_{k+1|k} \sum_{\mathbf{z} \in \mathbf{Z}} f_{k+1|k}[p_D L_{\mathbf{z}}] \cdot \frac{\kappa(Z - \{\mathbf{z}\})}{\kappa(Z)} \end{array} \right) && \text{(G.412)}\\
&= \kappa(Z) \cdot \left(\begin{array}{c} 1 - p_{k+1|k} \cdot f_{k+1|k}[p_D] \\ + p_{k+1|k} \sum_{\mathbf{z} \in \mathbf{Z}} f_{k+1|k}[p_D L_{\mathbf{z}}] \cdot \frac{\kappa(Z - \{\mathbf{z}\})}{\kappa(Z)} \end{array} \right). && \text{(G.413)}
\end{aligned}
$$

Thus (14.181) tells us that when $X = \emptyset$, (G.401) becomes

$$f_{k+1|k+1}(\emptyset) \tag{G.414}$$

$$= \frac{\kappa(Z) \cdot f_{k+1|k}(\emptyset)}{\kappa(Z) \cdot \left(\begin{array}{c} 1 - p_{k+1|k} \cdot f_{k+1|k}[p_D] \\ +p_{k+1|k} \sum_{\mathbf{z} \in Z} f_{k+1|k}[p_D L_{\mathbf{z}}] \cdot \frac{\kappa(Z-\{\mathbf{z}\})}{\kappa(Z)} \end{array} \right)} \tag{G.415}$$

$$= \frac{1 - p_{k+1|k}}{\begin{array}{c} 1 - p_{k+1|k} \cdot f_{k+1|k}[p_D] \\ +p_{k+1|k} \sum_{\mathbf{z} \in Z} f_{k+1|k}[p_D L_{\mathbf{z}}] \cdot \frac{\kappa(Z-\{\mathbf{z}\})}{\kappa(Z)} \end{array}}. \tag{G.416}$$

By definition, $f_{k+1|k+1}(\emptyset) = 1 - p_{k+1|k+1}$. So we get

$$p_{k+1|k+1} \tag{G.417}$$

$$= 1 - \frac{1 - p_{k+1|k}}{\begin{array}{c} 1 - p_{k+1|k} \cdot f_{k+1|k}[p_D] \\ +p_{k+1|k} \sum_{\mathbf{z} \in Z} f_{k+1|k}[p_D L_{\mathbf{z}}] \cdot \frac{\kappa(Z-\{\mathbf{z}\})}{\kappa(Z)} \end{array}} \tag{G.418}$$

$$= \frac{\begin{array}{c} p_{k+1|k} - p_{k+1|k} \cdot f_{k+1|k}[p_D] \\ +p_{k+1|k} \sum_{\mathbf{z} \in Z} f_{k+1|k}[p_D L_{\mathbf{z}}] \cdot \frac{\kappa(Z-\{\mathbf{z}\})}{\kappa(Z)} \end{array}}{\begin{array}{c} 1 - p_{k+1|k} \cdot f_{k+1|k}[p_D] \\ +p_{k+1|k} \sum_{\mathbf{z} \in Z} f_{k+1|k}[p_D L_{\mathbf{z}}] \cdot \frac{\kappa(Z-\{\mathbf{z}\})}{\kappa(Z)} \end{array}} \tag{G.419}$$

$$= \frac{1 - f_{k+1|k}[p_D] + \sum_{\mathbf{z} \in Z} f_{k+1|k}[p_D L_{\mathbf{z}}] \cdot \frac{\kappa(Z-\{\mathbf{z}\})}{\kappa(Z)}}{p_{k+1|k}^{-1} - f_{k+1|k}[p_D] + \sum_{\mathbf{z} \in Z} f_{k+1|k}[p_D L_{\mathbf{z}}] \cdot \frac{\kappa(Z-\{\mathbf{z}\})}{\kappa(Z)}}. \tag{G.420}$$

Similarly, (14.182) tells us that when $X = \{\mathbf{x}\}$, (G.401) becomes

$$f_{k+1|k+1}(\{\mathbf{x}\}) = f_{k+1|k}(\mathbf{x}) \cdot \frac{1 - p_D(\mathbf{x}) + p_D(\mathbf{x}) \sum_{\mathbf{z} \in Z} L_{\mathbf{z}}(\mathbf{x}) \cdot \frac{\kappa(Z-\{\mathbf{z}\})}{\kappa(Z)}}{\left(\begin{array}{c} p_{k+1|k}^{-1} - f_{k+1|k}[p_D] \\ + \sum_{\mathbf{z} \in Z} f_{k+1|k}[p_D L_{\mathbf{z}}] \cdot \frac{\kappa(Z-\{\mathbf{z}\})}{\kappa(Z)} \end{array} \right)} \tag{G.421}$$

Since

$$f_{k+1|k+1}(\{\mathbf{x}\}) = p_{k+1|k+1} \cdot f_{k+1|k+1}(\mathbf{x}) \tag{G.422}$$

then from (G.420) this becomes

$$f_{k+1|k+1}(\mathbf{x}) \tag{G.423}$$

$$= \frac{1}{p_{k+1|k+1}} \cdot f_{k+1|k}(\mathbf{x}) \tag{G.424}$$

$$\cdot \frac{1 - p_D(\mathbf{x}) + p_D(\mathbf{x}) \sum_{\mathbf{z}\in \mathbf{Z}} L_{\mathbf{z}}(\mathbf{x}) \cdot \frac{\kappa(Z-\{\mathbf{z}\})}{\kappa(Z)}}{p_{k+1|k}^{-1} - f_{k+1|k}[p_D] + \sum_{\mathbf{z}\in \mathbf{Z}} f_{k+1|k}[p_D L_{\mathbf{z}}] \cdot \frac{\kappa(Z-\{\mathbf{z}\})}{\kappa(Z)}} \tag{G.425}$$

$$= f_{k+1|k}(\mathbf{x}) \cdot \frac{1 - p_D(\mathbf{x}) + p_D(\mathbf{x}) \sum_{\mathbf{z}\in \mathbf{Z}} L_{\mathbf{z}}(\mathbf{x}) \cdot \frac{\kappa(Z-\{\mathbf{z}\})}{\kappa(Z)}}{1 - f_{k+1|k}[p_D] + \sum_{\mathbf{z}\in \mathbf{Z}} f_{k+1|k}[p_D L_{\mathbf{z}}] \cdot \frac{\kappa(Z-\{\mathbf{z}\})}{\kappa(Z)}}. \tag{G.426}$$

G.25 P.G.FL. FORM OF THE MULTITARGET CORRECTOR

We are to establish (14.280):

$$G_{k+1|k+1}[h] = \frac{\frac{\delta F}{\delta Z_{k+1}}[0,h]}{\frac{\delta F}{\delta Z_{k+1}}[0,1]} \tag{G.427}$$

where

$$F[g,h] \triangleq \int h^X \cdot G_{k+1}[g|X] \cdot f_{k+1|k}(X)\delta X \tag{G.428}$$

$$G_{k+1}[g|X] \triangleq \int g^Z \cdot f_{k+1}(Z|X)\delta Z. \tag{G.429}$$

From the Radon-Nikodým theorem for functional derivatives, (11.251), we know that

$$\frac{\delta G_{k+1}}{\delta Z}[g|X] = \int g^Z \cdot f_{k+1}(Y \cup W|X)\delta W. \tag{G.430}$$

Thus

$$\frac{\delta F}{\delta Z}[g,h] \tag{G.431}$$

$$\triangleq \int h^X \cdot \left(\int g^Z \cdot f_{k+1}(Z \cup W|X) \delta W \right) \cdot f_{k+1|k}(X) \delta X \tag{G.432}$$

$$= \int g^Z \cdot \left(\int h^X \cdot f_{k+1}(Z \cup W|X) \cdot f_{k+1|k}(X) \delta X \right) \delta W \tag{G.433}$$

and so, substituting $g = 0$,

$$\frac{\delta F}{\delta Z}[0,h] = \int h^X \cdot f_{k+1}(W|X) \cdot f_{k+1|k}(X) \delta X. \tag{G.434}$$

Substituting $h = 1$ into this,

$$\frac{\delta F}{\delta Z}[0,1] = \int f_{k+1}(W|X) \cdot f_{k+1|k}(X) \delta X = f_{k+1}(W|Z^{(k)}). \tag{G.435}$$

Thus, as claimed,

$$\frac{\frac{\delta F}{\delta Z_{k+1}}[0,h]}{\frac{\delta F}{\delta Z_{k+1}}[0,1]} = \frac{\int h^X \cdot f_{k+1}(W|X) \cdot f_{k+1|k}(X) \delta X}{f_{k+1}(W|Z^{(k)})} \tag{G.436}$$

$$= \int h^X \cdot \frac{f_{k+1}(W|X) \cdot f_{k+1|k}(X)}{f_{k+1}(W|Z^{(k)})} \delta X \tag{G.437}$$

$$= \int h^X \cdot f_{k+1|k+1}(X) \delta X \tag{G.438}$$

$$= G_{k+1|k+1}[h]. \tag{G.439}$$

G.26 INDUCED PARTICLE APPROXIMATION OF PHD

We are to prove (15.78): If $\theta_X \triangleq \sum_{\mathbf{x} \in X} \theta(\mathbf{x})$ where $\theta(\mathbf{x})$ is any unitless function then

$$\int \theta_X \cdot f_{k|k}(X) \delta X = \int \theta(\mathbf{x}) \cdot D_{k|k}(\mathbf{x}) d\mathbf{x}. \tag{G.440}$$

To see this, note that

$$\int \theta_X \cdot f_{k|k}(X)\delta X \tag{G.441}$$

$$= \sum_{n=0}^{\infty} \frac{1}{n!} \int \theta_{\{\mathbf{x}_1,...,\mathbf{x}_n\}} \cdot f_{k|k}(\{\mathbf{x}_1, ..., \mathbf{x}_n\}) \tag{G.442}$$

$$\cdot d\mathbf{x}_1 \cdots d\mathbf{x}_n \tag{G.443}$$

$$= \sum_{n=0}^{\infty} \frac{1}{n!} \int (\theta(\mathbf{x}_1) + ... + \theta(\mathbf{x}_n)) \tag{G.444}$$

$$\cdot f_{k|k}(\{\mathbf{x}_1, ..., \mathbf{x}_n\}) d\mathbf{x}_1 \cdots d\mathbf{x}_n \tag{G.445}$$

$$= \sum_{n=0}^{\infty} \frac{1}{(n-1)!} \int \theta(\mathbf{x}) \cdot f_{k|k}(\{\mathbf{x}, \mathbf{x}_1, ..., \mathbf{x}_{n-1}\}) \tag{G.446}$$

$$\cdot d\mathbf{x} d\mathbf{x}_1 \cdots d\mathbf{x}_{n-1} \tag{G.447}$$

$$= \sum_{n=0}^{\infty} \frac{1}{(n-1)!} \int \theta(\mathbf{x}) \cdot f_{k|k}(\{\mathbf{x}, \mathbf{x}_1, ..., \mathbf{x}_{n-1}\}) \tag{G.448}$$

$$\cdot d\mathbf{x} d\mathbf{x}_1 \cdots d\mathbf{x}_{n-1} \tag{G.449}$$

$$= \int \theta(\mathbf{x}) \sum_{j=0}^{\infty} \frac{1}{j!} \int f_{k|k}(\{\mathbf{x}, \mathbf{x}_1, ..., \mathbf{x}_j\}) d\mathbf{x}_1 \cdots d\mathbf{x}_j d\mathbf{x} \tag{G.450}$$

$$= \int \theta(\mathbf{x}) \cdot D_{k|k}(\mathbf{x}) d\mathbf{x}. \tag{G.451}$$

Consequently, as claimed,

$$\int \theta(\mathbf{x}) \cdot D_{k|k}(\mathbf{x}) d\mathbf{x} = \int \theta_X \cdot f_{k|k}(X)\delta X \cong \sum_{i=0}^{\nu} w_{k|k}^i \cdot \theta_{X_{k|k}^i} \tag{G.452}$$

$$= \sum_{i=0}^{\nu} w_{k|k}^i \cdot \sum_{\mathbf{x} \in X_{k|k}^i} \theta(\mathbf{x}). \tag{G.453}$$

G.27 PHD COUNTING PROPERTY

We are to prove (16.15). That is, we are to show that

$$\int_S D_\Psi(\mathbf{y})d\mathbf{y} = \mathrm{E}[\,|\Psi \cap S|\,]. \tag{G.454}$$

To prove this, note that

$$\begin{aligned}
\mathrm{E}[\,|\Psi \cap S|\,] &= \int |X \cap S| \cdot f_\Psi(Y)\delta Y && \text{(G.455)}\\
&= \int \left(\sum_{\mathbf{x}\in X} \mathbf{1}_S(\mathbf{x})\right) \cdot f_\Psi(Y)\delta Y && \text{(G.456)}\\
&= \sum_{n=0}^\infty \frac{1}{n!} \int (\mathbf{1}_S(\mathbf{y}_1) + ... + \mathbf{1}_S(\mathbf{y}_n)) && \text{(G.457)}\\
&\quad \cdot f_\Psi(\{\mathbf{y}_1, ..., \mathbf{y}_n\})d\mathbf{y}_1 \cdots d\mathbf{y}_n && \text{(G.458)}\\
&= \sum_{n=0}^\infty \frac{n}{n!} \int \mathbf{1}_S(\mathbf{y}_n) \cdot f_\Psi(\{\mathbf{y}_1 ..., \mathbf{y}_{n-1}, \mathbf{y}_n\}) && \text{(G.459)}\\
&\quad \cdot d\mathbf{y}_1 \cdots d\mathbf{y}_{n-1} d\mathbf{y}_n && \text{(G.460)}\\
&= \sum_{n=0}^\infty \frac{1}{(n-1)!} \int_S f_\Psi(\{\mathbf{y}_1 ..., \mathbf{y}_{n-1}, \mathbf{y}_n\}) && \text{(G.461)}\\
&\quad \cdot d\mathbf{y}_1 \cdots d\mathbf{y}_{n-1} d\mathbf{y}_n. && \text{(G.462)}
\end{aligned}$$

Thus, as claimed,

$$\begin{aligned}
&\mathrm{E}[\,|\Psi \cap S|\,] && \text{(G.463)}\\
&= \int_S \left(\sum_{j=0}^\infty \int \frac{1}{j!} f_\Psi(\{\mathbf{y}_1 ..., \mathbf{y}_{n-1}, \mathbf{y}_n\})d\mathbf{y}_1 \cdots d\mathbf{y}_{n-1}\right) d\mathbf{y}_n && \text{(G.464)}\\
&= \int_S \sum_{j=0}^\infty \left(\int f_\Psi(Y \cup \{\mathbf{y}_n\})\delta Y\right) d\mathbf{y}_n && \text{(G.465)}\\
&= \int_S D_\Psi(\mathbf{y}_n)d\mathbf{y}_n. && \text{(G.466)}
\end{aligned}$$

G.28 GM-PHD FILTER PREDICTOR

We are to prove (16.253)-(16.260). From (16.95) and (16.96) the PHD predictor equations are

$$D_{k+1|k}(\mathbf{x}) = b_{k+1|k}(\mathbf{x}) + \int F_{k+1|k}(\mathbf{x}|\mathbf{x}') \cdot D_{k|k}(\mathbf{x}')d\mathbf{x}' \quad \text{(G.467)}$$

$$F_{k+1|k}(\mathbf{x}|\mathbf{x}') \triangleq p_S \cdot f_{k+1|k}(\mathbf{x}|\mathbf{x}') + b_{k+1|k}(\mathbf{x}|\mathbf{x}'). \quad \text{(G.468)}$$

Substitute into these equations: (16.243) for $b_{k+1|k}(\mathbf{x})$, (16.241) for $f_{k+1|k}(\mathbf{x}|\mathbf{x}')$, (16.244) for $b_{k+1|k}(\mathbf{x}|\mathbf{x}')$, and (16.252) for $D_{k|k}(\mathbf{x}')$. Then

$$D_{k+1|k}(\mathbf{x}) = \sum_{i=1}^{a_k} \beta_k^i \cdot N_{B_i}(\mathbf{x} - \mathbf{x}_b^i) \quad \text{(G.469)}$$

$$+p_S \cdot \int N_Q(\mathbf{x}' - F\mathbf{x}') \quad \text{(G.470)}$$

$$\cdot \left(\sum_{i=1}^{n_{k|k}} w_{k|k}^i \cdot N_{P_i}(\mathbf{x}' - \mathbf{x}^i) \right) d\mathbf{x}' \quad \text{(G.471)}$$

$$+ \int \left(\sum_{j=1}^{b_k} \gamma_k^j \cdot N_{G_j}(\mathbf{x} - E_j\mathbf{x}') \right) \quad \text{(G.472)}$$

$$\cdot \left(\sum_{i=1}^{n_{k|k}} w_{k|k}^i \cdot N_{P_i}(\mathbf{x}' - \mathbf{x}^i) \right) d\mathbf{x}' \quad \text{(G.473)}$$

or

$$D_{k+1|k}(\mathbf{x}) = \sum_{i=1}^{a_k} \beta_k^i \cdot N_{B_i}(\mathbf{x} - \mathbf{x}_b^i) \quad \text{(G.474)}$$

$$+p_S \cdot \sum_{i=1}^{n_{k|k}} w_{k|k}^i \quad \text{(G.475)}$$

$$\cdot \int N_Q(\mathbf{x}' - F\mathbf{x}') \cdot N_{P_i}(\mathbf{x}' - \mathbf{x}^i)d\mathbf{x}' \tag{G.476}$$

$$+\sum_{j=1}^{b_k}\sum_{i=1}^{n_{k|k}} \gamma_k^j \cdot w_{k|k}^i \tag{G.477}$$

$$\cdot \int N_{G_j}(\mathbf{x} - E_j\mathbf{x}') \cdot N_{P_i}(\mathbf{x}' - \mathbf{x}^i)d\mathbf{x}'. \tag{G.478}$$

From the fundamental identity for Gaussian distributions (D.1),

$$D_{k+1|k}(\mathbf{x}) = \sum_{i=1}^{a_k} \beta_k^i \cdot N_{B_i}(\mathbf{x} - \mathbf{x}_b^i) \tag{G.479}$$

$$+p_S \cdot \sum_{i=1}^{n_{k|k}} w_{k|k}^i \cdot N_{P_{k+1|k}^i}(\mathbf{x} - \mathbf{x}_{k+1|k}^i) \tag{G.480}$$

$$+\sum_{j=1}^{b_k}\sum_{i=1}^{n_{k|k}} \gamma_k^j \cdot w_{k|k}^i \tag{G.481}$$

$$\cdot \int N_{G_j+E_iP_iE_i^T}(\mathbf{x} - E_j\mathbf{x}^i) \tag{G.482}$$

$$\cdot N_{P_i}(\mathbf{x}' - \mathbf{x}^i)d\mathbf{x}'. \tag{G.483}$$

From this we conclude that

$$D_{k+1|k}(\mathbf{x}) = \sum_{i=1}^{a_k} v_{k+1|k}^i \cdot N_{B_i}(\mathbf{x} - \mathbf{b}_{k+1|k}^i) \tag{G.484}$$

$$+\sum_{i=1}^{n_{k|k}} w_{k+1|k}^i \cdot N_{P_{k+1|k}^i}(\mathbf{x} - \mathbf{x}_{k+1|k}^i) \tag{G.485}$$

$$+\sum_{j=1}^{b_k}\sum_{i=1}^{n_{k|k}} \gamma_k^j \cdot w_{k|k}^i \cdot N_{P_{k+1|k}^{i,j}}(\mathbf{x} - \mathbf{x}_{k+1|k}^{i,j}) \tag{G.486}$$

where

$$
\begin{aligned}
v^i_{k+1|k} &= \beta^i_k, \quad \mathbf{b}^i_{k+1|k} = \mathbf{x}^i_b, \quad B^i_{k+1|k} = B_i & \text{(G.487)}\\
w^i_{k+1|k} &= p_S \cdot w^i_{k|k}, \quad \mathbf{x}^i_{k+1|k} = F\mathbf{x}^i & \text{(G.488)}\\
P^i_{k+1|k} &= Q + FP_iF^T & \text{(G.489)}\\
w^{i,j}_{k+1|k} &= \gamma^j_k \cdot w^i_{k|k}, \quad \mathbf{x}^{i,j}_{k+1|k} = E_j\mathbf{x}^i & \text{(G.490)}\\
P^{i,j}_{k+1|k} &= G_j + E_iP_iE_i^T. & \text{(G.491)}
\end{aligned}
$$

G.29 GM-PHD FILTER CORRECTOR

We are to prove (16.270)-(16.277). The PHD corrector equations (16.108) and (16.109) are

$$D_{k+1|k+1}(\mathbf{x}) \cong L_{Z_{k+1}}(\mathbf{x}) \cdot D_{k+1|k}(\mathbf{x}) \tag{G.492}$$

where for any measurement set Z,

$$L_Z(\mathbf{x}) \triangleq 1 - p_D + p_D \sum_{\mathbf{z} \in Z} \frac{L_{\mathbf{z}}(\mathbf{x})}{\lambda c(\mathbf{z}) + p_D D_{k+1|k}[L_{\mathbf{z}}]}. \tag{G.493}$$

The integral $D_{k+1|k}[L_{\mathbf{z}}]$ can be evaluated in closed form as follows. Substitute $f_{k+1}(\mathbf{z}|\mathbf{x})$ from (16.242) and $D_{k+1|k}(\mathbf{x})$ into the formula for $D_{k+1|k}[L_{\mathbf{z}}]$. We get, from an application of the fundamental identity for Gaussian distributions—see (D.1):

$$\begin{aligned}
& D_{k+1|k}[L_{\mathbf{z}_j}] & \text{(G.494)} \\
= & \int f_{k+1}(\mathbf{z}_j|\mathbf{x}) \cdot D_{k+1|k}(\mathbf{x})d\mathbf{x} & \text{(G.495)} \\
= & \int N_R(\mathbf{z}_j - H\mathbf{x}) \cdot \left(\sum_{i=1}^{n_{k+1|k}} w^i_{k+1|k} \cdot N_{P_i}(\mathbf{x} - \mathbf{x}^i) \right) d\mathbf{x} & \text{(G.496)} \\
= & \sum_{i=1}^{n_{k+1|k}} w^i_{k+1|k} \int N_R(\mathbf{z}_j - H\mathbf{x}) \cdot N_{P_i}(\mathbf{x} - \mathbf{x}^i) d\mathbf{x} & \text{(G.497)} \\
= & \sum_{i=1}^{n_{k+1|k}} w^i_{k+1|k} \int N_{R+HP_iH^T}(\mathbf{z}_j - H\mathbf{x}^i) \cdot N_{C_i}(\mathbf{x} - \mathbf{c}_i) d\mathbf{x} & \text{(G.498)} \\
= & \sum_{i=1}^{n_{k+1|k}} w^i_{k+1|k} \cdot N_{R+HP_iH^T}(\mathbf{z}_j - H\mathbf{x}^i) & \text{(G.499)}
\end{aligned}$$

where

$$\begin{aligned}
C_i^{-1} &= P_i^{-1} + H^T R^{-1} H & \text{(G.500)} \\
C_i^{-1} \mathbf{c}_i &= P_i^{-1} \mathbf{x}^i + H^T R^{-1} H \mathbf{z}_j. & \text{(G.501)}
\end{aligned}$$

Abbreviate

$$\begin{aligned}
\Lambda_j &= \lambda c(\mathbf{z}_j) + p_D \cdot D_{k+1|k}[L_{\mathbf{z}_j}] & \text{(G.502)} \\
&= \lambda c(\mathbf{z}_j) + p_D \sum_{i=1}^{n_{k+1|k}} w^i_{k+1|k} \cdot N_{R+HP_iH^T}(\mathbf{z}_j - H\mathbf{x}^i). & \text{(G.503)}
\end{aligned}$$

Substituting this and (16.242) into (G.492) we get:

$$\begin{aligned}
D_{k+1|k+1}(\mathbf{x}) &= (1 - p_D) \sum_{i=1}^{n_{k+1|k}} w^i_{k+1|k} \cdot N_{P_i}(\mathbf{x} - \mathbf{x}^i_{k+1|k}) & \text{(G.504)} \\
&\quad + p_D \sum_{i=1}^{n_{k+1|k}} \sum_{j=1}^{m_{k+1}} \frac{w^i_{k+1|k}}{\Lambda_j} & \text{(G.505)} \\
&\quad \cdot N_R(\mathbf{z}_j - H\mathbf{x}) \cdot N_{P_i}(\mathbf{x} - \mathbf{x}^i_{k+1|k}). & \text{(G.506)}
\end{aligned}$$

Using our previous results for evaluating $D_{k+1|k}[L_{\mathbf{z}}]$, this becomes

$$D_{k+1|k+1}(\mathbf{x}) = (1-p_D)\sum_{i=1}^{n_{k+1|k}} w^i_{k+1|k}\cdot N_{P_i}(\mathbf{x}-\mathbf{x}^i_{k+1|k}) \quad \text{(G.507)}$$

$$+p_D\sum_{i=1}^{n_{k+1|k}}\sum_{j=1}^{m_{k+1}} \frac{w^i_{k+1|k}}{\Lambda_j} \quad \text{(G.508)}$$

$$\cdot N_{R+HP_iH^T}(\mathbf{z}_j - H\mathbf{x}^i)\cdot N_{C_i}(\mathbf{x}-\mathbf{c}_i). \quad \text{(G.509)}$$

Or, rewritten in standard Kalman form, see (2.9)-(2.11),

$$\mathbf{x}^{i,j}_{k+1|k+1} = \mathbf{x}_{k+1|k} + K_i\left(\mathbf{z}_j - H\mathbf{x}^i\right) \quad \text{(G.510)}$$

$$P^{i,j}_{k+1|k+1} = (I - K_iH)\,P_i \quad \text{(G.511)}$$

$$K_i = P_iH^T\left(HP_iH^T+R\right)^{-1} \quad \text{(G.512)}$$

$$w^{i,j}_{k+1|k+1} \quad \text{(G.513)}$$

$$= \frac{w^i_{k+1|k}\cdot p_D\cdot N_{R+HP_iH^T}(\mathbf{z}_j - H\mathbf{x}^i)\cdot N_{C_i}(\mathbf{x}-\mathbf{c}_i)}{\lambda c(\mathbf{z}_j) + p_D\sum_{e=1}^{n_{k+1|k}} w^e_{k+1|k}\cdot N_{R+HP_eH^T}(\mathbf{z}_j - H\mathbf{x}^e)}. \quad \text{(G.514)}$$

G.30 EXACT PHD CORRECTOR

We are to demonstrate (16.293). That is,

$$D_{k+1|k+1}(\mathbf{x}) = \ell_Z(\mathbf{x})\cdot D_{k+1|k}(\mathbf{x}) \quad \text{(G.515)}$$

where

$$\ell_Z(\mathbf{x}) = \frac{1}{p_{k+1|k}}\cdot\frac{1-p_D(\mathbf{x})+p_D(\mathbf{x})\sum_{\mathbf{z}\in Z}\frac{L_{\mathbf{z}}(\mathbf{x})}{\lambda c(\mathbf{z})}}{p^{-1}_{k+1|k} - f_{k+1|k}[p_D] + \sum_{\mathbf{z}\in Z}\frac{f_{k+1|k}[p_D L_{\mathbf{z}}]}{\lambda c(\mathbf{z})}}. \quad \text{(G.516)}$$

For from the JoTT corrector, (14.202) and (14.203), we get

$$\ell_Z(\mathbf{x}) \tag{G.517}$$
$$\triangleq \frac{p_{k+1|k+1}}{p_{k+1|k}} \tag{G.518}$$
$$\cdot \frac{1 - p_D(\mathbf{x}) + p_D(\mathbf{x}) \sum_{\mathbf{z}\in Z} L_{\mathbf{z}}(\mathbf{x}) \cdot \frac{\kappa(Z-\{\mathbf{z}\})}{\kappa(Z)}}{1 - f_{k+1|k}[p_D] + \sum_{\mathbf{z}\in Z} f_{k+1|k}[p_D L_{\mathbf{z}}] \cdot \frac{\kappa(Z-\{\mathbf{z}\})}{\kappa(Z)}} \tag{G.519}$$
$$= \frac{1}{p_{k+1|k}} \tag{G.520}$$
$$\cdot \frac{1 - p_D(\mathbf{x}) + p_D(\mathbf{x}) \sum_{\mathbf{z}\in Z} L_{\mathbf{z}}(\mathbf{x}) \cdot \frac{\kappa(Z-\{\mathbf{z}\})}{\kappa(Z)}}{p_{k+1|k}^{-1} - f_{k+1|k}[p_D] + \sum_{\mathbf{z}\in Z} f_{k+1|k}[p_D L_{\mathbf{z}}] \cdot \frac{\kappa(Z-\{\mathbf{z}\})}{\kappa(Z)}}. \tag{G.521}$$

Thus the exact PHD update equation is nonlinear in $D_{k+1|k}(\mathbf{x})$:

$$D_{k+1|k+1}(\mathbf{x}) \tag{G.522}$$
$$= \frac{\left(\begin{array}{c} 1 - p_D(\mathbf{x}) \\ +p_D(\mathbf{x}) \sum_{\mathbf{z}\in Z} \frac{L_{\mathbf{z}}(\mathbf{x}) \cdot \kappa(Z-\{\mathbf{z}\})}{\kappa(Z)} \end{array} \right) \cdot D_{k+1|k}(\mathbf{x})}{1 - D_{k+1|k}[p_D] + \sum_{\mathbf{z}\in Z} \frac{D_{k+1|k}[p_D L_{\mathbf{z}}] \cdot \kappa(Z-\{\mathbf{z}\})}{\kappa(Z)}} \tag{G.523}$$

and the updated expected number of targets is

$$N_{k+1|k+1} \tag{G.524}$$
$$= \int D_{k+1|k+1}(\mathbf{x}) d\mathbf{x} \tag{G.525}$$
$$= \frac{N_{k+1|k} - D_{k+1|k}[p_D] + \sum_{\mathbf{z}\in Z} \frac{D_{k+1|k}[p_D L_{\mathbf{z}}] \cdot \kappa(Z-\{\mathbf{z}\})}{\kappa(Z)}}{1 - D_{k+1|k}[p_D] + \sum_{\mathbf{z}\in Z} \frac{D_{k+1|k}[p_D L_{\mathbf{z}}] \cdot \kappa(Z-\{\mathbf{z}\})}{\kappa(Z)}}. \tag{G.526}$$

Finally, let the false alarm process be Poisson: $\kappa(Z) = e^{-\lambda} \prod_{\mathbf{z}\in Z} \lambda c(\mathbf{z})$. Then we get

$$\ell_Z(\mathbf{x}) = \frac{1}{p_{k+1|k}} \cdot \frac{1 - p_D(\mathbf{x}) + p_D(\mathbf{x}) \sum_{\mathbf{z}\in Z} \frac{L_{\mathbf{z}}(\mathbf{x})}{\lambda c(\mathbf{z})}}{p_{k+1|k}^{-1} - f_{k+1|k}[p_D] + \sum_{\mathbf{z}\in Z} \frac{f_{k+1|k}[p_D L_{\mathbf{z}}]}{\lambda c(\mathbf{z})}}. \tag{G.527}$$

G.31 GM-CPHD FILTER PREDICTOR

We are to prove (16.370)-(16.375). To see this, note that the CPHD predictor equations, (16.311) and (16.312), are

$$G_{k+1|k}(x) \cong B(x) \cdot G(1 - p_S + p_S \cdot x) \quad \text{(G.528)}$$

$$D_{k+1|k}(\mathbf{x}) = b(\mathbf{x}) + p_S \cdot \int f_{k+1|k}(\mathbf{x}|\mathbf{x}') \cdot D_{k|k}(\mathbf{x}')d\mathbf{x}'. \quad \text{(G.529)}$$

Into the second equation, substitute (16.245) for $D_{k|k}(\mathbf{x}')$ and (16.241) for $f_{k+1|k}(\mathbf{x}|\mathbf{x}')$ and (16.243) for $b(\mathbf{x})$. Then we get

$$D_{k+1|k}(\mathbf{x}) = \sum_{i=1}^{a_k} \beta_k^i \cdot N_{B_k^i}(\mathbf{x} - \mathbf{x}_b^i) \quad \text{(G.530)}$$

$$+p_S \cdot \int N_Q(\mathbf{x} - F\mathbf{x}') \quad \text{(G.531)}$$

$$\cdot \left(\sum_{i=1}^{n_{k|k}} w_{k|k}^i \cdot N_{P_{k|k}^i}(\mathbf{x} - \mathbf{x}_{k|k}^i) \right) d\mathbf{x}'. \quad \text{(G.532)}$$

From the fundamental identity for Gaussian distributions (D.1), this becomes

$$D_{k+1|k}(\mathbf{x}) = \sum_{i=1}^{a_k} \beta_k^i \cdot N_{B_i}(\mathbf{x} - \mathbf{x}_b^i) \quad \text{(G.533)}$$

$$+p_S \sum_{i=1}^{n_{k|k}} w_{k|k}^i \cdot \int N_Q(\mathbf{x} - F\mathbf{x}') \cdot N_{P_i}(\mathbf{x} - \mathbf{x}^i)d\mathbf{x}' \quad \text{(G.534)}$$

$$= \sum_{i=1}^{a_k} \beta_k^i \cdot N_{B_i}(\mathbf{x} - \mathbf{x}_b^i) \quad \text{(G.535)}$$

$$+p_S \sum_{i=1}^{n_{k|k}} w_{k|k}^i \cdot N_{Q+FP_iF^T}(\mathbf{x} - F\mathbf{x}^i). \quad \text{(G.536)}$$

From this we conclude that

$$D_{k+1|k}(\mathbf{x}) = \sum_{i=1}^{a_k} b^i_{k+1|k} \cdot N_{B^i_{k+1|k}}(\mathbf{x} - \mathbf{b}^i_{k+1|k}) \quad \text{(G.537)}$$

$$+p_S \sum_{i=1}^{n_{k|k}} w^i_{k|k} \cdot N_{Q+FP_iF^T}(\mathbf{x} - F\mathbf{x}^i). \quad \text{(G.538)}$$

where

$$b^i_{k+1|k} = \beta^i_k \quad \text{(G.539)}$$

$$\mathbf{b}^i_{k+1|k} = \mathbf{x}^i_b, \quad B^i_{k+1|k} = B_i \quad \text{(G.540)}$$

$$w^i_{k+1|k} = p_S \cdot w^i_{k|k} \quad \text{(G.541)}$$

$$\mathbf{x}^i_{k+1|k} = F\mathbf{x}^i, \quad P^i_{k+1|k} = Q + FP_iF^T. \quad \text{(G.542)}$$

G.32 GM-CPHD FILTER CORRECTOR

We are to prove (16.380)-(16.396). Under our current assumptions the CPHD corrector equations are

$$G_{k+1|k+1}(x) \cong \frac{\sum_{j=0}^m x^j \cdot C^{(m-j)}(0) \cdot \hat{G}^{(j)}(q_D x) \cdot \sigma_j\left(Z_{k+1}\right)}{\sum_{i=0}^m C^{(m-i)}(0) \cdot \hat{G}^{(i)}(q_D) \cdot \sigma_i\left(Z_{k+1}\right)} \quad \text{(G.543)}$$

$$D_{k+1|k+1}(\mathbf{x}) \cong L_{Z_{k+1}}(\mathbf{x}) \cdot D_{k+1|k}(\mathbf{x}) \quad \text{(G.544)}$$

where

$$L_Z(\mathbf{x}) \triangleq \frac{\sum_{j=0}^m C^{(m-j)}(0) \cdot \hat{G}^{(j+1)}(q_D) \cdot \sigma_j\left(Z\right)}{\sum_{i=0}^m C^{(m-i)}(0) \cdot \hat{G}^{(i)}(q_D) \cdot \sigma_i\left(Z\right)} \cdot (1 - p_D) \quad \text{(G.545)}$$

$$+p_D \sum_{\mathbf{z} \in Z} \frac{L_{\mathbf{z}}(\mathbf{x})}{c(\mathbf{z})} \quad \text{(G.546)}$$

$$\cdot \frac{\sum_{j=0}^{m-1} C^{(m-j-1)}(0) \cdot \hat{G}^{(j+1)}(q_D) \cdot \sigma_j\left(Z - \{\mathbf{z}\}\right)}{\sum_{i=0}^m C^{(m-i)}(0) \cdot \hat{G}^{(i)}(q_D) \cdot \sigma_i\left(Z\right)} \quad \text{(G.547)}$$

and where

$$\sigma_i(Z) \triangleq p_D^i \cdot \sigma_{m,i}\left(\frac{D_{k+1|k}[L_{\mathbf{z}_1}]}{c(\mathbf{z}_1)}, \ldots, \frac{D_{k+1|k}[L_{\mathbf{z}_m}]}{c(\mathbf{z}_m)}\right) \quad \text{(G.548)}$$

$$\hat{G}^{(i)}(x) \triangleq \frac{G^{(i)}(x)}{G^{(1)}(1)^i} = \frac{G^{(i)}(x)}{N_{k+1|k}^i}. \quad \text{(G.549)}$$

From the derivation of the corrector equations for the GM-PHD filter, we know—see (G.499)—that

$$D_{k+1|k}[L_{\mathbf{z}_j}] = \sum_{i=1}^{n_{k+1|k}} w_{k+1|k}^i \cdot N_{R+HP_iH^T}(\mathbf{z}_j - H\mathbf{x}^i) \quad \text{(G.550)}$$

$$\cdot \int N_{C_i}(\mathbf{x} - \mathbf{c}_{i,j})d\mathbf{x} \quad \text{(G.551)}$$

$$= \sum_{i=1}^{n_{k+1|k}} w_{k+1|k}^i \cdot N_{R+HP_iH^T}(\mathbf{z}_j - H\mathbf{x}^i) \quad \text{(G.552)}$$

where

$$C_i^{-1} = P_i^{-1} + H^T R^{-1} H \quad \text{(G.553)}$$

$$C_i^{-1}\mathbf{c}_{i,j} = P_i^{-1}\mathbf{x}^i + H^T R^{-1} H \mathbf{z}_j \quad \text{(G.554)}$$

Abbreviate

$$\Lambda_0 \overset{\text{abbr.}}{=} \frac{\sum_{j=0}^m C^{(m-j)}(0) \cdot \hat{G}^{(j+1)}(q_D) \cdot \sigma_j(Z)}{\sum_{i=0}^m C^{(m-i)}(0) \cdot \hat{G}^{(i)}(q_D) \cdot \sigma_i(Z)} \cdot (1 - p_D) \quad \text{(G.555)}$$

$$\Lambda_e \overset{\text{abbr.}}{=} \frac{p_D \sum_{j=0}^{m-1} C^{(m-j-1)}(0) \cdot \hat{G}^{(j+1)}(q_D) \cdot \sigma_j(Z - \{\mathbf{z}_e\})}{c(\mathbf{z}_e) \cdot \sum_{i=0}^m C^{(m-i)}(0) \cdot \hat{G}^{(i)}(q_D) \cdot \sigma_i(Z)}. \quad \text{(G.556)}$$

Then

$$D_{k+1|k+1}(\mathbf{x}) = \Lambda_0 \cdot D_{k+1|k}(\mathbf{x}) + \sum_{e=1}^{m_{k+1}} \Lambda_e \cdot L_{\mathbf{z}_e}(\mathbf{x}) \cdot D_{k+1|k}(\mathbf{x}). \quad \text{(G.557)}$$

Substituting (16.242) for $L_{\mathbf{z}}(\mathbf{x})$ and (16.245) for $D_{k+1|k}(\mathbf{x})$, we get

$$D_{k+1|k+1}(\mathbf{x}) = \Lambda_0 \cdot \left(\sum_{i=1}^{n_{k+1|k}} w_{k+1|k}^i \cdot N_{P_i}(\mathbf{x} - \mathbf{x}^i) \right) \tag{G.558}$$

$$+ \sum_{e=1}^{m_{k+1}} \Lambda_e \cdot N_R(\mathbf{x} - H\mathbf{z}_e) \tag{G.559}$$

$$\cdot \left(\sum_{i=1}^{n_{k+1|k}} w_{k+1|k}^i \cdot N_{P_i}(\mathbf{x} - \mathbf{x}^i) \right). \tag{G.560}$$

From the fundamental identity for Gaussian distributions (D.1),

$$D_{k+1|k+1}(\mathbf{x}) = \Lambda_0 \sum_{i=1}^{n_{k+1|k}} w_{k+1|k}^i \cdot N_{P_{k+1|k}^i}(\mathbf{x} - \mathbf{x}^i) \tag{G.561}$$

$$+ \sum_{e=1}^{m_{k+1}} \sum_{i=1}^{n_{k+1|k}} w_{k+1|k}^i \cdot \Lambda_e \tag{G.562}$$

$$\cdot N_R(\mathbf{x} - H\mathbf{z}_e) \cdot N_{P_i}(\mathbf{x} - \mathbf{x}^i) \tag{G.563}$$

$$= \Lambda_0 \sum_{i=1}^{n_{k+1|k}} w_{k+1|k}^i \cdot N_{P_{k+1|k}^i}(\mathbf{x} - \mathbf{x}^i) \tag{G.564}$$

$$+ \sum_{e=1}^{m_{k+1}} \sum_{i=1}^{n_{k+1|k}} w_{k+1|k}^i \cdot \Lambda_e \tag{G.565}$$

$$\cdot N_{R+HP_iH^T}(\mathbf{z}_j - H\mathbf{x}^i) \cdot N_{C_i}(\mathbf{x} - \mathbf{c}_i). \tag{G.566}$$

From this we conclude that

$$D_{k+1|k+1}(\mathbf{x}) = \sum_{i=1}^{n_{k+1|k}} w_{k+1|k+1}^i \cdot N_{P_{k+1|k+1}^i}(\mathbf{x} - \mathbf{x}_{k+1|k+1}^i) \tag{G.567}$$

$$+ \sum_{j=1}^{m_{k+1}} \sum_{i=1}^{n_{k+1|k}} w_{k+1|k+1}^{i,j} \tag{G.568}$$

$$\cdot N_{P_{k+1|k+1}^i}(\mathbf{x} - \mathbf{x}_{k+1|k+1}^{i,j}) \tag{G.569}$$

where

$$w^i_{k+1|k+1} = \Lambda_0 \cdot w^i_{k+1|k} \tag{G.570}$$

$$\mathbf{x}^i_{k+1|k+1} = \mathbf{x}^i, \quad P^i_{k+1|k+1} = P^i_{k+1|k} \tag{G.571}$$

and where

$$w^{i,j}_{k+1|k+1} = w^i_{k+1|k} \cdot \Lambda_e \cdot N_{R+HP_iH^T}(\mathbf{z}_j - H\mathbf{x}^i) \tag{G.572}$$

$$\mathbf{x}^{i,j}_{k+1|k+1} = \mathbf{c}_{i,j}, \quad P^i_{k+1|k+1} = C_i. \tag{G.573}$$

Or, expressed in standard Kalman filter form, (2.9)-(2.11),

$$\mathbf{x}^{i,j}_{k+1|k+1} = \mathbf{x}^i_{k+1|k} + K_i \left(\mathbf{z}_j - H\mathbf{x}^i\right) \tag{G.574}$$

$$P^i_{k+1|k+1} = (I - K_i H)\, P_i \tag{G.575}$$

$$K_i = P_i H^T \left(H P_i H^T + R\right)^{-1}. \tag{G.576}$$

G.33 MEMBER FILTER TARGET NUMBER

We are to prove (17.72)-(17.75) for the MeMBer filter:

$$N_{k|k} = \sum_{i=1}^{\nu} q_i \,, \quad \sigma^2_{k|k} = \sum_{i=1}^{\nu} q_i(1-q_i) \tag{G.577}$$

$$\hat{n}_{k|k} = \arg\max_n \; B_{q_1,\ldots,q_\nu}(n) \tag{G.578}$$

$$\hat{\sigma}^2_{k|k} = \sigma^2_{k|k} + N^2_{k|k} - n^2_{k|k}. \tag{G.579}$$

Equation (17.75) follows from the fact that

$$\hat{\sigma}^2_{k|k} = \sum_{i=1}^{\nu} (i - \hat{n}_{k|k})^2 \cdot B_{q_1,\ldots,q_\nu}(i) \tag{G.580}$$

$$= \sum_{i=1}^{\nu} i^2 \cdot B_{q_1,\ldots,q_\nu}(i) - n^2_{k|k} \tag{G.581}$$

and

$$\sigma^2_{k|k} = \sum_{i=1}^{\nu}(i - N_{k|k})^2 \cdot B_{q_1,...,q_\nu}(i) \tag{G.582}$$

$$= \sum_{i=1}^{\nu} i^2 \cdot B_{q_1,...,q_\nu}(i) - N^2_{k|k}. \tag{G.583}$$

The easiest way to demonstrate (17.72) and (17.73) is to look at the p.g.f. $G_{k|k}(x)$ of $G_{k|k}[h]$ and use (11.164) and (11.165). The p.g.f. of $G_{k|k}[h]$ is

$$G_{k|k}(x) = (1 - q_1 + q_1 \cdot x) \cdots (1 - q_\nu + q_\nu \cdot x).$$

Its first and second derivatives are

$$G'_{k|k}(x) = \sum_{i=1}^{\nu}(1 - q_1 + q_1 \cdot x) \cdots q_i \cdots (1 - q_\nu + q_\nu \cdot x) \tag{G.584}$$

$$G''_{k|k}(x) = \sum_{1 \leq i \neq j \leq \nu} (1 - q_1 + q_1 \cdot x) \cdots q_i \cdots q_j \tag{G.585}$$

$$\cdots (1 - q_\nu + q_\nu \cdot x). \tag{G.586}$$

Thus

$$N_{k|k} = G'_{k|k}(1) = \sum_{i=1}^{\nu} q_i \tag{G.587}$$

$$G''_{k|k}(1) = \sum_{1 \leq i \neq j \leq \nu} q_i \cdot q_j \tag{G.588}$$

and so

$$\sigma^2_{k|k} = G''_{k|k}(1) - N^2_{k|k} + N_{k|k} \tag{G.589}$$

$$= \sum_{1 \leq i \neq j \leq \nu} q_i \cdot q_j - \left(\sum_{i=1}^{\nu} q\right)^2 + \sum_{i=1}^{\nu} q_i \tag{G.590}$$

$$= \sum_{1 \leq i \neq j \leq \nu} q_i \cdot q_j - \sum_{i,j=1}^{\nu} q_i \cdot q_j + \sum_{i=1}^{\nu} q_i \tag{G.591}$$

$$= -\sum_{i=1}^{n} q_i^2 + \sum_{i=1}^{\nu} q_i = \sum_{i=1}^{\nu} q_i(1 - q_i). \tag{G.592}$$

G.34 PARA-GAUSSIAN FILTER PREDICTOR

We are to prove (17.93)-(17.97). From (17.128)-(17.131) we know that the predicted p.g.fl. is

$$G_{k+1|k}[h] = G_B[h] \cdot \prod_{i=1}^{\nu'} (1 - q_i + q_i \cdot f_i[h]) \tag{G.593}$$

where

$$\begin{aligned} G_B[h] &= \prod_{i=1}^{e} (1 - b_i + b_i \cdot b_i[h]) &\text{(G.594)} \\ q_i &= q'_i \cdot f'_i[p_S] &\text{(G.595)} \\ f_i(\mathbf{x}) &= \frac{f'_i[p_S M_\mathbf{x}]}{f'_i[p_S]} &\text{(G.596)} \\ M_\mathbf{x}(\mathbf{x}') &\triangleq f_{k+1|k}(\mathbf{x}|\mathbf{x}'). &\text{(G.597)} \end{aligned}$$

If p_D is constant then $q_i = q'_i \cdot f'_i[p_S] = p_S \cdot q'_i$ follows trivially.

As for $f_i(\mathbf{x})$, from the fundamental identity for Gaussian distributions, (D.1), we get

$$\begin{aligned} f_i(\mathbf{x}) &= f'_i[M_\mathbf{x}] = \int f_{k+1|k}(\mathbf{x}|\mathbf{x}') \cdot f'_i(\mathbf{x}') d\mathbf{x}' &\text{(G.598)} \\ &= \sum_{e=1}^{n_i} w'_{i,e} \cdot \int N_Q(\mathbf{x} - F\mathbf{x}') \cdot N_{P'_{i,e}}(\mathbf{x} - \mathbf{x}'_{i,e}) d\mathbf{x}' &\text{(G.599)} \\ &= \sum_{e=1}^{n_i} w'_{i,e} \cdot N_{Q + F P'_{i,e} F^T}(\mathbf{x} - F\mathbf{x}'_{i,e}). &\text{(G.600)} \end{aligned}$$

G.35 PARA-GAUSSIAN FILTER CORRECTOR

We are to prove (17.105)-(17.112). From (17.145)-(17.151) we know that the data-updated p.g.fl. is, approximately,

$$G_{k+1|k}[h] \cong \prod_{i=1}^{\nu'} (1 - q_i + q_i \cdot f_i[h]) \quad \text{(G.601)}$$

$$\cdot \prod_{j=1}^{m} (1 - Q_i + Q_i \cdot F_i[h]) \quad \text{(G.602)}$$

where

$$q_i = \frac{q_i' f_i'[q_D]}{1 - q_i f_i[p_D]} \quad \text{(G.603)}$$

$$f_i(\mathbf{x}) = \frac{1 - p_D(\mathbf{x})}{1 - f_i'[p_D]} \cdot f_i'(\mathbf{x}) \quad \text{(G.604)}$$

$$Q_j = \frac{\sum_{i=1}^{\nu} \rho_i \cdot f_i'[p_D L_{\mathbf{z}_j}]}{\lambda c(\mathbf{z}_j) + \sum_{i=1}^{\nu} \rho_i \cdot f_i'[p_D L_{\mathbf{z}_j}]} \quad \text{(G.605)}$$

$$F_j(\mathbf{x}) = \frac{p_D(\mathbf{x}) \cdot L_{\mathbf{z}_j}(\mathbf{x}) \cdot \sum_{i=1}^{\nu} \rho_i \cdot f_i'(\mathbf{x})}{\sum_{i=1}^{\nu} \rho_i \cdot f_i'[p_D L_{\mathbf{z}_j}]} \quad \text{(G.606)}$$

$$\rho_i \triangleq \frac{q_i'}{1 - q_i' f_i'[p_D]} \quad \text{(G.607)}$$

Now p_D is constant. Thus

$$\rho_i = \frac{q_i'}{1 - q_i' f_i'[p_D]} = \frac{q_i'}{1 - p_D q_i'} \quad \text{(G.608)}$$

Also,

$$q_i = \frac{q_i' f_i[h q_D]}{1 - q_i f_i[p_D]} = \frac{q_i' \cdot (1 - p_D)}{1 - p_D \cdot q_i'} \quad \text{(G.609)}$$

and

$$f_i(\mathbf{x}) = \frac{1 - p_D}{1 - p_D} \cdot f_i'(\mathbf{x}) = f_i'(\mathbf{x}). \quad \text{(G.610)}$$

As for the measurement-generated tracks,

$$Q_j = \frac{\sum_{i=1}^{\nu'} p_D \rho_i \cdot f_i'[L_{\mathbf{z}_j}]}{\lambda c(\mathbf{z}_j) + \sum_{i=1}^{\nu'} p_D \rho_i \cdot f_i'[L_{\mathbf{z}_j}]}. \quad \text{(G.611)}$$

However,

$$f'_i[L_{\mathbf{z}_j}] \tag{G.612}$$

$$= \int L_{\mathbf{z}_j}(\mathbf{x}) \cdot f'_i(\mathbf{x}) d\mathbf{x} \tag{G.613}$$

$$= \sum_{e=1}^{\nu'} w'_{i,e} \int N_R(\mathbf{z}_j - H\mathbf{x}) \cdot N_{P'_{i,e}}(\mathbf{x} - \mathbf{x}'_{i,e}) d\mathbf{x} \tag{G.614}$$

$$= \sum_{e=1}^{\nu'} w'_{i,e} \int N_{R+HP'_{i,e}H^T}(\mathbf{z}_j - H\mathbf{x}'_{i,e}) \cdot N_{C_i}(\mathbf{x} - \mathbf{c}_{i,e,j}) d\mathbf{x} \tag{G.615}$$

$$= \sum_{e=1}^{\nu'} w'_{i,e} \cdot N_{R+HP'_{i,e}H^T}(\mathbf{z}_j - H\mathbf{x}'_{i,e}) \tag{G.616}$$

and so

$$Q_j = \frac{\sum_{e,i=1}^{\nu'} p_D \rho_i w'_{i,e} N_{R+HP'_{i,e}H^T}(\mathbf{z}_j - H\mathbf{x}'_{i,e})}{\lambda c(\mathbf{z}_j) + \sum_{e,i=1}^{\nu'} p_D \rho_i w'_{i,e} N_{R+HP'_i H^T}(\mathbf{z}_j - H\mathbf{x}'_{i,e})} \tag{G.617}$$

where

$$\mathbf{c}_{i,e,j} = \mathbf{x}'_{i,e} + K_{i,i}(\mathbf{z}_j - H\mathbf{x}'_{i,e}) \tag{G.618}$$

$$C_{i,e} = (1 - K_{i,e}H)P'_{i,e} \tag{G.619}$$

$$K_{i,e} = P'_{i,e}H^T \left(HP'_{i,e}H^T + R\right)^{-1}. \tag{G.620}$$

As for $F_j(\mathbf{x})$,

$$F_j(\mathbf{x}) \tag{G.621}$$

$$= \frac{p_D \cdot \sum_{i=1}^{\nu} \rho_i \cdot f'_i(\mathbf{x}) \cdot L_{\mathbf{z}_j}(\mathbf{x})}{p_D \cdot \sum_{i=1}^{\nu} \rho_i \cdot f'_i[L_{\mathbf{z}_j}]} \tag{G.622}$$

$$= \frac{\sum_{e,i=1}^{\nu'} \rho_i w'_{i,e} \cdot N_{P'_{i,e}}(\mathbf{x} - \mathbf{x}'_{i,e}) \cdot N_R(\mathbf{z}_j - H\mathbf{x})}{\sum_{e=1}^{\nu'} \sum_{i=1}^{\nu} p_D \rho_i w'_{i,e} N_{R+HP'_{i,e}H^T}(\mathbf{z}_j - H\mathbf{x}'_{i,e})} \tag{G.623}$$

$$= \frac{\left(\begin{array}{c} \sum_{e,i=1}^{\nu_{k+1|k}} \rho_i w'_{i,e} \cdot N_{R+HP'_{i,e}H^T}(\mathbf{z}_j - H\mathbf{x}'_i) \\ \cdot N_{C_{i,e}}(\mathbf{x} - \mathbf{c}_{i,e,j}) \end{array} \right)}{\sum_{e,i=1}^{\nu'} p_D \rho_i w'_{i,e} N_{R+HP'_{i,e}H^T}(\mathbf{z}_j - H\mathbf{x}'_{i,e})} \tag{G.624}$$

and so

$$= \frac{\left(\begin{array}{c}\sum_{e,i=1}^{\nu'} \rho_i w'_{i,e} \cdot N_{R+HP'_{i,e}H^T}(\mathbf{z}_j - H\mathbf{x}'_{i,e}) \\ \cdot N_{C_{i,e}}(\mathbf{x} - \mathbf{c}_{i,e,j}\end{array}\right)}{\sum_{e,i=1}^{\nu'} p_D \rho_i w'_{i,e} N_{R+HP'_{i,e}H^T}(\mathbf{z}_j - H\mathbf{x}'_{i,e})} \tag{G.625}$$

$$= \sum_{e,i=1}^{\nu'} w_{i,e,j} \cdot N_{C_{i,e}}(\mathbf{x} - \mathbf{c}_{i,e,j}) \tag{G.626}$$

where

$$w_{i,e,j} \tag{G.627}$$

$$= \frac{\rho_i \cdot w'_{i,e} \cdot N_{R+HP'_{i,e}H^T}(\mathbf{z}_j - H\mathbf{x}'_{i,e})}{\sum_{e'=1}^{\nu_{k+1|k}} \sum_{i'=1}^{\nu} p_D \rho_{i'} w'_{i',e'} N_{R+HP'_{i',e'}H^T}(\mathbf{z}_j - H\mathbf{x}'_{i',e'})}. \tag{G.628}$$

Appendix H

Solutions to Exercises

Solution 1 *to Exercise 1: We are to prove (2.114):*

$$f_{k+1}(\mathbf{z}|\mathbf{x}) = f_{\mathbf{W}_{k+1}}(\eta^{-1}_{k+1,\mathbf{x}}(\mathbf{z})) \cdot J_{\eta^{-1}_{k+1,\mathbf{x}}}(\mathbf{z}) \tag{H.1}$$

To see this, note that

$$\begin{aligned}
\int_T f_{k+1|k}(\mathbf{z}|\mathbf{x})d\mathbf{z} &= \Pr(\mathbf{Z}_{k+1} \in T|\mathbf{x}) && \text{(H.2)}\\
&= \Pr(\eta_{k+1,\mathbf{x}}(\mathbf{W}_{k+1}) \in T|\mathbf{x}') && \text{(H.3)}\\
&= \Pr(\mathbf{W}_{k+1} \in \eta^{-1}_{k+1,\mathbf{x}}(T) && \text{(H.4)}\\
&= \int \mathbf{1}_{\eta^{-1}_{k+1,\mathbf{x}}(T)}(\mathbf{w}) \cdot f_{\mathbf{W}_{k+1}}(\mathbf{w})d\mathbf{w} && \text{(H.5)}\\
&= \int \mathbf{1}_S(\eta_{k+1,\mathbf{x}}(\mathbf{w})) \cdot f_{\mathbf{W}_{k+1}}(\mathbf{w})d\mathbf{w} && \text{(H.6)}\\
&= \int \mathbf{1}_S(\mathbf{y}) \cdot f_{\mathbf{W}_{k+1}}(\eta^{-1}_{k+1,\mathbf{x}}(\mathbf{y})) \cdot J_{\eta^{-1}_{k+1,\mathbf{x}}}(\mathbf{y})d\mathbf{y} && \text{(H.7)}\\
&= \int_S f_{\mathbf{V}_k}(\eta^{-1}_{k+1,\mathbf{x}}) \cdot J_{\eta^{-1}_{k+1,\mathbf{x}}}(\mathbf{y})d\mathbf{y}. && \text{(H.8)}
\end{aligned}$$

Solution 2 *to Exercise 2: We are to prove (2.149):* $K(f_{k|k}; f_0) = \log V - \varepsilon_{k|k}$. *This follows from*

$$K(f_{k|k}; f_0) = \int_D f(\mathbf{x}) \cdot \log\left(\frac{f_{k|k}(\mathbf{x}|Z^k)}{V^{-1}}\right) d\mathbf{x} \tag{H.9}$$

$$= \int_D f_{k|k}(\mathbf{x}|Z^k) \cdot \log f_{k|k}(\mathbf{x}|Z^k) d\mathbf{x} \tag{H.10}$$

$$+ \log V \cdot \int_D f_{k|k}(\mathbf{x}|Z^k) d\mathbf{x} \tag{H.11}$$

$$= \int_D f_{k|k}(\mathbf{x}|Z^k) \cdot \log f_{k|k}(\mathbf{x}|Z^k) d\mathbf{x} + \log V \tag{H.12}$$

$$= \log V - \varepsilon_{k|k}. \tag{H.13}$$

Solution 3 *to Exercise 3: We are to prove the corrector equations for the Gaussian sum filter, (2.196)-(2.200). That is, we are to show that*

$$f_{k+1|k+1}(\mathbf{x}|Z^{k+1}) = \sum_{j=1}^{L_{k+1}} \sum_{i=1}^{n_{k+1|k}} w_{k+1|k+1}^{i,j} \cdot N_{E_{k+1|k}^{i,j}}(\mathbf{x} - \mathbf{e}_{k+1|k}^{i,j}) \tag{H.14}$$

where

$$(E_{k+1|k}^{i,j})^{-1} \mathbf{e}_{k+1|k+1}^{i,j} = (P_{k+1|k}^{i})^{-1} \mathbf{x}_{k+1|k}^{i} \tag{H.15}$$

$$+ (H_{k+1}^{j})^T (R_{k+1}^{j})^{-1} \mathbf{z}_{k+1} \tag{H.16}$$

$$(E_{k+1|k}^{i,j})^{-1} = (P_{k+1|k}^{i})^{-1} + (H_{k+1}^{j})^T (R_{k+1}^{j})^{-1} H_{k+1}^{j} \tag{H.17}$$

and

$$w_{k+1|k+1}^{i,j} \tag{H.18}$$

$$= \frac{\lambda_k^j \cdot w_{k+1|k}^i \cdot N_{C_k^{j,i}}(\mathbf{z}_{k+1} - H_k^j \mathbf{x}_{k+1|k}^i)}{\sum_{e=1}^{L_{k+1}} \sum_{l=1}^{n_{k+1|k}} \lambda_k^e \cdot w_{k+1|k}^l \cdot N_{C_k^{j,i}}(\mathbf{z}_{k+1} - H_k^e \mathbf{x}_{k+1|k}^l)} \tag{H.19}$$

and where

$$C_k^{j,i} \triangleq R_k^j + H_k^j P_{k+1|k}^i (H_k^j)^T. \tag{H.20}$$

By assumption, from (2.176) and (2.178)

$$f_{k+1}(\mathbf{z}|\mathbf{x}) = \sum_{j=1}^{L_{k+1}} \lambda_k^j \cdot N_{R_k^j}(\mathbf{z} - H_k^j \mathbf{x}) \tag{H.21}$$

$$f_{k+1|k}(\mathbf{x}|Z^k) = \sum_{i=1}^{n_{k+1|k}} w_{k+1|k}^i \cdot N_{P_{k+1|k}^i}(\mathbf{x} - \mathbf{x}_{k+1|k}^i). \tag{H.22}$$

The Bayes normalization factor is

$$f_{k+1}(\mathbf{z}|Z^k) = \int f_{k+1}(\mathbf{z}|\mathbf{x}) \cdot f_{k+1|k}(\mathbf{x}|Z^k) d\mathbf{x} \tag{H.23}$$

$$= \int \left(\sum_{j=1}^{L_{k+1}} \lambda_k^j \cdot N_{R_k^j}(\mathbf{z} - H_k^j \mathbf{x}) \right) \tag{H.24}$$

$$\cdot \left(\sum_{i=1}^{n_{k+1|k}} w_{k+1|k}^i \cdot N_{P_{k+1|k}^i}(\mathbf{x} - \mathbf{x}_{k+1|k}^i) \right) d\mathbf{x} \tag{H.25}$$

$$= \sum_{j=1}^{L_{k+1}} \sum_{i=1}^{n_{k+1|k}} \lambda_k^j \cdot w_{k+1|k}^i \tag{H.26}$$

$$\cdot \int N_{R_k^j}(\mathbf{z} - H_k^j \mathbf{x}) \cdot N_{P_{k+1|k}^i}(\mathbf{x} - \mathbf{x}_{k+1|k}^i) d\mathbf{x}. \tag{H.27}$$

By the fundamental identify for Gaussian distributions (D.1), this becomes

$$f_{k+1}(\mathbf{z}|Z^k) = \sum_{j=1}^{L_{k+1}} \sum_{i=1}^{n_{k+1|k}} \lambda_k^j \cdot w_{k+1|k}^i \tag{H.28}$$

$$\cdot N_{R_k^j + H_k^j P_{k+1|k}^i (H_k^j)^T}(\mathbf{z} - H_k^j \mathbf{x}_{k+1|k}^i). \tag{H.29}$$

The denominator of Bayes' rule is

$$
\begin{aligned}
& f_{k+1}(\mathbf{z}|\mathbf{x}) \cdot f_{k+1|k}(\mathbf{x}|Z^k) && \text{(H.30)} \\
= & \left(\sum_{j=1}^{L_{k+1}} \lambda_k^j \cdot N_{R_k^j}(\mathbf{z} - H_k^j \mathbf{x}) \right) && \text{(H.31)} \\
& \cdot \left(\sum_{i=1}^{n_{k+1|k}} w_{k+1|k}^i \cdot N_{P_{k+1|k}^i}(\mathbf{x} - \mathbf{x}_{k+1|k}^i) \right) && \text{(H.32)} \\
= & \sum_{j=1}^{L_{k+1}} \sum_{i=1}^{n_{k+1|k}} \lambda_k^j \cdot w_{k+1|k}^i && \text{(H.33)} \\
& \cdot N_{R_k^j}(\mathbf{z} - H_k^j \mathbf{x}) \cdot N_{P_{k+1|k}^i}(\mathbf{x} - \mathbf{x}_{k+1|k}^i) && \text{(H.34)} \\
= & \sum_{j=1}^{L_{k+1}} \sum_{i=1}^{n_{k+1|k}} \lambda_k^j \cdot w_{k+1|k}^i && \text{(H.35)} \\
& \cdot N_{R_k^j + H_k^j P_{k+1|k}^i (H_k^j)^T}(\mathbf{z} - H_k^j \mathbf{x}_{k+1|k}^i) && \text{(H.36)} \\
& \cdot N_{E_{k+1|k}^{i,j}}(\mathbf{x} - \mathbf{e}_{k+1|k+1}^{i,j}). && \text{(H.37)}
\end{aligned}
$$

If we substitute these results into Bayes' rule

$$
f_{k+1|k+1}(\mathbf{x}|Z^{k+1}) = \frac{f_{k+1}(\mathbf{z}_{k+1}|\mathbf{x}) \cdot f_{k+1|k}(\mathbf{x}|Z^k)}{f_{k+1}(\mathbf{z}_{k+1}|Z^k)} \tag{H.38}
$$

then the result immediately follows from inspection.

Solution 4 *to Exercise 4: We are to prove the formula for the expected a posteriori (EAP) estimator for the Gaussian mixture filter—see (2.204). We are to show that*

$$
\hat{\mathbf{x}}_{k+1|k+1}^{EAP} = \sum_{j=1}^{L_{k+1}} \sum_{i=1}^{n_{k+1|k}} w_{k+1|k+1}^{i,j} \cdot \mathbf{e}_{k+1|k}^{i,j}. \tag{H.39}
$$

From the definition of an EAP estimate and from (2.196) we get

$$\hat{\mathbf{x}}^{EAP}_{k+1|k+1} = \int \mathbf{x}\cdot f_{k+1|k+1}(\mathbf{x}|Z^{k+1})d\mathbf{x} \quad \text{(H.40)}$$

$$= \int \mathbf{x}\cdot \sum_{j=1}^{L_{k+1}}\sum_{i=1}^{n_{k+1|k}} w^{i,j}_{k+1|k+1}\cdot N_{E_{i,j}}(\mathbf{x}-\mathbf{e}^{i,j}_{k+1|k})d\mathbf{x} \quad \text{(H.41)}$$

$$= \sum_{j=1}^{L_{k+1}}\sum_{i=1}^{n_{k+1|k}} w^{i,j}_{k+1|k+1}\cdot \int \mathbf{x}\cdot N_{E_{i,j}}(\mathbf{x}-\mathbf{e}^{i,j}_{k+1|k})d\mathbf{x} \quad \text{(H.42)}$$

$$= \sum_{j=1}^{L_{k+1}}\sum_{i=1}^{n_{k+1|k}} w^{i,j}_{k+1|k+1}\cdot \mathbf{e}^{i,j}_{k+1|k}. \quad \text{(H.43)}$$

Solution 5 *to Exercise 5: We are to prove that the operator '⊗' of Example 7 in Section 3.5.3 is commutative and associative. From (3.62) it is easy to see that*

$$\{(C_1,\mathbf{c}_1)\otimes(C_2,\mathbf{c}_2)\}\otimes(C_3,\mathbf{c}_3) \quad \text{(H.44)}$$

$$= (C,\mathbf{c}) \quad \text{(H.45)}$$

$$= (C_1,\mathbf{c}_1)\otimes\{(C_2,\mathbf{c}_2)\otimes(C_3,\mathbf{c}_3)\} \quad \text{(H.46)}$$

where

$$C = (C_1^{-1}+C_2^{-1}+C_3^{-1})^{-1} \quad \text{(H.47)}$$

$$\mathbf{c} = (C_1^{-1}+C_2^{-1}+C_3^{-1})^{-1}(C_1^{-1}\mathbf{c}_1+C_2^{-1}\mathbf{c}_2+C_3^{-1}\mathbf{c}_3) \quad \text{(H.48)}$$

Solution 6 *to Exercise 6: We are to prove (4.24)-(4.25):*

$$\Sigma_A(f\wedge f') = \Sigma_A(f)\cap\Sigma_A(f') \quad \text{(H.49)}$$

$$\Sigma_A(f\vee f') = \Sigma_A(f)\cup\Sigma_A(f'). \quad \text{(H.50)}$$

For (4.24), note that the following statements are respectively equivalent: (1) $u\in\Sigma_A(f\wedge f')$, (2) $A\leq\min\{f(u),f'(u)\}$, (3) $A\leq f(u)$ and $A\leq f'(u)$, (4) $u\in\Sigma_A(f)$ and $u\in\Sigma_A(f')$, (5) $u\in\Sigma_A(f)\cap\Sigma_A(f')$. Thus $u\in\Sigma_A(f\wedge f')$ if and only if $u\in\Sigma_A(f)\cap\Sigma_A(f')$. From this, $\Sigma_A(f\wedge f')=\Sigma_A(f)\cap\Sigma_A(f')$ follows. Similar reasoning applies to (4.25).

Solution 7 *to Exercise 7: We are to prove that the Hamacher fuzzy conjunction of Example 9 is (a) such that $0\leq a\wedge a'\leq 1$ for all a,a', (b) associative, and (c) a* *copula. The first two properties are easy to prove. To prove that the conjunction is a copula we must show that*

$$\frac{ab}{a+b-ab}+\frac{a'b'}{a'+b'-a'b'} \overset{?}{\geq} \frac{ab'}{a+b'-ab'}+\frac{a'b}{a'+b-a'b} \tag{H.51}$$

for all $a, a', b, b' \in [0,1]$ *with* $a \leq a', b \leq b'$. *We first prove it for* $a = 0$, *which will also establish it for* $b = 0$. *In this case we can write*

$$\frac{1}{x'+y'-1} \overset{?}{\geq} \frac{1}{x'+y-1} \tag{H.52}$$

where

$$x' = \frac{1}{a'}, \quad y = \frac{1}{b}, \quad y' = \frac{1}{b'} \tag{H.53}$$

and thus $y \geq y'$. *Since* $x'+y'-1 \geq 0$ *and* $x'+y-1 \geq 0$ *we get*

$$x'+y-1 \overset{?}{\geq} x'+y'-1, \tag{H.54}$$

which is identically true. In general we are to show that

$$\frac{1}{x+y-1}-\frac{1}{x'+y-1} \overset{?}{\geq} \frac{1}{x+y'-1}-\frac{1}{x'+y'-1} \tag{H.55}$$

where $x \geq x'$, $y \geq y'$, *and* $x, x', y, y' \in [1, \infty)$. *Then*

$$\frac{x'+y-1-x-y+1}{(x+y-1)(x'+y-1)} \overset{?}{\geq} \frac{x'+y'-1-x-y'+1}{(x+y'-1)(x'+y'-1)} \tag{H.56}$$

$$\frac{x'-x}{(x+y-1)(x'+y-1)} \overset{?}{\geq} \frac{x'-x}{(x+y'-1)(x'+y'-1)} \tag{H.57}$$

$$\frac{1}{(x+y-1)(x'+y-1)} \overset{?}{\leq} \frac{1}{(x+y'-1)(x'+y'-1)} \tag{H.58}$$

$$(x+y'-1)(x'+y'-1) \overset{?}{\leq} (x+y-1)(x'+y-1) \tag{H.59}$$

$$xx'+(x+x')(y'-1)+(y'-1)^2 \tag{H.60}$$

$$\stackrel{?}{\leq} xx' + (x+x')(y-1) + (y-1)^2 \tag{H.61}$$

$$(x+x')(y'-1) + (y'-1)^2 \stackrel{?}{\leq} (x+x')(y-1) + (y-1)^2 \tag{H.62}$$

$$0 \stackrel{?}{\leq} (x+x')(y-y') + (y-1)^2 - (y'-1)^2 \tag{H.63}$$

$$0 \stackrel{?}{\leq} (x+x')(y-y') + (y-y')(y+y'-2) \tag{H.64}$$

$$0 \stackrel{?}{\leq} (x+x'+y+y'-2)(y-y'). \tag{H.65}$$

This last inequality is identically true since, by assumption, $x, x', y, y' \geq 1$ *and since* $y \geq y'$.

Solution 8 *to Exercise 8: We are to prove that the FGM conjunction of (4.57) is associative. After expansion of terms it can be shown that*

$$
\begin{aligned}
& a \wedge (a' \wedge a'') && \text{(H.66)}\\
= \; & a\,[a'a'' + \theta(1-a')(1-a'')] && \text{(H.67)}\\
& +\theta(1-a)(1-[a'a'' + \theta(1-a')(1-a'')]) && \text{(H.68)}\\
& -aa'a'' + \theta a(1-a')(1-a'') && \text{(H.69)}\\
& +\theta(1-a)(1-a'a'' - \theta(1-a')(1-a'')) && \text{(H.70)}\\
& (2aa'a'' - aa' - aa'' - a'a'')\,\theta && \text{(H.71)}\\
& +(-1+a+a'+a-aa'-aa''-a'a''+aa'a'')\,\theta^2 && \text{(H.72)}\\
& -aa'a''. && \text{(H.73)}
\end{aligned}
$$

This formula is invariant under permutation of the a, a', a'' *and so associativity is established.*

Solution 9 *to Exercise 9: We are to verify that (4.61) and (4.62) are true:*

$$\Sigma_\alpha(f) \cap \Sigma_\alpha(g) = \Sigma_\alpha(f \wedge g) \tag{H.74}$$

$$\Sigma_\alpha(f) \cup \Sigma_\alpha(g) = \Sigma_\alpha(f \vee g) \tag{H.75}$$

where $\alpha(u)$ *is a uniform random scalar field, and* $\Sigma_\alpha(f) \triangleq \{u|\ \alpha(u) \leq f(u)\}$. *To see this note that*

$$\begin{aligned}
\Sigma_\alpha(f) \cap \Sigma_\alpha(g) &= \{u|\ \alpha(u) \leq f(u)\} \cap \{u|\ \alpha(u) \leq g(u)\} && \text{(H.76)}\\
&= \{u|\ \alpha(u) \leq f(u),\ \alpha(u) \leq g(u)\} && \text{(H.77)}\\
&= \{u|\ \alpha(u) \leq \min\{f(u), g(u)\}\} && \text{(H.78)}\\
&= \{u|\ \alpha(u) \leq (f \wedge g)(u)\} && \text{(H.79)}\\
&= \Sigma_\alpha(f \wedge g) && \text{(H.80)}
\end{aligned}$$

The other equation follows similarly.

Solution 10 *to Exercise 10: We are to prove (4.71) and (4.72),* $W_f \cap W_g = W_{f\wedge g}$ *and* $W_f \cup W_g = W_{f\vee g}$, *but that* $W_f^c \neq W_{f^c}$. *To see this, recall from (4.81) that* $W_f = \{(a,u)|\ a \leq f(u)\}$. *Thus*

$$\begin{aligned}
W_f \cap W_g &= \{(a,u)|\ a \leq f(u)\} \cap \{(a,u)|\ a \leq g(u)\} && \text{(H.81)}\\
&= \{(a,u)|\ a \leq f(u),\ a \leq g(u)\} && \text{(H.82)}\\
&= \{(a,u)|\ a \leq \min\{f(u), g(u)\}\} && \text{(H.83)}\\
&= \{(a,\mathbf{y})|\ a \leq (f \wedge g)(u)\}\} = W_{f\wedge g}. && \text{(H.84)}
\end{aligned}$$

The proof of $W_f \cup W_g = W_{f\vee g}$ *is similar. On the other hand, note that* $W_f^c = \{(a,u)|\ a > f(u)\}$ *but that*

$$W_{f^c} = \{(a,u)|\ a \leq 1 - f(u)\} = \{(a,u)|\ 1 - a \geq f(u)\} \qquad \text{(H.85)}$$

Solution 11 *to Exercise 11: We are to prove (4.76)-(4.79):*

$$\begin{aligned}
\Sigma_A(V \cap W) &= \Sigma_A(V) \cap \Sigma_A(W) && \text{(H.86)}\\
\Sigma_A(V \cup W) &= \Sigma_A(V) \cup \Sigma_A(W) && \text{(H.87)}\\
\Sigma_A(W^c) &= \Sigma_A(W)^c && \text{(H.88)}\\
\Sigma_A(\emptyset) &= \emptyset, \quad \Sigma_A(\mathfrak{U}^*) = \mathfrak{U}. && \text{(H.89)}
\end{aligned}$$

Recall that $\Sigma_A(W) \triangleq \{u|\ (u,A) \in W\}$. *First,*

$$\begin{aligned}
\Sigma_A(V) \cap \Sigma_A(W) &= \{u|\ (u,A) \in V\} \cap \{u|\ (u,A) \in W\} && \text{(H.90)}\\
&= \{u|\ (u,A) \in V,\ (u,A) \in W\} && \text{(H.91)}\\
&= \{u|\ (u,A) \in V \cap W\} = \Sigma_A(V \cap W). && \text{(H.92)}
\end{aligned}$$

Likewise for $\Sigma_A(V \cup W) = \Sigma_A(V) \cup \Sigma_A(W)$. *Note that*

$$\Sigma_A(W^c) = \{u|\ (u,A) \in W^c\} = \{u|\ (u,A) \notin W\} \quad \text{(H.93)}$$
$$= \{u|\ (u,A) \in W\}^c = \Sigma_A(W)^c. \quad \text{(H.94)}$$

And so on.

Solution 12 *to Exercise 12: We are to prove (4.116): Voorbraak probability is consistent with Dempster's combination. Note that*

$$\nu_{m_1 * m_2}(u) \propto \sum_{U \ni u} (m_1 * m_2)(U) \quad \text{(H.95)}$$
$$\propto \sum_{U \ni u} \sum_{V \cap W = U} m_1(V) \cdot m_2(W) \quad \text{(H.96)}$$
$$= \sum_{V \ni u} \sum_{W \ni u} m_1(V) \cdot m_2(W) \quad \text{(H.97)}$$
$$= \left(\sum_{V \ni u} m_1(V) \right) \left(\sum_{W \ni u} m_2(W) \right) \quad \text{(H.98)}$$
$$\propto \nu_{m_1}(u) \cdot \nu_{m_2}(u) \propto (\nu_{m_1} * \nu_{m_2})(u). \quad \text{(H.99)}$$

Thus $K \cdot v_{m_1 * m_2}(u) = (\nu_{m_1} * \nu_{m_2})(u)$ *for some constant* K. *Summing both sides over* u, *we get* $K = 1$, *from which (4.116) follows from the definition of Dempster's combination—see (4.87).*

Solution 13 *to Exercise 13: We are to prove (4.121): Pignistic probability is consistent with modified Dempster's combination. Note that*

$$\pi_{m_1 *_q m_2}(u) = q(u) \sum_{U \ni u} \frac{(m_1 *_q m_2)(U)}{q(U)} \quad \text{(H.100)}$$
$$\propto q(u) \sum_{U \ni u} \sum_{V \cap W = U} \frac{m_1(V) \cdot m_2(W) \cdot \alpha_q(V,W)}{q(U)} \quad \text{(H.101)}$$

and so

$$= q(u) \sum_{V \ni u} \sum_{W \ni u} \frac{m_1(V) \cdot m_2(W)}{q(V) \cdot q(W)}$$

and so

$$= \left(q(u)\sum_{V\ni u}\frac{m_1(V)}{q(V)}\right)\left(q(u)\sum_{W\ni u}\frac{m_2(W)}{q(W)}\right) \quad \text{(H.102)}$$

$$\cdot q(u)^{-1} \quad \text{(H.103)}$$

$$= \pi_{m_1}(u)\cdot\pi_{m_2}(u)\cdot q(u)^{-1} \propto (\pi_{m_1} *_q \pi_{m_2})(u) \quad \text{(H.104)}$$

from which follows the assertion.

Solution 14 *to Exercise 14: We are to prove (4.128): The probabilistic interpretation of Dempster's combination. Dempster's combination can be expressed as a conditional probability defined in terms of random subsets. Note that if* $U \neq \emptyset$*, then*

$$\Pr(\Sigma_{m_1}\cap\Sigma_{m_2} = U, \Sigma_{m_1}\cap\Sigma_{m_2}\neq\emptyset) \quad \text{(H.105)}$$

$$= \sum_{V,W}\Pr(\Sigma_{m_1} = V, \Sigma_{m_2} = W, V\cap W = U, V\cap W\neq\emptyset) \quad \text{(H.106)}$$

$$= \sum_{V\cap W = U\neq\emptyset}\Pr(\Sigma_{m_1} = V, \Sigma_{m_2} = W) \quad \text{(H.107)}$$

$$= \sum_{V\cap W = U\neq\emptyset}\Pr(\Sigma_{m_1} = V)\cdot\Pr(\Sigma_{m_2} = W) \quad \text{(H.108)}$$

$$= \sum_{V\cap W = U\neq\emptyset} m_1(V)\cdot m_2(W) \quad \text{(H.109)}$$

whereas

$$\Pr(\Sigma_{m_1}\cap\Sigma_{m_2}\neq\emptyset) = \sum_{V\cap W\neq\emptyset}\Pr(\Sigma_{m_1} = V, \Sigma_{m_2} = W) \quad \text{(H.110)}$$

$$= \sum_{V\cap W\neq\emptyset} m_1(V)\cdot m_2(W). \quad \text{(H.111)}$$

Solution 15 *to Exercise 15: We are to prove that fuzzy Dempster combination (4.129) reduces to ordinary Dempster's combination (4.87) when all focal sets are crisp. By (4.129)*

$$(m_1 *_{FDS} m_2)(f) \triangleq \alpha_{FDS}^{-1}\sum_{f_1\cdot f_2 = f} m_1(f_1)\cdot m_2(f_2) \quad \text{(H.112)}$$

where

$$0 \neq \alpha_{FDS} = \sum_{f_1 \cdot f_2 \neq 0} m_1(f_1) \cdot m_2(f_2). \tag{H.113}$$

By assumption, $m_1(f) = 0$ *except when* f *has the form* $\mathbf{1}_S$ *for some subset* S*; and likewise for* m_2*. Write* $\tilde{m}_1(S) = m_1(\mathbf{1}_S)$ *and* $\tilde{m}_2(S) = m_2(\mathbf{1}_S)$*. Then* $f_1 \cdot f_2 = \mathbf{1}_{S_1} \cdot \mathbf{1}_{S_2} = \mathbf{1}_{S_1 \cap S_2}$ *and so*

$$\alpha_{FDS} = \sum_{S_1 \cap S_2 \neq \emptyset} m_1(\mathbf{1}_{S_1}) \cdot m_2(\mathbf{1}_{S_2}) \tag{H.114}$$

$$= \sum_{S_1 \cap S_2 \neq \emptyset} \tilde{m}_1(S_1) \cdot \tilde{m}_2(S_2) = \alpha_{DS}. \tag{H.115}$$

Likewise,

$$(m_1 *_{FDS} m_2)(f) = \alpha_{FDS}^{-1} \sum_{S_1 \cap S_2 = S} m_1(\mathbf{1}_{S_1}) \cdot m_2(\mathbf{1}_{S_2}) \tag{H.116}$$

$$= \alpha_{DS}^{-1} \sum_{S_1 \cap S_2 = S} \tilde{m}_1(S_1) \cdot \tilde{m}_2(S_2) \tag{H.117}$$

$$= (m_1 *_{DS} m_2)(S). \tag{H.118}$$

Solution 16 *to Exercise 16: We are to show that* $T \cap Y = (S_1 \cap X_1) \cup (S_2 \cap X_2)$*, where* T *and* Y *are defined in (4.149) and (4.150). By (4.149) and (4.150) we know that*

$$Y \triangleq (S_1 \cap X_1) \cup (S_2 \cap X_2) \cup (X_1 \cap X_2) \tag{H.119}$$

$$T \triangleq S_1 \cup S_2. \tag{H.120}$$

Thus

$$T \cap Y = [S_1 \cup S_2] \cap [(S_1 \cap X_1) \cup (S_2 \cap X_2) \cup (X_1 \cap X_2)] \tag{H.121}$$

$$= [(S_1 \cup S_2) \cap S_1 \cap X_1] \cup [(S_1 \cup S_2) \cap S_2 \cap X_2)] \tag{H.122}$$

$$\cup [(S_1 \cup S_2) \cap X_1 \cap X_2] \tag{H.123}$$

$$= [S_1 \cap X_1] \cup [S_2 \cap X_2] \cup [(S_1 \cup S_2) \cap X_1 \cap X_2] \tag{H.124}$$

$$= [S_1 \cap X_1] \cup [S_2 \cap X_2] \cup [S_1 \cap X_1 \cap X_2] \tag{H.125}$$

$$\cup [S_2 \cap X_1 \cap X_2] \tag{H.126}$$

$$= [S_1 \cap X_1] \cup [S_1 \cap X_1 \cap X_2] \cup [S_2 \cap X_2)] \tag{H.127}$$

$$\cup [S_2 \cap X_1 \cap X_2] \tag{H.128}$$

$$= [S_1 \cap X_1] \cup [S_2 \cap X_2]. \tag{H.129}$$

Solution 17 *to Exercise 17: We are to show that*

$$f(W_1, ..., W_m|\mathbf{x}) = f(W_1 \cap ... \cap W_m|\mathbf{x}) \tag{H.130}$$

for any generalized fuzzy subsets $W_1, ..., W_m$ *of* $\mathfrak{Z}_0$. *It is enough to prove the statement for* $m = 2$. *In this case*

$$\begin{aligned} f(W_1, W_2|\mathbf{x}) &\triangleq \Pr(\eta(\mathbf{x}) \in \Sigma_A(W_1),\ \eta(\mathbf{x}) \in \Sigma_{A'}(W_2)) & \text{(H.131)} \\ &= \Pr(\eta(\mathbf{x}) \in \Sigma_A(W_1) \cap \Sigma_{A'}(W_2)) & \text{(H.132)} \\ &= \Pr(\eta(\mathbf{x}) \in \Sigma_A(W_1 \cap W_2)) & \text{(H.133)} \\ &= f(W_1 \cap W_2|\mathbf{x}) & \text{(H.134)} \end{aligned}$$

where the third equation follows from (4.71).

Solution 18 *to Exercise 18: We are to generalize (5.119)-(5.123) to the case when the focal sets of b.m.a.s and generalized b.m.a.'s are fuzzy:*

$$f(o|\mathbf{x}) = |o| \cdot f(m_o|\mathbf{x}), \quad m_{o \cap o'} = m_o * m_{o'} \tag{H.135}$$

where $m_o(g) \triangleq |o|^{-1} \cdot o(g)$ *and* $m_o(0) = 0$ *otherwise; and where* $|o| \triangleq \sum_{g \neq 0} o(g)$. *For the first equation, from (5.73) we get*

$$\begin{aligned} f(m_o|\mathbf{x}) &= \sum_{g \neq 0} m_o(g) \cdot g(\eta(\mathbf{x})) & \text{(H.136)} \\ &= \sum_{g \neq 0} |o|^{-1} \cdot o(g) \cdot g(\eta(\mathbf{x})) & \text{(H.137)} \\ &= |o|^{-1} \cdot \sum_{g} o(g) \cdot g(\eta(\mathbf{x})) & \text{(H.138)} \\ &= |o|^{-1} \cdot f(o|\mathbf{x}). & \text{(H.139)} \end{aligned}$$

As for the second equation, from the definitions of unnormalized and normalized DS combination, (4.95) and (4.87),

$$\begin{aligned} m_{o \cap o'}(g'') &= \frac{(o \cap o')(g'')}{\sum_{g'' \neq 0}(o \cap o')(g'')} = \frac{\sum_{g \cdot g' = g''} o(g) \cdot o'(g')}{\sum_{g \cdot g' \neq 0} o(g) \cdot o'(g')} & \text{(H.140)} \\ &= (o * o')(g''). & \text{(H.141)} \end{aligned}$$

Solution 19 *to Exercise 19: We are to prove (5.147) and (5.148) and, from them, that conversion* $o \mapsto m_o$ *of generalized fuzzy b.m.a.s to fuzzy b.m.a.s is Bayes-invariant. For (5.147:) From (5.73) we get*

$$f(o|\mathbf{x}) = \sum_g o(g) \cdot g(\eta(\mathbf{x})) = |o| \cdot \sum_{g \neq 0} |o|^{-1} \cdot o(g) \cdot g(\eta(\mathbf{x})) \quad \text{(H.142)}$$

$$= |o| \cdot \sum_{g \neq 0} m_o(g) \cdot g(\eta(\mathbf{x})) = |o| \cdot f(m_o|\mathbf{x}). \quad \text{(H.143)}$$

For (5.148): From (4.87) we get

$$m_{o \cap o'}(g'') = |o \cap o'|^{-1} \cdot (o \cap o')(g'') \quad \text{(H.144)}$$

$$= \frac{\sum_{g \cdot g' = g''} o(g) \cdot o'(g')}{\sum_{g \cdot g' \neq 0} o(g) \cdot o'(g')} \quad \text{(H.145)}$$

$$= (m_o * m_{o'})(g''). \quad \text{(H.146)}$$

Solution 20 *to Exercise 20: We are to prove (5.150) and (5.151) and, from them, that conversion* $g \mapsto o_g$ *of fuzzy membership functions to fuzzy b.m.a.s is Bayes-invariant. For (5.150), from (5.73) and (5.29) we get*

$$f(o_g|\mathbf{x}) = \sum_{g'} o_g(g') \cdot g'(\eta(\mathbf{x})) = g(\eta(\mathbf{x})) = f(g|\mathbf{x}). \quad \text{(H.147)}$$

For (5.151): If $g \cdot g' \neq 0$ *then from (4.87)*

$$o_{g \cdot g'}(g'') = \begin{cases} 1 & \text{if} \quad g'' = g \cdot g' \\ 0 & \text{if} \quad \text{otherwise} \end{cases}. \quad \text{(H.148)}$$

$$= \sum_{\gamma \cdot \gamma' = g''} o_g(\gamma) \cdot o_{g'}(\gamma') = (o_g * o_{g'})(g'') \quad \text{(H.149)}$$

where

$$\alpha_{DS}(o_g, o_{g'}) = \sum_{\gamma \cdot \gamma' \neq 0} o_g(\gamma) \cdot o_{g'}(\gamma') = o_g(g) \cdot o_{g'}(g') = 1. \quad \text{(H.150)}$$

Solution 21 *to Exercise 21: We are to prove (5.153) and (5.154) and, from them, that the conversion* $o \mapsto \mu_o$ *from fuzzy b.m.a.s to fuzzy membership functions is*

Bayes-invariant. For (5.153), from (5.73) and (5.29) we get

$$f(o|\mathbf{x}) = \sum_g o(g) \cdot g(\eta(\mathbf{x})) = \mu_o(\eta(\mathbf{x})) = f(\mu_o|\mathbf{x}). \tag{H.151}$$

For (5.154): From (4.87),

$$\mu_{o*o'}(\mathbf{z}) = \sum_{g''} (o * o')(g'') \cdot g''(\mathbf{z}) \tag{H.152}$$

$$= \alpha^{-1} \sum_{g''} \sum_{g \cdot g' = g''} o(g) \cdot o'(g') \cdot g(\mathbf{z}) \cdot g(\mathbf{z}) \tag{H.153}$$

$$= \alpha^{-1} \sum_{g,g'} o(g) \cdot o'(g') \cdot g(\mathbf{z}) \cdot g'(\mathbf{z}) \tag{H.154}$$

$$= \alpha^{-1} \left(\sum_g o(g) \cdot g(\mathbf{z}) \right) \left(\sum_{g'} o'(g') \cdot g'(\mathbf{z}) \right) \tag{H.155}$$

$$= \alpha^{-1} \cdot \mu_o(\mathbf{z}) \cdot \mu_{o'}(\mathbf{z}) = \alpha^{-1} \cdot (\mu_o \cdot \mu_{o'})(\mathbf{z}). \tag{H.156}$$

Solution 22 *to Exercise 22: We are to prove (5.161) and (5.162) and, from them, that the conversion* $o \mapsto \varphi_o$ *of fuzzy b.m.a.s to probability density functions is Bayes-invariant. For (5.161), from (5.160) and (5.73) we get*

$$f(\varphi_o|\mathbf{x}) = \frac{\varphi_o(\eta(\mathbf{x}))}{\sup_{\mathbf{z}} \varphi_o(\mathbf{z})} = \frac{\sum_g o(g) \cdot g(\eta(\mathbf{x}))}{\sup_{\mathbf{z}} \sum_g o(g) \cdot g(\mathbf{z})} = K \cdot f(o|\mathbf{x}) \tag{H.157}$$

where $K = \sup_{\mathbf{z}} \sum_g o(g) \cdot g(\mathbf{z})$. *For (5.162), from (4.129) and (5.158)*

$$\varphi_{o*o'}(\mathbf{z}) \propto \sum_{g''} (o * o')(g'') \cdot g''(\mathbf{z}) \tag{H.158}$$

$$= \sum_{g \cdot g' = g''} o(g) \cdot o'(g') \cdot g(\mathbf{z}) \cdot g'(\mathbf{z}) \tag{H.159}$$

$$= \sum_{g,g'} o(g) \cdot o'(g') \cdot g(\mathbf{z}) \cdot g'(\mathbf{z}) \tag{H.160}$$

$$= \left(\sum_g o(g) \cdot g(\mathbf{z}) \right) \left(\sum_{g'} o'(g') \cdot g'(\mathbf{z}) \right) \tag{H.161}$$

$$\propto \varphi_o(\mathbf{z}) \cdot \varphi_{o'}(\mathbf{z}). \tag{H.162}$$

Thus $\varphi_{o*o'}(\mathbf{z}) = K \cdot \varphi_o(\mathbf{z}) \cdot \varphi_{o'}(\mathbf{z})$ *for some constant* K *and so, integrating,* $1 = K \cdot \int \varphi_o(\mathbf{z}) \cdot \varphi_{o'}(\mathbf{z}) d\mathbf{z}$. *From this it follows that*

$$\varphi_{o*o'}(\mathbf{z}) = \frac{\varphi_o(\mathbf{z}) \cdot \varphi_{o'}(\mathbf{z})}{\int \varphi_o(\mathbf{w}) \cdot \varphi_{o'}(\mathbf{w}) d\mathbf{w}} = (\varphi_o * \varphi_{o'})(\mathbf{z}). \tag{H.163}$$

Solution 23 *to Exercise 23: We are to prove (5.164) and (5.165) and, from them, that the conversion* $\varphi \mapsto \mu_\varphi$ *of probability density functions to fuzzy membership functions is Bayes-invariant. For (5.164): From (5.29) and (5.160) we get*

$$f(\mu_\varphi|\mathbf{x}) = \mu_\varphi(\eta(\mathbf{x})) = \frac{\varphi(\eta(\mathbf{x}))}{\sup_{\mathbf{w}} \varphi(\mathbf{w})} = f(\varphi|\mathbf{x}). \tag{H.164}$$

For (5.165): From (5.160) we get

$$\begin{aligned}
\mu_{\varphi*\varphi'}(\mathbf{z}) &= \frac{(\varphi * \varphi')(\mathbf{z})}{\sup_{\mathbf{w}} (\varphi * \varphi')(\mathbf{w})} = \frac{\varphi(\mathbf{z}) \cdot \varphi'(\mathbf{z})}{\sup_{\mathbf{w}} \varphi(\mathbf{w}) \cdot \varphi'(\mathbf{w})} && \text{(H.165)} \\
&= K \cdot \frac{\varphi(\mathbf{z})}{\sup_{\mathbf{w}} \varphi(\mathbf{w})} \cdot \frac{\varphi'(\mathbf{z})}{\varphi'(\mathbf{w})} && \text{(H.166)} \\
&= K \cdot \mu_\varphi(\mathbf{z}) \cdot \mu_{\varphi'}(\mathbf{z}) = K \cdot (\mu_\varphi \cdot \mu_{\varphi'})(\mathbf{z}) && \text{(H.167)}
\end{aligned}$$

where

$$K = \frac{(\sup_{\mathbf{w}} \varphi(\mathbf{w})) \cdot (\sup_{\mathbf{w}} \varphi'(\mathbf{w}))}{\sup_{\mathbf{w}} \varphi(\mathbf{w}) \cdot \varphi'(\mathbf{w})}. \tag{H.168}$$

Solution 24 *to Exercise 24: We are to verify that (6.11) reduces to (5.29):* $f(g|\mathbf{x}) = g(\eta(\mathbf{x}))$. *From (6.11) we know that*

$$f(g|\mathbf{x}) = \sup_{\mathbf{z}} \min\{g(\mathbf{z}), \eta_{\mathbf{x}}(\mathbf{z})\}. \tag{H.169}$$

If $\eta_{\mathbf{x}}(\mathbf{z}) = \delta_{\eta(\mathbf{x}),\mathbf{z}}$ *then*

$$f(g|\mathbf{x}) = \sup_{\mathbf{z}} \min\{g(\mathbf{z}), \delta_{\eta(\mathbf{x}),\mathbf{z}}\} = g(\eta(\mathbf{x})) \tag{H.170}$$

as claimed.

Solution 25 *to Exercise 25: We are to prove (6.14). Begin by noting that the quantity* $\min\{g(z),\ h_x(z)\}$ *will achieve its supremal value at an intersection point of* $g(z)$ *and* $h_z(z)$. *Setting* $g(z) = h_x(z)$ *implies that*

$$\frac{(z-z_0)^2}{\sigma_0^2} = \frac{(z-z_x)^2}{\sigma_x^2} \tag{H.171}$$

and hence

$$\frac{\sigma_x^2}{\sigma_0^2} = \left(\frac{z-z_x}{z-z_0}\right)^2 = \left(1 + \frac{z_0 - z_x}{z-z_0}\right)^2 \tag{H.172}$$

and so

$$z = z_0 + \frac{z_0 - z_x}{\pm\frac{\sigma_x}{\sigma_0} - 1} = z_0 + \sigma_0 \cdot \frac{z_0 - z_x}{\pm\sigma_x - \sigma_0}. \tag{H.173}$$

Substituting into $g(z)$ *yields*

$$\exp\left(-\frac{(z_0 - z_x)^2}{2(\pm\sigma_x - \sigma_0)^2}\right). \tag{H.174}$$

This is maximized if $\pm\sigma_x - \sigma_0$ *is maximized. That is, it is maximized if we choose the positive sign. Thus the supremal value is, as claimed,*

$$f(g|x) = \exp\left(-\frac{(z_0 - z_x)^2}{2(\sigma_0 + \sigma_x)^2}\right). \tag{H.175}$$

Solution 26 *to Exercise 26: We are to show that (6.18):* $f(W_g|\mathbf{x}) = f(g|\mathbf{x})$. *To see this, note that by (4.76) and (4.71),*

$$f(W_g|\mathbf{x}) = \sup_{\mathbf{z}} \int_0^1 \mathbf{1}_{W_g \cap W_{\eta_\mathbf{x}}}(\mathbf{z}, a) da = \sup_{\mathbf{z}} \int_0^1 \mathbf{1}_{W_{g \wedge \eta_\mathbf{x}}}(\mathbf{z}, a) da. \tag{H.176}$$

However, by (4.74)

$$\sup_{\mathbf{z}} \int_0^1 \mathbf{1}_{W_{g \wedge \eta_\mathbf{x}}}(\mathbf{z}, a) da = \sup_{\mathbf{z}} (g \wedge \eta_\mathbf{x})(\mathbf{z}) = \sup_{\mathbf{z}} \min\{g(\mathbf{z}), \eta_\mathbf{x}(\mathbf{z})\} \tag{H.177}$$

which is just $f(g|\mathbf{x})$ *by (6.11).*

Solution 27 *to Exercise 27: We are to prove that (6.19) reduces to* $f(o|\mathbf{x}) = Pl_o(O_{\mathbf{x}})$ *(the plausibility of* $O_{\mathbf{x}}$*) when the model* $\sigma_{\mathbf{x}}(O)$ *has* $O_{\mathbf{x}}$ *as its unique focal set. Show that (6.19) further reduces to (5.58) when* $O_{\mathbf{x}} = \{\eta(\mathbf{x})\}$*. To see this, note that*

$$f(o|\mathbf{x}) = \alpha_{DS}(o, \sigma_{\mathbf{x}}) = \sum_{O \cap O' \neq \emptyset} o(O) \cdot \sigma_{\mathbf{x}}(O') \tag{H.178}$$

$$= \sum_{O \cap O_{\mathbf{x}} \neq \emptyset} o(O) = Pl_o(O_{\mathbf{x}}). \tag{H.179}$$

In the special case $O_{\mathbf{x}} = \{\eta(\mathbf{x})\}$, $f(o|\mathbf{x}) = \sum_{O \ni \eta(\mathbf{x})} o(O)$, *which is just (5.58).*

Solution 28 *to Exercise 28: We are to prove (8.73) and (8.74). From (8.72) we know that the likelihood of a DS state-estimate* $s(f)$ *is*

$$f(s|\mathbf{x}) \triangleq \left(\sum_f \frac{s(f)}{p_0(f)} \right)^{-1} \left(\sum_f \frac{s(f)}{p_0(f)} \cdot f(\mathbf{x}) \right). \tag{H.180}$$

The Bayes normalization constant is, therefore,

$$f(s) = \int f(s|\mathbf{x}) \cdot p_0(\mathbf{x}) \tag{H.181}$$

$$= \left(\sum_f \frac{s(f)}{p_0(f)} \right)^{-1} \left(\sum_f \frac{s(f)}{p_0(f)} \cdot \int f(\mathbf{x}) \cdot p_0(\mathbf{x}) dx \right) \tag{H.182}$$

$$= \left(\sum_f \frac{s(f)}{p_0(f)} \right)^{-1} \left(\sum_f \frac{s(f)}{p_0(f)} \cdot p_0(f) \right) \tag{H.183}$$

$$= \left(\sum_f \frac{s(f)}{p_0(f)} \right)^{-1} \left(\sum_f s(f) \right) = \left(\sum_f \frac{s(f)}{p_0(f)} \right)^{-1}. \tag{H.184}$$

Consequently the posterior distribution is

$$f(\mathbf{x}|s) = \frac{f(s|\mathbf{x}) \cdot p_0(\mathbf{x})}{f(s)} = p_0(\mathbf{x}) \sum_f \frac{s(f)}{p_0(f)} \cdot f(\mathbf{x}) \tag{H.185}$$

which is (8.73). On the other hand, from (8.63) we know that

$$(\eta^{\sim 1}o)(f) \triangleq \frac{\left(\sum_{\eta^{-1}g=f} o(g)\right) \cdot p_0(f)}{\sum_{g'} o(g') \cdot p_0(\eta^{-1}g')} \tag{H.186}$$

where $(\eta^{-1}g)(\mathbf{x}) \triangleq g(\eta(\mathbf{x}))$. *Substituting* $s = \eta^{\sim 1}o$ *into (H.185) we get*

$$f(\mathbf{x}|\eta^{\sim 1}o) \tag{H.187}$$

$$= \frac{p_0(\mathbf{x}) \sum_f \frac{1}{p_0(f)} \left(\sum_{\eta^{-1}g=f} o(g)\right) \cdot p_0(f) \cdot f(\mathbf{x})}{\sum_{g'} o(g') \cdot p_0(\eta^{-1}g')} \tag{H.188}$$

$$= \frac{p_0(\mathbf{x}) \sum_f \left(\sum_{\eta^{-1}g=f} o(g)\right) \cdot f(\mathbf{x})}{\sum_{g'} o(g') \cdot p_0(\eta^{-1}g')} \tag{H.189}$$

and so

$$f(\mathbf{x}|\eta^{\sim 1}o) = \frac{p_0(\mathbf{x}) \sum_g o(g) \cdot (\eta^{-1}g)(\mathbf{x})}{\sum_{g'} o(g') \cdot p_0(\eta^{-1}g')} \tag{H.190}$$

$$= \frac{p_0(\mathbf{x}) \sum_g o(g) \cdot g(\eta(\mathbf{x}))}{\sum_{g'} o(g') \cdot p_0(\eta^{-1}g')}. \tag{H.191}$$

Also, from (5.73) the likelihood for a fuzzy DS measurement $o(g)$ *is*

$$f(o|\mathbf{x}) = \sum_g o(g) \cdot g(\eta(\mathbf{x})). \tag{H.192}$$

The Bayes constant for such a measurement is

$$f(o) = \int f(o|\mathbf{x}) \cdot p_0(\mathbf{x}) d\mathbf{x} \tag{H.193}$$

$$= \sum_g o(g) \cdot \int g(\eta(\mathbf{x})) \cdot p_0(\mathbf{x}) d\mathbf{x} \tag{H.194}$$

$$= \sum_g o(g) \cdot p_0(\eta^{-1}g). \tag{H.195}$$

So, the corresponding posterior distribution is

$$f(\mathbf{x}|o) = \frac{f(o|\mathbf{x}) \cdot p_0(\mathbf{x})}{f(o)} \tag{H.196}$$

$$= \frac{\left(\sum_g o(g) \cdot g(\eta(\mathbf{x}))\right) \cdot p_0(\mathbf{x})}{\sum_{g'} o(g') \cdot p_0(\eta^{-1}g')} \tag{H.197}$$

which agrees with (H.191).

Solution 29 *to Exercise 29: We are to prove (8.75) and (8.76):*

$$\eta^{\sim 1}(o * o') = \eta^{\sim 1}o_1 *_0 \eta^{\sim 1}o' \tag{H.198}$$

$$f(\mathbf{x}|o_1, ..., o_m) = f(\mathbf{x}|\eta^{\sim 1}o_1, ..., \eta^{\sim 1}o_m) \tag{H.199}$$

In the first case we know from (8.63) that

$$(\eta^{\sim 1}o)(f) \triangleq \frac{\left(\sum_{\eta^{-1}g=f} o(g)\right) \cdot p_0(f)}{\sum_{g'} o(g') \cdot p_0(\eta^{-1}g')}. \tag{H.200}$$

Thus on the one hand, from the definition of DS combination (4.87),

$$(\eta^{\sim 1}(o * o'))(f) = \frac{\sum_{\eta^{-1}g''=f}(o * o')(g'') \cdot p_0(f)}{\sum_{g''}(o * o')(g'') \cdot p_0(\eta^{-1}g'')} \tag{H.201}$$

$$= \frac{\sum_{\eta^{-1}g''=f} \sum_{g \cdot g'=g''} o(g) \cdot o'(g') \cdot p_0(f)}{\sum_{g''} \left(\sum_{g \cdot g'=g''} o(g) \cdot o'(g')\right) \cdot p_0(\eta^{-1}g'')} \tag{H.202}$$

$$= \frac{\sum_{(\eta^{-1}g)\cdot(\eta^{-1}g')=f} o(g) \cdot o'(g') \cdot p_0(f)}{\sum_{g,g'} o(g) \cdot o'(g') \cdot p_0(\eta^{-1}g \cdot \eta^{-1}g'))}. \tag{H.203}$$

On the other hand, from the definition of fuzzy modified Dempster's combination (4.132),

$$\left[(\eta^{\sim 1}o) *_0 (\eta^{\sim 1}o')\right](f'') \tag{H.204}$$

$$= \frac{\sum_{f \cdot f'=f''}(\eta^{\sim 1}o)(f) \cdot (\eta^{\sim 1}o)(f') \cdot \alpha_0(f, f')}{\sum_{f,f'}(\eta^{\sim 1}o)(f) \cdot (\eta^{\sim 1}o')(f') \cdot \alpha_0(f, f')} \tag{H.205}$$

$$= \frac{\sum_{f\cdot f'=f''} \left(\sum_{\eta^{-1}g''=f} o(g'')\right) \left(\sum_{\eta^{-1}g''=f'} o'(g'')\right) p_0(f\cdot f')}{\sum_{f,f'} \left(\sum_{\eta^{-1}g=f} o(g'')\right) \left(\sum_{\eta^{-1}g'=f'} o'(g'')\right) p_0(f\cdot f')} \tag{H.206}$$

$$= \frac{\sum_{(\eta^{-1}g)\cdot(\eta^{-1}g')=f''} o(g)\cdot o'(g')\cdot p_0(f'')}{\sum_{(\eta^{-1}g),(\eta^{-1}g')} o(g)\cdot o'(g')\cdot p_0(\eta^{-1}g\cdot\eta^{-1}g')} \tag{H.207}$$

$$= \frac{\sum_{(\eta^{-1}g)\cdot(\eta^{-1}g')=f''} o(g)\cdot o'(g')\cdot p_0(f'')}{\sum_{g,g'} o(g)\cdot o'(g')\cdot p_0(\eta^{-1}g\cdot\eta^{-1}g')}. \tag{H.208}$$

which is the same as (H.203). So this proves (H.198). As for (H.199), by (5.120) we know that

$$f(\mathbf{x}|o_1,...,o_m) = f(\mathbf{x}|o_1 * ... * o_m). \tag{H.209}$$

To see this, note that by (8.74)

$$f(\mathbf{x}|o_1 * ... * o_m) = f(\mathbf{x}|\eta^{\sim 1}(o_1 * ... * o_m)) \tag{H.210}$$

and by (8.75)

$$f(\mathbf{x}|\eta^{\sim 1}(o_1 * ... * o_m)) = f(\mathbf{x}|\eta^{\sim 1}o_1 \ *_0 ... *_0 \ \eta^{\sim 1}o_m) \tag{H.211}$$

and finally by (8.77) we know that

$$f(\mathbf{x}|\eta^{\sim 1}o_1 \ *_0 ... *_0 \ \eta^{\sim 1}o_m) = f(\mathbf{x}|\eta^{\sim 1}o_1 \ ,..., \ \eta^{\sim 1}o_m) \tag{H.212}$$

as claimed.

Solution 30 *to Exercise 30: We are to prove (8.77):*

$$f(\mathbf{x}|s_1 *_0 ... *_0 s_m) = f(\mathbf{x}|s_1,...,s_m). \tag{H.213}$$

It is enough to show that

$$f(s *_0 s'|\mathbf{x}) = f(s|\mathbf{x}) \cdot f(s'|\mathbf{x}) \tag{H.214}$$

where $f(s_1|\mathbf{x})$ *was defined in (8.72):*

$$f(s|\mathbf{x}) \triangleq \frac{\sum_f s(f)\cdot p_0(f)^{-1}\cdot f(\mathbf{x})}{\sum_f s(f)\cdot p_0(f)^{-1}}. \tag{H.215}$$

To see this, note that in this case

$$
\begin{aligned}
f(\mathbf{x}|s *_0 s') &\propto f(s *_0 s'|\mathbf{x}) \cdot f_0(\mathbf{x}) && \text{(H.216)}\\
&= f(s|\mathbf{x}) \cdot f(s'|\mathbf{x}) \cdot f_0(\mathbf{x}) && \text{(H.217)}\\
&= f(s, s'|\mathbf{x}) \cdot f_0(\mathbf{x}) && \text{(H.218)}\\
&\propto f(\mathbf{x}|s, s_2). && \text{(H.219)}
\end{aligned}
$$

To demonstrate (H.214) note that

$$
\begin{aligned}
& f(s *_0 s'|\mathbf{x}) && \text{(H.220)}\\
=\; & \frac{\sum_{f''} (s *_0 s')(f'') \cdot p_0(f'')^{-1} \cdot f''(\mathbf{x})}{\sum_{f''} (s *_0 s')(f'') \cdot p_0(f'')^{-1}} && \text{(H.221)}\\
=\; & \frac{\left(\begin{array}{c} \sum_{f''} \sum_{f \cdot f' = f''} s(f) \cdot s'(f') \\ \cdot \alpha_0(f, f') \cdot p_0(f \cdot f')^{-1} \cdot (f \cdot f')(\mathbf{x}) \end{array} \right)}{\left(\begin{array}{c} \sum_{f''} \sum_{f \cdot f' = f''} s(f) \cdot s'(f') \\ \cdot \alpha_0(f, f') \cdot p_0(f \cdot f')^{-1} \end{array} \right)} && \text{(H.222)}\\
=\; & \frac{\sum_{f,f'} s(f) \cdot s'(f') \cdot p_0(f)^{-1} \cdot p_0(f')^{-1} \cdot f(\mathbf{x}) \cdot f'(\mathbf{x})}{\sum_{f,f'} s(f) \cdot s'(f') \cdot p_0(f)^{-1} \cdot p_0(f')^{-1}} && \text{(H.223)}\\
=\; & \left(\frac{\sum_f s(f) \cdot p_0(f)^{-1} \cdot f(\mathbf{x})}{\sum_f s(f) \cdot p_0(f)^{-1}} \right) && \text{(H.224)}\\
& \left(\frac{\sum_{f'} s(f') \cdot p_0(f')^{-1} \cdot f'(\mathbf{x})}{\sum_f s(f') \cdot p_0(f')^{-1}} \right) && \text{(H.225)}\\
=\; & f(s|\mathbf{x}) \cdot f(s'|\mathbf{x}) && \text{(H.226)}
\end{aligned}
$$

as claimed.

Solution 31 *to Exercise 31: We are to prove (9.31)* $\int_T f_C(Z)\delta Z = e^{\lambda p_c(T)-\lambda}$. *To see this, note that*

$$\int_T f_C(Z)\delta Z \tag{H.227}$$

$$= f_C(\emptyset) + \sum_{m=1}^{\infty} \frac{1}{m!} \int_{T\times\ldots\times T} f_C(\{\mathbf{z}_1,\ldots,\mathbf{z}_m\})d\mathbf{z}_1\cdots d\mathbf{z}_m \tag{H.228}$$

$$= e^{-\lambda} + \sum_{m=1}^{\infty} \frac{1}{m!} \int_{T\times\ldots\times T} e^{-\lambda}\cdot\lambda^m\cdot c(\mathbf{z}_1)\cdots c(\mathbf{z}_m) \tag{H.229}$$

$$\cdot d\mathbf{z}_1\cdots d\mathbf{z}_m \tag{H.230}$$

$$= e^{-\lambda} + e^{-\lambda}\sum_{m=1}^{\infty} \frac{\lambda^m p_c(T)^m}{m!} \tag{H.231}$$

$$= e^{-\lambda}\cdot e^{\lambda p_c(T)} = e^{\lambda p_c(T)-\lambda}. \tag{H.232}$$

Solution 32 *to Exercise 32: Let* $\mathbf{Y}$ *be a random vector with distribution* $f_{\mathbf{Y}}(\mathbf{y}) = \delta_{\mathbf{y}_0}(\mathbf{y})$. *We are to show that the p.g.fl. of* $\Psi = \{\mathbf{Y}\}$ *is* $G_\Psi[h] = h(\mathbf{y}_0)$. *To see this note that*

$$G_\Psi[h] = \int h^Y\cdot f_\Psi(Y)\delta Y = \int h(\mathbf{y})\cdot f_{\mathbf{Y}}(\mathbf{y})d\mathbf{y} \tag{H.233}$$

$$= \int h(\mathbf{y})\cdot\delta_{\mathbf{y}_0}(\mathbf{y})d\mathbf{y} = h(\mathbf{y}_0). \tag{H.234}$$

Solution 33 *to Exercise 33: Let* $\mathbf{w}_1,\mathbf{w}_2 \in \mathfrak{Y}_0$ *be fixed, and let* $\Psi = \{\mathbf{Y}_1,\mathbf{Y}_2\}$, *where* $\mathbf{Y}_1,\mathbf{Y}_2$ *are independent with respective distributions* $f_{\mathbf{Y}_1}(\mathbf{y}) = \delta_{\mathbf{w}_1}(\mathbf{y})$ *and* $f_{\mathbf{Y}_2}(\mathbf{y}) = \delta_{\mathbf{w}_2}(\mathbf{y})$. *We are to show that the probability density of* Ψ *is*

$$f_\Psi(Y) = \left\{\begin{array}{ccc} \delta_{\mathbf{y}_1}(\mathbf{w}_1)\cdot\delta_{\mathbf{y}_2}(\mathbf{w}_2) & \text{if} & \left\{\begin{array}{c} Y=\{\mathbf{y}_1,\mathbf{y}_2\} \\ |Y|=2 \end{array}\right. \\ +\delta_{\mathbf{y}_2}(\mathbf{w}_1)\cdot\delta_{\mathbf{y}_1}(\mathbf{w}_2) & & \\ 0 & \text{if} & |Y|\neq 2 \end{array}\right. . \tag{H.235}$$

To see this, note that

$$\frac{\delta G_\Psi}{\delta Y}[h] = \left\{\begin{array}{ccc} h(\mathbf{w}_1)\cdot h(\mathbf{w}_2) & \text{if} & Y=\emptyset \\ \delta_{\mathbf{y}_1}(\mathbf{w}_1)\cdot h(\mathbf{w}_2) & \text{if} & Y=\{\mathbf{y}_1\} \\ +h(\mathbf{w}_1)\cdot\delta_{\mathbf{y}_1}(\mathbf{w}_2) & & \\ \delta_{\mathbf{y}_1}(\mathbf{w}_1)\cdot\delta_{\mathbf{y}_2}(\mathbf{w}_2) & \text{if} & \left\{\begin{array}{c} Y=\{\mathbf{y}_1,\mathbf{y}_2\} \\ |Y|=2 \end{array}\right. \\ +\delta_{\mathbf{y}_2}(\mathbf{w}_1)\cdot\delta_{\mathbf{y}_1}(\mathbf{w}_2) & & \\ 0 & \text{if} & |Y|\geq 3 \end{array}\right. . \tag{H.236}$$

Solution 34 *to Exercise 34: We are to show that (11.121) defines a probability density function:* $\int f(Y)\delta Y = 1$. *We are to further show that the corresponding belief-mass function and p.g.fl. of a Poisson process are, respectively,* $\beta(S) = e^{\lambda p_f(S)-\lambda}$ *and* $F[h] = e\lambda^{f[h]-\lambda}$ *where* $p_f(S) \triangleq \int_S f(\mathbf{y})d\mathbf{y}$ *and* $f[h] \triangleq \int h(\mathbf{y})f(\mathbf{y})d\mathbf{y}$. *To see this note that*

$$\int f(Y)\delta Y \tag{H.237}$$

$$= f(\emptyset) + \sum_{n=1}^{\infty} \frac{1}{n!} \int f(\{\mathbf{y}_1, ..., \mathbf{y}_n\}) \tag{H.238}$$

$$\cdot d\mathbf{y}_1 \cdots d\mathbf{y}_n \tag{H.239}$$

$$= p(0) + \sum_{n=1}^{\infty} \frac{n!p(n)}{n!} \int f(\mathbf{y}_1) \cdots f(\mathbf{y}_n) d\mathbf{y}_1 \cdots d\mathbf{y}_n \tag{H.240}$$

$$= p(0) + \sum_{n=1}^{\infty} p(n) = 1. \tag{H.241}$$

Next,

$$\beta(S) = \int_S f(Y)\delta Y \tag{H.242}$$

$$= f(\emptyset) + \sum_{n=1}^{\infty} \frac{1}{n!} \int_{S\times...\times S} f(\{\mathbf{y}_1, ..., \mathbf{y}_n\}) d\mathbf{y}_1 \cdots d\mathbf{y}_n \tag{H.243}$$

$$= e^{-\lambda} + e^{-\lambda} \sum_{n=1}^{\infty} \frac{\lambda^n}{n!} \int_{S\times...\times S} f(\mathbf{y}_1) \cdots f(\mathbf{y}_n) d\mathbf{y}_1 \cdots d\mathbf{y}_n \tag{H.244}$$

$$= e^{-\lambda} \left(1 + \sum_{n=1}^{\infty} \frac{\lambda^n p_f(S)^n}{n!}\right) \tag{H.245}$$

$$= e^{-\lambda} \cdot e^{\lambda p_f(S)} = e^{\lambda p_f(S) - \lambda}. \tag{H.246}$$

Solution 35 *to Exercise 35: We are to show that the multiobject uniform density function of (11.126) is a probability density:* $\int f(Y)\delta Y = 1$. *For,*

$$\int f(Y)\delta Y \tag{H.247}$$

$$= f(\emptyset) + \sum_{n=1}^{\infty} \frac{1}{n!} \int f(\{\mathbf{y}_1, ..., \mathbf{y}_n\}) d\mathbf{y}_1 \cdots d\mathbf{y}_n, \tag{H.248}$$

and so

$$= \frac{1}{M+1} + \sum_{n=1}^{M} \frac{1}{n!} \int_{D\times\ldots\times D} \frac{n!}{|D|^n (M+1)} d\mathbf{y}_1 \cdots d\mathbf{y}_n \quad \text{(H.249)}$$

$$= \frac{1}{M+1} + \sum_{n=1}^{M} \frac{1}{M+1} = \frac{M+1}{M+1} = 1. \quad \text{(H.250)}$$

Solution 36 *to Exercise 36: We are to verify (11.125). That is, we are to show that* $\delta_{Y'}(Y)$ *satisfies the defining property of a Dirac delta density:*

$$\int \delta_{Y'}(Y) \cdot f(Y)\delta Y = f(Y'). \quad \text{(H.251)}$$

To see this, note that

$$\int \delta_{Y'}(Y) \cdot f(Y)\delta Y \quad \text{(H.252)}$$

$$= \sum_{j=0}^{\infty} \frac{1}{j!} \int \delta_{Y'}(\{\mathbf{y}_1, ..., \mathbf{y}_n\}) \cdot f(\{\mathbf{y}_1, ..., \mathbf{y}_n\}) d\mathbf{y}_1 \cdots d\mathbf{y}_n \quad \text{(H.253)}$$

$$= \frac{1}{n!} \int \left(\sum_{\sigma} \delta_{\mathbf{y}'_{\sigma 1}}(\mathbf{y}_1) \cdots \delta_{\mathbf{y}'_{\sigma n}}(\mathbf{y}_n) \right) \cdot f(\{\mathbf{y}_1, ..., \mathbf{y}_n\}) \quad \text{(H.254)}$$

$$\cdot d\mathbf{y}_1 \cdots d\mathbf{y}_n \quad \text{(H.255)}$$

$$= \int \delta_{\mathbf{y}'_1}(\mathbf{y}_1) \cdots \delta_{\mathbf{y}'_n}(\mathbf{y}_n) \cdot f(\{\mathbf{y}_1, ..., \mathbf{y}_n\}) d\mathbf{y}_1 \cdots d\mathbf{y}_n \quad \text{(H.256)}$$

$$= f(\{\mathbf{y}'_1, ..., \mathbf{y}'_n\}) = f(Y'). \quad \text{(H.257)}$$

Solution 37 *to Exercise 37: We are to verify (11.164) and (11.165):*

$$N_\Psi = G'_\Psi(1), \qquad \sigma^2_\Xi = G''_\Psi(1) - N^2_\Psi + N_\Psi \quad \text{(H.258)}$$

Abbreviate $p_n \overset{abbr.}{=} p_\Psi(n)$. *Since* $G_\Psi(x) = \sum_{n=0}^{\infty} p_n \cdot x^n$ *it follows that*

$$G'_\Psi(x) = \sum_{n=1}^{\infty} n \cdot p_n x^{n-1}, \quad G''_\Psi(x) = \sum_{n=2}^{\infty} n \cdot (n-1) \cdot p_n x^{n-2} \quad \text{(H.259)}$$

and so

$$G'_\Psi(1) = \sum_{n=1}^{\infty} n \cdot p_n, \quad G''_\Psi(1) = \sum_{n=2}^{\infty} n \cdot (n-1) \cdot p_n. \tag{H.260}$$

The first equation immediately tells us that $G'_\Psi(1) = N_\Psi$. *As for the second equation,*

$$G''_\Psi(1) = \sum_{n=2}^{\infty} n^2 \cdot p_n - \sum_{n=2}^{\infty} n \cdot p_n = \sum_{n=0}^{\infty} n^2 \cdot p_n - \sum_{n=0}^{\infty} n \cdot p_n \tag{H.261}$$

$$= \sum_{n=0}^{\infty} n^2 \cdot p_n - N_\Psi \tag{H.262}$$

$$= \sum_{n=0}^{\infty} n^2 \cdot p_n - \sum_{n=0}^{\infty} n^2 \cdot p_n - N_\Psi^2 + N_\Psi^2 - N_\Psi \tag{H.263}$$

$$= \sum_{n=0}^{\infty} (n - N_\Psi)^2 \cdot p_n + N_\Psi^2 - N_\Psi \tag{H.264}$$

$$= \sigma_\Xi^2 + N_\Psi^2 - N_\Psi. \tag{H.265}$$

Solution 38 *to Exercise 38: We are to verify (11.174):* $G_{Y'}[h] = h^{Y'}$. *To see this note that, on the one hand, if* $Y' = \emptyset$, *then*

$$G_\emptyset[h] \triangleq \int h^Y \cdot \delta_\emptyset(Y) Y \tag{H.266}$$

$$= \delta_\emptyset(\emptyset) \tag{H.267}$$

$$+ \sum_{n=1}^{\infty} \frac{1}{n!} \int h(\mathbf{y}_1) \cdots h(\mathbf{y}_n) \cdot \delta_\emptyset(\{\mathbf{y}_1, ..., \mathbf{y}_n\}) \tag{H.268}$$

$$\cdot d\mathbf{y}_1 \cdots d\mathbf{y}_n \tag{H.269}$$

$$= 1 + 0 = 1 = h^\emptyset. \tag{H.270}$$

On the other hand, if $\emptyset \neq Y' = \{\mathbf{y}'_1, ..., \mathbf{y}'_n\}$, *then*

$$G_{Y'}[h] \triangleq \int h^Y \cdot \delta_{Y'}(Y) Y \tag{H.271}$$

$$= \frac{1}{n!} \int h(\mathbf{y}_1) \cdots h(\mathbf{y}_n) \cdot \delta_{Y'}(\{\mathbf{y}_1, ..., \mathbf{y}_n\}) d\mathbf{y}_1 \cdots d\mathbf{y}_n \tag{H.272}$$

and so

$$= \frac{1}{n!} \int h(\mathbf{y}_1) \cdots h(\mathbf{y}_n) \tag{H.273}$$

$$\cdot \left(\sum_{\sigma} \delta_{\mathbf{x}'_{\sigma 1}}(\mathbf{x}_1) \cdots \delta_{\mathbf{x}'_{\sigma n}}(\mathbf{x}_n) \right) d\mathbf{y}_1 \cdots d\mathbf{y}_n \tag{H.274}$$

and so

$$= \int h(\mathbf{y}_1) \cdots h(\mathbf{y}_n) \cdot \delta_{\mathbf{x}'_1}(\mathbf{x}_1) \cdots \delta_{\mathbf{x}'_n}(\mathbf{x}_n) d\mathbf{y}_1 \cdots d\mathbf{y}_n \tag{H.275}$$

$$= h(\mathbf{y}'_1) \cdots h(\mathbf{y}'_n) = h^{Y'}. \tag{H.276}$$

Solution 39 *to Exercise 39: We are to verify (11.185):*

$$G[h] = \frac{1}{\mathring{n}+1} \sum_{n=0}^{\mathring{n}} \left(\frac{\mathbf{1}_D[h]}{|D|} \right)^n. \tag{H.277}$$

To see this note that

$$G[h] = \int h^Y \cdot f(Y) \delta Y \tag{H.278}$$

$$= \sum_{n=0}^{\infty} \frac{1}{n!} \int h(\mathbf{y}_1) \cdots h(\mathbf{y}_n) \cdot f(\{\mathbf{y}_1, ..., \mathbf{y}_n\}) d\mathbf{y}_1 \cdots d\mathbf{y}_n, \tag{H.279}$$

and so

$$= \sum_{n=0}^{\mathring{n}} \frac{1}{n!} \int_{D \times ... \times D} h(\mathbf{y}_1) \cdots h(\mathbf{y}_n) \tag{H.280}$$

$$\cdot \frac{n!}{|D|^n \cdot (\mathring{n}+1)} d\mathbf{y}_1 \cdots d\mathbf{y}_n, \tag{H.281}$$

and so

$$= \frac{1}{\mathring{n}+1} \sum_{n=0}^{\mathring{n}} \left(\frac{1}{|D|} \int_D h(\mathbf{y}) d\mathbf{y} \right)^n \tag{H.282}$$

$$= \frac{1}{\mathring{n}+1} \sum_{n=0}^{\mathring{n}} \left(\frac{\mathbf{1}_D[h]}{|D|} \right)^n. \tag{H.283}$$

Solution 40 *to Exercise 40: We are to prove (11.166):*

$$G_{\Psi_1 \cup ... \cup \Psi_s}[h] = G_{\Psi_1}[h] \cdots G_{\Psi_s}[h] \tag{H.284}$$

if $\Psi_1, ..., \Psi_s$ *are independent. For,*

$$G_{\Psi_1 \cup ... \cup \Psi_s}[h] = \int h^Y \cdot f_{\Psi_1 \cup ... \cup \Psi_s}(Y) \delta Y \tag{H.285}$$

$$= \int h^{Y_1 \cup ... \cup Y_s} \cdot f_{\Psi_1}(Y_1) \cdots f_{\Psi_s}(Y_s) \delta Y_1 \cdots \delta Y_s \tag{H.286}$$

$$= \int h^{Y_1} \cdots h^{Y_s} \cdot f_{\Psi_1}(Y_1) \cdots f_{\Psi_s}(Y_s) \delta Y_1 \cdots \delta Y_s \tag{H.287}$$

where the last equation results from the fact that $Y_i \cap Y_j \neq \emptyset$ *is a zero-probability event if* $i \neq j$. *Thus, as claimed,*

$$G_{\Psi_1 \cup ... \cup \Psi_s}[h] = \left(\int h^{Y_1} \cdot f_{\Psi_1}(Y_1) \delta Y_1 \right) \tag{H.288}$$

$$\cdots \left(\int h^{Y_s} \cdot f_{\Psi_s}(Y_s) \delta Y_s \right) \tag{H.289}$$

$$= G_{\Psi_1}[h] \cdots G_{\Psi_s}[h]. \tag{H.290}$$

Solution 41 *to Exercise 41: We are to prove the third chain rule (11.283). Let* $\tilde{F}[h] \triangleq F[s_h]$. *Then*

$$\frac{\delta \tilde{F}}{\delta \mathbf{y}}[h] = \lim_{\varepsilon \searrow 0} \frac{F[s_{h+\varepsilon \delta_\mathbf{y}}] - F[s_h]}{\varepsilon}. \tag{H.291}$$

However, to first order

$$s_{h+\varepsilon\delta_\mathbf{y}}(\mathbf{w}) = s(h(\mathbf{w}) + \varepsilon \delta_\mathbf{y}(\mathbf{w})) \tag{H.292}$$

$$\cong s(h(\mathbf{w})) + \varepsilon \delta_\mathbf{y}(\mathbf{w}) \cdot \frac{ds}{dy}(h(\mathbf{w})) \tag{H.293}$$

$$= s_h(\mathbf{w}) + \varepsilon \delta_\mathbf{y}(\mathbf{w}) \cdot \frac{ds}{dy}(h(\mathbf{y})). \tag{H.294}$$

Thus in the limit

$$\frac{\delta \tilde{F}}{\delta \mathbf{y}}[h] = \lim_{\varepsilon \searrow 0} \frac{F[s_h + \varepsilon \frac{ds}{dy}(h(\mathbf{y}))\delta_{\mathbf{y}}] - F[s_h]}{\varepsilon}. \tag{H.295}$$

$$= \frac{ds}{dy}(h(\mathbf{y})) \cdot \lim_{\varepsilon \searrow 0} \frac{F[s_h + \varepsilon \frac{ds}{dy}(h(\mathbf{y}))\delta_{\mathbf{y}}] - F[s_h]}{\varepsilon \frac{ds}{dy}(h(\mathbf{y}))} \tag{H.296}$$

$$= \frac{ds}{dy}(h(\mathbf{y})) \cdot \frac{\delta F}{\delta \mathbf{y}}[s_h]. \tag{H.297}$$

Solution 42 *to Exercise 42: We are to show that*

$$f_{\Psi'}(Y) = \int f_{\Psi}(Y - \mathbf{y}) \cdot f_{\mathbf{Y}}(\mathbf{y}) d\mathbf{y} \tag{H.298}$$

and that $f_{\Psi'}(\emptyset) = f_{\Psi}(\emptyset)$. *To see this, note that*

$$\beta_{\Psi'}(S) = \Pr(\Psi' \subseteq S) = \Pr(\Psi + \mathbf{Y} \subseteq S) \tag{H.299}$$

$$= \int \Pr(\Psi + \mathbf{y} \subseteq S) \cdot f_{\mathbf{Y}}(\mathbf{y}) d\mathbf{y} \tag{H.300}$$

$$= \int \Pr(\Psi \subseteq S - \mathbf{y}) \cdot f_{\mathbf{Y}}(\mathbf{y}) d\mathbf{y} \tag{H.301}$$

$$= \int \beta_{\Psi}(S - \mathbf{y}) \cdot f_{\mathbf{Y}}(\mathbf{y}) d\mathbf{y}. \tag{H.302}$$

Let $E_{\mathbf{w}}$ *be an arbitrarily small hypersphere centered at* $\mathbf{w}$. *Then*

$$\frac{\delta}{\delta \mathbf{w}} \beta_{\Psi}(S - \mathbf{y}) \tag{H.303}$$

$$= \lim_{|E_{\mathbf{w}}| \searrow 0} \frac{\beta_{\Psi}((S \cup E_{\mathbf{w}}) - \mathbf{y}) - \beta_{\Psi}(S - \mathbf{y})}{|E_{\mathbf{w}}|} \tag{H.304}$$

$$= \lim_{|E_{\mathbf{w}}| \searrow 0} \frac{\beta_{\Psi}((S - \mathbf{y}) \cup (E_{\mathbf{w}} - \mathbf{y})) - \beta_{\Psi}(S - \mathbf{y})}{|E_{\mathbf{w}}|} \tag{H.305}$$

$$= \lim_{|E_{\mathbf{w}}| \searrow 0} \frac{\beta_{\Psi}((S - \mathbf{y}) \cup E_{\mathbf{w}-\mathbf{y}}) - \beta_{\Psi}(S - \mathbf{y})}{|E_{\mathbf{w}-\mathbf{y}}|} \tag{H.306}$$

$$= \frac{\delta \beta_{\Psi}}{\delta(\mathbf{w} - \mathbf{y})}(S - \mathbf{y}). \tag{H.307}$$

Consequently,

$$\frac{\delta\beta_{\Psi'}}{\delta Y}(S) = \int \frac{\delta\beta_{\Psi}}{\delta(Y-\mathbf{y})}(S-\mathbf{y}) \cdot f_{\mathbf{Y}}(\mathbf{y})d\mathbf{y} \tag{H.308}$$

and thus, as claimed,

$$\begin{aligned} f_{\Psi'}(Y) &= \frac{\delta\beta_{\Psi'}}{\delta Y}(\emptyset) = \int \frac{\delta\beta_{\Psi}}{\delta(Y-\mathbf{y})}(\emptyset) \cdot f_{\mathbf{Y}}(\mathbf{y})d\mathbf{y} && \text{(H.309)} \\ &= \int f_{\Psi}(Y-\mathbf{y}) \cdot f_{\mathbf{Y}}(\mathbf{y})d\mathbf{y}. && \text{(H.310)} \end{aligned}$$

Solution 43 *to Exercise 43: We are to show that the probability density function of* $\Psi \cap T$ *is*

$$f_{\Psi\cap T}(Y) = \begin{cases} \frac{\delta\beta_{\Psi}}{\delta Y}(T^c) & \text{if} \quad Y \subseteq T \\ 0 & \text{if} \quad \text{otherwise} \end{cases}. \tag{H.311}$$

To see this, note that

$$\begin{aligned} \beta_{\Psi\cap T}(S) &= \Pr(\Psi \cap T \subseteq S) && \text{(H.312)} \\ &= \Pr([(\Psi \cap S) \cup (\Psi \cap S^c)] \cap T \subseteq S) && \text{(H.313)} \\ &= \Pr([\Psi \cap S \cap T] \cup [\Psi \cap S^c \cap T] \subseteq S) && \text{(H.314)} \\ &= \Pr(\Psi \cap S \cap T \subseteq S,\ \Psi \cap S^c \cap T \subseteq S) && \text{(H.315)} \\ &= \Pr(\Psi \cap S^c \cap T \subseteq S) && \text{(H.316)} \\ &= \Pr(\Psi \cap S^c \cap T = \emptyset) && \text{(H.317)} \\ &= \Pr(\Psi \subseteq (S^c \cap T)^c) && \text{(H.318)} \\ &= \Pr(\Psi \subseteq S \cup T^c) && \text{(H.319)} \\ &= \beta_{\Psi}(S \cup T^c). && \text{(H.320)} \end{aligned}$$

Thus

$$\begin{aligned} \frac{\delta\beta_{\Psi\cap T}}{\delta\mathbf{y}}(S) &= \frac{\delta}{\delta\mathbf{y}}\beta_{\Psi}(S \cup T^c) && \text{(H.321)} \\ &= \lim_{|E_{\mathbf{w}}|\searrow 0} \frac{\beta_{\Psi}(S \cup E_{\mathbf{w}} \cup T^c) - \beta_{\Psi}(S \cup T^c)}{|E_{\mathbf{w}}|} && \text{(H.322)} \\ &= \lim_{|E_{\mathbf{w}}|\searrow 0} \frac{\beta_{\Psi}(S \cup T^c \cup E_{\mathbf{w}}) - \beta_{\Psi}(S \cup T^c)}{|E_{\mathbf{w}}|} && \text{(H.323)} \\ &= \frac{\delta\beta_{\Psi}}{\delta\mathbf{y}}(S \cup T^c). && \text{(H.324)} \end{aligned}$$

The last equation results from the assumption $\mathbf{w} \in T$, *which implies* $\mathbf{w} \notin S^c \cap T$. *If on the other hand* $\mathbf{w} \notin T$, *then the last limit is zero. Repeating this for higher-order set derivatives, we get the desired result.*

Solution 44 *to Exercise 44: We are to show that the probability density function of* $\beta(S|T)$ *is*

$$f_{\Psi|T}(Y) = \begin{cases} \frac{f_\Psi(Y)}{\beta_\Psi(T)} & \text{if} \quad Y \subseteq T \\ 0 & \text{if} \quad \text{otherwise} \end{cases} . \tag{H.325}$$

To see this, note that

$$\frac{\delta \beta_{\Psi|T}}{\delta \mathbf{y}}(S) \tag{H.326}$$

$$= \frac{1}{\beta_\Psi(T)} \lim_{|E_\mathbf{w}| \searrow 0} \frac{\beta_\Psi((S \cup E_\mathbf{w}) \cap T) - \beta_\Psi(S \cap T)}{|E_\mathbf{w}|} \tag{H.327}$$

$$= \frac{1}{\beta_\Psi(T)} \lim_{|E_\mathbf{w}| \searrow 0} \frac{\beta_\Psi((S \cap T) \cup (E_\mathbf{w} \cap T)) - \beta_\Psi(S \cap T)}{|E_\mathbf{w}|}. \tag{H.328}$$

If $\mathbf{w} \notin T$ *this is zero. Otherwise, If* $\mathbf{w} \in T$ *but* $\mathbf{w} \notin S$, *for sufficiently small* $|E_\mathbf{w}|$ *this becomes*

$$\frac{\delta \beta_{\Psi|T}}{\delta \mathbf{y}}(S) \tag{H.329}$$

$$= \frac{1}{\beta_\Psi(T)} \lim_{|E_\mathbf{w}| \searrow 0} \frac{\beta_\Psi((S \cap T) \cup E_\mathbf{w}) - \beta_\Psi(S \cap T)}{|E_\mathbf{w}|} \tag{H.330}$$

$$= \frac{1}{\beta_\Psi(T)} \cdot \frac{\delta \beta_\Psi}{\delta \mathbf{y}}(S \cap T) \tag{H.331}$$

Repeating this for higher-order set derivatives, we get the desired result.

Solution 45 *to Exercise 45: We are to prove the fourth chain rule (11.285). Let* $\tilde{F}[h] \triangleq F[T[h]]$. *Then*

$$\frac{\delta \tilde{F}}{\delta \mathbf{y}}[h] = \lim_{\varepsilon \searrow 0} \frac{F[T[h + \varepsilon \delta_\mathbf{y}]] - F[T[h]])}{\varepsilon}. \tag{H.332}$$

However, to first order

$$T[h + \varepsilon \delta_\mathbf{y}](\mathbf{w}) \cong T[h](\mathbf{w}) + \varepsilon \cdot \frac{\delta T}{\delta \mathbf{y}}[h](\mathbf{w}) \tag{H.333}$$

or, abbreviated,

$$T[h + \varepsilon\delta_{\mathbf{y}}] \cong T[h] + \varepsilon\frac{\delta T}{\delta \mathbf{y}}[h]. \tag{H.334}$$

Thus in the limit

$$\frac{\delta \tilde{F}}{\delta \mathbf{y}}[h] = \lim_{\varepsilon \searrow 0} \frac{F[T[h] + \varepsilon\frac{\delta T}{\delta \mathbf{y}}[h]] - F[T[h]])}{\varepsilon} = \frac{\partial F}{\partial \frac{\delta T}{\delta \mathbf{y}}[h]}[T[h]] \tag{H.335}$$

where the rightmost expression is a gradient derivative at $g = (\delta T/\delta \mathbf{y})[h]$ *as defined in Appendix C. Note that*

$$\frac{\delta T}{\delta \mathbf{y}}[h](\mathbf{w}') = \int \frac{\delta T}{\delta \mathbf{y}}[h](\mathbf{w}) \cdot \delta_{\mathbf{w}}(\mathbf{w}')d\mathbf{w} \tag{H.336}$$

or, abbreviated,

$$\frac{\delta T}{\delta \mathbf{y}}[h] = \int \frac{\delta T}{\delta \mathbf{y}}[h](\mathbf{w}) \cdot \delta_{\mathbf{w}}d\mathbf{w}. \tag{H.337}$$

Then from (11.189) we get the claimed result:

$$\begin{aligned} \frac{\partial F}{\partial \frac{\delta T}{\delta \mathbf{y}}[h]}[T[h]] &= \int \frac{\delta T}{\delta \mathbf{y}}[h](\mathbf{w}) \cdot \frac{\partial F}{\partial \delta_{\mathbf{w}}}[T[h]]d\mathbf{w} && \text{(H.338)} \\ &= \int \frac{\delta T}{\delta \mathbf{y}}[h](\mathbf{w}) \cdot \frac{\delta F}{\delta \mathbf{w}}[T[h]]d\mathbf{w}. && \text{(H.339)} \end{aligned}$$

Solution 46 *to Exercise 46: Let* $\tilde{F}[h] \triangleq F[T[h]]$. *We are to show that*

$$\frac{\delta T}{\delta \mathbf{y}}[h] = \delta_{\mathbf{y}} \sum_{i=1}^{e} i \cdot h_i(\mathbf{y}) \cdot h(\mathbf{y})^{i-1} \tag{H.340}$$

and thus that

$$\frac{\delta \tilde{F}}{\delta \mathbf{y}}[h] = \left(\sum_{i=1}^{e} i \cdot h_i(\mathbf{y}) \cdot h(\mathbf{y})^{i-1} \right) \cdot \frac{\delta F}{\delta \mathbf{y}}[T[h]. \tag{H.341}$$

For any density function $g(\mathbf{y})$, *to first order,*

$$\begin{aligned}
T[h+\varepsilon g](\mathbf{w}) &= \sum_{i=0}^{e} h_i(\mathbf{w}) \cdot (h(\mathbf{w}) + \varepsilon g(\mathbf{w}))^i && \text{(H.342)} \\
&\cong \sum_{i=0}^{e} h_i(\mathbf{w}) \cdot \left(h(\mathbf{w})^i + \varepsilon i h(\mathbf{w})^{i-1} g(\mathbf{w})\right) && \text{(H.343)} \\
&= \sum_{i=0}^{e} h_i(\mathbf{w}) \cdot h(\mathbf{w})^i && \text{(H.344)} \\
&\quad + \varepsilon g(\mathbf{w}) \sum_{i=1}^{e} i \cdot h_i(\mathbf{w}) \cdot h(\mathbf{w})^{i-1} && \text{(H.345)}
\end{aligned}$$

or

$$\begin{aligned}
\frac{\delta T}{\delta \mathbf{y}}[h](\mathbf{w}) &= \lim_{\varepsilon \searrow 0} \frac{T[h+\varepsilon g](\mathbf{w}) - T[h](\mathbf{w})}{\varepsilon} && \text{(H.346)} \\
&= g(\mathbf{w}) \sum_{i=1}^{e} i \cdot h_i(\mathbf{w}) \cdot h(\mathbf{w})^{i-1}. && \text{(H.347)}
\end{aligned}$$

Setting $g = \delta_{\mathbf{y}}$, *we get the first result. Then from (11.285) we get, as claimed,*

$$\begin{aligned}
\frac{\delta \tilde{F}}{\delta \mathbf{y}}[h] &= \int \frac{\delta T}{\delta \mathbf{y}}[T](\mathbf{w}) \cdot \frac{\delta F}{\delta \mathbf{w}}[T[h]] d\mathbf{w} && \text{(H.348)} \\
&= \int \delta_{\mathbf{y}}(\mathbf{w}) \left(\sum_{i=1}^{e} i \cdot h_i(\mathbf{y}) \cdot h(\mathbf{y})^{i-1} \right) \cdot \frac{\delta F}{\delta \mathbf{w}}[T[h] d\mathbf{w} && \text{(H.349)} \\
&= \left(\sum_{i=1}^{e} i \cdot h_i(\mathbf{y}) \cdot h(\mathbf{y})^{i-1} \right) \cdot \frac{\delta F}{\delta \mathbf{y}}[T[h]. && \text{(H.350)}
\end{aligned}$$

Solution 47 *to Exercise 47: We are to show that the third chain rule (11.283) follows from the fourth chain rule (11.285). Let* $\tilde{F}[h] \triangleq F[T[h]]$. *In this case*

$T[h](\mathbf{y}) = s_h(\mathbf{y}) = s(h(\mathbf{y}))$. *For any density function* $g(\mathbf{y})$,

$$\frac{\partial T}{\partial g}[h](\mathbf{w}) = \lim_{\varepsilon \searrow 0} \frac{T[h+\varepsilon g](\mathbf{w}) - T[h](\mathbf{w})}{\varepsilon} \tag{H.351}$$

$$= \lim_{\varepsilon \searrow 0} \frac{s(h(\mathbf{w}) + \varepsilon g(\mathbf{w})) - s(h(\mathbf{w}))}{\varepsilon} \tag{H.352}$$

$$= g(\mathbf{w}) \cdot \lim_{\varepsilon \searrow 0} \frac{s(h(\mathbf{w}) + \varepsilon g(\mathbf{w})) - s(h(\mathbf{w}))}{\varepsilon g(\mathbf{w})} \tag{H.353}$$

$$= g(\mathbf{w}) \cdot \frac{ds}{dy}(h(\mathbf{w})). \tag{H.354}$$

Setting $g = \delta_{\mathbf{y}}$ *we get*

$$\frac{\delta T}{\delta \mathbf{y}}[h](\mathbf{w}) = \delta_{\mathbf{y}}(\mathbf{w}) \cdot \frac{ds}{dy}(h(\mathbf{w})) = \delta_{\mathbf{y}}(\mathbf{w}) \cdot \frac{ds}{dy}(h(\mathbf{y})). \tag{H.355}$$

So from (11.285) we get, as claimed,

$$\frac{\delta \tilde{F}}{\delta \mathbf{y}}[h] = \int \frac{\delta T}{\delta \mathbf{y}}[h](\mathbf{w}) \cdot \frac{\delta \tilde{F}}{\delta \mathbf{w}}[s_h] d\mathbf{w} \tag{H.356}$$

$$= \int \delta_{\mathbf{y}}(\mathbf{w}) \cdot \frac{ds}{dy}(h(\mathbf{y})) \cdot \frac{\delta \tilde{F}}{\delta \mathbf{w}}[s_h] d\mathbf{w} \tag{H.357}$$

$$= \frac{ds}{dy}(h(\mathbf{y})) \cdot \frac{\delta \tilde{F}}{\delta \mathbf{y}}[s_h]. \tag{H.358}$$

Solution 48 *to Exercise 48: We are to show that the second chain rule (11.282) follows from the fourth chain rule (11.285). Let* $\tilde{F}[h] \triangleq F[T[h]]$. *In this situation* $T^{-1}[h](\mathbf{y}) = h(T(\mathbf{y}))$, *and so*

$$\frac{\delta T^{-1}}{\delta \mathbf{y}}[h](\mathbf{w}) = \lim_{\varepsilon \searrow 0} \frac{h(T(\mathbf{w})) + \varepsilon \delta_{\mathbf{y}}(T(\mathbf{w})) - h(T(\mathbf{w}))}{\varepsilon} \tag{H.359}$$

$$= \delta_{\mathbf{y}}(T(\mathbf{w})) = \frac{1}{J_T(\mathbf{y})} \cdot \delta_{T^{-1}\mathbf{y}}(\mathbf{w}) \tag{H.360}$$

where the last equation follows from (B.7). Thus from (11.285),

$$\begin{aligned}\frac{\delta \tilde{F}}{\delta \mathbf{y}}[h] &= \int \frac{\delta T^{-1}}{\delta \mathbf{y}}[h](\mathbf{w}) \cdot \frac{\delta \tilde{F}}{\delta \mathbf{w}}[T^{-1}[h]]d\mathbf{w} && \text{(H.361)}\\ &= \int \frac{1}{J_T(\mathbf{y})} \cdot \delta_{T^{-1}\mathbf{y}}(\mathbf{w}) \cdot \frac{\delta \tilde{F}}{\delta \mathbf{w}}[T^{-1}[h]]d\mathbf{w} && \text{(H.362)}\\ &= \frac{1}{J_T(\mathbf{y})} \cdot \frac{\delta F}{\delta T^{-1}\mathbf{y}}[T^{-1}[h]]. && \text{(H.363)}\end{aligned}$$

Thus for $Y = \{\mathbf{y}_1, ..., \mathbf{y}_n\}$ *with* $|Y| = n$*, we get as claimed*

$$\begin{aligned}&\frac{\delta^n \tilde{F}}{\delta \mathbf{y}_1 \cdots \delta \mathbf{y}_n}[h] && \text{(H.364)}\\ &= \frac{1}{J_T(\mathbf{y}_1) \cdots J_T(\mathbf{y}_n)} \cdot \frac{\delta^n F}{\delta T^{-1}\mathbf{y}_1 \cdots \delta T^{-1}\mathbf{y}_n}[T^{-1}[h]] && \text{(H.365)}\\ &= \frac{1}{J_T^Y} \cdot \frac{\delta F}{\delta T^{-1} Y}[T^{-1}[h]]. && \text{(H.366)}\end{aligned}$$

Solution 49 *to Exercise 49: We are to show that the probability-generating function* $G_{n,q}(y)$ *of the binomial distribution* $B_{n,q}(m)$ *is*

$$G_{n,q}(y) = (1 - q + qy)^n. \qquad \text{(H.367)}$$

To see this, note that

$$\begin{aligned}G_{n,q}(y) &= \sum_{m=0}^{\infty} B_{n,q}(m) y^m && \text{(H.368)}\\ &= \sum_{m=0}^{n} C_{n,m} \cdot q^m \cdot (1-q)^{n-m} y^m && \text{(H.369)}\\ &= (qy + (1-q))^n && \text{(H.370)}\end{aligned}$$

where the final equation follows from the binomial expansion

$$(a+b)^n = \sum_{i=0}^{n} C_{n,m} a^m b^{n-m}. \qquad \text{(H.371)}$$

Solution 50 *to Exercise 50: We are to prove (12.264):* $\int f(Z|a, \mathbf{x})\delta Z = 1$. *To see this, note that*

$$\int f(Z|a, \mathbf{x})\delta Z \tag{H.372}$$

$$= f(\emptyset|a, \mathbf{x}) + \sum_{m=1}^{\infty} \frac{1}{m!} \int f(\{\mathbf{z}_1, ..., \mathbf{z}_m\}|a, \mathbf{x}) d\mathbf{z}_1 \cdots d\mathbf{z}_m \tag{H.373}$$

$$= B_{a,p_D(\mathbf{x})}(0) \tag{H.374}$$

$$+ \sum_{m=1}^{\infty} B_{a,p_D(\mathbf{x})}(m) \int f(\mathbf{z}_1|\mathbf{x}) \cdots f(\mathbf{z}_m|\mathbf{x}) d\mathbf{z}_1 \cdots d\mathbf{z}_m \tag{H.375}$$

$$= \sum_{m=0}^{\infty} B_{a,p_D(\mathbf{x})}(m) \cdot 1^m \tag{H.376}$$

$$= G_{a,p_D(\mathbf{x})}(1) = 1. \tag{H.377}$$

Solution 51 *to Exercise 51: We are to show that the p.g.fl. of a single extended target is*

$$G_{k+1}[g|\mathbf{x}] = e^{\lambda c[g] - \lambda} \prod_{\ell=1}^{L} \left(1 - p_D^\ell(\mathbf{x}) + p_D^\ell(\mathbf{x}) p_g^\ell(\mathbf{x})\right) \tag{H.378}$$

and that the p.g.fl. for multiple extended targets is

$$G_{k+1}[g|X] = e^{\lambda c[g] - \lambda} \prod_{\ell=1}^{L} \left(1 - p_D^\ell + p_D^\ell p_g^\ell\right)^X. \tag{H.379}$$

To see this, note that there is a direct analogy between the likelihoods $f_{k+1}^\ell(\mathbf{z}|\mathbf{x})$ *and probabilities of detection* $p_D^\ell(\mathbf{x})$ *for the sites of an extended target and the likelihoods* $f_{k+1}(\mathbf{z}|\mathbf{x}_i)$ *and the probabilities of detection* $p_D(\mathbf{x}_i)$ *for the targets in a target group. Thus (12.366) follows immediately from (12.151). From (11.166) we know that*

$$
\begin{aligned}
G_{k+1}[g|X] &= e^{\lambda c[g]-\lambda} \prod_{\mathbf{x}\in X} \prod_{\ell=1}^{L} \left(1 - p_D^\ell(\mathbf{x}) + p_D^\ell(\mathbf{x}) p_g^\ell(\mathbf{x})\right) &\text{(H.380)}\\
&= e^{\lambda c[g]-\lambda} \prod_{\ell=1}^{L} \prod_{\mathbf{x}\in X} \left(1 - p_D^\ell(\mathbf{x}) + p_D^\ell(\mathbf{x}) p_g^\ell(\mathbf{x})\right) &\text{(H.381)}\\
&= e^{\lambda c[g]-\lambda} \prod_{\ell=1}^{L} \left(1 - p_D^\ell + p_D^\ell p_g^\ell\right)^X . &\text{(H.382)}
\end{aligned}
$$

Solution 52 *to Exercise 52: We are to verify that* $\overset{*}{f}_{k+1}(Z|X)$ *as defined in (12.193) is a likelihood function, that is,* $\int \overset{*}{f}_{k+1}(Z|X)\delta Z = 1$. *To see this, note that*

$$
\begin{aligned}
&\int \overset{*}{f}_{k+1}(Z|X)\delta Z &\text{(H.383)}\\
&= \overset{*}{f}_{k+1}(\emptyset|X) + \int_{Z\neq\emptyset} \overset{*}{f}_{k+1}(Z|X)\delta Z &\text{(H.384)}\\
&= 1 - \overset{*}{p}_T + \overset{*}{p}_T \cdot f_{k+1}(\emptyset|X) + \overset{*}{p}_T \int_{Z\neq\emptyset} f_{k+1}(Z|X)\delta Z &\text{(H.385)}\\
&= 1 - \overset{*}{p}_T + \overset{*}{p}_T \cdot f_{k+1}(\emptyset|X) + \overset{*}{p}_T \left(1 - f_{k+1}(\emptyset|X)\right) &\text{(H.386)}\\
&= 1. &\text{(H.387)}
\end{aligned}
$$

Solution 53 *to Exercise 53: We are to verify (12.194):*

$$
\overset{*}{G}_{k+1}[g|X] = 1 - \overset{*}{p}_T + \overset{*}{p}_T \cdot G_{k+1}[g|X]. \quad \text{(H.388)}
$$

To see this, note that

$$\mathring{G}_{k+1}[g|X] = \int h^Z \cdot \mathring{f}_{k+1}(Z|X)\delta Z \tag{H.389}$$
$$= \mathring{f}_{k+1}(\emptyset|X) + \int_{Z\neq\emptyset} h^Z \cdot \mathring{f}_{k+1}(Z|X)\delta Z \tag{H.390}$$
$$= 1 - \mathring{p}_T + \mathring{p}_T \cdot f_{k+1}(\emptyset|X) \tag{H.391}$$
$$+\mathring{p}_T \int_{Z\neq\emptyset} h^Z \cdot f_{k+1}(Z|X)\delta Z \tag{H.392}$$
$$= 1 - \mathring{p}_T + \mathring{p}_T \cdot f_{k+1}(\emptyset|X) - \mathring{p}_T \cdot f_{k+1}(\emptyset|X) \tag{H.393}$$
$$+\mathring{p}_T \int h^Z \cdot f_{k+1}(Z|X)\delta Z \tag{H.394}$$
$$= 1 - \mathring{p}_T + \mathring{p}_T \int h^Z \cdot f_{k+1}(Z|X)\delta Z \tag{H.395}$$
$$= 1 - \mathring{p}_T + \mathring{p}_T \cdot G_{k+1}[g|X]. \tag{H.396}$$

Solution 54 *to Exercise 54: We are to show that the p.g.fl. of the finite-mixture model likelihood of (12.353) is*

$$G[g|Q, \mathbf{p}] = G_{\mathbf{p}}(p_Q[g]) \tag{H.397}$$

and also that

$$G[g|Q] = e^{\lambda(Q)\cdot p_G[g] - \lambda(Q)}. \tag{H.398}$$

To see this, note that from (12.353) we get

$$G[g|Q, \mathbf{p}] = \int g^Z \cdot f(Z|Q, \mathbf{p})\delta Z \tag{H.399}$$
$$= \sum_{m=0}^{\infty} \frac{1}{m!} \int g(\mathbf{z}_1) \cdots g(\mathbf{z}_m) \cdot m! p_{\mathbf{p}}(m) \tag{H.400}$$
$$\cdot f(\mathbf{z}_1|Q) \cdots f(\mathbf{z}_m|Q) d\mathbf{z}_1 \cdots d\mathbf{z}_m \tag{H.401}$$
$$= \sum_{m=0}^{\infty} p_{\mathbf{p}}(m) \cdot \left(\int g(\mathbf{z}) \cdot f(\mathbf{z}|Q) d\mathbf{z} \right)^m \tag{H.402}$$
$$= \sum_{m=0}^{\infty} p_{\mathbf{p}}(m) \cdot p_Q[g]^m = G_{\mathbf{p}}(p_Q[g]). \tag{H.403}$$

If $p_{\mathbf{p}}(m) = e^{-\lambda(Q)} \lambda(Q)^m / m!$ *then* $G_{\mathbf{p}}(x) = e^{\lambda(Q)x - \lambda(Q)}$ *and we are done.*

Solution 55 *to Exercise 55: We are to show that the p.g.fl. for the simple virtual leader-follower model of (13.95) is*

$$G_{k+1|k}[h|X'] = p_h^{X'-\varphi_k(\bar{\mathbf{x}}')-\bar{\mathbf{x}}'} \tag{H.404}$$

where

$$p_h(\mathbf{x}') \triangleq \int h(\mathbf{x}) \cdot f_{\mathbf{V}}(\mathbf{x}+\mathbf{x}')d\mathbf{x} \tag{H.405}$$

and where, for any $\mathbf{x} \in \mathfrak{X}_0$,

$$X' - \mathbf{x} \triangleq \left\{ \begin{array}{ll} \{\mathbf{x}' - \mathbf{x} | \, \mathbf{x}' \in X'\} & \text{if} \;\; X \neq \emptyset \\ \emptyset & \text{if} \;\; X = \emptyset \end{array} \right. . \tag{H.406}$$

To see this, note that from (13.83)

$$
\begin{aligned}
& G_{k+1|k}[h|X'] && \text{(H.407)}\\
= & \int h^X \cdot f_{k+1|k}(X|X')\delta X && \text{(H.408)}\\
= & \frac{1}{n!}\int h(\mathbf{x}_1)\cdots h(\mathbf{x}_n) && \text{(H.409)}\\
& \cdot \sum_\sigma f_{\mathbf{V}}(\mathbf{x}_1 - \varphi_k(\bar{\mathbf{x}}') - \bar{\mathbf{x}}' + \mathbf{x}'_{\sigma 1}) && \text{(H.410)}\\
& \cdots f_{\mathbf{V}}(\mathbf{x}_n - \varphi_k(\bar{\mathbf{x}}') - \bar{\mathbf{x}}' + \mathbf{x}'_{\sigma n}) && \text{(H.411)}\\
& \cdot d\mathbf{x}_1 \cdots d\mathbf{x}_n && \text{(H.412)}\\
= & \frac{1}{n!}\left(\begin{array}{c} \sum_\sigma \left(\int h(\mathbf{x}) \cdot f_{\mathbf{V}}(\mathbf{x} - \varphi_k(\bar{\mathbf{x}}') - \bar{\mathbf{x}}' + \mathbf{x}'_1)d\mathbf{x}\right) \\ \cdots \left(\int h(\mathbf{x}) \cdot f_{\mathbf{V}}(\mathbf{x} - \varphi_k(\bar{\mathbf{x}}') - \bar{\mathbf{x}}' + \mathbf{x}'_n)d\mathbf{x}\right) \end{array} \right) && \text{(H.413)}\\
= & p_h(\mathbf{x}'_1 - \varphi_k(\bar{\mathbf{x}}') - \bar{\mathbf{x}}') \cdots p_h(\mathbf{x}'_n - \varphi_k(\bar{\mathbf{x}}') - \bar{\mathbf{x}}') && \text{(H.414)}\\
= & p_h^{X'-\varphi_k(\bar{\mathbf{x}}')-\bar{\mathbf{x}}'}. && \text{(H.415)}
\end{aligned}
$$

Solution 56 *to Exercise 56: We are to verify (13.111). That is, we are to show that the p.g.fl. for the virtual leader-follower with target disappearance of (13.104) is*

$$G_{k+1|k}[h|X'] = (1 - p_S + p_S p_h)^{X'-\varphi_k(\bar{\mathbf{x}}')-\bar{\mathbf{x}}'} \tag{H.416}$$

where p_S *is constant and where* p_h *and* $X' - \varphi_k(\bar{\mathbf{x}}') - \bar{\mathbf{x}}'$ *are as defined in Exercise 55. To see this, note that from (13.104) we have*

$$
\begin{aligned}
& f_{k+1|k}(X|X') && \text{(H.417)}\\
= \; & f_{k+1|k}(\emptyset|X') && \text{(H.418)}\\
& \cdot \sum_{1\leq i_1\neq\ldots\neq i_n\leq n'} \frac{p_S^n f_{k+1|k}^{i_1}(\mathbf{x}_1|X')\cdots f_{k+1|k}^{i_n}(\mathbf{x}_n|X')}{(1-p_S)^n} && \text{(H.419)}
\end{aligned}
$$

where $f_{k+1|k}(\emptyset|X') = (1-p_S)^{n'}$. *Thus*

$$
\begin{aligned}
& G_{k+1|k}[h|X'] && \text{(H.420)}\\
= \; & \int h^X \cdot f_{k+1|k}(X|X')\delta X && \text{(H.421)}\\
= \; & f_{k+1|k}(\emptyset|X') \sum_{n=0}^{n'} \frac{1}{n!} \int h(\mathbf{x}_1)\cdots h(\mathbf{x}_n) && \text{(H.422)}\\
& \cdot \sum_{1\leq i_1\neq\ldots\neq i_n\leq n'} \frac{p_S^n \cdot f_{k+1|k}^{i_1}(\mathbf{x}_1|X')\cdots f_{k+1|k}^{i_n}(\mathbf{x}_n|X')}{(1-p_S)^n} && \text{(H.423)}\\
& \cdot d\mathbf{x}_1\cdots d\mathbf{x}_n && \text{(H.424)}\\
= \; & f_{k+1|k}(\emptyset|X') \sum_{n=0}^{n'} \frac{1}{n!} \sum_{1\leq i_1\neq\ldots\neq i_n\leq n'} && \text{(H.425)}\\
& \int \frac{\begin{array}{c} p_S^n \cdot h(\mathbf{x}_1) f_{\mathbf{V}}(\mathbf{x}_1 - \varphi_k(\bar{\mathbf{x}}') - \bar{\mathbf{x}}' + \mathbf{x}'_{i_1}) \\ \cdots h(\mathbf{x}_n) f_{\mathbf{V}}(\mathbf{x}_n - \varphi_k(\bar{\mathbf{x}}') - \bar{\mathbf{x}}' + \mathbf{x}'_{i_n}) \end{array}}{(1-p_S)^n} d\mathbf{x}_1\cdots d\mathbf{x}_n && \text{(H.426)}
\end{aligned}
$$

and so

$$= f_{k+1|k}(\emptyset|X') \sum_{n=0}^{n'} \frac{1}{n!} \sum_{1 \leq i_1 \neq \ldots \neq i_n \leq n'} \tag{H.427}$$

$$\left(\int \frac{p_S h(\mathbf{x}_1) f_{\mathbf{V}}(\mathbf{x}_1 - \varphi_k(\bar{\mathbf{x}}') - \bar{\mathbf{x}}' + \mathbf{x}'_{i_1})}{1 - p_S} d\mathbf{x}_1 \right) \tag{H.428}$$

$$\cdots \left(\int \frac{p_S \cdot h(\mathbf{x}_1) f_{\mathbf{V}}(\mathbf{x}_n - \varphi_k(\bar{\mathbf{x}}') - \bar{\mathbf{x}}' + \mathbf{x}'_{i_n})}{1 - p_S} d\mathbf{x}_n \right) \tag{H.429}$$

$$= f_{k+1|k}(\emptyset|X') \tag{H.430}$$

$$\sum_{n=0}^{n'} \sum_{1 \leq i_1 < \ldots < i_n \leq n'} \frac{p_S p_h(\mathbf{x}'_{i_1} - \varphi_k(\bar{\mathbf{x}}') - \bar{\mathbf{x}}'))}{1 - p_S} \tag{H.431}$$

$$\cdots \frac{p_S \cdot p_h(\mathbf{x}'_{i_n} - \varphi_k(\bar{\mathbf{x}}') - \bar{\mathbf{x}}')}{1 - p_S}. \tag{H.432}$$

Using (11.141)

$$\sigma_{n',n}(x_1, \ldots, x_{n'}) = \sum_{1 \leq i_1 < \ldots < i_n \leq n'} x_{i_1} \cdots x_{i_n} \tag{H.433}$$

we can write this as

$$G_{k+1|k}[h|X'] \tag{H.434}$$

$$= f_{k+1|k}(\emptyset|X') \tag{H.435}$$

$$\sum_{n=0}^{n'} \sigma_{n',n} \left(\begin{array}{c} \frac{p_S p_h(\mathbf{x}'_{i_1} - \varphi_k(\bar{\mathbf{x}}') - \bar{\mathbf{x}}'))}{1 - p_S}, \ldots, \\ \frac{p_S p_h(\mathbf{x}'_{i_n} - \varphi_k(\bar{\mathbf{x}}') - \bar{\mathbf{x}}')}{1 - p_S} \end{array} \right). \tag{H.436}$$

From the fundamental identify for symmetric functions (11.142),

$$G_{k+1|k}[h|X'] \tag{H.437}$$

$$= f_{k+1|k}(\emptyset|X') \cdot \left(1 + \frac{p_S p_h(\mathbf{x}'_1 - \varphi_k(\bar{\mathbf{x}}') - \bar{\mathbf{x}}'))}{1-p_S}\right) \tag{H.438}$$

$$\cdots \left(1 + \frac{p_S p_h(\mathbf{x}'_n - \varphi_k(\bar{\mathbf{x}}') - \bar{\mathbf{x}}')}{1-p_S}\right) \tag{H.439}$$

$$= (1 - p_S + p_S p_h(\mathbf{x}'_1 - \varphi_k(\bar{\mathbf{x}}') - \bar{\mathbf{x}}')) \tag{H.440}$$

$$\cdots (1 - p_S + p_S \cdot p_h(\mathbf{x}'_n - \varphi_k(\bar{\mathbf{x}}') - \bar{\mathbf{x}}')) \tag{H.441}$$

$$= (1 - p_S + p_S p_h)^{X - \varphi_k(\bar{\mathbf{x}}') - \bar{\mathbf{x}}'}. \tag{H.442}$$

Solution 57 *to Exercise 57: We are to verify (14.230). That is, we are to establish the formula for the variance of target number for the single-target joint detection and tracking filter. To see this, note that substituting (14.215) into (14.229),*

$$\sigma^2_{k+1|k+1} \tag{H.443}$$

$$= \frac{1 - p_D + p_D \sum_{\mathbf{z}\in\mathbf{Z}} f_{k+1}(\mathbf{z}) \cdot \frac{\kappa(Z-\{\mathbf{z}\})}{\kappa(Z)}}{p^{-1}_{k+1|k} - p_D + p_D \sum_{\mathbf{z}\in\mathbf{Z}} f_{k+1}(\mathbf{z}) \cdot \frac{\kappa(Z-\{\mathbf{z}\})}{\kappa(Z)}} \tag{H.444}$$

$$\cdot \left(1 - \frac{1 - p_D + p_D \sum_{\mathbf{z}\in\mathbf{Z}} f_{k+1}(\mathbf{z}) \cdot \frac{\kappa(Z-\{\mathbf{z}\})}{\kappa(Z)}}{p^{-1}_{k+1|k} - p_D + p_D \sum_{\mathbf{z}\in\mathbf{Z}} f_{k+1}(\mathbf{z}) \cdot \frac{\kappa(Z-\{\mathbf{z}\})}{\kappa(Z)}}\right) \tag{H.445}$$

and so

$$= \frac{1 - p_D + p_D \sum_{\mathbf{z}\in\mathbf{Z}} f_{k+1}(\mathbf{z}) \cdot \frac{\kappa(Z-\{\mathbf{z}\})}{\kappa(Z)}}{p^{-1}_{k+1|k} - p_D + p_D \sum_{\mathbf{z}\in\mathbf{Z}} f_{k+1}(\mathbf{z}) \cdot \frac{\kappa(Z-\{\mathbf{z}\})}{\kappa(Z)}} \tag{H.446}$$

$$\cdot \frac{p^{-1}_{k+1|k} - 1}{p^{-1}_{k+1|k} - p_D + p_D \sum_{\mathbf{z}\in\mathbf{Z}} f_{k+1}(\mathbf{z}) \cdot \frac{\kappa(Z-\{\mathbf{z}\})}{\kappa(Z)}} \tag{H.447}$$

$$= \left(p^{-1}_{k+1|k} - 1\right) \tag{H.448}$$

$$\cdot \frac{1 - p_D + p_D \sum_{\mathbf{z}\in\mathbf{Z}} f_{k+1}(\mathbf{z}) \cdot \frac{\kappa(Z-\{\mathbf{z}\})}{\kappa(Z)}}{\left(p^{-1}_{k+1|k} - p_D + p_D \sum_{\mathbf{z}\in\mathbf{Z}} f_{k+1}(\mathbf{z}) \cdot \frac{\kappa(Z-\{\mathbf{z}\})}{\kappa(Z)}\right)^2} \tag{H.449}$$

and so

$$= \left(p_{k+1|k}^{-1} - 1\right) \cdot p_{k+1|k}^2 \tag{H.450}$$

$$\cdot \frac{1 - p_D + p_D \sum_{\mathbf{z} \in Z} f_{k+1}(\mathbf{z}) \cdot \frac{\kappa(Z - \{\mathbf{z}\})}{\kappa(Z)}}{\left(1 - p_{k+1|k} p_D + p_{k+1|k} p_D \sum_{\mathbf{z} \in Z} f_{k+1}(\mathbf{z}) \cdot \frac{\kappa(Z - \{\mathbf{z}\})}{\kappa(Z)}\right)^2} \tag{H.451}$$

$$= p_{k+1|k} \cdot \left(1 - p_{k+1|k}\right) \tag{H.452}$$

$$\cdot \frac{1 - p_D + p_D \sum_{\mathbf{z} \in Z} f_{k+1}(\mathbf{z}) \cdot \frac{\kappa(Z - \{\mathbf{z}\})}{\kappa(Z)}}{\left(1 - p_{k+1|k} p_D + p_{k+1|k} p_D \sum_{\mathbf{z} \in Z} f_{k+1}(\mathbf{z}) \cdot \frac{\kappa(Z - \{\mathbf{z}\})}{\kappa(Z)}\right)^2} \tag{H.453}$$

and so

$$= \sigma_{k+1|k}^2 \tag{H.454}$$

$$\cdot \frac{1 - p_D + p_D \sum_{\mathbf{z} \in Z} f_{k+1}(\mathbf{z}) \cdot \frac{\kappa(Z - \{\mathbf{z}\})}{\kappa(Z)}}{\left(\begin{array}{c} 1 - p_{k+1|k} p_D \\ + p_{k+1|k} p_D \sum_{\mathbf{z} \in Z} f_{k+1}(\mathbf{z}) \cdot \frac{\kappa(Z - \{\mathbf{z}\})}{\kappa(Z)} \end{array}\right)^2}. \tag{H.455}$$

Solution 58 *to Exercise 58: We are to show that if* $G[h] = e^{\mu s[h] - \mu}$, *then* $G^\rho[h] = e^{\mu^\rho s^\rho[h] - \mu^\rho}$, *with* $\mu^\rho = \mu \cdot s[\rho]$ *and* $s^\rho(\mathbf{x}) = \rho(\mathbf{x}) \cdot s(\mathbf{x}) / s[\rho]$. *For, from (14.296)*

$$G^\rho[h] = e^{\mu s[1 - \rho + \rho h] - \mu} \tag{H.456}$$

$$= e^{-\mu s[\rho] + \mu s[\rho h]} = \exp\left(-\mu s[\rho] + \mu s[\rho] \cdot \frac{s[\rho h]}{s[\rho]}\right) \tag{H.457}$$

$$= e^{\mu^\rho s^\rho[h] - \mu^\rho}. \tag{H.458}$$

Solution 59 *to Exercise 59: We are to show that if* $G[h] = G(s[h])$, *then* $G^\rho[h] = G^\rho(s^\rho[h])$, *where* $\mu^\rho = \mu \cdot s[\rho]$ *and* $s^\rho(\mathbf{x}) = s(\mathbf{x}) / s[\rho]$. *To see this, first notice that*

$$G^\rho[h] = G(s[1 - \rho + \rho h]) = G(1 - s[\rho] + s[\rho h]) \tag{H.459}$$

and thus that

$$G^\rho(x) \triangleq G^\rho[x] = G(1 - s[\rho] + s[\rho] \cdot x). \tag{H.460}$$

Consequently,

$$\begin{aligned} G^\rho[h] &= G\left(1 - s[\rho] + s[\rho h]\right) &\text{(H.461)}\\ &= G\left(1 - s[\rho] + s[\rho] \cdot \frac{s[\rho h]}{s[\rho]}\right) &\text{(H.462)}\\ &= G^\rho\left(\frac{s[\rho h]}{s[\rho]}\right) = G^\rho\left(s^\rho[h]\right). &\text{(H.463)} \end{aligned}$$

Solution 60 *to Exercise 60: We are to show that if* $G^\rho[h] = G[1-\rho+\rho h]$, *then* $D^\rho(\mathbf{x}) = \rho(\mathbf{x}) \cdot D(\mathbf{x})$ *and*

$$\begin{aligned} G^\rho(x) &= G(1 - s[\rho] + s[\rho] \cdot x) &\text{(H.464)}\\ p^\rho(n) &= \frac{1}{n!} \cdot s[\rho]^n \cdot G^{(n)}(1 - s[\rho]). &\text{(H.465)} \end{aligned}$$

To see this, note that by the chain rule for affine functional transformations (11.215),

$$\frac{\delta G^\rho}{\delta \mathbf{x}}[h] = \frac{\delta}{\delta \mathbf{x}} G[1-\rho+\rho h] = \rho(\mathbf{x}) \cdot \frac{\delta G}{\delta \mathbf{x}}[1-\rho+\rho h]. \qquad \text{(H.466)}$$

Thus

$$D^\rho(\mathbf{x}) = \frac{\delta G^\rho}{\delta \mathbf{x}}[1] = \rho(\mathbf{x}) \cdot \frac{\delta G}{\delta \mathbf{x}}[1] = \rho(\mathbf{x}) \cdot D(\mathbf{x}). \qquad \text{(H.467)}$$

On the other hand,

$$\begin{aligned} G^\rho(x) &\triangleq G^\rho[x] = G(s[1-\rho+\rho \cdot x]) &\text{(H.468)}\\ &= G(1 - s[\rho] + s[\rho] \cdot x). &\text{(H.469)} \end{aligned}$$

Thus

$$\frac{d^n G^\rho}{dx^n}(x) = s[\rho]^n \cdot G^{(n)}(1 - s[\rho] + s[\rho] \cdot x) \qquad \text{(H.470)}$$

and so

$$p^\rho(n) = \frac{1}{n!} \frac{d^n G^\rho}{dx^n}(0) = \frac{s[\rho]^n}{n!} \cdot G^{(n)}(1 - s[\rho]).$$

Solution 61 *to Exercise 61: We are to verify (17.132):*

$$G_{k+1|k}[h] = G_B[h] \cdot G_{k|k}[1 - p_S + p_S p_h] \qquad \text{(H.471)}$$

where

$$G_B[h] = \prod_{i=1}^{e}(1 - b_i + b_i \cdot b_i[h]) \tag{H.472}$$

is the p.g.fl. of the birth process. In Section 14.8.1, I showed that the multitarget predictor equation, (14.14), can be written in p.g.fl. form as follows:

$$G_{k+1|k}[h] = \int G_{k+1|k}[h|X'] \cdot f_{k|k}(X'|Z^{(k)})\delta X' \tag{H.473}$$

where $G_{k+1|k}[h]$ *is the p.g.fl. of* $f_{k+1|k}(X|Z^{(k)})$ *and where*

$$G_{k+1|k}[h|X'] \triangleq \int h^X \cdot f_{k+1|k}(X|X')\delta X \tag{H.474}$$

is the p.g.fl. of $f_{k+1|k}(X|X')$. *We must determine the form of* $G_{k+1|k}[h|X']$ *for the current multitarget measurement model. It has the form*

$$\Xi_{k+1|k} = \overbrace{\Gamma_k(X')}^{\text{persisting targets}} \cup \overbrace{B_k}^{\text{birth targets}}. \tag{H.475}$$

Since $\Gamma_k(X')$ *and* B_k *are assumed independent, the p.g.fl. has the form*

$$G_{k+1|k}[h|X'] = G_{\Gamma_k(X')}[h] \cdot G_B[h]. \tag{H.476}$$

The persisting targets have the same form as in the standard motion model with no target appearance (13.60):

$$G_{k+1|k}[h|X'] = (1 - p_S + p_S p_h)^{X'} \tag{H.477}$$

where

$$p_h(\mathbf{x}') = \int h(\mathbf{x}) \cdot f_{k+1|k}(\mathbf{x}|\mathbf{x}')d\mathbf{x}. \tag{H.478}$$

On the other hand,

$$G_B[h] = \prod_{i=1}^{e}(1 - b_i + b_i \cdot b_i[h]) \tag{H.479}$$

where

$$b_i[h] = \int h(\mathbf{x}) \cdot b_i(\mathbf{x})d\mathbf{x}. \tag{H.480}$$

References

[1] S. Ahlberg, P. Hörling, H. Kjellström, K. Jöred, C. Mårtenson, G. Neider, J. Schubert, P. Svenson, P. Svensson, K. Undén, and J. Walter, "An information fusion demonstrator for tactical intelligence processing in network-based defense," *Information Fusion*, Vol. 8, No. 1, pp. 84-107, 2007.

[2] S. Ahlberg, P. Hörling, H. Kjellström, K. Jöred, C. Mårtenson, G. Neider, J. Schubert, P. Svenson, P. Svensson, K. Undén, and J. Walter, "The IFD03 information fusion demonstrator," *Proc. 7th Int'l. Conf. on Inf. Fusion*, pp. 936-943, Stockholm, Sweden, June 28-July 1, 2004, Int'l. Soc. Inf. Fusion, Sunnyvale, CA.

[3] J. Albert, "Algebraic properties of bag data types," *Proceedings of the 17th International Conference on Very Large Data Bases*, pp. 211-219, Barcelona, Sept., 1991.

[4] S. Arnborg, "Robust Bayesianism: Imprecise and paradoxical reasoning," *Proc. 7th Int'l. Conf. on Inf. Fusion*, pp. 1-20, Stockholm, Sweden, June 28-July 1, 2004, Int'l. Soc. Inf. Fusion, Sunnyvale, CA.

[5] D. L. Alspach, "A Gaussian sum approach to the multi-target identification-tracking problem," *Automatica*, Vol. 11, pp. 285-296, 1975.

[6] C. Andrieu, A. Doucet, S.S. Singh, and V.B. Tadić, "Particle methods for change detection, system identification, and control," *Proc. IEEE*, Vol. 92, No. 3, pp. 423-438, 2004.

[7] R. T. Antony, *Principles of Data Fusion Automation*, Norwood, MA: Artech House, 1995.

[8] M.S. Arulamalan, S. Maskell, N. Gordon, and T. Clapp, "A tutorial on particle filters for online nonlinear / non-Gaussian Bayesian tracking," *IEEE Trans. Sign. Proc.*, Vol. 50 No. 2, pp. 174-188, 2002.

[9] B. Balakumar, A. Sinha, T. Kirubarajan, and J. P. Reilly, "PHD filtering for tracking an unknown number of sources using an array of sensors," *Proc. 13th IEEE Workshop on Stat. Sign. Proc.*, pp. 43-48, Bordeaux, France, July 17-20, 2005.

[10] D. J. Ballantyne, J. Hailes, M. A. Kouritzin, H. Long, and J. H. Wiersma, "A hybrid weighted interacting particle filter for multi-target tracking," in I. Kadar (ed.), *Sign. Proc., Sensor Fusion, and Targ. Recogn. XII*, SPIE, Vol. 5096, pp. 244-255, Bellingham, WA, 2003.

[11] Y. Bar-Shalom and X.-R. Li, *Estimation and Tracking: Principles, Techniques, and Software*, Norwood, MA: Artech House, 1993.

[12] O. E. Barnsdorff-Nielsen, W. S. Kendall, and M. N. M. van Lieshout (eds.), *Stochastic Geometry: Likelihood and Computation*, Boca Raton, FL: Chapman & Hall/CRC Press, 1999.

[13] M. Baudin, "Multidimensional point processes and random closed sets," *J. Applied Prob*, Vol. 21, pp. 173-178, 1984.

[14] E. Biglieri and M. Lops, "Multiuser detection in a dynamic environment," *Workshop of the Center on Information Theory and Applications*, La Jolla, CA, Feb. 6-10, 2006.

[15] J. A. Bilmes, "A gentle tutorial of the EM algorithm and its application to parameter estimation and hidden Markov models," Technical Report, Int'l Computer Science Inst., April 1998, 13 pages.

[16] S. S. Blackman, "Multiple hypothesis tracking for multiple target tracking," *IEEE Aerospace and Electr. Sys. Mag., Part 2: Tutorials*, Vol. 19 No. 1, pp. 1-18, 2004.

[17] S. S. Blackman, *Multiple-Target Tracking with Radar Applications*, Norwood, MA: Artech House, 1986.

[18] S. Blackman and S. Popoli, *Design and Analysis of Modern Tracking Systems*, Norwood, MA: Artech House, 2000.

[19] D. Blount, M. Kouritzin, and J. McCrosky, "Artificial learning approaches for multi-target tracking," in F. A. Sadjadi (ed.), *Automatic Targ. Recogn. XIV*, SPIE, Vol. 5426, pp. 293-304, Bellingham, WA, 2004.

[20] M. Bolić, P. M. Djurić, and S. Hong, "Resampling algorithms for particle filters: A computational complexity perspective," *EURASIP J. on Applied Sign. Proc.*, Vol. 15, pp. 2267-2277, 2004.

[21] W. G. Carrara, R. S. Goodman, and R. M. Majewski, *Spotlight Synthetic Aperture Radar: Signal Processing Algorithms*, Norwood, MA: Artech House, 1995.

[22] S. Challa, R. J. Evans, and D. Musicki, "Target tracking—a Bayesian perspective," *Proc. 14th Int'l Conf. on Digital Sign. Proc.*, Vol. 1, pp. 437-440, Santorini, Greece, July 1-3, 2002.

[23] S. Challa and B.-N. Vo, "Bayesian approaches to track existence—IPDA and random sets," *Proc. 5th Int'l Conf. on Information Fusion*, pp. 1228-1235, Annapolis, MD, July 8-11, 2002, Int'l. Soc. Inf. Fusion, Sunnyvale, CA.

[24] P. Cheeseman, J. Kelly, M. Self, J. Stutz, W. Taylor, and D. Freeman, "Autoclass: A Bayesian classification system," *Proc. Fifth Int'l. Conf. on Machine Learning*, June 12-14, 1988, University of Michigan at Ann Arbor, pp. 54-64, Morgan Kaufmann Publishers, San Mateo, CA,1988.

[25] P. Cheeseman, M. Self, J. Kelly, J. Stutz, W. Taylor, and D. Freeman, "Bayesian classification," *Proc. Seventh Nat'l. Conf. on Artificial Intelligence*, Vol. 2, Aug. 21-26, 1988, St. Paul MN, pp. 607-611, 1988.

[26] C.-Y. Chong, S. Mori, and K.-C. Chang, "Distributed multitarget multisensor tracking," in Y. Bar-Shalom (ed.), *Multitarget-Multisensor Tracking: Advanced Applications*, Norwood, MA: Artech House, 1990, Chapter 8.

[27] C. K. Chui and G. Chen, *Kalman Filtering with Real-Time Applications*, 3rd ed., New York: Springer, 1999.

[28] D. E. Clark and J. Bell, "Bayesian multiple target tracking in forward scan sonar images using the PHD filter," *IEE Radar, Sonar, and Nav.*, Vol. 152, No. 5, pp. 327-334, 2005.

[29] D. E. Clark and J. Bell, "Convergence results for the particle PHD filter," *IEEE Trans. on Sign. Proc.*, Vol. 54, No. 7, pp. 2652-2661, 2006.

[30] D. E. Clark and J. Bell, "Data association for the PHD filter," *Proc. Conf. on Intell. Sensors, Sensor Networks, and Info. Proc.*, pp. 217-222, Melbourne, Australia, Dec. 5-8, 2005.

[31] D. E. Clark and J. M. Bell, "GM-PHD filter multitarget tracking in sonar images," in I. Kadar (ed.), *Sign. Proc., Sensor Fusion, and Targ. Recogn. XV*, SPIE, Vol. 6235, Bellingham, WA, 2006.

[32] D. E. Clark, J. Bell, Y. de Saint-Pern, and Y. Petillot, "PHD filter multi-target tracking in 3D sonar," *Proc. IEEE OCEANS05-Europe*, pp. 265-270, Brest, France, June 20-23, 2005.

[33] D. E. Clark, K. Panta, and B.-N. Vo, "The GM-PHD filter multiple target tracker," *Proc. 9th Int'l Symp. on Information Fusion*, Florence, Italy, July 10-13, 2006, Int'l. Soc. Inf. Fusion, Sunnyvale, CA.

[34] D. Clark, and B. Vo, "Convergence analysis of the Gaussian mixture probability hypothesis density filter," *IEEE Trans. Sign. Proc.*, to appear, 2007.

[35] D. Crisan, "Particle filters—a theoretical perspective," in A. Doucet, N. de Freitas, and N. Gordon (eds.), *Sequential Monte Carlo Methods in Practice*, pp. 17-41, New York: Springer, 2001.

[36] D. Daley and D. Vere-Jones, *An Introduction to the Theory of Point Processes*, New York: Springer-Verlag, 1988.

[37] D. Daley and D. Vere-Jones, *An Introduction to the Theory of Point Processes, Volume 1: Elementary Theory and Methods*, New York: Springer-Verlag, 2003.

[38] F. Daum, "Nonlinear filters: Beyond the Kalman filter," *IEEE Aerospace & Electr. Sys. Mag.*, Vol. 20, No. 8, Part 2: Tutorials, Aug., 2005.

[39] D. Denneberg, *Non-Additive Measure and Integral*, Dordrecht: Kluwer Academic Publishers, 1994.

[40] J. Dezert, "Foundations for a new theory of plausible and paradoxical reasoning," *Information & Security, an Int'l J.*, Vol. 9, pp. 90-95, 2002.

[41] P. M. Djurić, J. H. Kotecha, J. Zhang, Y. Huang, T. Ghirmai, M. Bugallo, and J. Míguez, "Particle filtering: A review of the theory and how it can be used for solving problems in wireless communications," *IEEE Sign. Proc. Mag.*, pp. 19-38, Sept., 2003.

[42] R. Douc, O. Cappé, and E. Moulines, "Comparison of resampling schemes for particle filtering," *Proc. 4th Int'l Symp. on Image and Sign. Proc. and Analysis*, pp. 64-69, Zagreb, Croatia, Sept. 15-17, 2005.

[43] A. Doucet, N. de Freitas, and N. Gordon (eds.), *Sequential Monte Carlo Methods in Practice*, New York: Springer, 2001.

[44] O. E. Drummond, "On track and tracklet fusion filtering," in O. E. Drummond (ed.), *Signal and Data Processing of Small Targets 2002*, SPIE, Vol. 4728, pp. 176-195, Bellingham, WA, 2002.

[45] A. Doucet, B.-N. Vo, C. Andrieu, and M. Davy, "Particle filtering for multi-target tracking and sensor management," *Proc. 5th Int'l Conf. on Information Fusion*, pp. 474-481, Annapolis, MD, July 8-11, 2002, Int'l. Soc. Inf. Fusion, Sunnyvale, CA.

[46] D. Dubois and H. Prade, "Random sets and fuzzy interval analysis," *Fuzzy Sets and Systems*, Vol. 42, pp. 87-101, 1991.

[47] D. Dubois, H. Prade, and C. Testemale, "Fuzzy pattern matching with extended capabilities: proximity notions, importance assessment, random sets," in W. Bandler and A. Kandel (eds.), *Recent Developments in the Theory and Applications of Fuzzy Sets, Proc. 1986 Conf. of the N. American Inform. Proc. Soc.*, New Orleans, June 2-4, 1986, pp. 125-139, 1986.

[48] D. Dubois and H. Prade, "A set-theoretic view of belief functions: logical operations and approximations by fuzzy sets," *Int'l J. of General Systems*, Vol. 12, pp. 193-226, 1986.

[49] A. El-Fallah, M. Perloff, B. Ravichandran, T. Zajic, C. Stelzig, R. Mahler, and R. Mehra, "Multisensor-multitarget sensor management with target preference," in I. Kadar (ed.), *Sign. Proc., Sensor Fusion, and Targ. Recogn. XIII*, SPIE, Vol. 5429, pp. 222-232, Bellingham, WA, 2004.

[50] A. El-Fallah, R. Mahler, B. Ravichandran, and R. Mehra, "Adaptive data fusion using finite-set statistics," *Signal Processing, Sensor Fusion, and Target Recognition VIII*, SPIE, Vol. 3720, pp. 80-91, Bellingham, WA, 1999.

[51] A. El-Fallah, R. Mahler, T. Zajic, E. Sorensen, M. Alford, and R. Mehra, "Scientific performance evaluation for sensor management," in I. Kadar (ed.), *Sign. Proc., Sensor Fusion, and Targ. Recogn. IX*, pp. 183-194, SPIE, Vol. 4052, Bellingham, WA, 2000.

[52] A. El-Fallah, A. Zatezalo, R. Mahler, and R. K. Mehra, "Unified robust-Bayes multisource ambiguous data rule fusion," in I. Kadar (ed.), *Sign. Proc., Sensor Fusion, and Targ. Recogn. XIV*, pp. 277-287, SPIE, Vol. 5809, Bellingham, WA, 2006.

[53] A. El-Fallah, A. Zatezalo, R. Mahler, R. K. Mehra, and M. Alford, "Advancements in situation assessment sensor management," in I. Kadar (ed.), *Sign. Proc., Sensor Fusion, and Targ. Recogn. XV*, SPIE, Vol. 6235, Bellingham, WA, 2006.

[54] A. El-Fallah, A. Zatezalo, R. Mahler, R. K. Mehra, and M. Alford, "Regularized multi-target particle filter for sensor management," in I. Kadar (ed.), *Sign. Proc., Sensor Fusion, and Targ. Recogn. XV*, SPIE, Vol. 6235, Bellingham, WA, 2006.

[55] A. El-Fallah, A. Zatezalo, R. Mahler, R. K. Mehra, and M. Alford, "Unified Bayesian situation assessment sensor management," in I. Kadar (ed.), *Sign. Proc., Sensor Fusion, and Targ. Recogn. XIV*, SPIE, Vol. 5809, pp. 253-264, Bellingham, WA, 2005.

[56] O. Erdinc, P. Willett, and Y. Bar-Shalom, "A physical-space approach for the probability hypothesis density and cardinalized probability hypothesis density filters," in O. E. Drummond (ed.), *Signal Processing of Small Targets 2006*, SPIE, Vol. 6236, Bellingham, WA, to appear, 2007.

[57] O. Erdinc, P. Willett, and Y. Bar-Shalom, "Probability hypothesis density filter for multitarget multisensor tracking," *Proc. 8th Int'l Conf. on Information Fusion*, Philadephia, PA, July 25-29, 2005, Int'l. Soc. Inf. Fusion, Sunnyvale, CA.

[58] R. Fagin, "Combining fuzzy information from multiple systems," *Journal of Computer and System Sciences*, Vol. 58, pp. 83-89, 1999.

[59] L. Fisher, "A survey of the mathematical theory of multidimensional point processes," in P. Lewis (ed.), *Stochastic Point Processes: Statistical Analysis, Theory, and Applications*, New York: John Wiley & Sons, 1972.

[60] D. Fixsen and R. Mahler, "The modified Dempster-Shafer approach to classification," *IEEE Trans. Sys., Man & Cyber.–Part A*, Vol. 27, No. 1, pp. 96-104, 1997.

[61] T. E. Fortmann, "A matrix inversion identity," *IEEE Trans. Auto. Contr.*, Vol. 15, No. 5, p. 599, Oct., 1970.

[62] T. E. Fortmann, Y. Bar-Shalom, and M. Scheffe, "Sonar tracking of multiple targets using joint probabilistic data association," *IEEE Jour. Oceanic Engineering*, Vol. OE-8, No. 3, pp. 173-184, July, 1983.

[63] H. Gauvrit, J. P. Le Cadre, and C. Jauffret, "A formulation of multitarget tracking as an incomplete data problem," *IEEE Trans. on Aerospace and Electr. Sys.*, Vol. 33, No. 4, pp. 1242-1257, Oct., 1997.

[64] A. Gelb, J. F. Kasper, Jr., R. A. Nash, Jr., C. F. Price, and A. A. Sutherland, Jr., *Applied Optimal Estimation*, Cambridge, MA: M.I.T. Press, 1974.

[65] C. Geyer, "Likelihood inference for spatial point processes," in O. E. Barnsdorff-Nielsen, W. S. Kendall, and M. N. M. van Lieshout (eds.), *Stochastic Geometry: Likelihood and Computation*, pp. 79-140, Boca Raton, FL: Chapman & Hall/CRC Press, 1999.

[66] K. Gilholm, S. Godsill, S. Maskell, and D. Salmond, "Poisson models for extended target and group tracking," in O. E. Drummond (ed.), *Sign. and Data Proc. of Small Targets 2005*, SPIE, Vol. 5913, 2005.

[67] C. Goffman, *Calculus of Several Variables*, New York: Harper & Row, 1965.

[68] I. R. Goodman, "Fuzzy sets as equivalence classes of random sets," in R. Yager (ed.), *Fuzzy Sets and Possibility Theory*, New York: Permagon Press, pp. 327-343, 1982.

[69] I. R. Goodman, "Toward a comprehensive theory of linguistic and probabilistic evidence: Two new approaches to conditional event algebra," *IEEE Trans. Sys., Man & Cyber.*, Vol. 24, pp. 1665-1698, 1994.

[70] I. R. Goodman, R. P. S. Mahler, and H. T. Nguyen, *Mathematics of Data Fusion*, Dordrecht: Kluwer Academic Publishers, 1997.

[71] I. R. Goodman and H. T. Nguyen, *Uncertainty Models for Knowledge Based Systems*, Amsterdam: North-Holland, 1985.

[72] I. R. Goodman, H. T. Nguyen, and E. A. Walker, *Conditional Inference and Logic for Intelligent Systems*, Amsterdam: North-Holland, 1991.

[73] I. R. Goodman, R. P. S. Mahler, and H. T. Nguyen, "What is Conditional Event Algebra and Why Should You Care?"in I. Kadar (ed.), *Sign. Proc., Sensor Fusion, and Targ. Recogn. VIII*, SPIE, Vol. 3720, pp. 2-13, Bellingham, WA, 1999.

[74] N. Gordon, "A hybrid bootstrap filter for target tracking in clutter," *IEEE Trans. on Aerospace and Electr. Sys.*, Vol. 33, No. 1, pp. 353-358, 1997.

[75] J. Goutsias, R. Mahler, and H. T. Nguyen (eds.), *Random Sets: Theory and Applications*, New York: Springer-Verlag, 1997.

[76] M. Grabisch, H. T. Nguyen, and E. A. Walker, *Fundamentals of Uncertainty Calculi with Applications to Fuzzy Inference*, Dordrecht: Kluwer Academic Publishers, 1995.

[77] J. Graf, "A Radon-Nikodým theorem for capacities," *J. für Reine und Angewandte Mathematik*, Vol. 320, pp. 192-214, 1980.

[78] J. E. Gray and S. R. Addison, "Characteristic functions in radar and sonar," *Proc. 34th Southeastern Symp. on Sys. Theory*, pp. 31-35, Mar. 18-19, 2002.

[79] W. H. Greub, *Linear Algebra*, 2nd ed., Berlin: Springer-Verlag, 1963.

[80] M. M. Gupta and J. Qi, "Connectives (AND, OR, NOT) and T-operators in fuzzy reasoning," in I. R. Goodman, M. M. Gupta, H. T. Nguyen, and G. S. Rogers (eds.), *Conditional Logic in Expert Systems*, pp. 211-233, Amsterdam: North-Holland, 1991.

[81] R. Haenni, "Are alternatives to Dempster's rule of combination real alternatives? Comments on 'About the belief function combination and the conflict management problem,'" *Information Sciences*, Vol. 3, pp. 237-239, 2002.

[82] D. L. Hall, *Mathematical Techniques in Multisensor Data Fusion*, Norwood, MA: Artech House, Boston, 1992.

[83] D. L. Hall and J. Llinas, (eds.), *Handbook of Multisensor Data Fusion*, Boca Raton, FL: CRC Press, 2001.

[84] K. Hestir, H. T. Nguyen, and G. S. Rogers, "A random set formalism for evidential reasoning," in I. R. Goodman, M. M. Gupta, H. T. Nguyen and G. S. Rogers (eds.), *Conditional Logic in Expert Systems*, pp. 309-344, Amsterdam: North-Holland, 1991.

[85] T. L. Hill, *Statistical Mechanics: Principles and Practical Applications*, New York: Dover Publications, 1987.

[86] Y. C. Ho and R. C. K. Lee, "A Bayesian approach to problems in stochastic estimation and control," *IEEE Trans. Automatic Control*, Vol. AC-9, pp. 333-339, 1964.

[87] J. Hoffman and R. Mahler, "Multitarget miss distance via optimal assignment," *IEEE Trans. Sys., Man, and Cybernetics—Part A*, Vol. 34, No. 3, pp. 327-336, 2004.

[88] J. Hoffman, R. Mahler, R. Ravichandran, R. Mehra, and S. Musick, "Robust SAR ATR via set-V=valued classifiers: New results," in I. Kadar, (ed.), *Sign. Proc., Sensor Fusion, and Targ. Recogn. XII*, SPIE, Vol. 5096, pp. 139-150, Bellingham, WA, 2003.

[89] J. Hoffman, R. Mahler, and T. Zajic, "User-defined information and scientific performance evaluation," in I. Kadar (ed.), *Signal Processing, Sensor Fusion, and Target Recognition X*, SPIE, Vol. 4380, pp. 300-311, 2001.

[90] U. Höhle, "A mathematical theory of uncertainty: Fuzzy experiments and their realizations," in R. R. Yager (ed.), *Recent Developments in Fuzzy Set and Possibility Theory*, pp. 344-355, New York: Permagon Press, 1981.

[91] P. J. Huber, *Robust Statistics*, New York: John Wiley & Sons, 1981.

[92] C. Hue, J.-P. Le Cadre, and P. Pérez, "Sequential Monte Carlo methods for multiple target tracking and data fusion," *IEEE Trans. Sign. Proc.*, Vol. 50, No. 2, pp. 309-325, 2002.

[93] C. Hue, J.-P. Le Cadre, and P. Perez, "Tracking multiple targets with particle filtering," *IEEE Trans. Aerospace and Electr. Sys.*, Vol. 38, No. 3, pp. 791-812, 2002.

[94] R. A. Hummel and M. S. Landy, "A statistical viewpoint on the theory of evidence," *IEEE Trans. on Pattern Analysis and Machine Intelligence*, Vol. 10, No. 2, pp. 235-247, 1988.

[95] M. B. Hurley, "An information theoretic justification for covariance intersection and its generalization," *Proc. 5th Int'l Conf. on Information Fusion*, Vol. 1, pp. 505-511, July 8-11, 2002, Int'l. Soc. Inf. Fusion, Sunnyvale, CA.

[96] N. Ikoma, T. Uchino, and H. Maeda, "Tracking of feature points in image sequence by SMC implementation of PHD filter," *Proc. Soc. of Instrument & Contr. Engineers (SICE) Annual Conf.*, pp. 1696-1701, Hokkaido, Japan, Aug. 4-6, 2004.

[97] T. Inagaki, "Interdependence between safety-control policy and multiple-sensor schemes via Dempster-Shafer theory," *IEEE Trans. On Reliability*, Vol. 40, pp. 182-188, 1991.

[98] K. Ito and K. Xiong, "Gaussian filters for nonlinear filtering problems," *IEEE Trans. AC*, Vol. 45, No. 5, pp. 910-927, 2000.

[99] A. H. Jazwinski, *Stochastic Processes and Filtering Theory*, New York: Academic Press, 1970.

[100] A. Jøsang, M. Daniel, and P. Vannoorenberghe, "Strategies for combining conflicting dogmatic beliefs," *Proc. 6th Int'l Conf. on Information Fusion*, pp. 1133-1140, Cairns, Australia, July 8-11, 2003, Int'l. Soc. Inf. Fusion, Sunnyvale, CA, 2003.

[101] S. Julier and J. K. Uhlmann, "General decentralized data fusion with covariance intersection (CI)," in D. L. Hall and J. Llinas, (eds.), *Handbook of Multisensor Data Fusion*, Chapter 12, Boca Raton, FL: CRC Press, 2001.

[102] V. Kadirkamanathan, M. H. Jaward, S. G. Fabri, and M. Kadirkamanathan, "Particle filters for recursive model selection in linear and nonlinear system identification," *Proc. 39th IEEE Conf. on Decision and Control*, pp. 2391-2396, Sydney, Australia, Dec., 2000.

[103] E. W. Kamen and C. R. Sastry, "Multiple target tracking using products of position measurements," *IEEE Trans. Aerospace and Electr. Sys.*, Vol. 29, pp. 476-493, 1993.

[104] A. F. Karr, *Point Processes and Their Statistical Inference*, 2nd ed., New York: Marcel Dekker, 1991.

[105] Y. Kharin, *Robustness in Statistical Pattern Recognition*, Dordrecht: Kluwer Academic Publishers, 1996.

[106] S. Kim, M. A. Kouritzin, H. Long, J. McCrosky, and X. Zhao, "A stochastic grid filter for multi-target tracking," in I. Kadar (ed.), *Sign. Proc., Sensor Fusion, and Targ. Recogn. XIII*, SPIE, Vol. 5429, pp. 245-253, Bellingham, WA, 2005.

[107] T. Kirubarajan and Y. Bar-Shalom, "Probabilistic data association techniques for target tracking in clutter," *Proc. IEEE*, Vol. 92, No. 3, pp. 536-557, 2004.

[108] J. H. Kotecha and P. M. Djurić, "Gaussian sum particle filtering," *IEEE Trans. on Sign. Proc.*, Vol. 51, No. 10, pp. 2602-2612, 2003.

[109] J. H. Kotecha and P. M. Djurić, "Gaussian particle filtering," *IEEE Trans. on Sign. Proc.*, Vol. 51, No. 10, pp. 2592-2601, 2003.

[110] M. A. Kouritzin, D. J. Ballantyne, H. Kim, and Y. Hu, "On sonobuoy placement for submarine tracking," in I. Kadar (ed.), *Sign. Proc., Sensor Fusion, and Targ. Recogn. XIV*, SPIE, Vol. 5809, pp. 244-252, Bellingham, WA, 2005.

[111] M. A. Kouritzin, H. Long, X. Ma, and W. Sun, "Non-recursive and recursive methods for parameter estimation in filtering problems," in I. Kadar (ed.), *Sign. Proc., Sensor Fusion, and Targ. Recogn. XII*, SPIE, Vol. 5096, pp. 585-596, Bellingham, WA, 2003.

[112] V. Kreinovich, "Random sets unify, explain, and aid known uncertainty methods in expert systems," in J. Goutsias, R. Mahler, and H. Nguyen (eds.), *Random Sets: Theory and Applications*, pp. 321-345, New York: Springer-Verlag, 1997.

[113] R. Kruse and K. D. Meyer, *Statistics with Vague Data*, Dordrecht: D. Reidel Publishing Co., 1987.

[114] R. Kruse, E. Schwencke, and J. Heinsohn, *Uncertainty and Vagueness in Knowledge-Based Systems*, New York: Springer-Verlag, 1991.

[115] C. Kwok, D. Fox, and M. Meilă, "Real-time particle filters," *Proc. IEEE*, Vol. 92, No. 3, pp. 469-484, 2004.

[116] J.-R. Larocque, J. P. Reilly, and W. Ng, "Particle filters for tracking an unknown number of sources," *IEEE Trans. on Sign. Proc.*, Vol. 50, No. 12, pp. 2926-2937, 2002.

[117] E. Lefèvre, O. Colot, and P. Vannoorenberghe, "Belief function combination and conflict management," *Information Fusion*, Vol. 3, pp. 149-162, 2002.

[118] M. D. Levine, "Feature extraction: A survey," *Proc. of IEEE*, Vol. 57, No. 8, pp. 1391-1407, 1969.

[119] Y. Li, "Probabilistic Interpretations of Fuzzy Sets and Systems," Doctoral Dissertation, Dept. of Elec. Eng. and Comp. Sci., Massachusetts Institute of Technology, July, 1994.

[120] L. Lin, Y. Bar-Shalom, and T. Kirubaragan, "Track labeling and PHD filter for multitarget tracking," *IEEE Trans. Aerospace and Electr. Sys.*, Vol. 42, No. 3, pp. 778-795, 2006.

[121] L. Lin, T. Kirubarajan, and Y. Bar-Shalom, "Data association combined with the probability hypothesis density filter for multitarget tracking," in O. E. Drummond (ed.), *Sign. and Data Proc. of Small Targets 2004*, SPIE, Vol. 5428, pp. 464-475, Bellingham, WA, 2004.

[122] W. K. Ma, B. Vo, S. Singh, and A. Baddeley, "Tracking an unkown time-varying number of speakers using TDOA measurements: A random finite set approach," *IEEE Trans. Sign. Proc.*, Vol. 54, No. 9, pp. 3291-3304, 2006.

[123] R. Mahler, "Bayesian cluster detection and tracking using a generalized Cheeseman approach," in I. Kadar (ed.), *Sign. Proc., Sensor Fusion, and Targ. Recogn. XII*, pp. 334-345, SPIE, Vol. 5096, pp. 334-345, Bellingham, WA, 2003.

[124] R. Mahler, "Bayes-invariant transformations of uncertainty representations," in I. Kadar (ed.), *Sign. Proc., Sensor Fusion, and Targ. Recogn. XV*, SPIE, Vol. 6235, Bellingham, WA, 2006.

[125] R. Mahler, "Bayesian tracking with Dempster-Shafer Measurements," in O. E. Drummond (ed.), *Sign. and Data Proc. of Small Targets 2005*, SPIE, Vol. 5913, Bellingham, WA, 2004.

[126] R. Mahler, "Bayesian vs. 'plain-vanilla Bayesian' multitarget statistics," in I. Kadar (ed.), *Sign. Proc., Sensor Fusion, and Targ. Recogn. XIII*, pp. 1-12, SPIE, Vol. 5429, Bellingham, WA, 2004.

[127] R. Mahler, "Can the Bayesian and Dempster-Shafer approaches be reconciled? Yes," *Proc. 8th Int'l Conf. on Information Fusion*, Philadephia, PA, July 25-28, 2005, Int'l. Soc. Inf. Fusion, Sunnyvale, CA.

[128] R. Mahler, "Combining ambiguous evidence with respect to ambiguous a priori knowledge, I: Boolean logic," *IEEE Trans. Sys., Man & Cyber.-Part A*, Vol. 26, pp. 27-41, 1996.

[129] R. Mahler, "Combining ambiguous evidence with respect to ambiguous a priori knowledge, II: Fuzzy logic," *Fuzzy Sets and Systems*, Vol. 75, pp. 319-354, 1995.

[130] R. Mahler, "Detecting, tracking, and classifying group targets: A unified approach," in I. Kadar (ed.), *Signal Processing, Sensor Fusion, and Target Recognition X*, pp. 217-228, SPIE, Vol. 4380, Bellingham, WA, 2001.

[131] R. Mahler, "An extended first-order Bayes filter for force aggregation," in O. E. Drummond (ed.), *Signal and Data Processing of Small Targets 2002*, pp. 196-207, SPIE, Vol. 4729, Bellingham, WA, 2002.

[132] R. Mahler, "Formal vs. heuristic modeling for multitarget Bayesian filtering," in O. E. Drummond (ed.), *Sign. and Data Proc. of Small Targets 2004*, pp. 342-353, SPIE, Vol. 5428, pp. 342-353, Bellingham, WA, 2004.

[133] R. Mahler, "A general theory of multitarget extended Kalman filters," in I. Kadar (ed.), *Sign. Proc., Sensor Fusion, and Targ. Recogn. XIV*, SPIE, Vol. 5809, pp. 208-219, Bellingham, WA, 2005.

[134] R. Mahler, *An Introduction to Multisource-Multitarget Statistics and Its Applications*, Lockheed Martin Technical Monograph, Mar. 15, 2000, 114 pages.

[135] R. Mahler, "Measurement models for ambiguous evidence using conditional random sets," in I. Kadar (ed.), *Sign. Proc., Sensor Fusion, and Targ. Recogn. VI*, SPIE, Vol. 3068, pp. 40-51, Bellingham, WA, 1997.

[136] R. Mahler, "Multitarget filtering via first-order multitarget moments," *IEEE Trans. Aerospace and Electr. Sys.*, Vol. 39 No. 4, pp. 1152-1178, 2003.

[137] R. Mahler, "Multitarget moments and their application to multitarget tracking," *Proc. Workshop on Estimation, Tracking, and Fusion: A Tribute to Y. Bar-Shalom*, pp. 134-166, Naval Postgraduate School, Monterey, CA, May 17, 2001.

[138] R. Mahler, "Multitarget motion models," in I. Kadar (ed.), *Sign. Proc., Sensor Fusion, and Targ. Recogn. VIII*, pp. 47-57, SPIE, Vol. 3720, Bellingham, WA, 1999.

[139] R. Mahler, "Multitarget sensor management of dispersed mobile sensors," in D. Grundel, R. Murphey, and P. Paralos (eds.), *Theory and Algorithms for Cooperative Systems*, Singapore: World Scientific, 2005.

[140] R. Mahler, "Optimal/robust distributed data fusion: A unified approach," in I. Kadar (ed.), *Sign. Proc., Sensor Fusion, and Targ. Recogn. IX*, SPIE, Vol. 4052, pp. 128-138, Bellingham, WA, 2000.

[141] R. Mahler, "Point target clusters and continuous-state multitarget statistics," in I. Kadar (ed.), *Sign. Proc., Sensor Fusion, and Targ. Recogn. XI*, SPIE, Vol. 4729, pp. 163-174, Bellingham, WA, 2002.

[142] R. Mahler, "PHD filters of higher order in target number," *IEEE Trans. Aerospace and Electronic Systems*, Vol. 43, No. 3, 2005.

[143] R. Mahler, "PHD filters of second order in target number," in O. E. Drummond (ed.), *Signal Proc. of Small Targets 2006*, SPIE, Vol. 6236, Bellingham, WA, 2006.

[144] R. Mahler, "Random sets: Unification and computation for information fusion—a retrospective assessment," *Proc. 7th Int'l. Conf. on Inf. Fusion*, pp. 1-20, Stockholm, Sweden, June 28-July 1, 2004, Int'l. Soc. Inf. Fusion, Sunnyvale, CA.

[145] R. Mahler, "Random set theory for target tracking and identification," in D. L. Hall and J. Llinas (eds.), *Handbook of Multisensor Data Fusion*, Chapter 14, Boca Raton, FL: CRC Press, 2002.

[146] R. Mahler, "Representing rules as random sets, I: Statistical correlations between rules," *Information Sciences*, Vol. 88, pp. 47-68, 1996.

[147] R. Mahler, "Representing rules as random sets, II: Iterated rules," *Int'l J. Intell. Sys.*, Vol. 11, pp. 583-610, 1996.

[148] R. Mahler, "'Statistics 101' for multisensor, multitarget data fusion," *IEEE Aerospace & Electr. Sys. Mag., Part 2: Tutorials*, Vol. 19 No. 1, pp. 53-64, 2004.

[149] R. Mahler, "Sensor management with non-ideal sensor dynamics," *Proc. 7th Int'l Conf. on Inf. Fusion*, Stockholm, June 28-July 1, 2004.

[150] R. Mahler, "Target preference in multitarget sensor management: A unified approach," in I. Kadar (ed.), *Sign. Proc., Sensor Fusion, and Targ. Recogn. XIII*, SPIE, Vol. 5429, pp. 210-221, Bellingham, WA, 2004.

[151] R. Mahler, "A theory of PHD filters of higher order in target number," in I. Kadar (ed.), *Sign. Proc., Sensor Fusion, and Targ. Recogn. XV*, SPIE, Vol. 6235, Bellingham, WA, 2006.

[152] R. Mahler, "Unified Bayes filtering with fuzzy and rule-based evidence," in I. Kadar, (ed.), *Sign. Proc., Sensor Fusion, and Targ. Recogn. XIV*, SPIE, Vol. 5809, pp. 265-276, Bellingham, WA, 2005.

[153] R. Mahler, "Unified Bayes multitarget fusion of ambiguous data sources," *IEEE Int'l Conf. on Integration of Knowledge Intensive Multi-Agent Systems (KIMAS)*, pp. 343-348, Cambridge, MA, Sept. 30-Oct. 4, 2003.

[154] R. Mahler, "A unified foundation for data fusion," in F. A. Sadjadi (ed.), *Selected Papers on Sensor and Data Fusion*, pp. 325-354, SPIE, Vol. MS-124, Bellingham, WA. Reprinted from *Proc. 7th Joint Service Data Fusion Symp.*, vol 1, pp. 153-173, Johns Hopkins Applied Research Laboratory, Laurel, Maryland, Oct. 25-28, 1994.

[155] R. Mahler, P. Leavitt, J. Warner, and R. Myre, "Nonlinear filtering with really bad data," in I. Kadar (ed.), *Sign. Proc., Sensor Fusion, and Targ. Recogn. VIII*, SPIE, Vol. 3720, pp. 59-70, Bellingham, WA, 1999.

[156] R. Mahler and M. O'Hely, "Multitarget detection and acquisition: A unified approach," in O. E. Drummond (ed.), *Signal and Data Proc. of Small Targets 1999*, pp. 218-229, 1999, SPIE, Vol. 3809, Bellingham, WA, 1999.

[157] R. Mahler, S.-H. Yu, R. Mehra, B. Ravichandran, and S. Musick, "Applications of unified evidence accumulation methods to robust SAR ATR," in I. Kadar (ed.), *Sign. Proc., Sensor Fusion, and Targ. Recogn. VIII*, SPIE, Vol. 3720, pp. 71-79, 1999.

[158] R. Mahler and T. Zajic, "Multitarget filtering using a multitarget first-order moment statistic," in I. Kadar (ed.), *Signal Processing, Sensor Fusion, and Target Recognition X*, SPIE, Vol. 4380, pp. 184-195, Bellingham, WA, 2001.

[159] R. Mahler and T. Zajic, "Probabilistic objective functions for sensor management," in I. Kadar (ed.), *Sign. Proc., Sensor Fusion, and Targ. Recogn. XIII*, SPIE, Vol. 5429, pp. 233-244, Bellingham, WA, 2004.

[160] G. Mathéron, *Random Sets and Integral Geometry*, New York: John Wiley, 1975.

[161] A. McCallum and K. Nigam, "A comparison of event models for naive Bayes text classification," *Proc. AAAI-98 Workshop on Learning for Text Categorization*, Madison, WI, July 28-30, 1998.

[162] G. J. McLachan and T. Krishnan, *The EM Algorithm and Extensions*, New York: John Wiley, 1997.

[163] E. Miranda, I. Couso, and P. Gil, "Relationships between possibility measures and nested random sets," *Int'l J. of Uncertainty, Fuzziness & Knowledge-Based Systems*, Vol. 10 No. 1, pp. 1-15, 2002.

[164] M. Mizumoto, "Pictorial representation of fuzzy connectives I. Cases of T-norms, T-conorms and averaging operators," *Fuzzy Sets and Systems*, Vol. 31, pp. 217-242, 1989.

[165] I. Molchanov, "Random closed sets," in *Space, Structure, and Randomness*, Springer Lect. Notes on Stat., Vol. 183, New York: Springer, pp. 135-149, 2005.

[166] I. Molchanov, *Theory of Random Sets*, New York: Springer-Verlag, 2005.

[167] R. Moore, *Interval Analysis*, New York: Prentice-Hall, 1966.

[168] M. Moreland and S. Challa, "A multi-target tracking algorithm based on random sets," in *Proc. 6th Int'l. Conf. on Information Fusion*, pp. 807-814, July 8-11, 2003, Cairns, Australia, July 8-11, 2003, Int'l. Soc. Inf. Fusion, Sunnyvale, CA, 2003.

[169] J. E. Moyal, "The general theory of stochastic population processes," *Acta Mathematica*, vol 108, pp. 1-31, 1962.

[170] D. J. Muder and S. D. O'Neil, "The multi-dimensional SME filter for multitarget tracking," SPIE, Vol. 1954, pp. 587-599, Bellingham, WA, 1993.

[171] C. K. Murphy, "Combining belief functions when evidence conflicts," *Decision and Support Systems*, Vol. 29, pp. 1-9, 2000.

[172] D. Musicki, R. Evans, and S. Stankovic, "Integrated probabilistic data association," *IEEE Trans. Auto. Contr.*, Vol. 39, No. 6, pp. 1237-1241, 1994.

[173] D. Musicki and R. Evans, "Joint integrated probabilistic data association," *IEEE Trans. Aerospace and Electronic Sys.*, Vol. 40, No. 3, pp. 1093-1099, 2004.

[174] C. Musso, N. Oudjane, and F. Le Gland, "Improving regularized particle filters," in A. Doucet, N. de Freitas, and N. Gordon (eds.), *Sequential Monte Carlo Methods in Practice*, pp. 247-271, New York: Springer, 2001.

[175] S. Naimpally, "What is a hit-and-miss topology?" *Topological Commentary*, Vol. 8, No. 1, 2003.

[176] J. Naylor and A. Smith, "Application of a method for the efficient computation of posterior distributions," *Applied Statistics*, Vol. 31, No. 3, pp. 214-225, 1982.

[177] R. B. Nelson, *An Introduction to Copulas*, New York: Springer, 1999.

[178] R. B. Nelson, "Properties and applications of copulas: A brief survey," in J. Dhaene, N. Kolev, and P. Morettin (eds.), *Proc. 1st Brazilian Conf. on Statistical Modeling in Insurance and Finance*, pp. 10-28, Ubatuba, Brazil, Sept. 1-6, 2003, Institute of Mathematics and Statistics, University of São Paulo, 2003.

[179] H. T. Nguyen, "On random sets and belief functions," *J. Math. Anal. and Appl.*, Vol. 65, pp. 531-542, 1978.

[180] A. I. Orlov, "Relationships between fuzzy and random sets: Fuzzy tolerances," *Issledovania po Veroyatnostnostatishesk. Modelirovaniu Realnikh System*, Moscow, 1977.

[181] A. I. Orlov, "Fuzzy and random sets," *Prikladnoi Mnogomerni Statisticheskii Analys*, Moscow, 1978.

[182] K. Panta, B. Vo, A. Doucet, and S. Singh, "Probability hypothesis density filter versus multiple hypothesis testing," in I. Kadar (ed.), *Sign. Proc., Sensor Fusion, and Targ. Recogn. XIII*, SPIE, Vol. 5429, pp. 284-295, Bellingham, WA, 2004.

[183] K. Panta, B. Vo, and S. Singh, "Improved probability hypothesis density (PHD) filter for multitarget tracking," *Proc. Int'l Conf. on Intelligent Sensing and Inform. Proc.*, pp. 213-218, Bangalore, India, 2005.

[184] K. Panta, B. Vo, and S. Singh, "Novel data association schemes for the probability hypothesis density filter," *IEEE Trans. Aerospace and Electronic Sys.*, to appear, 2007.

[185] A. Pasha, B. Vo, H. D. Tuan, and W.-K. Ma, "Closed-form PHD filtering for linear jump Markov models," *Proc. 9th Int'l Symp. on Information Fusion*, Florence, Italy, July 10-13, 2006, Int'l. Soc. Inf. Fusion, Sunnyvale, CA.

[186] A. Pasha, B. Vo, H. D. Tuan, and W.-K. Ma, "A Gaussian mixture PHD filter for jump Markov system models," submitted to *IEEE Trans. Electr. and Aerospace Sys.*, 2006.

[187] A. B. Poore, S. Lu, and B. J. Suchomel, "Data association using multiple frame assignments," in D. L. Hall and J. Llinas, (eds.), *Handbook of Multisensor Data Fusion*, Chapter 11, Boca Raton, FL: CRC Press, 2001.

[188] G. W. Pulford, "Taxonomy of multiple target tracking methods," *IEE Proc. on Radar & Sonar Nav.*, Vol. 152, No. 5, pp. 291-304, Oct., 2005.

[189] K. Punithakumar, T. Kirubarajan, and A. Sinha, "A distributed implementation of a sequential Monte Carlo probability hypothesis density filter for sensor networks," in I. Kadar (ed.), *Sign. Proc., Sensor Fusion, and Targ. Recogn. XV*, SPIE, Vol. 6235, Bellingham, WA, 2006.

[190] K. Punithakumar, T. Kirubarajan, and A. Sinha, "A multiple model probability hypothesis density filter for tracking maneuvering targets," in O. E. Drummond (ed.), *Sign. and Data Proc. of Small Targets 2004*, SPIE, Vol. 5428, pp. 113-121, Bellingham, WA, 2004.

[191] P. Quinio and T. Matsuyama, "Random closed sets: A unified approach to the representation of imprecision and uncertainty," in R. Kruse and P. Siegel (eds.), *Symbolic and Quantitative Approaches to Uncertainty*, pp. 282-286, New York: Springer-Verlag, 1991.

[192] L. Råde and B. Westergren, *Mathematics Handbook for Science and Engineering*, Springer-Verlag, Berlin, 2004.

[193] D. B. Reid, "An algorithm for tracking multiple targets," *IEEE Trans. Auto. Contr.*, Vol. AC-24, No. 6, pp. 843-854, 1979.

[194] A. W. Rihaczek, *Principles of High-Resolution Radar*, Norwood, MA: Artech House, 1996.

[195] B. Ripley, "Locally finite random sets: Foundations for point process theory," *Annals of Prob.*, Vol. 4, No. 6, pp. 983-994, 1976.

[196] B. Ristic, S. Arulampalam, and N. Gordon, *Beyond the Kalman Filter: Particle Filters for Tracking Applications*, Norwood, MA: Artech House, 2004.

[197] C. P. Robert and G. Casella, *Monte Carlo Statistical Methods*, New York: Springer, 2004.

[198] J. A. Roecker, "A class of near optimal JPDA algorithms," *IEEE Trans. Aerospace and Electr. Sys.*, Vol. 30, No. 2, pp. 504-510, Apr. 1994.

[199] L. H. Ryder, *Quantum Field Theory*, 2nd Ed., Cambridge, U.K.: Cambridge University Press, 1996.

[200] D. J. Salmond and N. Gordon, "Group and extended object tracking," in O. E. Drummond (ed.), *Sign. and Data Proc. of Small Targets 1999*, SPIE, Vol. 3809, pp. 284-296, Bellingham, WA, 1999.

[201] D. J. Salmond and N. J. Gordon, "Group tracking with limited sensor resolution and finite field of view," *Sign. and Data Proc. of Small Targets 2000*, SPIE, Vol. 4048, pp. 532-540, Bellingham, WA, 2000.

[202] S. Särkkä, A. Vehtari, and J. Lampinen, "Rao-Blackwellized particle filter for multiple target tracking," *Proc. 9th Int'l Conf. on Information Fusion*, Florence, Italy, July 10-13, 2006, Int'l. Soc. Inf. Fusion, Sunnyvale, CA.

[203] A. H. Sayed, "A state-space approach to adaptive RLF filtering," *IEEE Signal Proc. Mag.*, Vol. 11, No. 3, pp. 18-60, July, 1994.

[204] D. C. Schleher, *MTI and Pulsed Doppler Radar*, Norwood, MA: Artech House, 1991.

[205] G. Shafer and R. Logan, "Implementing Dempster's rule for hierarchical evidence," *Artificial Intelligence*, Vol. 33, pp. 271-298, 1987.

[206] S. Shi, M. E. C. Hull, and D. A. Bell, "A new rule for updating evidence," in Z. W. Ras and M. Zemankova (eds.), *Proc. 8th Int'l Symp. on Methodologies for Intell. Sys.*, pp. 95-104, New York: Springer, 1994.

[207] D. A. Shnidman, "Expanded Swerling target models," *IEEE Trans. Aerospace and Electronic Sys.*, Vol. 39, No. 3, pp. 1059-1069, 2003.

[208] H. Sidenbladh, "Multi-target particle filtering for the probability hypothesis density," *Proc. 6th Int'l Conf. on Information Fusion*, pp. 800-806, Cairns, Australia, July 8-11, 2003, Int'l. Soc. Inf. Fusion, Sunnyvale, CA, 2003.

[209] H. Sidenbladh, P. Svenson, and J. Schubert, "Comparing multitarget trackers on difference force level units," in I. Kadar (ed.), *Sign. Proc., Sensor Fusion, and Targ. Recogn. XIII*, SPIE, Vol. 5429, Bellingham, WA, 2004.

[210] H. Sidenbladh and S.-L. Wirkander, "Tracking random sets of vehicles in terrain," *Proc. 2003 IEEE Workshop on Multi-Object Tracking*, Madison, WI, June 21, 2003.

[211] I. M. Skolnik (ed.), *Radar Handbook*, New York: McGraw-Hill, 1990.

[212] P. Smets, "The combination of evidence in the transferable belief model," *IEEE Trans. on Pattern Analysis and Machine Intelligence*, Vol. 12, No. 5, pp. 447-458, 1990.

[213] P. Smets and R. Kennes, "The transferable belief model," *Artif. Intell.*, Vol. 66, pp. 191-234, 1994.

[214] D. L. Snyder and M. I. Miller, *Random Point Processes in Time and Space*, 2nd ed., New York: Springer, 1991.

[215] E. Sorensen, T. Brundage, and R. Mahler, "None-of-the-above (NOTA) capability for INTELL-based NCTI," in I. Kadar (ed.), *Sign. Proc., Sensor Fusion, and Targ. Recogn. X*, SPIE, Vol. 4380, pp. 281-287, Bellingham, WA, 2001.

[216] H. W. Sorenson, "Recursive estimation for nonlinear dynamic systems," in J. C. Spall (ed.), *Bayesian Analysis of Statistical Time Series and Dynamic Models*, New York: Marcel Dekker, 1988.

[217] H. W. Sorenson and D. L. Alspach, "Recursive Bayesian estimation using Gaussian sums," *Automatica*, Vol. 7, pp. 465-479, 1971.

[218] W. D. Stanley, *Digital Signal Processing*, Reston, VA: Reston Publishing/Prentice-Hall, 1975.

[219] D. Stein, S. Theophanis, W. Kuklinski, J. Witkoskie, and M. Otero, "Mutual information based resource management applied to road constrained target tracking," *Proc. 2006 MSS Nat'l Symp. on Sensor & Data Fusion*, McLean, VA, June 6-8, 2006.

[220] D. Stoyan, W. S. Kendall, and J. Meche, *Stochastic Geometry and Its Applications*, 2nd ed., New York: John Wiley & Sons, 1995.

[221] J. J. Sudano, "A generalized belief fusion algorithm," *Proc. 6th Int'l Conf. on Information Fusion*, pp. 1126-1132, Cairns, Australia, July 8-11, 2003, Int'l. Soc. Inf. Fusion, Sunnyvale, CA, 2003.

[222] P. Svenson, "Capabilities-based force aggregation using random sets," *Proc. 8th Int'l Conf. on Information Fusion*, pp. 872-878, Philadelphia, PA, July 25-29, 2005, Int'l. Soc. Inf. Fusion, Sunnyvale, CA.

[223] Y. Tang, S. Sun, and Z. Li, "The generalized combination of evidence involved a priori knowledge," *Proc. 2003 IEEE Int'l Conf. on Systems, Man and Cybern.*, Vol. 5, pp. 4998-5003, Washington, D.C., Oct. 5-8, 2003.

[224] M. Tobias and A. D. Lanterman, "Probability hypothesis density-based multitarget tracker using multiple bistatic range and velocity measurements," *Proc. 36th Southeast Symp. on System Theory*, pp. 205-209, Alanta, GA, March 14-16, 2004.

[225] M. Tobias and A. D. Lanterman, "Probability hypothesis density-based multitarget tracking with bistatic range and Doppler observations," *IEE Proc. Radar, Sonar, and Navig.*, Vol. 152, No. 3, 2005.

[226] M. Tobias and A. D. Lanterman, "Multitarget tracking using multiple bistatic range measurements with probability hypothesis densities," in I. Kadar (ed.), *Sign. Proc., Sensor Fusion, and Targ. Recogn. XIII*, pp. 296-305, SPIE, Vol. 5429, Bellingham, WA, 2004.

[227] J. K. Uhlmann, "Algorithms for multiple-target tracking," *American Scientist*, Vol. 80, pp. 128-141, 1991.

[228] J. K. Uhlmann, "An introduction to the algorithmics of data association in multiple-target tracking," in D. L. Hall and J. Llinas, (eds.), *Handbook of Multisensor Data Fusion*, Chapter 3, Boca Raton, FL: CRC Press, 2001.

[229] J. K. Uhlmann, "An introduction to the combinatorics of optimal and approximate data association," in D. L. Hall and J. Llinas, (eds.), *Handbook of Multisensor Data Fusion*, Chapter 9, Boca Raton, FL: CRC Press, 2001.

[230] B. Van Fraasen, "Probabilities of conditionals," in W. L. Harper and C. A. Hooker (eds.), *Foundations of Probability Theory, Statistical Inference, and Statistical Theories of Science*, Dordrecht: D. Reidel, Holland,1976, pp. 261-300.

[231] H. L. van Trees, *Detection, Estimation, and Modulation Theory, Part I: Detection, Estimation, and Linear Modulation Theory*, New York: John Wiley & Sons, 1968.

[232] M. Vihola, "Random set particle filter for bearings only target tracking," in I. Kadar (ed.), *Sign. Proc., Sensor Fusion, and Targ. Recogn. XIV*, SPIE, Vol. 5809, pp. 301-312, Bellingham, WA, 2005.

[233] M. Vihola, "Random Sets for Multitarget Tracking and Data Fusion," Licentiate Thesis, Department of Information Technology, Tampere University of Technology, Tampere, Finland, 2004.

[234] M. Vihola, "Rao-Blackwellized particle filtering in random set multitarget tracking," *IEEE Trans. Aerospace and Electronic Sys.*, to appear, 2007.

[235] B. Vo, and W.-K. Ma, "A closed-form solution to the probability hypothesis density filter," *Proc. 8th Int'l Conf. on Information Fusion*, pp. 856-863, Philadelphia, PA, July 25-29, 2005, Int'l. Soc. Inf. Fusion, Sunnyvale, CA.

[236] B.-N. Vo, and W.-K. Ma, "The Gaussian mixture probability hypothesis density filter," *IEEE Trans. Sign. Proc.*, Vol. 54, No. 11, pp. 4091-4104, 2006.

[237] B.-N. Vo, S. Singh, and A. Doucet, "Random finite sets and sequential Monte Carlo methods in multi-target tracking," *Proc. Int'l Conf. on Radar*, pp. 486-491, Adelaide, Australia, Sept. 3-5, 2003.

[238] B.-N. Vo, S. Singh, and A. Doucet, "Sequential Monte Carlo methods for multi-target filtering with random finite sets," *IEEE Trans. Aerospace and Electronic Sys.*, Vol. 41, No. 4, pp. 1224-1245, Oct., 2005.

[239] B.-N. Vo, S. Singh, and W. K. Ma, "Tracking multiple speakers using random sets," *Proc. 2004 IEEE Int'l Conf. on Acoustics, Speech, and Signal Proc.*, Vol. II, pp. 357-360, Montreal, Canada, May 17-21, 2004.

[240] B.-T. Vo, and B.-N. Vo, "Performance of PHD based multi-target filters," *Proc. 9th Int'l Conf. on Information Fusion*, Florence, Italy, July 10-13, 2006, Int'l. Soc. Inf. Fusion, Sunnyvale, CA.

[241] B.-T. Vo, B.-N. Vo, and A. Cantoni, "Analytic implementations of the cardinalized probability hypothesis density filter," *IEEE Trans. Sign. Proc.*, to appear, 2007.

[242] B.-T. Vo, B.-N. Vo, and A. Cantoni, "The cardinalized probability hypothesis filter for linear Gaussian multi-target models," *Proc. 40th Annual Conf. on Info. Sciences & Systems*, Princeton, New Jersey, March 22-24, 2006.

[243] B.-N. Vo, B.-T. Vo, and S. Singh, "Sequential Monte Carlo methods for static parameter estimation in random set models," *Proc. 2nd Int'l Conf. on Intelligent Sensors, Sensor Networks, and Information Processing*, pp. 313-318, Melbourne, Australia, Dec. 5-8, 2004.

[244] F. Voorbraak, "A computationally efficient approximation of Dempster-Shafer theory," *Int'l J. Man-Machine Studies*, Vol. 30, pp. 525-536, 1989.

[245] E. Waltz and J. Llinas, *Multisensor Data Fusion*, Norwood, MA: Artech House, 1990.

[246] Y.-D. Wang, J.-K. Wu, A. A. Kassim, and W.-M. Huang, "Tracking a variable number of human groups in video using probability hypothesis density," *Proc. 18th In'l Conf. on Pattern Recognition*, Hong Kong, Aug. 20-24, 2006.

[247] M. J. Waxman and O. E. Drummond, "A bibliography of cluster (group) tracking," in O. E. Drummond (ed.), *Sign. and Data Proc. of Small Targets 2004*, SPIE, Vol. 5428, pp. 551-560, Bellingham, WA, 2004.

[248] N. Wilson, "Decision-making with belief functions and pignistic probabilities," in M. Clarke, R. Kruse, and M. Serafin (eds.), *Symbolic and Quantitative Approaches to Reasoning and Uncertainty*, pp. 364-371, New York: Springer-Verlag, 1991.

[249] J. Witkoskie, W. Kuklinksi, S. Theophanis, D. Stein, and M. Otero, "Feature-aided random set tracking on a road constrained network," *Proc. 2006 MSS Nat'l Symp. on Sensor and Data Fusion*, McLean, VA, June 6-8, 2006.

[250] J. Witkoskie, W. Kuklinksi, S. Theophanis, D. Stein, and M. Otero, "Random set tracker experiment on a road constrained network with resource management," *Proc. 9th Int'l Conf. on Information Fusion*, Florence, Italy, July 11-13, 2006, Int'l. Soc. Inf. Fusion, Sunnyvale, CA.

[251] R. R. Yager, "On the Dempster-Shafer framework and new combination rules," *Information Sciences*, Vol. 41, pp. 93-137, 1987.

[252] Y. Yen, "Generalizing the Dempster-Shafer theory to fuzzy sets," *IEEE Trans. Sys., Man & Cyber.*, Vol. 20, No. 3, pp. 559-570, 1990.

[253] L. Zadeh, review of G. Shafer, *A Mathematical Theory of Evidence*, in *The AI Magazine*, Fall, 1984.

[254] T. Zajic and R. Mahler, "A particle-systems implementation of the PHD multitarget tracking filter," in I. Kadar (ed.), *Sign. Proc., Sensor Fusion, and Targ. Recogn. XII*, SPIE, Vol. 5096, pp. 291-299, Bellingham, WA, 2003.

[255] T. Zajic, B. Ravichandran, R. Mahler, R. Mehra, and M. Noviskey, "Joint tracking and identification with robustness against unmodeled targets," in I. Kadar (ed.), *Sign. Proc., Sensor Fusion, and Targ. Recogn. XII*, SPIE, Vol. 5096, pp. 279-290, Bellingham, WA, 2003.

[256] L. Zhang, "Representation, independence, and combination of evidence in the Dempster-Shafer theory," in R. R. Yager, J. Kacprzyk, and M. Fedrizzi (eds.), *Advances in the Dempster-Shafer Theory of Evidence*, New York: John Wiley, pp. 51-69, 1994.

[257] Y. Zhiu and X.-R. Li, "Extended Dempster-Shafer combination rules based on random set theory," in B. V. Dasarathy, (ed.), *Multisensor, Multisource Information Fusion: Architectures, Algorithms, and Applications 2004*, SPIE, Proc. Vol. 5434, pp. 112-120, Bellingham, WA, 2004.

[258] B. P. Zeigler, "Some properties of modified Dempster-Shafer operators in rule based inference systems," *Int'l J. General Systems*, Vol. 14, pp. 345-356, 1988.

About the Author

Dr. Ronald P. S. Mahler is a staff scientist at Lockheed Martin MS2 Tactical Systems, Eagan, Minnesota, where he has been employed since 1980. From 1974 to 1979, he served as an assistant professor of mathematics at the University of Minnesota in Minneapolis, Minnesota. He has a B.A. in mathematics from the University of Chicago, a B.S. in electrical engineering from the University of Minnesota and a Ph.D. in mathematics from Brandeis University, Waltham, Massachusetts. He is a recipient of Lockheed Martin MS2 quality circle and author of the year awards.

Since 1991 Dr. Mahler's research has focused on data fusion, expert systems theory, multitarget tracking, multitarget nonlinear filtering, sensor management, random set theory, point process theory, and conditional event algebra. He has been a Lockheed Martin principal investigator for several U.S. Department of Defense (DoD) research and development contracts, including DARPA, the Air Force Research Laboratory (AFRL), the Air Force Office of Scientific Research (AFOSR), the Army Research Office (ARO), the Missile Defense Agency (MDA), and SPAWAR Systems Center. He has also served on technology advisory meetings for ARO, AFRL, DARPA, and SPAWAR Systems Center.

Since 1995 he has authored or co-authored 60 papers, an earlier book (*Mathematics of Data Fusion*, Kluwer, 1997), and one monograph, including nine journal papers. He was principal organizer, cochair, and proceedings coeditor for an August 1996 international scientific workshop on random sets (*Random Sets: Theory and Applications*, Springer, 1997). He is currently a session chair for the annual SPIE Defense & Security Symposium in Orlando, Florida. He has been invited to speak at many conferences, universities, and DoD labs, including Harvard, Johns Hopkins, the University of Massachusetts, Amherst, the University of Wisconsin, Madison, the U.S.A.F. Institute of Technology, the IEEE Conference on Decision and Control, the IEEE Workshop on Multiobject Tracking, and the International Conference on Information, Decision, and Control. He was a plenary speaker at the International Conference on Information Fusion.

Index

AGA measurement
defined, 214
Dempster-Shafer, 219
fuzzy, 215
fuzzy Dempster-Shafer, 220
generalized fuzzy, 219
simple examples of, 211
Agreement
Dempster-Shafer, 137
Fixsen-Mahler, 138
fuzzy Dempster-Shafer, 144
modified Dempster-Shafer, 138, 277, 282
AGU measurement
defined, 249
simple example of, 249
Alpha-beta filter, 33
and PHD filter, 568
as a primitive SMC filter, 547
Approximation
first-order moment
multitarget, 569
single-target, 567
fixed-grid
multitarget, 544
single-target, 68
Gaussian-mixture (GM)
filter, 668
CPHD filter, 646
PHD filter, 623
single-target, 72
i.i.d. cluster process
for CPHD filter, 635, 636
multi-Bernoulli
multitarget, 657
Poisson
multitarget, 579
SMC
JoTT filter, 523
multitarget, 545, 551
single-target, 79
Taylor series, 71, 696
Association
and multitarget likelihoods, 422
as a one-to-one function, 318, 329
as a permutation, 326
Bayesian analysis, 340
distance, 316
global nearest neighbor, 317
in general, 332
likelihood, 318
global, 318, 327
global, for fuzzy DS, 342
total, 323, 328, 330
nearest neighbor, 317
non-Bayesian?, 422
of track with fuzzy DS measurement, 342
optimal, 317, 320
partial, 335
Associativity, 121
Asynchronous random set representation
of fuzzy set, 131, 373

b.m.a., 134
fuzzy, 106, 144
measurement, 173
state, 204
measurement, 171

random, 108
Bag, 705
Basic mass assignment (b.m.a.), 134
fuzzy, 106, 144
magnitude of, 204
generalized, 138
random, 108
Bayes
and expert systems, 152
Bayes' rule, 39, 62
combination operator, 111
formal modeling
defined, 109
multitarget, 110
single-target, 110
unconventional measurements, 110
implementation, deceptive simplicity of, 68
normalization factor, 42
multitarget, 484
single-target, 62
parallel combination, 137, 186, 272
recipe, 341
resolution of Zadeh's paradox, 141
risk, 63
robust classifier, 223, 240, 244, 265
Bayes-invariant data conversion, 113
Bayes-invariant measurement conversion, 114, 190
Bayes-optimal state estimator, 63
Bearing-only measurement, 448, 547, 611
Belief function, 136
Belief-mass function, 296, 360
as special case of p.g.fl.s, 373
conditional, 397
importance of, 357
Binomial distribution, 436
and false alarms, 300
continuous, 437
p.g.f. of, 438
Birth target
defined, 313
Boolean algebra
and generalized fuzzy logic, 133
of events, 121
Bootstrap SMC filter, 83
Capacity of a random set, 713
Cardinality distribution, 363, 497
multitarget SMC approximation, 561
of CPHD filter, 633
of MeMBer filter, 666
of SMC multitarget filter, 553
Cartesian product, 56
Censored random finite set, 397
Central entropy
multitarget, 513
single-target, 66
Central formula of multitarget measurement modeling, 400
Central formula of multitarget motion modeling, 464
Chain rule
affine functional transformations, 379
first, 390
fourth, 391
second, 390
third, 390
Choquet functional, 713
Choquet-Mathéron theorem, 713
Clutter measurement, 291
Coherent passive localization (PCL), 614
Combination
Bayes operator, 112
Bayes parallel, 137, 142, 143, 186, 193, 272
Dempster's, 136
fuzzy Dempster's, 144
fuzzy modified Dempster's, 145, 282
modified Dempster's, 138, 273, 280
unnormalized Dempster's, 137
unnormalized fuzzy Dempster's, 145
Commonality function, 136
Commutativity, 121
Complementation
in conditional event algebra, 148
of fuzzy sets, 122
of multisets, 706
Computational complexity
CPHD filter, 640
fixed grid
multitarget, 544
single-target, 68

ideal MHC, 338
PHD filter, 592
SMC
multitarget, 545
single-target, 69
Conditional event algebra, 148
Conjunction
in conditional event algebra, 149
of fuzzy sets, 122
of multisets, 706
Consonant random set, 127, 163, 256, 736
Constant rule, 387
Continuity
of point cluster likelihoods, 444
Conversion of measurements, 191
Convolution
multiobject
fundamental theorem, 385
of ordinary functions, 344
Copula, 129
fuzzy logic, 130
equivalence to Bayes' rule, 186
Sklar's theorem, 129
Correlation
composite-hypothesis, 335
multihypothesis, 321
single-hypothesis, 319
Correlation operators, 195
Cost function, 63
Covariance
intersection
Julier-Uhlmann, 271
matrix, 27
and association, 315
and hybrid spaces, 65
and unification, 182
as a second-order moment, 567
definition of, 65
EKF corrector, 72
EKF predictor, 71
Kalman corrector, 30
Kalman predictor, 29
multitarget track distribution, 523
simple example, 65
single-target SMC filter, 86
standard multitarget measurement model, 312
standard multitarget motion model, 314
multitarget track, 509
CPHD filter, 573, 633
compared with JoTT filter, 641
computational complexity, 640
corrector, 636
derivation of, 652
Gaussian-mixture (GM) approximation, 646
initialization, 633
physical interpretation of, 642
predictor, 634
derivation of, 650
state estimation, 639

Data clustering, 452
EM-algorithm, 623
hierarchical, 623
k-means, 623
De Morgan laws, 121
Dead reckoning motion model, 480
Death target
defined, 313
Dempster's combination, 136
and Voorbraak probability, 142
as equivalent to Bayes' rule, 203
fuzzy, 144
modified, 138, 273, 280
modified fuzzy, 145
unnormalized, 137
unnormalized fuzzy, 145
Zadeh's paradox, 138
Dempster-Shafer
agreement, 137
basic mass assignment (b.m.a.), 134
generalized, 138
fuzzy, 144
measurement, 163
Density
covariance, 572
Dirac delta
and the PHD, 577
caution using, 376
in functional derivatives, 375
in Janossy densities, 707

        multitarget, 366, 374
        properties of, 693
        SMC filters, 80
    filtering, 42
    intensity, 366, 569, 577
    Janossy, 707
    Markov transition
        multitarget, 464, 467
        single-target, 51
    multiobject
        defined, 362
        simple example of, 364
    multitarget factorial moment, 586
        inversion formula, 587
    multitarget first-moment, 577
    probability hypothesis (PHD), 569, 577
    proposal
        multitarget, 557
        single-target, 83
Derivative
    directional, 375
    Frechét, 375
    functional
        and physics, 376, 382
        and the PHD, 582
        and units of measurement, 377
        as Mobius transform, 383
        definition of, 375
        iterated, 376
        linear-continous property, 376
        rules for, 386
        with respect to finite set, 376
    gradient, 71, 72, 375, 695
    partial, 696
    set, 296
        and functional derivatives, 381
        and PHD, 581
        definition of, 381
        is not measure theoretic, 716
        rules for, 386
Detection
    clutter, 291
    false, 289
    measurement, 288
    probability of, 289
Digital voltmeter, 97
Dirac
    delta density
        and the PHD, 577
        caution using, 376
        in functional derivatives, 375
        in Janossy densities, 707
        multitarget, 366, 374
        PHD of multitarget, 585
        properties of, 693
        single-target, 354
        SMC filters, 80
    measure, 694, 706
    sum, 706, 708
Discrete random set, 127
Disjunction
    in conditional event algebra, 149
    of fuzzy sets, 122
    of multisets, 706
Distance
    association, 316
    Euclidean, 315
    global association, 317
    Hausdorff, 511
    Mahalanobis, 317
    Wasserstein, 511
Distribution
    binomial, 300, 436
    cardinality, 363
        multitarget SMC approximation, 553, 561
    Gaussian, 34
        normalized, 166, 202
    initial, 61
    multi-Bernoulli, 369
    multinomial, 85
    posterior, 39
        multimodal, 51
    uniform, 143
Distributivity, 121
Double counting, 271
Dynamic prior, 83
Dynamic state space model, 56, 109

EAP estimator, 79
    GM filter, 573
    KEF filter, 208

PHD filter, 639
single-target, 63
SMC filter, 86
EKF, 71
Elementary symmetric functions, 369
and CPHD filter, 641
ELINT, 222, 240
EM algorithm, 86, 505, 510, 562, 595, 613, 614, 623
Entropy
central
multitarget, 513
single-target, 66
single-target, 66
Epanechnikov kernel, 86
Error bar
fuzzy, 256
random, 252
Error covariance matrix, 27
Euclidean
distance, 315
measurement space, 57
state space, 56
Event
antecedant, 147
consequent, 147
crisp, 120
fuzzy, 131
Expected a posteriori (EAP) estimator, 63
Expected value
and p.g.f.s, 372
multitarget
indirect, 576
posterior, 63
Extended Kalman filter (EKF), 71
Extended target, 292
measurement model, 429
motion model, 474
Eyeball track labeling, 508

False alarm, 289
state-dependent, 301, 424
Feature, 94
active-acoustic, 613
as a crisp subset, 120
datalink, 238
frequency, 215
and model bases, 215
example of filtering, 225
from EO/IR image, 292
from HRRR signature, 615
in expert-system theory, 120
in images, 292, 614
operator-mediated, 211, 238, 292
fuzzy model of, 167
random set model of, 157
SAR image, 240
statistically correlated, 195
passive-acoustic, 211
FGM family of fuzzy logics, 131
Filter, 668
alpha-beta, 33
as a primitive SMC filter, 547
as first-moment filter, 567
bootstrap
in SMC approximation, 83
CPHD, 573, 633
Gaussian-mixture (GM) approximation, 646
SMC approximation, 644
Gaussian-mixture
single-target, 72
IPDA, 515
multitarget generalization of, 667
JoTT, 515
SMC implementation of, 524
target prioritization, 535
Kalman, 27
as second-order moment filter, 567
constant-gain, 567
Kalman evidential (KEF), 199
multitarget
general Bayes, 357, 422, 453, 484
generalized IPDA, 667
JIPDA, 667
JPDA, 335
multihypothesis, 321
single-hypothesis, 319
multitarget Bayes
appropriateness of, 546
multitarget multi-Bernoulli (MeMBer), 657
para-Gaussian, 658

PDA, 334, 337
probability hypothesis density (PHD), 569
GM implementation of, 623
SMC implementation of, 615
strengths and weaknesses of, 571
single-target Bayes, 42
appropriateness of, 70
with fuzzy measurements, 221
with random imprecise data, 222
filter
initialization, 669
Filtering density, 42
Finite set
measurement, 287
random, 349
state, 404
Finite-mixture model, 108, 453
Firm track, 326
Fixed-grid approximation
multitarget
intractability of, 544
single-target
intractability of, 68
Focal set, 136
fuzzy, 144
Frame of discernment, 120
Function
belief, 136
belief-mass, 296, 357, 360
conditional, 397
generalized, 373
characteristic, 344
commonality, 136
cost, 63
elementary symmetric, 369
and CPHD filter, 641
fuzzy membership, 121
likelihood
single-target, 37, 51
moment generating, 344
multiobject density
defined, 362
simple example of, 364
one-to-one, 318
plausibility, 136
plausibility-mass, 360
probability-generating, 372
probability-mass, 359
set derivative of, 381
set, 359
induced by p.g.fl., 380
set indicator, 167
sigmoidal, 437
tactical importance (TIF), 358, 533
test, 370
Functional
definition of, 370
derivative
definition of, 375
rules for, 386
integral, 709
linear, 371
derivative of, 377
probability-generating, 370
importance of, 358
quadratic
derivative of, 378
transformation, 358, 379, 382
Fundamental convolution formula, 385
Fundamental Gaussian identity, 699
Fundamental theorem
of multiobject calculus, 384
of ordinary calculus, 383
Fuzzy
complementation, 122
conjunction, 122
Dempster's combination, 144
modified, 145, 282
unnormalized, 145
disjunction, 122
error bar, 256
event
definition of, 131
logic
copula, 130
definition of, 122
FGM family of, 131
generalized, 133
Hamacher, 131, 196
nilpotent, 123, 131
prodsum, 122, 130
Zadeh, 133

measure, 360
measurement, 162
simple example of, 166
membership function, 121
and the PHD, 579
finite-level, 126
random, 126
rule, 148
random interpretations, 150
set
asynchronous random set representation, 131, 373
data cluster as, 458
generalized, 133, 145
level set of, 123
normal, 122
synchronous random set representation, 123
state, 275

g.b.m.a., 138
fuzzy, 145
Gaussian
distribution, 34
normalized, 166, 204
fundamental identity, 699
General data modeling, 94
the four steps of, 95
Generalized b.m.a., 138
Generalized fuzzy logic, 133
Generalized fuzzy set, 133, 145
Generalized likelihood, 100
AGA, 214
Dempster-Shafer, 219
fuzzy, 217
fuzzy Dempster-Shafer, 220
generalized fuzzy, 219
AGU, 254
naive, 249
for SAR image, 264
joint, 100
nonlinearity of, 101
of Dempster-Shafer state-estimate, 278
of NOTA target type
AGA measurements, 232
UGA measurements, 197
UGA, 164
composite fuzzy rule, 179
Dempster-Shafer, 171
first-order fuzzy rule, 174
fuzzy, 166
fuzzy Dempster-Shafer, 173
generalized fuzzy, 169
second-order fuzzy rule, 180
statistical, 165
Generalized measurement, 100
imprecise and slightly random, 101
in general, 107
random and slightly imprecise, 97
statistically correlated, 195
uncertain and nonrandom, 103
vague and nonrandom, 102
Global
association distance, 317
association likelihood, 318
mean deviation, 510, 562
GM-CPHD filter, 646
corrector, 647
initialization, 646
predictor, 646
state estimation, 649
GM-PHD filter, 599, 623
assumptions, 625
corrector, 628
initialization, 626
predictor, 626
pruning and merging, 630
state estimation, 630
strengths and weaknesses, 624
Gradient derivative
and EKF corrector, 72
and EKF predictor, 71, 696
and functional derivative, 375
defined, 695
Group PHD filter, 615
Group target, 461, 565, 615

Hamacher fuzzy logic, 131, 196, 241
Hard data cluster, 453
Hausdorff distance, 511, 562
Heuristic model, 53
Hit-or-miss topology, 711

Hybrid
measurement space, 57, 287
state space, 56
measures of dispersion, 65
Hyperspace, 110, 153, 291, 349, 404, 407, 711
defined, 295
Hypothesis
Dempster-Shafer, 136
null, 105, 136
fuzzy Dempster-Shafer, 144
probability, 322, 330, 333
definition, 326
simplest case, 327
single multitarget, 320
table, 321

i.i.d. cluster process, 366
for CPHD filter, 635, 636
state estimation, 502
Implementations
CPHD filter
survey of, 643
fixed-grid
multitarget, 544
single-target, 68
Gaussian mixture
single-target, 72
multitarget filter
bearing-only tracking, 547
multiuser detection in comms, 550
tracking in road networks, 549
tracking in terrain, 549
PHD filter
active sonar tracking, 613
active sonar tracking-3D, 613
bistatic RF tracking, 614
direction of arrival tracking, 612
distributed tracking, 611
multiple-model tracking, 611
tracking in images, 614
tracking in terrain, 610
SMC
multitarget, 545
single-target, 69, 79
Taylor series
EKF, 71
Importance
sampling, 83
weights, 80
Imprecise measurements, 97
Innovation
in Kalman filter, 30, 336
Integral
change of variables, 60, 166
for Dirac densities, 694
functional, 709
in general Bayes modeling, 109
Lebesgue
generalized to finite sets, 715
measure theoretic, 712
on a hybrid space, 158
set, 296, 297
and PHD, 580
formally defined, 361
general definition of, 714
is not measure theoretic, 714
nonadditivity of, 362
Stieltjes, 497
Intensity density, 569, 577
of Poisson process, 366
Interval
argsup, 254
arithmetic, 254
measurement, 223
MLE, 254
IPDA filter, 515, 667
multitarget generalization of, 667

Jacobian determinant, 60, 166, 382, 390
and Dirac delta densities, 694
Janossy measure, 707
JIPDA filter, 667
Joint generalized likelihood, 100
Joint probabilistic data association, 335
JoM multitarget state estimator, 498
for JoTT filter, 521
JoTT filter, 515
and target prioritization, 535
compared with CPHD filter, 641
SMC implementation of, 524
JPDA filter, 335

k-means clustering, 623
Kalman filter, 27
as second-order moment filter, 567
Bayes interpretation of, 45
constant-gain, 32
as first-order moment filter, 567
corrector, 29
Bayes formulation of, 37
derivation of, 30
data fusion using, 32
derivation of, 30
extended (EKF), 71
information form, 32
inverse, 271
predictor, 28
Bayes formulation of, 35
derivation of, 31
state and error estimation, 40
top-down perspective, 34
KEF filter, 199
compared to Gaussian-mixture filter, 208
corrector
conventional measurements, 207
fuzzy DS measurements, 205
predictor, 205
state estimation, 208
Kullback-Leibler discrimination
multitarget, 513
and the PHD, 579
single-target, 66
relationship with central entropy, 67
relationship with entropy, 66

Label
table, 626
track, 319
Likelihood
as limiting case of generalized likelihood, 100, 102
bearings-only sensor field, 449
constructive, 302
data clustering, 456
extended target, 430
Poisson approximation, 431
for generalized measurement, 100
for SAR image
datalink, 241
generalized
AGA measurement, 214
Dempster-Shafer state-estimates, 278
of NOTA target type, 197
in general Bayes modeling, 109
interval-valued, 252
joint
multitarget, 445
single-target, 67
multiple point clusters, 442
continuity of, 444
multitarget
and associations, 422
bearing-only sensor field, 452
extended targets, 430
multisensor, 445
multisource, 447
simple example of, 400, 418
true, 297
of association, 318
global, 318
single point cluster, 435
single-target, 51
true
multitarget, 294, 296, 302, 303, 357, 400
single-target, 53, 60
UGA measurements, 254
Linear rule, 387
for set integrals, 362
Lumpy pillow syndrome
in algorithm development, 54

MaM multitarget state estimator, 497
for JoTT filter, 521
MAP estimator
GM filter, 79
single-target, 63
SMC filter, 86
Markov density
in general Bayes modeling, 109
multitarget
multiple point clusters, 477
simple example, 464
single point clusters, 476

single-target, 51
true
multitarget, 358, 464
single-target, 53, 59
Mathe-magic, 51, 299
Mathéron topology, 711
Matrix
completion of the square, 30
covariance, 27
and association, 315
and hybrid spaces, 65
and unification, 182
as a second-order moment, 567
EKF corrector, 72
EKF predictor, 71
general definition, 65
Kalman corrector, 30
Kalman predictor, 29
multitarget track distribution, 523
simple example, 65
single-target SMC filter, 86
standard multitarget measurement model, 312
standard multitarget motion model, 314
measurement matrix, 29
norm, 33, 203
positive-definite, 34
state transition, 29
transportation, 511
transpose, 35
Maximum a posteriori (MAP) state estimator, 63
Maximum likelihood state estimator, 64
Measure
counting, 706
Dirac, 706
fuzzy, 360
Janossy, 707
Lebesgue
generalized to finite sets, 715
plausibility, 360
possibility, 360
Sugeno, 360
uncertainty, 360
Measurement
AGA, 115
defined, 214
simple example of, 211
AGU, 115
defined, 249
simple example of, 249
bearing-only, 448
defined, 97
Dempster-Shafer, 163
determinism versus uncertainty in, 106
finite-set, 287
general, 293
fuzzy, 162
simple example of, 166
fuzzy and nonrandom, 102
generalized, 100
and multitarget filtering, 493
imprecise, 97, 162
imprecise and slightly random, 101
interval, 223
lossless conversion of, 113
matrix
in Kalman filter, 29
natural-language, 94, 102, 103
random and slightly imprecise, 97
sensor transform model, 115
single-source, single-target
conventional, 57
UGA, 115, 157
defined, 164
statistical, 165
UGU, 114
uncertain, 163
uncertain and nonrandom, 103
unresolved, 289
vague, 102, 162
Measurement model
bearing-only sensor field, 450
extended target, 428
finite mixture, 453
in general Bayes filter, 109
multisource, 445
multitarget
central formula, 400
standard, 311, 408
single-target
Kalman filter, 29

with false alarms, 414
unresolved targets, 432
Measurement space
for generalized measurements, 163
hybrid, 158
in general Bayes modeling, 109
multisensor-multitarget, 408
single-sensor, multitarget, 407
MeMBer filter, 657
and generalized IPDA filter, 667
corrector, 662
derivation of, 677
interpretation of, 664
initialization, 661
predictor, 661
derivation of, 675
pruning and merging, 665
state estimation, 666
MLE estimator
and AGU measurements, 254
and target identification, 225
interval, 254
multitarget, 501
single-target, 64
MMLE multitarget state estimator, 500
Mobius transform, 136
and set derivative, 383
Model
Bayes, 54
dynamic state space, 56, 109
general data, 94
heuristic
lumpy pillow syndrome, 55
single-target, 53
Model-based approaches, 196
Modified Dempster's combination, 138, 280
and pignistic probability, 143
and unification, 283
Modus ponens, 147, 187
generalized, 149
Monomial rule, 387
Motion model
at most one target, 469
constant-velocity, 33, 480
dead reckoning, 480
extended target, 474
in general Bayes modeling, 109
multitarget
birth and death, 314
birth, no death, 314
central formula, 464
no birth or death, 314
standard, 313, 467
single-target
additive-nonlinear, 58
constant-velocity, 43, 48
deterministic, 28
general nonlinear, 60
in Kalman filter, 28
Multi-Bernoulli process, 368
Multidimensional extrema, 696
Multidimensional second-derivative test, 697
Multidimensional Taylor series, 696
Multinomial
distribution, 85
resampling, 85
Multiobject process
binomial i.i.d. cluster, 370
clutter, 292
clutter only, 412
Dirac, 374
p.g.fl. of, 374
PHD of, 585
Dirac delta, 366
i.i.d. cluster, 366
p.g.fl. of, 374
PHD of, 584
i.i.d. cluster process
state estimation, 502
multi-Bernoulli, 368
for multitarget filtering, 657
p.g.fl. of, 374
PHD of, 584
Poisson, 366, 373
p.g.fl. of, 373
PHD of, 583
uniform, 367
p.g.fl. of, 375
PHD of, 585
state estimation, 503
Multiset, 705
and point processes, 707

as generalized fuzzy membership function, 706
intersection of, 706
union of, 706
Multitarget
central entropy, 513
density function, 362, 484
unitless, 496
factorial moment density, 586
inversion formula, 587
i.i.d. cluster process, 366
Kullback-Leibler discrimination, 513
likelihood
and associations, 422
bearing-only sensor field, 452
extended targets, 430
multiple point clusters, 442
multisensor, 445
simple example of, 400
measurement model
standard, 408
motion model
standard, 467
multi-Bernoulli approximation, 657
multi-Bernoulli filter, 657
particle system, 551
equal weighting, 553
Poisson process, 366
state
definition of, 404
rigorous representation of, 405
state estimation
PHD visualization, 505
state estimator
JoM, 498
MaM, 497
state space, 404
track covariances, 509
track labeling, 508
Multitarget multi-Bernoulli (MeMBer) filter, 661
Multitarget probability density, 357
and target prioritization, 535
importance of, 357
simple example of, 364
Multitarget uniform process, 367
state estimation, 503
Multiuser detection (MUD) for comms, 550

Natural-language measurement, 102, 103
Nilpotent fuzzy logic, 123, 131
None-of-the-above (NOTA), 245
for AGA measurements, 232
for UGA measurements, 197
Normalized Gaussian, 166, 202
Notation
and generalized measurements, 158
set derivatives and physics, 382
set versus vector, 364
transparent, 683
Null hypothesis
in Dempster-Shafer theory, 105

One point covering function, 123
One sigma containment region, 65
One-to-one function
as an association, 318

p.g.f., 344
and expected value, 372
and p.g.fl.s, 372
and variance, 372
p.g.fl., 358
and fuzzy sets, 132
and p.g.f.s, 372
as generalized belief-mass function, 373
data clustering, 459
definition of, 370
extended target, 458
for a tactical importance function, 533
multiple extended targets, 458
of Dirac process, 374
of multi-Bernoulli process, 374
of multitarget corrector, 530
used for PHD, 651
of multitarget predictor, 528
PHD predictor, 650
of Poisson process, 373
of standard measurement model, 421
state-dependent false alarms, 426
transmission drop-outs, 427
of uniform process, 375

of virtual leader-follower model, 482
properties of, 371
Para-Gaussian filter, 658, 668
corrector, 671
predictor, 669
pruning and merging, 673
state estimation, 675
Particle impoverishment, 85
Particle system
for PHD, 616
JoTT filter, 524
multitarget, 551
equal weighting, 553
single-target, 80
and equal weighting, 80
Particle system approximation, 523
PDA filter, 334, 337
Performance
CPHD versus PHD, 643
MHC versus SHC, 339
PHD versus MHC, 596, 598
PHD
as first-order moment, 569
filter, 569
functional derivative formula, 582
multitarget SMC approximation
visualization tool, 561
set derivative formula, 581
set integral formula, 580
simple example of, 569
visualization, 505
PHD filter, 569
and target ID, 599
computational complexity, 592
corrector, 590
derivation of, 651
interpretation of, 592
multisensor, 594
physical interpretation of, 603
derivation of, 649
error estimation, 595
for group target tracking, 615
Gaussian-mixture (GM), 599, 623
initialization, 587
limitations of, 570
target number, 631
MHC-PHD hybrid, 596
predictor, 588
derivation of, 650
physical interpretation of, 600
sensor management using, 614
SMC approximation, 615
state estimation, 595
track labels, 596
Pignistic probability, 142
Bayes interpretation of, 279
Plausibility function, 136
Plausibility-mass function, 360
Point process, 707
and information fusion, 710
in Dirac sum form, 709
integrals with respect to, 709
probability distribution of, 709
simple, 707
Point target cluster, 434
definition of, 434
likelihood of, 435
Markov density, 476
multicluster Markov density, 477
Poisson
distribution, 43, 366, 457
and binomial distribution, 301
p.g.f. of, 345
false alarm model, 43, 300, 301, 329, 410, 412
and MeMBer filter, 664
and p.g.fl., 422
and point clusters, 442
multidimensional process, 300, 366
and the PHD, 579, 590, 605, 651
initialization of multitarget filter, 486, 555
p.g.fl. of, 373
PHD of, 583
target birth model, 314, 472
and p.g.fl., 474
and point clusters, 477
Possibility measure, 360
Posterior expectation, 63
Power rule, 388
Probabilistic data association, 334, 337
Probability

copula, 129
density
Gaussian, 34
multitarget, 357
event defined, 120
of a hypothesis, 322
of detection, 289
transmission, 426
of hypothesis, 326, 330, 333
of track, 326
pignistic, 142, 279
total probability theorem, 61
Voorbraak, 142, 173, 186, 279
Probability hypothesis density (PHD)
as first-order moment, 569
Probability-generating function
and p.g.fl.s, 372
Probability-generating functional, 344
definition of, 370
properties of, 371
Probability-mass function, 359
set derivative of, 381
Prodsum fuzzy logic, 122, 130, 131
Product rule
general
for ordinary functions, 389
two factors, 389
multiple factors, 389
simple, 388

Radar
as source of nonmeasurement data, 97
bistatic, 613
GMTI, 240, 291
high range-resolution (HRRR), 53, 57, 615
millimeter-wave, 292
surveillance, 288
synthetic aperture (SAR), 53, 57
datalink data, 240
robust classifier, 260
Radon-Nikodým derivative
definition of, 713
Radon-Nikodým theorem, 715
of multiobject calculus, 385
generalized, 385
of ordinary calculus, 383
Random
finite set (RFS), 349
integer, 348
number, 349
vector, 349
as random set, 349
twinkling, 350, 355
Random disk, 103
Random finite set, 349
censored, 397
definition of, 712
probability measure of, 712
Random set, 97
consonant, 127, 163, 256, 736
definition of, 712
discrete, 127, 171
one point covering function, 123
probability measure of, 712
representation
of composite rule, 151
of DS evidence, 143
of first-order fuzzy rule, 150
of first-order rule, 150
of fuzzy DS evidence, 145
of fuzzy evidence, 123
of fuzzy evidence, general, 131
of generalized fuzzy evidence, 134
of second-order fuzzy rule, 152
of second-order rule, 152
of UGA measurements, 252
uncertainty representation, 119
uniformly distributed, 152
Random set modeling
basic example of, 99
Receiver operating characteristic (ROC) curve, 289
Regularization
in SMC approximation, 85
Rejuvenation
in SMC approximation, 85
Resample-move
in SMC approximation, 85
Resampling, 84
Robust Bayes classifier, 223, 240, 244, 265
Robust statistics, 250
Roughening

in SMC approximation, 85
Rule
composite, 147
random set representation, 151
firing of, 147
first-order, 147
random set representation, 150
fuzzy, 148
partial firing of, 188
second-order, 148
random-set representation, 152

Sensor management
PHD-based approach, 614
tracking in road networks, 549
Sensor tag, 406
Sensor transform model, 115, 157, 164
Set
crisp, 121
derivative, 296, 716
rules for, 386
focal, 136
fuzzy, 144
function, 359
induced by p.g.fl., 380
fuzzy, 121
finite-level, 126
indicator function, 167
integral, 296, 297
formally defined, 361
general definition of, 714
is not measure theoretic, 714
nonadditivity of, 362
integrals
units of measurement and, 362
notation
versus vector notation, 364
SIGINT, 240
Sigma-count
of a fuzzy set, 580
Single-target Bayes filter, 42
computational complexity, 68
corrector
derivation of, 62
fixed-grid implementation
intractability of, 68
fusing data with, 67
initialization, 61
interpretation of, 45
predictor
derivation of, 61
simple example of, 43
with fuzzy measurements, 221
with random imprecise data, 222
Single-target Gaussian-mixture filter, 72
compared to KEF filter, 208
corrector, 76
predictor, 76
state estimation, 79
Single-target SMC filter
and generalized measurements, 101
corrector, 83
initialization, 82
predictor, 82
state estimation, 86
Sklar's theorem, 129
SMC approximation
multitarget
definition of, 551
importance of proposal density, 557
Rao-Blackwellized, 548
strengths and limitations, 545
of JoTT filter, 523
single-target, 79
SMC-CPHD filter, 644
corrector, 645
initialization, 644
predictor, 644
state estimation, 646
SMC-PHD filter, 615
corrector, 621
and regularization, 623
initialization, 618
particle PHD, 616
predictor, 618
appearing targets, 619
persisting targets, 619
spawned particles, 620
state estimation, 623
Smoothing
in SMC approximation, 85
Soft data cluster, 453, 457

Space
Euclidean, 56
hybrid, 56, 65, 287, 349
hyper, 110, 153, 291, 295, 349, 407, 711
measurement
in general, 109
multisensor, 408
single-sensor, 407
single-target, 57
single-target, 56
state
in general, 109
multisensor, 406
multitarget, 403
Spawned target, 313, 461, 472
multitarget SMC approximation, 559
SMC-PHD filter, 620
Standard multitarget measurement model, 408
Standard multitarget motion model, 467
State
as data, 97
Dempster-Shafer, 275
and UGA measurements, 279
fusion, 272
fuzzy, 275
fuzzy Dempster-Shafer, 281
fuzzy DS, 204
generalized, 274
imprecise, 274
multitarget
definition of, 404
sensor, 406
single-target, 56
transition matrix, 29
unresolved target, 434
State estimator, 63
Bayes-optimal, 63
EAP
GM filter, 79
KEF filter, 208
PHD filter and target number, 573, 639
single-target, 63
SMC filter, 86
MAP
GM filter, 79
single-target, 63
SMC filter, 86
multitarget, 494
discrete MAP, 499
JoM, for JoTT filter, 521
MaM, 497
MaM, for JoTT filter, 521
MMLE, 500
PHD visualization, 505
multitarget EAP
failure of, 494
multitarget MAP
failure of, 494
practical, 64
single-target
MLE, 64
State space
hybrid, 158
in general Bayes modeling, 109
multisensor, 407
multitarget, 404
of generalized states, 275
single-target, 56
unresolved targets, 434
State-dependent false alarm, 424
defined, 301
Statistical moment
first-order
multitarget, 568, 576
single-target, 40, 567
higher-order multitarget, 586
second-order
multitarget, 572
single-target, 40, 567
Stieltjes integrals, 497
Sufficient statistic, 111, 202
fuzzy Dempster-Shafer, 203
Sugeno measure, 360
Sum rule
for derivatives, 388
Synchronous random set representation
of fuzzy set, 123

Table
hypothesis, 321
label, 626
track, 319

Tactical importance function (TIF), 533
Target
    birth
        defined, 313
    closely spaced, 565
    confuser, 292
    death
        defined, 313
    density
        and the PHD, 577
    extended, 292
        approximate likelihood, 431
        likelihood, 430
        measurement model of, 428
        motion model, 474
    group, 461, 565, 615
    identification
        robust-Bayes classifier, 223
        unknown target types, 196, 232
    mode, 461
    of interest (ToI), 358, 531
    partially existing, 437
    point, 432
    prioritization, 358, 531, 565
        and the PHD, 614
    signal detection, 57
    spawning, 313, 461, 472
    unresolved, 289, 432
Taylor series approximation, 71, 696
Test function, 370
ToI, 358, 531, 565
Total probability theorem, 61
Track
    continuity, 505
    firm, 326
    initialization
        multitarget, 321
    label, 319, 505
        and PHD filter, 596, 598
        multitarget SMC filter, 563
        table, for GM-PHD filter, 628
    probability, 326
    table
        for MeMBer filter, 657
        for MHC filter, 322
        for SHC filter, 319
Tracking
    active sonar
        with PHD, 613
    air traffic control, 33
    bearing-only, 547
    bistatic RF
        with PHD, 614
    direction of arrival
        with PHD filter, 612
    distributed
        with PHD filter, 611
    group and cluster, 566
    group target, 615
    in images
        with PHD filter, 614
    in road networks, 549
    in terrain, 549
        with PHD filter, 610
    multiple-model
        with PHD filter, 611
    with HRR radar
        with the PHD, 615
Tracklets, 271
Transform
    and p.g.fl.s, 372
    Fourier, 343
    geometric, 344
    Laplace, 343
    Mobius, 136
        and functional derivatives, 383
    sensor, 157
    utility of, 345
    z-transform, 343
Transmission probability of detection, 426
Transparent notation, 683
Transportation matrix, 511
Transpose of a matrix, 35
True likelihood function
    and associations, 424
    multitarget, 302
        and set derivatives, 303, 357, 400
        of standard model, 294, 296, 410
    single-target, 53, 294
True Markov transition density
    multitarget, 464
        and set derivatives, 358

of standard model, 467
single-target, 53
derivation of, 59
Twinkling random vector, 349, 350, 355, 391

UGA measurement
absence of opinion in, 163
and Dempster-Shafer state-estimates, 281
composite fuzzy rule, 179
defined, 164
Dempster-Shafer, 171
first-order fuzzy rule, 174
fuzzy, 166
fuzzy Dempster-Shafer, 173
generalized Dempster-Shafer, 171
generalized fuzzy, 169
generalized likelihood
defined, 164
joint generalized likelihood
defined, 164
lossless measurement conversion, 190
second-order fuzzy rule, 180
simple example of, 157
statistical, 165
unified measurement fusion
copula fuzzy conjunction, 186
Dempster's combinations, 185
in finite case, 188
of rule-firing, 187
Uncertainty
due to imprecision, 97
measure, 360
sources of, 92
Unified
data modeling, 95
measurement fusion, 182
in finite case, 188
of copula fuzzy conjunctions, 186
of Dempster's combinations, 185
of rule-firing, 187
multitarget state representation, 404
multitarget-multisource integration, 493
target detection and estimation, 498
theory of data modeling, 55
Uniform distribution
and Dempster's combination, 138
and entropy, 66
and multitarget central entropy, 67
and the MLE, 64
discrete multitarget, 367
in filter initialization, 45, 61
multiobject, 367
on power set of a universe, 150
on unit interval, 123
and copulas, 129
and fuzzy logic, 123
Uniform random scalar field, 132
Units of measurement
and functional derivatives, 377
and generalized Lebesgue measures, 715
and multiobject densities, 362
and multitarget likelihoods, 299
and set integrals, 362
Universe of discourse, 120
Unknown target type, 196, 232
Unnormalized Dempster's combination, 137
fuzzy, 145
Unresolved target, 289, 432

Variance
and fuzzy filtering, 221
and JoM estimator, 501
and p.g.f.s, 372
of target number
JoTT filter, 523
MeMBer filter, 666
Vieta's theorem, 641
Virtual leader-follower motion model, 478
Voorbraak probability, 142, 173, 186, 193
Bayes interpretation of, 279

Wasserstein distance, 511, 562

Zadeh's paradox, 138
as a criticism of Bayes' rule, 140
Bayes resolution of, 141
impact on Dempster-Shafer theory, 89

The Artech House Information Warfare Library

Electronic Intelligence: The Analysis of Radar Signals, Second Edition, Richard G. Wiley

Electronic Warfare for the Digitized Battlefield, Michael R. Frater and Michael Ryan

Electronic Warfare in the Information Age, D. Curtis Schleher

Electronic Warfare Target Location Methods, Richard A. Poisel

EW 101: A First Course in Electronic Warfare, David Adamy

Information Warfare and Organizational Decision-Making, Alexander Kott, editor

Information Warfare Principles and Operations, Edward Waltz

Introduction to Communication Electronic Warfare Systems, Richard A. Poisel

Knowledge Management in the Intelligence Enterprise, Edward Waltz

Mathematical Techniques in Multisensor Data Fusion, Second Edition, David L. Hall and Sonya A. H. McMullen

Modern Communications Jamming Principles and Techniques, Richard A. Poisel

Principles of Data Fusion Automation, Richard T. Antony

Stratagem: Deception and Surprise in War, Barton Whaley

Statistical Multisource-Multitarget Information Fusion, Ronald P. S. Mahler

Tactical Communications for the Digitized Battlefield, Michael Ryan and Michael R. Frater

Target Acquisition in Communication Electronic Warfare Systems, Richard A. Poisel

www.ingramcontent.com/pod-product-compliance
Lightning Source LLC
Chambersburg PA
CBHW071252180926
51613CB00001B/18
9781596930926